Air Transport in Canada

Air Transport in Canada

Larry Milberry

VOLUME 2

CANAV Books

Canadian Cataloguing in Publication Data

Milberry, Larry, 1943-
   Air transport in Canada

Includes bibliographical references and index.
ISBN 0-921022-09-3

1. Aeronautics, Commercial - Canada - History. I. Title.

HE9815. A3M54 1997   387.7'0971   C97-931133-0

Design: Robin Brass Studio, Toronto
Maps: Ralph Clint
Photo retouching: Stephen Ng/ML Studio, Toronto
Proof reading: Lambert Huneault, Ralph Clint

Printed and bound in Canada by Friesen Printers Ltd., Altona, Manitoba

Published by
CANAV Books
Larry Milberry, Publisher
51 Balsam Avenue
Toronto, Ontario M4E 3B6
Canada

Front endpaper
A pair of Canadian Forces C-130 Hercules delivering food relief to Jijiga, Ethiopia on December 2, 1991. They were operating from Djibouti on behalf of the UN. (Larry Milberry)

Half title
Canadian's Boeing 747-475 "Maxwell W. Ward" departs Hong Kong (Kai Tak Airport) for Vancouver on November 19, 1995. For the late 1990s Canadian was focusing on its historically strong trans-Pacific markets and new Canada-US Open Skies opportunities, while downsizing domestically and in Europe. With this plan, the troubled carrier hoped to survive into the 21st century. (Anthony J. Hickey)

Title page
Then and now over the Montreal waterfront in June 1997. Canadair's photography department arranged for this "photo op", which put Air Canada's famous Lockheed 10 together with one of the airline's Canadair CRJ-200s. Lucio Anodal took this view from a CRJ-200 flown by Bruce Campbell and Jacques Thibodeau. (Canadair Ltd.)

Above
The RCAF started modestly in 1924, using hand-me-down WWI types. It came a long in the next 40 years, by when it had such types as the C-130 Hercules, one of which is shown using JATO—jet assist takeoff. (DND IE69-102)

Back endpaper
Midnight at Yellowknife on June 25, 1993. A Beech 18 is finally being put to bed after a long day's flying. In winter it's another story in the Arctic for such VFR operations. Having four or five hours for flying makes for a good day. (Larry Milberry)

Contents

Air Transport in Colour

Finding useful photos from the early years of colour film was a challenge while assembling this book. Few were shooting colour film in the 1940s-50s. Certainly, good photos must exist, but most seem buried away. No doubt many others have been lost forever, but enough were scrounged up to make a good beginning here in Chapter 32. Some of these photos are from individuals who took the occasional airplane picture for general interest. Others were shot by dedicated enthusiasts like John Davids and Al Martin. By the early 1960s a handful of younger fellows appeared who started regularly using colour film. From that period on there is no shortage of decent colour photos. By the 1990s aviation photography was a sizeable hobby—hardly an airplane moves these days without someone photographing it.

(Left) PWA's Junkers W.33 CF-AQW in Yellowknife in the mid-1950s. Pilot Al Roach was heading south after some DEW Line work. An ex-RCAF air gunner and POW, Roach later was killed in Beech 18 CF-RRE. This Junkers had come to Canada in 1931. After years with Canadian Airways it served Central BC Airlines, PWA, and in May 1959 went to Skyways Air Services. It crashed at Kootenay Lake on August 10 that year. Then (below), AQW ready to go at Peace River, Alberta. The engineer is adjusting his heater used to warm the cowling tent, where he has work to do on the engine. (Reg Phillips, Provincial Archives of Alberta/John Davids Col. D.1069)

(Below) O. J. Wieben of Superior Airways operated Canada's last Bellanca 31-55 Skyrocket. It's seen at his Kaministiquia River base outside Fort William on September 3, 1961. CF-DCH retired in 1966, then was a training aid in a New Hampshire school until 1989. The Reynolds Aviation Museum returned it to Canada. Once restored, it went on display in Wetaskiwin—the only complete Canadian-built Bellanca. The remains of CF-DCE, another old Wieben Skyrocket, were added to the CMFT of Surrey, BC in 1978. (Larry Milberry)

(Above) Skyrocket CF-EQQ at Fort MacKay on the Athabasca River. Associated bought EQQ in June 1949. It was sold 10 years later to John Midget of Meadow Lake, Saskatchewan, then to F.M. Clark of the same town. On September 6, 1961 it was destroyed by fire. (Provincial Archives of Alberta/John Davids Col. D1037)

(Left) A dandy shot by bush pilot Ed McIvor of Transair's famous Bellanca 66-70 in front of the HBC post at Cross Lake, Manitoba in February 1966. Edwin Johnson was flying it on this occasion. This "jumbo" bushplane came to Canada in 1941 for Mackenzie Air Service. It became the property of CPA in 1943, was sold to Central Northern Airways in 1947, stayed on through CNA's transformation to Transair in 1956, then went to Hooker Air Service when Transair left the bush in 1967. Barney Lamm acquired it and sold it in the US in the late 1980s.

(Left) Norseman IV No. 3537 on the Ottawa River at Rockcliffe after WWII. It had come on strength in January 1942. It's shown in the standard RCAF colours of the period, but little is known of its particular history. (Howard Levy)

(Below) A Norseman always looks attractive in red and white. CF-BDF of St. Felicien Air Service is seen in 1975 at home base on the western shore of Lac St-Jean, Quebec. The 13th Norseman, it first flew in July 1937. From 1938-46 it was with Canadian Airways and CPA. Post-1947 its flew mainly in Quebec with operators like Gagnon, Dolbeau and Cargair. It was lost September 7, 1977—it stalled on landing at Lac Sacacomie, but those aboard survived. (Hugh Halliday)

Northward Aviation of Edmonton operated Norseman CF-INN. Here it is at Yellowknife in 1967. It was ex-RCAF No. 789, a 1943 UC-64A that Noorduyn took back in 1956, refurbished as a Mark V, then re-sold to Lamb Airways. It moved to Northward in 1964, thence to Northland Airlines of Winnipeg. On June 23, 1967 it crashed in Sharpe Lake, Manitoba. The pilot died of exposure while trying to swim ashore. (Provincial Archives of Alberta/John Davids Col. D2912)

(Below) Bradley's Norseman CF-HAD while supporting exploration for Newmont Mining (a Sherritt Gordon company) around Ferguson Lake, Eskimo Point and Padley in the NWT. Reg Phillips was the pilot and John Jamieson, fishing off the float, was his helper. Jamieson later became a pilot and engineer. Along with Dick de Blicquy and Dan Kirkonnell he later ran Bradley. This Norseman had come from Cuba, where it had belonged to the dictator Batista. (Reg Phillips)

(Left) In 1954 Jeff Wyborn was in Port Harrison, Quebec with this Austin Airways Norseman. He carried in supplies and mail, then took out Eskimo TB patients for treatment in Moosonee. This was always traumatic for a Northern community—the people knew that some who got on the Norseman would not be coming home. (Geoff Wyborn)

(Below) Around 1960 Austin switched to this colour scheme. Here Norseman GSR was at Moosonee on August 31, 1965. It still was active in 1997. (Larry Milberry)

Norseman HQD of Ontario Central Airlines taxis at Kenora in June 1964, wearing the most common colour scheme at OCA. This Mark VI had been USAAF 43-5357, then NC88760. In 1954 it came to Canada for Rainy Lake Airways of Fort Frances, but was sold quickly to Warren Plummer of Flin Flon. He was busy developing a sport fishing operation on Great Bear Lake. Chukuni Airways of Kenora was the next owner (1959), then OCA (1961), Norell and Chyk of Sioux Lookout (1967), Slate Falls Airways of Sioux Lookout (1968), then Aircraft Technical Services of Kakabeka Falls. Reportedly it then was cannibalized to help in rebuilding CF-OBG. (Ed McIvor)

Parsons Airways Norseman CF-ECE had trouble one October day in 1966. Ed McIvor snapped it making a dicey landing and recalled in 1996: "ECE had broken its left float on rough water at Granville Lake while hauling fish. Ed Dulonavich flew it back to Lynn Lake. He told me later that his big fear was that the float would break off on landing and flip the Norseman. That's why he ran ashore with a high power setting." Here ECE roars landward, one float askew. Beyond are the hangars of La Ronge Aviation. Then, ECE sits ashore. The overall view shows the general setting with the Parson's base.

The Stinson Reliants were among Canada's best-liked bushplanes. CF-OAW was an SR-9 bought by the OPAS in July 1937. It was sold later to Bob Dale of Sault Ste. Marie and photographed at the Air-Dale dock in August 1963. Next it went to Ptarmigan Airways in Yellowknife. Resplendent in red, it's next shown at Red Lake on July 15, 1995, following restoration by Gerry Arnold of Winnipeg. SR-9 CF-IBR was at Toronto Island Airport in February 1961. When new in 1937 it had been an executive plane with Gulf Oil (NC49628). In March 1955 it came to Canada for Northern Skyways of Little Current, Ontario. When that company failed, IBR went to Leavens Brothers, then to Robert Page of Quebec City. Finally, Ron Hayashi, who ran a television repair company in Toronto, bought it for $6,000. In 1964 he took it seal hunting to Grindstone in the Magdalen Islands. There IBR was wrecked on the ice on April 6. (Larry Milberry)

(Left) CF-DJB (at Kenora September 6, 1961) was owned by prospector J.A. Edwards and was Canada's last working Fox Moth. After it fell into disrepair, Max Ward bought it in 1972 for $5,000 and restored it to flying condition. Wardair pilot Garth Martin was flying DJB in the CNE airshow September 5, 1976, when mechanical failure caused it to dive into Lake Ontario. Martin was OK, but engineer George Benedik was seriously hurt. Again DJB was restored, this time with parts from CF-BNO, which Allan Coggon had preserved. Its final home was the National Aviation Museum. (Larry Milberry)

(Below) Sherritt Gordon Air Transport's Husky EIR in a rare colour view from May 25, 1958. This mining company earlier had bought Husky EIO from Fairchild, but in 1950 it was wrecked by a twister. Like all who flew the Husky professionally, Sherritt Gordon's pilot Ralph Shapland thought highly of the rugged, practical bushplane. Then, Boreal's Husky EIL in a Quebec scene from about 1950. (Al Martin, Stults Col.)

(Left) Husky SAQ and Norseman GSR of Austin Airways at Moosonee August 31, 1965. GSR was active in 1997 with Stewart Lake Airways in Northwest Ontario. SAQ ended on the West Coast with Island Airlines, eventually was wrecked, then went to the CMFT. (Larry Milberry)

(Right) Pilot-photographer Grant Webb entitled this action shot "I'm outta here!" C-FEIM of Northcoast Air Services was the last working Husky. In 1979 owner Jack Anderson wrote of it to Bob Halford of the *Canadian Aircraft Operator*: "Our Super Husky has an Alvis Leonides 514-8A radial engine producing 560 bhp for take off. It swings a three-bladed de Havilland propeller with 11-foot diameter. It is presently on Edo 6500 floats and certified for 12 passengers plus pilot, giving a 6,300 pound certified gross weight. The aircraft is going strong, making money and a long way from being a museum piece." EIM later was wrecked at Prince Rupert, but Grant Webb dove to recover it. It went to the CMFT, thence to *La Fondation Aérovision Québec* at St. Hubert. In the second view EIM, this time shot by Ken Swartz, is seen near Prince Rupert on June 28, 1980.

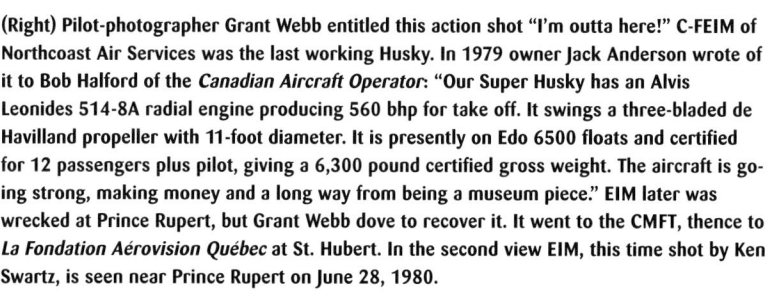

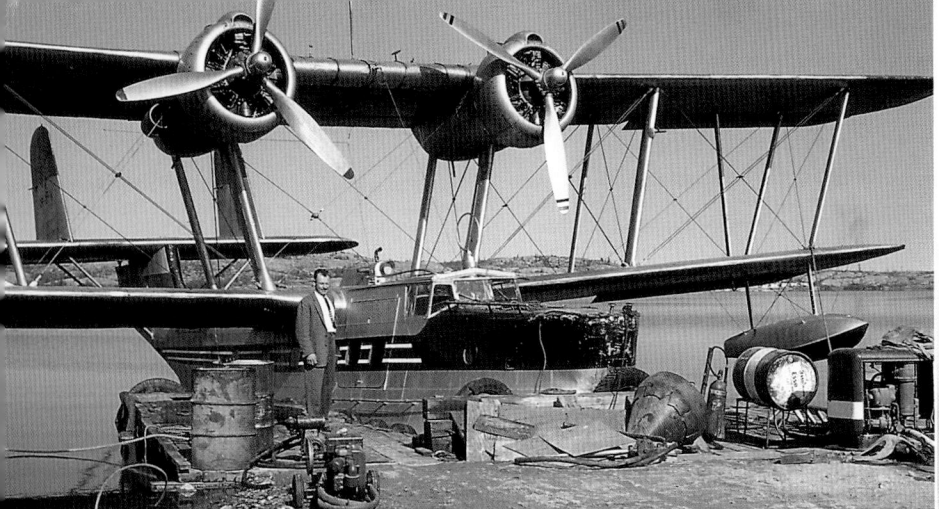

(Left) A rare colour view of PWA Stranraer BYM at Yellowknife; then (below) BXO departing there. BXO was still working for a living in 1963 and may be seen today in the RAF Museum at Hendon, UK. (Len McHale, Provincial Archives of Alberta/John Davids Col. D2685)

(Left) In the mid-1950s Don Routhier photographed Fleet 50 Freighter CF-BXP on the edge of Sandgirt Lake, Labrador, where it had piled up in June 1944. Through the efforts of M.L. "Mac" McIntyre these ruins later were salvaged for the National Aviation Museum.

(Left) Rapide DIM and Stinson 108 EDJ ready to head home on a crisp winter's morning in 1951 at Malton. Matane Air Service had just acquired DIM from Carl Millard. (Sirois Col.)

(Below) A classic view of Matane's Electra CF-TCC. For TCA's 25th anniversary in 1962, it was leased from Matane to make a nostalgic trans-Canada tour. TCC had been purchased by TCA in October 1937 for $49,585. It went to the DOT in 1940 on VIP and airways inspection duties. Surplus in December 1954, it went in trade, along with CF-BXE, to Trans Aircraft of Hamilton for three new Apaches worth $126,939. TCC then went to Matane. (Sirois Col.)

(Below) One of the scores of Cessna T-50 Cranes sold after the war by Ottawa. Most were used by small local carriers and did solid work into the mid-1950s, by when more modern types were taking over. This example (ex-RCAF 8170) went to Superior Airways in September 1945. After a few months it was sold to Leavens Brothers with whom it operated till July 11, 1958. That day it was spraying near Chute des Passes in Quebec, when it crashed doing a low turn. In this view BRK was at Toronto Island on May 16, 1957. (Al Martin)

(Left) EPA's 1947 Seabee CF-DLS in a classic Newfoundland setting. (A.J. Lewington)

(Below) While Beavers and Huskys were the glamorous postwar bushplanes, Canada also introduced the Fleet 80 Canuck. More than 200 were sold, but the market quickly became saturated and production ceased in 1946. CF-GAW, at Toronto Island Airport on April 15, 1966, was one of the Central Airways Canucks on which young students learned. Hundreds went on to careers with Canada's airlines. Another of Central's Canucks (CF-EBE) went to the National Aviation Museum. (Larry Milberry)

(Below) Beavers JGV and HJW warming up at Cache Lake, near Chibougamau on a spring morning in 1959. Arthur Fecteau had put a $5,000 deposit on JGV a few weeks before it was delivered. Reg Phillips collected it August 27, 1956 at the factory after handing a $35,000 cheque to Buck Buchanan, DHC's head of sales. (Phillips Col.)

(Left) Fecteau's Chibougamau base in the summer of 1953 with two Beavers waiting. (Routhier Col.)

(Right) When Arthur Fecteau saw Otter No. 3 on the production line in December 1952, his reaction was, "I'll take this one." Phil Garratt told him it was already sold to the OPAS. On checking, he found that the OPAS would let it go and take a later machine. ODH was in a bad prang at Lake Mistassini in April 1974, but rose from the scrap heap to soldier on. (Phillips Col.)

(Left) One of PWA's amphib Beavers. By the mid-1950s PWA had 22 Beavers, the world's largest commercial fleet. (Len McHale)

(Below) A 1957 overview of the Mid Canada Line station near Hopedale, Labrador, illustrating how sites were placed on the highest terrain. Hopedale was the most easterly in a chain that extended west to Dawson Creek, BC. (Routhier Col.)

Those who fly over the Canadian north enjoy spectacular sights. Above is Chubb Crater, the most famous feature on the Ungava Peninsula. Formed by a meteor strike, it lies about half way between Wakeham Bay on Hudson Strait and Mosquito Bay on Hudson Bay, and is near the headwaters of the Povungnituk River. (Routhier Col.)

(Below) Grand Falls, high on the Hamilton River in interior Labrador. The Hamilton drains the great Labrador plateau. Beginning in 1961 dams and generating stations were built on it to produce electricity for mining centres and for export to the US. This reduced the 300-foot falls to a trickle, so this wonderful sight no longer exists. (Routhier Col.)

(Above) Laurentian's famous Goose CF-BXR at Uplands August 15, 1970. It came to Canada in 1942 as RCAF 797 and was one of 345 built. In 1944 it joined LAS, where it served for decades till sold in BC. On October 8, 1979 it struck deadheads while taking off at Seal Cove. The pilot ran ashore and all 10 aboard got off. BXR was carried away and sank. Retrieved, as hundreds of BC coast wrecks have been, it was restored. On May 31, 1992 trouble again hit—BXR capsized in Rose Harbour. Thereafter it was sold in the US. (Larry Milberry)

(Left) MacMillan Bloedel had one of the oldest corporate flight departments, founded in 1948 on behalf of the Powell River Co. Goose IOL was put in service in 1950, followed by Goose HUZ. These were used for the next 44 years in managing over a million hectares of forest. On February 19, 1992 IOL, flown by Capt Larry Veitch and FO Dale Douglas, turned 25,000 hours. On April 3 HUZ turned 21,000. In September 1992 the department completed 70,000 accident-free hours. In the mid-1990s IOC and HUZ were sold to Pacific Coastal Airlines. (Christopher Buckley)

(Above) Otter CZP loads at Cambridge Bay. Originally with CPA, it went to PWA, then Northward (1966-73), which had taken over PWA's Edmonton-based VFR operation. CZP still was at work in the late 1990s. (Provincial Archives of Alberta/John Davids Col. D2886)

(Left) In this typical John Davids photo a PWA Beaver is ready for a flight from Inuvik in 1961. The Otter beyond is in PWA colours. (Provincial Archives of Alberta/John Davids Col. D2102)

Imports that made good bush planes in Canada were the Swiss Pilatus Porter and German Dornier Do.28. Turbo Porter CF-ZIZ (right) was at Ross River, Yukon in August 1979. Porter HB-FAL (below left) was in Toronto Harbour in August 1964 on a sales tour. Do.28 CF-WAN was at Lac à la Tortue in August 1969. (Kenneth I. Swartz, Larry Milberry, Hugh Halliday)

The Barkley-Grow company was named for its top men: Archibald S. Barkley and Harold B. Grow. Their T8P-1 first flew at Detroit in the spring of 1937. A sturdy, sleek and fast machine, it deserved greater success, but only 11 were built. Seven spent their working years in Canada, where they were a favourite. In these views by Les Corness are Associated's T8P-1 CF-BMW; and BQM in PWA colours, then, after its resurrection. BQM had served with Yukon Southern, CPA, Sioux Narrows Airways, Parsons Airways and Northland Airlines. In 1971 it was restored by a Montreal doctor, who flew it in the 1973 Burlington, Ontario Seaplane Race. Thereafter, BQM's future was uncertain. Calgarian Roy Staniland, an old Barkley-Grow pilot, heard of its plight. He organized its rescue with the help of old-time AME Jim Dick and pilot Art Bell. Looking a little battered, BQM is seen below on Chestermere Lake near Calgary on July 9, 1979 soon after its westward ferry.

(Right and below) Lockheed 10 BXE was at Gaspé in August 1966. The tough and reliable L.10 had a long history in Atlantic Canada and Quebec. BXE started in Canada in April 1943 with MCA as a replacement for its Barkley-Grow BMV. It was sold to the DOT in 1951, moved to EPA in 1955, to Trans Gaspésian Air Lines in 1958, and to Air Gaspé in 1962. Next, Carter Air's Lockheed 10 at Edmonton October 3, 1971. It recently had had a cargo door installed. One season Jim McAvoy of Yellowknife had hauled fish in HTV on the 300-mile Hottah Lakes-Hay River run. Each load was 4,600 pounds of premium whitefish. HTV came to its end in a crash while hauling fish in the NWT. (Hugh Halliday, Les Corness)

(Below) These two ex-TCA planes were at Malton in 1961. Lockheed 14 Super Electra CF-TCN recently had begun with an up-start charter company, Execaire. Then, Noranda Mines' L.18 Lodestar. Considerably faster than the DC-3, Lockheeds were favoured in the corporate world. Soon, however, they faded, replaced by turboprops like the Gulfstream and jets like the D.H.125. (Larry Milberry)

(Left) Ex-TCA Lodestar CF-TDE newly refurbished at Canadair on June 1, 1948. It soon was delivered to BA Oil of Toronto. (Canadair Ltd.)

(Right) The combination of solid airframe and unbeatable P&W engines guaranteed long life to the Lockheed airliners. Here Southern Provincial's Lodestar, an old TCA plane, visits Frobisher Bay April 29, 1958. (Les Corness)

(Below) Pacific Petroleum's Lodestar at Edmonton October 13, 1970. It had joined the company in 1955 and later served with Sterling Air Services of Red Deer. (Les Corness)

(Right and below) The life cycle of a Lodestar: CF-TDD (c/n 2246) joined the TCA fleet in October 1942. Seven years later it was sold in the US as N9949H and is seen as such at Malton on August 30, 1959. In 1965 it came to Montreal for Royalair as CF-SEQ. In 1971 it was abandoned in a vacant lot, then became a *patates frites* drive-in north of Montreal, where it is shown "under new management". (Merlin Reddy, Hugh Halliday, Larry Milberry)

This highly modified Howard 250 was one of the last Lodestars in executive service. It was refuelling at Toronto on October 31, 1975. (Larry Milberry)

(Above) The Lockheed PV-1 Ventura was popular on the Canadian civil scene—surveying with Spartan, serving corporate operators and speeding priority packages city to city. This Howard 500 conversion (at Vancouver in February 1981) was one of Kelowna Flightcraft's first trans-Canada courier planes. On July 9, 1981 it was taking off at Toronto when it stalled and crashed, killing the three aboard. It had been improperly loaded, with a dangerously aft C of G. (Gary Vincent)

(Above) Massey Ferguson's impressive Howard 500 at Malton about 1960. It began as USAAF B-34 Ventura 41-38020, went to the Cuban army, then was bought in 1951 by the Babb Co. of Newark, NJ. It next went to Dee Howard in Texas as N1489V for executive conversion and was sold to Massey Ferguson of Toronto. In 1962 it was replaced by a Gulfstream and sold to Dominion Tar and Chemicals, then to Canadian Inspection and Testing. It returned to the US in 1971. In 1996 it was listed on the USCR to Florida-based World Photography as N8GW. (Al Martin)

(Left) Starting about 1960 exec planes like the Ventura were replaced by the first generation of jets and turbo-props. Domtar's CF-DOM (Malton, March 22, 1971) was one of Canada's first D.H.125s. In 1996 it was C-GXTP of Pacific Jet Charters. Lockheed Jetstar N12R was at Toronto Island on June 4, 1966. The Jetstar and North American Saberliner used the P&WA JT12 engine. Something rarely noted is how the JT12 was designed by a team of young engineers from Canadian Pratt & Whitney. In 1983 pressure from Toronto City Council led to most jets being banished from Toronto Island. Only medevac Citations were allowed. (Larry Milberry)

(Below) Canada's first Lockheed Jetstar was CF-ETN, owned by the Eaton family of Toronto. The Eatons had made their fortune in department stores. ETN was delivered in 1962 and later went to the Department of Transport. (Al Martin)

(Left) This famous Boeing 247, which resided for decades in Calgary with the Canadian branch of a US oil concern, ended in the National Aviation Museum. Another Canadian B.247 that ended in a museum was ex-RCAF No. 7839 and ex-MCA CF-BTB. It is exhibited at the Boeing Museum of Flight in Seattle. (Les Corness)

(Left) Although Ansons and Cranes served well after WWII, their days were restricted by wooden wing spars. Within a decade most had been scrapped, but several excellent types appeared as options. One of the best turned out to be the Piper PA-23 Apache, first flown in 1952 at Lock Haven, Pennsylvania. Production terminated in January 1962 with the 2,047th plane. CF-KFX came to Canada in 1957 for Central Airways at Toronto Island Airport. Many young pilots won their multi-engine IFR tickets on it, then went on to the airlines. KFX was at Toronto Island on April 21, 1973. (Larry Milberry)

(Left) One of the last Cessna T-50s airworthy in Canada was CF-FGF. It was at Rockcliffe on June 10, 1973. FGF had been RCAF 7862 and came on the CCAR in 1948. It spent many years with Matane Air Service and in later years was in the aerial survey business in Ottawa (registered to D. Wardle). FGF eventually was acquired by the CWH and restored to RCAF markings. (Larry Milberry)

(Below) Aztec CF-RGI, at Malton May 14, 1961, was owned by a Toronto construction company. An offshoot of the Apache, the Aztec flew in 1959 and production (4,929 examples) continued to November 1981. Aztecs found a hundred and one uses in Canada over the decades and remained popular through the 1990s. Beyond in this view are CF-LEU, one of Canada's early (1959) Ce.310s; a vintage Ce.150; and International Air Freighter's North Star CF-TFC, by then retired from its Cuban adventures. (Larry Milberry)

481

(Left) No transport influenced the world as much as the DC-3. After more than 60 years hundreds still were at work at the turn of the century. Near the end of WWII Canadair refurbished many for airline, military and corporate customers. These were bought at surplus depots in the US and overseas, then ferried to Cartierville, generally looking like this example on arrival. (Canadair Ltd.)

(Below) DC-3s left Canadair like new. Imperial Oil's refurbished CF-ESO was shot at Cartierville in 1946. Its bears company titles, a practice which faded in the 1960s, when corporations began showing more sensitivity about their private affairs. (Canadair Ltd.)

(Right) CF-DJT on a fine Cartierville morning. It was recorded on early Kodak color 4 x 5 film. DJT had been a Canadair plane at first, appeared briefly in USCAN markings, then went to Maritime Central Airlines in 1950. In 1953 it became Avro Canada's VIP plane, then moved to the Robert Simpson Co. of Toronto in 1959. In 1966 it went to Sears Roebuck of Chicago as N34110, thence to a series of owners, last being heard of in Colombia. (Canadair Ltd)

A nice close-up of a newly refurbished TCA DC-3 at Cartierville in 1946. Then, an overall view at Canadair as CF-TDJ, TCA's first DC-3, taxis out. This aircraft later made its way into the National Aviation Museum in the markings of Goodyear Tire and Rubber. (Canadair Ltd.)

This brightly-coloured DC-3 rolled out at Canadair in 1946. Charter airline Transair formed in New Jersey in January 1945. It operated down the Atlantic coast as far as Havana. Col W. Deering Howe headed Transair, but his death soon after the war led to its quick demise. N88801 moved to Mexico as XB-JUX and later was in the Mexican Air Force. (Canadair Ltd.)

(Right) This DC-3 was done for Northeast Airlines. It had operated in India and the Middle East during WWII, then served NEA 1944-67. (Canadair Ltd.)

Canadair refurbished NC86597 for Eastern Air Lines. It later flew for the Aluminum Company of America as N200A. On January 24, 1980 it crashed at Bogota as HK-2214X. (Canadair Ltd.)

(Left and below) ZONDA's LV-ABX (ex-CF-DYK) and Aeropostal's LV-ADG (ex-CF-DXT). These cleanly-attired DC-3s were at Canadair in 1946 awaiting their ferry flights to Argentina. (Canadair Ltd.)

(Below) CR-LBM of the Angolan company Direccao de Exploracao dos Transportes Aereos. It survived to Angola independence in 1973, then joined the new airline TAAG as D2-LBM. (Canadair Ltd.)

(Left) Few DC-3s ever looked this clean. SE-BBO was running up prior to delivery from Cartierville in 1946. Later it served SAS, Swedish Transair and Faroe Airways of Denmark. It was scrapped at Kastrup, Denmark in 1968. (Canadair Ltd.)

(Right) HUT Dakota CF-HGD in a setting around 1950. It later flew with a host of Canadian companies, finishing with Nordair in 1968, then going to the US as N341W. (Willis McLaughlin)

(Below) One of MCA's early DC-3s. FKQ was originally a US Marine Corps R4D-1. In June 1952 Spartan was set to buy it, but it went instead to MCA later that year. It was wrecked at Moncton on April 15, 1961. (Stults Col.)

(Left) A CPA DC-3 at DEW Line Site 22 on April 13, 1957. This classic colour scheme was known across Canada from the 1940s till CPA adopted its gaudy orange scheme in the 1960s. This DC-3 served CPA 1946-60, went on to Cities Service Oil, Columbian Carbon, Syncrude, Great Northern Airways, International Jet Air, Northward Aviation, Gateway Aviation, Southern Frontier Airlines, then disappeared south of the border, where it was last heard of in the early 1990s. (Les Corness)

(Below) This PWA Dak came to Canada in 1955 as CF-INB (Associated Airways). It later flew as CF-JWP with such outfits as Gateway Aviation, Banff Oil, GNA and Northward Aviation. Les Corness shot it at Edmonton September 17, 1960 just before it departed on its sked to Prince Albert.

(Left) A Wheeler DC-3 at Frobisher Bay on June 25, 1958. HTH had served in the RAF and SAAF during the war. In 1953 it was rebuilt in South Africa from scrap. In February 1955 it joined MCA, then went to Wheeler, Canadian Aircraft Renters, EPA and Nordair. On February 6, 1973 it was severely damaged when hit by a truck at Dorval, and never flew again. (Les Corness)

(Left) Spartan's globe-spanning DC-3 CF-ICU at Frobisher Bay June 17, 1958. (Les Corness)

(Below) Eastern Canada Stevedoring's DC-3 at Frobisher September 11, 1958. ILW was an early C-47 (built in April 1942). After the war it operated in Mexico, then was bought by ECS. Among other things it was used moving marine crews to and from the Arctic. It later served with EPA, Field Aviation and Kenting before returning to the US as N16625, where it served with a parachute club and other owners into the 1980s. (Les Corness)

(Below) A detail of the famous World Wide DC-3 "Arctic Rose" on a visit to the DEW Line site at Foxe Main on June 17, 1957, and an overall view of the aircraft at Dorval on May 11, 1959 (Les Corness, Al Martin)

(Below) A lovely setting in DEW Line days. A DC-3 of Arctic Wings (Spartan's air transport division) is shown supplying a remote camp. (Dyson Webb)

(Left) This 1942-vintage C-47 spent its early years with the USAAF in the South Pacific and postwar was VH-INE in Australia. In 1956 it came to Canada with Timmins Aviation as CF-JIP. In February 1957 it was sold to Canadian Aircraft Renters of Toronto, becoming CF-CAR. In 1965 it went to South Africa as ZS-EDX. Here it was at Malton in the late 1950s. (Al Martin)

(Left) In longevity the DC-4 kept pace with the DC-3. This MCA DC-4 was off-loading construction material at an Arctic site. The DC-4 carried an 18,000-pound payload compared to the DC-3's 6,500. (Stults Col.)

(Right) DC-4 CF-PWB on a 1962 sked at Norman Wells. After serving in the USAAF, it had been with American Airlines, Eastern Air Lines and Flying Tiger Line. It joined PWA in 1957 and served till 1972, then returned to the US. (Provincial Archives of Alberta/John Davids Col. D2286)

(Left) The Carvair was a curious DC-4 conversion with a bulbous nose and hinged nose door designed for the English Channel auto transport trade. This EPA Carvair was at Summerside, PEI March 10, 1969. Three examples flew for EPA, mainly supporting iron ore and hydro developments in Labrador. (Andy Graham)

(Below) The Canso was another long-serving Canadian bush plane. Here CF-EPX refuels at Thicket Portage, Manitoba around 1950. (Harry Mochulsky)

(Above) EPA's Canso CF-CRP at Gander in June 1962. (R.F. Gaudet Col.)

(Left) Water-bombing Canso CF-PQP (ex-RCAF 11079) at Quebec City August 24, 1967. On July 18, 1987 it bounced and crashed while practising landings on Lac Caché. The pilot was thrown clear and lived; the co-pilot died. In 1994 Quebec disposed of its Cansos. PQL and PQM were taken over by Conifair, while PQK went to *La Fondation Aérovision Québec*. In May 1996 PQM was fighting fires with Buffalo Airways. (Larry Milberry)

(Right) A number of Canadian PBYs operated around the world as survey aircraft. This ex-US Navy machine was the longest-serving (more than 30 years). Here it is at Malton on May 26, 1958 soon after arriving from California to become CF-JMS. For most of its career (which included tours in Australia) it served Selco Exploration Co. of Toronto. (Al Martin)

(Below) Great Lakes Paper of the Lakehead was operating this executive PBY in 1971. It was ideal for VIP fishing trips and the side blisters were retained for sightseeing. (Gary Vincent)

(Bottom) Canso HGE of HUT at Menihek on September 10, 1953. The mishap followed a flight from Mile 253. HGE came perilously close to the fuel dump. This Canso was built by Canadian Vickers in 1943 as an OA-10A for the US Army. Postwar it served Bahamas Airways as VP-BAB. HUT purchased it for $53,000 in 1953. In April 1955 it was sold to Charlie Hoyt and Jack Scott (Trans Labrador Airlines). On March 27, 1957 George H. Moore was captain and Kenneth H. Moore was co-pilot aboard HGE flying from Lac Jeannine to Seven Islands. When the engines quit, HGE landed in the St. Lawrence. C-46 CF-FBJ quickly arrived and dropped a liferaft. Canso CF-FZJ, DC-3 CF-FST and Beaver CF-GBF monitored the scene. This impromtu SAR operation started coming together within a few minutes of Moore's Mayday call. Those aboard soon were rescued by a small boat. HGE drifted downstream, sinking about 30 miles off Shelter Bay. The engine failure was caused when one of the crew, attempting to open the nose-wheel doors, which were jammed, accidentally hit an emergency fuel shut-off valve, causing both engines to stop. (Willis McLoughlin)

(Left) The Canso proved a superb water bomber; some still were active in the late 1990s. Here one taxis at Malton April 29, 1972. Unlike earlier examples with twin 350-gallon external tanks, CF-NLF (ex-RCAF 11005) had 800-gallon internal tanks developed by Knox Hawkshaw of Field Aviation. Water bombing is tough work, with long hours and a high stress rate. W.J. "Bill" Brady bombed with Cansos for Transair. On one day in 1972 he and FO Richard Cousineau flew two sorties for eight flying hours, making 120 drops with Canso CF-GLX. (Larry Milberry)

(Below) A York arrives at Churchill to join an RAF Hastings and an Arctic Wings DC-3. The York proved a useful DEW Line freighter, although most ended in mishaps. (Dyson Webb)

(Above) A rare colour view of the last Canadian Avro York. CF-HAS was shot in the old hangar at The Pas in July 1964. Soon it was pushed outside and dumped beside the highway. Some of the local bozos soon found a great use for it—they set it alight. (Ed McIvor)

Hard working C-46 Commandos carried much of the load on the DEW Line. Above World Wide's CF-IQQ is at home at Dorval July 26, 1959. It later worked in South America and the Caribbean. Then, DAT's CF-FBJ—an overall view at Dorval on May 10, 1959 and a close-up at Foxe Main on February 11, 1958. (Larry Milberry, Al Martin, Les Corness)

Two other Arctic workhorses. The Bristol Freighter (left) proved its worth on many airlifts. Associated, Central Northern and MCA introduced it on the DEW Line, but Wardair made it famous in the region. Here CF-TFX, which Ward bought from Transair in 1958, refuels at Edmonton March 31, 1960. Years later it became a monument in Yellowknife. Then, Nordair's L.1049H Super Constellation CF-NAM at Dorval February 21, 1966. The type attained "bush plane" status with Nordair by operating in the Arctic off gravel and ice strips. Originally delivered to National Airlines in 1957, NAM came to Nordair in December 1964 for overseas charters and Arctic skeds. It went to Canair Relief in April 1969. On one flight to Uli airstrip it was damaged by a bomb. After the Biafran civil war it was abandoned in São Tomé. (Les Corness, Larry Milberry)

(Right) The P-38 was an important high-altitude photo survey plane in Canada. Its excellent performance coupled with ready availability from US surplus dealers made it a good bet. Weldy Phipps of Spartan brought in several P-38s. Here CF-HDI is seen after a landing gear failure. (Len McHale)

(Below) In July 1956 Survey Aircraft of Vancouver purchased this sharp-looking P-38L (CF-JJA, ex-USAAF 44-27205, N21765, N34992) from Fairchild Aerial Surveys of Burbank for $35,000. Fairey Aviation of Victoria modified it for surveying. Here JJA runs up at Fort William on September 3, 1961, while en route Vancouver-Toronto for Kenting to be readied for a South American contract. It went to Argentina in November 1961 as LV-HIV. While taking off at Buenos Aires one day, it had an engine failure. It crash landed back on the runway, injuring the Argentine crew. So ended this famous Lightning. (Larry Milberry)

(Left) P-38L CF-NMW at Malton in March 1964. Originally USAAF 44-53193, it was an F-5G PR version. After the war WASP ferry pilot Nadine Ramsey bought it from Kingman, Arizona as N33993. From 1951 it was with Aero Service Corp. Weldy Phipps brought it to Canada in 1961. That season he, Russ Bradley, and Bob Bolivar crewed it from Watson Lake and Yellowknife. Bob recalled that he earned $8,000 that summer and was able to pay cash for a new car. In 1965 NMW was sold to the Age of Flight Museum in Niagara Falls, Ontario. When the museum went broke, NMW returned to the US as N3005. It was owned by museums in Delaware and Pennsylvania, but in 1977 Pete Sherman of Florida acquired it. He restored it, but while flying from Florida to Oshkosh in 1978, crashed, killing himself and his wife. (Al Martin)

(Left and below) When a higher-flying photo platform was needed, Spartan switched from the P-38 to the Mosquito, buying 10 in the UK. Kenting, meanwhile, already had Mossies GKK and GKL. Here they are at North Bay and Oshawa respectively. (Dyson Webb, A.L. Humphreys)

(Left) A Spartan Mossie at an unknown Arctic site. The sand suggests Pelly Lake. (Len McHale)

(Below) Heavyweight surveyor: Spartan's Lancaster CF-IMF. It was used mainly in establishing SHORAN transmitting sites across the Arctic. (Len McHale)

(Right) A Photo Air Laurentide Beech AT-11 Kansan at La Tuque, Quebec in July 1966. A derivative of the Beech 18, the Kansan served for years on the Canadian scene. (Hugh Halliday)

490

(Above) Aerial Photography Co.'s Lodestar CF-CPA with a Kansan at Quebec City over the summer of 1959. (Routhier Col.)

(Left and below) CF-CPA where it ended in a northern Quebec bog. Another aero survey Lodestar was Photo Air Laurentide's CF-SYV (c/n 2273). It was imported in 1966 and is seen in 1973. (Routhier Col., Gary Vincent)

(Right) A Lockheed 14 in Kenting colours at Malton in September 1965. The following year it was sold in the US as N14126. (Al Martin)

(Below) Aerial ice patrol has been on-going in Canada since the 1920s, the most famous operation being the Hudson Strait Expedition. Since then ice has been tracked from aircraft and satellites on behalf of maritime transportation. In the 1960s Kenting converted DC-4s CF-KAD and KAE for government ice patrol contracts. The DC-4s were equipped with dual Doppler and astro compass systems, weather radar and closed-circuit TV. The crew included three pilots, two engineers, an electronics tech, a navigator and four ice observers (two in side blisters, one under an F-86 canopy, one spare). Here KAD fires up at Malton on May 17, 1971. It had come to Canada as CF-PWA, went to Transair as CF-TAM, then joined Kenting. 1975-78 it was freighting with Worldways of Toronto. Last heard of it was in storage in Texas in 1982. (Larry Milberry)

Field Aviation of Toronto developed an international reputation converting aircraft for surveying. This Venezuelan Beech 50 Twin Bonanza was at Field's Malton hangar November 24, 1971. Then, a German-registered Sikorsky S-58 there on April 27, 1972 to be converted for a Brazilian survey contract. When it left Field, it flew to St. Hubert, where it was converted to the PT6 TwinPac. In October it left for Brazil under German pilot Heiko Plessow. Years later D-HAGB returned to Canada to work in the heli-logging field. (Larry Milberry)

In the 1950s-60s the Department of Transport fleet wore its distinctive grays and yellows. Here are the famous Lockheed 12 CF-CCT (above), which was the regional superintendent's plane in Edmonton for many years. Les Corness shot it there in the late 1950s. The L.12 first flew in June 1936; CCT was delivered the following May. In September 1963 it joined the National Aeronautical Collection. Then, a DOT Aztec shot in 1966 by Hugh Halliday, and two Beech 18s shot by him at Malton in September 1964.

(Left) A DOT DC-3 at Edmonton on October 17, 1963. DTH had served earlier with the RCAF (KG479) and TCA (CF-TEB). In the mid-1990s it was flying with the Canadian Coast Guard. (Les Corness)

(Left and below) Two of the DOT's prime executive planes. Viscount TGP (ex-TCA) and Jetstar DTZ were at Uplands in June 1965. In the 1970s the DOT abandoned these classy colours for mundane red-and-white. (Larry Milberry)

(Below) North Star 17516 at Dakar in June 1949, while 426 Squadron was working up on the new type. This example was so fresh from Canadair that it had not yet been painted. (R.M. Edwards)

North Stars at Trenton in May 1952 during an exercise moving the Army to Goose Bay. Then, North Star 17504 loading for departure at Wake Island in February 1951. (R.M. Edwards)

(Below) As his North Star flew low, F/O Bob Edwards took an overview of the Winnipeg flood of April 1950. Then, in contrast, a photo he took while approaching Narsarssuak, Greenland in February 1956.

(Clockwise from above) Aircrew from 426 Squadron relaxing during crew rest at Shemya in August 1951: F/O John Lindgren (RO), F/O Paul Camire (RO), F/O Monty Montgomery (pilot) and F/O Jack Egan. F/O "Spoof" Logan and F/L Mickey Queale in the crew rest area of their 426 North Star en route Haneda-Ashiya in April 1951. Then, Cpl Harper (FE) and F/L Forbes Nellis (pilot) on a November 1950 flight. F/L John Watt (nav) and F/L Ed Boland (nav) pose at the entrance to the weather shack at the USAF base at Thule, Greenland during a resupply trip of March 1950. (R.M. Edwards)

Fine views of TCA North Star CF-TFN, RCAF North Star 17511 and BOAC Argonaut G-ALHD. This type put Canadair on the map as a manufacturer of fine transport planes. Fifty years later it still was a leader in the field, building the superb Challenger, CRJ and Global Express family. (Canadair Ltd.)

(Below) North Star TFE delivers passengers at Vancouver in November 1959, then (inset) runs up at Malton on a blustery day in February 1960. TFE left TCA in 1961 and spent its declining years with LEBCA of Venezuela. It was bashed up at Miami by Hurricane Cleo in August 1964, then scrapped. (via Gary Vincent, Larry Milberry)

(Left) North Star TFM arrives at Malton on an October 1959 flight. After leaving TCA it had a career working for shady weirdos, and carrying phony registrations. On October 11, 1966 it crashed in the Cameroons while smuggling guns. (Merlin Reddy)

(Right) North Star CF-TFD at Fort William in 1953. It was lost December 9, 1956 with all 62 aboard. Facing engine trouble and terrible weather, it got off track returning to Vancouver and smacked into Mount Slesse, near Hope, BC. (A.L. Humphreys)

(Right and below) North Star CF-TFC in two of its post-TCA guises: first in Cubana markings at Malton on July 13, 1961; then in those of International Air Freighters at Malton some time later. (Al Martin)

(Right) Ex-RCAF North Star CF-UXB served the Caribbean from its base in Sarasota for several years. Once a year it returned to Field Aviation in Toronto to have its certificate of airworthiness renewed. Here it was being rolled out at Field on August 4, 1968. In the often murky world of aircraft ownership, UXB at this time was registered to a company called Cavalier Aircraft of Canada. (Al Martin)

CF-TFP (above) ended in this sad state at Dorval. From here it was off to the recycler's. This scene was recorded February 19, 1966. Then (right), CF-TFG as a café in Mexico City, where it opened for business in 1965. In the early 1990s it was taken away as scrap. So ended the last of the world's civil North Stars/Argonauts. (Larry Milberry, Paul Duffy Col.)

496

(Left) CF-TDT (ex-RCAF FZ558) went through Canadair for TCA in 1945. It later served Matane Air Service and was last seen derelict in the Bahamas in 1971. It's shown at Winnipeg May 14, 1961 . (Larry Milberry)

(Below) Two fine Super Constellation shots. First, CF-TGF has just shut down at Malton on September 7, 1959. The stewardesses await as ground crew roll up the stairs. This was a typical scene looking westward from the public viewing area atop the old Malton terminal. Then, another Connie taxies in. (Merlin Reddy)

497

The Viscount was the foundation of TCA/Air Canada intercity operations in the 1950s-60s. The company couldn't have had a better plane. Here THZ was waiting at Malton in March 1965. THS was landing on R23L on July 22, 1972, while THV was boarding on April 23, 1973 (both at Malton). Although the Viscount mostly led a quiet life, there were moments of excitement, like the day Capt Harry Holland had a propeller and part of an engine fall off soon after takeoff from Malton. Then, on September 11, 1968 TIB was hijacked in Saint John, NB. Capt Ron Hollet and FO Bob Bromley were directed to fly to Havana. Instead, they went to Dorval, telling the hijacker that they needed fuel. There the incident was brought to a peaceful end. (Larry Milberry)

(Left) CF-THP was one of many Viscounts scrapped at Winnipeg. Here it was being reduced to scrap in August 1984. THP had been delivered to TCA in January 1958. (Jan Stroomenbergh)

(Left) In 1956 TCA chose the Vickers Vanguard to replace its North Stars. On July 3, 1959 Vanguard No. 2 visited Toronto on a PR tour. TCA ticket agent Al Martin, always on top of local developments, took this shot soon after the Vanguard arrived. Among the crowd were several schoolboy aviation buffs. Always keen to promote aviation, Al had alerted them the day before.

(Above) TCA Vanguard CF-TKI awaits its morning departure from Malton September 3, 1965. It served in Canada 1961-69, then went to the UK with Air Holdings to be scrapped. (Larry Milberry)

(Left) Vanguard TKR in a nice pose at Quebec City in May 1962. It served to late 1969, then had a career in the charter world with the French operator Europe Aero Service, finally being retired in 1981. (Merlin Reddy)

(Right) TKT, its mighty Rolls-Royce Tynes idling, lands at Toronto July 2, 1971. The Vanguard was developed as a 100-seater capitalizing on the 50-seat Viscount. It flew in January 1959, was ordered by BEA and TCA, but only 43 were built. With four 5,500-shp Tynes, this propliner beauty cruised at over 400 mph and offered top passenger appeal. Pilots loved it, but interest faded as operators and travellers began focusing on jets. The Vanguard, along with the exotic Electra, soon was demoted to the second-level carriers, TKT to Europe Air Service. The world's last Vanguard flight was on October 17, 1996. That day an ex-BEA aircraft was delivered to the Brooklands Museum in the UK. (Larry Milberry)

The DC-8 came to TCA in March 1960, quickly replacing the Super Constellation. Travellers now had the ultimate in air travel—luxurious, quiet cabins, high speed and long range. Here DC-8 CF-TJH waits at Edmonton International on January 7, 1964. It was sold in 1977 and was active in the 1990s as a freighter in Zaire. Then, CF-TJB landing at Prestwick in 1960. It served to 1977, then was sold for scrap in Switzerland. TCA/Air Canada operated 44 DC-8s. They logged 1.765 million hours, and 744,057 cycles. Aircraft CF-TJL (f/n 812) tallied the most hours and cycles: 55,859 and 23,354. The last Air Canada DC-8 passenger flight was AC071, Tampa-Halifax on April 24, 1983. The airline retired its last cargo DC-8 on April 1, 1994. (Les Corness, Wilf White)

(Below) Two other types of great interest to TCA were the Sud Caravelle and the Douglas DC-9. The Caravelle (inset) had interested TCA since the mid-1950s, but it was the early 1960s before it was evaluated seriously. That was the era when TCA was studying replacements for its Viscounts and Vanguards. The B.727, BAC 1-11, Caravelle, DC-9 and DH Trident all were considered. In 1963 TCA selected the DC-9, in spite of overblown lobbying by Quebec and France to oblige TCA to buy Caravelles. France's meddling head of state, Charles de Gaulle, pressured Ottawa on this score; but TCA, renowned for its world-class engineering, knew the right choice. Prime Minister Pearson allowed TCA to make its pick objectively and the choice was right. The DC-9 remained in service for more than 30 years, long after most Caravelles had gone for scrap. Here Caravelle No. 2 was on a sales promotion visit to Malton on June 21, 1957. Then, TCA's first DC-9 on its inaugural visit to Toronto on April 1, 1966. The DC-9 never wore TCA colours. Through an Act of Parliament the company had become Air Canada on June 1, 1964. (Al Martin)

(Left and below) The DC-6B remained in service with CPA to 1961, when it was edged out by the Britannia. CF-CZQ "Empress of Santiago" was at Malton on July 26, 1961. It ended with the El Salvador Air Force in 1976 numbered FAS301. CF-CZV "Empress of Lima" was at Malton in November 1961, the month it was sold to Transair Sweden. It was earning revenue in the early 1990s as ZS-MUL of the South African carrier Avia Air. (Al Martin, Merlin Reddy)

Britannia CF-CZX at Vancouver in November 1959, then in a later colour scheme. Note how the empress name changed from "Madrid" to "Santa Maria", something that was common practice. CZX served CPA 1958-65, then went to Caledonian Airways and other charter companies. It finished its days at Biggin Hill, UK, where it was scrapped in 1974. (via Gary Vincent, Al Martin)

(Above) CPA took delivery of its first DC-8 (CF-CPF) on February 22, 1961. The type was introduced on Vancouver-Honolulu March 25, 1961. On April 21 it went into trans-Canada service. Here CF-CPG "Empress of Buenos Aires" departs Dorval August 22, 1967. Capt Ralph Leslie had delivered it to Vancouver from Douglas in Santa Monica on November 15, 1961. CPG was the first jetliner to exceed Mach 1. CPG was scrapped at Opa Locka, Florida in 1981. (Larry Milberry)

CPA's one-of-a-kind Boeing 707 N791SA. It had begun as a Qantas machine in 1959. It joined CPA as "Empress of Sydney" in October 1967 from Standard Airways, but was wrecked at Vancouver February 7, 1968. Wilf White took this shot at Vancouver.

In 1968 Canadian Pacific adopted a new corporate image. For its airline it moved to this garish orange. DC-8 CF-CPT was typical—it was landing at Malton in October 1971. CPT had begun in 1966 with PanAm, then served on lease with Braniff and Seaboard and Western till joining CPA in 1967. Its last known activity was as 5N-ATY of Liberia World Airways in 1993. (Larry Milberry)

(Right) By 1960 the major world carriers were visiting Canada daily with the new jetliners. These rapidly replaced the best propliners of the time like this TWA L.1649 "Jetstream" at Gander in 1957. This example served to February 1963, then was scrapped soon afterwards in Seattle. (A.L. Humphreys)

The Lockheed L.188 Electra was one of the great propliners in the transition years between the decline of the piston airliners and the domination of the pure jets. American's N6103A was on a media visit to Malton January 16, 1960. This Electra later served in Brazil and Zaire. Eastern's Electra N5526 was at Malton on March 19, 1968. The following year it was sold in Ecuador, then was scrapped in Miami in 1975. (Al Martin, Larry Milberry)

(Below) Beginning with the RCAF's pair, the Comet was the first jetliner from abroad in Canada. Comet IVC G-APDB was at Dorval July 26, 1959. It joined BOAC in September 1958 to serve seven years. It was later with Malaysia-Singapore Airlines and Dan-Air London, and finished in the Duxford Aviation Museum. (Merlin Reddy)

(Right and below) The USSR's Tu.104 was the second foreign jetliner in Canada. This example was at Goose Bay in 1958. Then, one of the first visits to Canada of Aeroflot's awesome Tu.114. It was at Dorval August 22, 1967 during the Montreal World's Fair (Expo 67). (Frank Reddy, Larry Milberry)

(Below) One of the most graceful-looking of the early jetliners. BOAC's VC-10 G-ASGG departs Dorval August 22, 1967. It served till 1981, when it went to the RAF as ZD235. (Larry Milberry)

(Below) A Northeast Convair CV880 at Dorval May 22, 1961. Delta had introduced 880 service in May 1960. Although the 880 was faster than the DC-8 and 707, it carried no more than 110 passengers. This limited marketability and only 65 were built (37 of the larger CV990 followed). Nordair was the only Canadian operator of Convair jetliners. In 1968 it leased two 990s for the summer tourist trade between Dorval and European centres. (Larry Milberry)

(Below) Aero Mexico's DC-8-51 XA-SID on final at Malton on July 22, 1972. After its Mexican days it flew as C-FFSB in 1989 with the Canadian operator Holidair. It went into storage at Waco, Texas when Holidair folded with $5 million in debts. (Larry Milberry)

(Below) G-APFM gleams as it lands at Toronto the afternoon of March 28, 1972. One of BOAC's original 707s, it served till 1976, then returned to Boeing, probably on a trade. It later was scrapped at Kingman, Arizona. (Larry Milberry)

(Above) A Qantas 707 at Vancouver in July 1968, a few months after it joined the fleet as "City of Sydney". It flew on lease with British Caledonian in 1975, then with Ontario Worldair 1978-81 as C-GRYN and Worldways 1981-83, after which it went to the RAAF as A20-623. (J.E. Vernon)

(Left) American Airlines always was quick to put new equipment on its Toronto runs. Here 707-323C N7559A unloads cargo at Malton in January 1966. In 1977 this aircraft went to Trans Mediterranean Airways and later was destroyed at Beirut during a shoot-out between Israeli and Lebanese forces. American introduced the 727 in early 1964; this example (N1901) was at Malton February 20, 1968. It served American till 1990. American still operated 727s in 1997, by when it had the biggest share of US domestic traffic—some 20%. Its affiliation with Canadian Airlines and the general freedom under the Canada-US Open Skies agreement allowed American to solidify its cross border links. In early 1996, for example, it commenced daily New York-Vancouver and Miami-Vancouver non-stops. By then American and Canadian had fully integrated their route systems. (Al Martin)

(Left) While the jets came in, the old piston pounders were scrapped or sold. In Canada they were a bonanza for the regionals, which were crowding into the charter market on the coat-tails of Wardair (Canada's air charter trend-setter). Seen first (at Prestwick) in this series of "old props" is ex-KLM DC-6B CF-PCI. In 1963 it got Wardair into the transpolar charter business between Canada and Europe. It left Wardair in 1973. In a later guise it was C-FKCJ working for CIDA in Africa. In 1980 it was XB-BQO in Mexico. (Wilf White)

(Right) DC-7C CF-TAY at Winnipeg in June 1966. After a career with Northwest Airlines of Minneapolis, it spent 1965-70 with Transair, then knocked around the US before ending as HI-524C in the Dominican Republic in 1989. The DC-7 had gotten its start with a $40-million order for 25 from American Airlines. Its gross weight was 122,000 pounds compared to 107,000 for the DC-6B, and it carried more passengers farther. The first example flew in May 1953. The longer-range DC-7B was introduced by Pan Am about two years later, but the ultimate was the DC-7C, the first truly intercontinental airliner (6,000-mile range). Pan Am introduced it in June 1956, obliging competitors like Alitalia, BOAC and Swissair to order the luxurious new type. Overall, 338 DC-7s were built. (Andy Graham)

(Left) Originally with KLM, this PWA DC-7C was at Prestwick in the mid-1960s. It was scrapped in Texas in 1973, but not before helping PWA get established in the international charter market. (Wilf White)

(Above) Nordair made good use of hand-me-down Super Connies in the 1960s. Here NAJ, originally with National Airlines, was at Prestwick. On August 3, 1969, while with Canair relief, it was lost near Uli airstrip in Biafra. (Wilf White)

(Right) World Wide's Super Connie CF-PXX at Malton in July 1964. PXX was an old Qantas machine (VH-EAA). It later served in Argentina, where it finally was scrapped. (Al Martin)

(Left) A nice study of Otter 9411 in June 1962, while with No. 102 CU at Trenton. It was lost in a crash May 19, 1974. (Merlin Reddy)

(Below) A pristine air force Dakota at Namao on June 9, 1973. It earlier had been KP224 and eventually was sold by CADC. From 1977-93 it was C-GSCB of Skycraft. (Les Corness)

(Left) Dakota target tug MH-660 at an unknown stopover on June 14, 1958. It later was renumbered 12941 and remained on strength to 1975. It last was heard of as spray plane N64767 in Florida in the late 1980s. (Al Martin)

(Left) VIP Mitchell 5220 of 412 Squadron at Uplands in June 1958. On April 19, 1960 it crashed while making an emergency landing at Mitchell Field, Milwaukee. Those aboard died: AC J.G. Stevenson, W/C G. Kusiar, F/L D.E. Dyck, F/O R.P. Howard, Sgt L. J. Bisson and LAC N.A. Porteous. (Turbo Tarling)

(Right) The Beech Expeditor was the most ubiqutous RCAF twin of the 1950s and it served a wide variety of uses, e.g. pilot and nav trainer, light transport, VIP, SAR. This spiffy example was at Dorval on May 10, 1961. (Al Martin)

(Left) Canso 11024 was at RCAF Station Sea Island (Vancouver) on August 21, 1957. It later became Field Aviation water bomber CF-UAW. (Al Martin)

(Right) Albatross 9309 at Kinaskan Lake, northern BC for a fall SAR training exercise. Ten of the versatile Albatrosses served the RCAF 1966-70. This example was sold later to Mexico. (Maxwell Col.)

(Above) Caribou 5320 in UN colours at London (Gatwick) in September 1962. It had been CF-LVA before joining the RCAF in 1960. It was damaged at Kokoropar, Kashmir on October 27, 1965, while on UN duty with 117ATU. It was repaired using the nose of 5324 (destroyed in India in 1965 by the Pakistan Air Force). In 1971 No. 5320 was gifted to Tanzania. It later was purchased by John Woods Inc. as N1016P and remained on the USCAR in the 1990s. (Tenby Col.)

(Right) Caribou 5325 in standard RCAF domestic colours in February 1967. It served 1964-71, then went to Tanzania as 9006. (Andy Graham)

A February 1960 air-to-air of 435 Squadron Box Car 22115 over Alberta farmland. Then, a typical Arctic scene with 22127. (Evans Col.)

(Above) No. 22135 between trips at an Arctic site. The C-119 proved admirably suited for RCAF needs, whether operating at minus 30°C in the Arctic or plus 30° in the tropics. (Len McHale)

(Left) Several ex-RCAF C-119s served as water bombers with Hawkins and Powers Aviation of Greybull, Wyoming. They had J34 jets added to boost performance in the mountains. Tanker 140 is seen March 4, 1987 still in its RCAF colours. (Christopher Buckley)

(Right and below) Lancaster 10 FM104 of 107 Search and Rescue Unit (Torbay, Newfoundland) and two Lancaster 10ARs of 408 Squadron at Downsview in April 1964 for the phase-out of this type from RCAF service. (Al Martin)

(Right) Lancaster 10AR KB976 of 408 Squadron visiting Uplands from Rockcliffe in the mid-1950s. (Turbo Tarling)

(Above) North Star 17525 with unusual "Canada" markings. It was visiting an RAF Shackleton base, perhaps Kinloss or St. Mawgan. (Ralph Heard)

(Left) The RCAF's last North Stars were these ex-107 Search and Rescue Unit machines. They were at Mountain View, Ontario waiting to be sold by CADC. No. 507 ended with the El Salvador Air Force, where it was delivered in 1968. No. 510 became CF-UXB. It was scrapped at Sarasota following a landing accident in September 1971. (Al Martin)

RCAF Comet 5301 at Gander in June 1953, a few weeks after entering service. Then, 5202 awaiting the scrap man's torch at Mountain View in June 1964. (A.L. Humphreys, Al Martin)

(Below) The RCAF's first C-130B Hercules on Arctic ops with 435 Squadron in 1964. It came on strength in October 1960, then was traded to Lockheed early in 1967 in a deal that brought the C-130E to Canada. It next went to the Chilean Air Force as 1003 and was lost in the Atlantic off Cape May, New Jersey on October 16, 1982. (Evans Col.)

This sight greeted many an RCAF crew visiting Resolute Bay in postwar times. "Rez" also was where the RCAF conducted winter survival courses for many years. (Evans Col.)

The Canadair CC-106 Yukon brought an important new era to Air Transport Command. It had unprecedented speed, range and payload, and quickly replaced the North Star. Here Yukon 929 prepares to depart Marville for Gatwick and Trenton. The 10th Yukon built, it was delivered in February 1959. This once stately propliner ended its days ignominiously—cannibalized and abandoned at Kinshasa in 1978. Then, a Yukon arriving at Lahr in April 1969. (Jack Wilkinson, W.H. Meaden)

NAE North Star CF-SVP-X with Yukon 106932 at Trenton on September 22, 1973 for the 25th anniversary of Air Transport Command. SVP ended as a Caribbean drug runner. 106932, still with its luxurious VIP interior, recently had been sold by CADC from Saskatoon. It ended as HC-AZH of the Ecuadorian carrier Andes Airlines, and worked into the mid-1980s. (Larry Milberry)

(Right) The smallest of Canada's military transports in the 1950s-60s was the Cessna 182. This example was at Malton about 1960. In an all-too-Canadian move the DND purchased its US-built 182s, while Found Brothers Aviation of Toronto was desperately trying to make its initial FBA-2C sales—Found had demonstrated its rugged, equally-priced but less attractive-looking plane to the DND. Perhaps overcome by the Cessna's good looks, and the feeling of importance in buying foreign (while putting down a small Canadian company), the DND went for the 182. (Al Martin)

US military aircraft have visited Canada since earliest post-WWI days. The peak for volume was WWII. The Cold War also was busy, when the Americans controlled bases like Goose Bay and Argentia, and had major interests at others like Namao. Although photography was forbidden, some enthusiast always managed to squeeze off a few shots when nobody was looking. A.L. Humphreys of Kenting caught this SAC KC-97 tanker (right) and C-123J Provider (below) at Frobisher in the mid-1950s.

(Above) No. 54-2956 was a YC-123H from the USAF Air Research and Development Command at Wright Field, Ohio, shot at Malton by Merlin Reddy on September 12, 1959. Len McHale of Spartan photographed the USN Piasecki HUP at a sandy spot somewhere in the North.

Two other DEW Line heavy-haulers as shot by Al Martin at Malton. The C-124 was there on May 28, 1957 to collect two fire trucks for the DEW line. The C-130 (right) visited on June 26, 1958 to collect an S-55 for another Arctic site. Kenting pilot and engineer Charlie Parkin accompanied the flight. No. 55-0001 was only the 30th of more than 2,000 Hercules built. It ended its days scrapped in Viet Nam due to corrosion. (Al Martin)

(Above) This indigenous Canadian single-seat helicopter design of about 1950 attained at least the pre-flight running stage. Little is known about this project. (Al Martin)

(Above) This Sikorsky H-5 (S-51) was on SAR stand-by during Ex. Checkerboard run by Air Defence Command at RCAF Station Chatham in April 1959. CF-100s of 428 Squadron are in the background. The RCAF received seven H-5s in 1947 and four survived into retirement in 1965. No. 9601 may be seen in the National Aviation Museum at Rockcliffe. 9603 is in the American Helicopter Museum in West Chester, Pennsylvania. (Turbo Tarling)

(Right and below) The initial RCAF order for six Vertol H-21As had a value of $2.5 million. Al Martin photographed H-21A No. 9614 at Trenton on February 21, 1959. Gary Vincent shot ADC Vertol H-21B No. 9639 (equipped with floatation bags) at Mount Hope on September 7, 1969. It later was CF-GMO of Nahanni, Tundra and Transwest. It disappeared in northern BC in August 1975. Then, No. 9642 shot by Hugh Halliday in Rockcliffe in 1971. This H-21 had an unusual tasking on July 30, 1969, when it was used to tow glider CF-URE at Bagotville. The H-21 initially made its name in the RCAF as a workhorse on the 2,600-mile Mid Canada Line. With the MCL complete most RCAF H-21s and H-19s were transferred to civil contractors to support the system. The first civil contracts were let in 1957 to Okanagan for the western half of the MCL and Spartan in the east. Dominion took over from Spartan in April 1961; Autair took over from Okanagan in 1962. In 1963 the MCL used 22 helicopters—Autair in the west—14; Dominion in the east—8. That year they carried some 15,000 passengers and 4,200 tons of cargo. Advances in technology led to the closure of the MCL beginning in 1964 with the stations at Bird, Cranberry Portage, Stoney Mountain and Dawson Creek. These were replaced by longer-range Pinetree Line stations at Gypsumville, Yorkton, Dana, Alsask and Penhold. Meanwhile, Ballistic Missile Early Warning System (BMEWS) stations were opened in the early 1960s at Thule, Greenland and Clear, Alaska. Kenting Helicopters had contracts supporting Thule.

(Left) A helicopter can land almost anywhere, but this was going a little far! Of this candid picture Jim Bell of Austin Airways noted: "The one and only helicopter to land on our Canso water bomber." The pilot was John Schultz, one of Canada's early helicopter pilots. After 44 years, John still was at work in 1997, flying an S-76 air ambulance in Toronto. (J.C. Bell)

(Below) A standard Bell 47D1 visits Malton in June 1964. After more than 40 years in Canada, the Bell 47 remained active into the 1990s, some with turbine engines. (Al Martin)

(Below) Two choppers at Air Alma's base in Quebec's Saguenay region on August 6, 1975. Closest is Bell 47J CF-AYH (ex-Heli Voyageur); then Bell 47G CF-NNG. In northern Quebec such machines were in demand on mineral exploration, water, fish and wildlife surveys, hydro line patrol, medevacs, etc. At this time there were some 300 Bell 47s on the CCAR. (Larry Milberry)

(Above) This Bell 47J2 came to Canada in 1961 for Inspiration Helicopters of Burlington. Here it's at Malton in June 1964, while with Pegasus. (Al Martin)

(Left) Operators found the helicopter indispensable in out-of-the-way places. This Bell 47 was gassing up in October 1955 as a Beaver waited at a remote drill camp near the Berland River in Alberta. (Provincial Archives of Alberta/ John Davids Col. D1172)

One of the rare Sikorsky S-52s operated briefly in Canada by Genaire before being sold in Florida. GWE, built in December 1951, was at St. Catharines on August 13, 1960. (Al Martin)

Aero Service Corporation had Sikorsky S-55 N753A on the Newmont exploration project in the NWT in 1957. Here it is following damage to its electronic gear. (Phillips Col.)

(Right) The S-55 was the first large helicopter in Canada. It made its name on the Alcan project in BC and the Mid Canada Line. This one was at Dorval on February 19, 1966. It was last heard of a decade later derelict in Costa Rica, where it worked shuttling guests from the mainland and an island resort. (Larry Milberry)

(Below left and right) Sometime around 1950 a USAF C-47 was forced down. One crewman was killed and the plane was abandoned. On September 25, 1954 Hudson Bay Air Transport's Ross Lennox and Bill Beveridge flew S-55 CF-HAB to the site–a 6,600-foot plateau north of Haines Junction and east of the lower end of Kluane Lake. According to historian J.M.G. Gradidge, in 1945 this C-47 (USAAF 45-1037) was assigned to Alaska. In 1966 it was listed in Texas as N7712B. This suggests a salvage–Ross Lennox noted in 1996 that it would have been relatively easy to get 45-1037 off the plateau on sledges. (W.H. Beveridge)

S-55 JJD at Malton on May 5, 1961; then, HNG prepares for lift-off at Toronto Island on June 26, 1958. Both machines flew in the Arctic. JJD worked for a time in Greenland on a DEW Line contract. (Al Martin)

(Left) After retirement this RCN HO4S became a gate guardian at CFB Shearwater, where it was seen in January 1987. The HO4S served various roles—training, transport, ASW and SAR. (Larry Milberry)

(Below) Dominion Helicopters' Vertol 44A CF-JJX (ex-RCAF 9596) slings 45s into a Mid Canada Line site on the Great Whale River in October 1964. (Don Marsh)

(Above) The four/five-place Fairchild Hiller FH-1100, powered by an Allison 250-C18, was the first civil production US turbine helicopter (June 1966). It was designed for the US Army Light Observation Helicopter competition. When Hughes took the LOH with its OH-6 Cayuse, the commercial FH-1100 was offered. A line was set up in Hagerstown, Maryland. Certified at 2,750 pounds all-up, the FH-1100 had a useful load of 1,355 pounds. Falconbridge Nickel, Hudson Bay Mining and Smelting, Klondike Helicopters, Okanagan Helicopters and Wescan Turbo Helicopters were early Canadian operators. The hot market of the late 1960s waned, however, not to forget the impact of the new Bell 206. FH-1100 sales died by 1971. In 1980 the project was sold to Hiller Aviation in California, then to Rogerson Aircraft in Washington. In 1994 Rogerson sold its UH-12/ FH-1100 interests to Stanley Hiller, Jr. Many of the approximately 250 FH-1100s remained in service in the 1990s. CF-DAL (c/n 23), at Malton October 7, 1971, was on lease from the factory to Toronto-based Geophysical Engineering and Survey. (Larry Milberry)

(Below) The S-58 found many uses in the Canadian market from 1970 onward. All-round medium lift slinging was important, as were fighting forest fires and heli-logging. Here Kenting's C-FNHJ is readied for an engine run at Calgary on August 19, 1976. (Larry Milberry)

(Below) An early Sud-Aviation Alouette II departs Fort Chimo. Powered by a 360-ehp Turbomeca Artouste, this type was the first widely used light turbine helicopter. First flown in 1955, more than 500 were in service by the time of this September 1959 photo. With an easy-to-start turbine, they were ideal in the North. Ontario Hydro was the first Canadian operator. (Routhier Col.)

An Air and Space 18A (a.k.a. Umbaugh) demonstrator at Toronto Island April 15, 1966. It was wrecked at Smiths Falls, Ontario on September 17 the same year when a banner tow rope snagged the undercarriage on take-off. (Larry Milberry)

The Merlin Reddy Collection

Merlin Reddy introduced Larry Milberry to the world of aviation. He and the Milberrys first met in 1953, when he was boarding with them at 27 Edgewood Avenue in east Toronto. Merlin was a renaissance man, with deep interests in many areas—religion, philosophy, all the arts, everything about nature, humour, astronomy, and the other sciences, language, photography and sports. Every hour in his company was a treat. What a difference compared to being trapped with those whose only interest involved airplanes.

Born in February 1918, Merlin spent his boyhood in Quebec City, where he enjoyed the occasional aviation event, including Lindbergh's visit in April 1928. In the late 1930s he was a teacher in downtown Toronto, then joined the RCAF early in the war. He trained as a radar tech at RCAF Station Clinton and was posted to the UK, serving at RAF Acklington, Bradwell Bay, Tarrant Rushton, Wethersfield, etc. He worked on airborne intercept radar (Beaufighters), and on Gee and IFF (Stirlings and Albemarles). After the war he studied broadcasting, attended the University of Toronto and began a career as a technical writer. He retired from Technical Economists in the late 1980s.

Merlin always was an aviation buff, but after the early 1960s he took few airplane photos. While active, however, he kept various twin lens reflex cameras clicking, eventually phasing them out for Nikon SLRs. These photos mostly were taken with his twin lens using Kodak Ektachrome. Unlike the durable Kodachrome, his Ektachrome transparencies faded over the decades, but they have been revived here via computer.

Merlin passed away June 22, 1995 in York Central Hospital, Richmond Hill. His long-forgotten airplane slides were passed along to CANAV. In these two snaps, Merlin is shown in RCAF uniform, then at one of the Reddy/Milberry haunts of the 1950s—the end of Clanton Park Road, which brought one to the east fence at RCAF Station Downsview. From there it was easy to shoot taxiing planes. (Reddy Col., Larry Milberry)

Super Connie CF-TGC runs up on the TCA ramp at Malton on November 7, 1959. The Toronto Flying Club shack sits across R28. The silhouettes of several of southern Ontario's once common (now almost extinct) elm trees appear in the distance. Then, a spectacular view of TGC swooshing across Airport Road to land on R23.

(Below) North Star CF-TFS at Malton in September 1959. By this time the North Star was a refined piece of equipment, reliably serving hemispheric centres from Vancouver to St. John's, New York, Chicago, Tampa, Bermuda and Kingston, Jamaica. Soon, however, it would retire—the Vanguard loomed. TFS left TCA in December 1961 having logged 36,070 hours—more than 3,000 yearly. It had earned handsome profits over those years. It was bought by International Air Freighters and scrapped at Malton.

(Right) While landing on R23 in a rain storm on October 3, 1959 Viscount TGY hit 3,400 feet short. It bounced off a water reservoir, skidded across Airport Road and crashed through the perimeter fence. There were no fatalities, but some among the four crew (Capt Harry Bell) and 36 passengers were hurt. TGY's next departure was slightly delayed!

(Left) MCA's C-54E CF-MCI taxis for takeoff on Malton's R28 after offloading Indian monkeys needed to produce anti-polio vaccine. MCI had earlier been with the USAAF, PanAm ("Clipper Malay") and CPA (CF-CUJ). Later it flew with EPA and Nordair.

N90754 joined American Airlines' DC-6B fleet in May 1951 as "Flagship Illinois". Here it is at Malton in September 1959 as "Chicago". In 1964 it went to Ecuador as HC-AJF, then served a number of other carriers in the US and Latin America. Last noted as N11VX of Execuflite, it was apparently damaged beyond repair in 1986. Then, another American Airlines beauty—N90758 "Flagship Boston" crosses Airport Road for R23. This was one of the best spots at Malton for action photography. "Boston" ended its days in a 1971 crash in Argentina.

(Below) Super Connie PH-LKL "Desiderius Erasmus" was at Malton on November 14, 1959 delivering Phillips tape recorders. Air cargo was growing slowly—Malton had several flights per week. LKL had come new to KLM in April 1958, then served various freight carriers as N45516 after July 1962. It operated in Alaska during the North Slope oil boom of the late 1960s, and ended its days in a crash at Mesa, Arizona in May 1975. (Merlin Reddy)

(Left) While the DC-6 offered excellent appeal, carriers were moving into the prop-jet age with the Viscount and Electra. There was no comparison in speed and comfort. Merlin Reddy caught this pair of Eastern Air Lines Electras at Dorval in 1959. They had been delivered only a few months earlier. N5525 still was at work in 1997, carrying night freight for Channel Express Air Services in the UK.

(Above) Flying Fortresses in peacetime... on August 15, 1959 the Reddy-Milberry duo made an airline spotting trip to Buffalo, flying from Malton on an American Airlines DC-6 and back on a Convair 240. When they returned in the evening, their exciting day was topped off by this Fairchild Aerial Surveys B-17G-110-VE converted to a photo plane by Pacific Airmotive. It was en route to the Middle East. Kenting's hard-working B-17E CF-ICB was on Field's ramp at Malton the same year. It had come to Kenting from Florida in April 1955.

(Above) This well-kept Spartan Anson overnighted at Toronto Island on July 6, 1960.

C-46 N1442V "City of Nashville" was hauling livestock from Malton on a rainy October 17, 1959. Then, Wheeler nostalgia—CF-ILJ on a bone-chilling Christmas Day at Dorval in 1959. It served various Canadian operators 1955-66, then was sold in Hawaii as N30046.

(Left) Merlin Reddy was especially keen on corporate aviation, enjoying its unusual types and unique colour schemes. Here is one of his great 1959 views—Home Oil's Lodestar CF-EAE with a second Lodestar beyond.

(Below) Another attractive Lodestar based at Malton was Massey Ferguson's CF-TDG, a much-modified speedster that was bought from TCA in 1947 for $50,000 and converted to nine-passenger VIP configuration. Periodically TDG was upgraded, e.g. in 1955 with mods available from Oakland Aeromotive, in 1957 with PacAero mods, and in 1960 with Howard mods. By then it was about as swish a Lodestar as ever flew and was valued at $250,000. In 1965, by when Massey Ferguson had its new Gulfstream MUR, TDG, with some 18,000 hours logged, was sold to Execaire.

A pair of attractive "exec" DC-3s at Malton in November 1959. Ford Motor's N310K (above) was at Malton November 21, 1959 sporting a radar nose. Originally C-53 No. 41-20073, it served in Hawaii during the war. It was sold by Ford in 1968 and later flew with operators like Atlantic Lobster and Seafood. Then, N124H (right), which carried VIPs to and from Malton, since International Harvester had a plant in Hamilton, about 30 miles away. N124H had joined IH in 1946 as NC24H, having come from the McCarthy Oil and Gas Co. Originally c/n 12510, it was so extensively rebuilt for the corporate market that it was assigned a new constructor's number—c/n 43076. N124H was last heard of in Mexico as XC-DOE in the early 1980s.

(Left) N58092 was one of several executive Douglas B-23s (C-67 Dragons) spotted at Malton in 1959-60. General Electric's N33310 and N33311 from White Plains, NY were regulars. This B-23 showed up on September 12, 1959, the same day DC-3s 981 (RCAF), N1942 ("Sky Mole" of Perini and Sons) , N1234X (Pillsbury Mills) and 43-16167 (USAF VC-47D) visited.

Mining man M.J. Boylen's attractive Goose at home base (Malton) in August 1959. Then, General Food's Mallard N2989 there on September 7, 1959. It was sold later to the Gold Bond Stamp Co. as N298GB. On May 3, 1967 it crashed taking off at Huron, North Dakota.

The Convair and Martin liners were contemporary lookalikes and rivals. Corporate Convairs owned by the likes of Shell Oil, Gulf Oil, General Motors and Arthur Godfrey often visited Malton in the late 1950s. Here Bethlehem Steel's beautiful Convair 440 (left) taxies in over the summer of 1959. It was last heard of about 1990 as N333TN (Citrus Air). Then (below), a Martin 404 of Kawanee Oil Co. (ex-TWA "Skyliner Bethlehem") on January 10, 1960. It ended with the George T. Baker Aviation School in Miami in the 1980s. The gate under the Martin's wing was always open, and photo buffs could use it any time for ramp access. These truly were the good old days of airplane photography.

(Left) One of the most exotic corporate planes of the fifties was the modified Douglas A-26 Invader. L.B. Smith conversion N600WB was parked at the Toronto Flying Club on September 10, 1959. In the period just before the corporate jets, the A-26s and Howard 500s were the hottest things in executive air transport.

(Left) Rare Canso colour photos. No. 9830 (note JATO bottles) was at Toronto Island for the CNE airshow September 10, 1960. It was retired a year later and became Quebec government water bomber CF-PQK. Then, a spectacular corporate visitor to Malton in July 1959. N19Q was owned by the Monsanto Chemical Co. Its mods included Wright R2600s, wing-mounted weather radar, underwing punts, and a "Super Cat" tail.

DHC-4 Caribou CF-LAN-X was caught on approach to Downsview over Wilson Heights Blvd. on September 11, 1959. Many Reddy shots show evening lighting. This is easily explained, for many of his airport visits had to be made in the remaining hours of daylight after he had put in a day's work. This Caribou was a test plane for years. One important program was evaluating the characteristics of lateral wing spoilers being developed for the Buffalo. Such work was conducted with great care. Test pilot Bob Fowler recalled in 1995 how they never would "try" anything. Every stage of a flight was discussed and planned among the flight test, engineering and production people. If someone on a flight got the idea to try such and such, he would have to "hold that thought" till back on the ground, then work up his idea with the team members.

Expeditor 1510 parked on "the flying club" side at Malton on January 9, 1960. It belonged to VC-920, the RCN air reserve squadron at nearby Downsview. VC-920 operated Avengers, Expeditors and Harvards. The Beeches provided general transport and were vital in multi-engine and instrument training. It was always a treat to see them at Malton.

One of Autair's superannuated S-51s dormant at Dorval over the winter of 1959-60.

(Above) US Navy Piasecki HUP-2 Retriever 129999 from NAS Grosse Ile (Detroit) was at Windsor, Ontario September 19, 1959, taking part in a salute to the flight of the Silver Dart 50 years earlier.

(Left) Vertol H-21B No. 9641 was at Toronto Island on September 12 1959 in SAR colours. It had joined the RCAF in January 1956 with 108 Communications Unit, where it served on the Mid Canada Line. Once the line opened, resupply and support were privatized, initially with Spartan Air Services and Dominion Helicopters.

(Below) Maryland Air National Guard Albatross 51-7160 at Toronto Island on September 10, 1960.

(Above) MATS Connie 0-80611 was delivered to the USAF as a C-121A in December 1948. It flew that year on the Berlin Airlift and in the early 1950s became a VC-121A, in which capacity it was at Malton on September 11, 1959 with General Curtiss Lemay. It was retired to Davis-Monthan AFB in 1968, then was sold in 1971 to become sprayer N611AS. After periods in storage it began freighting from Miami as HI-393 with Argo SA in 1981.

(Right) This Fairchild C-123J from Harmon AFB, Newfoundland was modified with skis, underwing tanks and wingtip J44 booster jets. It was on Malton's flying club ramp December 12, 1959 as part of the operation to support an unserviceable USAF C-47B 43-16266 (inset).

The Fairchild C-123 Provider (60,000 pounds AUW) was one of the USAF's key transports in the 1950s. It evolved from the Chase YG-18A glider of 1947, which progressed to the YC-122, then YC-123. In 1954 the project went to Fairchild, which manufactured 302 C-123Bs into 1958. In 1959 the USAF Thunderbirds visited Toronto's CIAS with their F-100 Super Sabres. Their C-123B support plane was at Downsview September 11. Photo buff Larry Milberry is on the left. Then a view across the tarmac at Downsview showing the planes which the C-123 supported. It carried the ground crew who kept the F-100s and the T-33 working, their spares and their tools.

Sidekicks Reddy and Milberry spent a lot of time chasing DHC products, whether at Malton, where L-20 Beaver 58-2072 was arriving on December 5, 1959 to clear customs before delivery to the US; or at the ends of the runways at Downsview, where YAC-1 Caribou 57-3079 is seen landing on October 9, 1959. On these enjoyable occasions, Reddy used Ektachrome 120 in something like a Yashica twin lens, while Milberry shot black and white, also in 2 1/4 format.

New Products and Booming Times

The 1950s-60s were boom times in Canada's aviation industry. Much activity was generated by the Cold War and the military-industrial complex. Defence dollars were keeping the economy buoyant—when de Havilland Canada received orders, it was Washington doing most of the buying—Beavers were needed for the Korean War, and everywhere else the US and USSR were glaring at each other—from the Baltic to the Black Sea to the Far East. Next, the US Army bought the Otter. Once things got hot in Southeast Asia, it put down cash for the Caribou, then funded the Buffalo. In the same era Ottawa purchased the CL-66 Cosmopolitan and CL-44 Yukon. The latter was needed to support Canada's NATO forces, and for UN ops in Cyprus, the Congo, etc. Other projects driven by defence dollars

Four of the original Otters on the line at Downsview. Closest (No. 5) was the first for Wardair. Next is No. 4, the first Otter delivered to a customer—Hudson Bay Air Transport. (DHC)

included Canadair's sophisticated tilt-wing CL-84. Meanwhile, Avro Canada took the high road with the advanced CF-100 and CF-105 fighters, large jet engines, and a weird little "flying saucer". These were risky winner-take-all ventures. Avro lost. Its once booming Malton facility was all but empty in early 1959.

Commercial ventures in Canada's aviation industry were few, although the CL-44 sold to civil operators specializing in carrying military cargo to Southeast Asia. As to the CL-66, its Eland engines were so troublesome that it flopped completely with the airlines. Small companies like Avian Industries, Found Brothers, and Saunders, meanwhile, developed projects for the civil market; but all faced an uphill struggle against a wary and fickle market.

De Havilland Projects: The Otter

While Avro and Canadair developed glamorous projects, DHC played it low key after WWII, starting with the humble Fox Moth, the DHC-1 Chipmunk and the DHC-2 Beaver. Prudent

US Navy Otters 142425 and '426 in a pre-delivery photo of October 1956. They were delivered to VX-6 Squadron for Antarctic work. No. 142425 later served at El Centro, California with the Naval Aerospace Recovery Centre, and the National Parachute Test Range. It went into storage at Davis-Monthan AFB in 1978, was sold as N1037G the year after, then migrated to Alaska with 40 Mile Air. No. 142426 was lost with VX-6 on August 30, 1957. (DHC 2325)

DHC preferred to hedge with a diverse line that appealed to a more workaday market. In 1950, following a good start with the Beaver, it launched the follow-on DHC-3 "King Beaver". It would use a 600-hp geared R1340 engine (compared to the Beaver's 450-hp R985); have excellent short field performance, and carry more than twice a Beaver's load. On December 12, 1951, a year after project approval, George Neal flew the prototype. The program advanced quickly, mods being incorporated as dictated, e.g. the fin and rudder were enlarged to compensate for unexpected torque from the large-diameter propeller. Soon the plane was

The prototype Otter and a Beaver aloft on December 19, 1951. DYK still had its original small tail. For many years it was RCAF 3667 and was involved in many experimental projects. Later it was CF-SKX with Lambair, finally Geoterrex. On May 1, 1970 it crashed near Dunrobin west of Ottawa. It had been testing mods for carrying geophysical equipment. While in a high-speed dive it disintegrated, killing the pilot. (DHC KBX16)

Otter No. 2 (DHC's performance and systems test prototype). Here it is in flight over Toronto Island Airport. (DHC/Best Col.)

renamed "Otter" and orders started coming in. Certification came in November 1952; the first delivery was made that month to Hudson Bay Mining and Smelting. Others soon followed, most going to historic operators of DHC products—Imperial Oil, the OPAS, RCAF, RCMP, Wardair, etc. In June 1954 Widerøe of Norway became the first foreign customer. The US Army began taking delivery of the first of some 200 U-1 Otters early in 1955.

DHC built 466 Otters, the last (CF-VQD) going in May 1967 to Laurentian Air Services. The Otter followed the Beaver to all continents—they became symbols of Canadian technology and Canada's ability to deliver products that served important human needs. Canada's largest Otter operator was the RCAF/

CF, which had 69 over the years. CF Otter 9408 went to the National Aviation Museum in September 1983. When the careers of the Beaver and Otter came to an end in the US military, dozens of those machines were repatriated in the 1970s-80s. Others arrived from more distant places, including several Indian Air Force Otters that reached Canada in the early 1990s to be refurbished for bush operators.

In the 1970s there was interest in converting the Otter to turbine power. The first example (1972) was N3904 (c/n 54) in which a Garrett TPE331 was tried, but that project faded. Next, CF-MES was modified. While with Gateway Aviation it had crashed at Cambridge Bay in August 1973. Salvage specialist Ray Cox of Cox Air Resources in Edmonton recovered the

wreck, restored it and installed a PT6A-27. As such it flew in September 1978. Cox soon moved his project to Washington state, where little more was heard of it. Some years later another PT6 installation was offered by Washington-based Vasar. More than 20 of these conversions were at work by 1997.

In 1983 an Otter upgrade was offered by AirTech of Peterborough, Ontario. It used the Polish PZL piston engine, first of 600, then of 1,000 hp. Raecom Air of Yellowknife tried the 600-hp PZL in Otter CZP. The first engine failed, but was replaced by AirTech; then CZP threw a propeller. Raecom lost interest, as did Labrador Airways. It had converted several Otters, then deconverted them after disappointing results. The 1,000-hp PZL, however, suc-

Otter Potpourri

(Right) Otter CF-RWU, seen on tundra tires at Resolute Bay, had gone new to Weldy Phipps' Atlas Aviation. It was lost in a crash at Browne Island, NWT July 17, 1971. (Halford Col.)

(Below) Qantas and the Australian military took 12 Otters. VH-EAW, delivered in 1958, was written off in an accident in New Guinea on August 14, 1961. (Peter R. Keating)

(Above and left) CF-EYO (c/n16) served as DHC's demonstrator from 1953, then went to EPA in 1960. In the second view (with Newfoundland Labrador Air Transport of Goose Bay) it's shown after being re-engined with a small (600 hp) Polish PZL engine. The PZL Otter was first offered in 1983 by AirTech of Peterborough, a company formed in 1977 by Bogdan Wolski. The 600-hp PZL, it was claimed, boosted take-off performance, lowered gas consumption, and had a longer TBO. Experience showed this conversion to be foolhardy, and the 600-hp conversions returned to P&WA power. A 1000-hp PZL proved more viable, but difficult to sell. (NAC/McNulty Col. PA191695)

(Right) The first Otter delivered to a customer was CF-GBX (c/n 4), which went to Hudson Bay Air Transport in November 1952. Here it is in its yellow, blue and white colour scheme at Jeff Lake in the Yukon, where pilot W.H. "Bill" Beveridge was supporting HBAT's S-55 at a drill camp. An entry in his log for June 6, 1958 notes "CF-GBX completely destroyed by fire." In 1996 he explained what happened: "GBX was being unloaded by a driller at Channing Lake near Flin Flon, while I sat on the shore catching up on my paperwork. During unloading, a roll of copper screen inadvertently was pulled over the posts of a six-volt battery. That ignited some spilled fuel and GBX went up." Later, a salvager fished one wing and the rudder out of the lake. In time a new Otter bearing the registration CF-GBX appeared. This was unusual, since such projects require the rebuilder to own the original aircraft data plate. That, however, was in Beveridge's possession—he had taken it as a souvenir after the fire. The second view (below) shows the counterfeit GBX at Pickle Lake on August 22, 1979 in a red, white and black scheme. On May 24, 1980 it crashed near Carling Lake, Ontario (between Sioux Lookout and Armstrong) and again was destroyed by fire. The eight aboard escaped with some injuries. (Larry Milberry)

(Right) CF-HXY was the 67th Otter and served DHC 1955-61, when it was sold to EPA. Because of some test work being done, it carried an "X" for "experimental" on the tail. In the 1990s HXY was operating with Lac Seul Airways of Ear Falls, Ontario. (DHC)

CF-IGM, the original Otter in aero-magnetic survey work, was at Toronto Island in 1960 while with Rio Canadian Exploration. Later it flew for Survair, Canadian Aero Service and Spartan, then became a bushplane with Norcanair. (Larry Milberry)

Shell Canada operated Otters JFH and KLC on exploration. Here is one of them in a mid-1950s scene with a leased Okanagan Bell 47. KLC went to Omineca Air Services in 1967 and to TPA in 1970. (Shell Canada 2993)

(Below) Shell Canada's Otter JFH, a Beaver and Spartan Anson GSB (ex-RCAF 11993) at Peace River December 1, 1956. JFH later went to Laurentian Air Services and was destroyed at Fort McPherson, NWT on July 13, 1966. A January 20, 1965 note in the old DOT files for GSB states: "We do not intend to reactivate this aircraft and it is only a matter of time until we turn it over to the RCAF Fire Fighting School at Uplands." (Provincial Archives of Alberta/John Davids Col. D171)

(Right) These Labrador Air Safari planes, shot November 18, 1992, were stored at Baie Comeau for the winter. After a US Army career (1956-74) C-FBEU operated in Quebec, where it survived at least two accidents. The forestry town of Baie Comeau was a significant aviation hub from the time of its founding in 1937. (Larry Milberry)

Over the winter of 1970-71 Austin Airways leased QEI from Bradley. Here it's seen between Moosonee and Fort Albany. Originally in the Norwegian AF, QEI came to Bradley in 1970, then went west in 1977 for Island Airlines, then Air BC. (Roland Brandt)

US Army Otters that survived their years in uniform invariably had later careers. This one was photographed at Malton, while clearing customs before delivery in 1960. 59-2210 came to Laurentian Air Services in August 1974 as C-GLAB. (Larry Milberry)

(Left) They'll try anything. In this US Army trial, Otter 55-2350 was an aerial tanker refuelling an H-21. (Halford Col.)

(Below) RCAF Otter No. 3671 was caught by Gordon McNulty crossing the finish line at Burlington, Ontario on September 15, 1973 during the Great Burlington Centennial Air Race. It was crewed by Maj Ron Pierce, Capt Pat Rieschi and WO Ron Wylie of the Toronto air reserve wing. M.L. "Mac" McIntyre shot 9415 after a minor "incident" at Buttonville, Ontario on October 22, 1966.

The Cox Air Resources PT6-powered Otter at Abbotts-ford in August 1981. Originally CF-MES, it had begun in 1961 with McMurray Air Services of Uranium City. After a crash at Cambridge Bay in 1973 it was resurrected and became the first turboprop Otter conversion. (Larry Milberry)

DHC in 1958 in a northward view with the main plant in the centre. The cafeteria is the small building just below the line of parked Otters, while the flight test hangar is to the right. Across the field is the wartime Mosquito plant, used at this time by DHC and 436 Squadron (C-119s). In the smaller view, taken four years earlier, there still were open fields right up to the plant gate, but they soon were gobbled up for parking. (DHC)

Key DHC men from the company's booming 1950s-60s era: P.C. Garratt–chairman and managing director, C.H. "Punch" Dickins–vice-president of sales, and Russ Bannock–vice-president military sales. (DHC 17010-12-63)

ceeded—about 15 Otters had adopted it by 1997. The turbo Otter, however, was more widely accepted in spite of being more costly than the PZL. By the turn of the century, in spite of being 50 years old, the Otter still held an important place in air transportation, especially in Canada, Alaska, Oregon and Washington.

Enter the Caribou

While the Otter was making its mark, DHC was eyeing a most ambitious concept—a twin-engine STOL in the DC-3 category. Through 1956 Russ Bannock, who kept in touch with the US Army, was receiving encouragement in Washington about the need for such a plane. The go-ahead came in January 1957 and the prototype, designated DHC-4 Caribou, flew July 30, 1958 with George Neal, Dave Fairbanks (pilots) and Hans Brinkman (engineer). Several pre-production machines soon were flying, mainly in US Army colours. The program advanced, but on February 24, 1959 Caribou 57-3079, operating as CF-LKI-X, was lost northeast of Toronto. Pilots George Neal and Walter Gadzos escaped by parachute. The elevator had failed due to uncontrollable flutter in a high speed dive. Mods were added to make sure that such an accident never recurred. On October 8, 1959 the US Army accepted its first three Caribous (US Army designation: AC-2). New orders came in, letting DHC

finally breathe easily. In the end 164 Caribous went to the US Army, becoming a mainstay in the war in Southeast Asia. The Caribou could use short, unprepared strips; get troops quickly to (or from) a hot spot, and lift 3 $^1/_2$ tons. Its rear ramp was ideal, especially when LOLEX (low level extraction) was required. With LOLEX, as the pilot flew a few feet over his drop zone, the loadmaster would deploy

palletized cargo by parachute extraction off the rear ramp.

In a change of policy at the Pentagon the Army was obliged to transfer its larger fixed-wing aircraft to the USAF in 1967. The Army's AC-2 became the USAF's C-7 Caribou. The Caribou proved a valuable asset in the Southeast Asia theatre till war's end in 1975. Thereafter, most returned to the US to serve in a va-

(Above) The prototype Caribou at Downsview on August 8, 1958 before the nose was lengthened by 45". George Neal, Dave Fairbanks and Hans Brinkman first flew KTK on July 10, 1958. On an early test flight with pilots Neal and Fowler, KTK had a runaway propeller at nearly maximum design speed (270 K). It went to 4,000 rpm in the ensuing dive. The plane vibrated so badly that it shook the paint off cockpit panels. It later was found to have stretched propeller blades and one engine was so badly damaged that it was scrapped. (Inset) An artist's rendition of the proposed DHC-4 from early 1957. (DHC)

(Left) KTK on March 23, 1959, by when a full paint scheme had been applied. In July 1960 KTK became RCAF No. 5303. It went to Tanzania in 1971 as JT9011. After that career it was bought by John Woods Ltd. as N1016N. (Paul J. Regan)

(Left) The second prototype Caribou in a portrait from November 1958. DHC sold LAN in 1971 as N6080. As such its work was mainly with CIA-related carriers like Air America. It was last noted with the Environmental Institute of Michigan in 1977. Many Caribous sold to foreign militaries turned up on the civil market in the 1980s-90s. (DHC 9044)

riety of important roles. The Caribou proved useful in other militaries. It served from Africa to Australia, India, Malaysia, and the Persian Gulf. The RCAF operated nine Caribous 1960-71, using them on UN operations in areas like the Middle East and India-Pakistan. Surviving aircraft were transferred to Tanzania. In time many ex-military Caribous were retrieved by brokers and converted to civil use. In Canada they found work on projects like the James Bay hydro development, on mineral exploration, and on DEW Line cleanup. The Caribou had been DHC's biggest gamble—had the Americans not pursued it, it would have ruined DHC. By the turn of the century, however, the tough and versatile transport still had years of work ahead.

A typical US Army AC-1 Caribou at Malton about 1960. This was the seventh example and is seen in the red-and-white used on cold weather trials. No. 57-3082 later was the jump plane for the Army's Golden Knights parachute team. (John Kerr)

(Left) Caribou CF-OYE (c/n 40) during a 1965 demo flight at Camp Borden with Turbo Beaver CF-PSM. OYE went to Shell Oil for exploration work in northern Alberta. It later served in Ecuador and Oman, but returned to Canada in 1978 as C-GVGU of Quebec's La Sarre Aviation. (Howard Levy)

(Right) Caribou CF-LVA (c/n 9) in a publicity photo from November 1959. It recently had returned from a 50,000-mile world sales tour covering 40 countries. In August 1960 LVA joined the RCAF with tail number 5320. It was damaged on UN duty in October 1968 and repaired using the nose of 5324 (destroyed in India in 1965 by the Pakistan Air Force). In 1971 it was gifted to Tanzania. It later was purchased by John Woods Inc. as N1016P and still was listed on the USCAR in 1990. (DHC)

(Left) GE T64 Caribou test bed RCAF 5303 (c/n 1) first flew September 22, 1961 with Bob Fowler in command. He later noted that it would do the Caribou's design dive speed (270 Kts) in level flight. (DHC 13917)

(Above) A typical classroom scene at de Havilland in October 1959. Bill Kavanagh was instructing a US Army group in the workings of the Caribou undercarriage. (DHC 11304)

(Left) A DHC/RCAF group pictured at course end in May 1964. Behind are Jim Gilmore, Bob Irving, Bert Beasley, Russ Burnham, Bill Calder, Bill Kavanagh, Ben Nugent, Gerald Healey and Ben Cox. In front are Donna Shirley, Charlene MacCormack, LAC Charles Jebb, Cpl Paul Dagley, Cpl Glen Nowe, Cpl Lawrence Burt, FSgt Howard Wilson, FSgt Frank Johnson, Sgt Revel Lewis, Sgt Malcolm Doherty, Cpl Don Pollen, LAC Oscar Ikert, Denise Hall and Sharon Thomson. Caribou 5325 (c/n 115) was new at the time. It was gifted to Tanzania by Canada in 1971, but years later cropped up on the USCAR as N3262Y. (DHC 17886)

Keeping The Caribou Flying

R.W. "Dick" Gleasure was born in 1913 in Tralee, Ireland, where he grew up on a farm. After his father passed away, Dick apprenticed at the Ford Motor Co., but wages were scant and he sought better opportunities. This led to the RAF, which was looking for recruits in view of Germany's growing militancy. Gleasure joined in 1937, but his mechanical know-how from Ford didn't interest the RAF. It wanted recruits who knew nothing about mechanics, so they could be taught from a clean slate. Gleasure felt lucky to be kept on to train as an airframe fitter. His first posting was to No. 98 Squadron (Fairey Battles), which he accompanied to France. When France fell in May 1940, many from his squadron were lost in the bombing of HMS *Lancastria* at St. Nazaire on

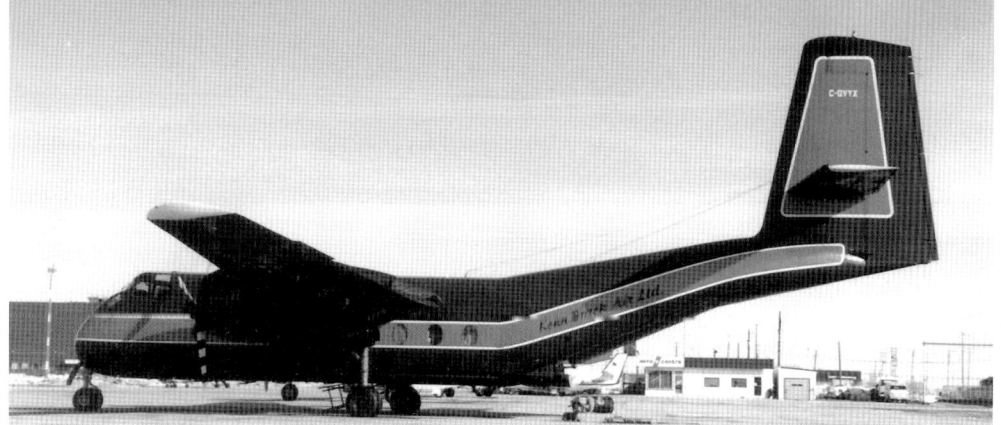

(Left) Caribou VYX at Winnipeg in March 1987. Later that year it crashed in the Yukon. (Jan Stroomenbergh)

(Below) The first of several Caribous for Ghana. After years in West Africa G401 was sold as N90567, then went to the Indian Air Force in 1979 as M2169. (Peter R. Keating)

(Below) RCAF Caribou 5321 (c/n 10, ex-CF-LWM) served on UN duty in places like Yemen and Kashmir. In 1971 it went to Tanzania as JW9013/5-HA, then (1982) became N1016S of Jim Woods Inc. It ended its days derelict in Djibouti after landing heavily and breaking its back. Its only use now was to shelter airport workers—a kind of flop house with wings. This photo dates to November 30, 1991. (Larry Milberry)

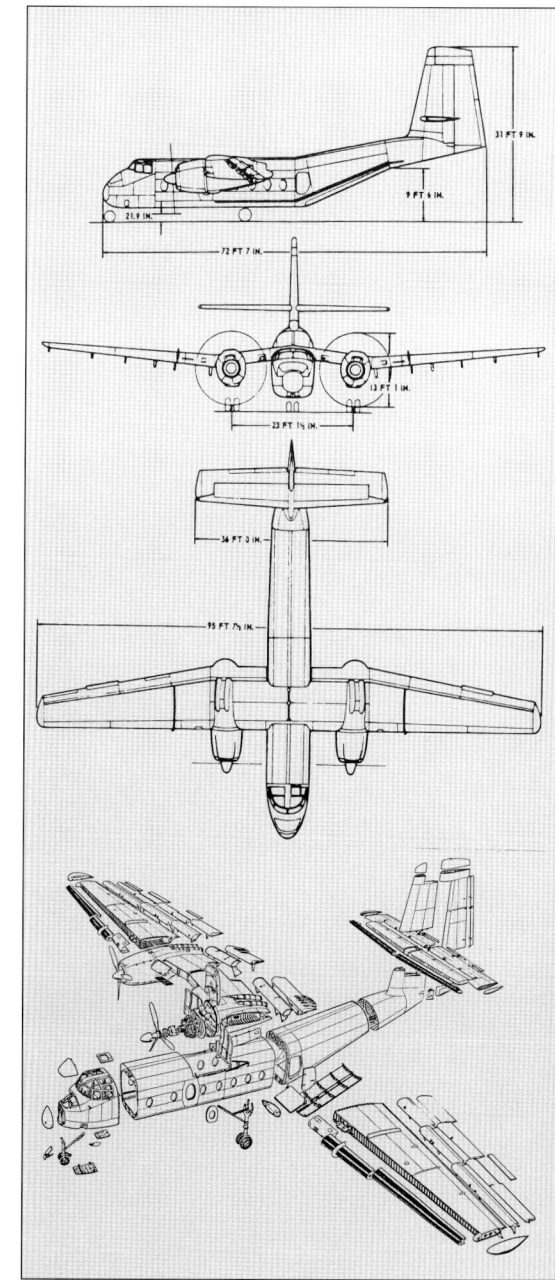

June 17, 1940; but Gleasure escaped by returning to England by air. For the rest of the war he was in North Africa and Europe in maintenance, overhaul, and the recovery of downed aircraft. In Italy he was with No. 322 Wing (Spitfires), which operated alongside a USAAF P-47 wing. This was a favourite period for Gleasure, who found that the Americans ran a first-class operation. (Years later, when his career at DHC put him back to work with the Americans, Gleasure could not have been happier.)

After Spitfires, Gleasure went to Algiers to work on a damaged B-24, then returned to the UK in an Avro York and was discharged. What he wanted now was to emigrate to Canada. After much red tape he caught a Pan American DC-4 for New York via Gander, then took the train to Montreal, arriving on June 6, 1946. From there he made his way to Toronto, where his sister lived. Initially he worked in an auto garage, but in 1950 was hired by DHC as a mechanic. He worked on Vampires, North Stars, Norsemen, etc., then was promoted to MRP (mobile repair party) duties. This took him to stations like St. Hubert, Chatham and Bagotville mainly on Vampires. He also worked on 412 Squadron's Comets in Ottawa. When he passed his engineer's exams, Gleasure became a DHC tech rep. Meanwhile, he also was serving with 411 (Aux) Squadron at Downsview, which flew Vampires and Harvards.

The US Army defined the role of a manufacturer's tech rep as someone providing "on site proficiency training, technical guidance

and assistance in the evaluation of unusual field problems", with the aim of elevating "the technical skills and abilities of AF personnel... to a level of self sufficiency." Gleasure's first tech rep posting was to the US Army's 202nd Aviation Company in Verona, Italy. He arrived in April 1957 to instruct tech staff, and assist in repair, overhaul and modification to Beavers and Otters. He returned to Downsview in July 1959, then went to Malmstrom AFB, Montana to help the USAF develop a crosswind landing gear for its Beavers.

Now Gleasure's career sent him all over the world. He spent from January to October 1964 training the US Army on the Caribou. This took him first to Fort Orde, California (17th Aviation Company) for field trials in the Mojave Desert. There a number of problems such as leaking fuel cells appeared. Gleasure noted: "These activities were ideal for training... The desert manoeuvres brought a great deal of experience in the new 'air mobile' concept, and everyone worked hard at it. The Army pilots adapted easily to the Caribou, but anyone who didn't pull his weight was soon posted out." Next, Gleasure moved to Fort Benning, Georgia and Fort Gordon, South Carolina. The Army now formed a Caribou unit for duty in Viet Nam—the 92nd Aviation Company, and Gleasure was invited to join. He jumped at the chance and flew to Saigon on a commercial flight to meet the 92nd ferrying its first 12 aircraft from the US to Qui Nhon, Viet Nam. This was the beginning of an exciting, demanding period. Gleasure was to serve from

Three-views and a basic exploded view of the Caribou. (DHC)

Dick Gleasure on duty in Viet Nam in the 1960s; then, at home in the Toronto suburb of Weston in 1994. The office was well-equipped and constantly used as Dick kept his archives up-to-date. Most of the photos in this section are from his files.

Tech rep scenes in Italy. U-1 Otters, an H-34 and an L-19 on exercise at Brunico in July 1957. Then, Dick Gleasure on an engine change at Bosco Mantico. Otter 55-3324 later returned to Canada as CF-FSU with Labrador Airways. On September 4, 1979 it collided at Squaw Lake, Quebec with Cessna 180 C-FHWR as they were landing on converging courses. The left float of FSU struck HWR. FSU flipped when it touched down. There were no injuries on HWR, but five of the eight aboard FSU drowned. In 1981 FSU was sold as N9895B to Tyee Airlines of Ketchikan, Alaska.

October 1964 to November 1965 (at Qui Nhon), then from September 1966 to July 1968 (An Khe). Crawford Byers, Jack Loriaux, Jack McGrogan and Bill Tyrell were other DHC tech reps in Viet Nam in this period.

In October 1964 Gleasure was on the beach when the 11th Air Assault Division came ashore at Qui Nhon, a port about half way between Cam Ranh Bay to the south and Da Nang to the north. Here he met some of the troops he had trained a few months earlier. No sooner was the 11th ashore than the boys were loaded aboard Caribous of the 92nd (part of the 14th Aviation Battalion) and ferried a short distance inland to An Khe, a fortified base. A big task for the Caribou, working with CH-47 and CH-54 helicopters, was to lift equipment from Qui Nhon. As An Khe developed, the Caribous were joined by C-123s and C-130s.

The airfield at An Khe could be unbearably dusty, or a muddy quagmire. In the rainy season it was difficult starting the Caribous, but a solution was found—injecting compressed nitrogen onto damp magnetos (prior to start) evaporated moisture and starts were again smooth. As to daily life, Dick Gleasure recalled, "We lived in tents, and drew water for washing from a tank truck. To shower we used 45-gallon drums braced by wooden beams up on trees. There was a makeshift pull string to dowse you when you were ready. Some of us also used the Song Ba River for washing." The isolation and the absence of conveniences was eased by occasional visits by celebrities like Bob Hope, Phyllis Diller and Billy Graham.

Caribou 63-9733 during training at Tonapah in the California desert in June 1964. Intensive evaluation, then full-scale exercises gave the Army the knowledge and experience it needed to make a battlefield success of Caribou ops in Viet Nam.

Crews of the 92nd soon got used to flying at least 100 hours a month, and surviving under enemy fire while going in and out of strips barely 1,000 feet long. One Caribou came home one day with 21 bullet holes. On another occasion a single bullet hole had to be patched—that bullet had killed a crewman. The 92nd's Caribou did LOLEX (low level extraction, known later as LAPES, low altitude parachute extraction system) and paradrops. Aircraft routinely logged 30-40 cycles a day. Gleasure noted that this wear and tear easily was handled by the Caribou: "Max gross loads was the criterion of the day. Low flying in poor weather into jungle strips was a constant challenge." When crashes occurred, they further demonstrated the Caribou's ruggedness, and men often walked away unhurt. Post-crash fires were rare. Daily ops were described in a nutshell by Gleasure: "Aircraft fly from dawn to dusk, usually 0500 to 1700 hours. On return to base, maintenance begins. There is very little flying at night unless in emergency. Not-

withstanding their long hours, the crews enjoy it."

The 92nd flew daily milk runs throughout the Central Highlands. One stop was the town of Dalat, a place described by Gleasure: "The climate and the beautiful surroundings with the right combination of mountains and trees, with an abundance of brilliant flowers, made Dalat a paradise." Daily runs operated to and from Saigon (250 miles south) carrying mail, personnel and supplies. On one occasion Gleasure went on an MRP to Dalat to change an engine on Caribou 63-9725: "With tools and a quick-change kit we were flown there by helicopter, removed and replaced the engine and were back in Qui Nhon that evening. The Caribou returned to Qui Nhon after a flight test." While the main base was at Qui Nhon, where heavy maintenance was done by the 51st Aviation Detachment, the 92nd had detachments at Danang, Pleiku and Nha Trang. One detachment carried 90,000 lb in 24 hours during a big push to rout Viet Cong from around Bong Son and Plaeme.

Crew chief reported that this was a continuing problem without refilling the reservoir...

61-2396 (1,484:55 hours): This aircraft was damaged on the ground [at Pleiku on February 6, 1965] in a VC mortar attack. Charges were placed between the undercarriage struts and inner wheels and tires...

63-9726 (720 hours): A heavy landing reported on this aircraft. Inspection of the nose leg indicated excessive oil leaking from the strut... As there are no piston seals at this echelon, the leg was replaced by a new one...

In 1967 Dick Gleasure investigated an unusual mishap at Ha Thanh, a Special Forces camp near the DMZ (demilitarized zone). On August 3 Caribou 62-4161 was landing with 2.4 tons of 155mm artillery shells. While at 200 feet and about 1,000 feet from the threshold, it was struck in the tail by an outgoing 155-mm round—it lost its tail and crashed fatally. A drawback to smooth operations was the annual rotation of personnel. As Dick Gleasure explained:

(Above) The maintenance crew runs up a Caribou at Qui Nhon. The engines are Pratt & Whitney Aircraft R2000s of 1,450 hp; the propellers, Hamilton Standards. R2000s for new Caribous were supplied by the US Army to Pratt & Whitney Canada in Longueuil, Quebec for conversion and shipment to DHC.

(Left) An Australian AF Caribou loads livestock at Qui Nhon for delivery to outlying camps that had no refrigeration. Livestock were kept at these camps until needed, then were slaughtered.

(Below) Caribou 62-4161 in its final moments after being hit by 155mm artillery fire at Ha Thanh; then, the aftermath.

An L-20 Beaver coming in to Qui Nhon in May 1965. Beavers and Otters did important work throughout Southeast Asia, from supply and leaflet dropping, to transporting casualties, mail and VIPs.

By this time the 17th Aviation Company arrived from Fort Orde to replace the 92nd.

The job of a tech rep included every imaginable challenge, from simple repairs, to recovering a badly smashed plane, to investigating an accident where nobody had survived. Every piece of work had to be recorded, and excerpts from Gleasure's Viet Nam notes for February 1965 list typical cases:

61-2395 (with 725 hours flown): Pilots complained of aircraft falling off on one wing at the stall. The cause of this problem is due to components removed and reinstalled, particularly flaps and ailerons... It is possible that these components were not checked for correct rigging after installation.

60-5438 (1,763:55 hours): No. 1 engine reported having fluctuating RPM and would not stabilize at cruise... It was reported also that the oil was overflowing from the vent line.

(Above) Caribous at Cam Ranh Bay in February 1967. The air force recently had taken over the Caribou from the army.

(Right) The R2000 engine test stand used at Cam Ranh Bay.

May 1968 the 483rd logged 10,885 hours. Aircraft were 94.5% operationally ready (compared to the USAF Pacific Air Force requirement of 75%). Sortie effectiveness was rated at 99.3%. For each Caribou hour, 3-4 maintenance hours were required (compared to 13 for the C-123, more for the C-130). Average time between overhauls for Caribou engines at this time was 800 hours.

On January 11, 1968 Caribou 63-9730 ran off the strip at Quang Tra Bong while taking off. Next day a recovery team travelled from Cam Ranh Bay to Quang Nhi (near Quang Tra Bong), but was delayed as napalm drops were being made on suspected VC positions. Eventually, the team had to fly from Quang Nhi to overnight at Phu Cat until things settled. On January 13 a Caribou returned the recovery team to Quang Tra Bong. A ramp had to be dug down to the aircraft for dragging it onto the runway, then, as Dick Gleasure wrote: "Using

A year in theatre, regardless of experience, was not enough. After a year skills were only becoming apparent. Our trained mechanics then returned to the US and technical and on-the-job training (OJT) started all over again for the newcomers.

The Caribou, because of its unique design, dual flaps, flap/aileron configuration, and flap/horizontal stabilizer combination, was time-consuming to teach before all was properly understood. Correct rigging of the combination control surfaces was absolutely essential. Mechanics had to be continually involved in the chores and maintenance. With repair and recovery, where components and assemblies were broken down and dismantled, rigging of the aircraft could get out of whack. Therefore, constant training and supervision were essential. This also applied to the hydraulic system, main landing gear, and nose gear door drag strut mechanism. Engine controls and the design of the streamlined cowlings were other sophisticated areas. These were all covered in the maintenance manuals. However, manuals rarely are consulted in an operational theatre until there are serious problems. Then the panic begins, resulting in a poor job one way or the other.

Steel roofing being laid down as the recovery team prepares to haul 9730 from the ditch where it had ended on January 11, 1968. DHC tech rep Dick Gleasure is on the right.

(Right) Dick Gleasure with a crowd of little "helpers" at Quang Tra Bong.

For his second Viet Nam tour Dick Gleasure joined the 17th at An Khe, 20 minutes flying time from Qui Nhon. An Khe was home to the 1st Air Cavalry and the 11th Air Assault divisions. The place was simply described by Gleasure: "Everyone was housed in tents. There was no running water, and rats and snakes were plentiful." On January 1, 1967 the USAF took over Caribou operations from the Army. The 483rd Tactical Airlift Wing at Cam Ranh Bay became the overall Caribou unit. It had six squadrons with 90 to 96 Caribous, two squadrons each at Cam Ranh Bay, Vung Tan and Phu Cat. This contrasted with the Army's way of dispersing Caribous throughout the countryside in small detachments. That system had given great service to the "end users", but was a nightmare when it came to aircraft servicing. With the USAF, any level of repair/over-

haul could be done at Cam Ranh Bay, including engine build-up. There was even a special mobile stand designed by DHC for testing R2000s. The product support given by DHC in Viet Nam was superb. Gleasure recalled: "The extended duration of the war proved a tough, exacting testing ground for the equipment involved. The Caribou met this test and passed with honours. De Havilland engineering design, combined with excellent maintenance management and corrosion control, enhanced the reputation of the Caribou. In Viet Nam it is a household word."

The 483rd TAW amassed impressive statistics as noted in a report of August 1967: January 1-July 30: 90,000 sorties, 600,000 passengers, 100,000 tons, and 50,000 flying hours. In

air bags we raised the nose so we could build a suitable support to hold the fuselage in a safe position, while completing repairs to the nose gear and forward fuselage." Beams (4 x 10s) were used to raise the fuselage and the work continued in stages till dusk. Then the men retired inside the defended compound to spend the night in uncomfortable bunkers. The camp CO advised that there were VC within a

Sampling of Caribou Mishaps in Viet Nam			
Aircraft	Date	Location	Details
61-2396	6-2-65	Pleiku	Sabotage (Viet Cong infiltration)
63-9732	"	"	"
63-9724	"	"	"
63-9732	2-6-65	Ba To	Power approach, too low, downdraft
62-4148	28-7-65	Ta Bat	Failed to lift off, wet & muddy
63-9751	4-10-66	An Khe	Crashed on radar controlled approach
62-4178	15-3-67	Litts	Landed short
60-1405	28-10-66	An Khe	Hit mountain top
63-9759	18-4-67	Dong Trei	Failed to take off, muddy, C of G
63-9747	12-6-67	Ban Hin Dong	Landed short
62-4161	3-8-67	Ha Thanh	Shot down on approach
63-9718	21-8-67	Dinh Quan	Hit ground on low level ops
63-9752	4-67	Vung Tau	Recovery by helicopter
63-9730	1-68	Quang Tra Bong	Failed to get off in muddy conditions
62-4182	23-2-68	Long Hai	Failed to get off in downwind departure necessitated by enemy fire
62-4176	7-5-68	Camp Evans	Failed to get off
60-5439	10-5-68	Phu Cat	Severely damaged by rocket attack. *Also 60-5442, 61-2391, 62-4155, 62-4190, 63-9728, 63-9731, 63-9758
61-2384	15-5-68	Ben Thuey	Damaged by helicopter collision
62-5430	12-66	Dak Pek	Landed short
62-4189	29-5-68	Dak Pek	Landed short
62-4192	26-6-68	Chiang Klang	Hit load of gravel left on runway
61-2399	25-4-68	Vung Tau	Direct hit by rocket
61-2391	3-68	Phu Cat	Nose gear collapse

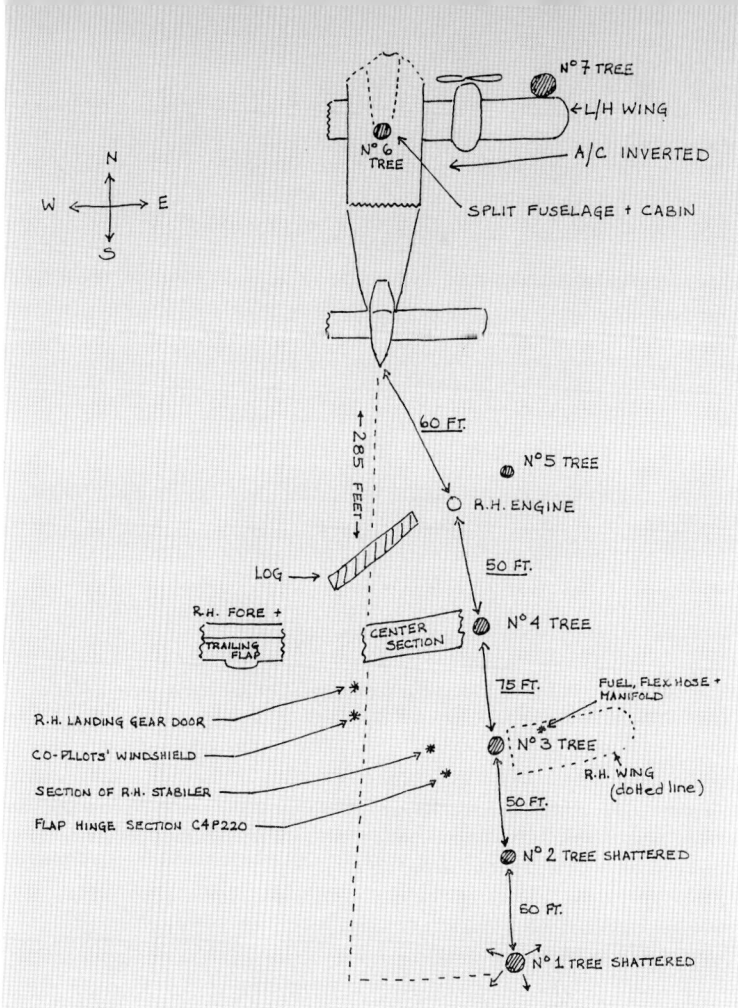

Dick Gleasure's on-site sketch showing the path of Caribou 63-9751.

kilometer. Next day the repairs continued, including changing No. 2 propeller. Gleasure had to go to Cam Ranh Bay for parts. On January 15 he returned by helicopter and the heavy repair work continued. Earth had to be dynamited from under the aircraft, and this was expertly done. South Viet Namese troops at the base proved poor helpers, while the local tribesmen were excellent. More parts, such as a wing leading edge, kept arriving, and repairs went on nonstop. Late on the 19th the landing gear was functioning and the Caribou was ready to haul up the ramp. By this time the crew was worn out with long days of work and miserable nights in the bunkers. On the 20th the Caribou was towed onto the runway and the engines run up. It rained on the 20th and 21st, so the airstrip was a mess. Next day a spare wing leading edge arrived, was fitted, and at 1530 the Caribou got safely airborne for Phu Cat. The recovery crew happily left Quang Tra Bong next day, another routine job completed.

In the case of 60-1405, it was on a daily milk run to An Khe from the 57th Aviation Company at Vung Tau. The approach was in heavy rain. Suddenly the aircraft hit a grassy slope and slid to a stop on its belly. It was badly damaged, but only one of the 26 aboard was lost. Dick Gleasure immediately visited the site and reported: "Searching around the area of the crash site, we spotted wrecks in close proximity, some dating to the days of the French." The crash of 62-4178 at Litts was apparently caused when the aircraft, flaring at 70 knots, hit turbulence from a departing CH-47. The runway was 1,300 x 90 feet, so there was little room to manoeuvre once the right wing dropped suddenly. Nobody was seriously hurt, but the Caribou was wrecked. No. 63-9759 crashed at Doing Trei in an over-gross-weight and forward C of G configuration. The 1,800-

foot strip was muddy, and ambient temperature high in zero wind. The pilot took off using 1,200 feet when he had the option of using more. He piled up in the local village killing several on the ground. No. 63-9747 crashed without injuries at Ban Hin Dong, Thailand when the main gear clipped the end of its 1,531-foot strip.

Caribou 63-9718 crashed on a LOLEX when the port wing struck the ground. LOLEX trips were never a piece of cake—the pilot had to hold the airplane a few feet off at nearly stall speed as parachutes wrenched loads from the aircraft. The demise of 61-2387 at Benh Thuey south of Saigon occurred at the end of a milk run from Vung Tau. About 12 miles out, No. 2 engine quit. The pilot quickly noted his fuel near zero; No. 1 then cut out. The cargo of C-rations was jettisoned and a successful crash landing was made in a rice paddy. Investigation showed that the fuel cross-feed manifold for No. 2 engine had been hit by ground fire. This led to a rapid fuel loss as the pilot, unaware of damage, switched to cross feed when No. 2 died. For the crash of 63-9751 that killed 12 of 32 aboard on October 4, 1966 no certain cause could be found. Dick Gleasure recalled the tragedy years later:

An Khe is located half way to Pleiku, inland from Quin Hon in the Vietnamese highlands along Route 17—the Street without Joy. Overlooking An Khe is Hong Kong Mountain, 2,360 feet high and covered with heavy jungle—tall mahogany trees, elephant grasses and thorny-type vines with poisonous needles. About 300 feet down from the summit a helicopter pad had been blasted and levelled. It was here that the crash occurred.

At 1630 hours everyone on the base heard

the loud roar of aircraft engines as a Caribou approached, then, abrupt silence. This was serious. Eye witnesses stated that the Caribou approached in cloudy conditions, then turned right into the heavily-wooded mountain. I was working with the maintenance crews at the time. At first we could not see where the Caribou had crashed, but a sharp-eyed fellow pointed to an area showing a slight gap in the trees with smoke wafting through it.

A rescue was organized immediately. Crews were assembled. Even though the mountain was within the camp area, I was kept from going, since the area had been infiltrated by the Viet Cong. The following morning at 0700 Major Anderson (commander of the 17th Aviation Company) and I, with a crew of two, started out on foot to the crash site to see about recovering the aircraft. We arrived several hours later totally exhausted and suffering the scourges of the thorn-infested slopes. At the site we found that several bodies had not been removed.

We investigated the site, making notes about the point of initial impact, and sketches of all the parts scattered around. Official pictures were taken to accompany the accident report. The day was dull and cloudy, as it was monsoon season. The humidity was excessive. I kept my M1 rifle and camera on a fallen log within easy reach. Normally the jungle is full of sounds of all kinds—the chattering of birds, the sounds of animals, the rustling of trees. Snakes were around, but rarely bothersome. All this was taken for granted this day.

We were busy checking all aspects of the accident when, all of a sudden, there was total silence, not the sound of a bird, nor the usual

(Above) The demise of 63-9718, which had a double engine failure during LOLEX ops near Dinh Quan.

(Right) Caribou 63-9752 being recovered at Vung Tau southeast of Saigon, by CH-54 Tarhe 57-8460. The CH-54 excelled as a heavy lifter in Viet Nam. Some remained in Army service into the 1990s. Many surplus CH-54s later found work in heli logging and fire fighting.

(Below) Two views of Caribou 63-9747, which cracked up on the strip at Ban Hin Dong on June 10, 1967. The aerial view shows the typical nature of strips in the boonies of Viet Nam.

Caribou 62-4182 shows damage sustained at Long Hai, then some of the temporary repairs made by the field party. Dick Gleasure, Sgt Gibb and Sgt Donaldson pose with their handiwork. Finally, 62-4182 back at Cam Ranh Bay, where it received a new nose. This Caribou served till 1983, later becoming N60NC of NewCal.

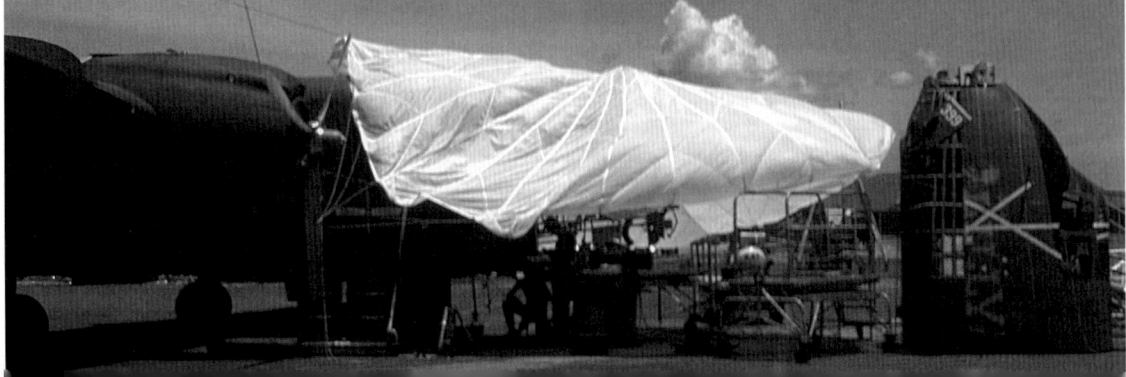

What the Caribou did best was carry troops and their gear. Here members of the 1st Air Cavalry in Viet Nam prepare to board. The Caribou was the biggest of the US Army's fixed-wing aircraft: wing span 95' 7¹/₂", length 72' 7", height 31' 9". Empty weight was 17,630 pounds, maximum takeoff weight was 28,500.

The USAF had an IRAN (inspect and repair as necessary) corrosion control facility in Manila. War-weary Caribous cycled through, emerging "like new". Dick Gleasure wrote: "This entailed complete dismantling of the aircraft including engines, control surfaces and treating control cables. Hydraulic components were removed, bench checked and re-installed. The airframe interior and exterior were stripped to the bare metal and repainted. The area beneath the floor was carefully inspected for corrosion, the worst of which had been caused by carrying live animals."

rustling. Someone said, "The VC are coming." I was petrified and looked around for my M1. It was a hundred feet away. I was about to make a dash for it, when we all were startled by trees rustling and a sudden movement in the tallest branches. A huge colony of monkeys was swinging at top speed from branch to branch. Within minutes they had disappeared, while we remained speechless and relieved. The normal sounds of the jungle resumed.

When Dick Gleasure's time came to leave Viet Nam, his superiors expressed their appreciation. In July 1968 LCol William P. Gilmartin of the 483rd TCW wrote to Russ Bannock:

I am taking this opportunity to say thanks to Dick Gleasure for his excellent service to the Air Force and the 483rd Troop Carrier Wing. I am completing my tour in Southeast Asia and the fine service that Dick has given, I feel, should be recognized. Military personal we can recognize through the efficient rating, commendation award, etc. The technical representative we can only recognize for a job well done by advising their company.

Dick Gleasure has more than fulfilled his duties as a technical representative. He has become a friend of airmen, flight crews and staff personnel, and his service is welcomed and truly respected. He is considered the authority on the C-7A Caribou. Airmen and officers of all ranks seek him out for information and I know, as Chief of Maintenance for the wing, that I have relied heavily on his knowledge and advice. As you no doubt know, when we accepted the Caribou from the US Army, we were expected to support the combat mission without any loss of effectiveness. This period of transition was exceedingly difficult and Dick's advice and timely guidance played no small part in the fine record that we have now attained. The Air Force has a need for people

Caribou 63-9729 at Hong Kong where it had taken troops for a weekend of R&R. In November 1972 it joined the South Vietnam Air Force.

Occasionally the troops in the back country got to Saigon for R&R. Dick Gleasure took this photo of rush hour there in April 1968.

Caribous 63-9724, 61-2396 and 63-9728 were damaged at Pleiku by the Viet Cong, who infiltrated the camp in February 1965. No. '724 was destroyed June 9, 1970 when dropped from a helicopter during recovery. Also seen here is Beaver 54-1726.

A Caribou ready to roll on the airstrip at Qui Nhon in March 1965.

like Dick and I truly hope that his services continue with the Air Force Caribou program for a long period of time.

Twenty Caribous were lost to enemy action or accident in Viet Nam. Beginning in 1971 most returned to the US; but 40 were left behind for the South Vietnamese. By late 1972 South Viet Nam had three Caribou squadrons.

Several old Caribous wended their ways back to Canada including C-GVYX (c/n 292). It had gone new to Guyana Airways, but later returned to Canada for Propair, then Air Inuit, Wardair, Kelowna Flightcraft and Air North. It came to a miserable end on November 10, 1987, while staging from Calgary to Ross River in the Yukon to haul ore concentrate. Aboard were two pilots, two maintenance men and 5,500 pounds of equipment. They refuelled at Fort Nelson, then continued to the mining town of Faro. There the weather was duff so VYX proceeded to its alternate, Watson Lake. En route it lost the right engine, so the captain decided to land at Ross River. On approach, warning lights indicated a faulty landing gear. On the overshoot VYX crashed on the bank of the Pelly River, killing the pilots. Cause of the accident involved such things as shaky maintenance practices and the use of unapproved parts.

Keeping the Caribou Alive

Over the years, New Jersey-based NewCal Aviation corralled dozens of Caribous. It refurbished many for resale, but also invested in a project to re-engine one with PT6A-67R engines. NewCal realized that there was always a market for an affordable freighter in this class, but there was a worsening problem—the old "round" engines were ever more costly to maintain and it was not always easy to obtain avgas. Jet fuel, however, was everywhere. Working from Gimli, NewCal completed a prototype PT6 Caribou (DHC-4T N400NC); it flew November 16, 1991. Basic weight for the "T" was 16,500 pounds compared to 17,630 for the standard DHC-4. The major weight saving came from the much lighter engines. Although take-off weight remained at 28,500 pounds, payload rose. The program advanced

N400NC was NewCal's PT6 proof-of-concept DHC-4T Caribou. It crashed at Gimli before the project could be completed. (Jan Stroomenbergh)

in spurts as time and money permitted. By mid-1992 N400NC was proceeding towards supplementary type certification. It had logged about 100 test hours by August 27. Just after takeoff at Gimli that day it nosed into the ground. All aboard died: pilots Mike Quirk and Perry Niforos of NewCal, and Winnipeg engineer Gordon Hagel. The crash was blamed on the elevator gust locks being left in place, making control impossible once airborne. The New-Cal project now was taken over by R. "Goby" Gobalian of Penn Turbo Aviation in Cape May, New Jersey. His prototype (N600NC, c/n 37) made its first flight August 22, 1996 with pilots Al Baker of Calgary and Stan Kereliuk of Ottawa. Gobalien estimated that as many as 200 good Caribou airframes remained of the 307 built. His vision was to convert many to the PT6, producing an excellent cargo plane with a five-ton payload compared to three and a half for the DC-3.

By 1990 only 14 US military Caribous remained active, the largest unit being the California National Guard at Fresno. The Caribou soon left the US military, most surplus examples going to Spain. In 1995 the Spanish Air Force still had 15. Many companies, such as Alenia of Italy with its G222, were vying to replace these veterans.

By the mid-1990s Caribou remained on the go with the Australian, Indian and Spanish militaries. Others worked on commercial jobs such as hauling relief in East Africa, or working on DEW Line clean-up in the Arctic (in the summer of 1992 Aklak Air of Inuvik leased N95NC from NewCal for three months of DEW Line flying). Others did clandestine flying with the CIA during the Contra-Sandanista

conflict in Central America and (probably some of the same machines) worked in the illicit drug trade in the same region.

In its normal life cycle an airplane's presence slowly recedes—10,000 DC-3s of 1945 became 3,000 by 1970, 1,000 by the turn of the century. As the years passed, the price of a DC-3 dropped. The same went for the Caribou. In the July 1993 *Trade-a-Plane* five ex-Union Flights Caribous were offered for sale as a package with spares, tools and ground support equipment. The advertisement noted, "Excellent maintenance records and history. Will help train your flight and maintenance personnel. Will deliver anywhere in the world. Will operate on your behalf." The asking price was US$1.7 million for the package:

c/n	Original Reg'n	Current Reg'n	Airframe Hours
98	62-4160	US Army N800NC	13,604
119	62-4178	US Army N5463	12,034
126	62-4184	US Army N700NC	13,231
212		N900NC	6,617
241	N544Y CIA	N544Y	13,231

US Army Caribous in 1990

Serial	c/n	National Guard
60-3767	18	California
60-5430	20	Missouri
61-2389	46	Mississippi
61-2392	50	Missouri
62-4146	81	Puerto Rico
62-4149	85	Connecticut
62-4188	130	Connecticut
63-9718	149	California
63-9732	172	Alabama
63-9737	182	Mississippi
63-9747	201	Puerto Rico
63-9757	220	Utah
63-9760	224	Alabama
63-9765	232	California

Development of the PT6

No sooner had DHC launched the Otter than its marketing men began musing about a light, twin-engine bush plane. This was an old concept in Canada dating to the days of the Vanessa, Vancouver, Sekani and Fleet 50, none of which succeeded commercially. The main problem was that, by the time the airframe designers finished their work, the planes were too heavy. There never seemed to be an engine with the required margin of safety and economy. Thus, when DHC considered twin R1340s, the engineers determined that its payload wouldn't be much better than a single Otter. Besides, such a plane fully loaded likely

wouldn't maintain level flight on one engine. If bigger engines were used, extra weight and fuel requirements brought the designers back to problem No. 1—no payload. For the early postwar years, then, there would be no viable small twin for bush operators. Everything depended on the state of engine technology.

For airframers a light now appeared at the end of the tunnel. The first small turboprop engines were beginning to appear. France, for example, announced the 400-shp Artouste, which flew successfully in the Alouette helicopter. It had lots of power, was easy to maintain, and was light. Its big drawback was high fuel consumption. In Canada there also was interest—CP&W was discussing a project. In 1958, following experience designing the JT12 turbojet, and after analyzing industry requirement, it formulated specs for a 450-shp, free turbine-type engine designated DS-10. A sales engineering report from K.H. Sullivan and E.L. Smith of CP&W (Report No. KHS & ELS-490R of March 3, 1959) indicated some company thinking, and hinted at possible markets:

DH is actively studying Beaver and Otter replacement aircraft powered by turboprop engines. To date they have visualized a single-engine Beaver replacement and a twinned Otter replacement. The latter would be a true STOL and, as such, could obtain DRB support. However, DH believes the commercial market potential to be much greater for the Beaver replacement. Their studies do not seem to have progressed too far. They are still fairly vague as to power requirements and ultimate aircraft size and payload. The only engines studied to date for these applications appear to be the [Lycoming] T53 and the [DH] Gnome (T58). Flat rating of the engine at sea level and 100°F was mentioned as a desirable feature. In the coupled version of the PT6 this can be comfortably accomplished at about 800 eshp. DH expects to complete aircraft studies and select a design within two months. Production aircraft would be available in four years. Availability of prototype engines in 30 months would be quite acceptable to their planning.

Although such a DHC project does not seem to have gone much farther, the PT6 was approved by P&WA in Hartford. It would be centred in Longueuil and be financed by CP&W with profits from its lucrative piston engine parts manufacturing. Development would be aided by a small cash infusion ($1.2 million) from Ottawa. The government could see potential in the project—the DS-10 was no Orenda Iroquois (the engine designed for the

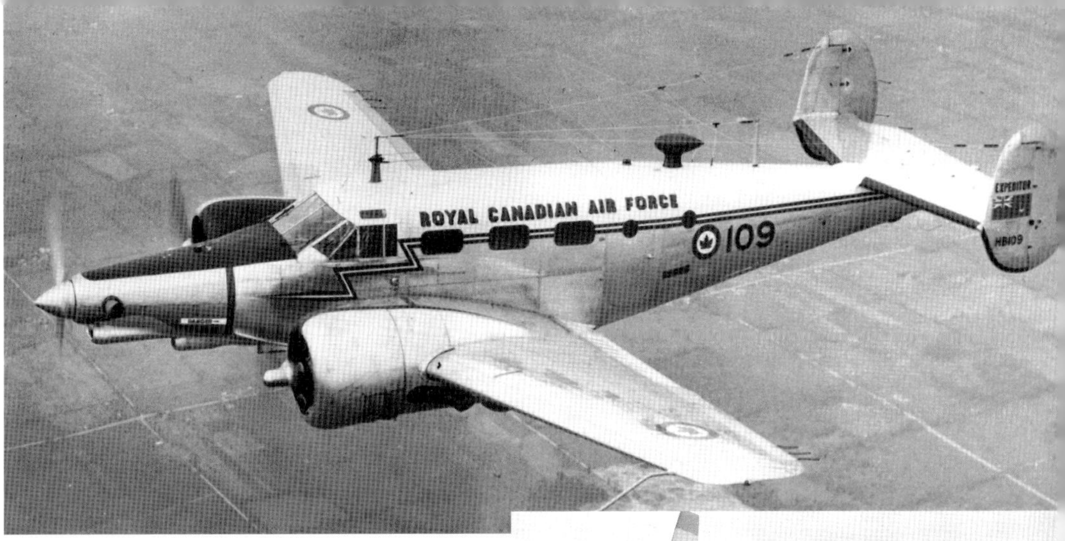

(Above) RCAF Expeditor HB109 was the original PT6 flying test bed. It first flew in this configuration in May 1961. (P&WC EP580)

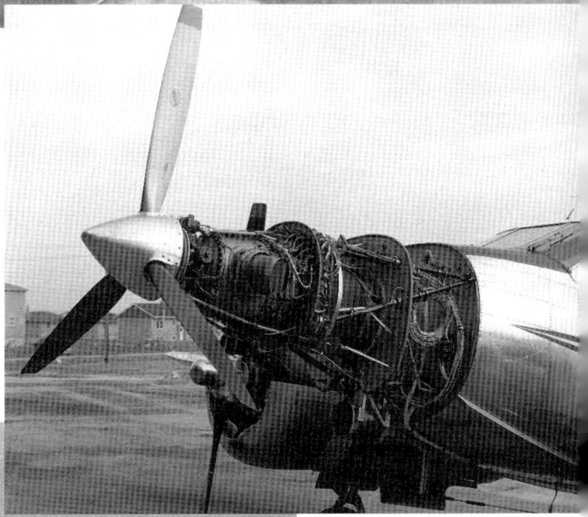

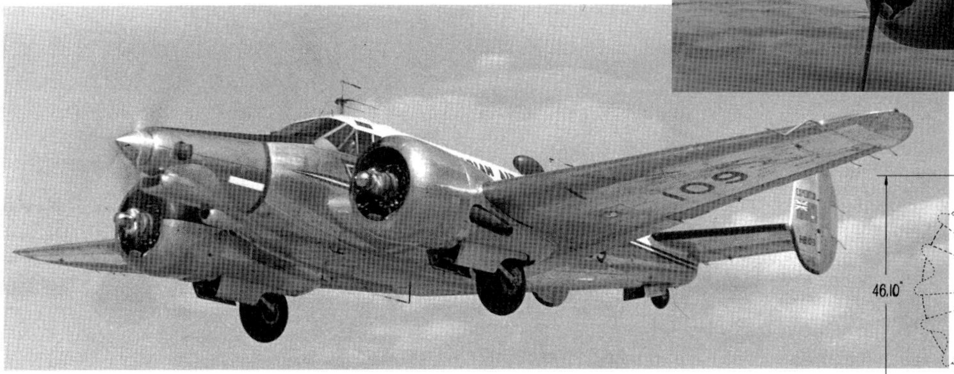

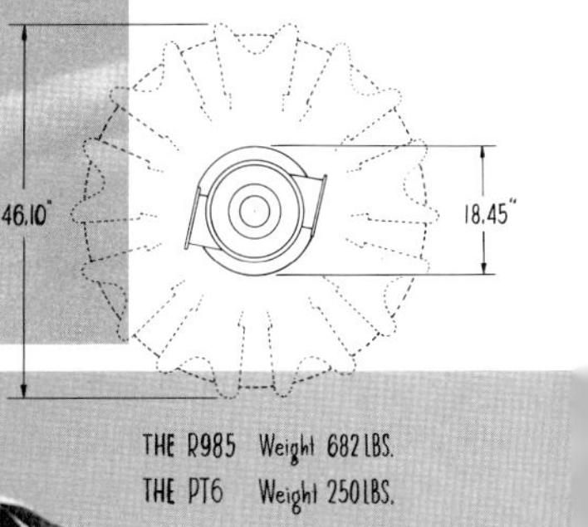

Further views of the test bed Beech 18, and a size comparison between the PT6 and its popular predecessor, the R985 Wasp Junior. While delivering about the same power, the PT6 was much smaller. Such features soon endeared airframers to the little engine. Beech HB109 later was P&WC's CF-ZWY-X. Before its retirement to *l'Ecole aéronautique* at St. Hubert it had logged 1,068 flying hours on 719 test flights. (DHC, P&WC)

THE R985 Weight 682 LBS.
THE PT6 Weight 250 LBS.

46.10"   18.45"

The first King Air (c/n LJ-1, later N925B and N26CH) takes to the air at Wichita on its first flight on January 20, 1964; then in a nice in-flight portrait. In 1997 LJ-1 was in Salina, Kansas—not airworthy, but with restoration a prospect. The proof-of-concept NU-8F (c/n LG-1) was in the Army Aviation Museum at Fort Rucker awaiting restoration. (Raytheon Aircraft BL17410A52, BL17410M)

The Little Engine that Revolutionized Aviation

(Right) The Hiller Ten99 was the first single-engine aircraft to get airborne with PT6 power. It was developed to meet a USMC requirement for an assault support helicopter, but lost to the Bell UH-1, even though the UH-1 was outside the weight limits specified in the bid. The sole Ten99 (N3776G) survives in the Hiller Northern California Aviation Museum in San Carlos, near San Francisco. It was opened in 1996 by 72-year-old helicopter pioneer Stanley Hiller, Jr. He flew his first design, the XH-44, on July 4, 1944. His company built helicopters till forced out of business in 1966. This followed a controversial competition for a light helicopter for the US Army which Bell won with the OH-58. (CANAV Col.)

(Left and below) The Lockheed XH-51 and Model 286 (two of each built) were exotic rotary-wing craft. Between 1962-1968 they logged more than 1,000 PT6 flying hours. No. 1262 is seen in flight. Then, Larry Milberry examines No. 1263 which also carried a P&W JT12 turbojet. It was used for experimental trials relating to an advanced attack helicopter project. This resulted in the Lockheed AH-56A Cheyenne, but it lost the US Army competition to the AH-64 Apache. Both XH-51s went to the US Army Aviation Museum at Fort Rucker. The model 286 were lost in a hangar fire in 1988. (P&WC, Mike Valenti)

(Above) The Piasecki 16H Pathfinder flew with a PT6 in March 1962. The experimental machine reached 173 mph in August 1964, but the was developed no further. (P&WC)

(Right) The Kaman K-1125 was the first helicopter powered by twin PT6s. It logged 70:25 hours before being abandoned. Although such machines never reached production, they provided P&WC with a great deal of PT6 experience. (P&WC)

Avro Arrow at horrendous cost). Unlike the Iroquois the DS-10 (soon redesignated "PT6") had practical applications. It went ahead early in 1959 with a budget of $21.1 million to see it to completion of the demanding 150-hour test phase.

Under the leadership of Thor E. Stephenson, who took over CP&W upon the death of Ron Riley in 1959, the PT6 moved quickly. The design team included Jack Beauregard, Fernand Desroches, Ken Elsworth, Fred Glasspoole, Dick Guthrie, Gordon Hardy, Hugh Langshur, Bruce Moss, Allan Newland, Pete Peterson, Elvie Smith and Thor Stephenson—"the dirty dozen" as they became known. The first PT6 ran on November 16, 1959. Developments thereafter were rapid, but

The 5,000th King Air was this eight-passenger, 1,500-mile-range Model 350 with P&WC PT6A-60A engines. The winning Beech-P&WC combination made the King Air the world's most successful turbine-powered business plane. (Raytheon Aircraft 960200-5)

Hiller UH-12 N5323V was used as a PT6 ground test bed at Longueuil. Hiller had hoped to sell PT6-powered UH-12s to the Canadian Army, but the sale turned out to be for piston-powered machines. (P&WC)

P&WC had many industrial and marine PT6 applications, from heavy trucks to speedy race cars, the spectacular Turbo Train, a wood chipper and this luxury cruiser. *Rimfakse*, first operated with a PT6 in 1962, was the original marine application. Through it P&WC received certification for the ST6—the marine version of the PT6. *Rimfakse* was used by P&WC during Expo 67, carrying VIPs up and down the St. Lawrence on sightseeing cruises and ferrying them to and from the Expo site. (P&WC)

People think of the DHC Dash 7 as the first four-engine PT6 application, but years before it the attractive French Potez P841 had four PT6A-6s. Potez envisioned the P841 as a commuter and corporate plane, but such thinking was premature when it appeared in 1964. Only two were built, one for the West German government, the other for a department store chain. (CANAV Col.)

Many a Grumman Goose, like G21T CF-BCI seen at Vancouver in the 1970s, also adopted the PT6. Keith McMann flew the Fletcher Challenge G21T for several years. This version had a 12,500-lbs gross weight compared to 8,500 for the tradional model; carried two pilots, nine passengers and 500 pounds of baggage. Most G21Ts had de-icing boots and electrically-heated props. (CANAV Col.)

Many old designs adapted the PT6. Most numerous were Beech 18s. Condor's Beech C-GBST (above) was shot at Mount Hope about 1980. (Jack McNulty)

(Right) One of the earliest King Airs was this 1968-model B90 (c/n 64-373) owned by Pratt & Whitney of Longueuil. It was among the first batch of 91 B90s built in 1968. This model had the PT6A-20, an improvement over the PT6A-6 of the original 295 King Airs. CF-PAW was at Toronto Island Airport on July 20, 1971. In 1997 it was listed as N275LE with Rockford Motors of Rockford, Illinois. (Larry Milberry)

two big problems arose: costs ran away, and marketing could find no buyers. James Young, CP&W's founding father, became restless and asked Jack Horner, head of P&WA, to terminate the PT6. Instead, in January 1961 Horner, to bring budget problems under control, despatched a six-man team from Hartford led by Bruce Torell to reorganize the project. Originally a Winnipeger, Torell had been at the NRC and Whittle's Power Jets Corp. before moving to P&WA. His team put the PT6 on track, then returned control to Stephenson. On May 30, 1961 Bob Fowler of DHC and John MacNeil of P&WC flew the PT6 for the first time—in the nose of a Beech 18. The next historic step was for the PT6 to power an airplane on its own.

The PT6-Beech Connection

In 1961 P&WC and Beech agreed to co-operate in a development program to place 500-shp PT6A-6s in a US Army L23F (civil Queen Air, normally with 380-hp Lycomings). A test aircraft, designated NU-8F, was sold for a dollar to the Army by Beech and UAC. On May 14 Beech pilot Steve Tuttle, P&WA's Bill Lee and UAC's Mick Saunders went aloft on the first flight of the NU-8F. This was inadvertent—the plane went over a bump during a high-speed ground run. Next day Tuttle and James D. Webber made the first official flight. An intensive flight test program ensued, with the Army putting the NU-8F through months of rugged trials at Fort Rucker, China Lake, etc. This is summed up in *The History of Beechcraft*: "It

was the first of an almost daily series of test flights that would take place over a six-month interval, providing data for evaluation of power plant performance that would firm up an eventual choice of engines for a wholly new Beechcraft."

Soon the US Army wanted a fleet of Beech turboprops. Encouraged by this, Beech decided to go ahead with a civil version (King Air 90), foreseeing an annual market for about 10 planes. Since the new plane evolved directly from the pressurized Queen Air, tooling and other manufacturing details cost a minimum, keeping selling price under control. The first King Air flew at Wichita on January 20, 1964. So important was this that Beech recorded: "The date was January 20, 1964—a date that

Otter 3682 was delivered to the RCAF in May 1954. It served with 121 Communications and Rescue Flight in Vancouver before being loaned for a joint DHC-DRB research program launched in 1956. It was modified progressively to investgate STOL parameters. Special features, as seen in this October 1959 photo, included an enlarged vertical tail, large flaps and new landing gear. A J85 turbojet was installed in the cabin to test the effects of thrust vectoring. The special undercarriage absorbed the impact of a rakishly steep approach and subsequent heavy arrival on the runway. The high, upswept elevator was clear of the turbulent slipstream generated by the flaps and J85 exhaust. George Neal made the first test flight (1:15 hours) on August 27, 1959. He logged 40:10 hours by the time of his last flight on May 10, 1961. Included was a demonstration flight for officials in Ottawa on June 8, 1960. Bob Fowler flew the thrust vectored phase in 1960-61, attaining speeds as low as 48 mph. Later (1963-1965), 3682 flew with two PT6 turboprops. As such it became the first twin PT6 application. The J85 had yet to be installed in the small photo taken at Downsview on September 11, 1959 of George Neal taxying. (DHC 11437, Larry Milberry)

(Below) Bat-winged Otter 3682 during test bed trials at Downsview in March 1958. Beyond is Otter 3692, which was used to test a series of large, wooden propellers. (DHC 7712)

(Below) Otter 3682 configured with twin PT6s. Although by no means a Twin Otter, this oddball machine must have set some wheels spinning in the minds of the engineering and marketing people at Downsview. (DHC 11437)

would stand out in the company's annals beside an earlier event, when on January 15, 1937, the first Model 18 Beechcraft made its initial flight..."

The first examples were re-engined Beech 65-80 Queen Airs. The new type was announced to the press in August 1963. First flight was at Wichita on January 20, 1964. On March 12 the US Army accepted its first production U-8F.

On May 27, 1964 the King Air won its FAA type certification. The first delivery to a corporate customer took place on July 7, when UAC chief pilot John MacNeil accepted King Air CF-UAC on behalf of his firm. By now Beech had gotten over its initial King Air sales jitters. These had pestered company brass for some time, but now the orders were rolling in—the King Air had taken the corporate aviation world by storm, dozens being sold in the first year. The 100th came off the line early in 1966. New versions began to appear, invariably coupled with improved versions of the PT6—the 550-shp -20, 750-shp -34, etc. In December 1971 the 2,000th PT6 was delivered to Beech; these sales already were worth $72 million to P&WC.

The UAC/P&WC-Beech relationship continued unabated. In 1978 P&WC turned out its 15,000th PT6. On November 15 it presented this engine (c/n 80138) to Mrs. Beech at P&WC's 50th anniversary celebrations in Montreal.

Beech installed the engine in a King Air for its longest-standing customer, Marathon Oil of Ohio. This was the 85th Beech plane ordered by Marathon over 40 years. The King Air rolled on and on. June 24, 1996 marked the delivery of No. 5,000, which went to JELD-WEN, a wood products giant in Klamath Falls, Oregon. At a time when P&WC was on the ropes and Beech eager for something new, the two had come together in one of aviation's most fortuitous and long-lasting partnerships.

Recalling the PT6-Beech Connection
One the great characters in Canadian aviation was J.C. "Jack" Charleson of Ottawa. He had started flying in 1929 at International Airways in Hamilton, then spent his life in aviation. He barnstormed and did long-distance flying in the US; and had careers with the DOT, CP&W,

and Okanagan Helicopters. In the 1930s he became acquainted with Walter and Olive Ann Beech of Wichita. When United Aircraft of Canada (*i.e.* P&WC) was struggling to find a launch customer for its revolutionary PT6 engine, it was (as Charleson related) Mrs. Beech who came to the rescue. At the same time she saved her company, which needed a vibrant, new product. It was building progressive versions of the Queen Air, but seemed stalled. UAC was trying to sell Beech on the PT6, but wasn't making headway. This is how Jack Charleson explained the details in a talk to the CAHS of November 1977.

The first lady of aviation had a great deal to do with the success of the PT6. Mrs. Beech was behind it all, this very sensible marriage of a Canadian engine to an American airframe.

While the last Otters were being built at Downsview, DHC was producing the first DHC-2T Turbo Beavers. The prototype (CF-PSM-X) flew first December 31, 1963. It's seen (inset) in shadow form about to alight at Downsview June 10, 1973. George Neal had just returned from Rockcliffe, where he had been flying some of the NAM's WWI planes at an air display. PSM crashed July 3, 1988 on Ahmic Lake, Ontario. It flipped when the pilot made a water landing with the wheels down—a dumb sort of *faux pas* that occasionally takes place. Here a new Turbo Beaver is seen before delivery to the Manitoba government. The Toronto skyline forms the background. CF-MAB was red with white trim. (Larry Milberry, DHC 25056)

Since then Beech has bought over a half-billion dollars worth of PT6s. I met a man who was presnt at the meeting where this significant decision was made. Our man, Thor Stephenson, a big, attractive Icelander, was making one more presentation in a long series. Beech argued that it wouldn't pay to put an engine as expensive as the PT6 in the King Air; the cost of the airplane would become prohibitive. They were competing head-on with Cessna, and their answer was still No.

At this point Olive Beech stepped in. She and Thor had developed a mutual respect and regard. According to my informant, Mrs. Beech stood up and said: "Listen. I don't normally interfere with engineering or management decisions, but I still control this outfit. I'm telling you to take those engines, that are costing us nothing, and put them in the airframe. Just try it." [CP&W was so anxious to get its engines into any plane that it offered them free.] Now that was the most important event in the history of Canadian aviation... We never had another aviation export that came near the PT6 in terms of money earned, not even the Beaver.

Other PT6 applications

Another early role for the PT6 was in the helicopter. P&WC teamed with Hiller of Palo Alto, California to place the PT6 in the experimental Model Ten99, which flew in July 1963. Hereafter the PT6 ran experimentally in a series of

aircraft, boats, trucks, even Indy 500 racing cars. It powered Lockheed's XH-51 rigid-rotor test helicopters. On a June 1967 flight an XH-51, augmented with a JT12, became the world's fastest rotorcraft (302.6 mph). The Piasecki 16H, Kaman K-1125 and Hiller UH-12 also were PT6 test beds. At DHC an Otter was fitted with two PT6s and a variable thrust J85 jet engine for STOL experimentation. Funded by the federal Defence Research Board, this was the first twin PT6 installation and proved an excellent opportunity for P&WC to further its program. Not long after, DHC decided on the PT6-powered Turbo Beaver. The prototype flew December 31, 1963, a few days after the PT6 won civil certification after 11,000 running hours and more than 1,000 flights.

The PT6 and the Twin Otter

By the early 1960s Otter and Caribou production was declining at DHC, and a run of 100 Grumman Trackers for the RCN ended. Something new and bold was needed. Russ Ban-

nock, always with his ear to the ground, was mulling over what US Army pilots in Viet Nam had been telling him—"If only the Otter had two engines." DHC always had felt that a rugged STOL twin was marketable; but for something in the 10,000- to 12,000-pound range there was no suitable engine till the PT6. DHC had gained valuable PT6 experience during the DRB Otter program, so the timing seemed right—the DHC-6 Twin Otter was launched. This involved stretching the Otter fuselage, extending the wings, and installing PT6s. There would be nothing fancy, so the plane would have a fixed gear. On April 29, 1965 Sir Roy Dobson officiated at the rollout of the prototype, CF-DHC-X. Four additional pre-produc-

With the pilot keeping an eye on the photo plane, DHC's demo Twin Otter CF-SUL (c/n 3) lands at Toronto Island in December 1965. In 1967 it went to Alaska, where it worked till 1974, then became C-GKAZ in Canada with Kimba Air, Ptarmigan Airways and Carter Air Services. In May 1995 it was sold in the US. (DHC 22777)

The prototype Twin Otter at Toronto Island during float trials on August 31, 1966. CF-DHC-X served DHC to 1981 and was later donated to the NAM. (Larry Milberry)

tion machines were coming down the line, and DHC committed to 10 production planes. It was sink or swim once again at de Havilland. Bob Fowler and Mick Saunders flew CF-DHC-X on May 20. Less than a year later the type was certified. The timing was perfect, for business was about to boom. Even though DHC lost a US Army competition to the King Air, this was just when US commuter airlines were about to boom. In 1965 there were only about two dozen of these so-called third-level air taxi operators filling small markets that didn't interest the big airlines. The commuters operated under Part 135 of the CAB's regulations, a special category in which an operator could fly skeds with aircraft not heavier than 12,500 pounds. Pilots did not have to be airline

rated, but only needed a commercial VFR licence.

Typical among the commuter firms was TAG (Taxi Air Group). It began in the mid-1950s with two Beavers and an Otter operating seasonally on floats between Toledo and Cleveland. In 1957 Ross Miller bought TAG, added Doves and Aztecs and inaugurated hourly downtown-to-downtown service between Detroit and Cleveland. Success led TAG to expand to Columbus and Cincinnati. More than 83,000 passengers were boarded in 1965; but the CAB steadfastly refused TAG permission to operate larger planes. The entry of Wright Airlines into its market, and the crash of one of its Doves in 1970 threw TAG off its game enough to force it to close. So it often went with such "mom and pop" carriers. Many were shaky financially, and rarely were they administered in sophisticated fashion.

By the time TAG folded, the commuter business had changed greatly—there were more than 200 operators and most parts of the country were being served. This explosion, it may be argued, resulted from the appearance of several new designs like the Cessna 402, Piper Navajo, Beech 99 and Twin Otter. These had passenger appeal and lower seat-mile operating costs. At the same time the CAB approved a policy allowing major airlines to opt out of unprofitable markets in favour of commuter carriers taking over, so long as service did not deteriorate. Thus did

small centres give up Convair, DC-6 or Electra service for more cost-effective commuter planes. This trend had nowhere to go but up. Soon upstart airlines were ordering fleets of Twin Otters.

Selling at about $300,000 the Twin Otter made money wherever it flew. It may have been slow and dowdy-looking compared to a Beech 99, but operators knew an opportunity when it stared them in the face. Air Wisconsin was the first US commuter airline to take delivery of a Twin Otter—N4043B (c/n13) in October 1966. Air New England, Command Airways, Executive Airlines, Hawaii Jet Aire, Pilgrim Airlines and Suburban Airlines all opted for Twin Otters. By the end of 1967 DHC had recorded 90 deliveries, most to US commuters. Others went to customers throughout Canada and around the globe. The Twin Otter was on the road to success. Meanwhile, P&WC was straining to keep on top of orders. Not only did it have hundreds of PT6s to deliver for Twin Otters, but also for Beech 99s, King Airs, and a host of other types including helicopters and ag planes; and airframe converters like Jack Conroy, Fred Frakes, Dee Howard, Ed Swearingen and Ed West were calling for engines for planes like the Aero Commander, Beech 18, DC-3, Goose and Mallard. Before long the PT6 had grown passed 1,000-shp, engendering the next series of commuter planes—Dash 7, Short 330/360, etc.

More Tech Rep Assignments

Between Viet Nam tours DHC tech rep Dick Gleasure was busy on other work, largely sales tours and deliveries. In early 1965 he accompanied a team demonstrating the Buffalo in Venezuela (which later ordered several aircraft). He described another busy time:

The year 1966 was one of intense activity as de Havilland came out with the Twin Otter powered by P&WC turboprop engines driving Hartzell propellers and carrying 20 passengers. This was also a STOL aircraft, and the most versatile we had built. It was easy to maintain and a pleasure to accompany on tour. One could carry in a portable kit all the tools necessary to service and maintain it. Best of all one could wear his best suit without getting it all messed up with grease and oil. We toured the Caribbean and the US, getting home for the occasional weekend with the family. Fortunately, as my wife Nicky taught school, and the children were busy with their own studies, everyone was too involved to make an issue of it. On one occasion, however, Nicky asked why I didn't get a job like everyone else and settle down with my family.

A Twin Otter (CF-SJB) sales tour was flown in 1966 throughout California with pilot Dave Fairbanks and salesman L.L. "Slim" Jones. Gleasure considered them wonderful gentlemen and professionals. Fairbanks had been a leading WWII fighter pilot (DFC and two Bars). Jones was an early Sunderland pilot, who had a DFC for action in Norway. Gleasure once asked what rank he had attained, but

In the 1950s-60s commuter airlines were little-known operations serving obscure markets. Typical was TAG, which started as a Beaver and Otter air taxi service. TAG grew to include the D.H. Dove—N640H was at Detroit on May 21, 1966. The explosion of the US commuter airline business in the late1960s was caused largely by the advent of affordable mini-airliners powered by the PT6. (Larry Milberry)

Company demonstrator Beech 99 (c/n 2) at Mount Hope, Ontario September 7, 1968. The 15-passenger plane first flew in December 1965 with 550-shp P&WC PT6A-27s. The first delivery was made to Commuter Airlines in May 1968. A version with the 680-shp -28 was introduced (1972), then one with the 715-shp -36 (1980). Production continued till 1987 by when some 239 had been built. It was several years before the Beech 99 caught on in Canada. By the mid-1970s, however, it was well entrenched. (Larry Milberry)

Twin Otter demonstrator CF-DHA at the terminal at Addis Ababa, Ethiopia in September 1972. (Gleasure Col)

Jones was evasive. Gleasure later discovered that his friend had been Wing Commander Jones. The California tour took SJB to airports large and small, and totalled 66 flying hours. Other demos in 1966 were to Air Wisconsin and Pilgrim Airlines, both of which purchased aircraft. In July and August a tour took Gleasure to the Caribbean with pilot Tom Appleton and salesman Bill Lynn. For 1966 he spent 126 days on US and South American sales tours.

In January 1972 Gleasure was with a Twin Otter in Afghanistan helping with maintenance and doing some training. Next he accompanied demo pilot George Northrop on an extensive tour with CF-DHA. It began on April 16 with a 2,800-mile, 13:15-hour nonstop flight from Gander to Hamburg. Four interconnected 120-gal. auxiliary cabin tanks feeding into the main fuel system under the floor provided the endurance. The plane was packed with personal luggage, survival kit, cans of engine oil and spares, but also with sales brochures and the usual sales tour giveaways, which added considerable weight. The crossing was not all routine, for one engine had a faulty fuel/oil heater that let all its oil leak out by the time DHA reached Hamburg.

Other places visited on this trip were the Hanover Airshow, Switzerland and Yugoslavia. In July Gleasure was on another trip in DHA, this time with Don Rogers. They visited the UK, France, Lebanon, Iran, India, and Malaysia. In September he was with George Northrop on a swing from Farnborough through France, Algeria, Cameroon and Sudan. In October he was with Dave Fairbanks with DHA in Ethiopia. Demonstration flying sometimes leads to trouble. On May 4, 1976 Twin Otter C-GDHA (c/n 428, CF-DHA was c/n 337) was on an African tour with DHC's Amos J. Pudsey in charge. On a Zambian flight with a local pilot at the controls DHA crashed, killing Pudsey and 10 of the other 11 aboard.

Another interesting assignment for Dick Gleasure was with pilot A.W. "Mick" Saunders delivering Twin Otter 6V-ADD to Senegal October 12-16, 1973. They began by flying from Toronto to Sept-Îles and Goose Bay. For Gleasure the next leg (to Greenland) was the most interesting, with its approach in perfect weather up the iceberg-filled gorge to a landing on Narsarssuaq's uphill runway. Next came

Iceland (with a view of one of its volcanoes); then Jersey, and on to Agadir and Dakar. The trip took 37:55 flying hours with Agadir-Dakar being the longest leg (8:40). In Dakar Gleasure and Saunders met with officials involved in the new Twin Otter. Included was Archie Vanhee, a veteran Canadian on a contract with CIDA to train Senegalese pilots. He had been flying since 1926—in the bush, on wartime anti-submarine patrol, and with CPA after the war. Later in November Gleasure was off again with Saunders to Senegal with Twin Otter 6V-ADE. They departed November 29: Toronto-Charlottetown-St. John's-Santa Maria-Las Palmas-Nouadibou-Nouakchott-Dakar, arriving December 5. This route saved considerable time—only 24:60 hours were logged. Once in Dakar, Gleasure had the long-range fuel tanks removed and freighted home. The interior was then fitted with seats, carpet, etc. and handed over to Air Senegal. Gleasure spent a few days training groundcrew, while Saunders made some pilot famil trips. A typical delivery flight is tough work, but there is usually some time to rest at the other end. On the second Senegal delivery Gleasure and Saunders visited the home of a local pilot: "Mr.

The missile tracking base at Meck in the remote South Pacific Kwajalein Atoll where Dick Gleasure assisted Global Associates in keeping its fleet of Caribous in running order. Then, a scene with one GA Caribou being readied as another departs. (Gleasure Col.)

Jenot was from France and had had a long career in aviation. Mick and I enjoyed a pleasant evening at his apartment. We also visited Archie Vanhee, who gave a brief history on Dakar and French involvement in the country, and told stories about the pioneering days of aviation across the South Atlantic." While most could only dream of a trip to Senegal, such deliveries were not always welcomed by DHC's crews. They were gruelling, especially one like the delivery of CF-DHA to Japan from March 15-27, 1974. This time Gleasure was crewing with George Northrop. They started on the 15th flying Toronto-Goose Bay. Then came stops at Reykjavik, Shannon, Hurn, Nice, Damascus, Dubai, Bombay, Bangkok, Hong Kong, Kagoshima and Osaka. At 10:35 hours Bombay-Bangkok was the longest leg. Along the way CF-DHA was shown off whenever possible; interest always was high.

Once in Japan, Gleasure again worked with the customer (Nippon Kingkori Airways), helping familiarize mechanics. He put on some lectures: "Seven students were present. Engine construction was covered. Component location and adjustment of controls was taught in conjunction with the maintenance manual, charts and blackboard... Notes detailing instructions how to compute engine TBO with aviation gas instead of JP fuel were also introduced... On April 2 starting and engine run-up procedures were carried out." Later that day Gleasure left for Vancouver. April 19-May 4, 1974 he was again with Mick Saunders, this time delivering YA-GAZ to BAA in Kabul. Gleasure also made trips to San Francisco and the remote Kwajalein Atoll in the South Pacific. There Global Associates of San Francisco was operating Caribous between islands where the

Twin Otters around the World

(Right) A great Twin Otter formation shot set up by DHC's keen photo department. 5X-UVN went to African Airways in December 1968, then became CF-GQK with Bannock Aerospace of Toronto. In 1977 it reassumed its original registration, this time with Uganda Airways. It went to Kenya in 1985 as 5Y-BEK. "Flying clothes line" CF-INB went to Survair of Ottawa, which operated it for Inco on airborne survey work. In 1976 it was exported to Indonesia as PK-KBD for Inco's subsidiary there. T-84 went to the Argentine Air Force in December 1968. (DHC 31146)

(Below) In 1971 DHC teamed with Field Aviation to develop a water bombing Twin Otter that used a centre-line tank compared to the usual system of water-dropping floats. Later in the year the aircraft was delivered to Bell Canada. In 1992 it was sold in Indonesia. (Larry Milberry)

(Above) The Canadian Forces accepted its first of nine Twin Otters on March 29, 1971. Two were lost over the years—13808 was destroyed on the ground in Kashmir by Indian Air Force Hawker Hunters in 1971 and 13807 crashed during a search in the foothills of the Rockies June 1986. Three others were sold in 1994, the year 440 Squadron moved with the remaining four aircraft from Namao to Yellowknife. (DHC 35347)

(Left) Series 300 N142SA began in 1969 as N385EX of Executive Airlines. It was later with Air New England, then went to the Grand Canyon tourist operator Scenic Airlines in 1982, where it was fitted with panoramic windows. (Halford Col.)

USAF had missile tracking stations (missiles would be launched from Vandenburg AFB in California to target areas around Kwajalein). One of Dick Gleasure's last jobs was in March 1978 investigating an accident (air force Otter 9424, which had pranged north of Montreal). He then worked in the plant for a short time and retired in June 1979. His career had been full of challenge, hard work and satisfaction. He took away many memories and among his awards was the United States Bronze Star, received in June 1985 for his work in Viet Nam.

Afghan Venture

Even in the early 1960s landlocked Afghanistan (251,000 sq. mi., population 15,000,000) was living in the past. Camels and donkeys were common for transportation; the country's few trucks and buses always were overloaded if not broken down. There were few modern roads. Although Afghanistan had been the site of one of the earliest airlifts, when the British evacuated civilians from Kabul during unrest

(Above) Geologist Thomas M. O'Sullivan recorded this typical Twin Otter scene in the NWT. The first Twin Otters to operate regularly in the NWT were Pan Arctic Oil's CF-PAT (c/n 2, January 1966), Atlas Aviation's CF-WWP (c/n 12, October 1966) and Shell Oil's CF-SCA (c/n 17, November 1966). Shown here is C-FWAB. O'Sullivan notes of this photo: "This Twin Otter supplied our camp with the basic necessities over the summer of 1987. We were situated due west of Bathurst Inlet, northeast of Contwoyto Lake, and south of the Burnside River. A very lonely piece of land." WAB had gone new to Wardair in 1973, thence to Ptarmigan in 1980. It was wrecked taking off at Thistle Lake, NWT on June 6, 1990. In spite of accidents Twin Otters will be serving deep into the 21st century in countries all over world.

(Right) Three new Twin Otters (c/ns 7, 10 & 11) for Chile in a nice formation shot taken in October 1966. In 1997 these veterans still were at work—with 15 Twin Otters, Chile had the world's third largest fleet, next only to those of Scenic Airlines in the US (21 aircraft), and Kenn Borek/Harbour Air in Canada (17). (DHC 25206)

(Above) Ontario Provincial Air Service Twin Otters and Turbo Beavers at Sudbury in the summer of 1993. These types were the mainstay of the OPAS for decades. (Larry Milberry)

British Antarctic Survey Twin Otter and Turbo Beaver near Toronto before departure to the Falkland Islands in late 1968. Ron Nunney, who earlier had been photographing CF-100s and CF-105s at Avro, took this shot. The Twin Otter was VP-FAO, which later served in France and Djibouti. (DHC 33301)

Argentine Twin Otter 200 T-82 at El Salvador while being delivered to Argentina in 1968. (Shrive Col.)

after WWI, the airplane was still a rarity. Even though it was an original signatory of the 1944 Chicago Air Convention (the start of ICAO), by 1965 Afghanistan had only nine civilian airports. The only air carrier was Ariana Afghan Airlines, formed with the help of Pan American in 1955. It was moving slowly toward national management and by 1961 had two

DC-4s and four DC-3s. There was also a small, Russian-dominated air force.

By 1965 His Excellency Sultan Ghazi, head of civil aviation, sought to improve air transport. He had been impressed by the Twin Otter at the 1965 Farnborough Air Show. Through ICAO, DHC was asked to help, and sent pilot Fred Hotson on a fact-finding mission in 1966. Ac-

companied by Harry Hunter of DHC's service department, he analysed how the Twin Otter might be used to bring more of the amenities of life to the average Afghan. After studying its history, geography, economics and culture, Hotson realized that introducing any modern airplane in Afghanistan would be a challenge. There were almost no landing strips, aids to

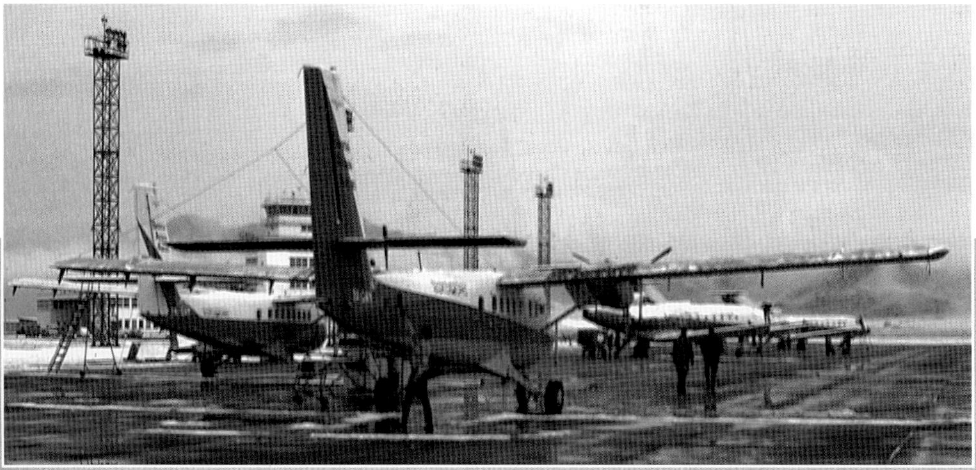

(Left) To support its products and expand markets, DHC sent tech reps and sales people to the remotest corners of the world. Here Afghan Twin Otters GAT and GAY, with YAK 40s, wait for their day's work to begin at Kabul in December 1971. For several years DHC men served in this isolated domain. (Gleasure Col)

(Below) Twin Otter YA-GAS attracted a crowd at this stopover at a remote Afghan village. (DHC 36964)

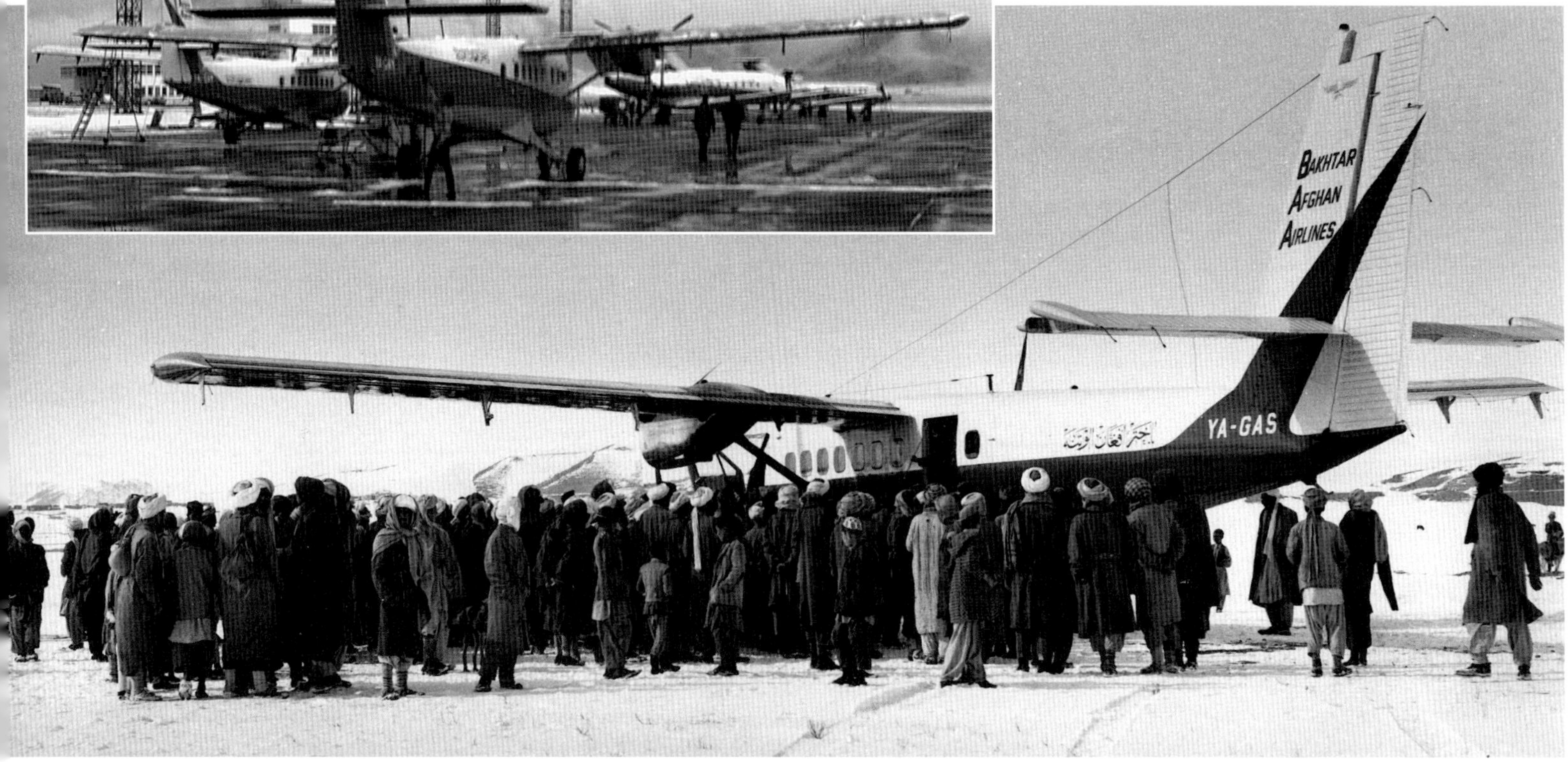

navigation or weather stations; few indigenous pilots, mechanics or other specialists needed to run an airline; and no hotels, taxis, fuel, etc. in the back country. There was a barter economy—there was no way the average Afghan could purchase an airline ticket. In short, running an internal air service could not be profitable.

Were DHC to become involved in setting up a Twin Otter operation, Hotson recommended that it be run as a self-contained aid program (distinct from Ariana Afghan Airlines, which had its own problems) with full support from the sponsoring body. Thus did ICAO become the overseeing authority for what lay ahead. As Hotson envisioned the program, it would be introduced in stages throughout five geographic areas. Kabul would be headquarters. Initially, Tom Appleton and Armand Hollinsworth took a Turbo Beaver and Twin Otter to Afghanistan to do some survey flights and familiarize some local pilots. Sultan Ghazi now raised the funds for two Twin Otter 100s. The first was delivered in December 1967 for newly-formed Bakhtar Afghan Airlines. It set to work on routes recommended earlier by Hotson. ICAO provided the expertise to make the operation work, but it wasn't long before there were problems. ICAO representative T.R. Nelson was concerned not only with unserviceabilities, but with single-engine "hot and high" capabilities of the Twin Otter. He asked DHC for assistance, so pilot Tony Shrive and tech rep George Kelly went out in July 1969 to help.

Shrive spent two months appraising and training pilots, and reporting on aircraft and airstrip conditions. Although one aircraft was

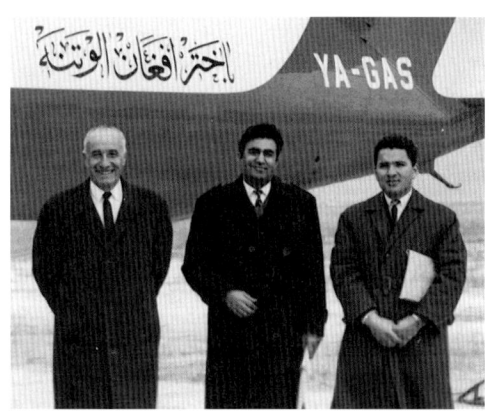

(Above) Fred Hotson (left) with two BAA men and the first Twin Otter delivered to BAA. (Hotson Col.)

(Right) A Bakhtar Twin Otter is refuelled from four-gallon cans trucked from hundreds of miles away. Then, pilot Tony Shrive with a local tribesman. (Shrive Col)

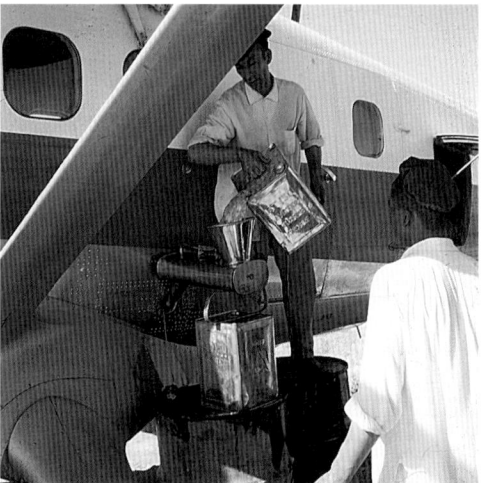

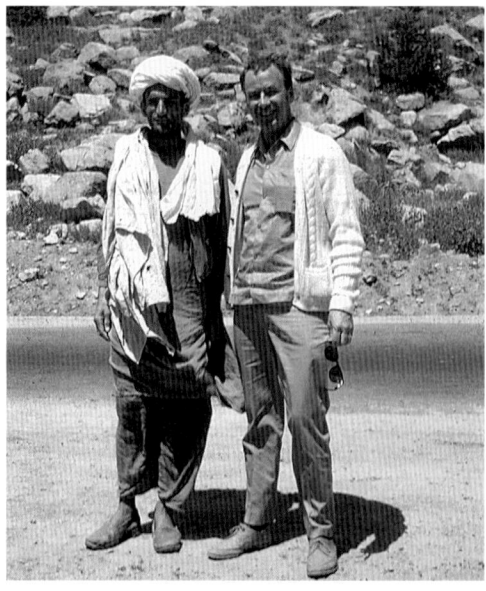

unserviceable when he reached Kabul, he was able to squeeze in the necessary training and evaluation. He wrote a brief report for BAA's seven pilots. Of Capt Mustafa Tayer he noted, "He is a competent pilot with no outstanding faults. He tends to be a little flamboyant, but not at the expense of safety." Of co-pilot Mir Omar he said, "He tends to over control and is reluctant to exploit the aircraft... His general flying improved considerably toward the end of the test. I suggest that he be given every opportunity to fly the airplane as a co-pilot and, when conditions permit, to fly in the left seat at the captain's discretion."

In general, Shrive was impressed with BAA's pilots. He wrote to Dave Fairbanks on

Afghanistan's Twin Otters			
Registration	c/n	Delivered	Notes
YA-GAS	77	15-12-67	Became C-GDQY of St. Andrews Airways 1-76; to Laurentian Air Services 8-76, to Labrador Airways 6-78
YA-GAT	111	25-3-68	Crashed Bamian 18-4-73
YA-GAX	331	11-10-71	To Ariana Afghan Airlines 2-88, active 1996
YA-GAY	332	9-11-71	Crashed Bamian 8-1-85
YA-GAZ	395	16-4-74	Crashed in mountains near Shashgow Ghanzi 10-3-83

July 8: "I have flown a few routes, when space was available... I give the pilots a lot of credit—10-hour days over the most rugged and inhospitable terrain that I have ever seen. If all they had to do was fly, they would certainly earn their $130 a month, but that's the easy part. There are three captains, about 10 co-pilots (one who flies regularly, because he shows up for work), and two who seem to spend their time in Beirut consistently failing their licence tests. Five more have been in India for 18 months learning to fly. Three more have never been up in an airplane."

As training progressed, Shrive started airstrip appraisal. He evaluated Bamian (7,400 feet ASL). He also worked to clarify technical details e.g. Could the Series 100 operate above 15,000 feet in icing with the fine mesh intake screen installed on the PT6, and was it feasible to convert the BAA aircraft from PT6A-20s to higher performance engines? Another strip was at Darwaz. The report noted that it was below requirements—only 550 metres long and needing levelling and compacting. Shrive suggested that no flight be made to Darwaz any time the weather en route fell below VFR.

Trials at Bamian showed that should an engine fail on climb-out at maximum takeoff weight in hot conditions, the aircraft would have to crash land. This was determined on July 20, when three simulated emergencies were flown at Bamian, one early in the day at 20°C, another in early afternoon at 28°C (these attempts were dicey, but the aircraft was able to climb marginally, since the weights were moderate), and the third when the day had warmed to 30°C. On that occasion the aircraft was heavy (10,530 pounds). As takeoff flap was removed and one engine was throttled back, power had to be restored immediately to avoid crashing. The crew then

Afghan memories: a camel train in Kunduz and some typical mountain scenery. (Gleasure Col.)

climbed to 9,000 feet and feathered an engine. The plane would not maintain height. Shrive's report concluded: "The trials at Bamian have clearly demonstrated that if an engine failure occurs during takeoff, before a minimum height of 300 feet above the runway has been achieved, a return to the runway (even a dumbbell turn with a downwind landing) would be so hazardous that it would probably be safer to make the best possible crash landing straight ahead." If Twin Otter 100s were to use this strip and maintain single engine go-around capability, the report recommended, they should be restricted to 10,000 pounds in temperatures not higher than 20°C. Shrive's notes covered every aspect of air transport in the country. He travelled extensively overland, observed the people and took photos. In a letter to Dave Fairbanks he described typical operations:

At one place we landed, as people tried to leave the aircraft, others were crawling between their legs to get in... Fights broke out among would-be passengers trying to attain priority. Departure had to be delayed and at one time we had 25 people on board. Some were on the floor under the seats, three were in the baggage compartment hiding under bed rolls. Incredibly, fuel is trucked hundreds of miles in four-gallon cans to some of the outlying strips. It takes an hour to put in 600 pounds... At one strip they built a brick building last fall to house items of minor equipment. The strip was abandoned for the winter months due to heavy snow. The first flight in the spring revealed that the building had been stolen, brick by brick...

As to maintenance, Shrive wrote of tech rep George Kelly, who accompanied him on the posting: "George is making valiant efforts to straighten out some of their maintenance administrative problems and lack of records. They do not even keep log books. They have no idea when the last inspections were done on the airplanes." In 1971 Bakhtar received three Twin Otter 300s with the PT6A-27. These had much-improved performance. At the same time

the Russian Yak 40 was introduced, only adding to the confusion.

In October 1971 and April 1974 Dick Gleasure accompanied Mick Saunders on Twin Otter deliveries to Afghanistan, first with Series 300 YA-GAX, then with YA-GAZ. Each time Gleasure remained several weeks to help BAA get the aircraft on line. He consulted with various department heads, and the two CIDA men involved—project manager Joe McElrea and flying instructor G. Reynolds. Gleasure enjoyed his visits to Afghanistan, a country whose people and natural beauty he found very special. The 1974 trip was from Downsview (leaving April 19) via Gander, Goose Bay, Narsarssuaq, Reykjavik, Jersey, Brindisi, Ankara and Tehran to Kabul (arriving April 25 after 52:25 flying hours). While in Kabul, Saunders trained four BAA captains in STOL techniques and gave a famil flight to several trainees, who had yet to fly in the Twin Otter. At this stage the BAA fleet comprised two Yak 40s, Twin Otter 100 YA-GAS (not in use) and Twin Otter 300s YA-GAX, 'Y and 'Z. GAS by this time had logged 3,495 hours since delivery in December 1967. A later DHC report noted that BAA was still struggling:

...lack of management, planning and scheduling still plague the airline, and have not improved in the last three years... The Yak 40s are costing so much to operate, and are so poorly supported by the Russians, that Bakhtar want to get rid of them and use only Twin Otters. However, as with all decisions, particularly political ones, this will be talked about for a long time before anything is done. As a result of poor reliability and the two accidents, the airline's reputation amongst tourists is not good. Their loads are poor, amounting at present to about 25 seats a day... as CIDA is in the process of allotting another $1/4 million for spares support, it would seem that the CIDA programme will continue. This money for spares will provide long term support, but BAA are unable to get spares for a hot-end inspection, due now, because of their lack of credit. CIDA were approached to put aside some proportion of the allotment to provide a credit against which BAA could order AOG items [urgently needed spares] without delay, but this apparently cannot be done. CIDA quite reasonably suggested that BAA should provide a small deposit for this purpose, but this suggestion so far has drawn a blank... the pilots

(Right) One of the pre-production Buffalos at Downsview in September 1964. Dave Fairbanks was doing an engine run. (Larry Milberry)

(Below) Ecuadorean Buffalo FAE064. The Buffalo proved ideal in developing nations, where barebones airstrips were common. (DHC 43597)

provide the only encouraging note in an otherwise distressing picture. They are still conscientious and do a good job under difficult conditions, where communications, scheduling and met forecasting are almost nonexistent.

Afghanistan's efforts to bring a few 20th century benefits to its people ultimately were thwarted. Tribal rivalries built through the 1970s and in December 1979 the Soviets invaded to support a communist faction. So began an all-out war, which didn't end till 1988, when the Soviets admitted defeat and withdrew. The war had cost them 50,000 dead. Inter-tribal warring continued through the 1990s, making impossible something so simple as running a small airline. By 1997 only one of the five Twin Otters delivered remained in Afghanistan and the country was in the midst of another brutal war.

"Have Shipped Everyone Out..."
While Tony Shrive was in Afghanistan in 1969, he received a letter dated July 16 from his boss, Dave Fairbanks. It included various instructions and gossip, and is a good general snapshot of life at de Havilland:

We presently have a full week in progress and have shipped everyone out except for Tom Appleton, who is in residence. Grant Davidson is off to Frobisher, for the Nordair Twin Otter operation, where they are having difficulty in producing the desired airspeeds in relation to the aircraft flight manual. George Northrop departed on Monday with YFT [Twin Otter], the waterbomber, and is not due back till late August. Rip Kirby is up trying to run the Georgian Bay airline with our own Twin Otter, due to late delivery of their 300 series. Bob Wilhelm is in Columbus, Ohio to train Trans Central Twin Otter crews before they proceed

to Texas. Due to a shortage of aircraft, we now have commandeered Twin Otter No. 176, CF-QDM. This belongs to TABA and we are operating it on a flight permit till the end of August. Twenty aircraft are scheduled for delivery this month and you can imagine how the chaos is going to mount by the end of the month so people can get away for vacation.

Naturally, the South American people, Aerochaco and YTF, are all going to descend on us for training during the last week of July... they probably expect that we should look after them for the first week of August. We are going to avoid this at all cost by putting everybody on flight training for these two customers. We will keep our fingers crossed and see if it will work. Money is tight and domestic and international sales aren't very good these days. Due to this the Twin Otter production rate has been cut to six or seven per month. Rumour has it that the Peruvian Air Force are just about to sign for 12 Buffalo aircraft. The negotiations with North American Rockwell appear to be coming to a head and we expect an announcement within the next two weeks.

In some cases the news is good, in others it's dragging. A bad piece of news concerns the New York Airways Twin Otter crash at Kennedy Airport. It seems that on the morning of July 15 the aircraft proceeded to take off from an intersection behind a Trans Continental Boeing. At 50 feet the wing went down and the aircraft hit. Two crew and one passenger were killed, and 11 passengers injured. Engineering have a team on the investigation board, and an initial report [indicates] wake turbulence. In addition to tight money and slow sales, this doesn't help ... if any good can come from such a tragedy, it might spur the FAA to get little aircraft off big jet runways. Time will tell. Let's hope that by the end of your two month period on location you have accom-

plished everything humanly possible under a difficult set of circumstances.

The Buffalo
In the late 1950s General Electric, DHC, the US Navy and the Canadian government collaborated in the flight test development stage of the 3,000-shp GE T64 turbine engine. DHC was to convert Caribou 5303, borrowed from the RCAF, to take the T64, then carry out flight tests. The plane first flew September 22, 1961. With extra power compared to a piston Caribou, it had high performance, and care had to be taken not to exceed airframe limitations. Flying totalled 220 hours, then GE continued T64 development in the US. The T64 later powered the Sikorsky CH-53 and Fiat G.222, but DHC was not going to let all its experience go to waste, especially since the US Army was interested in a larger version of the Caribou for which a 3,000-shp engine would be ideal.

The Army specified its requirement for a fixed-wing transport to lift a payload similar to a CH-47 Chinook (seven tons including such typical items as a 105-mm field piece). US manufacturers submitted proposals, but DHC won with its entry, the DHC-5 Buffalo. A contract was awarded for four pre-production aircraft. DHC, the US Army and Ottawa each became one-third risk sharers in the $22.5-million venture. Bob Fowler, Mick Saunders and Bob Dingle crewed on the first flight on April 9, 1964. About a year later Ottawa ordered 15 Buffalos for the RCAF. This was a rare case of Ottawa supporting a major DHC project by jumping in as the first serious customer. True, it had bought the Chipmunk, but never took any Beavers. The RCAF's first Otter was the seventh off the line; Ottawa also dilly-dallied over the Caribou and Twin Otter.

The four pre-production Buffalos underwent trials in the US and Viet Nam. They were required to prove themselves in carrying troops and cargo, conducting LOLEX, standard paradropping, etc. As much as the US Army wanted the Buffalo, however, the program was cut short—the USAF took over the Army's heavier fixed-wing capability at this time, leaving it with only light planes and helicopters. The USAF, well-supplied with C-123s and C-130s in Viet Nam, was not interested in the Buffalo. It now fell to DHC's seasoned salesmen to find other markets. Good sales were made in Brazil (18), Peru (16), and elsewhere

(Left) World tour Buffalo 115460/ CF-LAQ on the ramp at Pago Pago. (Shrive Col)

(Below) CF-LAQ makes an equipment drop during one of its Australian demos. (Polkinghorn & Stevens Photography)

Good Buffalo country—a typical New Guinea airstrip used on the world tour. (Shrive Col)

This group photo was taken the day the Buffalo arrived at Downsview after its world tour. Shown are Mrs. W.B. Boggs (wife of Bill Boggs, president of DHC), John Ferriera (DHC FE), J.R. McGrogan (DHC FE), G. Capellanis (GE tech rep), L.T. Trotter (DHC international sales), E.E. McCullough (DHC military sales), Mrs. McCullough, Mrs. Shrive, Tony Shrive (DHC pilot), G.H. Northrop (DHC Buffalo captain), Mrs. Northrop and the Northrop children. (DHC 29362)

Round-the-World Buffalo Trip

Date	Route	Time
March 10	Downsview–Omaha	4:45
11	San Francisco	7:15
13	San Francisco	:20
14	Honolulu	12:20
14	Pago Pago	12:20
15	Fiji	9:40
17	Sydney	8:00
17	Bankstown	:10
19	Camden–Richmond	3:10
22	Sydney	:20
24	Richmond	:20
25	Richmond	1:15
25	Bankstown	:20
26	Williamstown	1:45
26	Bankstown	:45
27	Richmond	1:50
27	Bankstown	:50
April 1	Melbourne	6:05
2	Canberra	1:30
3	Canberra	4:20
4	Canberra	1:40
5	Sydney	:30
7	Port Moresby (PM)	7:30
8	Davao–Kabuna–Tapini	5:05
9	Lae–Boana–Waugo–PM	2:50
10	Wewak–Davao	9:15
11	Singapore	7:00
12	Kuala Lumpur	1:15
16	Kuala Lumpur	2:10
16	Bangkok	3:30
18	Bangkok	3:50
19	Vientiane	1:30
20	Bangkok	2:30
22	Calcutta	4:15
22	Jorhat	3:30
30	New Delhi–Srinagar	2:20
May 2	Fukche	2:20
2	Srinagar–New Delhi	4:10
3	Dubai	8:00
4	Kuwait	2:30
5	Kuwait	1:00
7	Tehran	2:10
9	Beirut	7:00
12	Rome	4:45
13	Rome	1:15
15	Pisa	:50
18	Pisa	:20
20	Rome	1:30
20	Madrid	4:30
21	Madrid	1:30
22	Gatwick	3:30
24	Prestwick–Reykjavik	3:45
25	Goose Bay	6:15
26	St. Hubert	3:15
26	Downsview	1:20

around the world, e.g. Ecuador, Egypt, Kenya, Sudan, Togo, Zaire and Zambia. Efforts were made to market the Buffalo commercially, but few operators had either the requirement or the dollars to support the exotic transport (the T64 had proven to be a costly "maintenance hog"). Buffalo production terminated in 1984 at 123 machines.

Buffalo Sales Tour

In the spring of 1968 DHC despatched a Buffalo on a world sales tour. For the purpose aircraft 460 was borrowed from the RCAF and temporarily registered CF-LAQ. Pilots George Northrop and Tony Shrive; DHC tech rep Jack McGrogan, GE tech rep George Capellanis and air force personnel LCol J.W. Fitzsimmons (OC 429 Squadron), F/L Sam Morrell (nav), F/L Ron Dunmall (pilot) and WO Roscoe comprised the crew (Cpl Diamond and Cpl Steers joined the tour later). They departed Downsview on March 10, 1968 heading to San Francisco, thence across the Pacific. At every opportunity the airplane was shown off. The first major demos were in Australia. Hopes abounded for sales—the Australian military had been flying Caribous for years. Everyone was impressed and the plane operated well. Next came New Guinea where there were lots of chances to show off the Buffalo's STOL capabilities.

By mid-April the tour was in Southeast Asia. A demo was done for the CIA in Laos. India was next, where trips were flown for the army, including one to the remote mountain strip at Fukche. There one engine refused to spool up and a single-engine takeoff had to be made. This was dicey, for the strip was at

14,000 feet ASL (the Indians, however, had learned to build proper mountain strips— Fukche was nearly three miles long). After takeoff, the landing gear malfunctioned and wouldn't retract. Northrop fought to get the plane around. Shrive watched for a spot for a forced landing, but Northrop managed to get around OK. As soon as the wheels touched, Capelianis noticed that the duff engine started spooling up. The crew had a cup of tea, then left for Srinigar. The Buffalo pushed on to the Emirates, the Middle East and back to Canada. It had been on the road for nearly 12 weeks. The trip was a great education for all involved, but did not result in any sales.

The air force point of view of the trip was

551

expressed by LCol Fitzsimmons in his report to DHC of June 21, 1968. In part he noted:

The 429 technicians generally agreed that servicing of the aircraft is quite simple and no special equipment was required except for an eight-foot ladder... A varied stock of spares was carried... other items shipped by air by commercial means were received in good time and the aircraft managed to meet a very strict schedule of demonstrations and displays ... we have here a fine aircraft well-suited to its role of worldwide mobility, operation in austere and primitive conditions and maintainable with reasonable logistic support.

Later in 1968 Tony Shrive delivered a Twin Otter to the Argentine Air Force. In March

Caribous 301 (crew: Tom Appleton and co-pilot) and 302 (George Northrop and Tony Shrive) were delivered to the Abu Dhabi Defence Force in February-March 1969. They routed Downsview-Gander-Santa Maria-Palma de Majorca-Alexandria-Riyadh-Abu Dhabi. Both these Caribous later crashed in Abu Dhabi—302 on July 2, 1970 and 301 on October 28, 1976. (Shrive Col)

1969 he and George Northrop ferried a Caribou to Abu Dhabi. The following month Shrive and Bob Wilhelm demonstrated the Buffalo in the Los Angeles and San Francisco areas, then in Peru. Following this, Shrive did a liaison assignment in Afghanistan. Thereafter he became a victim of shrinking aircraft sales and was furloughed from DHC. Later, he was recalled to deliver Sabre Air's Twin Otter 9V-BCE to Singapore and train the pilots there. In 1971 he had a short-term job flying Lambair's Twin Otter CF-AUS between Halifax and Sable Island on a Mobil Oil contract. In early 1974 he joined Transport Canada as an airways inspector. He worked in the development and administration of aircraft noise abatement procedures. His last big project was the installation of IFR beacons and associated communications in Northwestern Ontario and along the west coast of James Bay and Hudson Bay. Shrive retired in Oakville, Ontario in 1985.

DHC Around the World: Buffalo Expert
Like Dick Gleasure, Michel Pierre spent much of his career involved one way or the other with de Havilland Canada. He started in aviation in 1959 and came by his interest honestly—his father had flown Fairey Battles for the Belgians early in WWII. The son trained in Dinan, France and had a licence a year later. Several years followed of towing gliders, instructing, doing some ag flying. It was the

usual school of hard knocks—living in a room at the airport and earning a slave's wages. In February 1965 Pierre earned his commercial ticket—he already had 1,200 hours. More routine work followed, until September 1967 when a job came up in Rwanda flying a Ce.206. The following year Pierre moved to Zambia to fly Zamair's Aztec. At this time Nigeria was embroiled in its civil war with Biafra, and Pierre got involved with the owners of DC-3 TR-LOR. Their business? Running guns at night from Libreville, Gabon into Uli airstrip. This was short-term, then it was back to Zamair to fly an Islander, and across the Atlantic to work for Air Guadeloupe. This gave Pierre the chance to fly his first DHC product, Twin Otter CF-DHA/F-OGES. Eventually, he moved to DC-3 F-OGDZ. This was good enough work for a young pilot, but gruelling—one day he logged 17.5 flying hours. By the end of 1973 Pierre had over 4,900 hours.

This kind of job offered little promise for the future. On a whim Pierre sent a résumé to DHC in Toronto. To his surprise he heard back—Dave Fairbanks invited him for an interview. He arrived in town in September 1975, and was amazed that his interview lasted all week! Each morning Fairbanks would pick him up at the hotel; they would sit in the office, while Fairbanks fastidiously picked through Pierre's logbooks, asking him about each entry. The only flying done was a famil ride in Turbo Beaver CF-PSM. Pierre returned south with the promise that he'd hear something by February.

When February came and went, Pierre contacted DHC only to hear that Fairbanks had died in the meantime. He hurried to Toronto, where George Neal listened to his story. Nei-

ther he nor anyone else at DHC had ever heard of Pierre's job application. Even though there was little he could do to help, Neal steered Pierre to CIDA, which was looking for a francophone pilot to instruct on the Twin Otter in the Congo. Pierre got the job and spent the next 11 months with Lina Congo (Twin Otter C-GNWD) where he had a good time along with a number of other Canadians like Gerry Deluce. When he returned to Toronto, DHC had a job for Pierre. He joined the company in July 1976 and immediately checked out on the Buffalo—his first flight was in C-GGQF on August 16.

Pierre's first big assignment came in the fall. He and Grant Davidson ferried Buffalo 9T-CBA to Zaire. They left Downsview on October 13 and flew 5:20 hours to St. John's. Next day they reached Santa Maria in the Azores in 6:40, thence to Dakar on the 15th (7:10), Lomé (Togo) on the 16th (5:20) and Kinshasa on the 17th (1:30). Davidson returned to Toronto, but Pierre remained to instruct for the Zaire Air Force. His first session was on October 29. He stayed until August 27, 1977, having logged 311 hours on type since leaving Toronto.

On October 6 DHC again sent Michel Pierre travelling. The task was to deliver Buffalo 5T-MAW to Mauritania in Saharan Africa. This time he was with Bob Hayward. He stayed on till December 1977 training the Africans. This entailed some wartime trips flying federal troops and supplies to hot spots where the Polisario guerrillas were causing trouble. In December Pierre ferried a Twin Otter to Rotterdam, then worked in Toronto till the following June 10, when he and Hayward took Buffalo 811 to Sudan, routing via Goose Bay, Reykjavik, Shoreham, Iraklion and Luxor. They deliberately avoided the often unpleasant Cairo. Like many in aviation Pierre had his opinions about aviation in Egypt, and agreed when some commented, "We always used to admire the Israeli Air Force, that is until we met their enemies."

The Buffalo touched down in Khartoum on June 14. This was to be Pierre's base for six

July 1978 scenes in Mauritania... Crowds of locals are attracted by the arrival of a government Buffalo; then, DHC pilot Michel Pierre leaning on a tank as his Buffalo is readied beyond. (Pierre Col.)

months—he had become DHC's chief instructor among the African nations. Again, there were ops, for Sudan was fighting its own guerrillas. The last flight on this assignment was on December 31, 1977. Pierre left Sudan with a further 377 Buffalo hours. Next came a Buffalo delivery to Dar es Salaam with Barry Hubbard March 13-18, 1978. They took C-GTJV across, where it immediately became air force 9019. These aircraft were all "D" models with 49,200 pounds gross weight compared to 41,000-43,000 for the "A" used by the CanForces. More production flying followed, then Pierre got the keys to Dash 7 7O-ACL and pointed towards South Yemen. Some demo flights were made en route, so the trip was long— March 27 to May 7, 1980. In August and September there was another demo tour. He and Mick Saunders flew to Hong Kong and picked up Dash 7 C-GNBX. They finally reached Toronto on September 21.

On December 13, 1980 Pierre set off in Dash 7 4W-ACK for Sa'an, North Yemen. He arrived on the 18th, and instructed till the following March 19. A few months later he and Barry Hubbard took Buffalo C-GDAF to Douala. He stayed around to teach—the local boys needed plenty of help. All this would wear down the average person, but not a keen aviator. Those like Pierre couldn't get enough flying, loved the travel and always got along with the locals. Of all the places Pierre enjoyed, out-of-the-way Yemen was his favourite. He found it an enticing land where, when one stepped from the plane, he moved into a long forgotten past.

In the first days of Dash 8 flying at DHC, George Neal made Michel Pierre chief production test pilot. He tested nearly all the first 120 aircraft, learning a great deal about such things as the certification game. Then, in 1989 Pierre was despatched to Zaire with Buffalo 9T-CBA. He trained the locals for several months, and again became embroiled in local feuding. An action called the Shaba War was in progress, so Pierre ended up flying more combat. These were busy days with up to 14 daily rotations. The Buffalo operated without major snags for

three months. Once Pierre asked DHC president Russ Bannock about getting some danger pay for flying in one African country after another. Being an old fighter pilot, Bannock may have agreed off the record, but turned down the idea. If Pierre wanted to get embroiled on ops, that was his business. On the other hand, Pierre knew that if word got back that he wasn't giving de Havilland's customer his full support, he'd get his knuckles rapped later.

On January 26, 1990 Michel Pierre made a flight at Kinshasa in Buffalo 9T-CBA. He returned to a new job—flying the Challenger at Canadair. He made his first trip in aircraft 6002 on February 14. In July the following year he left Canadair—the organization there was not to

his liking. Soon he was back in his comfortable old Buffalo. Florida-based Avior hired him for Operation Lifeline Sudan and on August 13 he was in Calgary to collect Buffalo C-GDOB (c/n 108). It had a story of its own. It had started as a DHC demo aircraft, then a deal had been made to sell it to Ecuador. This fell apart soon after DOB flew south. Some functionary in Quito was able to squelch a good deal by reminding people that the government could not buy used aircraft. DOB, with 110 hours, was

about 10 hours passed being officially "new". Pierre was sent to retrieve it, his pockets bulging with $14,000 in cash needed to pay the parking fee at Quito. Aegis, a Florida company interested in some used Dash 7s, had a look at DOB while in Toronto, and bought it for a song. Soon it was leased to Avior and making an honest living. Lifeline Sudan was based at Lokichokio, Kenya. The ferry flight took DOB from Calgary to Thunder Bay, St. Jean, Burlington (Vermont), St. John's, Ponta del Gada (Azores), Cairo and Djibouti. On August 22 DOB pulled into its spot on the tarmac at Nairobi. It made its first trip from "Loki" two days later. Michel Pierre flew it through 1992.

The Buffalo was ideal in this environment. It carried a good load, had the range for the trips (often 400 nm legs), was fast enough, and handled Sudan's rainy season with ease. Full payload landings on muddy 1,500-foot strips were routine. Snags were manageable, but it

probably helped that DOB barely had 3,000 airframe hours. After his tour Pierre did other contract flying including instructing for Aegis on Dash 7 N726AG. In March 1993 he received a call from North Yemen. After many years the lads hadn't forgotten their old teacher. They needed someone for proficiency

(Above) Hard-working freighters at Loki: Avior's Buffalo, and a Southern Air Transport Hercules. Mostly veterans of the USAF and the CIA, SAT's people were a knowledgeable and hard-nosed bunch who freighted anything anywhere for anybody, so long as the price was right. If there was a famine or revolution, SAT was sure to be there. (Larry Milberry)

(Right) ET-AHJ departing Loki for Sudan on March 30, 1993. It remained active there into 1996. As C-GBXL this Buffalo was ferried to Addis Ababa May 29-June 4, 1981 by Wess McIntosh and Lou Sytsma of DHC. (Larry Milberry)

rides. Pierre spent a delightful three months renewing old friendships—his students of former years now were captains. It was fun getting back to work again on 7O-ABD and ACZ.

Another Buffalo stint followed North Yemen. This time Pierre was running his own company, Montreal-based Gwenair. There was more work in Sudan. Through an old contact, Pierre was able to lease 5V-MAG, one of Togo's two Buffalos. July 5, 1993 he had it at Loki with a mainly Canadian crew. They were soon into their routine, flying relief to places like Watt, Nassir, Kongar, Yuay, Kapoeta and Lafon, most being legs about 2:30 hours. There were dreadful stories of civil war during this period, like the case of one southern faction surrounding the city of Kongar. It was ruled by another southern group, but was overwhelmed by its rivals. Everyone in Kongar was exterminated—some 50,000, but there was little media coverage.

During this phase 5V-MAG needed an engine overhauled. It was shipped to Rolls-Royce in Montreal. Engines were never the Buffalo's big selling point. Its T64s started with TBOs under 1,000 hours. MAG's were legal by this time to 1,500. The overhaul bill was over a half-million US dollars. It was even worse news for Ethiopian Airlines which, at the same time, needed two engines overhauled. They had quit simultaneously—in flight! The day was saved because of the Buffalo's in-flight APU. The result was a repair bill of about US$1.5 million. By 1997 Michel Pierre was in Yemen, this time in a joint venture with Yemen Airways—Gwenair was operating Twin Otters Nos. 664 and 764 bought in Switzerland and Adu Dhabi. These were involved mainly in oil exploration.

The welcoming sign at Loki; then, one of the strangest sights there—a pile of canoes. Some schemer in the UN had requistioned them for the locals for a fishing project. Nobody in Sudan, however, seemed to be interested in the enterprise. (Larry Milberry)

Togo Buffalo 5V-MAG at Loki while on a half-year relief contract to Montreal-based Gwenair at Loki. A turbo DC-3 on UNICEF work is beyond. (Pierre Col.)

A Buffalo at its best. Captained by Dave Fairbanks, a pre-production example approaches to land at the East River Park in downtown New York. It was participating in Metro 66, a demonstration of STOL capabilities sponsored by the FAA in September 1966. (DHC 25653)

Bob Fowler and Seth Grossmith, two of the Canadians who flew the augmenter-wing Buffalo at NASA's Ames Flight Research Centre in California. (Halford Col.)

Two configurations of the joint NASA-Canadian government research Buffalo program, first with two engines (Trenton, September 1982), then with four (Abbotsford, July 1986). This program dated to Avro Canada days and was headed by former Avro engineer Don C. Whittley. On the first version, Rolls-Royce provided the nacelles/engines; Boeing modified the plane in Seattle. The new wing used an augmentor flap through which air was blown, leading edge slats, and rotatable, thrust-vectoring ducts on the engines. Thomas E. Edmonds of Boeing conducted the first flight at Seattle on May 1, 1972. In the next few years more than 250 hours were accumulated, proving that steep approaches could be made as low as 55 knots, and takeoff in as little at 350 feet. (Mike Valenti, Ken Swartz)

DHC – A Colour Overview

De Havilland types served in the Canadian bush from the late 1920s. Few remained by the time colour film became available after WWII, but some early transparencies survived, and a number of vintage D.H. types have been restored. Seen here is D.H.87A Hornet Moth CF-AYG. It came to Canada in 1936 for Consolidated Mining and Smelting. It was at Winnipeg on September 5, 1961, crammed into a corner in the flying club hangar. A helpful mechanic went out of his way to pump up the tires, push AYG outside, unfold its wings and dust it for a kid with a Kodak Pony camera who wanted a few shots. In 1964 AYG was sold in the US, but returned in 1992, when purchased by the Reynolds Aviation Museum of Wetaskiwin. Then, George Neil's 1936 D.H.87B at Mountain View, Ontario on June 26, 1971. George bought EEJ in England in 1969. By 1996 he had flown it for 177 hours. (Larry Milberry)

Fox Moth DJB (left) was at Kenora on September 6, 1961, still working in mineral exploration. The restored Tiger Moths (below) were at Trenton on September 10, 1983. This type also worked in the bush after WWII. (Larry Milberry)

The DHC-1 Chipmunk was de Havilland Canada's first major post-WWII venture. A little two-seater with a 145-hp DH Gipsy engine, it was to replace wartime trainers. It built on DHC's great experience with the Tiger Moth. Pat Fillingham of De Havilland (UK) flew the prototype Chipmunk at Downsview on May 22, 1946. Production went ahead with orders for the RCAF and other nations like Chile, Egypt, India and Thailand. Canadian production totalled 218, but 952 were built in England and another 60 in Portugal. The Chipmunk also equipped a number of Canadian flying clubs. Many members who learned to fly it later worked in the airlines. This ex-flying club Chipmunk was used for promotion by Commander Aviation and was at Toronto in June 1965. It had been bought for about $2,000 from CADC at Dunnville, Ontario, then was meticulously restored. John Matthews of Ingersoll, Ontario designed the Golden Hawk-type colour scheme, and Keith Hopkinson of Goderich did the paint job. The boys at Commander Aviation—Dan Seagrove, Gordon Schwartz, Irv Watson— had a ball with the plane till it was sold at a nice profit. The second Chipmunk, restored in RCAF colours, was at Downsview on June 2, 1974. Many Chipmunks survived into the 1990s. (Larry Milberry)

On the step... Northcoast Air Service's DHC-2 Beaver EYN bounces along at Seal Cove near Prince Rupert in a Grant Webb photo from 1985. The Beaver is one of Canada's most famous bushplanes. Its reputation is richly deserved—after 50 years most still were commercial workhorses, even though some already were museum pieces. But how famous is famous? In 1996 a new book appeared—*The Immortal Beaver: The World's Greatest Bush Plane.* Many would argue that this plane or that is just as great. While Beech 18, DC-3, Norseman and Otter come to mind, one type has no rival. With nearly three times the land mass (and "bush") of Canada, the USSR developed the Antonov An-2, a chunky, 12,000-pound, single-engine bushplane. First flown in 1947, the An-2 remained in production in the 1990s, 30 years after the last Beaver was rolled out. More than 12,000 An-2s were built in the USSR, unknown thousands in China (DHC built 1,692 Beavers). An-2s served around the world, even making North American inroads in the 1990s. While Canadians love the Beaver, the An-2 has no rival as a bushplane.

CF-GQD was the 76th Beaver and served the Hudson's Bay Company for may years. Later it went to La Ronge Aviation, with which it is shown on wheel-skis in May 1967 working in the area between Dubawnt and Great Slave lakes. (Ed McIvor)

(Left) CF-FHP of Austin Airways was Beaver No. 57. It was at Ogoki Post, north of Nakina, Ontario on March 3, 1975. (Larry Milberry)

(Below left) Air-Dale's Beaver QXI, an old US Army L-20, being flown at Sault Ste. Marie by Earle W. Richardson on July 12, 1991. (Larry Milberry)

(Below) Gordon Schwartz shot this freshly-painted Beaver waiting at Malton for delivery to the owner in Fort William. It had spent most of its years since new with Bowaters Newfoundland Ltd.

(Left) Every aircraft requires regular maintenance. Sometimes a few hours of work suffices. At other times days, weeks or months are needed. This Beaver was receiving some serious attention at Sault Ste. Marie in July 1991. Rebuilding old Beavers is an important activity. Some 3,000 hours of labour go into a complete rebuild. The finished product was selling in the mid-1990s as high as US$500,000—a new Beaver in 1950 was about Cdn$30,000. (Larry Milberry)

(Above) Beaver No. 51 CF-FHS of Atlas Aviation was on tundra tires on the southwest coast of Ellesmere Island when photographed over the summer of 1965. Tundra tires were developed by Welland Wilfred "Weldy" Phipps, one of Canada's great Arctic aviators. He passed away on October 29, 1996. (via Lorna Deblicquy)

(Left) Former Ontario Lands and Forests Beaver OCP riding high at Ignace, Ontario on August 24, 1976. (Larry Milberry)

This Nestor Falls Fly-In Outposts formation was shot over the Lake-of-the-Woods on August 9, 1994 by bush pilot Richard Hulina. Beech 18 WYR was flown by Derek Campbell, Beaver MDB by Mike McFayden.

Close in with Ignace Airways Beaver C-FOBV near Ignace in August 1994. Kent Van Vliet was flying OBV. (Richard Hulina)

(Above) Ignace Airways' cut-down Cummer van added some atmosphere in this Beaver shot from July 1995. (Larry Milberry)

Richard Hulina gasses up for a flight on July 11, 1995; then (left) gets his Ignace Airways Beaver going from home base. (Larry Milberry)

Beavers at Prince Rupert in the early light of August 2, 1993. (Larry Milberry)

559

(Left) Marg Watson of Sudbury Aviation in her famous red Beaver IUU. She was taking some of the Snowbird flight demonstration team on a trout fishing trip to Solace Lake on June 9, 1992. (Larry Milberry)

(Below) Pilots David Burns and Jerry Rowe service the Wasp Junior in one of Adlair's Beavers at Cambridge Bay on August 19, 1995. Their job through the brief 1995 float season was hauling Arctic char from river sites to a fish plant at Cambridge Bay, whence the prepared char went to market in Edmonton and Winnipeg. (Larry Milberry)

(Above) This Beaver was used in rescuing disabled Otter CF-GBY on Reindeer Lake in May 1969. See page 566. (Ed McIvor)

(Right) Ramsey Airways Beaver AEG near Sudbury on August 28, 1988. AEG disappeared from the CCAR in 1995. (Larry Milberry)

(Below) The light float plane and flying boat have been a lifeline on the BC coast for generations. Here coast pilot Sandy Parker steps off Beaver "Kynoc Chief" at Masset in the Queen Charlottes on August 2, 1993. His passengers all summer were the likes of fishing, forestry and environmental people, tourists, doctors, students and travelling sales people. Parker, a top coast pilot, later lost his life in the crash of a Goose. (Larry Milberry)

A Harbour Air Beaver (ex-OPAS) posing off Vancouver on June 20, 1992 for Henry Tenby's lens.

(Below) Beavers of Washington-based Northwest Aviation await passengers at Campbell River on August 10, 1993. They were heading north for sport salmon fishing. (Larry Milberry)

(Above) Coast pilot Peter Killin photographed Beaver FRL at Chamiss Bay, BC on January 23, 1984 after a rare fall of snow.

(Left) Otter LCP and Beaver FHT of Gulf Air near Campbell River on September 14, 1980. LCP started in 1961 as G305 in the Ghana Air Force. It made its way back to Canada in 1974 through a German agent. One of the earliest Beavers (c/n 55), FHT spent most of its career on the West Coast. (Kenneth Swartz)

Air BC's Beaver CF-JFQ at Shearwater in May 1982. Built by Don Braithwaite's Gulf Air, this was Bella Bella's first airport. A replacement strip later was built closer to the village. (Uwe Ihssen)

561

The Beaver is a photogenic subject whatever the season or setting. Here AEE prepares to depart from a lake near Atlin, BC in June 1981. (Kenneth I. Swartz)

Many government agencies used the Beaver over the decades. This photo from September 1968 shows a DOT Beaver coasting to its buoy at Toronto Island Airport. (Larry Milberry)

(Above) Beaver OCN about to get wet at Vancouver on April 11, 1992. Cut-down, front-wheel-drive vehicles that otherwise are ready for the junk yard live on when converted for launching float planes. (Larry Milberry)

(Above) Beaver GCY gets away from Tofino April 9, 1992. GCY joined Tofino Air in February 1987. The 216th Beaver, it served earlier with Northward Aviation of Edmonton, Leo Doucette (Magic Lake Estates), Burrard Air, and Shuswap Flight Centre of Salmon Arm. Tofino Air was a long-serving Vancouver Island operator catering to logging, fishing and the local Indians. Another activity was taking money away from tourists anxious to watch whales from the air. (Larry Milberry)

(Right) A classic bush flying scene—Beaver JXO departs with a canoe strapped to the float. (Andy Graham)

With break-up months away a tourist camp Beaver waits out Old Man Winter at Sioux Lookout in February 1992. Hundreds such bush planes have nothing to do all winter, so the economics of owning a plane must be studied carefully by camp operators. (Larry Milberry)

The US Army operated hundreds of Beavers from Korean War days into the 1980s. It was happy to give the Beaver a place of honour in its museum at Fort Rucker, Alabama. The last three US military Beavers belonged to the US Navy test pilot school at NAS Patuxent River, Maryland. They still were active in the 1990s. (Larry Milberry)

DHC hoped to revive Beaver production with the PT6 turbine engine, but only a few Turbo Beavers were sold. In the 1980s, however, West Coast conversion shops like Viking and Kenmore began doing a booming business converting old Beavers to the PT6. Here is the prototype Turbo Beaver at Toronto Island Airport on April 15, 1966. (Larry Milberry)

A trio of turbine Beavers. Labrador Airways' CF-GFS (PT6A-20) was at Northwest River (above) in April 1975 on contract to the Grenfell Mission. It was used mainly for medevac trips. C-FOER (left) of the Ontario Ministry of Natural Resources was at Sudbury on July 11, 1991. It had been up-graded from the -20 to the bigger -27 engine, and had the cabin extended by pushing back the rear fuselage bulkhead. N754 (below) was an unusual conversion with a Garrett turbine engine seen at Port Hardy in May 1980. (Larry Milberry, Uwe Ihssen)

(Right) CF-VQD, the 466th and last Otter, was bought new in May 1967 by Laurentian Air Services. It moved to Sabourin Lake Airways 10 years later. Here it was loading for Sabourin at Red Lake on July 15, 1991. Then (below), VQD's front office at the same time. (Larry Milberry)

(Right) Al McNeil of Sabourin aimed Otter PHD at the camera boat for this action shot. PHD had spent years at Goose Bay with the US Army. Bradley bought it in 1976, then it went to Sabourin in 1980. When Sabourin folded in 1996 PHD went to Cargair of Quebec. (Larry Milberry)

(Below left and right) A 1950s scene at a Newfoundland outport as EPA Otter CF-GCV does a few minutes of business before pushing on. The second Otter built, it had been DHC's performance and systems test prototype. It later served with PWA, Northern Thunderbird, Silver Pine Air Services and Walsten Air Services, with which it is seen at Kenora on July 16, 1991. On September 20, 1995 GCV crashed in Salveson Lake on a charter from Kenora. All six aboard died. (A.J. Lewington, Larry Milberry)

(Below) Otter maintenance at Sioux Lookout on September 17, 1996. Peter Breton was doing a 100-hour inspection on Slate Falls Airways' C-FNWX. Then, an Air North Otter roaring off a lake near Atlin, BC. (Richard Hulina, Kenneth I. Swartz)

Otter QRI at Vancouver on August 18, 1983. It had gone to the US Army in April 1959, but returned to Canada in 1972 with Trans Mountain Air Services. In 1976 it moved to Gulf Air, thence to Alert Bay Air Services and Air BC. On January 22, 1982 QRI suffered damage (below) in a start-up fire at Port Hardy, but soon was rebuilt. (Larry Milberry, Uwe Ihssen)

(Above) The 330th Otter was delivered to the US Army in June 1959. It appeared on the USCAR as N212NY in May 1981 and was shot at Watson Lake in October 1983. (Mike Valenti)

RCMP Otter CF-MPO departing Thompson, Manitoba on July 20, 1973. As No. 3686 it served the RCAF 1954-62. The last RCMP Otter, it was retired from Edmonton on April 14, 1992. It was donated to No. 825 Air Cadet squadron at Yellowknife; then was sold to local operator Direct North. In June 1996 MPO had a PT6 installed in Vancouver and was exported as N87AW. (Larry Milberry)

(Below) Coval Air's dock at Campbell River on April 10, 1992. Otter LCP was in the latest of numerous schemes it had worn since new in 1961. Companies like Coval got their start when Jim Pattison of Air BC sold his float operations in the 1980s. (Larry Milberry)

(Above) Otter BEW in Kuby's yard at Kenora in July 1991. Worse-looking wrecks regularly rise from the scrap heap, looking like new. This ex-US Army machine came back to Canada for White River Air Services in 1973. While serving Pickle Lake Air Service it crashed on February 4, 1986—it failed to get airborne while trying to take off from slush-covered Fressel Lake and ended in the trees. (Larry Milberry)

Strange place for a float plane... La Ronge Aviation's Otter CF-GBY blew a cylinder half way across Reindeer Lake one day in May 1969. In 1996 Ed McIvor recalled: "George Sewell had to set down on the ice, which was just about to disintegrate. I flew over in Beaver GYR and picked him up; then flew in a mechanic to make repairs." GBY (c/n 5) had been Wardair's first Otter. It went to La Ronge in 1967 and was destroyed there on February 25, 1974. (Ed McIvor)

Stan Deluce founded White River Air Services soon after WWII. Below is his Otter CF-ODT at Tukanee Lake, Ontario on July 23, 1974. ODT had started with the OPAS in April 1957. Then (right), White River's ex-US Army Otter CF-QMN at Dick Watson's dock in Wawa, Ontario on July 17, 1991. (Larry Milberry)

CF-AYR had begun in 1963 as UN No. 307, and later was N9744F. Bannock Aerospace brought it back to Canada in 1970, then it went to Geoterrex till sold in 1981 to Alkan Air of Whitehorse. Here it was at Malton June 11, 1972. (Larry Milberry)

(Left) An Ontario Lands and Forests water bombing Otter in Toronto Harbour on August 3, 1966. Its water bombing tanks were built into the floats. (Larry Milberry)

(Left) A Beaver-Otter duo at Sept-Îles on November 18, 1992. JZN started its days in 1957 in Quebec, where it stayed a decade with companies like Wheeler Northland, Northern Wings, Fecteau and Alexandair. (Larry Milberry)

Lashing a boat to an Otter at Yellowknife on July 17, 1992. CZP had gone new to CPA in 1955. Later careers were at PWA, Northward Aviation, Carter Air and Raecom Air. Like most aged bush planes, it has been through the mill. One incident was a crash at Nahanni Butte on January 30, 1976. In 1995 CZP moved to Williams Aero Service. Jim McAvoy (right), who started flying in 1948, is shown ready to cast off CZP for François Lake. He and his brother had set up McAvoy Air Service in the 1960s using the Fairchild 82, Fox Moth, Howard DGA 15 and T-50. Jim retired from bush flying in 1994. His brother had disappeared years earlier in the F.82. (Larry Milberry)

A Laurentian Air Services Otter (ex-US Army 57-6108) goes back on floats at Uplands in May 1977 after a winter on skis. LAS was formed in June 1936, its first base being at the Ottawa-New Edinburgh Canoe Club. Winter flying was from Uplands. For 1938 LAS operated two Wacos and an Aristocrat. It logged 1,807 hours and carried 495 passengers. LAS ran into financial problems in the 1990s and closed its doors. In 1997 LAA was flying with Air Schefferville. (Larry Milberry)

Richard Hulina took this classic photo on Pine Lake near Nestor Falls on August 9, 1994, while Shane Pope and Will Hay were doing some training in the Otter. This machine was ex-US Army 57-6131 and ex-N1UW of the University of Wisconsin. In 1979 it became C-FYYS with Northwestern Flying Services of Nestor Falls, Ontario.

The Otter was widely used by the RCAF/CAF, especially by the Air Reserve. No. 3669 of No. 102 Communications and Rescue Flight was at Trenton on June 11, 1966, while Montreal air reserve Otter 3663 (below) was at Maxville, Ontario in July 1976. (Larry Milberry)

(Below) An Air Reserve Otter from Toronto beats up 411 Squadron's ops tent during 1971 summer training at Gagetown, New Brunswick. (Mike Valenti)

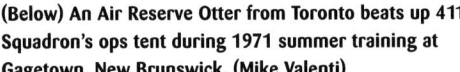

(Above) This US Army Otter was at Downsview on February 24, 1960. The colour indicates that it was destined for the Arctic or Antarctic, although this Otter's end came in a 1971 crash in Viet Nam. (Al Martin)

(Above) Museum Otters. No. 57-6135, delivered in August 1958, spent most of its career with the Army's Golden Knights parachute team. It went to the US Army Aviation Museum in May 1981. Then, No. 144672 at the Naval Air Museum at Pensacola, Florida in October 1994. It was delivered in 1956 and served in Antarctica with VX-6 Squadron for nearly a decade. Later it was with other specialized units including the Pacific Missile Range at Point Mugu, California. It joined the museum in 1975. (Larry Milberry)

(Left) The last US Army Otters (55-3251 and 58-1696) were at Goose Bay with Strategic Air Command. This example, seen March 18, 1975, went to Labrador Airways a year later as C-GLJH; the other joined the Alaskan Civil Air Patrol as N5323G. (Larry Milberry)

(Left) Over the winter of 1994-95 Otter CF-ODJ of Green Airways was converted in Peterborough, Ontario to the 1,000-hp PZL engine. This mod allows an Otter to get airborne quicker than a standard machine, but it vibrates more and burns more fuel (40 gph compared to 30). The added cost is partially offset by slightly higher speed, but the PZL's main feature is improved reliability, which customers and insurers favour. Here ODJ was supplying a fishing camp at Birch Lake on July 13, 1996. In 1996 Green Airways added Otter C-FLEA to its PZL roster. (Larry Milberry)

Warren and Chummy Plummer used two PZL Otters in the 1990s to serve their fishing camps on Great Bear Lake. Here C-GSMG leaves Plummer's Great Bear Lake Lodge on Dease Arm on August 8, 1996. Then, a view of Plummer's C-FKOA departing Yellowknife on July 3, 1993. (Larry Milberry)

(Left) Ray Baert flies SMG on August 7, 1996 from Great Bear Lake Lodge to a fishing spot across Dease Arm on Clearwater Bay. The Narakay Islands are beyond. Baert, who grew up on a farm north of Edmonton, learned to fly in 1966 at Slave Lake. His first job was flying Stinson Reliant CF-OAW. In 1976 he and Evelyn Comrie started Raecom Air in Yellowknife. In time Raecom grew to include eight aircraft, but eventually was sold and became Air Tindi. (Larry Milberry)

(Above) Rae Baert's first trip on August 7 was taking four guides and the day's equipment and supplies to Clearwater Bay. He then returned for a party of eight American fishermen. Here are guides Carl Rolf of Edmonton, Darren Anderson of Fort Saskatchewan, Alberta, Tate Pinder of Calgary and Shane Jonker of Winnipeg. (Larry Milberry)

(Left) Plummer's Otters usually wintered at Selkirk, Manitoba. (Larry Milberry)

The original turbine Otter was C-FMES, converted to a PT6A-27 (680 shp) in 1978 by Cox Air Resources of Edmonton. Turbo Tarling shot it at Cold Lake in December 1978.

The $1-million PT6 Otter conversion slowly was gaining in popularity in the 1990s. More operators seem willing to pay for a PT6 than a PZL, which was only one-fifth the cost to convert. The efficiency of a turbine, its quiet compared to an ear-shattering PZL, and an edge in speed over the piston Otter accounted for this. PT6A-135 (750 shp) conversions became commonplace. Air Tindi's (previously Central Mountain Airlines, converted by AeroFlite Industries of Vancouver) was shot over Yellowknife by Henry Tenby on April 24, 1994, then was seen at the dock there by Larry Milberry on August 18, 1995. Pilot Bryan Knutson was returning from a trip to an exploration site. XUY had been delivered to Philippine Air Lines in 1956 as PI-C54. It returned to Canada in 1968 for Lamb Airways, then went to Omineca Air Services, TPA, Gulf Air, Air BC and Central Mountain. Then, views by Milberry and Tenby of Harbour Air's C-GUTW: early in the morning at Vancouver on April 11, 1992; then over the Georgia Straits on June 20 that year.

C-GCMY was in Wolverine colours at Yellowknife on July 17, 1992. CMY later moved to Harbour Air. On August 18, 1996 it crashed into a mountain in the Queen Charlottes, while en route Tasu Sound-Alliford Bay in bad weather. The three aboard were lost: pilot Stanley Raymond Bray and passengers Tara A. Rowe and Desmond E. Caldwell. (Larry Milberry)

(Left) CF-EBX at Moh Bay on amphibious floats to pick up loggers on April 10, 1992, was ex-RCAF 3680, then became CF-MPK of the RCMP. It went to EB Exploration 1972-74, then returned to the RCMP, went to Brown's Air Service, Harbour Air and, finally, Western Straits in 1988, a company owned by Irvin and Thomas Olsen. In mid-1989 it underwent the Vasar PT6 conversion. At dusk on September 27, 1995, while approaching Campbell River from a camp at Triumph Bay, near Kitimat, it crashed in fog. Pilot Dan White and seven of nine passengers died. The Kenmore machine was at Campbell River August 13, 1995 on a sport fishing charter. In 1996 Washington-based Kenmore had five Turbo Otters and 10 Beavers. Included in their busy work each day were four Seattle-Victoria skeds. (Larry Milberry)

(Bottom left) C-FHPE of Athabaska Airways was at Stony Rapids January 7, 1993. With its PT6 there would be no trouble starting in -40° – no engine tent, no blow pots required. HPE was ex-Burmese Air Force UB651. (Larry Milberry)

(Below) Waweig Lake Outfitters' brightly-painted turbine Otter was at Geraldton, Ontario in May 1996. (Andy Graham)

The prototype Twin Otter at Toronto Island during float trials on August 31, 1966. CF-DHC-X served DHC to 1981 and later was donated to the NAM. (Larry Milberry)

The Twin Otter took naturally to the British Columbia coast, where it gradually replaced types like the Otter, Goose, Mallard and DC-3. This pair was in downtown Vancouver on August 3, 1974. AWF was lost at Bella Coola on September 22, 1976. Then, AWC in Air BC colours landing at Vancouver International in January 1987. Note the CPAir connector zapper. AWC had been ordered new in 1968 by Executive Airlines (N204E). It joined Air West in May 1972, Air BC in 1980, and was sold to North Caribou Air in 1989. Air BC phased out its last Twin Otters in late 1995, selling the operation to West Coast Air. (Larry Milberry, Maxwell Col.)

(Below) Twin Otter No. 2 (originally CF-SJB-X) at Vancouver July 31, 1993. In December 1967 DHC sold it to Air Commuter of Cleveland as N856AC. This deal was short-lived—within a few weeks the plane became CF-PAT of Pan Arctic Oil. A decade later it moved to Ptarmigan Airways, thence to Kenn Borek and Harbour Air, with various leases in between. (Larry Milberry)

(Left) One of the early Twin Otter sked operators was norOntair. Operated by the Ontario government, it linked small northern centres with larger ones like Sault Ste. Marie, Timmins and Sudbury, from where connections could be made with Air Canada. CF-TVO joined norOntair in 1971. After years of service it went to the Ontario Provincial Police. Here it is at Toronto Island on August 31, 1973 wearing the original norOntario stylized loon. (Larry Milberry)

(Left) CF-GON of norOntair in a later colour scheme. It had just landed at Timmins on August 31, 1982. GON was sold in 1985 to Labrador Airways. (Larry Milberry)

(Above) One of Canada's longest-serving pilots was Archie Vanhee, who served for a half-century in aviation. He began flying in the 1920s and continued into the 1980s. Along the way he flew for operators like Canadian Airways, the RCAF and CPA. Here he was at Timmins as a Twin Otter captain with Austin Airways on September 5, 1982. (Larry Milberry)

(Below) The Twin Otter performed a host of tasks from commuter airliner to bush plane, air ambulance and military transport. Aero survey Twin Otter CF-YTH was with Survair when photographed training in Toronto's outer harbour July 6, 1972. In 1985 it was sold as N800LJ, wandered around the US with various operators, went to Aero Taxi in Colombia, then returned to Canada for Kenn Borek in 1993 as C-FQIM. (Larry Milberry)

(Above) Sabourin's Twin Otter DMR near Red Lake on March 26, 1992. Norm Wright was flying it, while Clark Griffith flew Ce.185 C-GDSJ as the photo plane. Norm was heading for Malahar, a camp 80 miles away, with seven passengers, a skidoo and other freight. By this time DMR was a 19,500-hour airplane. In June 1994 it went to work for Pacific Coastal Airlines. It was lost disastrously at Fish Egg Inlet, BC on September 17, 1994, when an elevator cable broke. Al Ross, an old-time coast pilot, was the sole survivor. Somehow he was thrown clear and was picked up with grave injuries. In 1995 the TSB reported that there were 1,432 seaplane accidents in Canada 1976-90. Of these 234 were fatal, taking 452 lives. (Larry Milberry)

(Right) Ontario's Ministry of Natural Resources operated de Havilland products since the 1920s. Later it purchased the Beaver, followed up with the Otter and Turbo Beaver, then the Twin Otter. Here Twin Otter OGB cruises near Sioux Lookout. It was shot from Glenn Tudhope's Found FBA-2C on July 14, 1991. (Larry Milberry)

(Below) Two famous DH products: Ontario's Twin Otter OEG and ex-RCAF Comet SVR at Mount Hope in October 1966. (Al Martin)

(Below) Sky Walker was a short-lived Twin Otter and Islander operation connecting the Niagara Peninsula with Toronto Island. It failed after its Islander ran out of gas and ditched fatally in Lake Ontario. Twin Otter C-FCSB is seen at Toronto Island. A 1967 model, it was operating in 1997 with the Sky-diving Centre of Greater Washington as N229YK. (Kenneth I. Swartz)

(Below) Port Menier on the west tip of Anticosti Island is the central community on the 200-mile long sand bar. Aviaton plays a key role linking it with the mainland. On November 18, 1992 La Ronge Aviation's Twin Otter VOG (ex-Wardair) was serving Anticosti deer hunting lodges. Pilot Robert Burns' run was 135 miles down island with five stops en route. The provincial government by this time had decided to let Anticosti's population fade—children would not be allowed to live on the island once they finished high school. Anticosti would revert to its natural state and be a haven for deer hunters and naturalists. (Larry Milberry)

(Right) One of the Arctic's many Twin Otters. This one of Bradley Air Services was at Resolute Bay on June 30, 1993. It joined the company nearly 20 years earlier, so had been everywhere across the top of Canada. Bradley was operating seven Twin Otters in early 1997. (Larry Milberry)

(Left) Twin Otter No. 37 with the same registration as Tom Lamb's first airplane. It was at Rankin Inlet on July 19, 1973. AUS served first with Golden West in California, then came to Lambair in 1971. It ended its days with Labrador Airways—on October 11, 1984 it crashed fatally in bad weather while en route St. Anthony's-Goose Bay. (Larry Milberry)

(Below) C-GTNO at Whitehorse in October 1983. It went new to Trans North Turbo Air in May 1980. Bannock Aerospace bought it in 1985, then re-sold it in Australia as VH-XSW. (Mike Valenti)

Views of Twin Otter C-FGOG at Yellowknife: first, taxying in the channel near the Old Town on July 17, 1992; then on skis in fresh livery at the Air Tindi base on May 6, 1996. GOG's original owner in 1972 was Gulf Oil Canada. (Larry Milberry)

Twin Otter C-GARW arrives at Hay River with school children on July 13, 1992. When new in 1973 it had served Downtown Airlines as N200DA. Ptarmigan bought it two years later. (Larry Milberry)

Further views of ARW. It's seen on its way from Yellowknife to MacKay Lake Lodge on July 2, 1995, shot from Matt Wasserman's Mooney. Len Robinson and Trevor Rod were on the flight deck. Then (below), ready for a May 6, 1996 Yellowknife-Little Discovery Mine charter with Kevin Elke and John Reimer (IZD had just pulled in from a Rae Lakes-Lac le Gras-Lac la Martre sked with Trevor Rod and Terry Kilcommons). (Henry Tenby, Larry Milberry)

(Above) Even after retiring, Max Ward operated Twin Otter C-FWAH from Yellowknife. Here he is July 17, 1992 at Yellowknife after a flight from his lodge. Then, he's seen taxiing on Back Bay on August 17, 1995. Bryan Currie of Great Slave Helicopters flew the photo ship—Bell 206B C-FGSD. Ward had purchased WAH new in 1969. It went to Chevron Petroleum in 1980, operating mainly in West Africa till Ward brought it home in 1985. (Larry Milberry)

(Right) C-FOEQ, the 44th Twin Otter, started with the Ontario government in 1967. Here it was in Calm Air colours at Toronto Island on June 20, 1975. (Larry Milberry)

C-FSCA was the 17th Twin Otter. In 1966 it began with Shell Canada. It was shot at La Ronge in 1986 while with Athabaska Airways. (Grant Webb)

The RCAF/CAF operated several CC-138 Twin Otters from 1971. By 1997 four remained, two having been destroyed and two sold. Here is 13804 on April 23, 1996 near Yellowknife, 440's home after it left CFB Edmonton. Then, 13805 landing at Abbotsford on August 6, 1993. (Henry Tenby, Larry Milberry)

(Left) De Havilland always needed a demonstrator for sales purposes. Here CF-DHA (c/n 337) was being demonstrated in Yugoslavia in May 1972. It later was delivered to NKA of Japan as JA8798, then went to various operators including Scenic Airlines, where it was N331SA. (Gleasure Col)

(Below) CF-DHA at Farnborough with the Concord and an Avro 748 in May 1972. (Gleasure Col)

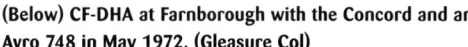

(Left) Series 100 Twin Otter YA-GAT and a second BAA Twin Otter at Kabul ready for a day's work in January 1972. Delivered in April 1968, GAT crashed at Bamyan, Afghanistan on April 18, 1973. (Gleasure Col)

A Twin Otter 300 at Goma, Zaire on August 8, 1994. Delivered initially to Air Illinois in 1980, it ended as 9Q-CBO with TMK Commuter in December 1989. The Twin Otter is ideal for countries like Zaire, where short, rough strips are more the norm than paved runways. (Larry Milberry)

Wounded Twin Otter CF-BEL (above) on a frozen New-foundland lake in the winter of 1974. It was delivered to Bell Canada in November 1969 and was operated for them by EPA. It later went to Labrador Airways, thence to Miami Aviation—1977 (N65308), Syd Aero—1978 (SE-GTU), Pilgrim Airlines—1983 (N126PM), Business Express—1986 and the Massachusetts Institute of Technology—1986. Then (right), Twin Otter QKZ come to grief at Red Lake on March 19, 1992. It stalled on takeoff—a case of questionable airmanship. Only the pilots were aboard; they were lucky to escape with only their pride damaged. (Gleasure Col., Ron Melville via Joe Sinkowski)

(Left) Series 300 HC-ASJ takes on 4,200 lbs. of fuel at the jungle strip of Havacaci, Ecuador in April 1971. ASJ was delivered in July 1970 to Transportes Aereos Orientales. Thereafter, its history is encyclopedic: 1973—N26TC, Omni Aircraft Sales; 1974—N288Z, Inter Mountain Aviation; 1975—Evergreen Helicopters; 1976—C-GFXJ, Patricia Air Transport; 1978—N288Z, Mountainwest Aviation; 1978—C-GFXJ, Air North; 1979—Commutair; 1980—Air North; 1981—Sierra Pacific Airlines; 1983—Banco de Pence, Crown Air, S.M. Miranda; 1987— KNP Air Lease, Crown Air; 1988—C-FCUS, 660 Syndicate Inc., Crown Air, Rocky Mountain Aircraft ; 1989—VH-TGC, Australian Regional Airlines, Vanair. (Gleasure Col)

(Right) C-GNZT-X was an aero survey Twin Otter for the Civil Aviation Admin-istration of China. It was at Downsview in April 1979 and was later registered No.510 in China. Canada had been eying the Chinese market since the 1960s. P&WC and DHC were among the first western aerospace companies with ties there. They made a few sales and assisted China in developing the indig-enous Y-12, a Twin Otter look-alike with PT6s. In time sales were made of Dash 8s in a market (1.2 billion people) that cried out for regional aircraft. (Gleasure Col.)

JA8798 (ex-CF-DHA) of NKA at work in Osaka in June 1974. Then, demo pilot Don Rogers with two NKA men. Rogers had started flying before WWII, became chief test pilot at Avro Canada, then held flying and training positions at DHC. (Gleasure Col.)

(Above) The Twin Otter has proven ideal for work in the savannah. These of Kenya Airways were at Nairobi's Wilson Airport on March 21, 1993. (Larry Milberry)

(Right) This Twin Otter began in April 1968 with Miami Aviation, but returned to Canada in 1972 to join the National Research Council. Andy Graham photographed it at home base (Uplands) in February 1996.

(Left) Twin Otter 300 LN-BNM went new to Widerøes in 1974 and served till lost in a crash at Namsos, Norway on October 27, 1993. (Anton Heuman)

(Right) The second prototype Caribou in its attractive colours at Toronto Island Airport on September 2, 1961. In 1971 it was sold in the US as N6080 where it remained active into the 1990s. (Larry Milberry)

(Left) There were five pre-production YAC-1 Caribous for the US Army. This was the first and also the first Caribou in South Viet Nam. It later spent several years at Fort Rucker and the Kwajalein Atoll missile range before retiring to the Army Aviation Museum at Fort Rucker in 1983. The "0" prefix on the tail number was the US Army/USAF way of indicating that an airplane had reached the 10-year point in its career. (Larry Milberry)

(Right) The US Army's fifth pre-production Caribou was 57-3083. Here it was visiting Downsview on August 30, 1975 with the Golden Knights parachute team. (Larry Milberry)

(Above) Kwajalein ramp scene. Caribou 66-254 (ex-N491GA, later N9984) sits alongside a MATS Douglas C-133 Cargomaster. While the Caribou did inter-island supply on the Kwajalein atoll missile test range, the C-133 flew skeds from the US mainland. (Gleasure Col.)

(Right) John Wegg, aviation historian and publisher of *Airways* magazine, took this photo of N9984 (ex-66-254) landing at Sacramento Metro Airport in May 1987.

Caribous on pierced steel planking at Qui Nhon in June 1965. (Gleasure Col)

Caribous of the 17th Aviation Battalion from Fort Orde at Tonapah, California. Aircraft 9730 later pranged at Quang Tra Bong while 9728 was lost to mortar fire at Phu Cat. (Gleasure Col)

(Right) La Sarre Air Services/ Propair operated Caribou VGX on the James Bay Hydro Development. Here it was at La Grande, Quebec on August 21, 1979 to collect a load for Paint Hills on the James Bay coast. VGX started in 1961 as Swedish Air Force No. 50001. It was in Australia 1965-69 as VH-BFC, then worked in Ecuador and Oman. It reached Quebec in 1977; after nine years it was exported to El Salvador, where it flew on CIA operations. (Larry Milberry)

(Left) Caribou VYX at Winnipeg in July 1984. It began with Guyana Airways, then came to Canada for a series of operators. On November 10, 1987 it crashed disasterously at Ross River in the Yukon. (R.W. Arnold)

No. 5326 in one of the schemes worn by RCAF Caribous over the years. Considering their frequent UN assignments, the basic white scheme prevailed. While at home, some red often was added. This Caribou went to Tanzania in 1971, but was later noted as N1016P. In these views it was at Trenton on June 11, 1966. (Larry Milberry)

(Left) NewCal Aviation's ill-fated turbo Caribou touches down at Gimli after its first flight in November 1991. (R.W. Arnold)

(Below) One of the first US Army Buffalos (63-13688) visiting Oshawa, Ontario in June 1965. (Larry Milberry)

(Right) Gary Vincent caught Buffalo 115453 landing at Brampton, Ontario on September 17, 1972. This was the original colour scheme for the 15 CF Buffalos that were accepted in 1967. By 1997 only 442 Squadron at Comox operated the type in the CF.

582

Other Buffalo schemes. Nos. 454 and 458 were at Comox August 6, 1974 in 442's overall white treatment of the day; 452 was at Ismailia April 20, 1978 in standard UN colours; and 458, doing a STOL landing, was at Sechelt, BC on February 24, 1987 in the overall yellow that typified SAR Buffalos in the 1980s-90s. (Larry Milberry)

(Below) This DHC-5 was the demonstrator for the proposed civilian "Transporter" version of the Buffalo. The project was abandoned when sales failed to materialize. Dave Thompson caught the Transporter at North Bay on May 30, 1980.

(Above) A 442 Squadron Buffalo just as the Snowbirds flight demonstration team swooshed past it near Comox. The scene was caught on April 6, 1992 from a Canadair CL-41 Tutor flown by "Snowbird 10"—Capt Réal Turgeon. (Larry Milberry)

The original US Army Buffalos had long careers at many specialized jobs, from the augmentor wing research project to this example used by the US National Centre for Atmospheric Research. It was at Downsview on August 31, 1973. (Larry Milberry)

The stately CL-44 was Canadair's product of the late 1950s. Here CL-44 15923 of 437 "Husky" Squadron poses over Niagara Falls in the summer of 1962. (DND PL21979)

Canadair's main plant at Cartierville at the height of the 1950s boom. The view is south towards Montreal along Laurentian (now Marcel Laurin) Blvd. The small hangars in the foreground belonged to Laurentide Aviation (Canadair Ltd. 18212)

Canadair Projects

No Canadian aircraft company in the 1950s-60s was as busy with so many programs as Canadair. Starting with DC-3 conversions and the North Star, it never seemed to rest. Korean War and NATO requirements brought F-86 and T-33 production. Concerns over the USSR's submarine fleet prompted the RCAF to order a new anti-submarine patrol plane—the Canadair CL-28 Argus, based on the Bristol Britannia. The Argus, in turn, developed into the CL-44 transport. The RCAF eventually needed a Sabre replacement. This led to licence manufacture of the F-104 Starfighter for the RCAF, USAF and various European nations. Missile R&D, and subcontracting, e.g. making components for the Bomarc missile, brought further growth and profits.

Meanwhile, Canadair developed its first indigenous design, the CL-41 jet trainer. It flew in January 1960, powered by the CP&W-designed JT12. Ottawa, however, decided against the JT12 so as to give GE J85 licence production to Orenda in Toronto—business there was in the doldrums following the Arrow cancellation. The JT12 went to P&W in the US, where it became a huge success. When the Northrop F-5 Freedom Fighter was adopted by the RCAF (designated CF-5), Canadair won the licence production contract. Work was extended when the Netherlands ordered a version. Meanwhile, Canadair was studying a twin-engine utility plane and doing VTOL research. Many non-aviation projects also kept Canadair active, everything from all-terrain vehicles to architecture and marine interests.

The CL-44

The RCAF gave the go-ahead for the CL-28 Argus in early 1954. The design team began with thousands of Bristol Britannia drawings, modifying and replacing them as required. While look-alikes, the two planes differed greatly. The Argus, for example, had the Wright R3350 piston engine; the Britannia had the Bristol Siddeley Proteus turboprop. The RCAF required piston engines for their rapid response to sudden power changes in a low-level combat scenario, compared to slow-responding turboprops. Besides, the R3350 was more fuel efficient at low level. The first Argus flew at Cartierville on March 28, 1958. Thirty-three were completed, the last rolling out on July 13, 1960. The Argus served till 1980, by when Canada was re-equipping with

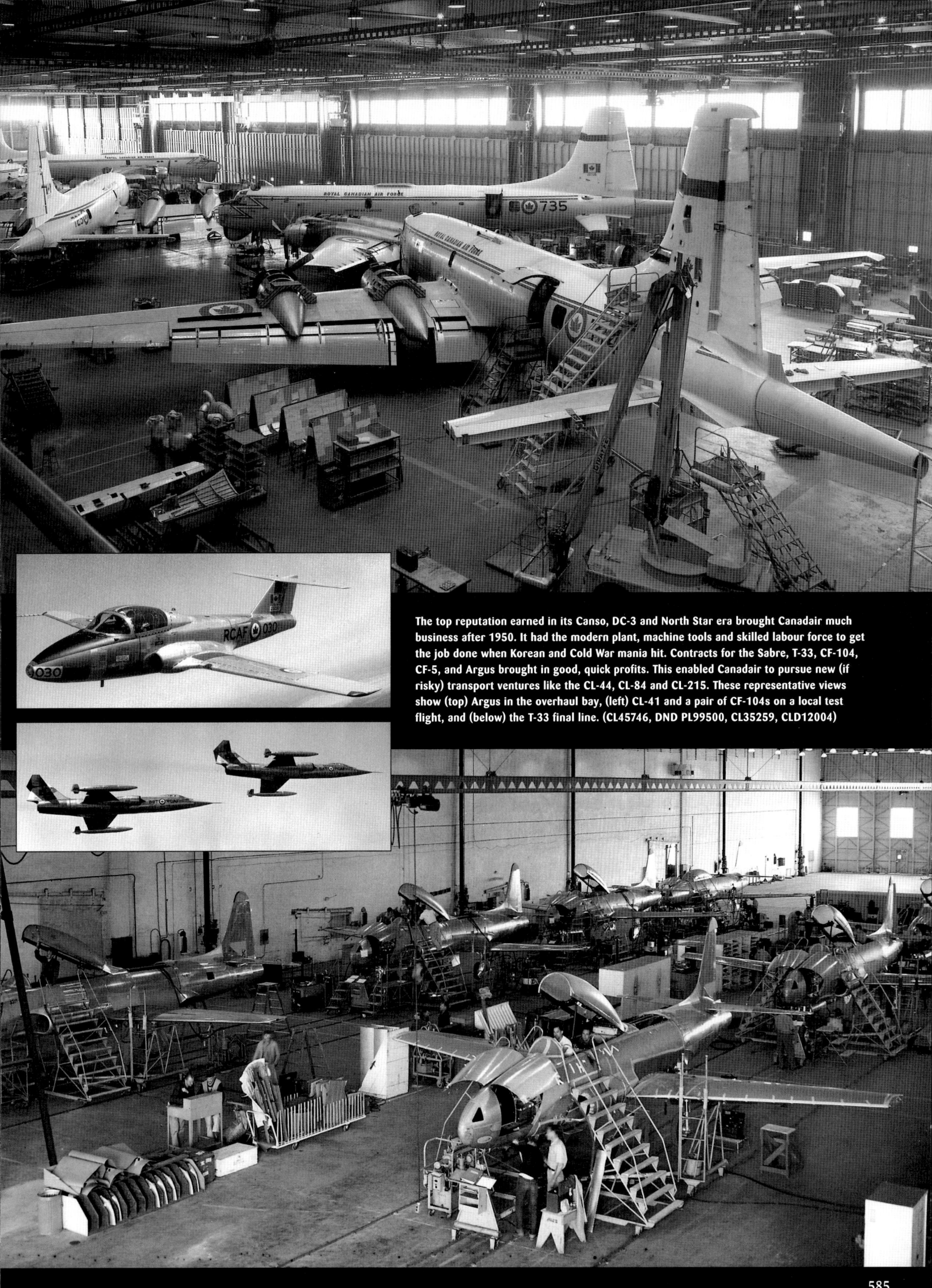

The top reputation earned in its Canso, DC-3 and North Star era brought Canadair much business after 1950. It had the modern plant, machine tools and skilled labour force to get the job done when Korean and Cold War mania hit. Contracts for the Sabre, T-33, CF-104, CF-5, and Argus brought in good, quick profits. This enabled Canadair to pursue new (if risky) transport ventures like the CL-44, CL-84 and CL-215. These representative views show (top) Argus in the overhaul bay, (left) CL-41 and a pair of CF-104s on a local test flight, and (below) the T-33 final line. (CL45746, DND PL99500, CL35259, CLD12004)

CL-44 Production List

Aircraft	c/n Operator	First Flight	Delivery	Remarks
15501	1 RCAF	15-11-59	1-5-62	Prototype, later 15921 and 106921, CF-DSY of Beaver Enterprises 18-11-71, YS-04C of TACA 7-74, returned Beaver, HK-1972 of Aerocondor, crashed near Medellin, Colombia 22-2-75.
15502	2 RCAF	29-3-60	20-3-62	Later 15922 and 106922, to Beaver 18-11-71, 9Q-CWN of SGA 12-11-73, to Tramaco 1978, withdrawn from use (WFU) and stored Kinshasa 1978.
15503	3 RCAF	20-5-60	19-7-60	Later 15923 and 106923, CF-CHC of Beaver 18-11-71, HC-AYS of Andes Airlines 7-9-73, stored Guayaquil 1986.
15924	4 RCAF	11-11-60	23-12-60	Later 106924, to Beaver 18-11-71, LV-LBS of Aeortransportes Entre Rios (AER) 8-72, WFU, stored Buenos Aires 1972.
15925	5 RCAF	25-3-61	23-5-61	Later 106925, to International Aerodyne 13-11-70, LV-JSY 13-11-70, crashed on takeoff Miami 27-9-75.
15926	6 RCAF	22-3-61	8-5-61	To Beaver 18-11-71, LV-JZR of AER 11-71, to OB-R-1005 Aeronaves del Peru 20-3-75, 9Q-CKQ of Vic Air Cargo (VAC) 11-82, Virunga Air Cargo 1-84.
15927	7 RCAF	21-1-61	3-3-61	Later 106927, Beaver 18-11-71, LV-JYR of AER 20-7-72, crashed near Santiago, Chile 20-7-72.
15928	8 RCAF	4-11-60	27-4-61	Later 106928, Beaver 18-11-71, LV-PRX and LV-JZB of Transporte Aereo Rioplatense (TAR) 11-71, LV-JZB of AER 1-76, CX-BKD of ALAS, crashed Montevideo 10-10-79.
CF-MKP-X	9 Loftleidir	16-11-60	25-5-65	Converted to CL-44J, TF-LLH Loftleider, later Cargolux 2-12-71, Aero Uruguay 4-78, Bab El Mandeb Airlines 28-8-78, EI-BGO AER Turas 12-5-79, WFU, scrapped Dublin 6-86.
15929	10 RCAF	21-1-61	5-2-61	Later 106929, C-GADY of Beaver 18-11-71, C-GADY of Batchair 28-10-74, 9Q-CWS of SGA 22-12-74, 9Q-CWS of Tramaco 1978, WFU, stored Kinshasa 1978.
15930	11 RCAF	10-1-61	6-2-61	Later 106930, Beaver 18-11-71, 9Q-CWK of SGA 12-11-73, 9Q-CWK of Tramaco 1978, 9Q-CWK of Katale Aero Transport 1983.
15931	12 RCAF	25-4-61	12-6-61	Later 106931, C-GACH of Beaver 18-11-71, C-GACH of International Air Lease 10-74, QB-R-1104 of Aeronaves del Peru 12-12-75. Crashed northern Peru 28-8-76.
15932	13 RCAF	11-7-61	30-3-62	Later 106932, CF-JSN of Beaver 18-11-71, HC-AZH of Andes Airlines 5-74, WFU, stored Guayaquil 1-86.
CF-MYO-X	14 Seaboard World	13-3-61	13-7-61	N124SW of Seaboard World, later G-AWUD of Transglobe, N124SW of Trans Meridian Airways (TMA) 5-69, TF-CLA of Cargolux 1-5-72, TR-LWF of SOACO 24-3-76, TR-LWF of Air Charters, 5A-DGE of United African Airlines, HC-BHS of AECA Carga 15-11-80, scrapped Miami 1982.
N446T	15 Flying Tiger Line	24-1-62	3-3-62	Later Conroy Aircraft 12-68, Mobil Oil, damaged beyond repair Anchorage 1-5-69.
N447T	16 Flying Tiger Line	24-7-61	16-8-61	Later Conroy Aircraft 12-68 and converted to guppy version (f/f 26-11-69), TMA Air Cargo, 8-7-70, British Cargo Airlines (BCA) 20-8-79, EI-BND of Heavylift Cargo Airlines 10-82.
N448T	17 Flying Tiger Line	4-2-62	16-3-62	Later G-AWWB of TMA 31-12-68, VR-HHC TMA 28-11-75, G-AWWB of BCA 20-8-79, N908L of Caribbean Air Express, N908L of Aeron International Airlines 11-88.
N449T	18 Flying Tiger Line	6-2-62	10-4-62	Later N449T and G-AXAA of TMA 13-1-69 and Express Flug Service, N123AE of Air Express International and Aeron 1982, N123AE of Wrangler Aviation 2-89.
CF-NBP-X	19 Flying Tiger Line	15-4-61	10-7-61	N450T, later G-AZIN of TMA 10-11-71, subsequently British Air Ferries, Express Flug Service, Limburg Air Cargo, BCA, PK-BAW of Bayu Indonesia 8-82, WFU, stored Jakarta.
CF-NND-X	20 Flying Tiger Line	10-5-61	1-6-61	N451T, later TF-LLJ of Loftleidir 30-4-68, TF-LLJ of Cargolux 4-1-70, OO-ELJ of Young Cargo and Cargolux from 4-3-75, TR-LVO of Affretair 25-8-75, subsequently Air Gabon Cargo, Royal Air Maroc as TR-LVO. Burned at Harare, Zimbabwe 5-2-82.
CF-NNE-X	21 Flying Tiger Line	17-5-61	4-7-61	N452T, later G-ATZH of Transglobe, TMA, British Air Ferries, BCA, crashed into sea off Hong Kong 2-9-77.
N453T	22 Flying Tiger Line	25-8-61	13-12-61	Crashed on landing at NAS Norfolk, Virginia 21-3-66 (first CL-44 crash).
CF-NNM-X	23 Seaboard World	24-6-61	7-7-61	N125SW, later G-AWDK of Transglobe, TMA and Tradewinds Airways, PK-BAZ of Bayu Indonesia 8-79, WFU, stored Jakarta.
N454T	24 Flying Tiger Line	8-11-61	25-1-62	Later G-AXUL of TMA (27-12-69), Tradewinds, Tropical Airlines, BCA, N104BB of Blue Bell Aviation (Wrangler) 1980.
N455T	25 Flying Tiger Line	11-7-61	23-7-61	Later Seaboard World 28-8-66 for one month, G-ATZI of TMA (24-4-70), British Air Ferries, HB-IEN of Transvalair 20-3-74, 5A-DHG (later DHJ) of United African Airlines (UAA) 13-2-81 and later Jamahiriyan Air Transport (JAT), N3951C of TRATCO 7-83, 9Q-CQU of Vic Air Cargo (9-8-83) and Virunga Air Cargo, Heavylift 6-85, N103BB of Wrangler 12-85.
N126SW	26 Seaboard World	31-8-61	7-10-61	Later leased to Flying Tiger Line, G-AWSC of Transglobe, N126SW of TMA 1969, G-AWSC of Tradewinds 31-10-70, DBR Lusaka 22-12-74.
N127SW	27 Seaboard World	20-11-61	22-12-61	Later G-AWGS of Transglobe (24-3-68), Tradewinds (6-3-69), N907L of Caribbean Air Express, Millon Air, Aeron, and Wrangler, scrapped in Dallas, Texas 1990.
CF-NYC-X	28 Slick Airways	28-9-61	17-1-62	N602SA, later Airlift International (1-7-66), TMA (1967), Intraco (12-73), Air Calypso (10-1-74), VP-LAT of Cairgo 15-4-74, N62163 of First Pennsyvania Leasing 3-10-74, G-BCWJ of Tradewinds (7-2-75), IAS Cargo Airlines, Tradewinds (1-12-77), Cyprus Airways (3-78), Tradewinds, DBR Nairobi 6-7-78 while hauling cement from Mombasa to Kigali (slid off runway after a precautionary landing).
N603SA	29 Slick Airways	26-12-61	5-2-62	Airlift International 1-7-66, TMA 1967, N100BB Blue Bell Aviation 7-73.
N123SW	30 Seaboard World	7-3-62	5-4-62	Later G-AWGT of Transglobe (29-5-68), Tradewinds (6-2-69), 5B-DAN of Cyprus Airways 1978, crashed Akrotiri, Cyprus 4-11-80.
CF-OFH-X	31 Seaboard World	16-4-62	18-6-62	N228SW Later BOAC (30-9-63), SW (31-10-65), Flying Tiger Line (3-11-65), crashed Da Nang, Viet Nam 24-12-66.
N229SW	32 Seaboard World	20-6-62	12-7-62	Later Flying Tiger Line (2-10-65), SW (1969), G-AWOV of Tradewinds 13-7-70, HB-IEO of Transvalair (19-12-77) and STAC (10-79), 5A-DGE of UAA (10-79) and JAT (1983).
N604SA	33 Slick Airways	11-7-62	25-9-62	Later Airlift International (1-7-66), TMA (11-67), LV-JZM of TAR 1972, derelict Buenos Aires 1986.
N605SA	34 Slick Airways	27-8-62	15-10-62	Later Airlift International (1-7-66), TMA 11-67, LV-JTN of TAR 10-4-75, shot down (or collided with pursuing Soviet fighter) Yerevan, USSR 18-7-81.
CF-PBG-X	35 Loftleidir	17-4-63	28-5-64	TF-LLF, later of Cargolux (8-2-72), Aero Uruguay (1978), N4993U of Cargosur (29-9-78), Air Charter (4-79), TL-AAL of STAC (1979), 5A-CVB and 5A-DGJ of UAA, 9Q-CQS of VAC 11-7-83, EI-BRP of Air Turas 4-1-86, Heavylift 17-5-89, stripped of useful parts at Southend 1992 .
TF-LLG	36 Loftleidir	25-3-64	17-10-64	CL-44J, later Cargolux (11-8-70), crashed Dacca 2-12-70.
CF-PZZ-X	37 Flying Tiger Line	27-11-64	31-12-64	Later G-AZKJ of TMA, Tropical Airlines (5-79), BCA (20-8-79), G-BRED of Redcoat Air Cargo 9-8-80, N106BB Blue Bell Aviation 7-12-83.
CF-RSL-X	38 Flying Tiger Line	10-12-64	14-1-65	N1002T, later G-AZML of TMA (28-1-72), BCA (20-8-79), N121AE of AEI (15-9-80), Aeron (10-85), Wrangler (2-89).
CF-SEE-X	39 Canadair	17-3-65	8-11-65	Originally a D4, converted to CL-44J. Later TF-LLI of Loftleidir 13-3-66, Cargolux 20-7-71, CX-BJV of Aero Uruguay 5-12-77, N4998S of Cargosur (1079), TAISA Peru (1981), HK-3148X of Aeronorte, crashed Barranquilla 6-8-88.

A CL-44 Gallery

(Above) The first CL44D swing tail flew in November 1960 as CF-MKP-X. It went to Loftleidir, which later had Canadair stretch it to a CL-44J. (Canadair Ltd. 24586)

Yukon 15921 (above) taxies at Lahr in a scene from April 1969. This was the first CL-44; it flew in November 1959. Then, 15926 landing at Trenton on August 24, 1970. When sold by CADC in 1971, it worked in South America, then ended its days in Zaire. (W.H. Meaden, Larry Milberry)

Flying Tiger Line, Seaboard World Airlines and Slick Airways were Canadair's key CL-44D customers. N124SW is shown with a specialized loader. The air cargo industry was beginning to appreciate the need for such equipment as business grew through the 1950s. (Canadair Ltd. CL26627, H136)

(Right) Seaboard's N228SW flew on lease to BOAC 1963-65. (Canadair Ltd. CL36775)

(Below) CF-SEE-X was the last of its type, first flying in March 1965. It became TF-LLI of Loftleidir. Here it fuels at Cartierville as an Argus and CL-41 await. (Canadair Ltd. 46446)

Tramp freighter EI-BRP of AER Turas. (Paul Duffy)

Canadair CL-44D Swingtail

Canadair CL-44's are powered by four Rolls-Royce Tyne Turbo-Prop engines.

6' 10 1/2"
10' 9"
Swingtail Opening

84' 0"
6' 10 1/2"

35' 11"   38' 6 1/2"   14' 1"

CL-44 SWINGTAIL CARGO COMPARTMENTS

LOCATION	MAX. LENGTH	MAX. WIDTH	MAX. HEIGHT	FLOOR AREAS	CAPACITY
MAIN CARGO COMPARTMENT	1008.25" (84'0.25")	137.0" (11'5")	82.5" (6'10.5")	924 SQ. FT.	5,528 CU. FT.
FWD UNDERFLOOR CARGO COMPARTMENT	431.0" (35'11")	54.75"	40.0"	162 SQ. FT.	463 CU. FT.
AFT UNDERFLOOR CARGO COMPARTMENT	462.5" (38'6.5")	54.75"	40.0"	173 SQ. FT.	524 CU. FT.
TAIL COMPARTMENT	169.0"	129.0"	81.5"	128 SQ. FT.	711 CU. FT.
TOTAL AREA & CAPACITY				1387 SQ. FT.	7,226 CU. FT.

SPECIFICATIONS

Maximum Takeoff Gross Weight......210,000 lbs.

Maximum Landing Gross Weight......165,000 lbs.

Maximum Payload......61,000 lbs.

Wingspan ....142'4"

Length (overall) ....136'10"

Height (overall) ....38'8"

Fuel Capacity...12,000 gallons

Speed ....350 mph

(Above) CL-44D dimensions and general specs.

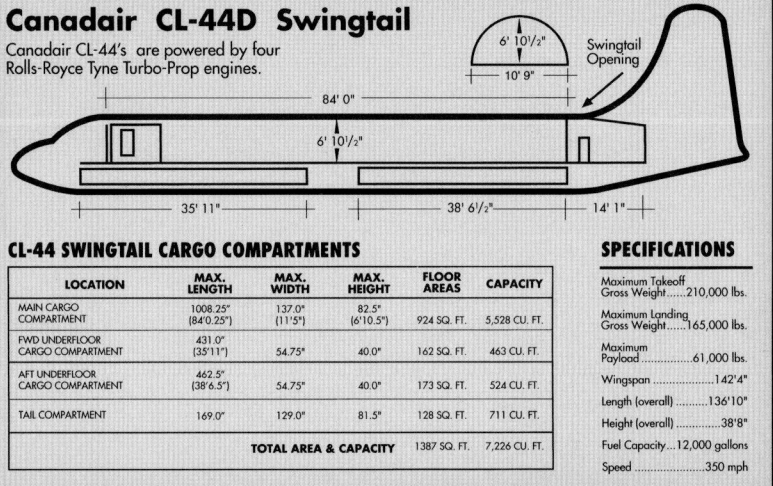

(Right) The CL-44-O "Skymonster" lands at Dorval in June 1994 to attend Canadair's 50th anniversary celebrations. (Canadair Ltd. C71223-19)

(Left) Fresh from storage at Saskatoon, Yukon CF-JSN visited Trenton on September 22, 1973 for ATC's 25th anniversary celebrations. It still carried full air force colours and had its 412 Squadron VIP interior. (Larry Milberry)

the Lockheed CP-140 Aurora (US Navy designation: P-3 Orion). Canadair became a subcontractor building major CP-140/P-3 components.

Even before the Argus flew, Canadair was involved in the Britannia-based CL-44. It would fill an RCAF requirement for a long-range transport to replace the ageing North Star. Ottawa ordered 12 CL-44s. The first CL-44 (RCAF designation: CC-106 Yukon) flew November 15, 1959, powered by the 5,730 ehp Rolls-Royce Tyne turboprop. The fleet entered service in October 1961, when 437 "Husky" Squadron re-formed at Trenton with 10 Yukons. Two others joined 412 Squadron as VIP planes. The Yukon would provide a decade of fine service in the RCAF, flying domestic, NATO and UN skeds; and serving on many emergency humanitarian airlifts—wherever there were floods, earthquakes or famines.

Canadair also built the CL-44D swing tail, which it hoped to sell to the USAF. A deal for 250 was rumoured, but this was not to be. Only 27 further CL-44s were built, mainly swing tails for US cargo carriers hauling military supplies to Southeast Asia. Eventually the CL-44Ds gave way to more modern freighters like the 707 and DC-8. In its bid to jump into the burgeoning trans-Atlantic travel market of the early 1960s, Loftleidir of Iceland ordered four stretched, 188-passenger CL-44Js. The first entered service from New York to Europe in June 1964.

Vagabond 44s
When the Canadian Forces received the first of its Boeing 707-320Cs in 1970, the Yukon retired. Prior to sale by CADC, interested parties were cautioned "... to inspect before bidding and to rely on their own inspection for the preparation of bids." Data quoted in the CADC sales brochure included:

Engines – R-R Tyne 12 Mk. 515 each of 5,730 EHP
Propellers – 16-foot de Havilland
Max weight – 205,000 pounds
Zero fuel weight – 155,000 pounds
Max payload – 60,763 pounds (39,480 in 134-passenger configuration)
Max payload range – 2,723 nm (3,700 nm in passenger configuration)
Range with max fuel – 4,800 nm (5,000 nm in passenger configuration)
Payload at max range – 31,593 pounds (23,918 for 112 passengers)
Passengers – 134, or 80 litters
Usable fuel – 10,150 Imp. gal.
Takeoff over 50 feet – 6,400 feet
Landing from 50 feet – 3,600 feet
Service ceiling – 26,000 feet

Following their RCAF and airline careers, CL-44 survivors passed through a series of tramp freighter operations. Most flew on ad hoc contracts. There could be a rush of work, then an aircraft could be idle for months. Others, such as ones in Zaire, had ongoing work, specializing in transporting live cattle and meat. CL-44D4 N447T was redesigned for outsize cargo up to 26 tons. The plane (CL-44-O) was converted by Conroy Aircraft in California and christened "Skymonster", though most referred to it as "the guppy". Initially it carried Rolls-Royce RB211 engines from the UK to Lockheed in California, where the L.1011 Tristar was being built. In 1982 it joined the Irish carrier HeavyLift as EI-BND to fly on global contracts. It carried auto parts from Cologne to Liverpool, a water purification plant from Cleveland, Ohio to Tunis, barges from Stansted to Burundi, oil field equipment from Brest to Saudi Arabia, race horses from Melbourne to Bangkok, etc. One never knew when a CL-44 might turn up; even crews could be surprised where they landed with their ageing freighters. One day in 1982 Yukon OB-R-1005 arrived unexpectedly in Shannon. Its picture, it goes without saying, soon appeared in *Propliner* magazine with the caption, "Exotic diversion to Shannon with engine problems on December 2... departed after engine change 20 days later. The Yukon is expected to join Zaire's motley collection of old prop freighters on the usual coffee and general cargo missions."

In October 1987 Canadair, under CL-44 technical expert Al Rankin, hosted the fifth CL-44 maintenance conference. Held in Greensboro, North Carolina, it was attended by 50 operators, maintainers and vendors. At the time Greensboro operator Wrangler was despatching 19 flights weekly using four CL-44s. Worldwide there were 14 serviceable CL-44Ds (no Yukons) and no shortage of spares (when one operator at the conference complained that he no longer could find a certain part, a vendor announced that he had 12 in stock). Eventually, only two CL-44 operators remained—Heavylift with Skymonster, and Wrangler/Bluebell of Greensboro. As of February 1992 Wrangler had N100BB and N121AE flying from Singapore. N103BB and N908L were grounded at Greensboro, and N106BB was flying from there. When a Caribbean carrier crashed its Constellation, Wrangler filled in with CL-44s on the busy Miami-Santo Domingo run. By 1996 the Greensboro CL-44s reportedly were scrapped, leaving only Skymonster, but its days were numbered in an era when hundreds of big, fast jet freighters were available at low prices.

The Crash of LV-JSY
Several CL-44s ended disastrously. LV-JSY of Aerotransportes Entre Rios crashed at Miami on September 27, 1975. Operating as F501/90, it was a cargo sked destined for Buenos Aires with stops at Panama City, Lima, Santa Cruz and Asuncion. Aboard were six crew and four passengers; and 13,021 pounds of aircraft engines, automobile and tractor parts, and perfume. Gross weight was 169,161 pounds, including 60,000 of Jet-A fuel. The C of G was within limits.

At 0555 JSY was cleared to taxi , then to take off on R27L. An air traffic controller watched it roll, then turned his attention elsewhere. Moments later he saw flames and alerted emergency services. JSY had failed to get airborne and ran into a canal beyond the runway. Six died. One survivor was the third pilot, who had been standing in the cockpit. He recalled that the takeoff run was normal until time to rotate. At that point the nose gear stayed on the runway. At 165 knots the captain yelled, "We are staying on the ground, we can't take off." He applied reverse thrust and brakes, trying to stop in the remaining third of the runway. The spare pilot fled to the rear of the plane as JSY cut a swath almost 1,000 feet beyond the runway. It finally hit a car, piled into the canal and burned.

The National Transportation Safety Board later reported "no evidence to indicate a failure of the aircraft's systems, structure or power plants". What could have caused the disaster? It was another case of carelessness. The inbound flight engineer had installed a makeshift gust lock on the right elevator on arrival in Miami. This was done since the primary hydraulic gust lock system was unserviceable. The outbound FE did not remove the temporary lock before departure, and did not twig to the fact that one elevator remained locked, even though he would have had warnings from his flight control position indicators (the NTSB noted that the pilots also should have noticed this). The partially burned temporary lock was found floating in the canal, a grim reminder of how human error can bring premature death.

The CL-66
Following WWII, D. Napier and Son Ltd. of Luton, England developed the Eland turboprop engine of about 3,000 shp. Napier envisioned the Eland as a replacement for piston engines like the R2800 in types like the Convair Liner and DC-6. The first Eland ran in 1952. Napier acquired CV340 N8458H to test its concept. Fitted with Elands, it flew at Luton in February 1956. It proved lighter and carried more payload at a higher speed. As the Eland developed, Napier canvassed Convair operators such as KLM and Swissair, but none seemed interested in conversions. Marketing now focused across the Atlantic. The prototype (re-registered N340EL) flew to the US in November 1957 to begin a US civil certification program. En route it stopped in Ottawa to be viewed by the RCAF and Canadair. Airframe N440EL now was converted by Eland's partner in California, PacAero Engineering.

Next came a marriage between Napier and Canadair. Although the RCAF sought a fleet of Viscounts, Ottawa gave Canadair an order for 10 Eland CV440s. Canadair, a subsidiary of General Dynamics (which also owned Convair) would acquire the manufacturing jigs, tools, etc. from Convair in San Diego, as soon as production terminated there. In May 1958 Canadair acquired two new 4-40s for conversion to what it termed the CL-66 (also called the Convair 540). These were CF-LMA and CF-MKO (CF-LMN came later). US type certification came in December 1958.

Canadair rolled out its first CL-66 (RCAF No. 11106, one of the pair bought from Convair) in January 1959. Bill Longhurst and Scotty McLean flew it on February 2. Tours of North and South America, and Europe ensued through 1959. No sales transpired, although Washington-based Allegheny Airlines leased N440EL for trials. Results were favourable and Allegheny established a fleet of seven 5-40s converted by AiResearch of California.

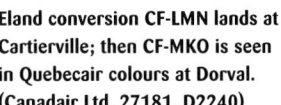

The first Canadair CL-66, seen receiving the final touches before completion, was a converted CV440. Pilots Bill Longhurst and Scotty McLean were on the flight deck for its first flight on February 19, 1959. (Canadair Ltd. 7195)

Eland conversion CF-LMN lands at Cartierville; then CF-MKO is seen in Quebecair colours at Dorval. (Canadair Ltd. 27181, D2240)

Canadair leased its three CL-66s to Quebecair. On January 7, 1960 Scotty McLean and Ian McTavish flew the first Canadair-built CL-66B —RCAF 11151. The tenth (11160—the last of 1,086 of the famed Convair series) was delivered to the RCAF on March 3, 1961. This model never received civil certification.

While the Eland initially operated satisfactorily, before long it became a costly maintenance hog. In one case RCAF 11152 had a catastrophic, in-flight engine failure that nearly resulted in disaster. Other than the RCAF and Allegheny, there were no other customers. In February 1962 Napier terminated the Eland and was absorbed by Rolls-Royce. Soon Elands became unsupportable; airline 5-40s

were "de-converted" to R2800s. Later, many became 5-80s with Allison 501 turboprops.

The CL-84 V/Stol

Canadair's most exotic project was the VTOL CL-84 Dynavert. Conceived in the early 1950s, it was a twin-engine, tilt wing, multi-role design. It was hoped that a demonstration prototype would attract enough attention to justify production. The National Research Council and Defence Research Board supported the idea. In August 1963 the Department of Defence Production gave it a green light, agreeing to fund three-quarters of development costs. Canadair would cover the rest, and a limit of US$10.7 million was stipulated. The US Army

loaned Canadair four Lycoming T53L-11 turbine engines for the prototypes (the T53 powered the Bell UH-1D helicopter).

As early as 1958 Canadair built a special rig for investigating low speed aerodynamic characteristics using models that would be free of the constraints of the wind tunnel. A special propeller test rig was built in 1961. A pilot/engineer simulator was built to study flying qualities, control system design, and ideas for cockpit layout. All this was done by a dedicated group within Canadair's engineering division headed by project manager Fred C. Phillips, an MIT graduate, whose experience included helicopter and convertiplane design at McDonnell in the US.

Many special features were built into the CL-84. There were fuel and hydraulic systems that functioned in either vertical or horizontal modes, a complex mechanism to tilt the wing and engines, a large wing (45 lb. per sq. ft. wing loading), full-span leading and trailing edge flaps, low horizontal tail (below the wing wake in cruise, always in the slipstream regardless of wing angle), triple vertical tails within the slipstream field giving steady directional stability throughout flight transition, nacelles placed low relative to the wing to assist in deflecting the slipstream to give high lift in the STOL mode, and propellers well ahead of the leading edge to reduce vibration loads. The fuselage had side-by-side pilot seating and some 200 cubic feet of cargo. There would be provision for nose-mounted 7.62mm guns and external ordnance.

The intricate part of the CL-84 was the harmonious integration of its triple control system encompassing fixed-wing flight, hovering, and transition between the two. A feature common in helicopters was adopted—SAS, or stability augmentation system, used during transition and hovering. The SAS was sensitive to pitch, yaw and roll and used gyro signals to operate actuators to automatically keep the aircraft stable. The propulsion system had a built-in capability lest an engine fail—the remaining engine would, by means of cross-shafting, give power to the failed propeller. A tail rotor provided sta-

Miscellaneous Canadair Projects

The CL-52 of 1956 was a specialized, one-off Canadair project—converting USAF TB-47E bomber No. 51-2059 as a test bed for the Orenda PS-13 Iroquois engine. Requirements included building mock-ups, the new engine nacelle and mounting pylon; protecting the fuselage and nacelle in case of test engine failure; strengthening the fuselage to take the new loads; modifying the rear (co-pilot) position to accommodate the Iroquois engine controls, and the nose (bombardier/navigator) position for a flight test engineer; installing fuel and hydraulic lines and controls to the test engine; installing test and recording equipment in the bomb bay; and adjusting the C of G by placing four tons of ballast in the nose. The B-47 is seen at Canadair for conversion. It spent its career at Malton, flying 31 flying hours with the Iroquois in 1957-58. (Canadair Ltd. F4057)

(Left) A model of the seven-passenger CL-69 proposed in 1958. A general-purpose or corporate amphibious turboprop, it never got off the drawing board. (Canadair Ltd. 16055-69)

(Left) Canadair became involved developing all-terrain vehicles for military and mineral exploration. Here one of its CL-91s goes aboard an RCAF Caribou. (Canadair Ltd. 51474)

(Right) In 1965-66 Canadair completed a run of 50 CL-218 transit buses needed for the forthcoming 1967 Montreal World's Fair. (Canadair Ltd. 47527)

(Above) Canadair CL-212 hovercraft CF-RSH-X is readied for a test run. Although Canadair spent millions on such projects, not one paid off. (Canadair Ltd. 42843)

In 1960 Canadair was getting interested in reconnaissance drones for target data, damage assessment information, etc. In 1963 Ottawa and the UK became partners with Canadair in this project and the CL-89 drone resulted. First flight was in March 1964 at the US Army range in Yuma, Arizona. The CL-89 entered service in 1972 and was a great success with Britain, France, Germany and Italy. A later development was the CL-289 for France and Germany. (Canadair Ltd. C5584)

CIL House rises in downtown Montreal in January 1961. The outer framework, or "curtain wall" (CL-92), was an innovation of Canadair's architectual department. The wall would transfer wind loads to the structure, making the building more impervious to winter weather. (Canadair Ltd. 23869)

From 1966-72 Canadair made components for the Lockheed C-5A Galaxy, delivering 81 shipsets. This diagram shows some of the parts involved. In June 1983 work began on the improved C-5B—aft cargo clamshell doors, aileron, leading edge slats, etc. The C-5A manufacturing jigs had been kept in outside storage for nearly a decade, but were ready for use after being hauled back inside, sandblasted and inspected. Jigs and tools not needed for the new C-5B work were shipped to the US in 77 truckloads and five boxcars. When C-5B work peaked in 1986, 700 at Canadair were involved. Such subcontracts are complex arrangements. In the case of the C-5B, for example, titanium extrusions were purchased in Los Angeles, travelled to St. Louis for forming, to Canadair for machining, to Texas for electron beam welding, and back to Canadair for final machining before being shipped to the C-5B assembly plant in Marietta, Georgia.

591

bility in transition and hover. Some 30 gears were needed to connect and operate all systems. Two T53s each put out 1,400 shp. The main propeller gearboxes, auxiliary gearboxes and shafting had growth potential to engines of 2,000 shp. Through 1964 testing was done with the prototype tied down to a steel I-beam rig, designed so that the CL-84 could swing as the wind changed. In part, a report from this phase read:

During February the engine-gearbox overrunning clutches malfunctioned and had to be rendered inoperative for some time while modification and additional testing were accomplished. Propulsion testing proceeded so that by mid-February the system was shown to have acceptably low vibration and stress levels throughout the operating range. During March the control system was completed and tested thoroughly without engines operating. Propeller governing tests showed that with wing up the scheduling cam arrangement relating power lever motion to propeller blade angle was not correct, as evidenced by inadequate RPM governing during throttle bursts.... In April, with the propulsion system operating reasonably well, attention turned to control system proving. This was a most important task in that cost and schedule considerations had denied to that program a control system rig or other means of testing the system prior to installation in the prototype... With the help of dynamic records from the strain-gauged tie downs on the rig, the control system was improved greatly for hover flight by a series of small modifications... By early May the wing tilt actuator and SAS were functioning dependably. It now appeared reasonable, after 36 hours of ground testing, to explore cautiously hover flight to determine if there were important flying qualities problems outstanding for the hover mode... After 37 minutes of successful [tethered] flight, ground testing was resumed in late May.

In June the prototype was in the hangar where the gearboxes, shafting and wing tilt actuators were removed and replaced after 56 operating hours. Both engines were inspected. Much additional testing of the control system followed in July. More tethered flights were logged in August 1964, bringing the prototype to about two hours of air time. On May 7, 1965 pilot Bill Longhurst completed the first free flight. Before this he had spent hours in a VTOL simulator, obtained his rotary wing rating, and piloted three different helicopters, including the National Aeronautical Establishment's variable-stability helicopter.

First flight for the CL-84 included four 10-second liftoffs to five feet with the aircraft at

slightly above its 10,600-pound gross weight. Longhurst was confident and in the next two weeks flew the prototype for 37 minutes. Eight more flights were made August 4-7, 1965, including one of 26 minutes. Progress included forward flight from hover (wing tilt 88°) to 33 knots (wing tilt 48°) and return to hover, demonstration that the CL-84 had good flying qualities even without the SAS, demonstration of adequate control in winds gusting to 25 knots, flight at 12,200 pounds, rearward and sideward flight, and turning in flight while in and out of ground effect.

Flying accelerated and included SAR trials in September 1966. Pilots from the Canadian Army, NAE, RAF and NASA flew the CL-84 that month. Later, a group of US military pilots checked out. All found the flying easy. The program had a setback when the prototype crashed September 12, 1967 after 145.5 flying hours. The following February Canadair received an order from Ottawa for three further machines (tail Nos. 8401-8403, designation CL-84-1). The 150 engineering changes incorporated included an increase in all-up weight to 14,500 pounds, lengthened fuselage, increased power (100 shp per engine) and hardpoints for weapons and external fuel. Bill Longhurst flew the first CL-84-1 on February 19, 1970. In February 1972 test pilot F.D. "Doug" Adkins flew a CL-84-1 to the US on a demo tour that included landings and takeoffs from the USN assault carrier *Guam*. Aircraft 8402 next went to NAS Patuxent River for a year of trials by CAF, RAF and USN pilots. On August 8, 1973 No. 8401 crashed following catastrophic failure in one of its propeller gearboxes. The two US pilots ejected safely. No. 8402 later completed trials aboard the carrier *Guadalcanal*.

Although the CL-84 showed great promise, no customer came forward. Development could not drag on forever, and in September 1974 the project was mothballed after 476 flying hours and 709 flights. Meanwhile, Canadair delved into the prospects for a VTOL commuter plane, using knowledge gained from the CL-84. Design of a 48-seat tilt-wing airliner soon was abandoned. Among other reasons, such a plane would require so much fuel as to be impractical. The basic concept survived, however. Beginning in 1981, Bell and Boeing started design of a tilt rotor (i.e. fixed wing) VTOL for the USAF, USMC and USN. This was the V-22 Osprey, which flew in March 1989. There were several evaluation aircraft by the mid-1990s, but full production still was pending. By mid-1996 Congress was considering that $7.1 billion could be saved by cancelling the V-22 and buying conventional helicopters. Meanwhile, a 9- to 20-passenger, PT6-powered version of the V-22 was studied into the late 1990s. The market would be corporate aviation and specialized areas such as SAR, air ambulance, and re-supply of offshore drill rigs. Selling the convertiplane remained as difficult as it had been in the early post-WWII years. As to the CL-84-1s No. 8402 was donated to the NAM; No. 8403, an unfinished airframe, went to the WCAM.

The CL-84 demonstrates its usefulness in the SAR role. Canadair pilot Bob Simmons volunteered to be the "dummy" for these trials. (Canadair Ltd. 51082)

(Below) Canadair board member Henry Marx, project leader Fred Phillips and chief test pilot Bill Longhurst during the early days of the CL-84. (Canadair Ltd. 45499)

No. 8401 was the first of three CL-84-1s in CAF colours. It later crashed. (Canadair Ltd.)

CL-84 No. 8401 was restored for display by AirTech, a Peterborough engineering company formed in 1977 by Bogdan Wolski. AirTech represented various Polish companies, and converted Otters and Beavers to PZL engines. After a year's work on the CL-84, AirTech returned it to the NAM in November 1994 via Calgary Gooseneck, a transport company specializing in aircraft moves, e.g. a Sea Fury from Florida to Alberta; Bandeirante N64GA from Oshawa to Miami; an ag plane (modified to resemble an ancient Potez for the movie *Wings of Courage*) from Calgary to Texas; S-58s from Arizona, California and Utah to BC for logging; and AH-1 Cobras from Fort Rucker, Alabama for overhaul at Bristol in Winnipeg. After the CL-84 job, André Therrien and his wife Gretchen (right) drove to Long Island for two Jet Rangers destined for Edmonton. A Montrealer, Therrien had flown for Photo Air Laurentide in Lodestar and AT-11 days, then with Aero Photo of Quebec. He moved to Calgary in 1977. In 1985 he started a motorcycle business. Along with one of his customers, Duncan Smith, he got into trucking. (Larry Milberry)

The CL-215 Waterbomber

Canadair's most enduring project was the CL-215 waterbomber. Such a project was in discussion around 1960—a twin-engine bush plane in the 12,500 pound range. Thus, Canadair and DHC were pursuing a similar project at the same time. While DHC went ahead, the result being the Twin Otter, Canadair digressed, turning its attention to a larger plane to replace Quebec's fleet of Canso waterbombers. At first a large float plane was envisioned, but it was a flying boat design—the CL-215—that finally emerged. It would carry 1,200 Imp. gal. of water compared to the Canso's 900, and use 2,100-hp R2800s compared to the Canso's 1,200-hp R1830s. In February 1966 Canadair launched the program. Bill Longhurst, Doug Adkins and Smoky Harris crewed the prototype on its first flight on October 23, 1967. Type approval followed in March 1969 and deliveries quickly began. Early production models went to Quebec (15), France (12), Spain (2) and Greece (1).

The CL-215 proved its worth in Quebec and throughout the Mediterranean, but stagnant sales caused a stop in production. In 1973, however, the line resumed with the improved CL-215 Series II, 20 of which were built. Sales tours were made in the US and as far away as Japan, but the market remained difficult. By 1979 Manitoba, Thailand, Venezuela and Yugoslavia had joined the list of operators; 15 Series III CL-215s were turned out, followed by 15 Series IVs. Meanwhile, sales efforts never slackened. When CL-215s normally would be

The CL-215 began with various earlier concepts, one being the CL-204 floatplane. The side views show evolution from the CL-204, to the parasol-wing CL-205, to CL-215 with blister cockpit, to the final layout. In the view below, design had moved along and wind tunnel models were being tested at the NRC in Ottawa. Finally, the first three airframes take shape early in 1967. (Canadair Ltd. 34954, 48765, 53583)

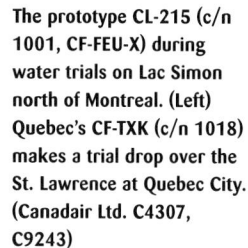

The prototype CL-215 (c/n 1001, CF-FEU-X) during water trials on Lac Simon north of Montreal. (Left) Quebec's CF-TXK (c/n 1018) makes a trial drop over the St. Lawrence at Quebec City. (Canadair Ltd. C4307, C9243)

(Right) Fred Kearns flew Spitfires during WWII. He joined Canadair in 1949 as a timekeeper, moved to accounting, then progressed step-by-step up the corporate ladder. In 1965 he became president, replacing Geoff Notman. Kearns led Canadair through exciting projects like the CL-84 and CL-215, then into the troubled early years of the Challenger. He retired in 1983 and passed away in November 1987. (Canadair Ltd. 48387)

One of Thailand's CL-215s departs on a test flight from Cartierville. (Canadair Ltd. C9456)

Some of Canadair's pilots through the 1950s-70s: (Top row) Al Lilly (Canadair's first chief pilot), Bill Kidd and Scotty McLean. (Bottom row) Ed Coe, Bud Scouten and Bruce Fleming. All these men came from solid aviation backgrounds. McLean, for example, flew in the RCAF in WWII, first as a flying instructor, then as a Mosquito pilot on 107 Squadron. Post-war he had a small flying service in partnershp with Angus Morrison, then flew for KLM. In 1951 he became a corporate pilot along with Ted Lawrence for Canada Packers on an ex-CPA Lodestar bought for $5,400. In 1953 Al Lilly wooed McLean to Canadair, where he stayed through many an interesting program. When he left flying, McLean headed Canadair's government lobby office in Ottawa. (Canadair Ltd. D9485, 4472, S1123, S1122, D9701, D9703)

dormant in Canada over the winter, some would go on southern sales tours or be leased out. In December 1980 CL-215 C-GUKM (c/n 1049) left Montreal for Argentina, where it spent the next few months fighting fires. Pilots Yves Mahaut, Larry Roluf and Mike Ross; and mechanics Maurice Langlois and José Ribiero; and tech rep Jack Forbes comprised the Canadair team. In the first five weeks the aircraft fought 30 fires, made 271 drops and logged 81 hours.

A boost came in 1983, when Ottawa, the provinces and the territories formed a consortium to purchase 29 CL-215 Series Vs. The general plan was that for each CL-215 ordered by Alberta, Quebec, the Yukon, etc., Ottawa would pay for a second. In December 1986 Bombardier purchased Canadair from Ottawa and began to revitalize the company. The first project authorized was to convert the CL-215 to 2,380-shp PW123AF turbines. Two of Quebec's Series Vs became CL-215Ts, the first of which flew June 8, 1989. This program had begun as a study in early 1986 under Ben Locke. After considerable development and modification, the CL-215T entered service with Quebec, France and Spain, but most operators opted to stay with the standard R2800-powered CL-215. Meanwhile, the final CL-215 was delivered to Greece in May 1990. From 1966, when metal had been cut for a prototype, 124 CL-215s were built.

DHC Moves Ahead: Dash 7 and Dash 8
In the late 1960s and early 1970s de Havilland Canada had a mix of activities. Caribou, Buffalo and Twin Otter production continued, and product support for the Beaver and Otter remained profitable; but a new project was on the horizon—the Dash 7 STOL airliner. In this era the concept of downtown-to-downtown air service became a fad. Governments were happy to fund companies for R&D in this field. DHC began market and engineering studies into a STOL commuter in the mid-1960s, but its parent company, Hawker Siddeley in the UK, was not supportive. DHC persevered, collaborating with the DOT and FAA regarding general operating standards. At the same time Ottawa supported the go-ahead at P&WC for a hybrid PT6 to power a 50-seat commuter plane. This became the 1,120-shp PT6A-50. It ran in May 1973 and took to the air on May 10, 1974 in the nose of P&WC's Viscount testbed. Meanwhile, Ottawa funded a sophisticated STOL experiment—Airtransit, an Ottawa (Rockcliffe)-Montreal (Dorval) operation using six modified Twin Otters. The experiment would provide information as to requirements for the Dash 7 and for operating profitable service between centres like Montreal and Toronto.

Airtransit's Twin Otters had the latest in nav and avionics equipment, and such mods as heavy-duty brakes, wing spoilers and propeller de-icing. Service began on July 20, 1974 with aircraft fitted for 14 passengers, not the usual 19. A $20 one-way fare was offered and as many as 30 flights per day operated. The service was well patronized and all the required data were gleaned by the time Airtransit shut down April 30, 1976. A key concept proven was that an interurban STOL service could

function independently and safely within the standard air traffic control system.

In early 1974 DHC, in partnership with Boeing, began marketing the Dash 7. On May 27 the Canadian government purchased DHC from Hawker Siddeley. Ottawa decided to put $77 million into the Dash 7, and expected to break even on its investment with 250 sales. The prototype rolled out at a gala affair on February 5, 1975. On March 27 pilots Bob Fowler and Mick Saunders, with flight test engineers Jock Aitken and Bob Dingle, flew the Dash 7 for the first time. No. 2 flew on June 26. The type certificate was awarded on May 2, 1977 and a European sales tour followed. On

February 3, 1978 Rocky Mountain Airlines of Denver became the first to operate the Dash 7, catering to the booming ski trade in Colorado. Various airlines followed, mainly ones with STOL requirements, or longtime DHC customers like Air Wisconsin, Wardair and the CanForces.

These were the early days of US airline deregulation, so a number of upstart carriers jumped in with Dash 7 orders. After all, they no longer were restricted to the 19-seat, 12,500-pound AUW category of plane like the Twin Otter. Air Pacific, Golden Gate, Henson, Metro, Ransome and Rio all ordered; but some of this action was illusory—the global reces-

Spantax of Spain was one of the early Dash 7 operators. It took delivery of its sole example (c/n 3) in May 1978. It later served on lease to Air BC as C-GFEL and in 1990 became N703WW of Worldwide Resources Ltd. Rocky Mountain's N27RM was the fourth Dash 7 and the first in scheduled use. It is seen accompanying one of the company's seven Twin Otters. After a decade N27RM went to Continental Express. Tyrolean of Austria also was an early Dash 7 operator. Over the years it had three aircraft. OE-HLS was the 22nd Dash 7 and was delivered in March 1980. (DHC 46543, 45807-11, 48697-3)

The prototype Dash 8-100 first flew on June 20, 1983. It later was stretched to become the first Dash 8-300 (below). Its useful days over, DNK was scrapped at Van Nuys, California early in 1990. The Dash 8 became a world leader in its class, and sales, once they got rolling, were steady. In August 1996 Horizon Airlines of Seattle ordered 70 Dash 8s in a move to build a one-plane fleet. Its Fairchild Metros, Do.328s and Dash 8-100s were sold once the Dash 8-300s started arriving. (DHC 52642-2)

(Below) Dash 8 No. 2 near the end of the line at Downsview. It first flew as C-GGMP on October 26, 1983. (DHC 52783-1)

(Left) A neat twosome over Toronto Harbour. Dash 8-100 C-GGTO (c/n 5) was the first airline delivery—it was turned over to City Express in September 1985. Here it flies with Dash 7 C-GGXS. (DHC 55073)

sion of the early 1980s erased much enthusiasm. Companies like Golden Gate and Rio went into bankruptcy as fast as they had appeared. Soon Dash 7s were lined up at storage depots in California and Arizona. DHC salesmen were scrambling in a buyers' market. Elsewhere, operators were finding the Dash 7 expensive to operate and limited on certain routes. The CanForces, for example, placed two Dash 7s on a traditional run between Lahr (West Germany) and Gatwick (London), then found that the Dash 7 could operate nonstop only with a fraction the normal payload. After a few years the planes were returned to DHC. Meanwhile, the STOL craze passed and airlines were effectively using Dash 7s as regular commuter planes.

In May 1979 DHC president John Sandford revealed that DHC was studying a fast, twin-engine, 36-passenger commuter . This was the Dash 8 and it got a green light in September 1980. P&WC had realized by now that, under US airline deregulation, commuter carriers soon would be looking for efficient new aircraft to carry as many as 60 passengers. It launched a new engine series—the PT7. The proof-of-concept prototype (TDE-1) ran first in

December 1977. Like the PT6, the PT7 would be flexible in power output—from less than 2,000 shp to above 4,000, depending on the airframe. DHC adapted the new engine for the Dash 8. After all, there were few options, the main one being the 30-year-old Rolls-Royce Dart, which was a noisy gas guzzler. Its days had passed as far as manufacturers of new airframes were concerned.

In 1980 P&WC introduced a new designation system; the PT7 became the PW100. It first flew on the Viscount on February 27, 1982. The timing couldn't have been better, for the airframers were in a design frenzy. The field included the Aerospatiale ATR42, BAe ATP, Dash 8, Embraer EMB120 Brasilia, Fokker 50 and Saab 340. All but Saab selected the PW100 and the first production engines (for the Dash 8 and ATR) were delivered in January 1984. The first Dash 8 flew with pre-production PW120s. At this time Dash 7 production was tailing off at about 100—far from the magic break-even point.

In April 1984 Prime Minister Trudeau told the House of Commons that his government would keep DHC in order to save jobs and preserve technology. This was in the face of the fact that the new Dash 8 was not yet selling and could be a failure. DHC had lost millions in recent years; Ottawa bolstered it with $1 billion. The press was up in arms, assailing Ottawa at every turn for squandering tax dollars. Such government support, however, was not unique to Canada. Governments in all Western nations were subsidizing their aerospace industries. In Britain, France, Germany and Italy much of the industry was government-owned in the first place; and huge military orders always kept US industry viable. In the 1970s France pumped $5 billion into its industry. It supported Airbus Industrie, when only 57 of its airliners could be sold in the first five years of production. Otherwise, the company would have failed. Instead, it flourished to rival Boeing as a giant in commercial aircraft production.

The reasons for government support to the aerospace industry were simple—hi-tech industry inspires and maintains valuable industrial, scientific and educational spin-offs; and provides jobs worldwide. For Ottawa to have backed planes like the Dash 8 and Challenger, when both had gloomy futures, was prescient. To the dismay of Canada's socialist losers, by the 1990s both companies were leading the world in their product lines.

To the Brink of Disaster

By the mid-1970s Canadair's glamorous projects of previous years had faded; some CL-215 orders, day-to-day product support, and a few subcontracts were all that remained. Employment fell from 9,000 in 1968 to as few as 1,400 in 1975 (employment across the industry in Canada had gone from 48,000 to 23,000). Work had to be found—and quickly. Fred Kearns, president of Canadair, sought a deal whereby Ottawa would buy his company as it had done DHC in 1974. In January 1975 Ottawa took an option to buy Canadair from General Dynamics. GD then sought a partner-

An artist's rendition of the LearStar 600 business jet as envisioned in 1977. Note how the engines exhausted over the horizontal tail, an unlikely configuration. The eventual configuration of the Challenger 600 is shown at right. (Canadair Ltd.)

ship with Ottawa, so long as Canada would buy its F-16 at the end of deliberations for a new fighter aircraft (this was not to be, for Ottawa chose the McDonnell Douglas F/A-18). Purchase offers were put on the table by other parties, but Ottawa, wary of Canadair leaving Quebec, decided to exercise its option. In January 1976, fearing that General Dynamics would not guarantee on-going support, and might let Canadair wither, Ottawa purchased Canadair for $48 million.

An opportunity for new work had arisen in 1975, when Fred Kearns met the famous inventor William P. Lear, the man behind the Lear bizjet. He had a new plane in mind, a long-range, up-scale executive jet of 30,000 pounds. Kearn's vice-president of engineering, Harry Halton, had talked earlier of such a project, so Canadair was interested. It sent a team to view Lear's set-up in Reno. The project was found to be little more than a mock-up and some rough diagrams. Nonetheless, Canadair liked Lear's idea of combining a new wing with fuel-efficient turbofan engines, and the Lear name would be an asset to any project. In April 1976 Canadair signed a deal for worldwide rights to the plane— the LearStar 600. Canadair described it as "an executive jet to accommodate 14... up to 30 passengers in the airliner configuration, or 7,500 pounds of high-priority cargo... non-stop range of over 4,500 miles... preliminary design, as already completed by Lear, incorporates the full supercritical wing and design technology that saves weight and reduces drag to the extent that its load-carrying and fuel-saving capabilities can exceed any aircraft in its category."

Detailed design began at Cartierville under project manager Jack Greeniaus. To head marketing Kearns hired James B. Taylor, who had experience selling Falcon and Citation bizjets. Taylor brought in some top US salesmen and convinced Kearns to set up a US-based operation (Canadair Inc.) in Westport, Connecticut. With orders and options for about 50 planes, Ottawa gave the project the green light on October 29, 1976. Minister of Industry, Trade and Commerce Jean Chrétien made the announcement at Canadair, promising that the project

CHALLENGER 600
Basic Dimensions

Two key men behind the Challenger were Fred Kearns (right) of Canadair and Jean Chrétien, Minister of Industry, Trade and Commerce. Kearns drove the project on every front; Chrétien cleared the way politically. Eventually, Kearns was forced out of Canadair, while Chrétien rose to be prime minister. This October 1976 photo is from the heady days when Canadair was getting the LearStar 600 "off the ground". (Canadair Ltd.)

Jim Taylor was hired by Fred Kearns to head LearStar sales. In 1997 Taylor still was involved with bizjets, consulting to VisionAire, builder of the single-engine, six-seat Vantage (JT15D-5 engine). (Canadair Ltd. 47172-29)

would boost employment there from 1,800 to 3,300. Design proceeded, wind tunnel models were tested (October 1976), the first order for parts was placed (February 23, 1977), a fuselage mock-up was completed (April 1977) and

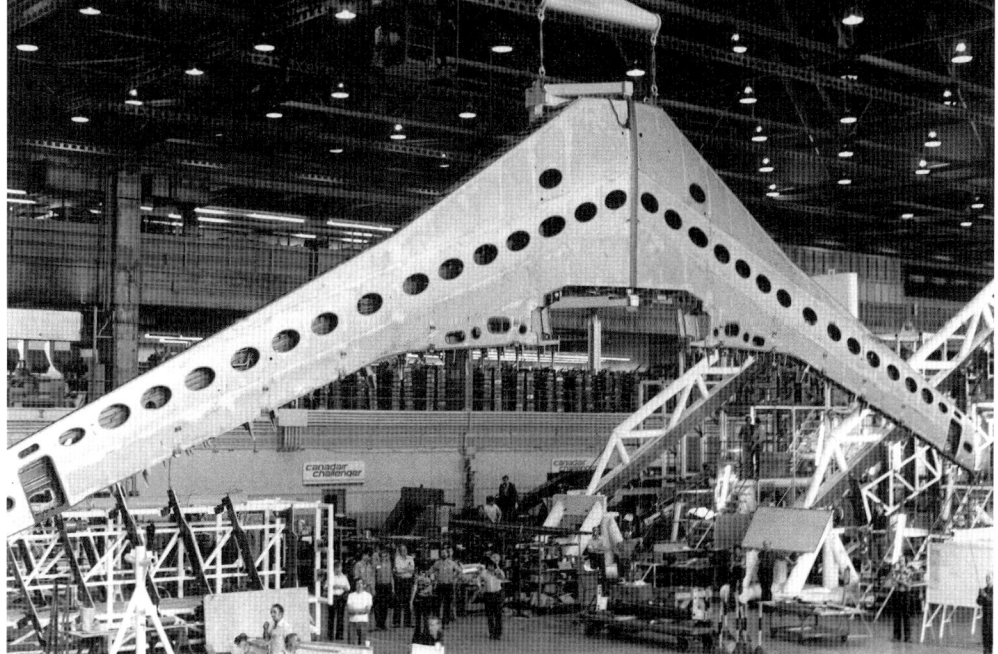

The wing for the first Challenger is moved from its jig on May 2, 1978 (top). Then, it is joined to the fuselage two days later. While much of the airframe was built at Cartierville, there was considerable subcontracting, e.g the rudder was made by Fleet Industries of Fort Erie and the horizontal stabilizer by Canadian Aircraft Products of Richmond, BC. (Canadair Ltd. 29337, C9916)

Bob Lingard, France Bouchard, Bruce Fleming and Bob Germain of Canadair double as models in this 1977 view aboard the Challenger mock-up. (Canadair Ltd. 12692)

things sounded upbeat. On May 11, 1977 Canadair announced that it was re-christening the new plane "Challenger". That month the mock-up went by sea to France for display at the Paris Airshow. In June Canadair announced that its order book showed orders for 80 Challengers. On July 8 an order came from TAG of Saudi Arabia for 21 LearStar 600s. In October the mock-up went to the 30th annual NBAA convention in Houston. At this time the price tag for one of the new planes was US$5,775,000.

By early 1978 Canadair and Bill Lear were at loggerheads. One issue was the concept that, to offer comfort on long flights, the LearStar 600 should have a fuselage diameter of 106" compared to Lear's proposed 88". To Lear this ruined the aesthetics of his design and he balked. He left to follow other projects, primarily his attractive Lear Fan 2100, a small business plane with a PT6 and a pusher propeller. Lear died on May 14, 1978, but his widow pursued the Lear Fan. It flew in January 1981, but no production ensued. Lear's estate later sued Canadair, claiming royalties for the Challenger and its offshoots. After years of litigation a 1995 decision favoured the Lears.

Decisions and controversy came rapidly at Canadair, e.g. which engine to use. One faction favoured the Lycoming ALF-502 of 6,750 pounds of thrust. Another favoured the GE CF34 of 8,660. The former was a fan version of the T55 helicopter turboshaft engine. The latter powered types like the Republic A-10 fighter and Lockheed SA-3 Viking anti-submarine plane. The ALF-502 won the contest on such arguments as being lighter, easier to maintain, and on the verge of growing in power. Going against GE was how FedEx, one of the first to order the LearStar 600 (25 aircraft, later cancelled), had had so much trouble with the GE CF700s in its fleet of Falcon 20s.

A piece of Canadair sales promotion from May 1977 noted of the Lycoming engine: "It's the best engine going... Because the high by-pass fans shuttle the airflow around the engine, the Challenger will be considerably more quiet than the competition... the fan portion is scheduled to carry a 10,000 hour TBO [time between overhaul], the compressor section 6,000 hours, the core section 4,000 hours." Lycoming, meanwhile, was going along with Canadair's pitch that the new plane could fly 4,000 nm. GE, however, stated early on that it could not assure such range with the CF34.

Harry Halton, driven by an ardor to complete the Challenger before Dassault, Gulfstream or Lockheed could introduce next-generation bizjets, established a blistering schedule for completing, flying and certifying the Challenger. Everyone involved was under pressure and something had to give, especially since the Challenger remained undefined—amazingly, from Kearns and Halton down, nobody had nailed down a basic definition to meet performance. Thus, as the first plane was assembled, to call the Challenger a crapshoot would be charitable, according to normal industry procedures. Starting in September 1977, for example, as components were completed

(Top) The CL-600 line looked impressive in 1978, but the bubble of Challenger enthusiam soon burst as setback after setback befell the project. Then, Challenger No. 1 during its first flight on November 8, 1978. (Canadair Ltd.)

Harry Halton (left) with Fred Kearns at the Challenger roll-out ceremony. (Canadair Ltd. C10196)

(Below) While the first Challenger was coming together, a great deal of experimental work was always underway. In this case the effects of bird strikes on the horizontal stabilizer were being evaluated. Carcasses were fired at the stab using a "bird cannon". (Canadair Ltd. 13858)

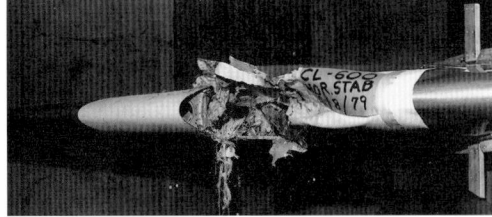

(often made by hand), they would suddenly be scrapped and replacements made to new drawings. An example was how the first wing was designed with a traditional structure. Suddenly this was thrown out and a new wing with a milled skin was introduced. This costly game kept pushing up the break-even point. Something else unorthodox was how Canadair was running two separate engineering departments, one under Halton, dealing only with the CL-600; the other under Frank Francis, cover-

ing all other projects. While this focussed usefully upon the CL-600, it created friction, e.g. Halton's group caused confusion throughout the company by ordering materiel directly from suppliers instead of routinely using Canadair's procurement department.

Challenger No. 1 rolled out May 25, 1978 in the presence of guests like Jack Horner (Minister of Industry, Trade and Commerce) and Jean Chrétien (Minister of Finance). At the gala Mansour Ojjeh of TAG (Canadair's Mid-

dle East sales representative) announced an order for a further 10 Challengers, upping TAG's total to 31. The Challenger flew on November 8 with the crew of Doug Adkins, Norm Ronaasen, Bill Greening and Jim Martin. At this time Canadair established a flight test centre at Mojave, California under J.W. Wood, an ex-USAF test pilot, to operate, maintain, modify, measure, record, control and analyze Challenger systems and performance. Mojave was chosen on account of its year-round flying weather, and strong aerospace infrastructure. The centre fell under Canadair Services Ltd., which was responsible for all operations outside St. Laurent. Mojave soon counted some 200 personnel, most of whom worked for Canadair's main contractor, Flight Systems. Within 2 1/2 years it tallied more than 1,000 Challenger flight test hours (Challenger No. 1 had gone to Mojave in December, 1978; No. 2 followed after its first flight of March 19, 1979.)

Besides activity at Mojave on system development and type certification; a third Challenger was flying from Cartierville. Eight Canadair and three DOT pilots were involved, along with FTEs (flight test engineers) and hundreds of support personnel from Canadair, Flight Systems, Lycoming, etc. A Canadair document from February 1980 noted: "The first aircraft (1001) was dedicated to establishing handling qualities and performance characteristics... The second (1002) was the systems aircraft, used to evaluate all of the basic operating on-board systems. Determination of flight loads and evaluation of the production fuel system were the prime functions of the third aircraft (1003). The fourth will fly the function and reliability tests." Special areas covered in flight test included airspeed calibration, stalls, stability, trim, engine performance and cooling, engine and APU fire extinguishers, avionics, minimum speeds, proving the auto pilot, water ingestion, flight in icing conditions, thrust reverser evaluation, noise control, and proving airframe integrity even after taking an eight-pound bird strike. Doug Adkins managed flight test at Mojave, assisted by fellow pilots Al Baker, Dave Gollings and Ron Houghton. To February 23, 1980 three Challengers had logged 509 flights and 941.5 hours (by January 30, 1981 these figures rose to 1072 and 2008). Meanwhile, a fatigue specimen at Cartierville completed more than 3,000 simulated flying hours.

Included in the work-up to certification were some 2,000 natural stalls, probably more than any new design ever had been required to perform. Transport Canada insisted on this intensive program, something that caused ill feelings between Ken Owen, who headed the program at TC, and Harry Halton. On one stall test flight (April 3, 1980) aircraft 1001 was lost. It entered a deep stall, a condition where there is little or no forward airspeed and where rudder and elevator control are neutralized. A tail drag chute, available for such an event, was used, pulling the tail up i.e. allowing the pilots to get the nose down to gain some speed. But the chute would not jettison and the crew decided to abandon the plane. Dave Gollings and FTE Bill Scott escaped, but Norm

Ronaasen died in the ensuing crash. In the US the FAA had the tightest and most practical certification regulations. These did not include anything resembling TC's stall proviso.

Approaching a stall, a sequence of automatic devices alerted the crew and gave time for corrective action. There was a light, a horn, a stick shaker (i.e. the control column shook in the pilot's hands) and a stick pusher (to shove the stick forward in order to gain speed), so any crew easily could avoid trouble. In spite of everything, TC would not relent about the need to ceaselessly demonstrate natural stalls (with the automatic warning system inoperative), and even obliged Canadair to stall-test each new plane. Even though the Challenger stalled normally, and a metal strip fitted on the port wing assured that it always fell off on the same wing. Jim Taylor suggested using US-based Canadair Inc to obtain an FAA STC to remove the stall strip. This was done using the regular crew at Mojave. Henceforth, each export aircraft went to a mod centre in the US to have its stall strip removed. The effect was to lower stall speed, permitting operations from shorter runways. Transport Canada initially would not approve this work being done in Canada, but this was changed after its own tests. An agreement was reached that the CL-601, in its re-defined form, would be certified by TC without the stall strip.

On August 11, 1980 the CL-600 earned Transport Canada type approval No. A131 (US FAA approval followed on November 7). Certification was limited, with take-off weights, flight in icing conditions, etc. restricted. No loads could be carried. So much had the Challenger grown through the design phase that a CL-600 with two pilots and full fuel was at maximum takeoff weight. Canadair, however, capitalized on the piece of paper without dwelling on its restrictions, then accelerated efforts to install mods to make the plane customer-acceptable. This put a load on the whole company. All this activity cost a great deal and Canadair was facing a steadily-worsening fiscal situation.

In order to bring some order to a situation that could end in the CL-600 failing, in late 1980 Fred Kearns brought R.D. Richmond back to Canadair. He had started there originally in 1947 and later held executive positions with Canadian Pratt & Whitney, McDonnell Douglas Canada, and Spar. Richmond assessed the problems facing Canadair. Some immediate relief was found by re-assigning certain personnel; e.g. Halton moved from an operational to a staff position, Francis went to procurement, Vince Ambrico became vice-president of manufacturing in Plant 1 (which included the Challenger), and Andy Throner became vice-president of manufacturing in Plant 2. To deal with the problem of project definition, Richmond held a series of weekend brainstorming meetings, where he obliged senior engineering managers to define, cost and schedule all outstanding items that needed to be designed or developed to complete certification. Simultaneously, sessions were held with manufacturing to contain costs and implement realistic schedules. To expedite the process, he made an arrangement with Boeing for technical assistance. On individual issues this

was costly, but overall was beneficial. In 1994 Ross Richardson, a veteran of the Challenger's early design pains and a keen aviation historian, recalled: "Dick Richmond sorted out the priorities and organized the problems in order of importance. He shook us all up initially, but got things back on the rails."

At this time several CL-600s were on the line, but no two were exactly the same. Mods were coming quickly, but could not be installed consistently airframe by airframe. An airplane could arrive at a certain station, where some component was to be fitted, but the parts might not be there, maybe hadn't even been ordered. The line would back up. As soon as possible a mod line was introduced at Throner's suggestion in Plant 2, and each new airframe was brought up to a standard before sending it for mods. Airplanes on the line gradually were upgraded until there was one that was completely on schedule and moving satisfactorily.

A problem with interiors was that Canadair had allowed only about 1,200 pounds for seats, cabinets and other furnishings. Customers, however, were fitting their planes more extravagantly—2,000 pounds, in one case 4,000. The latter plane would not get off the ground with full fuel and passengers. Canadair's salesmen were ordered to make it crystal clear to customers just how they could furnish their planes. Gradually, the allowance for furnishing rose to about 2,000 pounds. In January 1981 Canadair opened its first Challenger Service Centre at Hartford. A second opened in September at Oberpfaffenhofen, near Munich, West Germany.

Financing Your New Challenger in 1977	
Price	US$4.75 million*
Deposit on signing	5%
Deposit 12 months before delivery	30%
Deposit 6 months before delivery	30%
Payment on delivery	35%
*Original 50 aircraft	

Besides being overweight, there were a number of equipment problems with early Challengers. Nosewheel steering frustrated pilots due to design of the system having been split between Canadair and Dowty. This was only resolved by giving total responsibility to Dowty. Some aircraft were operated at weights above the recommended limits for braking. Since brakes were in short supply, this caused concern until the complete fleet had satisfactory brakes installed.

Of all the Challenger's troubles perhaps the worst was the Lycoming engine. Firstly, deliveries were slow and Canadair had to delay deliveries. There were so few engines that an airplane would be ferried to the centre in Hartford, where the engines would be removed and trucked back to Canadair so the next plane could fly. This set back deliveries, so Frank Francis was sent to Lycoming in Bridgeport to find out what was going on. He started tracking every part destined for a Canadair engine and deliveries slowly picked up. Before long, however, even worse troubles appeared—engines began failing. This was blamed on the rush to deliver extra thrust—Lycoming was turning out more powerful engines using the turbine blades, vanes, etc. from smaller-thrust versions. These engines were on progressively heavier Challengers and couldn't take the

strain. The Canadian Imperial Bank of Commerce's Challenger was one aircraft that had an in-flight engine failure. Dick Richmond had a call from the CIBC chairman saying that his officials refused to ride on their airplane. Richmond asked if the chairman wanted to tell that directly to the president of AVCO (Lycoming's parent). The latter came to the CIBC's head office in Toronto and had his eyes opened to the hazards that his engine might cause. Henceforth there was a concerted effort at Lycoming to correct problems, although this took time.

A coincident problem was cabin noise. The fans were inherently noisy. This was complicated by the use of hard engine mounts, which carried engine fan vibrations into the fuselage. While all Canadair's sales literature and press releases emphasized how quiet the Challenger was, it had one of the noisiest cabins among bizjets. A quick fix was to add thicknesses of lead-vinyl, but it was heavy and not very effective, considering the weight penalty. Eventually, it was determined that most of the noise was coming from small fan imbalances. A procedure was established whereby all engines were balanced during ground running. This brought marked improvements and was continued on the CL-601.

Another Challenger problem was range— early models could not fly as far as promised. The company had guaranteed 4,000 nm for the original 32,500-pound CL-600, but by the time the plane reached 38,500 it could fly only 2,850 nm with the specified load. This cost Canadair millions to rectify. The initial solution was to install a belly fuel tank that stretched range to 3,150-nm, but this meant returning planes to Canadair and having them out of service for two or three months. An alternative was offered. Winglets were designed that increased range about 4% (about 115 nm)—less than the extra range offered with the belly tank mod, but it only took about two weeks for installation. All this was done at Canadair's expense. Performance eventually was guaranteed and customers (more or less) were satisfied. Throughout this period Andy Throner oversaw the situation and became known in Plant 2 as "the mod king".

Through these difficult times all senior managers faced one tough job—placating irate customers. First, deliveries were delayed; then, planes had snags. Canadair was bound by contract to make good all deficiencies. Some customers would not be appeased. They cancelled orders or obliged Canadair to take back planes. One US corporation cancelled an order for five Challengers. Fortunately, Canadair had some support through its trials. Xerox, which had three CL-600s, remained positive. Its planes flew almost daily, including on a New York-Syracuse sked. Plenty of hours were logged and whatever snags occurred were expeditiously resolved. The chief pilot was an old hand in the corporate jet field, having experi-

enced the teething problems of Xerox's earlier Falcon 20s.

In March 1980 Fred Kearns had announced two new Challengers—the CL-601 with GE CF34s and the stretched CL-610 for FedEx, GE, TAG, etc. The CL-601 got the go-ahead, but in August 1981 the CL-610, which would have needed bigger engines and a bigger wing, was halted. This was recommended by Richmond, who realized that Canadair had enough going without starting down another uncertain road with a doubtful entry. The CL-601 was redefined, e.g. with winglets as standard equipment, and the specification changed to be able to meet new guarantees. Customers awaiting the CL-610 were offered $1 million off the price of a CL-601; 19 of 49 accepted. In October 1981 the first Challenger CL-601 aft fuselage was flown aboard a Belfast cargo plane from Cartierville to Mojave. There it was mated to a CL-600 (c/n 1003) to become the CL-601 prototype (c/n 3991). It flew initially on April 10, 1982 with pilots Doug Adkins and Jamie Sutherland. Canadian and US certification (to a maximum ramp weight of 42,250 pounds) were received in February and March 1983. Meanwhile, sales efforts continued. From September 18 to October 3, 1981 Challenger 600 No. 1005 completed a 35,270-nm tour (crew: director of flight ops Les McClelland, pilots Doug Glime and Hartwig Baier, mechanic Hugo Bartel). They departed Hartford for Singapore, where the Challenger was displayed at the Asian Aerospace Exhibi-

C-GCGT (top) was the prototype CL-601. It combined the forward fuselage of the third CL-600 with a new aft section and GE engines. N601CL, seen at its roll-out, was the first production CL-601 (c/n 3001). It later was N789DR of Dresser-Rand Co. in Elmira, NY. (Canadair Ltd. C48528-7)

(Above) The CDIC's Joel Bell and Gil Bennett and Canadair's Dick Richmond formed the trio that saved Canadair after the nearly disastrous events of the late 1970s and early 1980s. (Canadair Ltd. 49126)

(Left) CL-601-3As on the line. Nearest is c/n 3014, which was delivered in October 1983 as N14PN. In 1997 it was N292GA of A-OK Jets in Fort Lauderdale. (Canadair Ltd. C28809-5)

In 1984 the Luftwaffe purchased seven CL-601-1As for government VIP flying. This sale bolstered confidence in the Challenger at a time when Canadair was emerging from its worst doldrums. Aircraft 12+05 was at Glasgow in May 1995. Next, the 100th Challenger, a CL-601-1A, shown being handed over to General Electric in January 1984. Then, N602CC, the eighth CL-601-3R. It was delivered in February 1994 to Itochu Aviation of El Segundo, California. (Wilf White, Canadair Ltd. 51853, C48528-7)

tion. By the time it returned to Hartford it had made 47 flights and logged 78.8 hours. The CL-601 was an early success. In its May 16, 1983 issue *Aviation Week* commented: "Replacement of AVCO Lycoming ALF-502L turbofan engines with higher thrust General Electric CF34s on the CL-601 has transformed the Canadair Challenger into the corporate aircraft the manufacturer originally promised its customers in the late 1970s."

Once the program got on schedule, providing 4 1/2 aircraft per month, and the engine problem became more manageable, unsold Challengers began collecting on the tarmac. Some customers had refused to accept the planes they had ordered due to delivery being as much as a year late. In mid-1982 Richmond convinced Kearns that the only way to deal with this was with a temporary lay-off. This was done with concurrence of the union. As some of the people were being rehired in December 1982 Ottawa sent Joel Bell and Gil Bennett to take over Canadair on behalf of the Canadian Development Investment Corp. With the world in the depths of one of the worst modern-day recessions, Canadair was facing $1 billion in debts and needed a further $300 million to keep the Challenger on track. In January 1982 Fred Kearns admitted that, due to re-engineering, engine problems and other setbacks, break even on the Challenger would be 350 airplanes, not the 140 he had predicted earlier. Kearns envisioned a 900-plane market over 20 years, something that Gulfstream America Corp.'s president Allen Paulsen scoffed at (his company was selling the Challenger's main competitor, the Gulfstream III). Ottawa stood by Canadair.

By 1983 things were starting to look a lot brighter for Canadair. On August 23-24, 1983 Challenger 601 c/n 3002 flew non-stop Calgary to Heathrow, London, a distance of 3,792 nm. Time aloft was 9:04 hours. This bettered by 455 nm an FAI world bizjet record set by a Falcon 50 in 1980. No. 3002 was en route to Germany for demonstration to the Luftwaffe. On March 19, 1984 the 100th Challenger was handed over to GE in a ceremony at Boston. N375G (c/n 3019) was the fourth CL-601 for GE's executive fleet. The same month CDIC announce a corporate restructuring and a new entity—New Canadair (later Canadair Limited). The plan was to enable Canadair to stand on its own and, after years of losses, make a profit. New Canadair began with no debt—a $1.35 billion debt remained with the old company i.e. was buried by the government. Canadair soon began showing a profit—$6 million in 1984 on sales of $376 million; $27.6 million in 1985 on sales of $438 million.

Through the Challenger years Canadair suffered much ridicule in the media, especially from another federal institution, the Canadian Broadcasting Corporation (itself famous as a squanderer of tax dollars). Nonetheless, on March 20, 1983 the CBC released a documentary in its *Fifth Estate* series lambasting the Challenger, and questioning its safety. The program focused on the crash of a Challenger in Idaho. It suggested design problems, even though the US National Transportation Safety Board had not released any findings (it eventually exonerated Canadair). The *Fifth Estate* also implied that Xerox was dissatisfied with its Challengers. Xerox rebutted this and remained a loyal Challenger customer (years later, when the CRJ was offered in a corporate configuration, Xerox was the first to buy). The *Fifth Estate* said that down time for Challengers was excessive (it had been, but the mod program now was well advanced), but the program re-

Challenger Specifications (1983)

	CL-600	CL-601
Engine	Avco Lycoming ALF-502L	GE CF34-1A
Engine thrust (take-off)	7,500 lb.	8,600 lb.
Max ramp weight	41,250	42,250
Max takeoff weight	41,100	42,100
Max landing weight	36,000	36,000
Max zero fuel weight	31,000	31,000
Max fuel	14,900*	16,725
Outfitting allowance	4,100	4,100
Max payload	7,830	6,475
Payload with full fuel	3,180**	1,000
Range (5 passengers at M.74, no wind, NBAA flight reserves)	2,800 nm***	3,500 nm
Wing span	61.9'	64.3'
Length	68.5	68.5
Height	20.8	20.8
Wing area	450 sq.ft.	450 sq.ft.
Passenger cabin length	28.3	28.3
Cabin height	6' 1"	6' 1"
Cabin width	7' 2"	7' 2"

    * 16,725 with belly tank full
    ** 1,055 with belly tank full
    *** 3,150 with belly tank full

searchers were factoring in aircraft at completion centre having interiors installed. The CBC's implication was that these planes were down for mechanical reasons.

The *Fifth Estate* attack brought cheers from Gulfstream America Corp., which happily distributed copies of the program around the world. The dust finally settled and Canadair edged forward. Of the bruhaha CIDC president Joel Bell commented: "We've got to be very careful not to allow what is a financial problem to denigrate the product. We have a very good product, a sophisticated product. It's on the leading edge of technology." Bell would not renege on CIDC's goal regarding the Challenger. He would not let adverse criticism "cause us to lose our nerve to participate in technology development and in manufacturing." At the same time, as reported in the Toronto *Star* of February 9, 1982, Kearns predicted that Canadair could be sold within a few years to a private buyer for far more than Ottawa had paid for it. This prediction came true—in August 1986 Canadair was purchased by Montreal-based Bombardier for $120 million. At this time it had some 5,000 employees. (Gil Bennett was president of Canadair until 1985. He was replaced by board member Pierre Desmarais, who remained until the Bombardier takeover in 1986. At that time Donald C. Lowe became president.)

The Challenger 601 first flew in April 1982 with GE CF34-1A engines (8,660 pounds of thrust, but GE won a 9,140 "auto power restoration" rating that could be used if one engine failed on take-off) compared to the Lycoming ALF-502-L2s (7,500 pounds) in the Challenger 600. Respective ranges quoted (with belly tanks) were 3,500 and 3,150 nm. The first production 601 flew on September 17, 1982. A few days later Doug Adkins and Ian McDonald took it on a surprise visit to the NBAA annual convention in St. Louis, where Canadair touted it as a trans-oceanic bizjet. On September 28, 1986 the first CL-601-3A flew (c/n 5001). The 3A included such improvements as the CF34, Sperry and Collins avionics package ("glass cockpit"), power-assisted passenger door. For 1986 seven CL-601s were delivered to the German government and three to China. By this time Canadair was getting out of the woods with the Challenger—the program finally was on solid ground, and its reputation as a top product was gaining. Few airplanes had come through such a hell to end in the clear sky.

Now Canadair was confident enough to look to new ventures. It revived an idea from 1980—an airliner version of the CL-600. This would have been a low-density (about 20 passengers) commuter for busy, high-frequency, premium fare, inter-urban markets. After talks with carriers like CPAir, the notion faded, but in 1984 Fred Smith, chairman of FedEx, came to Canadair suggesting a new slant. He had two CL-600s that he wondered about stretching into package freighters. If this succeeded, he might try leasing a passenger version to commuter airlines for trials. Canadair talked prices with airframe modifier Jack Conroy. CL-600s were about US$4.5 million second

hand. With conversion each would be about US$11 million—too high. Next, Canadair considered a 601 freighter. Fedex looked at it, but passed. Richmond and consultant Eric McConachie of Aviation Planning Services now tried something else. Starting in 1986 they started talking to a select number of operators about a commuter concept. They visited American Airlines and Texas International Airlines on two occasions with Challengers borrowed from GE as representative of a stretched airplane based on the CL-601, and from Fred Smith, who had a CL-600 with airline-type seating used to move FedEx crews around. People were interested; one airline even talked about using a 29-seat standard CL-601 until a firm commitment could be made about a stretched airplane. Ultimately, nobody could come up with favourable enough numbers to make a commitment at that time. Undeterred, shortly after Bombardier assumed control, Richmond proposed to Bombardier chairman Laurent Beaudoin that Canadair consider a 38- to 40-seat stretched CL-601. It appeared as though there might be a market, even though acquisition price per seat could be 50% more than for a comparable turboprop like the Dash 8. He suggested that it would take $15-$20 million to define the aircraft and thoroughly test the market, including getting deposits. After he retired, Beaudoin gave the plan a green light and retained Richmond as an advisor through the program definition phase. Bob Wohl took charge of the program and in March 1989 it was launched. Air Canada, Delta and Lufthansa looked like early prospects for sales.

Challenger: Sample Production and Deliveries

c/n	Registration	Owner
3003	N500PC	Pepsico
3005	C-FAAL	Alcan Aluminum
3006	N372G	General Electric
3016	N1107Z	Pennzoil
3020	C-GCFG	Transport Canada
3026	N927A	Penn Central
3027	N17CN	Consolidated Natural Gas
3028	C-FBEL	Bell Canada Enterprises
3030	N34CD	Schering Realty
3031	12+01	Luftwaffe
3033	N601TJ	Northrop Corp.
3038	144616	Canadian Forces
3041	C-GRBC	Royal Bank of Canada
3046	B-4005	Government of China
3050	N62MS	Wells Fargo Leasing
3058	N125PS	Callendar Pie Shops
5006	C-FLPC	Canadian Pacific Corp
5012	N1868M	Metropolitan Life Inc.
5013	N711PD	Paul L. Deutz, Jr.
5017	N700KC	Kimberley-Clarke Aviation
5019	N915BD	Dillard Department Stores
5024	B-4010	Poly Technologies (Beijing)
5026	N601WM	Waste Management Inc.
5034	C-GIOH	Imperial Oil Ltd.
5037	N608CC	Gulfstream Aerospace Corp.
5047	N140CH	Chase Manhattan Inc.
5048	N2004G	Readers Digest Inc.
5053	N5PG	Proctor and Gamble Inc.
5067	N603CC	Government of Croatia
5100	N510	Bristol-Myers Squibb

Rather than being a stretched CL-601, something altogether new emerged—the Canadair Regional Jet. Within two years it would be airborne and quickly become the leader in its class.

Till 1968 DC-3 CF-DXU was Canadair's main corporate plane. In this 1946 scene Canadair vice-president of engineering Ken Ebel, president J.J. Hopkins and general manager Geoff Notman relax aboard DXU. It was furnished modestly, compared to bizjets of 40 years later. The Challenger interior shown below is from CL-601-3A XA-RZD of the Mexican company Consorcio Industrial Escorpion. (Canadair Ltd. C57693-3)

Douglas Aircraft

Douglas Aircraft Co. of Canada (Dacan, later McDonnell Douglas Canada) appeared in December 1965 from an earlier effort by de Havilland Canada to revitalize the Avro Canada factory at Malton by manufacturing DC-9 components. When DHC's activities at Malton faltered after 27 sets of DC-9 wings had been built, Dacan took over. Bill Baker, one of Avro Canada's pioneers and the flight engineer on the first flight of the C.102 Jetliner, became general manager. Dacan thrived—employment rose from 3,300 in 1965 to 6,600 in 1969, by when Dacan had delivered 600 sets of DC-9 wings. As well, it completed other components for the DC-8 and DC-9. These products, worth $300 million, were shipped by rail to Douglas in Santa Monica,

California. Interviewed in 1969, Bill Baker was asked about a company building commercial compared to military planes. He had been through the topsy-turvy era of the CF-100 and CF-105 and concluded that what Dacan was doing made more sense. While military contracts were short-term; something like the DC-9 provided stability. Baker foresaw work lasting into the 1980s on the DC-9 and the forthcoming DC-10. The first DC-10 contract was signed in May 1970 and included 300 93-foot wing sets. This saw employment rise in the early 1970s to 8,500. Baker's prediction was conservative—the DC-9 and its MD-80/90 developments; the DC-10/KC-10; and the MD-11 carried on into the late 1990s.

Throughout the McDonnell Douglas era several thousand well-paying jobs were always

available. Economic downturns caused by global phenomena sometimes struck the company. Business slumped, but rebounded. The bane of Canadian productivity—the union strike—caused setbacks, forcing the parent company to move some work to more productive and affordable areas in the US, even to places like China. By 1986 employment had fallen to 4,600. (At McDonnell Douglas in the US the layoffs were brutal. In a 1989 incident, company president Robert H. Hood called his 5,200 managers into a hangar at Long Beach, told them they were all fired and left them an option of reapplying—for 4,200 jobs.) Into the late 1990s the operation was still turning out high-quality aircraft components; it also had a few military contracts, e.g. manufacturing F/A-18 Hornet components.

(Left) An overview of the McDonnell Douglas Canada complex at LBPIA. Here famous types like the Lancaster, Jetliner, CF-100 and CF-105 were built. Since the 1960s the plant has manufactured wings and other components for the DC-8, DC-9, DC-10, MD-11 and MD-80. (McDonnell Douglas Canada Ltd.)

(Below) MD-80 wings ready to be wrapped and loaded; then, MD-80 and MD-11 wings set for shipping to California. (McDonnell Douglas Canada Ltd.)

(Above) The wings for this Wardair DC-10 were manufactured at LBPIA by McDonnell Douglas Canada Ltd. (Larry Milberry)

(Left) McDonnell Douglas built 976 sets of DC-9 wings, then switched the MD-80 wings. Here the 524th set, i.e. the 1500th set for the DC-9/MD-80, is ready to leave for the US. (McDonnell Douglas Canada Ltd.)

Canadair and DHC in Colour

CL-44 No. 1 soon after rollout in October 1959. It wore a special promotional colour scheme, but soon switched into RCAF garb. (Canadair Ltd.)

(Below) The first CL-44, this time in RCAF colours, is seen in a fine set-up shot looking north on Runway 33. Included is some classic Cartierville background. On the distant right are the old Curtiss-Reid Flying Service hangars (later burned). Further north is Canadair Plant 4. Below is Yukon 15924 on a shakedown flight before it was delivered to the RCAF in December 1960. (Canadair Ltd.)

(Left) Yukon 15921 flies across the Toronto waterfront during the Canadian International Air Show on August 31, 1967. It wore the special centennial-year colour scheme, including red-and-white roundels. (Larry Milberry)

(Right) Dick Gleasure snapped this Yukon-Argus formation at Uplands in November 1962. The Yukon was on RCAF strength 1959-71. The Argus served 1956-80. While the Yukon went on to a career as a tramp freighter, 24 Argus were sold for scrap to Bristol Metal Industries of Toronto for $72,000. The remaining survivors were set aside for museums. In the end the Yukons were worked to death, none surviving for museum use.

(Left) One of Flying Tiger's CL-44Ds in pre-delivery times at Cartierville in 1961. (Canadair Ltd.)

(Below) The speedy, long-range CL-44D played an important role supporting US military operations in Southeast Asia in the 1960s. Here CF-MYO-X of Seaboard World taxis at Cartierville before delivery. It departed as N124SW in July 1961. (Canadair Ltd.)

607

Views of the CL-44-O, first at Malton on July 17, 1971 while it was in long-term storage for Transmeridian; then at Valencia, Spain with HeavyLift on June 30, 1990. (Larry Milberry, Kenneth Swartz)

(Left) EI-BRP of AER Turas over Irish countryside. It had begun in 1964 with Loftleidir, then worked for years in Africa. It joined Aer Turas in 1986, then moved to HeavyLift. It was scrapped at Southend in 1992. (Paul Duffy)

(Below) Canadair's photographers captured a million great scenes over the years. This view of an RCAF Cosmopolitan is typical. Such photos usually were taken with a press camera from a T-33 or CL-41. Tight quarters for such a big camera, but the staff could handle any situation. (Canadair Ltd.)

"Cosmopolitician" 11152 lands at Uplands in June 1965. Then, 109159 at Trenton in October 1978. (Larry Milberry)

(Above) A scene at Gander on November 19, 1986. Cosmo 109156 awaits it passengers while a Tradewinds 707 arrives from a trans-Atlantic crossing. (Falk Foto)

(Above) The protoype CL-84 in level flight near Cartierville. (Canadair Ltd.)

(Right) The two CL-84-1s during a photo shoot at Cartierville. The third example did not get a chance to fly before the program was shelved. (Canadair Ltd C7173)

(Left) One of Quebec's CL-215s demonstrates a water drop at Quebec City airport August 18, 1991. (Larry Milberry)

(Below) CL-215s of the NWT on standby at Yellowknife on August 17, 1995. The Cessna 337s on the ramp are the bird dog (spotter) aircraft. These CL-215s were part of a 29-plane run whereby Ottawa shared the cost of equipping Newfoundland, Ontario, Manitoba, Alberta, the NWT and the Yukon with CL-215s. The first federal-provincial aircraft was handed over at Dorval on October 10, 1985. (Larry Milberry)

(Above) In winter Canada's CL-215s are usually dormant. Occasionally, however, some are leased out in places like Chile, where fires are a constant menace. These Ontario water bombers were at Sault Ste. Marie on February 4, 1992. (Larry Milberry)

(Left) While most customers took their CL-215s in long-familiar yellow, Newfoundland adopted a colour scheme all its own. C-FAYU was overwintering at Gander on February 28, 1992. (Larry Milberry)

(Right) One of Ontario's CL-215s in a nice action shot on a northern lake. (Canadair Ltd.)

(Left) The scene at the rollout of the Dash 7 at Downsview on February 5, 1975. The Dash 7 flew on March 27, 1975. Below it is at the Canadian International Air Show in Toronto on August 30 that year. (Larry Milberry)

Richard Beaudet caught this City Express Dash 7 landing at Toronto Island in June 1988. Ken Swartz shot C-GHRV departing from there in March 1987. HRV served Henson in the US as N903HA from 1982 before coming back to Canada in 1985. In 1991 it was sold in Abu Dhabi as A6-ADA.

The PT6A-50 was the hybrid engine developed by P&WC for the Dash 7. (Larry Milberry)

Dash 7s in the west. Time Air's TAD was at Vancouver on September 16, 1982; TAZ was there on April 4, 1990 wearing Canadian/Time Air colours. The two Air BC Dash 7s (HSL closest) were at Vancouver January 2, 1986; then Air BC's YMC at Comox April 5, 1990 along with a Time Air Short 360. By this time Air BC had switched to the Air Canada connector colours. TAD was sold in the US in 1990 as N726AG. TAZ became N341DS in 1991. HSL was last heard of in 1996 with Paradise Airlines as N780MG. YMC, which had been Air Wisconsin N707ZW 1981-87, returned to the US in 1993 as N59AG. (Kenneth I. Swartz Col., Larry Milberry)

(Right) This Dash 7 was converted for ice patrol in eastern Canada. It was at Summerside, PEI on January 14, 1987. At this time Bradley Air Services was under contract to operate and maintain it. (Larry Milberry)

(Left) This Alyemda (Yemen Airlines) Dash 7 was at Djibouti on November 30, 1991. It was delivered new to Yemen in October 1980. (Larry Milberry)

(Right) C-GJCB "James C. Bell" of norOntair at Timmins on January 24, 1992. The sixth Dash 8, it entered service in the fall of 1984. It served till norOntair disappeared in 1996. (Larry Milberry)

(Left) Air Canada's Quebec partner was Air Alliance. In 1997 it was operating 10 Dash 8-102s. Here C-FVON/808 alights at Quebec City on August 18, 1991. (Larry Milberry)

(Right) Passengers board an Air BC Dash 8 at Vancouver on April 4, 1991. C-FACD was one of 20 Dash 8-102s and -311s in the fleet in 1997. (Larry Milberry)

(Left) N907HA was the 11th Dash 8 and was delivered to Henson Airlines in May 1985. Later it was with Piedmont Airlines, a US Air connector. Here it was at Baltimore on December 9, 1989. (Larry Milberry)

(Right) Austria's Tyrolean Airways, based in Innsbruck, had 21 Dash 8s and five Canadair Regional Jets by 1997, and earlier had been an important Dash 7 operator. An important niche market for Tyrolean was the skiing crowd. (DHC)

(Left) Several Dash 8s were converted for special duties. 84-047 was one of a pair of E-9s delivered in the summer of 1986 and operated by the USAF in missile telemetry relay work at the Air Defence Weapons Center at Tyndall AFB in Florida. It was seen there on October 19, 1994. (Larry Milberry)

613

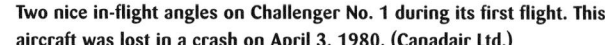
Two nice in-flight angles on Challenger No. 1 during its first flight. This aircraft was lost in a crash on April 3, 1980. (Canadair Ltd.)

Challenger prototype No. 2 was used in certification trials. It flew initially on March 17, 1979 with pilots Doug Adkins and Dave Gollings and flight test observer Jim Martin. It later became CanForces CX-144 No. 144612, working a few years with the Aerospace Engineering and Test Establishment at CFB Cold Lake. In 1993 it was retired and given a place of honour in Air Command's aviation heritage park at CFB Winnipeg. In this photo it has a spin chute instead of the usual tail cone. Should the plane get into a flat spin, the chute would be popped. This would yank up the tail, thus pointing the nose down. The pilots would jettison the chute and regain flying speed as they dove. A malfunction in this system led to the crash of Challenger No. 1. (Canadair Ltd. C12279)

An example of the art turned out by Canadair illustrator R. Dalabona. In this case, his scene depicted the proposed Challenger package freighter. (Canadair Ltd.)

(Above) The CanForces acquired a fleet of CL-600s (military designation CC-144) including 144608 (ex-C-GBLN), the 15th aircraft built. Except for No. 612, these went to 412 Squadron to replace the Falcon 20 on VIP duties. A few CanForces Challengers later were used for maritime patrol (CP-144) and electronic warfare training (CE-144). This 412 Squadron CC-144, seen at Gander on October 16, 1986, later became a CE-144. (Falk Photo)

(Left) CC-144 No. 602 of 412 Squadron in an angle showing the Challenger's pleasing lines. (Tony Cassanova)

(Below) A "green" (i.e. incomplete) Challenger 601-3A before going to the completion centre. It was owned by Household International and became N601CH. (Larry Milberry)

(Left) This view towards the west up Runway 28 shows Cartierville airport and the various Canadair plants about 1980. Plant 1 is centre front and Laurentian Blvd. running left to right. Plant 4 is centre right at the intersection of Laurentian and Bois Franc Rd., while Plant 2 (the old Noorduyn facility) is top centre. The growth of Canadair may be measured by comparing this view with the one in Chapter 17 (page 223). Cartierville was closed in the early 1990s and quickly turned into a vast housing estate, although Plant 1, much modernized, remained. (Canadair Ltd.)

Big Ideas, Small Budgets

Found Brothers Aviation

In each era there have been aircraft for Canada's unique needs. This began in earnest with the Vedette, and continued with planes like the Norseman, Husky, and Beaver. In a smaller but similarly Canadian way came the Found Brothers Aviation FBA-2C. The Founds were Alberta farm boys, but with the meagre 1930s the family moved to Edmonton. Here they took an interest in the local aviation scene. S.R. "Mickey" Found became an air engineer and commercial pilot. He flew for Mackenzie Air Service, then TCA. Grey became a pilot with TCA; Dwight worked briefly at Aircraft Repair, as did N.K. "Bud" Found. Dwight later went into mining in Yellowknife.

With their experience and interest the Founds discussed a concept for a small utility bush plane to fill a niche—while types like the Norseman or Reliant were popular, sometimes they were too much for the job. An operator's overhead could soar if his plane created excessive overhead compared to the revenue coming in. There were smaller bush planes in the mid-1940s, but they often were antiquated and makeshift. By 1944 the Founds had begun thinking about a new bushplane. They developed wind tunnel models for testing under Prof. T.A. Louden at the University of Toronto. In 1946 they established Found Brothers Aviation and began work on their FBA-1 proof-of-concept prototype to test a constant-taper cantilever wing with a deep-section, low-drag airfoil. The prototype was begun in a root cellar on the Agincourt, Ontario farm of Fred Staines, a former chief engineer for Mackenzie Air Service. Work later moved to the Toronto Flying Club hangar at Malton. The prototype was a fabric-covered, steel tube structure with a Gipsy Major engine and Tiger Moth propeller. Mickey Found flew the FBA-1 at Malton on June 27, 1949. A carefully monitored flight test program began, and about 20 hours were

logged by year's end. By that time the Founds had spent about $12,500 on the venture.

The focus now turned to a pre-production prototype, the all-metal, five-seat FBA-2A. It would have a modified airfoil with a constant-chord centre section. Various components would be easy to install and be interchangeable, e.g. rudder and elevators. A cantilever wing meant that no strut obstructed loading. The cowling would be hinged for ready engine access. The FBA-2A would operate on wheels, skis or floats; and have tricycle or tail-wheel configuration. The fuselage would have a spacious cabin with four doors to handle bush loads like drill rods, lumber, stretchers and fuel drums.

Progress with the FBA-2A was intermittent, as funds were sparse in the tight postwar economy. The Founds augmented their finances by recycling war surplus material—everything from Lancaster bombers to Merlin engines, aircraft fittings, instruments, even sparkplugs. Dwight and Grey managed this operation at Vulcan, Alberta. Meanwhile, Mickey flew the line for TCA. To build capital, Found Brothers Aviation diversified into aircraft servicing, support for the auto industry, etc. Between 1952-56 the FBA-2A inched along. Unfortunately, this was just when the four-place, all-metal Cessna 180 entered the market. First flown in January 1952, it had many attractive features e.g. steerable tail wheel, and steel spring main gear. With its 225-hp Continental O-470-A, the 2,250-pound (gross) aircraft had a lot to offer at an opening price of $12,950. The Ce.180 roared into production. For its first year (1953) 641 were built. In the next four years 2,467 more were delivered. A great many crossed the border, swamping Founds' target market and earning a favourable reputation. Even though it lacked many of the FBA-2A's tailored features, the Ce.180 was there when the customers needed it. Bush, corporate and private operators in Canada were enthusiastic.

Found's program was reactivated in 1958. Cecil Grainger was hired to fabricate components and L.M. Treleman became chief engineer, but it was June 1960 before the Founds completed the FBA-2A. Former Avro test pilot Stan Haswell made the first flight on August 11. Important support was obtained at this point from the National Research Council under Dr. F.R. Thurston. The NRC offered Found structural, aerodynamic and flight test analysis. Through 1960-62 the FBA-2A and the first FBA-2C (2,950-pounds AUW) were tested on wheels at Malton, and on floats (by Ernie Skerrett) in Toronto harbour. The aircraft met all expectations. There were no negative characteristics. Through no fault of the aircraft, the first FBA-2C (CF-NWT-X) crashed on June 26, 1962 at Brampton, a few miles from Malton. Pilot John Temple's engine quit as he made a high-performance takeoff with only a small quantity of fuel left in his tanks.

Aircraft CF-OZV-X was used for type certification. This was granted on February 2, 1964 (DOT Type Certificate No. 67). The first sales were of two aircraft for Toronto entrepreneur John David Eaton, owner of Georgian Bay Airways in Parry Sound, Ontario. He also purchased shares and became a key source of funding for the program. By this time development costs had reached $240,000. In March 1964 the first aircraft was delivered. The basic fly-away price was $17,000, a thousand dollars more than that year's Ce.180. About 25 employees were involved as production got under way at Founds' Rexdale, Ontario factory, not far from Malton. Through 1964 production reached $2^{1}/_{2}$ aircraft per month. The future looked good and various minor improvements were envisioned for 1965, when FAA certification was granted. The FBA-2C now had 10% of the annual Canadian market for its category. At this level Ottawa imposed a 15% federal import duty on foreign competitors. This an-

(Above) Stan Haswell takes the FBA-2A on its first flight at Malton. In early 1959 he was one of thousands at Avro Canada laid off when the CF-105 was cancelled. He had been a production test pilot on CF-100s and was airborne when he heard that he was fired! Haswell flew the first 25 hours on the FBA-2A. In 1996 he recalled that, in general, it was normal, except for poor elevator control. To rectify this, the second machine had a somewhat longer fuselage. Soon after his FBA contract Haswell left aviation for a career in the financial world. Then, Bud Found (left) and Stan Haswell discussing flight test matters at Malton. (Larry Milberry, Halford Col.)

(Right) The FBA-2A during a photo shoot in Toronto Harbour on May 28, 1961. Bud Found was at the controls. Publisher Bob Halford featured the plane on the July 1961 cover of *Aircraft* magazine. This was the first photo for which Larry Milberry received payment. The $25 from Halford paid for Milberry's film for the next year. (Larry Milberry)

(Right) The first FBA-2C climbs out on a demo flight. The pilot is holding the manual flap lever with his right hand. (Halford Col.)

noyed Canadian dealers of types like the Ce.180. They lobbied Washington for a reprisal duty against the FBA-2C. The Founds themselves were opposed to either duty.

By early 1966 aircraft No. 22 was sold to Roland Simard, and production of Nos. 23-26 was under way. Expenditures so far were $680,000. Plans now were made for a series of improvements to the FBA-2C. About this time, however, the company faced a new financial emergency. Eaton withdrew his support and the Founds were forced out of the company and new management took over. It shelved ideas for the improved FBA-2D in favour of a larger model. This became the Centennial 100, first flown April 7, 1967. The company moved to Grand Bend, Ontario, and geared for production. The Centennial 100 did not meet expectations. Although it was in the 3,500-pound gross weight range, there was no improvement in payload. It won type approval in August 1968, but sales did not materialize. Only four production machines were built, then the company folded. Some $2 million had been spent on the

CF-OZV-X was used for type certification of the FBA-2C. Later, OZV was donated to Centennial College in Scarborough, Ontario. It served for years as an instructional airframe. Here it is (wearing a bogus registration) doing a practice engine run at the college in October 1974. A few days later it was trucked to Ottawa to join the National Aviation Museum. (Halford Col., Larry Milberry)

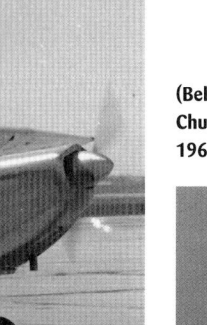

(Below) CF-WFN-X while it had work as a utility plane at the rocket range near Churchill, Manitoba. It was visiting the airshow at Brantford, Ontario on June 8, 1968. (Robert Finlayson)

(Above) The prototype Centennial 100 at Malton in the summer of 1967. Although it had a 290-hp Lycoming (FBA-2C: 250 hp), it had no better performance. On floats (Edo 3430s) it was a real dog. Only five were built: CF-IOO-X, WFN, WFO, WFP and WFQ. They found little use as workaday bush planes. (Jack McNulty)

CF-WFP takes on a load of parachute enthusiasts at Baldwin, Ontario in a view from June 1974; then it roars off with them, heading for 6,000 feet. At this time Murray Coursey ran the club and owned the plane. WFP was the last Centennial 100 known active—it was flying in 1997. (Larry Milberry)

Centennial. The company assets were sold at auction in March 1969.

The Founds built 26 FBA-2Cs for buyers across Canada— B.C. Airlines, Mahood Logging, Ocean Air, PWA, and Northern Mountain Airlines on the West Coast; Calm Air in Manitoba; Georgian Bay Airways, Northwestern Flying Services and Starratt Airways in Northern Ontario; and Air Alma in Quebec. Operators were pleased with the rugged, no frills plane. Roland Simard of Air Alma used his aircraft (CF-SOQ) mainly during the tourist and sport fishing season. For him the five-seat Found was perfect. As he put it, "People fish two-by-two." At one time he replaced his 250-hp Lycoming with a 290-hp version and a three-bladed prop to gain takeoff performance. This worked, but there was a price—an hour's less endurance due to added fuel consumption. SOQ reverted to its normal configuration.

Inevitably, FBA-2Cs began changing hands, others were deactivated, and there was the occasional crash. Missionary and free trader Dale Tozer of Moosonee operated SDB and SOT in the early 1980s, using bits and pieces of a wreck (OZW) for spares. In 1980 he noted, "The Founds are good, tough airplanes. We haven't had any trouble with them. They're quick on takeoff and easier to get off than a Cessna 185. We've logged over 500 hours carrying fishermen, moose hunters and freight. We haul a lot of 45-gallon drums and these are easily handled. The Found is roomy enough to carry snow machines. My only complaint is the lack of wing struts. How do you push an airplane around in slush or deep snow if it has no struts?" Another complaint was the FBA-2C's

two-notch flap handle. Veteran Found pilot Hugh MacCallum reported: "For long climb-outs, when heavily loaded, 5 to 10 degrees of flap had to be held in an unlocked position with the right hand for as long as five minutes. This could get tiring." Arnold Morberg of Calm Air in Thompson, Manitoba was a long-time fan of the FBA-2C: "It never let me down. With it I was able to move forward to where we became a Twin Otter operation flying skeds and charters." Bush pilot/author Robert S. Grant flew FBA-2Cs with GBA: "It came as a surprise to find that the machine I flew, CF-SDB, would outperform any 180 or 185 in the area." On the other hand he found that CF-OZU was the opposite—it didn't have any get-up-and-go. Grant also recalled: "A mishap occurred one day at Parry Sound. An inexperienced pilot was being checked out on CF-OZW. While landing on glassy water, the plane crashed and was a write-off; but the cabin structure remained intact. Neither pilot was hurt, a tribute to the Found's solid construction."

After 30 years a number of FBA-2C were still in use. In 1996 Fawnie Mountain Outfitters of Anahim Lake, BC was flying two. Owner John Blackwell corralled several hulks and a horde of original spares, so had no trouble maintaining his planes. John Vandene was one of his pilots and in 1992 wrote to Bud Found: "Your initial design concepts are just as valid today as they were when the plane first saw service. Our planes are used hard to move people and equipment every day of the six-month season. In late fall one is stored for the winter, a second is readied for wheel-ski opera-

tions. Due to the innovative main gear design, it can be changed from floats to wheels in less than two hours. Wheel performance differs little from floats and rough field handling is good... The Founds require little maintenance and most repairs can be made with little difficulty. Our main base is at 3,500 asl and the aircraft regularly operate at gross weight under standard conditions. A turbocharger would help the 250-hp Lycoming in cutting takeoff distance. Once off the water, however, the Found climbs at a respectable 600-800 fpm.... Few planes are easier to load, maintain and talk about. For a design that few remember and fewer can identify, the FBA-2C is one of the most interesting Canadian bushplanes ever built."

After WWII Bob Starratt's sons, Dean and Don, established Starratt Airways in Hudson. Dean was the pilot and Don the mechanic. In 1965 they purchased FBA-2C CF-SDC at the factory in Toronto. Dean took delivery and made his first flight with Found pilot George Ayerhart on August 13. Two days later he ferried SDC to Hudson via Parry Sound and Wawa and on August 16 put it to work earning revenue. That day he flew four trips for 2.5 hours. This was the beginning of a phenomenal career for the little plane. Each year hereafter it logged about 400 hours, flying on 2870 floats in summer and Federal skis in winter. Most often it was flying supplies to Indian reserves, but in summer served the tourist trade.

In 1984 the Starratts sold their operation to Glenn Tudhope, and retired. SDC still had its 1965 paint, but had graduated to bigger floats—2960s. Tudhope was just the second

Found Aircraft Earn Their Keep

Pilot Hugh MacCallum and helper John Robinson load boxes of pheasants aboard a GBA Found at Parry Sound in July 1969. (Robert S. Grant)

Found Brothers Aviation FBA-2 Series

C/n	Registration	Original Owner	Misc.
1	CF-GMO-X	Found Brothers Aviation	To Ken Gamble for parts
2	CF-NWT-X	Found Brothers Aviation	Crashed Brampton
3	CF-OZU	Georgian Bay Airways	To Jim McAvoy, Mini-Lab, V. Dikaitis, Lishman & Kroocmo. Crashed in Yukon 16-10-81 (4 killed). To Fawnie Mountain Outfitters (parts only)
4	CF-OZV-X	Georgian Bay Airways	To Centennial College, National Aviation Museum
5	CF-OZW	Georgian Bay Airways	Crashed Parry Sound 9 6 65
6	CF-RJV	Mahood Logging	To Industrial Acceptance Corp., Ocean Air, Nootka Air, Calm Air 1971. Accident at Cameron Lake, MB 23-2-68
7	CF-RRQ	Calm Air	
8	CF-RWV	PWA (lease)	
9	CF-RXA	BC Airlines	To Calm Air 1971
10	CF-RXB	BC Airlines	Hit trees en route Port Hardy-Seymour Inlet 11-4-66 (4 killed)
11	CF-RXC	BC Airlines	To Ocean Air. Crashed UchuckLake, BC 29-10-68 (1 killed)
12	CF-RXD	Michael P. Carr-Harris	To Bull Moose Aircraft, Ocean Air, Calm Air, L. Hohle, FMO
13	CF-RXI	Northern Mountain Airlines	To Arctic Air, Ocean Air, Hans Grunwald, T.J. John Logging, Shantymen's Christian Association, FMO
14	CF-RXJ	Corporate Plan Leasing	To Northwestern Flying Service, Don Boyd, Tinker's Places
15	CF-RXK	Great Northern Airways	Crashed Moose Lake, Yukon 7-7-69
16	CF-SDB	Georgian Bay Airways	To Gregco Aviation, Dale Tozer (Northern Missionary Fellowship), crashed Moosonee, 29-5-80.
17	CF-SDC	Starratt Air Service	To Tudhope Airways
18	CF-SON	Dominion Air Lease	Crashed 24-6-66 at Riverton, Man.
19	CF-SOO	Dominion Air Lease	To Garth Morritt, L. Hohle, FMO
20	CF-SOP	Chiupka Airways	To Calm Air 1971, R.A. Williams
21	CF-SOQ	Air Alma	To B.D. Kelly, L.C. Albey
22	CF-SOR	St. Felicien Air Service	To Air Alma
23	CF-SOT	Kipawa Air Service	To W. Newport, Dale Tozer, B.D. Kelly, J. Kivilahti
24	CF-SVB	EPA (lease)	Crashed 15-7-67 Williamsport, Nfld.
25	CF-SVC	Found Brothers Aviation	To Midwest Airlines, St. Andrews Airways, crashed Island Lake 9-12-74
26	CF-SVD	Northland Airlines	To Midwest Airlines, St. Andrews Airways, T. Fortin, RND Aviation, FMO, Found Brothers Aviation 1996 as proof-of-concept for new FBA-2.
27	CF-SVE	Found Brothers Aviation	To Northland Airlines, Midwest Airlines, McCully Aviation, L.G. Scruggs and P. Leach, R. Carey

FBA-2C CF-RJV (above left) of Mahood Logging starts its takeoff in Toronto harbour over the summer of 1965. The FBA proved ideal as a timber cruiser and on fire patrol. Then (right), an Ocean Air FBA-2C at Tofino, BC in 1968. RXA earlier had served BC Airlines. (NA/McNulty Col. PA191781, Wilf White)

FBA-2Cs served Georgian Bay Airways for about 15 years. Bob Finlayson saw CF-SDB (left) at Moosonee in the early 1970s. Then (below left), SDB shot there by Larry Milberry on August 24, 1979. By this time it was with Dale Tozer's operation. Tozer went to Moosonee in 1960 with the Northern Missionary Fellowship. His first plane was a Fleet Canuck. While preaching and running summer Bible camps for the Cree, he built up a small "buy and sell" operation with light planes and began doing ad hoc charters, mostly hauling groceries for the Indians. He also ran hunting and fishing camps, and did some trapping.

(Below) CF-OZU, which later was wrecked, at Parry Sound in August 1967. (Robert S. Grant)

(Above) The sole FBA-1 is pre-flighted at Malton over the summer of 1949. N.K. "Bud" Found is by the nose wheel, S.R. "Mickey" Found is by the prop. At this time Bud was running day-to-day operations, while Mickey was flying North Stars on the Atlantic for TCA. (Howard Levy)

(Right) This excellent 1960 view by Howard Levy speaks of the FBA-2's built-in ruggedness.

(Above left and right) The ill-fated CF-NWT-X and the plane that replaced it for certification trials, CF-OZV-X. (Al Martin)

(Below) Roland Simard of Air Alma liked the FBA-2C for fishing trips. Here SOQ has a three-blade prop and a 290-hp Lycoming compared to the usual 250 hp. This scene is at Simard's dock at Alma, Quebec. Pilot plus four passengers were heading out to fish. (Robert S. Grant)

(Above) Georgian Bay Airways made good use of the FBA-2C through the 1960s-70s. SDB was at Parry Sound on August 16, 1973. Those who knew this tough bushplane always found it a reliable revenue earner. Jim McAvoy of Yellowknife once operated CF-OZU. In 1992 he recalled, "The Found was better than a Cessna. It was roomier, had more doors and carried more cargo. I could carry two 45s and a fellow to help me with them. The Found was made for that kind of work and on 2960 floats was a very good performer." (Larry Milberry)

(Right) Ocean Air's FBA-2C CF-RXD at Vancouver in November 1967. Thirty years later it was active in BC with Fawnie Mountain Outfitters. (Tenby Col.)

Fred Chiupka's FBA-2C fresh from the paint shop in Toronto on November 18, 1965; then, earning revenue at Loon Narrows, Manitoba in February 1966. Chiupka earlier operated types like the Norseman, but considered the Found "tops". In a letter of November 13, 1996 his former pilot Ed McIvor recalled: "Loon Narrows, where Fred had a fish cooling shed, was about two-thirds up the shore of South Indian Lake on the west side. The fishermen kept their fish there until an airplane came to fly it to Lynn Lake. I made many a trip there when I flew for Fred." In 1995 Uwe Ihssen, chief pilot for Coval Air, recalled: "Four old bosses I'd rate in the 'super' category were Fred Chiupka, Barney Lamm, Gene Storey and Porky Wieben. They always kept the equipment in good condition and treated their employees with respect. They'd never ask a pilot to do a job they wouldn't do themselves." Chiupka sold to Calm Air in the early 1970s and lost his life in a car accident about 20 years later. (Al Martin, Ed McIvor)

The most famous FBA-2C is CF-SDC. At left it is at Hudson on June 25, 1974; then at the same dock with Tudhope Airways on July 14, 1991. SDC flew from here starting in 1965. George Ayerhart of Found Brothers Aviation, Dean Starratt and Glenn Tudhope were the only pilots to fly it over more than three decades. By early 1997 SDC had recorded 116,000 takeoffs and landings. (Larry Milberry)

Glenn Tudhope pushes paper in his office in July 1991. Like any small operator he would rather have been flying, but huge amounts of government red tape demand many hours of work each month. (Larry Milberry)

Forest Helseth, operator of Tinker's Places lodge, operated this FBA-2C since it was new in 1965. Here it sits at base near Sioux Narrows on July 16, 1991. RXJ had accumulated 4,600 hours by this time, mainly on camp runs of 50-60 miles. (Larry Milberry)

Found Centennial 100 CF-WFP at the parachute strip near Baldwin, Ontario in June 1974. WFP remained serviceable in 1997. (Larry Milberry)

(Above) Calm Air's FBA-2C CF-SOP at Eldon Lake, Manitoba, stripped and ready for fresh paint. Behind is Chiupka's Norseman CF-EPZ. (Hugh MacCallum)

(Left) Not every day is worth writing home about. Here Calm Air's CF-RJV is salvaged at Wheatcroft Lake near Lynn Lake, Manitoba. It sank when the floats filled after three days of high winds. (Hugh MacCallum)

CF-SOT was a company demonstrator in 1966. Here it was at Kipawa, Quebec. In 1968, by when it had only 319.5 flying hours, SOT was offered with floats for $24,000. It still was active in 1997. (CANAV Col.)

An FBA-2C on experimental wheel-skis at St. Jovite, Quebec on October 3, 1965. Fawnie Mountain Outfitters was operating SOO in the late 1990s. (Geoff A. Rowe)

Fawnie Mountain Outfitters/Moose Lake Lodge had C-FRXI in 1995. Its trim was blue and yellow. (John Vandene)

CF-RXJ while being used by Don Boyd Insurance in Northwestern Ontario. (Robert S. Grant)

Glenn Tudhope's Found waits out the weather at Hudson, Ontario on February 6, 1992. Then, a summer scene at Hudson (July 14, 1991) showing SDC at the dock and the historic Starratt/Tudhope base. (Larry Milberry)

After more than 20 years in the Quebec northland SOQ migrated west. It was seen at Fort Langley, BC on August 9, 1995. (Larry Milberry)

pilot to fly it. The company became Tudhope Airways, but little else changed. SDC sat at the same dock, although it did get new paint in 1988—after 24 years its orange had faded. Tudhope, who had flown for Bearskin earlier, quickly came to prefer SDC to his 185 or Beaver. It had a solid feel compared to the 185, and operating costs were less. As far as Tudhope was concerned the FBA-2C with its 918-pound useful load (on floats—1,300 on wheels) was ideal for his market. Fuelled with 55 Imperial gallons, it had a four hour endurance. This was ideal for summer forestry contracts. The Lycoming IO-540 always gave outstanding performance. Its 2,000-hour TBO only added to the economy of running a Found. As of July 14, 1991 SDC had accumulated 11,065 hours, and an astounding 112,000 cycles (takeoffs and landings), mostly on short hops to nearby Lac Seul Indian Reserve.

While FBA-2Cs continued giving good service to the turn of the century, CF-OZV-X made it into the National Aviation Museum in Ottawa. Following the rigorous program it went through for type certification, then a period in commercial service, it spent years as an instructional airframe in an aviation technology program at Centennial College in Scarborough, Ontario. In April 1979 OZV was donated to the NAM, assuring that at least one of these unique Canadian workhorses would be preserved.

FBA Mishaps

Like any active fleet, the FBA-2C had a number of accidents. The worst involved CF-SVB of EPA on July 15, 1967. That day it was flying from Williamsport, Newfoundland to Gander with the pilot and three passengers. Soon after takeoff from Fourché Harbour it plunged to earth from about 300 feet. The temperature was high (80°F) and the plane loaded to gross weight, so SVB would not give peak performance. It stalled when attempting a sharp turn, but without the room to clear nearby terrain. The pilot, new in the business, had only 344 flying hours (75 on SVB in the previous 90 days).

Another crash involved CF-SON at Riverton, Manitoba on June 24, 1966. It had been flying from Gods River to Selkirk when it hit deteriorating weather. DOT accident report No. 2990 described what happened: "A power-on approach to land was made due to reduced visibility and lack of depth perception [the sun had set]. The navigation lights were on, however the aircraft was not equipped with landing lights. It was reported that the approach was normal, and additional power was applied as the aircraft was settling. The heels of the floats touched and there was a loud bang and jolt, and the aircraft swung to the right. The pilot thought the float had struck something and applied power for an overshoot. After he got airborne, the flaps were raised gradually, but the airspeed did not increase. The flaps were again lowered to the first notch and a turn to the right was made in an attempt to complete a circuit and land. The aircraft did not respond normally and airspeed could not be maintained. The nose was raised as far as possible as the aircraft descended into

the trees under power." It appears that the left float had been badly damaged and offset by a hard landing. All four occupants walked away, but SON was destroyed.

The Return of Found Brothers Aviation

For years following the demise of Found Brothers Aviation, Bud Found contemplated a comeback for the FBA-2. By 1996 this seemed an especially good idea, for there was a dearth of new light utility planes. This was mainly a result of Cessna having withdrawn from the light plane market in 1986 following a flood of costly lawsuits. Courts were awarding plaintiffs millions for what really were rip-off cases; e.g., a pilot would take off drunk, crash, and die. His estate would sue Cessna, charging that it really was Cessna's fault. Time after time the courts agreed with the estates. Cessna, Beech and Piper could do nothing but cease building light planes—the alternative was bankruptcy. After a decade certain sectors of the market were starving for new equipment. About this time the US government decided to make it harder to win such frivolous lawsuits. In the mid-1990s Cessna returned to the market with the Ce.172 and Ce.206, promising that other lines soon would reappear.

In 1996 Found acquired mothballed FBA-2C C-FSVD from John Blackwell of BC and had it refurbished at Parry Sound, Ontario. A new 260-hp Lycoming IO-540D4A5 and 86-inch McCauley propeller were installed. On November 27, 1996 Elton Townsend took SVD on a 45-minute test flight. Found then began modifying SVD in anticipation of producing the improved FBA-2E Bush Hawk at Parry Sound airport.

The Avian 2/180

Cancellation of the CF-105 had a variety of spin-offs. While some skilled engineers headed to the US to work in exciting areas like the Apollo man-in-space project, others stayed home to develop original ideas. One ex-Avro engineer produced a small, all-terrain vehicle. Known as the Jigger, it was a success, but the engineer was not able to keep his life on track. He ended living with his worldly possessions in the back of a derelict car. Other engineers found good jobs with Canadair, DHC and CP&W. On another tangent Peter R. Payne led

While Found was developing the FBA-2, others had entries for the Canadian market. Fleet of Fort Erie made a handful of STOL Helios in the early 1950s, but the project fizzled. In 1960 the Helio appeared again in Canada, but only as a blip. Here CF-NNC, then a factory fresh H395A demo plane, does a STOL departure at Waterloo-Wellington Airport on July 9, 1961. The Lockheed LASA-60 was another bush type from the 1960s. This demonstrator, the sole Canadian example, was at Hamilton on September 7, 1968. Such types are reminders that it takes more than a good airplane to challenge the domination of Beavers and Cessnas in the Canadian bush. (Larry Milberry)

four ex-Avro men in the design of the Avian 2/180 gyroplane ("2" for the number of people carried, "180" for the horsepower). It was based on the principle discovered by Juan de la Cierva in the 1920s that a set of blades would rotate in an airflow and provide lift. A pusher propeller provided forward speed to keep the blades whirling. Since the main blades rotated at all times it was impossible to stall a gyroplane. The blades would rotate steadily in forward flight; level flight could be maintained as low as 25 mph. Even at 20 mph, the aircraft would not drop out of the sky, since air still passed through the rotor, and the aircraft would sink gently to the ground. Hovering was not possible.

The ex-Avroites' new company, formed in March 1959, was Avian Industries, based in the McNally Wood Products building at 1 Elgin St. in Georgetown. A few miles west of Malton, Georgetown had been home to about 500

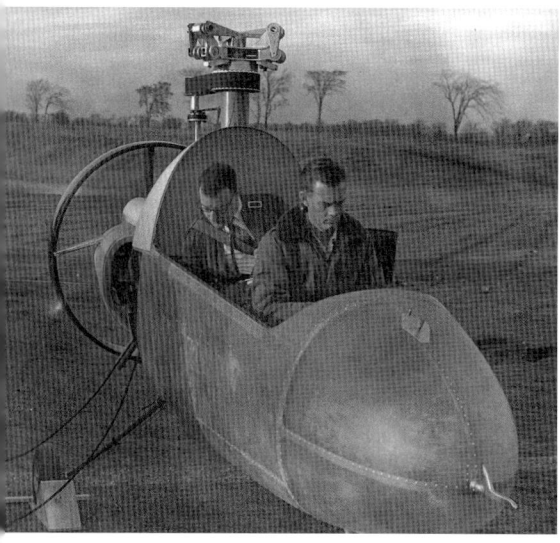

The prototype is seen above running up behind the plant in Georgetown. Ron Peterson (a TCA pilot) is in front, while Avian president and project manager Peter Payne is behind. Next, the plane, this time painted maroon with white, taxies at Waterloo-Wellington airport with the same crew. This machine later was wrecked. Then, one of the Avian 2/180s under construction in Georgetown. (Halford Col.)

Avroites. Avian felt that by establishing there it could help the community through rough times following cancellation of the CF-105. There were five original employees: Peter Payne (president and project manager), Gordon Sampson (engine, propeller and duct installation), John Sandford (airframe stress analysis), Howard Smith (airframe design) and Gordon Hunt (stress analysis of hub, blades and controls). At Avro Payne had worked in the preliminary design office studying futuristic rotary-wing concepts, including a twin-engine, jump-start gyroplane. He and his partners applied such knowledge in designing the Avian, which they envisioned as a flying machine for the ordinary person. It would be safe and easy to fly, execute a jump start (get airborne in a short distance), and be affordable—about $10,000. This was based on a company claim "that the original market estimate of 5,000 aircraft per year will prove to be too low by a substantial margin."

Initial funding for Avian Industries came from selling shares to Georgetown citizens. This kept Avian going until January 1960. It also approached the military, whose support would provide some cash to get rolling, but there was no interest. (Avian was able to arrange a cost sharing grant from Ottawa worth about $300,000, but this didn't come until late 1964.) By November 7, 1959, the day the prototype rolled out, Avian had about 20 people on the payroll. Initial prototype testing was done at Waterloo-Wellington airport, a short distance west of Georgetown, in 1959. In February 1960 Fred Walter of Thermo Electric Inc. became president of Avian and backed it until it closed in December 1969.

At first there were disappointments with the Avian. It proved impossible to track its tab-controlled rotor. In early 1960 new blades with conventional controls at the root were installed

and precise tracking proved a simple task. At Waterloo-Wellington airport Ron Peterson, Reg Jaworski and Emil M. Zuber each tried to get the plane airborne, but the four-foot, fixed-pitch propeller lacked the thrust for flight. On March 18, 1960 the prototype was involved in an accident. It became airborne for a few seconds, then crashed. Zuber was injured and the project was set back. In October 1961 the Avian 2/180 made its first sustained flight with an enlarged propeller and duct. Several other developmental machines were built (CF-JTO, MTV, NWS, OTC and PEV) with some of the test flying done from the company's parking lot. In the final version the rotor had grown to 37 feet in diameter, the propeller to six feet.

Harold Koehler headed flight test for Avian. Grant Davidson did all the flight testing of the sixth aircraft. After extensive fatigue testing, including several weeks of testing the main rotor blades in ground test rigs, Dick Bentham of the DOT put the Avian through exhaustive flight tests, including takeoffs and landings in high 90° crosswinds. These tests always seemed to take place in midwinter, in this case

The small Avian factory in Georgetown in 1960. (Halford Col.)

at Guelph in February 1968. The Avian won DOT certification the same year and promptly applied for FAA certification. This usually is automatic, but the FAA required some hot and high flight testing, done by Grant Davidson and Dick Bentham at Warm Springs Mountain, Virginia in May 1969. Based on air density, this was done at the equivalent of 5,000 feet. Davidson did takeoffs and landings and explored all the critical areas including engine failure at various points during takeoff. The thinner air caused the rotor to spin faster, so there was no difference in control, but the aircraft dropped a lot faster, although always was

(Left) Avian 2/180 CF-PEV-X lands at Georgetown in a scene from August 1965. (Gordon J. Hunt)

(Below) Pilot Grant Davidson aboard Avian CF-JTO-X. The spring-type main undercarriage legs had great strength. The nose wheel castored and differential braking was used with it to steer on the ground. The Avian 2/180 was 16' 2" long with a rotor diameter (originally) of 33'. It used a four-cylinder, 180-hp Lycoming O-360-A, and carried 26 Imperial gallons of fuel and two gallons of oil. Empty weight was 1,090 pounds; all-up weight 1,720. Cruise speed was 130 mph, stall speed 25 mph, and normal range 400 miles. The 2/180 was another good idea that proved difficult to market, as did the Umbaugh and other gyroplanes. (Halford Col.)

Avian CF-MTV-X on a local test flight with pilot Harold Koehler. A small tab on the extremity of each 60-pound blade made moving it a finger-tip affair. For jump starts, rotation of the main blades was achieved using 3,000 psi airflow from an on-board bottle of compressed air (capacity: one minute), and, later, by an engine-driven pump. The air was piped through each blade to a nozzle at the blade tip, and the blades quickly rotated. Although a gyroplane can leap into the air, it cannot hover, except briefly if flown into a stiff headwind. (Halford Col.)

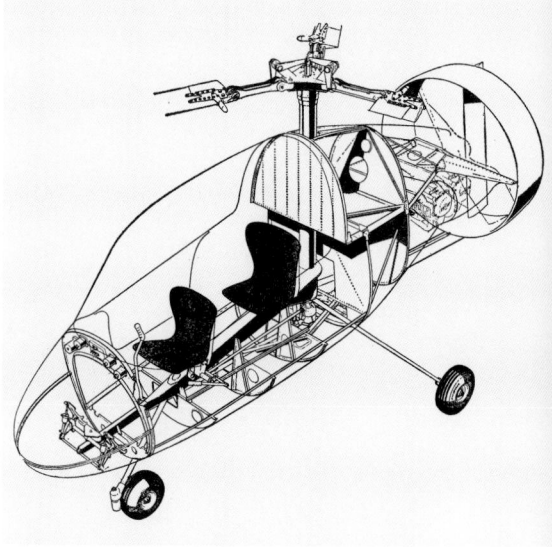

A hefty steel keel, around which the fuselage was assembled, took the Avian's landing gear load. Fuselage skin was reinforced glass fiber.

The Avian 2/180 was an attractive machine. Here the third propotype flies near Georgetown. It was powered by a Lycoming 200-hp IO-320. Take-off was simple with the Avian. During rotor spin-up, the rotor was set at zero pitch and the constant-speed propeller was also at shallow pitch until the rpm built up. When the rotor was up to speed, the pilot increased rotor pitch to a cruising angle of 4˚. At the same time the main rotor was disconnected from the engine and all engine power was used to turn the propeller, which automatically increased pitch. The result was a jump take-off (no ground roll) with maximum thrust from the ducted propeller. (Howard Levy)

controllable. FAA certification was obtained in 1969 but no market appeared. In December Avian closed its doors. Harvey Krotz of Listowel, Ontario was part of a small group interested in bringing the project to their town, but this did not transpire. In 1996 Gordon Hunt looked back on his involvement with the Avian 2/180: "I can truly say that this was the most interesting project I ever worked on. I saw it develop from a blank sheet of paper to first flight, extensive flight and fatigue testing, and DOT and FAA certification. It was a rare privilege to be involved and I am pleased to have been part of it." Test pilot Dick Bentham, who had flown other gyroplanes, also had some special words: "The Avian was far and away the Cadillac in a field of Yugos." No Avian is known to exist, although rumours in the 1990s suggested that one or two may be in storage.

(Above) Gordon Schwartz gassing up his plane about 1960, then (below) boarding a Super V at Goderich. (Schwartz Col.)

(Right) An in-flight angle of N4530V showing its pleasant lines. (via Claude Fournier)

The Super V

In June 1961 George Clark, president of Fleet, asked Toronto aviator Gordon Schwartz if he would be interested in managing a new light twin concept called the Super V. Clark wanted to get Fleet back into aircraft production—since Fleet Canuck days it had done only subcontract work. The Super V was conceived in 1955 by Dave Peterson, chief pilot for Sinclair Oil in Tulsa. His idea was to get a supplementary type certificate (STC) to turn the standard Beech Bonanza into a light twin, replacing the 205-hp Continental E185-11 with two 150-hp Lycoming O-320s. In 1958 he sold his project to Oakland Airmotive in California, which decided to upgrade to the 180-hp IO-360. Full type certification was awarded in June 1960. To attract publicity, Oakland Airmotive sent Pan Am Captain Chuck Banfe around the world in a Super V. The trip took only 8 1/2 days (216 hours aloft).

Fleet obtained manufacturing rights for the Super V and exclusive sales rights in Canada and the northeast US. Gordon Schwartz took over the project and became general manager. He started by appraising the airplane— listing its shortcomings, then taking it to San Francisco for discussions with Bay Aviation (formerly Oakland Airmotive). Schwartz found that Bay Aviation had little knowledge of manufacturing. Fleet purchased the whole operation and moved it to Fort Erie, where it was set up early in 1962. A number of essential improvements were made. A prototype was developed from N4530V, Fleet's demonstrator. In a letter to prospective customers, Schwartz described how a Super V was built: "The airframe enters our plant at which time it is completely dismantled for inspection. The first operation is to bring all components back to zero time, not just repair... Our aircraft then proceeds down the assembly line in the normal accepted manner. Well over 1,000 components

are built into the aircraft to support the new requirements of a high performing twin. New engines, propellers, bearings, cowlings, accessories, electrical system, vacuum system, engine controls, etc. are added as the aircraft proceeds towards finish. Upon completion and factory flight checks, the aircraft is painted, sound proofed and upholstered." In August the Super V received Canadian Type Certificate No. 59 after Walter Gadzos of the DOT, flying with Schwartz, put it through its paces. By year's end two US dealerships had been established. The first customer for a Super V was the president of the US Aircraft Owners and Pilots Association, "Doc" Hartranft. The aircraft listed at $36,700.

It soon was clear that Fleet would have to commit to 50 Super Vs if the venture was to be profitable. By this time the Beech 65 Travelair and Piper PA-30 Twin Comanche had swept the light twin executive market, so Fleet dropped its project. In early 1964 it was sold to Paul Mitchell of Savannah, after which the Super V faded. Ten aircraft had been made in California and five new ones by Fleet. Some were still in use in the 1990s. Fleet carried on as a subcontractor. In 1984 a new corporate entity appeared—Fleet Aerospace Corporation. Work through the 1980s-90s was greatly varied including component manufacture for types like the 707 (fin and rudder, engine nacelles) and 747SP (fairings); Twin Otter, Buffalo, Dash 7 and Dash 8; Challenger (rudder),

L1011 (landing gear doors, fairings), P-3C (flight deck assemblies) and satellite structural bodies (ANIK "C"). Fleet acquired Orenda in the 1990s and in 1996 changed its name to Magellan Aerospace

Bristol Projects

Through the 1960s Bristol of Winnipeg had a great variety of work. Much of this was overhaul and modification for RCAF aircraft like the CF-100 and CF-101. Another contract was to modify 24 UH-12Es for the DND. Bruce Jones of Hiller delivered the first of these from Hiller in May 1961, then Bristol carried out the work and handed the Hillers over to the Army. In 1968 and 1971 Bristol became the overhaul centre for the Army's Bell CUH-1H Hueys and CH-135 Twin Hueys. It later converted the Hueys for search and rescue, and in 1974 added the Bell CH-136 Kiowa. In 1977 the DND asked Bristol to develop a means to protect its helicopters from wire strikes during NOE (nap-of-the-earth) flying. By early 1978 Bristol, under engineer Nelson Chan, was ready for a simulated wire strike test program at Gimli. Kiowa airframe 136266 was fitted with trial wire cutters, mounted on a truck, then driven through various wires at different speeds. By 1980 the CanForces was equipping its Kiowas and Twin Hueys with Bristol's WSPS (wire strike protection system). The US military also adopted the WSPS as did many commercial operators. In 1980 Chan was hon-

One of Bristol's projects was developing wheel-skis for the ubiquitous Beech 18. This was done under engineer Haakon Kristansen. CF-OWU-X was the trial aircraft and is seen near Winnipeg in 1963. (WCAM 19808)

After the Arrow was cancelled, Avro continued with the experimental, US-funded Avrocar. Les Baxter photographed one of the two Avrocars on November 16, 1959. Eighteen feet in diameter, this proof-of-concept air-cushion vehicle was a step towards a high-speed "flying saucer". The Avrocar, which flew a few feet off the ground, was cancelled when funding ceased in 1961. One was saved for the US Army Transportation Museum at Fort Eustis, Virginia. (CANAV Col.)

Superannuated C-119s (No. 22119 closest) await disposal at the Bristol-run RCAF storage base at Saskatoon in May 1964. (Norm Malayney)

oured for this new technology—he received the Romeo Vachon Award from the Canadian Aeronautics and Space Institute.

Another important activity started in 1964, when Bristol took over facilities at RCAF Station Saskatoon as a storage base for surplus CanForces aircraft. These were received and stored; and many were disposed of through Crown Assets Disposal Corporation. Initially, Harvards and Expeditors were flown in—114 Harvards and 48 Expeditors eventually were sold. Of the Dakotas processed, 70 were returned to service (some more than once) and 78 sold. In 1967 30 C-119s were sold, many becoming fire bombers in the US. Twenty-four P2 Neptunes arrived in 1965, but only four were sold "fly-away". The rest were melted down. From 1970-74 ATC's 12 Yukons passed through Saskatoon for disposal, the last leaving in 1974 for Zaire. Other types handled included: Albatross (9), Argus (2), CF-5 (8), Chipmunk (13), H-34 (3), L-19 (13), T-33 (61), Tracker (37), UH-12E (12). Most were sold to brokers or private buyers, although the Argus went as scrap (one later reached the WCAM);

and 14 T-33s were crated and shipped to Turkey. The last aircraft to leave the site was T-33 No. 133275. It flew away to the air force on September 26, 1977. A total of 558 aircraft were processed at Saskatoon by Bristol. As well, its staff did field work on CF-104s at Cold Lake.

Bristol continued manufacturing floats, its original product line. Model 4580s for Beavers were built till 1969 and the popular 2870s were steadily turned out for the Ce.180, Found FBA-2C, etc. The 7170, which Bristol had first made in 1935 for the Norseman, and which later flew on the Otter, was another steady line. From 1955-66 some 57 sets of amphibious Model 348 Beaver floats and 68 Model 324 Otter floats were made. In 1961 the 7850 was developed for the Beech 18, and a retractable wheel-ski gear for the Beech 18 also was tested. The 7850s became popular; the wheel-ski gear found its place on the Twin Otter (55 sets built). The 4930 for the Turbo Beaver eventually was adapted to the straight Beaver, boosting its water performance. Some 300 sets were built, mainly for surplus US Army

Beavers that migrated to the Canadian bush in the 1970s.

Like many Canadian companies Bristol had subcontracted on the Avro Arrow program—it manufactured five sets of Arrow tail cones and nozzle boxes for the Iroquois engine. Bristol was hit when the Arrow was cancelled in 1959, but a note in a company document shows that diversified Bristol was able to survive: "While the sudden termination caused drastic changes in program planning, and curtailed anticipated increases in employment, the company gained some experience, such as in processing of titanium, and was left with equipment obtained for the Arrow project, which has seen good service through the succeeding years." Besides work on the Arrow, Bristol had previous experience in special manufacturing for Orenda (CF-100 and F-86) and Nene engines (T-33). It also made tail pipes for the Bomarc missile, jet pipes and muffs for the CL-44; and parts for such other engines as the PT6, J79, J85, and F404. Throughout the 1950s-80s Bristol also had an active nuclear research program associated with electric power generation.

In 1968 Bristol won a contract from Lockheed to manufacture the "S" duct for the centre engine of the L1011 Tristar. The 7' 1"-diameter, 1,200-pound duct was built of aluminum and titanium alloys. New tooling and processing

Northwest Industries Ltd. was another busy contractor in the post-WWII era. It had started with a run of new Bellanca bushplanes, then specialized in heavy maintenance for the RCAF, overhauling types like the Dakota, C-119, then C-130. One contract in the 1960s was overhauling USAF Republic RF-84F Thunderflash reconnaissance fighters. On this occasion anti-Viet Nam War zealots had sabotaged several Thunderflashes at NWI's base at Edmonton Industrial Airport. (CANAV Col.)

The prototype Trigull was CF-TRI-X. It is shown taxiing from the Fraser River at Vancouver airport; then touching down on the runway there in February 1974. (Halford Col.)

(Right) Entrepreneur Dave Hazelwood was the driving force behind the Trigull concept. In this photo he was at Fort Langley, BC in August 1995. (Larry Milberry)

(Left and below) The second Trigull shown first at Abottsford on August 14, 1976. Test pilot Norm Ronaasen was at the controls; then, in a dramatic takeoff view. (Larry Milberry, Canadair Ltd. C9727)

equipment was introduced for this 10-year program. Some 350 ducts were shipped to Lockheed in Palmdale by rail and truck. Other projects in the 1970s-80s included flaps, ailerons and wing leading edges for the Lockheed P-3/Aurora, and nacelles and cowlings for the Dash 7. Meanwhile, Bristol was steadily developing versions of the Black Brant research rocket from the late 1950s into the 1990s. Other involvement in rocketry included developing the booster for Canadair's CL-289 reconnaissance drone, and the CRV-7 air-to-ground rocket for attack aircraft. By the late 1970s Bristol employed more than 1,000.

Building a Better Seabee

A controversial hi-tech venture of the 1970s-80s in BC was the Trident Trigull, a five-seat amphibian originally promoted by Dave Hazelwood. The Trigull looked like a trimmed-up Seabee. Initially announced in the $90,000 range, it would be ideal for fisherman, hunter, and lodge operator and perfect for weekend getaways to remote spots. That the Trigull looked like a Seabee was no accident. The original layout was done in the US in the 1960s by Chuck Herbst, son-in-law of Seabee designer Percival Spencer. Herbst worked on the design with Bob Dent with a plan to market it as a kit plane. After lay-

ing a foundation and manufacturing a few components, Herbst and Dent let their brain child languish. Hazelwood and partner Paddy Newton learned of this. In 1969 they took over the project for $135,000. Herbst joined the project in Victoria and later became a partner. Several Boeing designers were hired and construction of a prototype began in 1971.

Hazelwood succeeded in interesting Canadian Aircraft Products—it would build the Trigull in its factory near Vancouver International Airport. A potential market for 1,000 Trigulls was envisioned. As often happens with a new airplane, however, costs soared—this was a sophisticated design and ate up millions in development dollars. There were some private funds, e.g. deposits from buyers (eventually $1.7 million), and some from the BC's Development Corporation ($2 million), but not enough for Trident to realize its ambitions. Meanwhile, the Trigull prototype (CF-TRI-X), powered by a 285-hp Teledyne Continental Tiara engine, was built at CAP, which had traded some of its invoices accrued in building the first machine, for Trigull manufacturing rights. Paul Hartman flew the prototype at Vancouver International Airport on August 5, 1973. The first production model (C-GATE) flew on July 2, 1976. The Trigull received Canadian certification on October 28, 1976. FAA certification followed. Due to a shortage of capital, however, production could not begin.

Relief finally came in 1977 when the federal Department of Industry, Trade and Commerce agreed to a $6 million loan guarantee. But ITC's money was not forthcoming. Layoffs began in late 1977 and the board of directors moved to force Hazelwood from the company. Bickering over financing dragged on for more than a year, this at a time when interest rates soared to 20%, severely hitting light aircraft sales. From 1978 to 1979 the price of a Trigull soared from $105,000 to $161,000. It wasn't until late 1980 that C-GATE officially was rolled out at Victoria (Trident had decided to cast aside its arrangement with CAP and open its own production facility).

The second Trigull built is seen in a nice pose by Canadair photographer Garth Dingman. (Canadair Ltd. C9721)

Trigull Specs (300-hp Lycoming)

Length	29' 3 1/2"
Wing span	41' 9" (floats up)
Height	12' 6"
Wing area	245.2 sq. ft. (floats up)
Empty weight	2,567 pounds
Max take-off weight	3,800 pounds
Max speed (300 hp)	148 knots
Max speed (350 hp)	172 knots

Federal small business minister Ron Huntington eventually decided that the project was too costly. He predicted that it would take at least $10 million more to keep Trident going—600 Trigulls would have to be sold to make the venture work, this at a time when there were letters of intent for only about 50. Trouble came in another form. The Tiara was pulled from production, forcing Trident to find a substitute. A 300-hp, fuel injected Lycoming IO-540 was selected, but this meant expensive re-engineering.

Trident became an issue in the campaign leading to the federal election of May 1979. The Conservatives under Joe Clark argued that the public should not fund high-risk private ventures, and mocked attempts to support Trident; but the Liberals under Pierre Trudeau touted aid to western hi-tech enterprises to encourage industrial growth outside central Canada. The Conservatives won the election. On December 13, 1981 the Clark government pulled the rug from under Trident. It closed its doors, putting 107 out of work, reportedly owing Ottawa $6 million and BC $2.4 million.

Don Cameron, president of CAP in the 1970s, summed up Trident's demise: "Trident got off track when it ceased to follow the original plan. When they decided to build their own parts, they had to spend money on a plant and equipment instead of on building airplanes. That's when it started not to make sense any more." In the end the Trigull and rights to build it went to Viking Aircraft of Victoria. Over the years it kept up an interest in the Trigull and as recently as 1993 conducted test flights.

Saunders Aircraft

In September 1964 Dave Saunders produced an original four-seat design, the Cheetah, but it didn't go beyond a prototype. A former Avro Canada engineer from the UK, and RCAF pilot, Saunders took on a bigger project. In May 1968 he established Saunders Aircraft Corp. in Montreal to produce a modernized version of the de Havilland Heron. Saunders stretched the Heron by 8' 6", modified the wing, and replaced four 250-hp Gipsy Queens with two 750-hp PT6A-34s. The new aircraft was the ST-27. It was promoted as "looking and feeling" like an airliner. The prototype was completed at Dorval, where it first flew May 28, 1969.

A number of Montreal interests were backing the project, including the Steinberg family and Prudential Insurance Co. The flight test program was completed in December 1970. At this time Manitoba was offering incentives to industries willing to relocate to recently abandoned CFB Gimli. Saunders made a deal to move there. Manitoba eventually took an 82% interest in the company. Dave Saunders, William Kelly, Thomas Ault and Reg Kersey were presidents of Saunders at various times. Others involved were J.E. Grandage (marketing), Lane Helms (engineering), George Henshaw (manufacturing), John Iverach (finance), F.W. Richter (product support), Gerry B. Smith (test pilot, marketing) and Bruce E. Thomson (administration). The first Manitoba-produced ST-27 flew in March 1971. Canadian type certification was awarded that May. Saunders appeared to have a bright future. A number of sales (basic price $425,000) and leases were booked, but FAA certification rules (FAR 25) meant that the ST-27 could not be sold in the US in the 12,500-pound commuter category. The all-new ST-28 (14,500 pounds) follow-on was built to FAR 25 specs and flew July 17, 1974.

The ST-27's market was smaller airlines linking urban areas. Joe Csumrik, a businessman and former mayor of Peterborough, Ontario, was developing Otonabee Airways to link Toronto, Ottawa, Montreal and points between (Kingston, Oshawa, Peterborough). To him the ST-27 was an affordable type with the capacity (19 seats) required. At the time there were few other choices other than the Twin Otter, which was slow and lacked passenger appeal. Csumrik bought his first ST-27 in April 1975.

Saunders' initial success was deceptive, for the company was on shaky ground. While

Pilot-engineer Dave Saunders' first all-metal design was the four-seat Cheetah. Saunders flew the prototype from St. Jean, Quebec on September 22, 1964, but was not able to organize financing for production. (Halford Col.)

Saunders ST-27 C-FCNX (above) at Mount Hope on October 5, 1974. It retired in 1988, but through its years provided excellent service and earned good revenue. Bought by Otonabee Airways (right) of Peterborough in 1975, it was later with Bearskin Lake Air Services of Thunder Bay, Voyageur Airways of North Bay and City Express of Toronto. (Larry Milberry, CANAV Col.)

Manitoba had invested, Ottawa was not enthusiastic, and provided only about 4 million over the years. As owner of de Havilland Canada it was pumping money into the Twin Otter and Dash 7. When Saunders had a sale to LAN Chile nearly sealed in 1973, Ottawa backed DHC in selling LAN Chile six Twin Otters. Through the Export Development Corp., a federal agency, DHC was able to offer LAN Chile payment terms which Saunders couldn't match. Unable to market the ST-27 in the US, and with the world economy in a slump, Saunders ultimately ran into a brick wall. It would cost millions to certify the $800,000, 23-seat ST-28. Manitoba, already having poured $32

million into the venture, would not hang in—in 1975 it pulled the plug. Saunders closed, leaving about 450 without work. Price Waterhouse Ltd. took over as receiver and put Saunders up for sale by tender in November 1978. Ten ST-27s and one ST-28 were offered, along with tooling, jigs, parts, engines, etc. Joe Csumrik purchased most of Saunders' chattels and moved them to Peterborough.

The demise of Saunders was a loss to Canada's aircraft industry. It had begun with a good concept and had the ST-28 nearly ready to produce. In the process it had trained skilled workers, and was diversifying. Although millions had been spent, Manitoba had something

it had lost in 1945—an aviation sector. A secondary goal at Saunders had been to bring back the Fairchild Husky. It can be argued that Manitoba showed a lack of vision and courage regarding Saunders. On the other hand maybe the province had been indulgent enough. In the end, 13 ST-27s and one ST-28 were completed. The ST-28 was donated to the WCAM; the ST-27s gave solid service throughout Canada into the 1990s, by when they had become unsupportable. CF-FZR, first delivered in May 1973 to ACES of Colombia, eventually joined the CMFT. C-GCML was donated by Max Shapiro of Voyageur Airways to the Canadian Bushplane Heritage Centre in Sault Ste. Marie.

(Right) This ST-27 spent a few months in 1976 in Barbados, then returned to fly as CF-HMQ for Otonabee, Northward and City Express. The ST-27 was an attractive 19-seater, but had one cabin feature that disconcerted passengers—the main wing spar passed through the cabin, obliging passengers to step over it when boarding or deplaning. (Halford Col.)

(Below) C-FCNT was the seventh of 12 ST-27s. It was delivered in May 1974 to Alberta-based Bayview Air Service. It later was with Otonabee, Northward and Labrador Airways. With the latter it was irreparably damaged on landing at St. John's, Newfoundland on August 25, 1984. Here it was on the ramp at Edmonton Municipal on July 31, 1974 with a 737, Electra and Lear. (Larry Milberry)

A nice in-flight view of YBM, the first ST-27, over Manitoba countryside. It had begun in 1954 as RAF Heron XG603 (c/n 14058). It had embassy postings in such places as Washington and Saigon before being sold to Saunders in 1969. In 1972-75 it served with ACES in Colombia as HK-1286 (right, at Gimli). Back in Canada it became the prototype ST-28, flying initially on July 17, 1977. With the collapse of Saunders it went to the WCAM. (Halford Col.)

(Left) Voyageur Airways of North Bay operated several ST-27s over the years. Here CML was at Lester B. Pearson International Airport as an Air Canada 727 departed on R24R on July 27, 1989. (Larry Milberry)

(Below) Most Saunders were cut up for scrap, but some were kept for museum purposes. C-GCAT, which sat on a trailer for years at Thunder Bay, finally went to the Canadian Museum of Flight and Transportation in Langley, BC. CAT was the last ST-27, having gone to On Air of Thunder Bay in May 1976. As for the ST-28 YBM, seen at Gimli in 1991, it went to the WCAM. (Larry Milberry, Jan Stroomenbergh)

YBM after being re-engineered as the first ST-28 in order to meet US Federal Aviation Administation requirements. Included in the mods were four-blade reversing propellers, stretched fuselage (22 passengers), larger vertical tail and increased fuel capacity. (Arnold Col.)

(Above) RAF Heron XH375 (c/n 14059) served the Queen's Flight 1955-64, then spent some years in storage. It was sold to Saunders Aircraft of Montreal, becoming CF-YAP. It was the fourth ST-27 (C-GYCR). In 1969 Al Martin shot it at Malton as a stock Heron still in Queen's Flight colours.

(Left) ST-27 CF-XOK in a nice December 1972 air-to-air. (CANAV Col.)

(Left) CF-LOL spent nearly five years serving the country north of Winnipeg for St. Andrews Airways. It later served Otonabee, Northward, Voyageur, Labrador Airways and City Express. Here it was at St. Andrews airport July 27, 1974. St. Andrews Airways was founded in 1971 by George and Barbara Brotherston, former store and lodge owners from Garden Hill, Manitoba. Needing more reliable air transport, they purchased the Midwest Airlines licence for the Island lake area, and three FBA-2C bushplanes. Within five years the company, based at Island Lake (290 miles north of Winnipeg) had 13 aircraft and 15 pilots. (Larry Milberry)

Otonabee's ST-27 CF-CNT at Kingston, Ontario on June 2, 1978 with (right) the crew of Capt Bill Dickey, FO Peter Scherm and FA Patty Brady (right). Their day's work included Toronto Island-Kingston-Ottawa-Kingston-Syracuse-Kingston-Ottawa. (Larry Milberry)

(Left) ST-27 C-FJFH at Winnipeg in December 1975. It recently had been delivered to Aero Trades, but in mid-1976 went to Otonabee. It later served City Express and was withdrawn from service in 1988. (John Kimberley)

(Below) A City Express ST-27 arrives at Toronto Island on a June 1988 flight from Peterborough. (Richard Beaudet)

Some of the Great Propliners

Douglas Propliners

Canada's most famous propliner is the Douglas DC-3. First flown in 1935, more than 10,000 were built. After so many years at work, a few DC-3s still were in Canadian service in the late 1990s. This gallery honours the DC-3 and its four-engine offspring. Air North's DC-3 C-GZOF is shown above at Abbotsford, BC on August 8, 1993 as a CF-18 departs; then, tail feather details. ZOF served the USAF as 43-16367 till bought by American Aero of Tucson in 1961. In 1975 it came to Canada for Contact Airways of Fort McMurray. With Air North in the 1990s it hauled fresh seafood on behalf of Waglisla Air. It came to a tragic end after an engine failed while departing Vancouver for Whitehorse on August 19, 1995. Capt Brad Wightman tried returning but ZOF hit a dike and burned. Wightman, FO Al Bandera of Air BC, and mechanic Lorenzo Roberte were badly injured. Wightman did not survive. By the late 1990s there was general concern in commercial aviation that planes like the DC-3 were getting too old. Where possible, operators were re-equipping. About the time that it lost a DC-4 in the summer of 1996, Air North of Whitehorse purchased two ex-Mount Cook Airline BAe 148s from New Zealand. Although 30 years old, they were advanced technology compared to a DC-3 or DC-4. (Larry Milberry)

(Right) Carl Millard's first DC-3 was this "maximized" beauty. It had served in North Africa, saw action over Normandy on D-Day, and was N1823 of Champion Spark Plug Co. when Millard bought it in 1966. WCM was photographed while clearing customs at Buffalo, NY on March 20 that year. It last was heard of in the US as N47FJ hauling for FedEx. (Larry Milberry)

(Above) Millardair's CF-WCO ("The Voyageur") viewed from atop Toronto's Aeroquay on January 4, 1974. A 1944 model (USAAF 43-15271), WCO was sold in 1946 to Remmert-Werner, a specialist in corporate airplanes. After being refurbished, it served Monsanto Chemical as N2005, then other US firms before coming to Canada in September 1967. In 1979 it became N9061H with a Florida operator, then migrated to Colombia. (Larry Milberry)

(Left) Dakota CF-AAM of Austin Airways departs Detour Lake, north of Cochrane, Ontario on August 31, 1982. The mine there depended on airlift, although a road later was built to Hwy 11. Traditionally, companies built towns for mine staff at such isolated sites. "Fly-in mining", however, is an option—miners commute by air, e.g. mines in the NWT at Lac de Gras, Little Cornwallis Island, Lupin and Nanisivik. Another reason companies like fly-in mining is that it avoids the trauma of closing a town once a mine is worked out. (Larry Milberry)

CF-AAM (ex-RCAF 10910) started in the RAF as FD941, then flew in BOAC as G-AGHO. It came to the US as N9993F in March 1951, but soon joined the RCAF. Jack Austin bought it surplus in 1970. It served Austin Airways till sold to Central Mountain Airlines in 1989. It crashed at Bronson Creek, BC on January 14, 1993. Here AAM is shown departing Smithers on April 3, 1992 heading to a remote mine with a heavy load of machinery. (Larry Milberry)

(Left) A detail of AAM's cockpit . (Larry Milberry)

(Below) A row of classic propliners at Edmonton on September 7, 1960. Transair's DC-3 CF-CPV is under the nose of the C-46. It served Canadian Pacific 1945-57 before joining Transair, was re-registered CF-TAR and crashed October 7, 1970. (Les Corness)

CF-CPY in CPA colours flying on forever on its pedestal at Whitehorse. CPA DC-3s served Canada from Seven Islands to Yellowknife and Port Hardy. (Mike Valenti)

CF-IAE went to the US military in August 1942, coming to Canada in 1955 for Shell Oil. The Alberta government bought it in 1969 for forestry work. IAE ended in the Reynolds-Alberta Museum in Westaskiwin, where it was photographed July 4, 1993. (Larry Milberry)

CF-DOT is thought to have been the last DC-3 converted at Canadair. After a career in Australia with the American 5th Air Force it reached Cartierville and was delivered to the DOT in February 1950. It's shown (right) over Toronto Island Airport on November 24, 1971 captained by Chris Oliver; then at Abbotsford on August 6, 1993 in Coast Guard markings. DOT remained in Canadian government service till sold in 1995 as N1XP. (Larry Milberry)

Transair's CF-TAU loading on a perfect July 19, 1973 at Rankin Inlet, NWT; then taking off for Baker Lake. TAU served years with Braniff Airways before coming to Transair in 1959 (originally as CF-LJS). It went to Lambair in 1973, then to the WCAM after it was cannibalized in 1981. (Larry Milberry)

(Below) CF-JRY at Malton on May 5, 1970, then on September 5 after being wrecked in a sudden storm. Built in 1942, JRY served with Western, TWA and Chicago and Southern. In 1957 it joined Shell Oil of Calgary. D.G. Harris Productions of Toronto bought it for movie making in 1969, but the wind ended that venture. (Larry Milberry)

(Left) CF-CSC "Arctic 7", which replaced JRY with Harris Productions, takes off at Oshawa in June 1971. It was NC75408 of Standard Oil after WWII, then came to Canada in 1961 for Chevron Oil of Calgary. Later operators were Air Brazeau and Nordair. CSC was destroyed by fire at La Grande, Quebec on November 15, 1975. (Larry Milberry)

Great Lakes Airline's CF-GLA landing at Malton on April 15, 1973. It was the oldest surviving DC-3 at the time, with some 80,000 flying hours since delivered to American Airlines in September 1936. Later it flew with Ozark Airlines, which traded it to Fairchild-Hiller for an F.27 in 1967. GLA joined Great Lakes in December that year, went to PemAir in 1971, then became XA-IOR of Aero California. It crashed in Mexico on January 29, 1986. Then, CF-GLB at Malton, where it and Wardair's 747 DJC were having engine changes. Great Lakes formed in January 1961. Besides DC-3s, it operated a Ce.310 and a Beech 18, before stepping up to Convairs. By 1968 it had a twice-daily Sarnia-Toronto service with tickets sold at $34 a round trip. (Larry Milberry)

(Below left and right) Air Brazeau's C-GABI departing Caniapiskau, Quebec on July 16, 1977 during construction of the James Bay hydro project. ABI returned to the US in 1985 as N140JR. C-FBJY of Val d'Or-based Air Creebec was at Winnipeg on May 23, 1988. It was lost in a crash at Pikangikum Lake, Ontario on November 1, 1988. (Larry Milberry, Jan Stroomenbergh)

(Above) Imperial Oil of Toronto operated CF-IOC as a VIP plane 1951-69. In this August 1979 photo it was at Timmins with Hudson Bay Air Transport on a mineral survey job. It later was N47CE of Condor Enterprises with which it crashed on May 22, 1989 at Dekalb, Illinois. (Larry Milberry)

(Above) In August 1986 ex-TCA, ex-DOT DC-3 C-FGXW made a 33,000-mile, round-the-world flight promoting Vancouver's Expo 86. It's seen at Abbotsford with crew and supporters. In 1990 GXW moved to the US as N173RD. (Kenneth Swartz)

(Right) CF-HGL at Sydney, Nova Scotia on August 15, 1975. An Eighth Air Force veteran, it was later one of Canadair's famous postwar conversions. Canadair sold it to Norway, where it served SAS. In 1953 Carl Burke bought it for MCA. It went to the US in 1978 as N37906 but soon was reported wrecked in Mississippi by Hurricane Frederick. (Larry Milberry)

(Left) This DC-3 was delivered in March 1945 to the RAF at Dorval. It was assigned to the RCAF as KN427 (later 12921), serving to June 1970. As C-FTVL it later operated with Bradley/Firstair, with whom it's seen at North Bay, Ontario in June 1980. In 1993 it became N16SA in the US with Saber Cargo Airlines. (Dave Thompson)

(Below) N4700C, converted to Rolls-Royce Darts, was at Edmonton on July 11, 1972. It was one of Jack Conroy's brainwaves and first flew on May 13, 1969. Later in life it was Conroy's "Tri-Turbo 3" with three PT6A-45s, first flown November 2, 1977. (Les Corness)

(Below) Robert S. Grant photographed C-FOOW near Kenora in February 1983. After servicing the RCAF/CF as 12954, OOW went to Atlas Aviation in 1971, then to Air-Dale. CF-IAX of Austin Airways was shot by Larry Milberry on the Albany River at Ogoki Post on March 8,1975. It was taking Toronto school children to Nakina to catch the transcontinental train. Veteran bush pilot Bill Ross was the captain. IAX was in Poland after the war as SP-LCD, then served Iranian Airways, the French Air Force, etc. till coming to Canada in 1972 with Ilford-Riverton Airways. It crashed at Fort George on November 10, 1976. John der Weduwen photographed CF-NNA supporting the caribou hunt inland from Great Whale River in March 1978. NNA crashed later at Big Trout Lake. Then, John's photo of CF-AAC at a survey camp at Inugsuin Fjord on Baffin Island over the spring of 1969.

(Left) Tragedy awaited CF-FOL of Toronto-based S-H Aviation Sales and seen at Malton on October 1, 1972. November 17 of that year it ran low on fuel 100 miles east of St. John's, Newfoundland. The crew ditched, but perished. (Larry Milberry)

(Right) CF-HPM at Saint John, New Brunswick on August 9, 1975. This was an original American Airlines DC-3 that was diverted to the USAAF as a C-49J in October 1942. In June 1944 it went to Chicago & Southern as NC38938, thence to Delta. L.B. Smith converted it to executive configuration and sold it in August 1954 to the Quebec textile company, Ayers Ltd. It went to Atlantic Airlines in Saint John in 1974, but was wrecked in a mishap there two years later. (Larry Milberry)

(Left) CF-ORD at Malton on April 19, 1969. It was delivered to the USAAF in March 1943 and later served Ford in Detroit as N303K. It joined Ford of Canada in June 1967 and finished its Canadian years in 1972, when sold by Northern Wings to a California company as N9011F. (Larry Milberry)

(Right) One of Millardair's classic C-117Ds on the company ramp in Toronto in December 1983. This type first appeared in 1950 as the Super DC-3, but Douglas over-estimated the postwar market for an improved DC-3. The airlines were not interested. The only decent order was 100 for the US Navy. These were conversions of R4D-6 and R4D-6 (i.e. C-47) airframes, the new aircraft being the R4D-8 (later redesignated C-117D). The C-117D had 1,425-hp Wright R1820s compared to 1,200-hp P&W R1830s in most versions of the DC-3/C-47. In 1995 Millardair sold C-117Ds C-GDIK, DOG, GKE and JGQ to Skyfreighters in BC. (Larry Milberry)

With as many as six on the flightline each day, Buffalo Airways was Canada's largest DC-3 operator in 1996. At right two of the fleet are seen at Yellowknife on August 18, 1995; the other (above) is undergoing maintenance the day after. (Larry Milberry)

Two grand Eldorado Aviation DC-4s shot at Edmonton Muni by John Kimberley in 1978. JRW came to Eldorado from Alaska Airlines in 1957. It later went to Aero Trades of Winnipeg and was wrecked at Spence Bay, NWT in December 1981. GNI came to Canada in 1969 after careers with the USAAF and American Airlines (Flagship "Illinois"). It served Great Northern Airways of Edmonton till GNA folded, then went to Eldorado (1971), Aero Trades (1980), Soundair (1986) and Air North (1988). On departing Bronson Creek, BC on August 14, 1996 headed for Wrangell for Prime Resources (operator of the Snip gold mine), GNI had an engine fire. The pilots tried putting down on a sandbar in the Iskut River, but landed in the water. The co-pilot and flight engineer Stewart Clark swam ashore, but Capt Don Bergren was swept away and lost.

(Left and below) A Conifair DC-4 lands at St. Jean in October 1983. Conifair specialized in forest spraying in Quebec with DC-4s, DC-6s and Constellations. This business gradually faded and the old crates were retired. DC-4 C-GDCH is shown rotting at St. Jean in September 1993. Conifair owner Michel Leblanc later branched into hunting charters with the DC-6 and Convair 580, then formed Royal Air, one of Canada's leading charter carriers in the 1990s. (Richard Beaudet, Larry Milberry)

Jim Smith and Graham Thoburn flying Buffalo Airways DC-4 C-FIQM near Yellowknife on August 29, 1992. IQM flew in Canada since 1956, the year it joined Wheeler Airlines. Later operators were Nordair, Pelly Bay Co-op, Aero Union and Kenn Borek, with whom it operated in the Antarctic in the late 1980s. In 1997 it was a water bomber with Buffalo Airways. (Henry Tenby)

(Left) In the 1980s CF-IQM was a water bomber in California with Aero Union. On March 29, 1990 its nose gear collapsed at Gun Barrel Inlet on Great Bear Lake. Seven mechanics took two weeks to change three props and two engines. Then, with a sad-looking patch on the nose, IQM flew to Calgary, where Chris Buckley photographed it in June. On March 1, 1992 it departed for Tucson for overhaul at Hamilton Aviation, then returned for a new career with Buffalo Airways.

(Right) Another fine John Kimberley DC-4 photo. QIX was at Winnipeg in the spring of 1979. It had served the USAAF, Colonial Airways, Eastern Air Lines, Southwest Airlines and Mercer Enterprises, then came north in 1969 for the Can-Arctic Co-op Federation. In 1979 it joined Aero Trades, with whom it was lost at Thompson that June 1. On take-off No. 1 engine caught fire. The crew came around and landed, but QIX burned. The investigation noted: "No. 7 induction pipe was found to be missing from the engine." It was believed to have become dislodged after start-up, "creating a torching effect which caused the magnesium alloy crankcase to burn."

(Left) Richard Beaudet shot this DC-4 at Churchill, Manitoba in July 1982. Calm Air did well freighting in the north with DC-4s, but eventually replaced them with 748s. In 1997 PSH still was earning a profit, by then with Buffalo Airways.

(Right) Down in the weeds... DC-4s turn up in the craziest places. N4989K of Contract Air Cargo visited the annual fly-in at Bob Spence's farm near Muirkirk, Ontario on the Labour Day weekend in September 1996. Aviation fan Gord McNulty noted: "It was a real show-stopper to see this heavy-weight fly into a 3,000-foot grass airstrip." N4989K served most of its USAF career in the medevac role and didn't retire till 1973. (David Frost)

(Left) Like the DC-4, the DC-6 lasted in Canada for a half-century. Veteran aviation buff Fred Prior of Winnipeg captured Conair's DC-6B "Tanker 47" ripping off at Abbotsford in August 1980. Besides working domestically, Conair often had international business, especially in the Mediterranean fighting fires or spraying. In 1988 five DC-6s and five Aerostar bird dogs were leased to the US Forest Service in Idaho. In 1990 three DC-6s fought fires in Mexico, while another worked in Alaska. After the *Exxon Valdez* disaster a Conair DC-6 dispersed oil throughout Prince William Sound, while another did reconnaissance. In 1996 Conair despatched a DC-6 to spray in New Zealand against the white-spotted tussock moth.

(Left) Ex-KLM, ex-Wardair DC-6B C-FKCJ. Assigned on CIDA humanitarian duty in the Congo, it was shot at Edmonton by John Kimberley in November 1974. Later it was XB-BQO in Mexico.

(Below left and right) Willy Laserich's colourful C-118 (DC-6) at Edmonton in March 1979. It was tied up in litigation for several years, then was sold to Northern Air Cargo of Alaska, becoming N99330. Then, Transair's DC-6B CF-CZS at Winnipeg in May 1971. This old CPA "Empress" later served in Greenland and the UK before being scrapped in 1985. (John Kimberley, Andy Graham)

Beech 18

First used in Canada by Starratt Airways, the Hudson's Bay Co. and CPA, the Beech 18 was the latest commuter type. Operators quickly came to appreciate its speed, comfort and economy. The Model 18A first flew at Wichita in January 1937. Six were built that year at a selling price of $37,500. They had 320-hp Wright R760s. The hotter Model 18S introduced in January 1939 had 450-hp P&W R985s. By September 1939 only 39 Model 18s had been sold, but the war changed that. By the time Model 18 production ceased in 1969, more than 7,000 had been delivered. Although not as popular on the West Coast as elsewhere in Canada, the incomparable Beech 18 was prized by certain operators, especially Vancouver Island Airlines. Here VIA's CF-CSN lands at home base on April 5, 1992; then is seen on its daily sked at the International Forest Products heli-logging barge on Simoom Inlet a few days later. (Larry Milberry)

Aboard CSN pilot Peter Killin chats in the cockpit, while his passengers enjoy the spectacular view coming down the coast to Campbell River. (Larry Milberry)

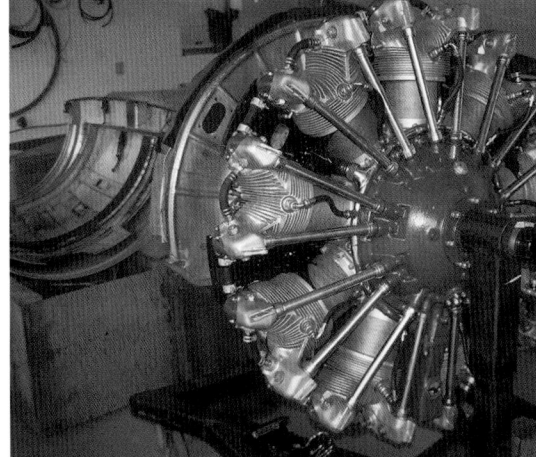

April 5, 1992... the first Seawind, a much-modified Beech, is being built up for VIA in Dave Nilson's Campbell River shop. Then, the heart of the matter—one of its zero-time R985s awaiting installation. (Larry Milberry)

(Right) Larry Langford pilots the Seawind on an early flight. Its floats were 8500s, suitable for the plane at its gross weight of 9,450 pounds. CSN used 7800s for a gross weight of 8,725 pounds. (Brian Kyle/VIA)

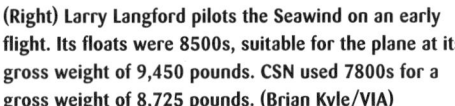

(Below) Northern Thunderbird Air operated the Beech 18 for many years, retiring its last in 1996. Here is CF-BCE (ex-BC government) in the hangar at Prince George on August 11, 1995; then a cockpit view. Note the crew door, faired-over photo nose panels, modified wingtips and stowage for a spare tail wheel. BCE was legal to 9,360 pounds. (Larry Milberry)

(Right) Beech CF-IMC at St. Albert on August 20, 1976. It was an old C-45G imported in 1964 for International Minerals and Chemicals of Esterhazy, Saskatchewan. Hopped up with extended wingtips and panoramic windows, it made a nice airliner when bought later by Athabaska Airways. (Larry Milberry)

Air Niagara's Super 18 CF-ANA (left) at Malton on May 15, 1971. It later served with White River and Laurentian air services. Then (below), Super E18 airliner "Snowbird" at St. Andrews, Manitoba on July 23, 1974. The extra headroom of the E18— a big selling feature—is apparent. Beech manufactured 460 E18s 1954-60. (Larry Milberry)

Rusty Myers Flying Service was the best-known bush operator in the Fort Francis-Rainy Lake region. Myers used many types over the decades, few of which came close to the Beech 18 for reliablity, safety or profitability. This grand air-to-air view of C-FERM was shot by bush pilot Richard Hulina, while the dockside scene was at Rainy River on the evening of July 16, 1991 by Larry Milberry.

(Left) Walsten's Beech 18 C-FCUK en route Kenora-Snowshoe Lake. Brent Wood was flying CUK, while Garth Tremblay flew the photo plane—Walsten's Ce.206. (Henry Tenby)

(Right) Beech EHX at Whitey Hostetler's Red Lake, Ontario maintenance base on July 15, 1991. (Larry Milberry)

(Left) Green Airways' Beech GNR at the dock at Red Lake as Otter ODJ ties up after its last trip of the day on July 15, 1991. (Larry Milberry)

(Right) Beech OII of Ignace Airways flew a long time from Ignace using "7850" floats built by MacDonald Brothers in 1939 for a Fleet Freighter. CF-OII was spotted in Kuby's scrap yard in Kenora on July 16, 1991. Over the years many wrecks rose from this yard to fly again. (Larry Milberry)

(Left) This beautiful Beech 18 of Northwestern Flying Services was at Nestor Falls on August 9, 1974. In the early 1990s it was derelict at Redditt, near Kenora, awaiting a new career. (Larry Milberry)

(Below) Wild Country's CF-BCC loads groceries at Red Lake for Indian reserves on March 25, 1992. With its extended cargo compartment BCC could carry up to seven full 45-gallon drums (AUW 10,100 pounds). Beyond, a Piper Navajo, a modern-day stalwart, awaits a trip, and a Sabourin Airways Otter chugs off. Then, pilot Marc Jaspar operating BCC Red Lake-Poplar Hills (75 mi.)-Deer Lake (45 mi.) on the same date. With its Garmin GPS 100, getting around was easy. Jaspar had learned to fly in 1968 at Central Airways in Toronto. Later he worked for OCA and Air Manitoba. Beech 18 C-FKEL replaced BCC after it crashed later in 1992. (Larry Milberry)

(Below) Bush pilot's office—Al Zaroski aboard TBH on July 12, 1995; then, his plane at Ear Falls, Ontario. Al began flying in 1968 and initially flew Ce.180A CF-IGU for R.J. Ball of Thunder Bay. He moved to George Theriault in Chapleau on the Ce.180 and Beaver, then to Rusty Myers of Rainy River in 1972 on Beaver OBY. For 1973 he flew Myers' Beech 18 CF-RVL; and in 1975 began a 10-year stint with OCA flying floats in summer, and DC-3s (BJE, CZG, XUS and YQG) in winter at places like Baker Lake, Churchill, Thompson and Yellowknife. After OCA folded, Zaroski flew the Beech 18 and Norseman for Red Lake Airways, spent a year as a mechanic at Red Lake Seaplane Service, and in 1987 moved to Northern Thunderbird Air on the Ce.337, Beech 18 and Navajo. Next came two seasons with Wild Country. By 1997 he had some 15,000 hours of which about 9,000 were on Beech 18s. (Larry Milberry)

CF-BCC details: the classic Beech 18 control wheel, and the most historic of aero engine company crests on the cowl of an R985. (Larry Milberry)

645

One of O.J. Wieben's Beech 18s on skis at Nakina on March 3, 1975. Wieben bought many of ex-RCAF Beech 18s from CADC in the mid-1960s. These were converted for bush service or cannibalized. Some still were operating 30 years later. As to WYR (ex-RCAF 1474), after years of dormancy it was listed to Saltwater West in 1996. (Larry Milberry)

One of Warren Plummer's Beech 18s in Toronto harbour while transitting west in September 1973. White with red trim, it was shot from Hughes 269 CF-WON. (Larry Milberry)

Many old designs were adapted the PT6. Most numerous were Beech 18s like CF-URS, shot at Mount Hope on September 6, 1968. URS was an old RCAF Expeditor with PT6s and tri-gear. With a top speed of 190K it was a bit of a rocket and easy winter starts were a boon; but one veteran pilot's conclusion was that a lot of money was spent "improving" the Beech 18 for what were marginal advantages. (Bob Finlayson)

(Right) CF-HZA at Baker Lake on June 28, 1967. Upon arrival, it landed on the ice when no open water could be found. HZA came to Canada in 1954 and served a list of owners—Cascade Drilling in Alberta, International Air Freighters in Ontario, La Ronge Aviation in Saskatchewan, Midwest Aviation and Warren Plummer's Sioux Narrows Airways. In 1997 HZA was with Ross Air in Northwestern Ontario. (Ed McIvor)

(Left) Air-Dale's Beech PSU at Sault Ste. Marie on July 23, 1974. By this time Air-Dale was expanding from its bush roots. It had won a contract to operate Twin Otters for norOntair. In October 1984, when norOntair added the Dash 8, Air-Dale became the first operator of that type. (Larry Milberry)

(Right) Selkirk, Manitoba. May 7, 1995. With pilot Ken Redding at the helm, Selkirk Air's C-FSFH "goes Beeching"—its initial flight of the season. Before long SFH was wrecked—taking off from Bradburn Lake for Selkirk on June 5 a float separated. The seven aboard made it ashore. SFH had been RCAF HB112 from 1943-65. (Larry Milberry)

(Left) CF-SUQ (ex-RCAF 2324) resplendent in the evening light at Thompson, Manitoba on July 18, 1974. The Beech 18 was an ideal aero survey platform. (Larry Milberry)

(Below) A few Beech 18s still served in Canada in the late 1990s. This example, at Hamilton on June 14, 1991, belonged to the Canadian Warplane Heritage. (Larry Milberry)

(Left) Bones to be picked clean. Clapped-out Beeches in the "back 40" at Sioux Lookout airport July 14, 1991. Included is Beech 18 CF-OLN with some fading nose art. The 1954 C50 Twin Bonanza (CF-SFB) had flown for Slate Falls in happier times. (Larry Milberry)

(Above) Manitoba ghosts. The back lot at the Gaffray brothers' strip on Highway 11 south of Pine Falls was home to all sorts of interesting old technology when visited on May 6, 1995. Included was the hulk of Beechcraft HVF with its Cameron Bay titles showing through after layers of paint had peeled in the prairie weather. HVF had been bought years earlier for its floats and other useful parts. The old Norcanair Beech 18, shot in the fall of 1967, had been wrecked on a northern lake. (Larry Milberry, Ed McIvor)

(Left) It's seen better days... Bradley's old Beech AT-11 Kansan CF-KJI at Carp airport near Ottawa on February 15, 1975. (Larry Milberry)

The Great Grummans

No amphibians lasted so long, nor made such a historic mark in aviation as those from Grumman of Long Island, New York. The Grumman Aircraft Engineering Co. was founded in 1929 by Leroy Grumman and partners, most of whom were from Loening. Beginning humbly, Grumman advanced to floatplane and fighter designs for the US Navy. The G-7, first flown in April 1933, was its first amphibian. It was ordered by the US Navy as the JF Duck. A family of amphibious twins followed. The first Canadian Grumman was G-21 Goose CF-BKE, registered on May 23, 1938; several G-21s still were flying here more than 60 years later. Although none was registered in Canada, the Duck made the occasional visit. This one, flown by California vintage plane collector Frank Tallman, was at Toronto Island Airport on August 29, 1975 for the CNE airshow. (Larry Milberry)

Trans-Provincial's Goose EFN (left) in the back channel at Seal Cove during the summer of 1985. EFN was wrecked at Sandspit in March 1991 and sold in the US. Then, Goose BXR (below) at Prince Rupert in the 1980s as a Jet Ranger departs and a 737 arrives. BXR spent more than 50 years at work, then was exported to the US in the 1990s. Like all types the Goose occasionally met disaster. Such a case occurred August 30, 1979: C-FUVJ of TPA crashed near Prince Rupert. Seven of 10 aboard perished. The pilot was approaching Runway 12 when both engines quit. The cause was a fuel shut-off valve left closed after maintenance—only fuel from the right tank could be used, even with the fuel selector set for both sides. UVJ ran the right tank dry a few moments before touchdown. (Kenneth Swartz, Grant Webb)

(Below) The most historic Canadian Goose was CF-MPG, which came to Canada in 1944 to serve as RCAF No. 391. After WWII it flew nearly 50 years with the RCMP. It's shown during a periodic overhaul in Transport Canada's Ottawa hangar on April 12, 1991; then arriving on the Ottawa River for delivery to the National Aviation Museum on August 17, 1995. MPG almost didn't make it here. The Crown initially planned to sell it for $350,000; but a Prince George group, "Save the Goose", lobbied to save MPG for posterity as the longest-serving aircraft in Canadian government service. A July 1995 statement from Canada's solicitor general noted: "In addition to the normal duties of ferrying RCMP staff, prisoners, supplies and equipment to remote settlements, the Goose also took part in surveillance work, drug busts and rescues, and escorted royalty, politicians and native chiefs." (Larry Milberry, Andy Graham)

(Right) Rainy Lake Airways' Goose CF-SBS far from home at Churchill in July 1972. With two 450-hp R985s the Goose had ample power to get off small lakes with a good load. In high-density configuration it could carry 8-9 passengers. All-up weight was 8,500 pounds. (Mike Valenti)

(Below) Pacific Coastal's Goose C-FUAZ en route Bella Bella-Port Hardy on September 7, 1988. (Henry Tenby)

Cliff Oakley's Goose C-GHAV (below) at Stuart Island, BC on May 25, 1990. In 1979 Oakley founded Windoak Air Services which, in 1982, became Harbour Air. Later he operated as Oakley Air, but lost his life in 1991 at Vancouver when HAV crashed on approach. (Bottom) C-FVFU of Forest Industries Flying Tankers at Port Alberni on April 9, 1992 with Reg Young at the controls. VFU logged about 500 hours yearly on timber cruising, VIP fishing trips, supporting Martin Mars water bombing operations, etc. (Christopher Buckley, Larry Milberry)

(Below) Beautifully finished G-44 Widgeon CF-JXX at Malton shortly before it was exported to the US around 1976. JXX came to Canada in 1957, serving Murfin Heating of Toronto and Chamcook Airways of Montreal. It then went to the West Coast for Harrison Airways. In this view Imperial Oil's hangar is in the background. The 4-5 passenger Widgeon was prompted by market pressure for a smaller version of the Goose. First flight was in June 1940. Grumman built about 200 wartime and 75 postwar Widgeons; 1949-52 a further 40 were made in France by SCAN. A US Coast Guard Widgeon sank the German submarine U-166 off Louisiana on August 1, 1942 using a 325-pound depth charge. (Gordon Schwartz)

(Above) The first PT6-powered G-21T Turbo Goose conversion was done in the mid-1960s by Fairey Aviation of Victoria for Alaska Coastal Ellis Airlines. Here Turbo Goose C-FBCI is at Kenora in June 1991. It was the first Canadian G-21T (PT6A-20s), having been converted at Victoria for the BC government by the king of Grumman converters, Angus McKinnon. The G-21T used the 550-shp PT6A-20 (10,000 pounds AUW) or 680-shp -27 (12,500 pounds AUW). It carried two pilots, nine passengers and 500 lbs baggage. Most had de-icing boots and electrically-heated props. Coast pilot Jim Soden commented in 1992: "The Turbo Goose handled like a bearcat, flew at an honest 200 mph and had four hours of gas. With reversing propellers it was a joy to handle on the water. Its big shortcoming was visibility, especially in a turn. With its PT6s protruding so far forward, it was hard to see." (Robert S. Grant)

(Left and below) SCAN Widgeon CF-ODR came to Ontario's Department of Lands and Forests in 1956. Later, real estate dealer Lorne Corley owned it. It always was based at Toronto Island, where it's shown in the water in September 1973, then landing on June 22, 1992. Redpath Sugar and Victory Soya Mills are beyond. The Victory silos later were demolished. (Larry Milberry)

(Right) Two lovely Grummans visiting Abbotsford for the airshow on August 7, 1993. (Larry Milberry)

(Right) Canada's first Grumman G-73 Mallard and its most historic corporate plane was CF-BKE. Toronto millionaire J.P. Bickell of McIntyre Porcupine Mines purchased it in September 1946. He sold it to Montreal tycoon Andrew P. Holt, who in turn sold it in 1950 to the Ontario Paper Company of Thorold, Ontario. Thereafter, BKE flew from Malton. Ken Irwin, Fred Hotson, Bill Henderson, Andy Gabura and others crewed it on business and fishing trips down the Quebec North Shore, to Labrador and the Caribbean. There also were trips to Chicago and New York to company associates, the *Tribune* and *Daily News*. Fifty years after coming to Canada BKE was still based in Toronto, but for the Quebec-based paper company Donohue. The crew was pilots Alan Bonnell and Patrick Meers, and engineer Martyn Boyle. The seven-passenger plane had about 14,000 hours logged and was being flown about 200 hours a year, mainly on fishing trips in Quebec. Here BKE was at Malton on April 29, 1972. In mid-1997 BKE was for sale. (Larry Milberry)

The Mallard first flew on April 30, 1946, but sales were disappointing. The last of 59 (21 of which had Canadian careers) was delivered six years later. This series of photos was taken at Vancouver: CF-HPU in November 1959, C-GENT of West Coast Airlines in August 1977, C-FHUB in June 1982, and C-GHDD "Patricia" in August 1981. HPU remained active in 1997 as the Mallard with the most hours—some 36,000. ENT, originally the executive plane of California-based Union Oil, was flying in Thailand in 1997. HUB, "Miss Daily News" of the New York *Daily News* when delivered in 1947, became VH-LAW of Australia's Air Whitsunday after leaving BC. HDD had been N2964. It later went to Wagalisla, then was sold in Australia. In 1997 CF-UOT was the only remaining Canadian Mallard. (Mel Lawrence via Henry Tenby, Kenneth I. Swartz, Larry Milberry, Kenneth I. Swartz)

CF-HPA at Malton on July 22, 1972. The seventh Mallard, it started with Powell Crossley, Jr. of Cincinnati, then came to Canada in 1954 to spend 16 years with BC Airlines. In 1970 it was bought by the Abitibi Power and Paper Co. of Toronto, which wanted it to serve a fishing camp in the Great Whale area. HPA was not ideal for this work due to shortcomings in range and payload. It was sold to Jack Anderson of North Coast Air Service. On March 5, 1974, while taking off at Prince Rupert, it crashed. Three of the 10 aboard died, including the pilot, Anderson's son. (Larry Milberry)

In the 1970s the first of several Mallards was adapted to the PT6. Here PT6-powered Mallard HUM ("The Hummer") is seen on its inaugural flight to Port Hardy on April 27, 1981. The first turbine Mallard (Northern Consolidated Airlines of Alaska, 1964) had a PT6A-6 on the starboard side and a standard R1340 on the port side. This trial suggested the viability of the PT6 on the Mallard. Next, Fred Frakes of California converted a Mallard to PT6A-27s and this became the standard turbine version. Frakes' Turbo Mallard was the first North American twin certified with the PT6. (Uwe Ihssen)

(Right) The largest of Grumman's flying boats was the Albatross, which gave a decade of good service in the RCAF. Here No. 9304 of 102 Composite Unit runs up at Trenton on June 11, 1966. (Larry Milberry)

(Below) This attractive Caribbean Albatross was at Abbotsford on August 7, 1993. N120FB was formerly USAF and USCG. It was sold back to Grumman in 1980, converted to civil use, then served in the Caribbean with Chalk's. In 1997 it was in storage. (Larry Milberry)

The Classy Convairs

(Left) The first of the renowned Convair Liners was the CV240, first flown at San Diego on March 16, 1947. American Airlines inaugurated service in June 1948. CPA brought the first Convairs to Canada in 1952—a fleet of CV240s. CF-CUU inaugurated service on January 24, 1953 flying Vancouver-Port Hardy-Sandspit. The CV240 developed into improved versions—the CV340 and CV440. Production totalled 541 civil and 520 military examples, all with the R2800 piston engine. Beginning in the mid-1950s, however, various turbine conversions appeared: CV540 (CL-66)—Napier Eland; CV580—Allison 501; CV600/640—Rolls-Royce Dart. By 1997 there still were many Canadian Convairs. Counting RCAF, airline and corporate examples, about 80 had been registered in Canada to 1997. Here CV440 CF-GLK is at Malton on January 4, 1974. Originally with SAS as OY-KPA it spent 1973-77 with Great Lakes Airlines, the predecessor of Air Ontario. In 1981 it joined Island Airlines in Hawaii. On January 17, 1982 it crashed into Pearl Harbor Channel soon after takeoff. (Larry Milberry)

(Right) Air Ontario's CV580 C-GGWJ departing Dorval in July 1988. It started in 1953 as a United Airlines CV340 (N73142 "Mainliner North Platt"). It went to Lake Central in 1960, and became a CV580. It moved to Allegheny, Commuter and Freedom airlines before coming to Air Ontario in 1984. In 1990 it started work with CanAir Cargo of Toronto. (Richard Beaudet)

(Left) CV440 Metropolitan C-GKFC starts its R2800s in a Vancouver scene from June 1980. KFC had started in 1958 with the Hughes Tool Co., then joined the Luftwaffe. In 1974 it came to Great Lakes Airlines as C-GOYO, returned to the US with various operators, then served Kelowna Flightcraft 1979-81. (John Kimberley)

(Right) Quebecair disembarks business travellers at Dorval March 19, 1987. In its earliest days this Convair was PanAm N11137, then went to National, Allegheny, LAN Chile and North Central before becoming C-GQBP in 1986. Like many a Canadian Convair, it moved around. After Quebecair it served Conifair and Air Inuit. (Kenneth I. Swartz)

(Left) CV580 NMQ at Ottawa on February 11, 1987 wearing the gaudy Nordair Metro scheme. NMQ had been United's "Mainliner Medford" in 1954. In 1960 Turbo-Prop Conversions fitted it with Elands. As such it flew with Allegheny as N544Z. Again it became a CV440, was sold to AVENSA as YV-C-AVA, then became a CV580. In 1986 it made its way to Nordair Metro, a short-lived commuter branch of Nordair bought later by Quebecair and re- named Inter-Canadien. (Larry Milberry)

(Right) Air Toronto CV580 C-FICA at LBPIA in May 1990. It started as United "Mainliner Boston" in 1953, then worked for other US companies, coming to Canada in 1984. Originally known as Commuter Express (a subsidiary of Soundair Corp.) Air Toronto had a diversified operation by 1989: Convairs for passenger charter, cargo and courier work; and Metros and Jetstreams on skeds to US points like Columbus, Harrisburg, Green Bay, and Madison. When Air Toronto went into receivership in April 1990, ICA moved to CanAir Cargo. On the night of September 18, 1991, while operating Moncton-Hamilton for FedEx, it crashed in Vermont with the loss of pilots John McDougall and Leonard Zilvytis. (Kenneth I. Swartz)

(Left) SEBJ CV580s at Bagotville July 7, 1977. They were on their daily skeds serving the vast James Bay hydro development along the Great Whale River. (Larry Milberry)

(Above) Can-Air Cargo was one of Canada's leading courier airlines in the 1990s, using a fleet of CV580s. This example started as a CV340 with United Air Lines ("Mainliner Des Moines") in 1952. It spent 1962-78 with Allegheny as N8424H. In 1988 it came to Canada for Soundair, then moved to CanAir in 1990. (Larry Milberry)

(Left) Allegheny CV580 N5834 lands at Toronto on July 22, 1972. The 24th Convair 340, it went to Hawaiian Airlines in 1952. In 1961 it joined Allegheny. By the mid-1970s Toronto was a key Allegheny base with daily CV580 service to Boston, Buffalo, Cleveland, Erie and Rochester. Eventually the DC-9, MD-80 and 737 took over. The Convairs then entered new markets with smaller local carriers and courier specialists. In 1987 N5834 became C-FAUF with Toronto-based Soundair. (Larry Milberry)

(Right) After five years with United ("Mainliner Portland") this CV580 spent its years in private or government hands. From 1958-70 it flew with Johnson and Johnson (N400J), then became CF-BGY with Great Lakes Paper of Thunder Bay. It returned to the US as N8EH, then (1974) was bought for the Canada Centre for Remote Sensing (Department of Mines and Resources) in Ottawa, and was converted for scientific research. It's seen at Uplands on February 11, 1995. (Andy Graham)

(Left) One of General Motors' long-serving CV580s at Toronto in June 1983. One of the last Convairs, it went to GM in October 1959, remaining till 1996, when it came to Canada as C-FTTA for Cypress Jetprop of Vancouver. This upstart no-frills carrier began service in the summer of 1996 to such places as Victoria, Cranbrook and Prince George. (Larry Milberry)

(Below) The Convair 5800 prototype at Abbotsford. It was rolled out at Kelowna Flightcraft on November 5, 1991. The company first discussed re-engineering the CV580 in 1985. General Dynamics and Allison agreed to assume product liability should the project go ahead. GD provided the data for modifying the airframe to take the -D22G—the latest version of the famous Allison 501 engine. The main airframe mod involved 2.1-meter plugs fore and aft of the wing. C-FKFS was produced as a freighter capable of an 11-ton payload. The "5800" flew its trials at Mojave, California and received its STC in December 1993. By 1997 only one other such conversion had been completed. (Larry Milberry)

(Right and below) For an August 1989 photo session Capt Gord Drysdale and FO Glen Hunter set up Convair 640 PWS near Saskatoon for the photographer. Besides sked flights, Time Air's Convairs flew countless charters, anything from trans-Canada prisoner transfers to West Coast and NWT fishing trips. C-FPWS began in 1957 with All Nippon Airways, returned to the US in 1965 and was converted. It did a stint with Hawaiian Airlines, then (below) was sold to PWA. It was scrapped at Springbank Airport near Calgary in the mid-1990s. (Henry Tenby, John Kimberley)

(Right) Worldways Convair 640s on a charter at Gimli on February 26, 1979. PWU went originally to American Arabian Oil as a CV340 in 1952. It joined PWA in April 1967, moved to Worldways in 1976 and returned to the US in 1982. (Bert Huneault)

The Hawker

The 52-passenger Avro 748, one of several postwar DC-3 replacements, first flew at Woodford, England in June 1960 and entered service in 1962 with the British carrier Skyways. Production continued to 1986 by when 377 had been delivered. Avro became Hawker Siddeley, which itself disappeared into the British Aerospace conglomerate; so the Avro 748 became the HS748, then the BAe748. Many of these excellent propliners served in Canada over the decades, remaining popular there well into the 1990s. Austin Airways 748 QSV is seen above at Fort Albany, Ontario on September 1, 1982, then (right) departing the same afternoon. It was originally with LAN Chile and came to Austin in June 1979. It was leased to Maersk Air of Denmark 1980-85 and still had the Maersk colours here. In one incident QSV was damaged at Sugluk. John der Weduwen, Chico Gonzales, Rick Hill, Serge Lavoie, Jack McCann and Raymond Ortez were some of those who went in to help with repair and salvage. Within 36 hours they changed an engine and both props, removed the gear doors and repaired the flaps. Then John Crocker and André Thibodeau flew it off the ice to Timmins for further work. QSV later went to Air Creebec and crashed at Waskaganish on December 3, 1988. (Larry Milberry)

(Right) John Morrison photographed Calm Air's C-FMAK at Winnipeg in May 1989. It had come new in 1969 to Midwest Aviation of Winnipeg, then moved to Gateway, Northward, Calm Air and, later, Air Creebec.

EPA used the 748 widely. This ex-Panamanian example was at Halifax in May 1981. It moved to Austin in 1986, then to the UK to become water bomber G-BNJK. (Gary Vincent)

Few planes are more at home in the Arctic than the rugged, reliable BAe748. Here First Air's C-GDUN (Capt Jim Merritt and FO Daryl Jackson) is shown near Yellowknife on August 9, 1992. Then (below) it's seen as Flight 842 at Coppermine on August 19, 1995, and (inset) departing Cambridge Bay later the same day for Gjoa Haven. (Henry Tenby, Larry Milberry)

(Above) Capt Mike Cahill (left) and FO Graham Hardy crewed F842, along with FA Ray Bayley and crewman Trevor Phypres. The following week First Air and the other NWT carriers were extra busy as some 3,600 students and 250 teachers headed back to school from points south. By this time First Air had a staff of about 35 in its Yellowknife operation. (Larry Milberry)

(Right) First Air 748 C-FYMX at Iqaluit on August 14, 1992. Air Inuit and Bradley Twin Otters and a Bell 206 are in the background. From here First Air served all Baffin Island and beyond to centres like Hall Beach. YMX had gone new in 1969 to Fiji Airways as VQ-FBK. It remained in South Pacific service till coming to the ice box of the world in 1980 for Bradley. (Larry Milberry)

The Viscount

The Rolls-Royce Dart powered some great propliners—the BAe748, Fokker/Fairchild F.27/227, Vickers Viscount and YS-11. Of these the leader was the incomparable Vickers Viscount, a type that came to Canada in 1954 and served more than 40 years. Here Air Canada Viscount CF-THL discharges passengers on May 4, 1971 at Toronto in a typical scene. Then (below) C-FTID, which went to P&WC in 1972 following its Air Canada career. It served P&WC well as a flying test bed till it retired in 1989. Here it was visiting DHC on February 5, 1975 for the rollout of the Dash 7. Its fifth engine is a Dash 7 PW100 series. Beyond is a Transport Canada Viscount. (Larry Milberry)

(Below) The DOT/Transport Canada and the Royal Bank were the first in Canada to operate the Viscount in the executive role. Here CF-GXK awaits at CFB North Bay in April 1980 as 414 Squadron CF-100s get ready to roll. Purchased in 1955, GXK served 27 years. (Dave Thompson)

(Above) Viscount CF-THQ started with TCA in 1958. It retired in 1971, then spent till 1987 on corporate duties for Wabush Mines at Dorval. In January 1988 THQ went to Zaire, the end of the line for many an old Canadian transport plane. (Richard Beaudet)

Friendship, YS-11 and Argosy

Like the Avro 748, the Fokker F.27 Friendship was a DC-3 replacement. First flown in 1955, it entered service with Aer Lingus in 1958. Fairchild began licence production in the US that year. It built more than 100, some of which served in Canada. Canada's first F.27 is shown at Dorval on September 5, 1960. It served Quebecair till sold to Horizon Airlines in 1981 as N273PH. It was traded to DHC in 1986 in a Dash 8 deal, then went to Airlift International. On August 24, 1992 it was wrecked at Miami by Hurricane Andrew. (Larry Milberry)

(Right) Another early F.27 in Canada was CF-LWN. It was bought in 1960 by Abitibi Power and Paper Co. of Toronto to replace DC-3 CF-ITQ. Here it was landing at Malton on July 1, 1969. LWN last was heard of in the early 1990s as N311RD with Airlift International. (Larry Milberry)

(Below) Norcanair F-27Js at Saskatoon on August 21, 1976. (Larry Milberry)

(Right) Great Northern Airways of Edmonton lost two F.27s. CF-GND crashed at Resolute June 12, 1968 less than a month after entering service. CF-GNG ran off the runway at Inuvik February 14, 1970. This one, fully stripped of anything useful, spent years close to Resolute as a beacon for aviators and a place for spray bomb artists to express themselves. (Richard Beaudet)

(Left) The F.27 was stretched in 1965 to become the F.227 or, in the US, the Fairchild Hiller FH-227. Seventy-eight FH-227s were built, seven of which operated in Canada. Nordair had three for southern skeds, but they eventually ended in the Arctic as vital equipment on DEW Line lateral resupply. Here CF-NAJ was at Hamilton on May 2, 1972. (Larry Milberry)

(Right) Many F.27s eventually ended as courier planes. Typical was this one of Time Air seen at Winnipeg in December 1988. (John Morrison)

(Left) Transair NAMC YS-11 CF-TAK (later dubbed "Norway House") at Churchill on July 20, 1973. The 46 to 60 seat YS-11 was a DC-3 replacement that flew initially in 1964; production totalled 182. It was welcomed by the travelling public at places like Thomspon and Red Lake, but considered too slow for longer routes like Thunder Bay-Soo-Toronto. Delivered in August 1968, TAK was sold in 1979 as N4989S and in the 1990s was a cargo plane with Mid Pacific. (Larry Milberry)

Argosy TAX (right) at Churchill on July 19, 1973. It served with BEA 1965-70, then came to Manitoba. The Argosy was supposed to be a great Arctic money-maker for Transair, but proved impractical. As Transair's senior man at the time Stan Wagner put it, "Goddamn Argosies, goddamn English airplanes!" Argosy N895U (below), on lease to PWA, was at Edmonton Industrial on February 4, 1970. It first flew in December 1960 and was delivered to Riddle as N6502R. As shown, it was with Universal Airlines. It later returned to the UK and in 1987 was pre-served at the East Midlands Aeropark. (Larry Milberry, Les Corness)

Some Other Famous Propliners

(Above) Vintage propliners at Abbotsford on August 4, 1986. Shown are Air Canada's Lockheed 10 CF-TCC, a Stout Bushmaster and a Ford Trimotor. The L.10 and Ford both had important careers in the earlier days of Canadian air transport. (Kenneth Swartz)

(Left) The Boeing 247 was introduced in 1933 as the world's first "three-mile-a-minute transport". United Air Transport took 60 of the 75 built and for a brief time wowed the US travel industry. Soon the 247 was eclipsed by designs like the Douglas DC-2 and Lockheed 10. With WWII a number of superannuated 247s were sold in Canada, giving excellent service to the RCAF, CPA and MCA. This 247 was owned by Charles McMaster of Ottawa, Kansas, who used it 1971-77 barnstorming at airshows. It was at Toronto Island Airport on September 4, 1972. N18E (ex-NC13340) had a long career, 14 years of which were spent after WWII with Island Airlines of Sandusky, Ohio. It ended with the Science Museum in London, England. Its delivery flight put N18E into the record book as the oldest plane (49 years) to fly the Atlantic. In 1997 four Boeing 247s still existed of which only the Boeing Museum of Flight's example was airworthy. (Larry Milberry)

659

Like the DC-3, the Consolidated PBY Canso easily lasted more than 50 years in Canadian use. Here C-FIZO was at Dryden, Ontario on water bombing duty August 9,1976. Some 40 Cansos still were active in Canada at this time. By 1997, however, the number had dwindled to about 10. (Larry Milberry)

The famous Canso C-FCRR of Avalon Aviation in its retirement at Parry Sound on July 11, 1991. Originally RCAF No. 9767, it sank the German submarine U-342 southwest of Iceland on April 17, 1944. In command that day was F/O T.C. Cooke, who later was a career bush pilot with the OPAS. (Larry Milberry)

If Torontonians were excited in 1937 when the great Short Empire flying boat "Cambria" visited, they were astounded on July 27, 1993. On that day the world's last airworthy Short flying boat landed. Here it is nearing Toronto after a trip from the UK via Iceland, Greenland and Gander Lake. G-BJHS (ex-RAF ML814) was one of 749 Short Sunderlands built during WWII to counter the German U-boat menace. Its wartime service included a stint with No. 422 (RCAF) Squadron. After the war it was refurbished for the RNZAF as NZ4108, then served 10 years with Ansett Airlines of Australia. Later owners were Antilles Air Boats in the US Virgin Islands and Edward Hulton of Chatham, England. In 1993 G-BJHS went to Florida aviation enthusiast Kermit Weeks. He engaged Ken Emmott, retired from British Airways, to fly it to Oshkosh, with an en route stop in Toronto. In the second view the Sunderland bobs in Toronto harbour before its departure on July 29. (Larry Milberry)

(Below) Another vintage warbird was Lancaster G-BCOH, ex-RCAF KB976. Northwestern Flying Services of St. Albert, Alberta converted it for water bombing in the 1960s. Then it was sold to the Strathallan Collection in Scotland. While being ferried to its new home on May 16, 1975, it stopped in Toronto, where Larry Milberry photographed it. Some years later G-BCOH was severely damaged in the collapse of a hangar.

(Left) The de Havilland Dove had a long career in Canada as a light business plane, but also as an air taxi. Here CF-GBE was at Malton in July 1964, while with Skyline Hotels of Toronto. By 1997 there were no active Doves in Canada. (Al Martin)

The Bristol 170 Freighter has earned a place of honour in Canadian air transport. Wardair's CF-WAD (right) was at Edmonton on July 31, 1974. It crashed at Hay River on November 20, 1977, killing one pilot and injuring the other. It had just taken off for Yellowknife with 12,000 pounds of corrugated steel when the load shifted, creating a deadly C of G situation. Freighter QWJ (below left) was at Thompson on July 20, 1973. It was built in 1951 and served 20 years in the UK before joining Lambair. It was lost at Rankin Inlet in March 1978, another case of shifting cargo. WAE (below right) , shown at La Ronge in October 1987, flew for the RCAF and Wardair before going to Norcanair. There it worked on cross-Canada jobs. Even after it retired to the Western Canada Aviation Museum in 1983, WAE was still in demand— TPA once tried buying it back! (Larry Milberry, John Kimberley)

(Right) The C-46 also carried the load in the Canadian northland. Here IBX of Nunasi Northland arrives at Winnipeg in July 1989. An Inuit corporation, Nunasi used some of its land claim millions to buy the OCA and Aero Trades licences in 1985. Besides the C-46, DC-3s and King Airs were used, but the airline folded in 1990. (R.W. Arnold)

(Below) IBX, by this time with Air Manitoba, unloads at Island Lake on November 16, 1990. It later served in East Africa with C-46s TXW and IXZ. IXZ was wrecked in Kenya, but IBX and TXW returned in October 1995 to join a new freight company, Commando Air. That day Capt Mark Mesdag and FO Dave Jackson operated Winnipeg-Island Lake (296 nm)-Gods River (74 nm)-Winnipeg (372 nm). (Larry Milberry)

C-GIBX at Winnipeg on January 11, 1993 as engine maintenance is done in typical Manitoba winter conditions. Then, A detail of the cockpit. (Larry Milberry)

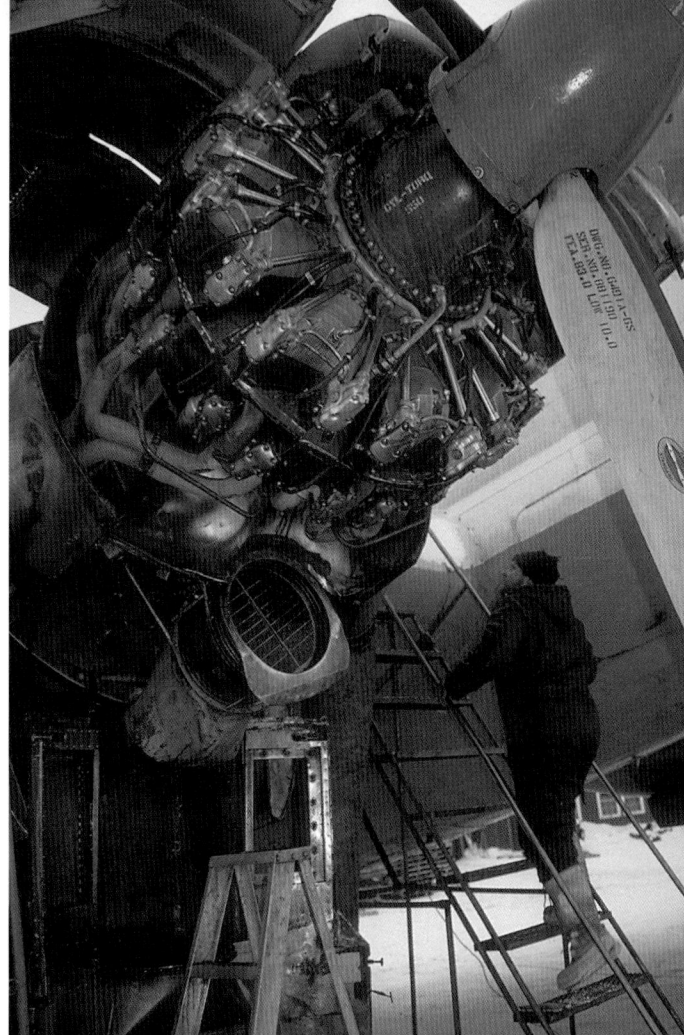

C-46 TPO languishing at Pickle Lake in February 1992, following an accident. (Larry Milberry)

The last Lockheed Constellations flying in Canada were these Conifair budworm-spraying beauties, seen at St-Jean, Quebec in June 1981. Closest is C-GXKO. In 1984 it was bought by Hollywood star John Travolta and ferried to Tucson to become N494TW. It was sold to the Constellation Group, Inc. of Scottsdale for $100,000. Enthusiasts could buy memberships in the group for US$3,995, which entitled them to a Connie type rating. Here N494TW was visiting Abbotsford, BC on August 8, 1993. (Hugh Halliday, Larry Milberry)

Formed in 1979 by Michel Leblanc, Conifair sprayed for many seasons in Quebec. For 1983 it covered 5,000,000 acres with Connies XKO and XKR; DC-4s BNV, BPA, BSK, XKN, N4994H and N62342; and DC-6As BYA, BYS and BZC. One DC-6 crashed when low-flying. Constellation XKS was lost on June 21, 1979 due to hydraulic trouble. The crew dumped fuel and returned to base at Rivière du Loup. On landing, the aircraft ran off the runway, the gear collapsed and all four engines were torn from their mounts. A few days later XKS burned to destruction during salvage work. Here XKR is seen at rest after the day's work in June 1984; then stored at Mont Joli in November 1992. It later was ferried to Arizona for restoration. (Richard Beaudet, Larry Milberry)

Boeing 377 "Pregnant Guppy" conversions, only a handful of which were made, rarely visited Canada. This example originally was PanAm's "Clipper Constitution". It was at Dorval on August 22, 1967, while freighting to Prague. The Pregnant Guppy, the first of which flew in California in September 1962, was the brainchild of Jack Conroy, one of aviation's great innovators. (Larry Milberry)

(Left) The Lockheed L.188 Electra first entered service with Eastern Airlines in January 1959. Production ended at 170 machines in 1962. Over the years 20 Electras worked in Canada for International Jet Air, Airspray, Nordair, Northwest Territorial Airlines, Pan Arctic Oils Ltd and PWA. For several years NWT Air Electras operated a transcontinental courier service for Air Canada. PWA Electra CF-PWG was seen at Edmonton Municipal on July 31, 1974. It started with National Airlines in 1959 as N5001K. In 1995 it was flying from Detroit with the cargo carrier Zantop International. (Larry Milberry)

(Right) The Electra was a boon to northern transportation. It was fast, rugged and carried a good load into marginal strips. It took an equally versatile plane, the Boeing 737, to finally unseat it as the leading NWT airliner of the 1970s-80s. Robert S. Grant caught Electra IJV at Coppermine on February 15, 1984. It had started with Western Airlines in 1961, then moved to International Jet Air in 1971, and NWT Air in 1976. Electras made money for NWT Air from start to finish. They were purchased for $250,000 to $350,000 a piece, and sold later at $2-$2.5 million.

Nordair/CAIL used Electras on ice patrol for nearly 20 years. Bases commonly were Frobisher, Inuvik and Churchill in summer; Gander and Summerside in the spring. Dorval was the maintenance base. Electra CF-NAZ is seen at Churchill on July 20, 1973. First delivered as N132US to Northwest Airlines, it came to Nordair in May 1972. On March 31, 1977 it was struck and severely damaged by an Argus attempting a three-engine approach at CFB Summerside. (Larry Milberry)

(Right) Ice patrol Electras C-GNDZ and C-GNAY with official and unofficial markings at Dorval on December 1, 1989. NDZ began as N128US of Northwest Airlines, flew later for Air Florida, then joined Nordair in February 1978. When NDZ was rebuilt in California, the fuselage of NAZ, flown there by Guppy, was used in the process. In the early 1990s NDZ/NAZ was operating in Zaire as 9Q-CRY. (Larry Milberry)

Tyne-powered classics of the fifties: a CL-44 (above) on a 1967 military charter at Danang, South Vietnam; then an Air Canada Vanguard (right) at Thunder Bay in November 1969. (Dick Gleasure, Andy Graham)

CPA – Evolution of an Airline

The Rush to Jets

Under Grant McConachie's gung-ho leadership CPA was Canada's first airline to order jets—Comet Is in 1950. Although the venture was brief and ended tragically with the infamous Karachi crash, CPA was determined to be a travel industry leader. It analysed types like the Comet IV, 707 and DC-8. McConachie insisted on trying the controls of each. It is claimed that he disliked the feel of the 707, but was impressed by the DC-8, which CPA ordered in 1959 at $6 million each. The first was accepted in February 1961 and introduced (Vancouver-Honolulu) on March 25, 1961. On April 21 Vancouver-Toronto, and Vancouver-Amsterdam (via Edmonton) were inaugurated. Service to the Orient began on October 8.

Passengers on long-haul runs felt spoiled by the DC-8—they wouldn't tolerate the old prop-liners much longer. In 1962 CPA sold seven DC-6Bs, bringing a much welcomed $3.5 million. Three went to Transair of Sweden, one to Trans-Caribbean, two to World Airways and one to Wardair. Another was leased to Cunard Eagle Airways. By 1969 CPA had disposed of its last Britannias and DC-6Bs—now the DC-8 was king. (TCA, also adding DC-8s, had its Super Connies up for sale at the same time for as little as $125,000.) For 1966 CPA's daily DC-8 utilization was 13.5 hours, the load factor, 55%. When DC-8 CF-CPK was lost at Tokyo on March 4, 1966, a replacement was leased from Trans International for a year at $4.15 million. So hard-to-find were DC-8s by this time that it took six months to find the TIA aircraft and get it into the fleet, prompting CPA's Ian Gray to comment, "I think we leased the last DC-8 in captivity."

One Airline or Two?

The argument about Canada supporting two major airlines began soon after WWII. In 1962 the cry went up that TCA and CPA should merge. After all, both were losing money—$6.45 million for CPA and $7.60 million for TCA in 1961. There was also the perennial argument that the two lines were duplicating services. Heading the one-airline faction was TCA president Gordon R. McGregor, who explained: "In this business, competition isn't funny. It's expensive and it's not efficient." As to merging, McGregor's view was simple—"It's a matter of pulling together or falling apart." In the August 1962 issue of TCA's in-house newsletter *Between Ourselves* he added: "In a nutshell it would seem that a properly-constituted merger could greatly improve the lot of both airlines and their respective employee bodies, but if the two dogs are asked to continue to feed simultaneously from the same platter, there will continue to be yappings from the smaller dog and wasteful duplication of effort." CPA didn't agree. In April 1962 Grant McConachie described the merger call as a "repetitious and plausible soap commercial which is riddled with fallacies." He accused the one-airline faction of being "too modest about Canada... It is perhaps because of this national inferiority complex that most Canadians outside aviation are astounded to learn that this country is the second largest air travel market in the free world." He also pointed out that Britain had 11 airlines in the international market, while Australia had three.

TCA and CPA were not really direct competitors. While TCA served the UK, Bermuda and the Caribbean, CPA flew to Hawaii, Holland and the Orient. The only face-to-face market was the transcontinental run on which CPA had only one daily Vancouver-Montreal return service. To McConachie this was a good arrangement (although he wanted a bigger share of the domestic market). Each airline was free to serve its domestic routes, and feed its overseas points-of-departure. He also claimed that present deficits would disappear as new jetliners appeared, causing an upsurge in travel. Even so, on April 27, 1964 Transport Minister Jack Pickersgill called in McGregor, McConachie and N.R. Crump of the CPR to talk merger. Crump, who was McConachie's boss, stated that there was room for competition and that the public wanted it. On May 26 TCA authorized McGregor to negotiate with CPA on such matters as international routes, but this led nowhere.

In 1965 Ottawa assigned CPA and Air Canada exclusive zones for their international operations. The plan had been worked out earlier between McGregor and McConachie. CPA got the Pacific region, Asia, Australia, New Zealand, southern and southeastern Europe, and Latin America, and kept its over-the-pole Edmonton-Amsterdam service. Air Canada kept its monopoly to the US and got Britain, western, northern and eastern Europe, and the Caribbean. McGregor, however, kept promoting merger.

On June 29, 1965 McConachie passed away and John C. Gilmer became CPA's president. Gilmer, who joined CPA in 1949, had been executive vice-president and comptroller. He had a bean counter's outlook. His simple explanation for the airline business was, "If you are right, you are right. If you are wrong, you are ruined." When he became president, he explained to the media:

There are no precedents for this airline business. Our product is the seat-mile, the most perishable commodity in the world. If we don't sell the product today we can't put it back on the shelf. It has gone. Our product changes from day to day, the development is staggering and the increase in passengers confounds the greatest experts and makes every estimate, even the wildest, look conservative.

The jet change-over saved the business, but it damned near killed it. Jets are more economical to run. You get more seats, more freight capacity, and it's cheaper to boot. But all the airlines got new jets at the same time, then threw them on the market at the same time. There was a glut of jet seats. Only the boom in travel saved us as the world's disposable income grew and travel restrictions dropped. This, and the reliable efficiency of jets, took us back into the black.

In October 1968 the merger question again was raised—on October 8, leader of the opposition John Diefenbaker asked the Liberal government in the House of Commons, "...whether representations have been made that Air Canada should merge with Canadian Pacific Air Lines..." Government spokesman James Richardson answered: "One of many alternatives which exists is a possible merger..." In January 1969 Air Canada president Yves Pratte described a merger as a reasonable proposition: "I believe that the present policy of dividing the globe between Air Canada and CP Air must be looked at anew, particularly since international carriers are penetrating farther and farther into Canada." At this time the two carriers were having informal talks about merging international operations, but CP Air would do so only if it held a majority position. The matter blew over. In April 1975 the airlines were back at the merger table. They had lost $20 million the year before; presidents Claude Taylor (Air Canada) and Ian Gray (CPAir) sat down. Gray proposed at least a pooling of maintenance operations, since there was some fleet commonality. These talks also petered out before anything serious transpired.

Looking at the SST
In 1966 CPA carried 738,857 passengers. Revenue was $82.6 million and operating income $8 million. On March 27, 1967 CPA was granted a second daily transcontinental sked, but still had a mere 9% of that market. It had 57,346 route miles in 1967 served by seven DC-8s, six DC-6Bs and three DC-3s. On order were four DC-8-63s, six 737s and three Boeing SSTs (supersonic transports). There were 3,200 employees, with projected growth to 6,000 in five to seven years.

It was no surprise that CPA should be one of the first airlines to order the SST. The concept for such a plane was big in the mid-1960s. Boeing first considered it in 1952. Although it focused mainly on designing a supersonic bomber for the USAF, it organized a continuous SST project in 1958. The bomber finally went to North American, which built the spec-

Air Canada and CPAir each ordered the Boeing SST, but the type was cancelled before it was built. (Air Canada)

tacular XB-70. Although Boeing lost another supersonic project to General Dynamics (the F-111), it learned a great deal from R&D. In 1963 the US government, wary about what the British, French and Soviets were doing in the field, decided to help industry fund an SST. Boeing won a competition with its proposal for the 350-seat Model 2707. It was 306 feet long and weighed 675,000 pounds all-up. A full-scale mockup was rolled out in September 1966—building it had consumed 42,800 board feet of lumber, 3,500 sheets of plywood and 50 tons steel!

The 2707 was to fly in 1970, and be in service within three years. With its V-swept wing, it would fly 4,000 miles at 1,800 mph. Twenty-six airlines placed deposits for 2707s, but in 1971 Washington withdrew financial support. Such funding itself had caused a public furor—it was viewed as an un-American way of doing business. Besides, environmental groups, which already had made things difficult for the Concorde, were bent on stopping the American SST; and the world was slipping into recession. The airlines seriously questioned the feasibility of the 2707. Except for the Concorde, the SST notion faded. Somewhat ironically, the Concorde was a success—late in the century it still was in daily service. At the same time manufacturers were regaining optimism about the SST.

CPAir's first 737 (CF-CPB) was accepted in October 1968. Seven were on order for $26,675,000. The first 737 flight operated on the "transcon" on April 1, 1969. The 737 soon

was operating from Vancouver to Prince George, Fort St. John, Fort Nelson, Watson Lake and Whitehorse. Greeted with excitement everywhere, it would prove costly on such local runs—it was too big for the business available. Also in 1968 four 240-seat DC-8-63 "Spacemasters" (37 feet longer than previous models) were delivered. Each cost $11-million. In 1968 CPA adopted a new corporate image, becoming CPAir and painting its planes a gaudy orange. That year revenue was $106.7 million and net income, $2.3 million. It carried 1,036,341 passengers, up 17% over 1967. The load factor was 50.4% (down from 56.6%, this being attributed to increased frequencies and the addition of the stretched DC-8). Cargo showed a 31% increase over 1967.

In 1969 CPAir won three more trans-Canada dailies, pushing its share of national capacity to 20%. International calls included Honolulu, Nandi, Sydney, Auckland, Tokyo, Hong Kong, Mexico City, Lima, Santiago, Buenos Aires, Santa Maria, Lisbon, Madrid, Amsterdam, Rome and Athens. CPAir's 1970 fleet numbered 11 DC-8s, two 727s, seven 737s as well as DC-3 CF-CRW (used for crew training, it remained till 1974, then was sold to Harrison Airways). CPAir's first two Boeing 727s were delivered early 1970. These were for trans-Canada service but, like the 737, covered the short-haul holiday market in the Caribbean, Mexico, etc. In November 1972 CPAir ordered two 747s at a package price of $77 million ($58 million for the planes). CPAir's first (C-FCRA) was delivered a year later.

(Left) Starting in 1968, CPAir built a large fleet of Boeing 737s. This one, delivered in 1982 as "Empress of Terrace", was at Vancouver on August 18, 1983. The 737 was ideal on CPAir's major domestic and holiday systems, but a money-loser on low-density, local services. (Larry Milberry)

(Below) "Stretched" DC-8-63 CF-CPO at Toronto in January 1978. This 240-passenger type, used by Air Canada and CPAir, was the biggest airliner prior to the Boeing 747. CF-CPO was with CPAir 1968-83, then served operators like Worldways, Burlington Air Express and American International Airways. (Gary Vincent)

(Right) The tri-jet Boeing 727 was another type that revolutionized air travel. First flown in February 1963, it entered service with Eastern and United airlines a year later. It joined CPAir in 1970 on the transcontinental run. CF-CUS was seen at Toronto on March 28, 1972, a year after it was delivered. It served till 1977 and later flew for Mexicana and SAN of Ecuador. (Larry Milberry)

(Above) CPAir's Boeing 747-217B C-FCRB "Empress of Canada" in a typical Toronto scene from the mid-1970s. The 747 gave travellers on CPAir the greatest in luxury, especially on the long haul to the Orient. CRB remained until 1986 when it went to Pakistan International Airlines in a swap for DC-10s. (Larry Milberry)

(Right) CPAir DC-10-30 C-GCPC came to the airline in March 1979. It was at Toronto on September 5, 1983 and remained in service into the late 1990s. The DC-10 first flew in August 1970. As many as 380 passengers could be accommodated. Including USAF KC-10 tankers, more than 500 DC-10s were built. (Larry Milberry).

million. On January 31, 1975 Gilmer retired and was replaced by Ian Gray, who had joined CPA as an engineer in 1943 and rose to be director of maintenance by 1961.

The Douglas DC-10

The original wide-body jetliners were the B.747, Douglas DC-10 and Lockheed L-1011. Air Canada, CPAir and Wardair adopted the 747, CPAir and Wardair the DC-10, and Air Canada the L-1011. In later years the L-1011 served Air Transat, Royal Air and Worldways. First flown in August 1970, the DC-10 won its C of A in July 1971 and entered service with American Airlines that August. CPAir chose it to complement the 747 on overseas operations. Its first was delivered in March 1979, its last in 1986. In 1987 CPAir and Varig of Brazil, each with 12 examples, had the world's largest DC-10-30 fleets. DC-10 production continued into 1989 by when 386 commercial and 60 military examples had been delivered. Through periodic upgrades, it kept its place with CPAir/Canadian Airlines International into the late-1990s.

All was not smooth for the DC-10, although CPAir had a good safety record with it. Cabin depressurization caused by badly-designed cargo door latches resulted in catastrophic

Boeing Jetliners with CPAir						
Type	Engine	Length	Span	Height	MTOW*	Max. Passengers
727-200	PW JT8D (14,500 lb.st)	153' 2"	108'	34'	184,800	189
737-200	PW JT8D (14,500 lb.st)	100' 2"	93'	37'	117,000	130
737-300	CFM56 (20,000 lb.st)	109' 7"	94' 9"	36' 6"	135,000	141
747-200	PW JT9D (54,750 lb.st)	231' 10"	195' 8"	63' 5"	800,000	447
747-400	PW4000 (56,750 lb.st)	231' 10"	211'	63' 4"	800,000	450
767-300	PW JT9D (50,000 lb.st)	180' 3"	156' 1"	52'	352.200	290
*max. takeoff weight						

Soon CPAir had four 747s and was an all-jet operation. CPA's best load factor in its early jet years came in 1972—63.5%. July 25 to September 22, 1973 it suffered a two-month machinists strike. At the same time fuel prices soared with an Arab oil embargo. For 1974 CPAir carried 2,284,000 passengers and had revenue of: $236.4 million—passengers, $22.1 million—cargo, $9.2 million —mail, $2.1 million—charters. Net income was $2.4 million. Revenue from the Orient rose by 33.4%, which was attributed to the 747. The airline employed 7,680 and had a fleet at year's end of four 747s, 12 DC-8-50s, six 727 and seven 737s.

Although it sought 17 new routes to the US, it won only Vancouver-LA.

In March 1973 CPAir had won rights for Peking and Shanghai. In return China was granted Vancouver. CPAir service did not begin immediately, and in 1975 still was on the back burner. In September that year it explained, "It's causing us no grief from the financial standpoint. We estimated it would be a loss operation for the initial development period. As we have other things already causing us losses, it's some consolation we are not starting China service immediately." For 1975 CPAir slipped into the red for the first time in 13 years, losing $6.4

damage for other carriers, no case worse than with the 1974 Turkish Airlines crash near Paris that killed 346. Even though an incident with American Airlines in 1972 had alerted Douglas to the potential for disaster, little remedial action was taken. The Paris crash exposed a system wherein manufacturer, airlines and government conspired to shield one another, rather than take action to prevent more accidents—a problem endemic in aviation.

In 1985 CPAir agreed to exchange four 747s for four Pakistan International Airliners DC-10-30s. The same year it began modifying several DC-10s with a 3,000-Imp. gal. belly tanks. This gave the aircraft Vancouver-Hong-Kong nonstop capability. The extended range DC-10-30 had an MTOW of 590,000 pounds, compared to 572,000 for standard aircraft. Range went from 6,700 nm to 7,140.

The Charter Scene

Canada's regional airlines were on the upswing by 1968 and starting the move to jets. PWA had two DC-7Cs, five DC-6s, two DC-4s, six DC-3s and three C-46s; but two 737s, four

One of aviation photo buff Wilf White's Prestwick photos of the mid-1960s that is so evocative of the propliner years. CF-PWF, an old Pan American plane from 1952, had just landed from Canada. Such types were used widely by Canadian charter airlines. PWF later went into water bombing with Conair; it was scrapped in 1986. PWA adopted this colour scheme with its first DC-6.

Convair 640s and a Hercules were on order—a $20-million investment. It planned to use one 737 on the Calgary-Edmonton air-bus run, the other in the North. Transair's fleet was aged—a Viscount, a DC-7C, a DC-6, three DC-4s, six DC-3s— but the company had an option on a Hercules freighter. Nordair was awaiting three 737s to replace its four Super Constellations. Other-wise it had four DC-4s, eight DC-3s and four C-46s. Quebecair had four Fairchild F.27s working alongside eight DC-3s. In the Maritimes EPA had three Dart Heralds, four DC-3s, a DC-4 and was considering a Hercules for freighting in Labrador.

The popularity of DC-6s and DC-7s with Canadian regionals was connected to the post-WWII growth of international passenger char-ters. Plans included the entity charter, used by sports teams, immigrant or military groups, ship's crews, etc. It was paid for by the organi-zation, and went to a specific destination. The pro rata charter was arranged by a recreation club, church or veterans group, etc. and was paid for by individuals making a trip to a spe-cific point. There also was the affinity charter. It was limited to persons with at least 6 months membership in an eligible club or organization. Airlines liked charters—they filled every seat. Canadian carriers had been missing a lot of business by not diving more aggressively into this market. Things changed in the early 1960s, when the less restrictive ITC appeared—the in-clusive tour, or vacation package charter. In 1962 there were 326 such flights to and from Canada carrying 35,700 passengers. Each year thereafter the percentage of charter (compared to sked) passengers mounted.

For 1963 of 438 Canadian ITCs, TCA oper-ated 149 with 131-seat DC-8s, CPA 111 with 140-seat DC-8s and 110-seat Britannias, and Wardair 34 with 89-seat DC-6Bs. The others were flown by Air France, Alitalia, BOAC Caledonian, El Al, Flying Tiger, KLM, Luft-hansa, Overseas National, Pan Am, Sabena, and SAS, using jets, DC-7s and Super Constel-lations. The older aircraft were still in good

Most Popular Canadian Charter Routes, 1969	
Route	Passengers
Toronto–London	107,811
Vancouver–London	39,925
Prestwick–Toronto	26,536
Toronto–Amsterdam	24,934
Toronto–Prestwick	24,105
Amsterdam–Toronto	22,363
Montreal–Paris	20,340
Manchester–Toronto	15,488
Montreal–London	12,405
Calgary–London	11,851
Toronto–Rome	10,840
Edmonton–London	10,262
Winnipeg–London	8,244
Belfast–Toronto	7,697
Vancouver–Amsterdam	7,428
Paris–Montreal	7,134
Toronto–Frankfurt	6,966

condition. They were free of debt, so operating costs were low. Passen-gers still would put up with flying on such rela-tively modern aircraft, especially since they were paying so little for tickets—96,000 bought charter seats to and from Canada in 1965. By 1967 most charters in Canada were still on foreign car-riers. Canadian airlines carried only 31.7%, but this rose to 41.9% in 1968. That year, however, Air Canada, Nordair and Wardair made only 150 of 433 Ontario-based charters. There was great potential yet to be developed. Although ITCs had their ups and downs (carriers alternately pulled out of, then re-entered the market, tour companies went bankrupt, etc.) they eventually developed into a key sector of Canada's travel industry.

PWA Takes Off

PWA was the first Canadian regional in ITCs, running some DC-6B trips to the Cayman Is-lands in 1964. It had taken the bold step in 1962 of spending $132,000 on its first DC-6B, acquired from International Aerodyne through its former vice-president, Duncan D. McLaren. A pressurized plane, the DC-6B brought PWA to the fringes of modern airline status. Be-sides doing ITCs, it also

flew north, reducing the Edmonton-Fort Smith-Norman Wells-Inuvik sked to 6:15 hours com-pared to 7:50 by DC-4. Further DC-6s and DC-7s followed from International Aerodyne, mainly for ITC work. They trail-blazed and taught PWA about the "big" airline business—training, maintenance, global routes, clear-ances and customs, advertising, catering and hotels, etc.

PWA's DC-7s were exchanged for bigger, faster, longer range DC-7Cs. These were classy, but passengers knew that they deserved jets by now—the 707 and DC-8 had been in mainline service for years. In October 1967 PWA acquired its first jet—a 138-seat ex-Qantas 707 bought for $4 million. Figures for 1969 illustrate the relentless growth in the charter business: 573,276 passengers flew the Atlantic to/from Canada compared to 345,632 in 1968. For 1968, Wardair led with 17% mar-ket share compared to 12% for Caledonian Air-ways, 11% for Air Canada, 8% for CP Air, 7% for PWA, 6% for BOAC. But this growth also left a few scars. For 1972 twelve Toronto travel agencies selling charter tickets folded, as did many others in Canada and Europe.

In April 1972 Ottawa replaced the pro rata charter with the ABC—advance booking char-ter. This was a handy arrangement whereby a traveller bought an airline ticket, but got a ho-tel and car in the same deal. In 1973 Ottawa re-duced the ABC booking period from 90 to 60 days, and reduced the deposit required from 25% to 10%. ABCs replace affinity charters. Pro rata charters remained in effect for groups organized, as stated by the CTC, "to attend a single event of a distinctive and special charac-ter and whose sole purpose is to get to and from that event".

Open Skies

PWA's 1966 order for turbines came as Minis-ter of Transport Jack Pickersgill released a pa-per that boded well for the regionals. Pickers-gill defined their role as providing services

PWA DC-6s and DC-7s

Registration	Type	Serial	Delivered	Original Owner
CF-PWA	DC-6B	44698	1962	Northwest/N569, del. 6-55
CF-PWF	DC-6B	43537	1964	Pan Am/N6537C del. 5-52
CF-PWD	DC-7	44136	1964	American/N315AA del. 2-54
CF-PWK	DC-7	44135	1964	American/N314AA del. 1-54
CF-PWP	DC-6	43127	1965	SAS/SE-BDI del. 12-48
CF-PWQ	DC-6	43128	1965	SAS/SE-BDK del. 9-48
CF-PWM	DC-7C	45181	1965	KLM/PH-DSB del. 4-57
CF-PCI	DC-6B	43555	1966	KLM/PH-TFN del. 7-52
CF-NAI	DC-7C	45129	1965	Mexicana/XA-LOD del. 4-57

Wilf White also took this great Prestwick action shot. PWA's DC-7C CF-NAI began in 1957 with Mexicana. It was with Nordair in 1964, then with PWA in 1965. NAI was scrapped in Miami in 1970.

CPA replaced its Britannias with DC-8s, the first four being delivered in 1961. CPA fought steadily for a bigger share of the trans-Canada sked market, but had to wait till 1967 before Ottawa, which tightly regulated such matters, granted it a second daily east-west sked. Here DC-8 CF-CPF "Empress of Rome" lands at Prestwick. It served CPA for 20 years before its retirement in 1981. (Wilf White)

The Boeing 737—one of air transport history's most important types. Its very appearance made passenger and revenue figures soar. With PWA the 737 brought the latest in air travel to many small BC, Alberta, Yukon and NWT centres and hastened the demise of older types like the DC-3, C-46 and DC-4. CF-PWW served PWA 1972-82 and 1987-92, being with US carriers in between. This colour scheme was introduced in 1977. (Boeing)

supplemental to the mainline carriers, and serving the North. PWA was assigned regional status in BC and northern Alberta. Pickersgill called this his Open Skies policy. PWA president Rusty Harris and his top men knew that modernization was essential. It was time to give up the cold weather idiosyncrasies of piston engines, and their costly overhead for the lower daily costs and improved public image of turbines. Meanwhile, PWA began selling single-engine VFR bases in BC, Alberta and the NWT. This was hastened by Open Skies, which allowed upstarts to compete in such areas as the Mackenzie basin. Disposing of VFR bases cleared PWA to focus on modernizing and establishing a new corporate culture. It also brought in some welcomed cash.

Soon the Air Transport Committee of the Canadian Transport Commission in Ottawa (established under Pickersgill) made more

news. In 1967 it had allotted a second daily trans-Canada sked to CPA, promising it 25% of this market. In November 1968 PWA was allowed to serve Vancouver-Calgary with a Kamloops stop. Kelowna, Penticton and Cranbrook soon were added, as were other interior centres. Vancouver-Victoria-Seattle were approved in 1970. The same year Air Canada abandoned Edmonton-Calgary, giving PWA a monopoly on Canada's third busiest city pair. In exchange PWA agreed to a CTC request to drop its application for Lethbridge-Calgary, leaving that promising route to upstart Time Air. PWA later entered the Lethbridge market, while Time Air pushed into PWA country, with Dash 7 service

all the way to Vancouver. Thus did several ambitious local carriers effectively become regionals. All the while the regionals sought mainline status.

After years of meetings and legal battles, PWA was awarded Vancouver-Edmonton in 1976, with a "regional" stop in Kelowna. Now it had an integrated system linking the coast and interior with Calgary, Edmonton and its northern operations. BC Airlines, which had been vying with PWA on coast and interior runs, was purchased by PWA from CAE Industries (which had acquired it in 1962 after purchasing Northwest Industries of which BC Air Lines was a subsidiary), giving further continuity to its new face. BCA had served places like Bella Coola, Castlegar, Quesnel, Terrace and Williams Lake. By 1970 PWA was serving 44 centres in BC, Alberta and the NWT, as far as Inuvik. Its had four 117-seat 737s, four 50-seat Convair 640s, two 159-seat 707s, three Hercules, four 24-seat Nord 262s and some DC-4s and DC-6s.

PWA kept up its applications for new services and hammered away at Lethbridge-Calgary. Time Air's Stubb Ross was equally adamant that PWA not tap his niche, and won a 1968 ATC decision to keep PWA out. Meanwhile, new players occasionally appeared e.g. Arrow Aviation, formed in Kamloops in 1970 by Percy Lotzer, served the southern interior centres of Kamloops, Kelowna, Castlegar,

(Above) The Convair served PWA well on passenger skeds and charters. CF-PWO served 1971-76 then went to Echo Bay Mines and North Caribou Air. It returned to the US in 1986 as N587CA. (John Kimberley)

(Right) With the 707 PWA became a modern international airline. Four were used 1967-79 in the holiday and cargo markets. The long-range jet also gave some extra bargaining power—PWA could negotiate with CPA and Air Canada by offering to get off long-range routes in exchange for coveted domestic routes. PWA had trouble with profits with its 707 freighters. There was a problem securing backhaul loads at points where cargo was delivered, competition from wide-body cargo planes was growing rapidly in the 1970s, and PWA did not have the volume to justify buying its own wide bodies. Here CF-PWJ (ex-Northwest Airlines N356US) was at Toronto in November 1973 delivering grapes from Chile. It ended its days in a 1992 accident in Nigeria. (Larry Milberry)

Penticton, Grand Forks and Cranbrook as a PWA partner, using PA-31s, then a Beech 99.

Everywhere in the world where it was introduced, jet travel caught on. Jets were customer and user friendly—fast, comfortable, safe and affordable. In Canada's remoter areas they beckoned the traveller—passenger statistics took off. In the case of Kamloops, for 1968, the last year for all-piston service there, 25,000 passengers used the airport on commercial domestic flights. Then the 737 appeared and five times that many used Kamloops by 1975. The 737 was especially welcomed down the Mackenzie, where it operated in cargo-passenger configuration. As oil and gas exploration in the Mackenzie Delta-Beaufort Sea expanded in the 1960s, and workers and equipment flowed into Inuvik, the 737 had a perfect market. It could carry a good load to Inuvik from Edmonton in just two hours. It shone everywhere, typically at Edmonton Municipal, where it operated handily, while Air Canada's DC-9, needing longer runways, was restricted to the distant international airport.

For several years PWA consolidated operations. Electras CF-PWG and PWQ were acquired in 1972 and proved excellent workhorses between Calgary and the Beaufort Sea. The company's six Convairs were ideal on BC milkruns and the Vancouver-Victoria shuttle, where they competed with Air Canada's

DC-9s. Through its years of coast, interior and Arctic flying PWA built an enviable safety record. This did not come easily, but gone were the days of VFR operations with shaky bush-planes and questionable operating procedures. February 11, 1978, however, brought PWA to the realization that it was not disaster-proof. That day one of its 737s was landing at Cranbrook from Calgary. Capt Christopher Miles, son of PWA vice-president Jack Miles, had just touched down when a vehicle loomed out of the blowing snow. A go-around was initiated, but the thrust reversers did not retract on one engine, leaving the 737 unflyable. It crashed beside the runway, killing 43. The scenario for this disaster began when Calgary ATC mis-estimated by 10 minutes the 737's arrival at Cranbrook; then the 737 failed to report to Cranbrook aeradio, when passing a beacon on final approach. An aeradio advisory about a snow blower on the runway was not acknowledged by the crew. In another incident Convair PWR crashed in Elk Lake Park, BC on September 18, 1969.

Alberta Buys PWA

On August 1, 1974 the airline industry was surprised at the news that the Province of Alberta had purchased PWA. The $36.5-million deal was made surreptitiously—the CTC was not consulted. Technically this was legal, consider-

ing the CTC's regulation that no "person" enter into such a deal without the CTC's involvement. Alberta simply argued that it was not a person, and won its case. Meanwhile, another PWA suitor, Federal Industries of Winnipeg, was left high and dry as it moved dutifully through regular Ottawa channels. Alberta, covetous of Edmonton's "Gateway to the North" title, had made its move (so it claimed) to prevent Federal from taking PWA, and changing its focus from Edmonton to Vancouver and Whitehorse, where Federal had its own transportation interests.

Although promising to let PWA run its affairs as usual, in March 1976 Alberta ordered head office moved to Calgary. President Don Watson was replaced by Roderick R. McDaniel. Harris quit a year later and a slate of young executives began running PWA. One was Rys Eyton, who had risen to head northern operations in 1970. He became president in time to preside over PWA's takeover of faltering Transair (the smallest of the regionals, with 11.6% of the five regionals' gross revenue). Transair seemed doomed to bankruptcy. Transport minister Otto Lang was anxious to avoid this, so supported PWA's bid.

PWA/Transair now had to placate Air Canada and CP Air, which feared PWA becoming a third transcontinental sked carrier. Air Canada asked that Transair not pursue its Regina-Brandon-Toronto application, nor seek direct Winnipeg-Calgary-Edmonton. It did not object to Transair having new Winnipeg-based Prairie routes, so long as they stopped east of the Rockies. This was fine by Transair, which received skeds from Winnipeg to Edmonton and Calgary, via Saskatoon and Regina. Meanwhile, PWA suspended Transair's Winnipeg-Thunder Bay-Toronto 737 service, assuming correctly that Ottawa would not allow a regional the run of Canada from Victoria to Toronto. CPAir, Great Lakes Airlines and Nordair now sought Winnipeg-Thunder Bay-Toronto. It went to Nordair. Great Lakes, at the time operating Convairs in Southern Ontario, lost on the argument that it was a "local" carrier, i.e. not entitled to broader regional privileges.

PWA inherited Transair's routes to such northern destinations as Thompson and Churchill, and took over the DEW Line vertical resupply to Resolute Bay and Cambridge Bay. Important new PWA routes opened in their turn—Vancouver-Prince George, Vancouver-Edmonton (nonstop), Vancouver-Whitehorse, etc. These came from hard-fought hearings where Air Canada argued that Vancouver-Calgary, etc. were mainline (not regional) markets. The government had to agree, but still favoured PWA on the basis that the routes were within its geographic domain. In 1980 PWA applied for Calgary-Brandon-Toronto. Brandon had long sought jet service, but TCA had dropped it in 1963. PWA was granted the route, which turned out to be the extension of a Vancouver-Calgary-Winnipeg service. The run enjoyed some popularity, temporarily forcing smaller Perimeter Airlines off Winnipeg-Brandon (Perimeter re-took the route, when PWA found that the 737 was impractical at Brandon).

A PWA 737 loads DEW Line cargo at Winnipeg over the summer of 1986. C-GOPW is in the colour scheme adopted in 1980. It remained in the fleet into 1997. (Jan Stroomenbergh)

PWA acquired two 767s in March 1983 to challenge Air Canada in domestic markets. The plan failed and both aircraft went to Air Canada two years later. C-GPWA was shot at Vancouver by John Kimberley, while Jan Stroomenbergh recorded the PWA signboard in Winnipeg.

When EPA won nonstop Halifax-Toronto service in 1980, the end of the regional airline policy was clear. PWA pushed its limits by introducing two 767s. Next year it took 24.5% of burgeoning Air Ontario (Air Canada took another 24.5%). The ultimate development in the rise of PWA came on February 1, 1987, when it purchased CPAir and gained control over what had been Nordair and EPA, both of which CPAir had absorbed earlier. PWA now began operations as Canadian Airlines International.

Inveterate Airline Pilot

After WWII Fred Lasby, who had instructed pilots, then flown with Ferry Command, left aviation and worked in construction on the prairies, flying nothing more than his own little Luscombe. Early in DEW Line days, however, he heard from Vern Simmonds of Associated Airways that Jack Crosby of PWA was looking for a pilot. Lasby went to see Crosby in Edmonton and took the job—flying a Bellanca on the DEW Line. A rush was on and there wasn't even time to take the Bellanca around for a circuit. Lasby soon was on his way to Hay River, then to Norman Wells and Cape Parry on Amundsen Gulf, where he served local sites. After two months he took some leave south, then was given an Anson and sent north for the summer.

Lasby progressed to types like the Bristol Freighter. His Bristol check-out was given by TCA at Winnipeg by Capt Bing Davis. Lasby thought highly of the Bristol in spite of adverse rumours following the Associated crash in Alberta. Before long another pilot put Lasby's Freighter through the ice north of Yellowknife (where it sat on the shore into the 1990s). Lasby switched to Avro Yorks, then DC-4s. In November 1958 he made the first cargo flight into the new Inuvik airport, landing from Whitehorse in a PWA DC-4 full of dynamite.

When PWA introduced the DC-6 in 1962 Lasby was one of the early captains. On May 21, 1963 he flew the inaugural PWA Edmonton-Calgary AirBus service. It had been scheduled for a DC-4, but it went unserviceable. The passengers boarded Lasby's DC-6, which had been due to fly Edmonton-Inuvik. In 1967 several pilots including Fred Lasby, W.A. "Wally" Crosson, Reg Scott, Dean MacLagen and Bob Bell went to Lockheed in Marietta for the Hercules technical course; then to Moses Lake, Washington for flight training. One day the Lockheed pilot there demonstrated stable flight with the Hercules on one engine! Lasby found Lockheed's training to be superb, and came to like everything about the Herc. His only criticism was that roll control was slow, but a pilot soon got used to that. PWA's first Hercules was delivered in May 1967. Early on it was stationed at Gatwick picking up whatever work was around. Typical was a charter to Djibouti with 45,000 pounds of cigarettes. This trip got dicey at Khartoum, where the local military was so menacing that Lasby fired up on the spot and blasted off, lest crew or plane come to harm. The day was sweltering, the Herc at max gross weight. Every bit of thrust was needed, so Lasby took off with engine bleed "off" i.e. with no air conditioning or other optional accessories running. In July 1967 Herc CF-PWO was subleased to Trans-Meridian Airways and based in Beirut for a twice weekly sked to Heathrow via Cyprus and Amsterdam. Failing more valuable cargo, it sometimes backhauled 20 tons of grapes from Cyprus to London. These, remembered by Lasby as the best he had savoured, were on sale in London within a couple of hours of arrival. Besides the sked, PWO flew some ad hoc charters to places like Tehran and Jedda.

PWA's first 737 arrived late in 1968 and was an instant hit. PWA tailored the 737 to airports in the BC interior, but especially to Edmonton Municipal. Fred Lasby was on PWA's initial 737 course, given in San Diego by Pacific Southwest Airlines, and spent many happy hours on this type. He also converted to the 707, taking the course with United Airlines in Denver. Now he was on the long haul with passengers to Honolulu, London, Frankfurt, etc. On January 2, 1972 PWA's 707 CF-PWZ went down. Two days earlier Lasby, being honoured that weekend on his retirement, had chatted with Capt Arthur E. Jung. He had joined the RCAF in 1943 and was one of the first Chinese Canadians to earn his wings. Posted overseas, he completed a tour on Lancs with 50 Squadron. He got into commercial flying soon after he got home to BC in 1945, flying for such coastal operators as QCA. Suddenly came bad news—Jung and his crew on PWZ were dead. This distressed Lasby, who knew Jung as a superior aviator, and recalled: "He was meticulous in every way and always had everything right on the numbers." The crash took place in heavy snow squalls; Lasby was convinced that the cause had to be something beyond even Jung's ability.

Upon retiring, Fred Lasby turned to the sea for a new career. He bought the 55-foot, Monk-designed cruiser *Pilgrim*, built in 1970 in Enno's boat yard in Vancouver. He sailed for Belize and the Caymans to get into the charter business. Eventually he retired in Fort Myers, Florida with the *Pilgrim*. Itchy for something else to do, in 1989 he bought a 1968 Piper PA-24 Comanche (N9250P) to fly for fun. This was a top model with the fuel-injected Lycoming IO-540 (260 hp). He would make long cross-country flights around the US and Canada, but was lured by a bigger challenge. In 1993 he began planning to fly solo around the world. He would arrive in Australia for the 1994 "Twenty-first World Comanche Congress" in Coolangatta, Queensland. He would take his Comanche across the Pacific to the convention, then keep going west till he got home. This required a great deal of preparation. To begin, he studied ways of improving his plane aerodynamically and mechanically. He ordered a new, cleaner engine cowling; and had numerous "speed mods" installed, such as cleaning up the nose gear cover, and fairing over gaps between the wing and the flight controls. This raised his cruise from 145 to 170 knots. A new alternator, new exhaust stacks, a vacuum pump and a digital fuel control system were some of the mods for improved engine performance and reliability. A three-axis auto pilot was installed, as were a Trimble 2000A GPS, a plug-in for a back-up Trimble hand-held GPS, HF radio and a WX900 Stormscope. A store of essential spares also was acquired—a tire and tube, fuel pump, plugs, battery, spare oil, etc. A major mod was twin long-range fuel tanks (150 US gallons) in the cabin. Along with the 90-gallon wing tanks, N9250P now had an 18-hour endurance. Lasby spent $30,000 getting his plane ready for the trip.

For flight planning and all en route and stop-over needs Lasby contracted

with Jeppesen Data Plan. This cost several thousand dollars, but proved its value later. When he arrived at each stop he would find that Jeppesen had faxed all the info needed for the next leg—contact people, clearances, etc. Lasby also carried let-down charts for every stop, low level charts for en route airways, and a set of GNC topographic charts for the whole route. He designed his own 16-point "howgozit" chart to monitor progress minute by minute e.g. rate of fuel consumption, speeds, position, winds, engine temperatures.

The great trip started Thursday, June 30, 1994 with a six-hour flight to Austin, Texas. Next day Lasby went 8:10 hours to Santa Barbara. The departure from Austin caused him some concern since the Comanche, 25% over-gross, was slow to unstick, reluctant to climb, and dicey in even a gentle turn. He soon learned to deal with these idiosyncrasies. On July 4 he departed on the Honolulu leg, the longest of all—2,152 nm. He noted: "En route was in the clear all the time at 8,000. Time: 13 hours, 25 minutes." He laid over several days, then heard from Jeppesen that he could not go to Tarawa in the Marshalls as planned—it was out of avgas. Majuro was the alternate, but he had to get there by 1630 local in order to avoid an exorbitant overcharge fee. This meant departing Honolulu by 0500. When 800 miles out, Lasby received an unusual transmission. A USAF controller called requesting that he hold for 2 ¹/₂ hours! A priority military flight was inbound and Johnson Island wanted full security in the area. They knew nothing of Lasby's flight plan. He simply replied, "I do not have holding fuel," and carried on.

Lasby reached Majuro at 1500 after 11:54 hours. The locals took two hours to hand-pump 90 gallons aboard the Comanche. Next day Lasby made for Honiara in the Solomons. On the 16th he reached Brisbane, then Coolongatta, where he was warmly welcomed by fellow Comanche owners. His plane later was judged best overseas visiter. Lasby departed on the 22nd on a 5:19-hour flight to Cairns. The next three legs were to Darwin, Singapore and Madras. On the last it was rough going: "This was not an easy trip, as communications were flat for eight hours. Weaved through a front for an hour with 35 knots on the nose. After getting clear of the front, winds decreased to 20 knots. Arrived Madras between intermittent rain showers and time en route was 12 hours 16 minutes. After arriving it took two hours to get cleared in, even with help from my handling agent."

On August 1 Lasby departed for Bahrain, a trip of 1,846 miles. For more than three hours he had headwinds up to 52 knots. At one point ground speed fell to 95 knots and Lasby noted, "I was able to get my planned airspeed of 160 knots nine hours into the trip... This set back my arrival time by more than two hours, and made this leg 13 hours, 52 minutes—my longest." Lasby departed for Luxor on August 4, thence to Rhodos and Rome, where he landed August 9. Next came Palma, about which he noted, "Made this leg VFR, but did not have special VFR arrival directions and the tower was very hostile. Although I was paying $150

per day at the hotel, the water was sulfurous and not fit to drink... not a good vacation destination. Fuel here was $4.84 a gallon ... but I had to top up for the ocean hop [to Santa Maria, Azores on August 14]." Weather delayed departure for St. John's, Newfoundland till the 17th. The 1,376-mile leg was uneventful except for headwinds. Lasby flew next day to Bangor to have his long-range tanks removed. On the 20th he went home via Richmond in rough weather. This day, and several times through the trip, he made good use of his B.F. Goodrich Stormscope, a device that pinpointed changes in the weather by sensing electrical impulses. It provided 360° coverage with readouts at 25, 50 or 100 miles. When indicating questionable weather on a certain course, he simply would deviate as required.

Besides the great personal satisfaction he enjoyed from his trip, Fred Lasby won a place in the Guinness Book of World Records. The Guinness plaque hanging in his TV room in 1996 noted him as "the oldest person to fly round the world solo, at age 82, leaving Fort Myers on 30 June 1994 and arriving back on 20 August 1994." He had covered 23,218 nm and visited 13 countries in 162:20 hours of flying. His engine operated beautifully and at no time did Lasby have to add more than two quarts of oil, even after the longest leg. Asked by a reporter how he had made out with language problems, he answered: "I had no trouble communicating with airport or hotel employees—anyone. People spoke English wherever I went. English is the universal language." Into 1997 Fred Lasby was as busy as ever. Two years earlier he had become a qualified scuba diver, and also qualified for a ham radio licence. Perhaps he best summed his life up when he said, "I never sit around... I have something to do every day."

The DC-3 of the Turboprop Age

The Lockheed L-100 (military designation: C-130 Hercules) is one of the world's greatest aviation success stories. It has been described justly as "the DC-3 of the turboprop age"; since it first flew in August 1954, more than 2,000 have been delivered. Early USAF C-130s visited Canada in the 1950s, but the first Canadian operator did not buy until 1960, when the RCAF ordered four C-130Bs. These were delivered to 435 Squadron and used to familiarize crews not only with flying a modern type, but with new cargo-handling concepts. Lockheed began offering the Herc to commercial operators in the late 1950s, but few had the funds or the work for such a specialized transport. Besides, there still were many cheap freighters on the market that could do the job—C-47s, C-54s, Connies, etc. It would take a major upswing in the resource sector before the commercial Herc gained a foothold. Beginning in the mid-1960s, there was an oil drilling rush in the Beaufort Sea along Alaska's North Slope, and on islands in the Canadian Arctic Archipelago. As there were no roads in the area, and only a few weeks of open water each summer for a sealift, transportation fell to the airplane. Millions of exploration dollars were available and the rush brought about one of the great postwar airlifts. Everything from Piper Cubs to L.1649 Starliners and civil C-133s were pressed into service hauling men and materiel from Alaska bases.

Across the border in Canada there was also exploration madness. Aircraft were needed to move drill rigs and all their infrastructure from one Arctic island to the next. PWA recognized this opportunity. Its interest in the Hercules was piqued by veteran bush pilot Stan McMillan, who headed the company's northern division. He pressed Jack Moul, head of con-

Wherever in the Third World PWA's Hercules went, they drew crowds. The 1972 scene below was with CF-PWR in Bangladesh during a humanitarian airlift. PWA loadie John Sproat (inset) is shown yukking it up with the Bangladeshi Army. PWR served 1969-80, when it was sold to Cargolux, which soon leased it in Libya. (Evans Col.)

Herc CF-PWN bringing supplies to Caniapiscau for the James Bay hydro project in July 1977. This Herc served PWA 1969-84. Its end came in a crash at Kamina, Angola on November 27, 1989, while flying as N9205T with Tepper Aviation. (Larry Milberry)

Capt George Chivers at work in his Canadian North 737 between Great Bear Lake Lodge and Winnipeg. Then, in his home workshop in Sherwood Park, Alberta, with a Yale trainer he was restoring in 1997. (Larry Milberry)

International Air. Ops were usually into marginal airstrips with no nav aids. Ice strips commonly were used, and blowing snow and ice crystals made flying an extra challenge.

In early 1969 PWA acquired Herc CF-PWN, ex-Zambia Air Transport 9J-RBW built in April 1966. Then two Hercs were leased from Interior. By now Chivers was a Herc instructor and involved in converting new pilots from Anchorage and Kenai, Alaska. In May 1969 a stretched L-100-20 (N7952S) was leased and Chivers flew it on a fuel resupply from Yellowknife to Pelly Bay. There were many long days on such jobs. May 22 was typical—15:20 hours of flying. As if flying the Herc 80-90 hours a month wasn't enough, Chivers kept up other aviation interests in his spare time. He flew his Tiger Moth (CF-CVZ) and a buddy's Cornell (CF-FDU) whenever possible. He built a two-thirds-scale P-51 Mustang (CF-XZI) with a 200-hp Ranger engine, flying it initially in October 1969.

On August 11, 1969 Chivers was landing PWN at Eureka on Ellesmere Island. The crew was unaware that the gravel strip there was saturated with water. The Herc's underside was severely damaged and the nose gear torn off. PWA rushed in an MRP and the Herc was ferried to Edmonton on the 22nd. The 13-hour flight was in a frigid, unpressurized airplane; the crew wore their parkas all the way. In November 1969 PWA took delivery of CF-PWR from Lockheed and the same month leased N7999S from Interior for drill rig moves from Resolute Bay. Before year's end CF-PWX also was delivered factory fresh. March 11, 1970 brought some unwelcome excitement for George Chivers. He had taken off from Resolute Bay with 5,000 gallons of fuel in bladders in the cargo bay (plus 4,000 in the wings). While they were en route to a drill site, fire broke out in wiring in the cargo area. The loadmaster couldn't quell the blaze, so the plane was taken to 22,000 feet where a centre hatch was blown to depressurize it and starve the fire of oxygen. The plan was successful.

While PWA's Hercs were busy in the Arctic, the sales department was finding contracts all over the world. July-September 1970 Chivers and crew were hauling cars and auto parts for Ford in West Germany, drill equipment to Lusaka, and cows between Norwich and Geneva. In 1971 there was a multi-trip contract beginning in May with PWN freighting military tanks from Manston in the UK to Colombo,

tracts, to take a look at the big freighter. A chance for this came in 1966 when PWA let Alaska Airlines use a Herc to haul fuel, pipe and other equipment along the 180 mile route in PWA territory from Peace River to the oil patch airhead at Rainbow Lake. Moul soon realized that the Herc, with its big payload, easy loading system, short field capability and high speed, would be a money-maker in the Arctic, where time was everything and money the least of the drill companies' worries. In 1966 Moul convinced PWA to order a Herc (CF-PWO), and immediately began hiring former RCAF aircrew. Veteran Air Transport Command captain Clare Agar became its first Herc chief pilot. Eventually, as many as seven Hercs were serving PWA.

Another of those involved in PWA's early Hercules operations was George Chivers. He had joined the RCAF as a teenager in 1957 and trained in electronics, then airframe. He served with 408 Squadron on Lancasters, then with 412 Squadron on the North Star and Comet. His next tour was in Toronto with RCAF and RCN air reserve squadrons. While there he earned his private pilot's licence at nearby Maple Airport and purchased a Tiger Moth. From 1959-68 he was at RCAF Station Marville on the Beech C-45, Bristol Freighter

and Dakota; then on C-119s and C-130s with 435 Squadron, a tour which took him on two UN Congo deployments.

Chivers won his commercial ticket in 1968, then left the RCAF. He worked briefly with the Edmonton Flying Club before joining PWA, where his first flight (a famil trip from Edmonton) was on DC-6 CF-PWQ on June 23, 1968. He went on the line on June 28 flying with Capt Bill May on DC-6B CF-PWA. They logged 12:55 hours that day flying Edmonton-Fort Smith-Yellowknife-Norman Wells-Inuvik return. Life on the DC-6 was a joy, but PWA had just taken delivery of its first Hercules. Chivers was picked to go on course. Lockheed at first loaned PWA N1130E, then delivered CF-PWO. Chivers' first flight in PWO was from Edmonton on December 1, 1968.

In February 1969 Chivers was promoted from second officer (occupying the flight engineer's seat) to the right seat. There he sat beside longtime captains like Bob Bell and Fred Lasby. From the start he was busy with trips into the North. PWO carried drummed fuel, vehicles and supplies of every kind into isolated Arctic outposts like Eureka, Isachsen and Mould Bay. There was lots of work on the North Slope in competition with US Hercs flown by Interior Airways and Alaska

Ceylon. PWN routed via Istanbul and Dubai. A typical trip to Colombo took 16:30 hours. Next, Chivers took PWN to Singapore for a series of flights into Katmandu with radio equipment, fire trucks, etc. He was back in Canada in August hauling insecticide from Saskatoon to Texas, where locust were on the rampage. Later that month came the first of several flights between Edmonton and Calgary, and Osaka via Cold Bay, Alaska carrying cattle and returning with women's fashions. Once the Herc had opened this route, PWA's 707 took it over. In November-December 1971, Chivers was involved in a number of rig moves in the Arctic. August 1972 found him in Bangladesh flying food aid. As happens with a lot of foreign aid, quantities of it were being siphoned off by UN and local officials. After one delivery, Chivers

PWA Lockheed Hercules Fleet Info

Reg'n	FN	c/n	Model	Enter Svc	Left Svc	Hrs	Miles	Tons
PWO	382	4197	L382B/L100-10	1967	1969	5,266	1,518,834	47,024
PWN	383	4129	L382F/L100-20	1969	1984	26,015	7,502,465	232,281
PWX	384	4361	L382E/L100-20	1969	1976	14,308	4,126,284	127,752
PWR	385	4355	L382E/L100-20	1969	1980	23,247	6,704,202	207,566
PWK	386	4170	L382E/L100-20	1973	1980	13,789	3,976,609	123,118
HPW	376	4799	L382G/L100-30	1978	1983	7,549	2,177,056	67,403
Totals						90,174	26,005,450	805,144

Streak", the Tiger Moth featured in one train scene was Chivers'—he was at the controls done up as a blond bombshell!

Herc Loadie

John Bonner worked nearly 40 years as a loadmaster on RCAF/CF and PWA Hercs. He joined the RCAF in 1951, a time when his trade was designated MCA—"Movement Control, Air". He then did six months of OJT with 426 Squadron at Dorval, before spending four years at Nos. 1 and 2 Wings in the RCAF Air Division. His aircraft there were the Beech 18, Dakota, Bristol Freighter and North Star. In 1959 Bonner converted to the C-119 at Trenton and was posted to 435 Squadron at Namao. The following year he spent six weeks at Lockheed in Marietta learning the cargo systems on the C-130B. His first C-130 flight was with F/L Reid Glenn going Edmonton-Calgary in 10302 January 4, 1961. He made his last trip in the C-119 with F/L Bill Flanders on March 12, 1965.

Bonner made his first Herc trip with PWA on CF-PWN July 17, 1969. Over his years with PWA he worked supporting oil rigs in the Arctic and in Chile; on UN food relief jobs in Bangladesh, Chad, Ethiopia and Sudan; on the James Bay project; and in Angola supporting diamond mining. On one long-haul job his Herc carried heavy machinery from London to Sydney, Australia. Equipment needed to salvage a ship was carried from Jacksonville, Florida to the southern tip of Chile; a load of

monkeys was taken from Singapore to Toronto; while cattle were carried from Calgary to Japan and Korea. In these early days a navigator was needed on oceanic trips, but he was soon replaced when PWA invested in Omega, then INS and GNS nav equipment.

Not all was rosy for the PWA Hercs. CF-PWO was lost, while landing downwind at Cauauya, Peru on July 15, 1969. There were no severe injuries. PWX (Capt Dean MacLagen) ran out of fuel and crashed near Kisangani, Zaire on November 21, 1976. It had arrived after a long trip from Ostend to find the region socked-in and all airport nav aids turned off for the day. All but one aboard died. The survivor was treated in hospital in Zaire. He contracted HIV from the local blood supply and died back in Canada. In 1981 John Bonner left PWA. He became involved with Island Jetfoil, a venture based in Victoria that planned to use a Boeing hydrofoil on passenger service. This plan failed and from 1986-91 he worked in operations for Helijet at Victoria. After retiring, Bonner was active in Winnipeg with the Western Canada Aviation Museum.

Time Air

In the early 1960s Lethbridge had a once-a-day Air Canada Viscount sked to Calgary. Local aviator Walter R. "Stubb" Ross felt the market was ill-served and offered a solution—his own airline. Ross, born on March 14, 1931, was the son of George Ross, a prosperous Lethbridge area rancher. He had flown in WWI, and took up flying again in the late 1920s—his ranch was so large that he needed an airplane to get around. He and Charles Elliot formed Southern Alberta Airlines, a flying school-cum-barnstorming business with a couple of Gipsy Moths. So it was in an aviation environment that the younger Ross grew up. He became an enthusiastic private pilot, and did a tour as president of the International Flying Farmers Association. To get a toehold in commercial aviation, he purchased a Piper Tri-Pacer from Walter Reich in 1963. For an extra $500 Reich included the charter licence for his company,

Lockheed L100 Specifications

	L100-10	L100-20	L100-30
Length	97' 9"	106' 1"	112' 9"
Height	38' 5"	38' 5"	38' 5"
Wingspan	132' 7"	132' 7"	132' 7"
Cargo Compartment Length	41'	49'	55'
Cargo Compartment Height	9'	9'	9'
Cargo Compartment Width	10'	10'	10'
Cargo Compartment Volume	4,155 cu.ft.	4,785 cu.ft.	6,000 cu.ft.
Zero Fuel Weight (lb)	118,000	127,000	127,000
Empty Operational Weight	73,500	74,500	76,500
Max Takeoff Weight	155,000	155,000	155,000
Max Payload	44,000	52,500	50,500
Max Fuel (lb)	66,000	66,000	66,000
Max Range Loaded (nm)	1,600	1,300	1,300
Max Range Empty	3,500	3,500	3,500

took it upon himself to have some sacks of grain handed to the needy right off the ramp of the Herc. One official took umbrage and next day presented the PWA crew with a basket of cobras, which he poured onto the ramp when the Herc shut down. One way or the other, flying the Herc was fun!

In off-time in 1971-72 Chivers rebuilt a Howard DGA-8 for a trader at Fort Rae. Over the summer of 1972 he flew DC-3 CF-VQV for Arctic Outpost Camps in northern BC. In 1973 he had PWR in Ethiopia moving drill rigs for Tenneco Oil. On August 1 he flew 11 shuttles, a record for PWA's Hercs. The Ethiopian job lasted till February 1974. The following month Chivers was hauling Rolls-Royce RB-211 engines for Air Canada L.1011s between East Midlands Airport in the UK and Dorval (two engines and one set of cowls per trip). He flew his last Hercules trip in PWK— a three-engine ferry from Resolute Bay to Edmonton. He had logged some 4,500 hours on type. He moved to the Boeing 737 and by 1996 had 9,000 hours on type. In more recent years his sidelines included rebuilding (with partner Len Weiss) CF-BFT, the oldest flying Norseman, a Stearman and a Yale. He still owned his Tiger Moth. For those who saw the Gene Wilder-Richard Prior movie "The Silver

The Twin Otter was ideal for short commuter runs in Time Air's early years. CF-DKK, which served the company for nearly a decade, was taxiing at Lethbridge in August 1974. In the 1990s it was operating in the Philippines as RP-C1154. (Larry Milberry)

Lethbridge Air Services. Encouraged by local business interests, Ross won a licence from the ATB to serve Lethbridge-Calgary. Air Canada and PWA opposed this. The ATB appeased PWA by giving it the lucrative Calgary-Edmonton route if it would withdraw its opposition. Ross acquired a nine-seat Beech 18; on May 16, 1966 he began operating Lethbridge-Calgary twice daily under the Lethbridge Air Services banner. A return ticket on the 45-minute run cost $23.20. The licence was for a year, but was renewable should Ottawa approve.

Time Air president Stubb Ross in the days when he was trying to get his new company established. (Halford Col.)

Ross built a regular clientele, even though only 2,703 passengers were carried between May and December 1966. He extended service to Edmonton with a stop at Red Deer. Now Ross was competing with PWA, paralleling its Calgary-Edmonton DC-4 air bus. He now changed his company name to Time Air. In early 1970 Air Canada sought to terminate Lethbridge-Calgary and supported PWA's application to take it over. Stubb Ross fought for exclusive rights on the route and won, provided he put larger aircraft into service. He leased his first Twin Otter at $5,650 monthly and added two Cessna 402s. A second Twin Otter was leased in September 1971.

Having good people at Time Air made the difference between success and failure. Ross hired engineer Don Ross, freshly graduated from SAIT, to take care of his Beech 18. Don went on to be Time Air's maintenance head and in 1994 celebrated 25 years of service. Richard Barton joined in 1970 as VP administration. He had worked in broadcasting and agriculture, and rose to be Time Air president in 1979. His skills got the company through some tough downturns. When Barton retired in 1993 he was executive VP and COO; he passed away in February 1994. One of Time Air's early stewardesses was Valerie Davis. She recalled the ups and downs of the early days, including

flights in the worst prairie weather: "I remember landing at Lethbridge, but we could not taxi to the terminal building because turning the aircraft might have flipped it! A bus was driven out to the runway, we exchanged passengers and took off for Calgary again." Such harum-scarum was short-lived, as Time Air rose in prominence.

In 1974 Time Air was confident enough to start 36-passenger F.27 operations—thrice daily from Lethbridge to Edmonton. A second F.27 was added in 1975. Three Short SD3-30s followed in 1976. The F.27s and a Twin Otter were sold the next year, and the first 48-seat Dash 7 was leased in 1979. The Twin Otters were shifted to communities like Grande Prairie and Pincher Creek. The Short SD3-30 carried its first fare-paying passengers for Time Air on August 24, 1976 on the Lethbridge-Calgary-Edmonton run. Time Air was the world's first airline to use the SD3-30 in sked service. The plane was good for Time Air—it was economical, had a comfortable cabin and was quiet in a "good-neighbourly" way. Its 30-seat capacity was symbolic of how the company had grown. By this time Time Air was operating 42 flights daily to Lethbridge, Calgary, Edmonton, Medicine Hat, Grande Prairie and Red Deer.

An application before the Air Transport Board in October 1974 illustrates how free enterprise and regulation worked. PWA sought to add Lethbridge under its license No. ATC 1851/69(S). PWA claimed that Time Air was not providing adequate service with the Twin Otter and stated that it "would not cope with the potential passenger traffic flow of the Lethbridge area." As Lethbridge grew there could be 55,000 passengers a year. PWA argued that what Lethbridge really needed was the 737. Its application was opposed by Time Air, Transair, International Jet Air, Air West, Air Canada, the province of Alberta and other parties, but sup-

ported by others. Time Air argued that if Ottawa would let it operate an F.27 on the run, the demands would be met for the immediate future. Meanwhile, it feared that should PWA win its case, Time Air would be squeezed out of the Lethbridge market, and forced to fold. Then small communities in the Lethbridge hinterland that Time Air served would lose their feeder flights. A third point was even more forceful. As stated in the ATC decision Time Air argued: "to permit a regional carrier to utilize a route developed by a third level or local service air carrier without first giving that carrier an opportunity to improve and upgrade its equipment or service to meet the alleged demand for increased air service, would defeat the initiative of small carriers, who have developed air traffic; impede the future development of air traffic by small carriers; and jeopardize the growth of communities which are dependent on the initiative of the smaller air carriers." Finally, Time Air put its case that there was no public interest to be served by PWA entering the Lethbridge market.

PWA argued that Lethbridge had potential far greater than Time Air could meet, but the ATB denied its application. It reminded PWA that even though Time Air was now allowed to operate aircraft like the HS748, it could not turn an Edmonton-Lethbridge aircraft around at Calgary to return to Edmonton, and so compete on the air bus route. The board then noted: "While the committee is not disputing the growth potential of Lethbridge, it must concern itself with traffic available to carriers in the present and foreseeable future and determine how that traffic is best carried in a manner that is in the interest of the public. Time Air with its eight daily return trips during the week and two daily return trips on Saturday and Sunday provides... adequate service at this time for the city of Lethbridge.... The committee is in agreement that the addition of the proposed service by PWA would create a situation where neither operation would be economically successful..."

In 1980 Time Air took over Edmonton-based Gateway Aviation and placed a 56-seat Convair 640 on Gateway's old Edmonton-Peace River-Rainbow Lake run. By 1981 Time Air had 50 daily flights and 200 staff. It pushed into BC in 1981 with Dash 7 service Medicine Hat-Lethbridge-Kelowna-Vancouver. Again it was head-to-head with PWA, but with more practical aircraft—PWA's 737s were too big for the market compared to a Dash 7 or Convair. Time Air's image changed further in

Time Air's Short SD3-30 C-GTAM (above) at Edmonton Municipal on August 5, 1977. It served till February 1984, then returned to the UK to fly with a series of carriers as G-BEEO. In 1992 it went to the Virgin Islands as VP-LVR, but was wrecked at Beef Island on May 2 the following year. Then, 3-60 C-GTAX at Campbell River on April 4, 1992. Straightforward and reliable, the PT6-powered Shorts were well-liked by operators. (Larry Milberry)

The Convair 640 was important to Time Air as it expanded in the 1980s. PWY, seen at the Muni on May 17, 1987 in a smart red-and-white scheme, came from Worldways of Toronto in 1984. Although a dinosaur compared to the forthcoming Dash 8, the CV640 filled a requirement. With a solid airframe and reliable Dart engines it carried its passengers comfortably and at a decent cruise speed. (Larry Milberry)

Gateway had used DC-3s, a Convair and this HS748 before its takeover in 1979 by Northward Aviation which, in turn, was absorbed by Time Air. HS748 MAK, seen taking off at the Muni on October 8, 1972, came from Midwest in 1972, then went to Calm Air upon Gateway's demise. (Les Corness)

September 1983 when it sold PWA Corp. 40% of its shares for $4.3 million—at this stage, partnership made more sense than perpetual competition. By now Time Air had dropped the Twin Otter and 3-30. It put Dash 7s on some PWA 737 routes. From 1985-87 further expansion saw Time Air takeover Southern Frontier of Calgary, Inter City Air Services of Kelowna, North Caribou Flying Service of Fort St. John and Norcanair of Saskatoon. With Norcanair came routes to Regina, Saskatoon and Minneapolis, three F.27s and two F.28 (Time Air's first pure jets). Several former PWA milk runs were taken over including Edmonton-Peace River-High Level, and Edmonton-Fort McMurray-Fort Chipewyan. From a management viewpoint these were better served by turboprops than 737s. Such expansion did much to help Time Air achieve an integrated operation in the West.

By late 1986 Time Air had nearly 400 employees. The fleet centred on Convair 580s and 640s, Dash 7s and Shorts. The latter were in Vancouver flying Air Canada Connector routes to places like Victoria and Campbell River. Numerous communities up the BC coast and in the interior also were added—Port Hardy, Kamloops, Prince George, Dawson Creek, etc. The 52-seat Convair was ideal for Time Air, but there were times when any 30-year-old plane will let you down. In one sales promotion Time Air boasted, "Passenger service is our pride... Our courteous, attentive flight attendants will pamper your group from start to finish of your charter." Flight attendant Anna Marello later recalled a Convair trip. One day she was unable to get the door open after a landing. The pilots couldn't budge it either. In the end the passengers were sent into the cargo hold from where each rode down on a baggage conveyor. Courtesy, with some old-fashioned innovation thrown in. Capt Peter Van Deursen had a memorable Convair charter: "One enthusiastic rugby team, rocking back and forth to the sound of 'stroke, stroke', actually got the aircraft's nose wheel bouncing up and down off the ground during the takeoff roll (subsequently aborted). And that was *before* the game!"

For its 1987 fiscal year PWA announced that it had carried 8,041,000 passengers at a 68% average load factor. That year it purchased CPAir, which it renamed Canadian Airlines International. Time Air was a leading partner in the new organization. For 1987 it showed revenue of $64.6 million, which included a profit of $3.4 million. By late 1987

Time Air had 28 aircraft working 175 flights daily. On September 14, 1987 Stubb Ross passed away. The following year the company phased out the F.27 and all but one Convair (which remained into 1993). In 1988 Time Air flew more than a million passengers. The first Dash 8-300 arrived in March 1989 and the Dash 7, of which the company eventually had seven, was retired. Although they had served only a few years, Captain Fred Kinniburgh logged 10,000 hours on type. For this DHC honoured him with a solid gold Dash 7 pin. Time Air now was an all-Dash 8 and F.28 fleet. In late 1990 all remaining Time Air shares were purchased by PWA Corporation.

Canadian Regional Airlines

For several years in the late 1980s the Canadian Regional Airlines concept had been gestating at PWA Corp. CRA was launched in January 1991 by when PWA had a 45% share each in Time Air, Calm Air International, and Air Atlantic, and 47% in Ontario Express (Canadian Partner). CRA hoped to encompass as much of this group as possible. In March 1993 Time Air, Ontario Express and Inter-Canadien of Quebec formally became CRA with a fleet of 17 Dash 8s and seven F.28s. CRA began introducing a list of improvements to make for more cost efficient operations e.g. improved operating procedures for the various fleets, reduced-power takeoffs to save fuel ($100,000 a year for the Dash 8 fleet, plus savings in engine maintenance), increasing F.28 passenger appeal by new soundproofing and more comfortable 737 seats.

With the advent of CRA, Duncan Fischer replaced Murray Sigler as president and CEO. Like many who eventually become airline executives, Fischer started his career in the lower echelons—as a customer agent for PWA in 1964. He became base manager at Cranbrook in 1968, then moved to marketing and scheduling. In 1983 he became VP of in-flight services

and, following PWA's takeover of Canadian in 1987, became VP of Canadian's Western Canada region. In 1990 he was tasked to set up Canadian North and next year went to Toronto as general manager and COO of Ontario Express. Fischer returned to Calgary in 1992 to reorganize CRA. The tough part was the deep cuts needed to bring costs under control. Ontario Express was especially hard hit. Hundreds were laid off and its 26-plane fleet of Brasilias and Jetstreams was phased out. By the summer of 1993 a fleet of seven ATR-42s was coming on line for Ontario with connections into Quebec, where Inter-Canadien also had ATRs. A joint ATR maintenance base was set up at Dorval. The Ontario-Quebec ATR-42 concept made sense. It was a mainline operation compared to the earlier Canadian Partner feeder system with its service to remote areas like James Bay. Such routes were soon grabbed by local carriers like Bearskin Air, which expanded to Red Lake, Sault Ste. Marie, Sudbury, North Bay and Ottawa. A new company, Visionair, set up in Timmins in 1992 with a Do.228 serving James Bay (although it soon folded).

In mid-1993 CRA continued downsizing, its 2,000 employees being slowly cut to 1,600. Fleet rationalization was in place. While CRA had 66 aircraft of nine types in January 1993, a year later there were 39 of four types: 14 Dash 8s, seven F.28s and three Short 3-60s in the West and 15 ATR-42s in Ontario-Quebec. First year savings following rationalization included: passenger revenue +1.9%, operating expenses -2.7%, revenue passengers +1.7% and break-even load factor -2.2%. These figures appear marginal, but in the airline industry such numbers were heartening in the 1990s. In March 1994 CRA added a new type—a Metroliner christened "City of Medicine Hat" leased from Calgary's Sunwest International Aviation Services for specialized Prairie routes. By 1995 CRA cut both the Short 3-60 and

ATR-42. This coincided with adding more Dash 8s.

By 1997 CRA had hundreds of daily flights. A typical one (still in Time Air days) was Flight 1260: Winnipeg-Regina on July 8, 1992. The aircraft was Fokker F.28 C-GTAH, one of eight in the fleet. Aboard were four crew and 26 passengers. Aircraft TAH taxied out to R31 weighing 57, 000 pounds and was airborne at 0716. It climbed first to FL240, then was cleared by ATC to FL310. The captain was Per Möller Warmedal, who had started flying in Norway and came to Canada in 1974 on a visit. Before he knew it, he was flying bushplanes for Jack Lloyd of Norcanair and had decided to make Canada his home. When Norcanair joined Time Air in April 1987, Warmedal stayed on. The first officer was Beth Moxley. She had started flying in 1974 as well. She instructed at the Yorkton Flying Club and with Skycraft in Oshawa, and went to work for Norcanair in 1979. She ran her own small flying operation for a while, before joining Time Air in 1980. Their route this day would be Winnipeg-Regina-Saskatoon, where they would overnight due to a long day's work the day before. At 0750, F1260 started down, landing at Regina 20 minutes later. On July 12 the continuation flight (F1262, C-FTAV) to Saskatoon (54 passengers) and Edmonton (38 passengers) was taken. The flight-deck crew was Capt Merv Andrew with FO Moxley.

Hearkening to CPA's early postwar years, a number of milk runs are operated on the Prairies and in the NWT. These remain a vital part of Canada's air transport infrastructure, connecting smaller centres with medium and major hubs. A typical milk run was CRA's F1142, operating on June 25, 1993. The aircraft was F.28 C-FTAV, Fleet No. 134, a veteran of 32,200 flying hours. It wore an odd-ball natural finish which explained its unofficial nickname, "The Silver Bullet". TAV had started its day several hours earlier in Victoria. It flew to Calgary and Edmonton, where it was being pushed back at 1125 local. Forty-nine passengers, Capt Pat Meek, FO Randy Hulkenberg and two FAs were aboard. As TAV taxied, a placard on the instrument panel was noted with some vital statistics: max brake release weight 66,500 lbs, max landing weight 59,000 lbs, max fuel 16,800 lbs, max altitude 35,000 feet. All good grist for the aviation history trivia mill. TAV got airborne at 1131 and was cleared to FL280 en route to Fort Smith on the NWT-Alberta border. Capt Meek was a graduate of one of the West's famous flight training institu-

tions, Mount Royal College. As a sprog pilot he flew with Ptarmigan, Simpson Air, Terra Mines and Wolverine Air, then joined Southern Frontier Airlines in the early 1980s. It was running two each of the Beech 99, Ce.404, Convair 440, DC-3 and King Air 90. Meek flew skeds (e.g. to Lloydminster and Cold Lake for the oil and gas industry), charters and "bag runs" (courier flights). In 1984 Southern Frontier was taken over by Time Air. Meek eventually became a senior Time Air captain with 5,000 hours flying Dash 7s in the Prairies and BC. He thought highly of the hybrid STOL, except that it was a slow climber and was not great in icing. Pilots used to argue whether it was propeller or wing icing that degraded Dash 7 performance. Meek recalled that once Time Air started applying an anti-ice substance on the props, performance in icing improved. He also reminisced about some of the old hands in Time Air including Capt R.J. "Pappy" Lundberg, who once pushed a DC-3 to the limits—14 hours aloft! He didn't quite reach his destination in the Arctic, and later described the "Lundberg way" to maximize a DC-3's range: "You fly it a while, you glide it a while, you slide it a while and you walk the rest of the way."

Meek and Hulkenberg sang the praises of the F.28, describing its "pilot friendly" features and how easy it was to check out on—just eight hours for initial captain training, six for an FO. At this time CRA was doing its F.28 simulator training every six months with US Air in Charlotte, North Carolina. At 1221 F1142 started down for Fort Smith. The airport was quiet. A Conair DC-6 was parked on fire standby, an old B-25 water bomber was rusting in the weeds. The town of 2,500 was a short ride down the road, sitting along the Slave River. Typical of the river communities of the region, Fort Smith had started as a Hudson's Bay Company post in 1874. It prospered with the gold rushes in the Territories, and was the first territorial seat of government, holding that honour until Yellowknife became capital in 1967.

The F.28 was on its way at 1312, blasting off from R29 at 53,500 lbs with 26 passengers (a crowd had gotten off at Fort Smith, including some students headed on a wilderness canoe holiday). This leg was no more than a hop—TAV's tires squeaked onto R31 at Hay River at 1336. Hay River also had its beginnings as an HBC post in the 1800s. A new site was built following floods in 1963 and the town of 3,000 prospered as a water transport

node. Tugs and barges that work the huge Mackenzie River system are based in Hay River, which has direct rail and highway connections with the south. This day the airport was quiet. Two Bell 204s on fire standby sat on the grass. Down a taxiway was Buffalo Airways' main base, with its compound full of DC-3s and tons of surplus parts and vehicles.

The last leg on F1142 was to Yellowknife, 105 nm across Great Slave Lake. Airborne at 1400, this time at 51,000 pounds and with 18 passengers, it took only 24 minutes to cross the lake and get on the ground. For the passengers, this was the end of the day's travels, but TAV still had a long way to go—back to Hay River, Fort Smith and Edmonton, then to Saskatoon. For Meek and Hulkenberg, the day would end late with 8:10 hours added to their logs.

The F28 Fellowship

One of the most successful small jetliners is the Fokker F.28 Fellowship. Fokker led its development in partnership with MBB and VFW-Fokker of Germany, and Short of Belfast. The F.28 was originally a 65-seater with two Rolls-Royce Spey 555s. The prototype flew on May 9, 1967 and LTU of Germany placed the first order. The F.28 quickly won the respect of the world's airlines. Production continued to 1986 with 241 built. Fokker by this time had determined to produce an advanced derivative of the F28—the F.100.

In 1972 Transair was the first Canadian airline with the F.28. It was ideal on such regional runs as Winnipeg-Thompson and Winnipeg-Thunder Bay. Norcanair acquired two F.28s in 1986 for its Edmonton-Minneapolis route, then flew them on Saskatoon-Regina-Calgary as a Canadian Partner. In March 1987 Time Air took 100% ownership of Norcanair and integrated its F.28s into the overall fleet. Before long Time Air discovered the benefits of the F.28 and began adding more. The F.28 found work on many new routes. It came east in November 1994 to begin three-times-daily service between Toronto and Quebec City. In 1994 CRA had seven F.28s but lacked enough work to keep its maintenance base in Calgary busy. Without more work it might have to lay off staff; instead, it began looking for outside F.28 overhaul work and found it. A contract was signed to do heavy maintenance ("C" checks) on nine of USAir's F.28s that had been mothballed in the California desert. A "C" check was done every 3,000 flying hours, or, about every 56 weeks an aircraft was on the line. Each F.28 required 40 working days involving

Time Air operated weekly skeds to northern Saskatchewan mines like Key Lake and Cuff Lake. Here southbound mine personnel arrive at Prince Albert on Time Air F.28 C-GTEO on January 8, 1993. By 1997 the company's fleet numbered 16 F.28s. These vintage airliners were in good condition. A company bulletin of September 1996 noted: "Fokker sees no reason why the F.28 cannot continue to fly as it could go as high as 100,000 cycles. Our highest aircraft right now has 55,000 cycles. As the aircraft ages there is more sampling and we look at different areas of the structure." (Larry Milberry)

110 personnel and 8,000-9,000 hours of labour. Instead of laying off, CRA hired 100 new people. Getting such work enabled CRA to offset overhead and generate extra profits. For the USAir job, it rented hangar space at Calgary from rival Air BC. The contract ran to late January 1995, when CRA was hopeful of more work from USAir.

CRA and the Fuel Picture

In March 1994 CRA reported on the important topic of fuel. Next to salaries, paying for fuel was its priciest item. By reducing types, careful route planning, etc., CRA saw the cost of fuel as a percentage of total expenses drop from 14.6% in 1991 to 12.2% in 1993. It explained that of the two commonly available fuel types (Jet A and Jet B), Jet A was preferred for the better distances it gave for the same quantity burned compared to Jet B. However, Jet B was preferred by the military since it ignites more easily below -20C, so fuel suppliers in more remote, northern places mainly kept Jet B. Since 1993 Canadian Airlines was the fuel supplier to Time Air and other Canadian partners. As CRA reported, "Due to their size and huge volume usage, Canadian Airlines are able to buy tanker loads of fuel in ports such as Toronto and Vancouver. The fuel is usually from the United States and is much cheaper; these price benefits flow through to Canadian Regional." Fuel prices change daily, so flights were planned to refuel wherever possible from the cheapest centre. Such planning is vital. As noted in early 1994, a one-cent per litre rise in the cost of fuel would cost CRA an extra $88,000 monthly. In 1994 CRA published these litres-per-block-hour figures for its types: F.28—2,299 litres, D8-300—689, D8-100—599, ATR42—655, Short 360—461, Metroliner—330, B.1900—300.

Canadian Airlines International Ltd.

Through the early 1980s various western carriers appeared, merged and failed. PWA of Calgary, which was privatized by Alberta in 1983, remained strong. In November 1986 CPAir lost its West Coast connector, when Jim Pattison sold Air BC to Air Canada. This was partially offset, however, when Air Canada connector Time Air was taken over by PWA in September 1986. That December PWA purchased CPAir for $300 million. The combined fleet now included 70 B.737s and 12 DC-10s. It served 82 North American and 17 international destinations. CP Ltd., parent company of CPAir, explained the sell-off in terms of its goal to strengthen its balance sheet (CPAir had lost $1.5 million in 1985) and build in areas like hotels. PWA raised much of the money needed for the purchase by selling, then leasing back, numerous 737s. From April 1987 the new company was known as Canadian Airlines International. It included Time Air, which became its new West Coast, prairie and territorial connector.

At first the employees and the industry in general were optimistic about the new company, but CAIL had trouble from the start. In 1989 it bought Wardair for a reported $241

million, taking on Wardair debt of $700 million. Much of this was retired by selling Wardair's A310s and an option on two Boeing 747-400s. The sale of five A310s to the DND brought $250 million. At the same time, CAIL took a long-term contract to support the DND fleet. There was a claim that the main reason CAIL acquired Wardair was to get rid of it as a rival—Wardair was by far Canada's most popular charter carrier and had a loyal following. CAIL quickly phased out the Wardair name.

Into the early-1990s CAIL lost millions every year. In 1992 its workers put $200 million into their company, buying shares at 80 cents that eventually slumped to 10 cents. The same year Ottawa loaned CAIL $50 million; Alberta and BC offered $70. CAIL now focused on arranging a partnership with American Airlines' parent company, AMR of Dallas. To achieve that, it needed freedom from Air Canada's Gemini reservation system (in which CAIL, Air Canada and Continental were partners). Air Canada opposed this. In April 1993 an Ontario court rejected CAIL's appeal to withdraw from Gemini. Air Canada claimed that Gemini, with 700 employees, would collapse if CAIL withdrew. The court decision was confirmed a few days later by the Competition Tribunal in Ottawa, but on May 27 the National Transportation Agency approved the PWA-AMR deal. Air Canada appealed to Cabinet, but lost on June 18. CAIL now joined AMR's Saber reservation system. This involved more than passenger bookings, e.g. fuel uplift data. Before Saber, CAIL made an educated guess as to how much fuel to tanker on a given flight. Saber began providing specific requirements for each station. Whereas fuelling had been done at major stops, it now was designated along the way, often at smaller stations. On paper this gave ideal quantities, although it also slowed turnarounds.

Meanwhile, CAIL tried rationalizing its balance sheet by curtailing service to many centres—a 15% reduction in domestic activity.

On November 9, 1992 NTA hearings had opened in Ottawa to discuss a CAIL—Air Canada merger. These were not pleasant talks. For example, when the topic of government aid to CAIL came up, Air Canada chairman Claude Taylor threatened that his company would take legal action should Ottawa give CAIL money without offering Air Canada the same. His argument was that Air Canada was losing more money than CAIL. President Rhys Eyton of CAIL countered: "Air Canada's focus is to put Canadian Airlines out of business. They intend to accomplish this by using their bigger bank account to outlast us... and by threatening government." Rhys Eyton stated that CAIL and Air Canada were losing a combined $1.5 million daily. "We're slowly driving one another out of business," he said, and suggested that CAIL and Air Canada put an end to over-capacity. At each phase one is reminded of earlier statements by McGregor and McConachie—the pros and cons of merger. In December 1993 Air Canada offered to buy CAIL's international routes for $250 million and assume an $800-

million debt on eight planes, but CAIL was not interested. It claimed that its international routes generated $14 billion annually. It also argued that restructuring as a domestic carrier would cost it $540 million and 6,000 jobs.

In a March 24, 1994 decision the federal government's Competition Tribunal ruled that CAIL should be allowed out of Gemini. This cleared the way after two years for CAIL to join AMR. On April 15, 1994 a youthful Harvard graduate, Kevin Jenkins, became CEO of CAIL. He replaced Rhys Eyton, who had spent 18 years as CEO, shaping a company from building blocks—CPAir, EPA, Nordair, PWA and Wardair . Jenkins set about to staunch the flow of red ink. He squeezed wage concessions from employees. Even though 1,200 were let go in 1994, CAIL kept losing. It took comfort, however, from Ottawa's belief in the concept of two strong Canadian air carriers; and by a partnership that gave AMR 33% ownership (25% voting stake) of Canadian in exchange for a $246 million cash infusion. The deal was finalized on April 27, 1994. AMR chairman Robert Crandall and vice-president of finance Donald Carty joined CAIL's 11-seat board (Carty had been president of CPAir at the time of the 1986 PWA takeover). In the deal CAIL agreed to buy services from AMR for a 20-year period.

Meanwhile, competition from Canada's charter companies grew fiercer. Air Transat, Canada 3000, Royalair, and several smaller carriers like Air Club, Greyhound, Skyservice and West Jet regularly added capacity, retired older planes, and built their image with the travelling public. Gone were the days when charter planes broke down all over the world, and had pieces falling off in flight. CAIL had to scramble to find new routes that might pay. On April 1, 1995 it began running the DND's domestic sked service with A320s. This included an extension to Comox from Vancouver. On March 6, 1995 CAIL inaugurated westbound service from Thunder Bay using the F.28. A summer 1995 sked between Dorval and Goose Bay was also introduced, with sport fishermen being targeted. Nonstop Toronto-Quebec service was reintroduced.

By 1996 the charter carriers had about 35% of Canada's domestic interurban business, and a big piece of the international pie. Air Canada had much of the domestic market's remaining 65%. By late 1996 its fleet was one of the newest in the world. On the whole, it was increasingly difficult for CAIL to hold on. On the surface it appeared that the Canada-US open skies agreement of the 1990s was helping CAIL, allowing it into many potentially lucrative US centres, but new routes alone were not going to save it. As Donald Carty stated in 1995, "One cannot grow out of a problem in the airline industry." Thus did CAIL aggressively pursue improved "cost performance"—getting the most bang for its bucks. This meant wiser spending down to the cost of paperclips. It also meant squeezing the employees. In 1995 CAIL was able to save millions by having its workers agree to do more for less e.g. pay cuts to 1,100 pilots totalled $41 million. The company was

Canadian's two stalwart regional jetliners through the years were the Boeing 737 (right) and Fokker F.28. By the 1990s the F.28 had taken over many routes previously operated by the larger 737 (which often represented wasted capacity). Canadian North 737 C-GOPW was at Iqaluit on August 13, 1992; while F.28 C-FCRU was at Toronto on September 5, 1996. (Larry Milberry)

(Left) The flight deck crew of Per Möller Wamedal and FO Beth Moxley aboard F.28 C-GTAH for the Winnipeg-Regina leg of F1260 on July 8, 1992. (Larry Milberry)

saying, "If you work harder and more efficiently, and don't press for raises, we will let you keep your jobs." This was the standard approach in an industry where jobs were scarce. In November 1995 CAIL announced that it was moving 330 maintenance jobs from Calgary to Vancouver and laying off or retiring early another 200 employees.

Constantly seeking new operating strategies, CAIL launched an aggressive 1995 summer schedule, adding 15% in capacity on busy domestic and international routes. The goal was to win back business lost to the charter lines. In January 1996 CAIL cut fares by 40% for a short period, seeking to fill otherwise empty seats. Return fares between Toronto and Florida, offered by the charter carriers for about $200, were even cheaper during CAIL's sale. Many CAIL routes, such as ones traditionally served with 737s, were no longer viable. Competitors such as Air BC were bleeding off CAIL's customers using smaller, more efficient planes. The 737 fleet gradually shrank, with local routes transferred to Canadian Regional. Internationally, in 1995 CAIL increased Pacific capacity by a third. It added a 747-400 for Vancouver-Nagoya and Vancouver-Taiwan; opened Toronto-Monterrey; and beefed up service to Frankfurt, Honolulu, Mexico City, Paris and Shanghai. It sought rights to Malaysia, the Philippines and Vietnam; and strengthened its Lufthansa alliance. CAIL still lost more than $100 million in the first half of 1995, about 10% more than Air Canada's loss for the period. Meanwhile, in order to keep pace with Air Canada's aggressive marketing, CAIL kept expanding—for 1995 it increased capacity by about 10% more over 1994, but carried only 4.6% more passengers. Excess capacity was dragging the company down.

In June 1996 Kevin Jenkins resigned. He had been under pressure for failing to pull CAIL out of its financial tail spin—the airline hadn't earned a penny since 1988 and for 1995 lost $194.7 million. Kevin Benson, formerly with Calgary-based real estate giant Trizec Corp, took over. By November 1996 CAIL faced $3 billion in debt and needed $180 million to pay bills. Meanwhile, rival Air Canada was making money—some $200 million in 1995. CAIL's employees, who numbered 16,400, were asked to take 10% pay cuts, more were laid off, and last-minute sources of funding were sought. Ottawa was asked to repeal a tax on aviation fuel. It agreed, extended CAIL's deadline on a $30-million loan payment, but refused cash handouts. BC offered a $3.1 million loan. Except for the recalcitrant Canadian Auto Workers, CAIL's unionized workers agreed to wage cuts, realizing that it was better saving most jobs than losing them all. The CAW proposed that Ottawa, the provinces and banks pour money into CAIL, and that Ottawa re-regulate the airline industry. The CAW also recommended that Ottawa allow AMR a bigger share of CAIL than Canadian law allowed. AMR was not in a rush to spend further—its $246 million in CAIL shares had dropped in value to $30 million, but it was willing to renegotiate certain CAIL contracts. CAIL worked on a plan to strengthen its Pacific business and reduce domestic overcapacity. It recently had lost millions on its Toronto-Montreal-Ottawa "Shuttle", which offered as many as 41 dailies between city pairs. These were flown by 737s and A320s that were too big for the market, so planes often were mostly empty. Having the wrong plane for the markets seems to have been an age-old problem at CPAir/PWA/CAIL.

On November 15 the CAIL board of directors resigned, fearing that they could be held liable should the airline collapse. Most of the 3,700 CAW members who worked for CAIL were demanding a free vote as to whether or not they would accept pay cuts to save their jobs. Union leaders were adamant that a vote not be allowed, but federal labour minister Alfonso Gagliano forced one on December 5. The travelling public began losing faith in CAIL and reservations fell, adding to the company's miseries. Finally, on December 7 the CAW agreed to a settlement and recommended that its members vote in favour of a deal that included wage cuts of 3.7% to last four years. The pressure was off and travellers' faith in CAIL rebounded. Even though CAIL now was assured of survival in the short term, its problems were not over. Air Canada, CAIL, Canada 3000, Greyhound, WestJet and a host of regional and local carriers still were competing with too many seats in a small market. Some sense had to be brought into the picture, but how? Early in 1997 transport minister Doug Anderson got the ball rolling by announcing a wide-ranging study to determine policies for preventing another major crisis.

The BC Scene

Vancouver International in a scene from February 23, 1995. Canadian Airlines International's Boeing 747-475 C-FBCA "G.W. 'Grant' McConachie" is arriving after a Pacific crossing. Beyond are Antonov An-124 freighter RA82047 and an Air Club 747. Vancouver is Canada's Pacific Rim gateway. In the 1990s it expanded greatly with further infrastructure like a new runway and control tower. (Jan Stroomenbergh)

The Origins of Air BC

Air BC was formed in 1980 from several historic coast carriers bought in the late 1970s by Vancouver entrepreneur James A. Pattison. These companies were Air West Airlines (Vancouver), Gulf Air Aviation (Campbell River), Haida Airlines (Vancouver), Island Airlines (Campbell River), Pacific Coastal Airlines (Nanaimo), West Coast Air Services (Vancouver) and Trans-Provincial Airlines (Prince Rupert). TPA was left to operate under its own name. The others came under Air BC, which began with a hodge-podge of more than 100 aircraft—Cessnas, Beavers, Mallards, etc. To head it all, Pattison appointed Iain Harris, a graduate in business from the London School of Economics and University of Chicago. Harris developed a plan to streamline operations, feeding passengers and cargo into hubs for the trunk carriers; but the plan faced difficulties. No sooner was it initiated than a recession hit. BC's primary industries were shaken. Airline traffic plummeted and fuel prices and interest rates soared. Another problem was that Air BC did not have the experienced personnel it needed. Old hands like Bob Langdon, founder of Island Airlines, had not come along with the buy-outs. Air BC suffered from some hasty decisions, e.g. buying the wrong planes, then having to dispose of them. It recorded a $3 million loss in 1981. Harris took drastic action. He cut overhead by selling float operations, giving rise to new coast operators like Coval Air and Air Nootka. There also were lay-offs—in early 1982 Air BC shrank from 585 to 185 employees and the hard-nosed Pattison was warning unionized employees to back off lest he shut down. In 1982 he won a three-year wage freeze.

Harris now set out to modernize and strengthen Air BC as a feeder line. As always on the coast, Vancouver-Victoria was the plum market, but CPAir was losing money on it, mainly because of the cost of running 737s on such a short leg. CPAir now teamed with Air BC, which placed a Dash 7 on the route in 1983 and took over completely for CPAir the following year. Air BC's marketing department also introduced incentives (cut-rate

fares, etc.) to attract passengers, whether from land, sea or air carriers. There was even a one-way Vancouver-Victoria 99¢ gimmick fare. By the end of 1983 Air BC carried 414,000 passengers compared to 214,000 the year before. It also took over such CPAir 737 destinations as Campbell River.

Air BC blossomed further when Ottawa de-regulated the airline industry in May 1984. This allowed carriers to decide where to fly, and with what frequency and equipment. Two Dash 7s were added for new destinations like Castlegar, Kamloops, Penticton and Smithers, but also some locations in Alberta. In 1985 Victoria-Seattle was added and eight Dash 8-100s were ordered. In 1986 Air BC carried 673,000 passengers, earning 2% on sales of $40 million. That year there were rumours of a PWA-CPAir amalgamation. Wary of this, in 1987 Air Canada bought 85% of Air BC stock from Jim Pattison, leaving the remainder with Iain Harris. With Air BC its largest connector Air Canada's West Coast revenue soared. New routes opened, and the 85-seat BAe146 was introduced on Vancouver-Terrace-Prince Rupert. Air BC began bombarding CAIL destinations with frequent turboprop service, forcing it to replace 737s with Time Air turboprops.

For 1987 revenue at Air BC climbed to $50 million and passengers to 804,000. In 1988 the company had its first million-passenger year. In 1989 six 19-seat BAe Jetstream 31s were added for smaller markets, and Dash 8s were placed on Vancouver-Portland, head-to head with Delta and Horizon of the US. Canadian eventually had to respond—in September 1996 it introduced daily Vancouver-Portland Dash 8s. Another strategy saw Air BC put the

BAe146 on CAIL's historic Calgary-Edmonton Municipal route, and by 1990 it was operating 18 Vancouver-Victoria dailies in peak season. By this time Air Canada recognized the wisdom in letting Air BC push east to Regina, Saskatoon and Winnipeg using economic Dash 8s (Air Canada had been flying high-overhead DC-9s and 727s). Air BC further enhanced its position in 1991 by ordering six 50-seat Dash 8-300s. In 1992 it phased out the Dash 7. By 1993 Air BC was Canada's largest regional with a fleet of Dash 8-100s and -300s, BAe146s, Jetstream 31s and Twin Otters. It had more flights per day from Vancouver than any carrier, operating more than 300 legs. From a bare bones operation it had 280 pilots, 175 flight attendants, 375 customer service and ground personnel, and more than 100 maintenance people. In August 1995 it reduced Jetstream service by subcontracting Calgary-Lloydminster and Calgary-Lethbridge to Sunwest International; and Calgary-Medicine Hat to Bar XH Aviation. In September Air BC Jetstream service was eliminated using extra Dash 8 sections.

A typical flight with Air BC was F1763 of April 1, 1992—Dash 8-300 C-FACV operating Winnipeg-Regina-Calgary. On the flight deck were Capt Rory Pleasants and FO Russell Brown, tending their 44 passengers were FAs Gary Mazurkewich and Kristin Babcock. Pleasants had come through the ranks, starting in Ontario with Big Trout Lake Air in April 1984, then flying for Alert Bay Air Services and Navair in BC. Brown, the son of a pilot, had spent his apprenticeship with Ptarmigan and Contact airways. F1763 taxied from the gate at 18:51 and was airborne from R31 at 38,328 lbs four minutes later. It climbed at 270 KIAS to FL240, then cruised easily over drab, late-winter prairies. After 287 nm and 1:13 hours it touched down on Regina's R36. Following a brief stop, the flight was on the move, this time with 24 passengers on the 357-nm/1:33-hour leg to Calgary. Here those pushing on waited for F1715, a BAe146 for Kelowna (246 nm) and Vancouver (168 nm). Aboard were Capts Dan Wenger and Les McAninch, seasoned men who had joined Air BC in 1979. Wenger had started gliding in Switzerland in 1965. He emigrated to Canada in 1970, flying

A Day on the Job: Air BC Progress Sheet, August 13, 1995		
Crew: Capt Glenn Langen, FO Victor Teply		
Duty Time: 10:50-21:35 hours (local)		
Flight No.	*Route*	*Flight Time*
642	Vancouver–Quesnel	1.3 hours
642	Quesnel–Williams Lake	.4
642	Williams Lake–Vancouver	1.0
593	Vancouver–Campbell River	.6
596	Campbell River–Vancouver	.6
644	Vancouver–Quesnel	1.3
644	Quesnel–Williams Lake	.4
644	Williams Lake–Vancouver	1.0

for Wilderness Air at a time when there was lots of business. One day his passenger was a timber cruiser spotting for a fortune in fir, the next day some high roller in herring roe with a suitcase stuffed with cash, hot to buy straight off the boats. With the recession in logging, Wenger was forced across the mountains in 1975 to fly an Otter for Buffalo Airways. McAninch, whose father was a longtime RCAF transport pilot, had trained at Rockcliffe, then worked for Laurentian Air Service and Kenn Borek.

F1715 crossed the mountains to Kelowna at FL260 indicating 370 knots. It brought its 51 passengers to the terminal after 46 minutes, then left 27 minutes later with 28 passengers for Vancouver. From here it was on to Terrace on BAe146 C-FBAO operating as F1599 with 29 passengers. Capt Kurt Miller and FO Dick Nassey were on the flight deck for the 390-mile run up the scenic Island and Coast ranges. As a junior pilot in the 1960s Miller had worked for Ocean Air at Tofino, flying the FBA-2C. Like most pilots who knew it, he liked the little Canadian-bred bush plane, describing it as "a marvellous aircraft in a lot of ways... I'd say that it did a better job than the 180." He also enjoyed the Fairchild Husky during a stint with Island Air, but Miller's favourite type was the Mallard on which he logged 4,000 hours with West Coast Air.

Dan Wenger and Les McAnnich at work for Air BC aboard their BAe146. (Larry Milberry)

(Above) The 50-seat Dash 7 replaced the DC-3 at Air BC. C-GFEL is seen on the inaugural Dash 7 service to Port Hardy in the summer of 1983. Then (right), some of those attending: Air BC president, Iain Harris (left) with Port Hardy and Port McNeill mayors John Davis and Gerry Furney, Air BC chairman Mel Cooper, and Capt Villi Douglas. (Uwe Ihssen)

(Left) Air BC's fleet began as a hodgepodge of types inherited from coast operators. Jim Pattison realized that a simplified fleet of turboprops was needed. In time modern Dash 7s, Dash 8s and BAe146s served the airline. Beaver OCJ was at the dock at Vancouver International on August 18, 1983. In 1997 it was on Harbour Air's roster. (Larry Milberry)

(Left) Otter UJM was at Dawson's Landing, a mail stop in Rivers Inlet (also the site of the only liquor store on the inlet) in May 1981. UJM was ex-US Army 55-3302, United Nations 302 and N12665 before coming to Ben Ginter Construction Co. in Prince George in 1965. After being damaged at Forbidden Plateau, BC in September 1971 it was restored for West Coast Air and later served Gulf Air, then Air BC. In 1997 UJM was sold in California. (Larry Milberry, Uwe Ihssen)

(Right) The reliable and low-overhead Britten-Norman Islander gave solid service in BC from the 1970s. Air BC's KAW is seen in June 1981 at Kimsquit, a logging camp at the head of Bean Channel. Islanders used to operate at the Campbell River spit from Fletcher Challenge's runway (turned into a camp ground around 1980). (Uwe Ihssen)

(Below) Passengers board Air BC's DC-3 WUG at Port Hardy in a scene from May 1983; the pilots this day were Capt Rolly Heinl (at one time Barney Lamm's chief pilot at OCA) and co-pilot Don Matheson. In early 1997 WUG was doing a coastal courier run for Kelowna Flightcraft. (Uwe Ihssen)

Capt Kurt Miller at the controls of a Turbo Mallard in September 1981. (Uwe Ihssen)

The Twin Otter proved ideal on the BC coast, where it replaced services offered by the Goose, Mallard and DC-3. Uwe Ihssen shot Twin Otter JAW at Port Hardy in May 1980. It had spent 1970-78 in Mexico, then came north for Air West. In 1997 it was listed to West Coast Air, a new float operation flying on Air BC's old Vancouver-Victoria run.

Air West

In the 1950s Norman A. Gold was a Powell River restaurateur with a private pilot's licence. He realized the need for improved local air service, so founded Powell River Airways in 1958 with a Cessna 172. He added a Seabee which, as he recalled, "carried three passengers, gingerly". Gold's first expansion (1960) was taking over PWA's Powell River-Comox run. PWA used to carry about 2,000 passengers a year in Beavers. Using an Apache, Gold carried 7,000. In 1964 he renamed his company Air West. In this period one carrier could not fly on routes served by another, so Gold had to purchase his competitors in order to expand. In 1965 he took over Nanaimo Air Lines with its four aircraft. Next year he added PWA's Vancouver, Kamloops and Nelson bases, and began taking over BC Airlines charter bases as that company expanded into sked operations. Using Beavers, Air West inaugurated twice daily Vancouver-Nanaimo service in 1966 and Vancouver-Victoria in 1968. In 1969 it bought BC Airline's Port Hardy base. Additional float bases were added as buy-out opportunities arose.

By 1970 Air West had 22 aircraft and was carrying nearly 60,000 passengers yearly. In February 1972 it bought its first Twin Otter—a

$450,000 investment—for the Vancouver (Coal Harbour)-Victoria (Inner Harbour) run. The Twin Otter replaced a nine-passenger Goose and a seven-passenger Beaver and flew its route in 28 minutes. The service was attractive to the many business people and government flunkies commuting between BC's two biggest cities. It was faster than the ferry boat, or flying airport-to-airport via TCA Viscount or PWA Convair. On April 29, 1972 Vancouver *Sun* reporter Phil Hanson compared the airline options, starting at the Hotel Vancouver and ending at the Empress Hotel in Victoria. Using taxis and buses to fill the gaps between terminals, the $16.75 trip on Air West took 1:05 hours. By PWA, it took 2:15 hours and cost

$17.00. PWA argued that if Ottawa would allow it to operate Vancouver-Victoria on an air bus basis it could match Air West in time and cost. Of course, the Air West concept had its limitations—it was strictly a day VFR operation.

In 1974 Air West carried 45,000 passengers downtown-to-downtown. In August 1975 it got approval from the Air Transport Board to operate Victoria-Seattle on floats; service commenced January 1, 1977. With each leg just 35 minutes, Air West had it all over the ferry. The same year Air West opened Vancouver-Comox and Vancouver-Powell River. By this time it had seven Twin Otters. Air West had its trials in this period, losing a Twin Otter and one passenger 50 miles from Bella Coola in September 1976. On December 16 Capt Ted Mitchell landed another Twin Otter hard in bad weather four miles off Victoria, buckling the floats. A US Coast Guard rescue helicopter arrived quickly, but struck its rotor blades on the sinking Twin Otter's wingtip. A number of boats came out to collect everyone, and the damaged chopper was towed ashore. Bad publicity from these incidents unnerved travellers. Another setback for Air West was a strike by 95 employees. Management kept five of 20 aircraft in service, but the trouble lasted from November 1977 to April 1978.

Meanwhile, West Coast Airlines of Vancouver also was growing, acquiring companies like Pacific Coastal of Nanaimo (1970), Staron Flights (1971), TPA's licence for Ocean Falls (1976) and Victoria Flying Services' licence for Vancouver-Victoria via the Gulf Islands (1977). Its Mallards and Twin Otters also operated Vancouver-Victoria. West Coast, however, was at a disadvantage, as its licence required every flight to stop in the Gulf Islands. In 1978 it was bought by Cromarty Holdings. Founder Al Michaud stayed on in management. Soon BC's main coast airlines were swept up by millionaire Jim Pattison, who started his spree by acquiring Air West in 1979. But operators like Tyee Air of Sechelt, and Northern Thunderbird of Prince George did not sell, so Pattison was unable to build a fully-integrated BC operation.

Coast Pilot

Don Thompson had a long, successful career on the BC coast. He had come to Canada in 1942 as an RAF student, arriving aboard the *Queen Elizabeth*. He recalled that among the 20,000 passengers were Hollywood stars Edward G. Robinson and David Fairbanks, Jr. Thompson learned to fly on Tiger Moths at No. 35 EFTS (Neepawa, Manitoba), then took his wings on Oxfords at No. 13 SFTS (North Battleford, Saskatchewan). Next came a tour instructing at No. 36 EFTS (Pearce, Alberta); a flying instructor's course at Trenton; instructing on Ansons at No. 33 SFTS (Carberry, Manitoba); and the Mitchell and Liberator OTU at Abbotsford and Boundary Bay, BC, before being posted to India on No. 99 Squadron. Life in England was difficult after the war. There were food, housing and job shortages. Thompson, who married a Canadian girl while at Carberry, flew Vickers Vikings for Britain, but felt that Canada had more to offer. In 1948

Don Thompson during his early days at Alert Bay. (Thompson Col.)

he emigrated and took a truck driving job in BC. He met Peter Deck, a flying instructor at Chilliwack, and this got him back into aviation. He soon was running a small flying school near Hope and in 1955 moved to Alert Bay on northern Vancouver Island. He planned to stay a year flying the Seabee and Fairchild 71 (CF-BXF) for BC Airlines. The deal was good—$250 a month, plus 3¢ a mile, his apartment and utilities. He enjoyed the work, the people, the way Bill Sylvester ran his company and the natural environment. Thirty years later, Thompson was still in Alert Bay.

Steamships still were carrying much of the trade to and from BC's isolated coast settlements, but aviation was making inroads. In 1955 a shipping strike lasted so long that the air carriers were able to win much of the steamship trade. The fliers never looked back, but coast aviation was changing. PWA bought QCA. Bill Sylvester sold to Paul Tak. Thompson watched for his own chance. In 1958 he and Bill Groth started Alert Bay Air Services Ltd., putting in $5,000 and securing a $4,000 note from Vic Youde of the Nimpkish Hotel in Alert Bay. Those were the days when an airline still could be launched with $5,000, so long as one had some experience, commitment and brashness. A J-3 Cub and Taylorcraft, plus a Seabee leased from Dr. Jack Pickup, made up the ABAS fleet. Bill Foyle came in as air engineer.

Times were lean, as Thompson relates: "We were so hungry in those days that on Sunday mornings we used to fly United Church minister Rev. MacKenzie to Port Hardy, go to his service, put a dollar in the collection plate and fly him back to Alert Bay... all for the grand sum of $15.68." In June 1959 ABAS' credit was good enough to borrow $25,000 for a new Cessna 180. The company now got rid of the Seabee which, as Thompson put it, "blew a pot a month and cost me $500 every time." By 1962 ABAS had bases in Kelsey Bay, Hardy Bay and Bella Bella. Two years later it won its first scheduled route—Alert Bay-Kelsey Bay, with intermittent stops. Port Hardy became a base, and in 1965 the company added its first Beaver. In 1971 Air West's Port Hardy amphibious operation was acquired. Business grew—ABAS became a PWA feeder—PWA had DC-3 skeds to Port Hardy.

ABAS's first twins (Goose CF-EFN and Beech 18 CF-ZMO) came in 1973. The nine-passenger Goose was bought for $120,000. Wheel operations started between Campbell River and Port Hardy. An IFR Navajo replaced the VFR Beech in 1977. Service between Port Hardy and Sandspit, 250 miles north, also opened in the 1970s. With a staff of solid people ABAS was a success. Its customers were the Indians, fishermen, loggers, miners, RCMP, etc. They were loyal toward the little airline and this kept it on a firm footing. At its peak in 1973 ABAS had 15 aircraft, a staff of 40 and was moving 30,000 passengers a year; but late summer that year, the economy went into a tailspin. Logging slowed and fuel prices soared. By December 1974 eight aircraft had been sold and only seven employees were left. ABAS was sold in 1978. By that time it was serving about 80 communities as a VFR float and IFR multi-engine operation. Its float licences went to Gulf Air of Campbell River, its IFR licence to Pacific Coastal Airlines.

In 1993 Don Thompson reminisced about Albert Bay Air Services. He remembered his staff as " a league of nations". He had Glen Rankine, an Aussie; Dave Hutcheon, a Rhodesian; Kayo Kawano, a Japanese; Willie Wong, a Chinese; Eric Andersen, a Dane and Henry Sweets, a Dutchman. Thompson mentioned one of his favourites as having "a great ability as a pilot, an enormous capacity for booze and an incredible sexual appetite... he was well-liked by everyone, especially the women." About 80 young pilots passed through ABAS. Many moved to successful careers with the airlines, but some had their careers cut short. Thompson's first two pilots were Norm Rogers and Dave Stronach, whom he admired for their flying skills and eccentric personalities. Norm was killed at Terrace in 1977 in a Twin Otter accident. Dave died in PWA's Hercules crash in Zaire.

Thompson gave the West Coast Indians a lot of credit for the success of small operators: "I loved the Indians. They were sensible, no bullshit people with a terrific sense of humour. Without their business and, particularly, their support during the 1964 coastal Air Transport Committee hearings, we wouldn't have gotten very far. It was hard enough for me to get up and speak before a battery of high-priced government officials with their counsels. For Indians from Bella Bella or Kingcome Inlet it was a real ordeal, but they did it for me in both Ocean Falls and Alert Bay." After Air BC bought out the coastal airlines, Thompson felt that service deteriorated. In earlier days the independents always gave exceptional service, and their pilots were held in high regard. Companies had mottos by which they lived or died (e.g. Island Airlines: "Instant Service"). The demise of the coast carriers led to more forestry companies setting up private flight departments to assure the service they required. Of course Air BC soon realized that its forte was in the regional market. Inevitably it sold its coast licences to a number of small, new companies and the cycle of coast flying began anew.

Trans-Provincial Airlines

Trans-Provincial Airlines formed in May 1960 with Beaver CF-MGS, bought new for $43,000. Its base was on Lakelse Lake near Terrace in northern BC. The first year was busy—there was plenty of work on a road building project between Stewart and Cassiar. From May to October MGS logged 800 hours and its engine was time-expired at season's end. In the fall a new Ce.180 on wheels was bought for $16,000. Beaver TPA came next. A Norseman was leased from OCA for the 1963 summer, then came Otters RHW and ROW. Doug Chappell and Cedric Mah were two early TPA pilots. Mah had flown in China and for QCA.

In 1965 founding member Lloyd Johnson, an ex-RCAF engineer, sold to Gary Reum and Bill McRae of Terrace. Reum, a wartime bomb aimer, had a private pilot's licence. TPA added bases—Ocean Falls, Prince George, Prince Rupert, Sandspit and Smithers. In 1968-69 it took over PWA's routes from Prince Rupert and Stewart; BC Airlines' VFR routes in the Queen Charlottes; and Omenica Air Services of Smithers. TPA's fleet—Beaver, Cessna, Goose and Otter—served all interests on the coast and in the interior. At its peak the Goose fleet alone numbered seven. DC-3s KAZ, PWI, PWH were acquired in 1971-72 as the BC economy faltered. The people had elected a socialist New Democratic Party government, and fishing, forestry and mining took a nosedive. Meanwhile, the price of insurance and avgas soared. An F.27 operation set up in 1975 to fly Prince Rupert-Terrace-Smithers-Prince George lost a small fortune. The plane was leased to Time Air for a year, then sold to Norcanair at a loss. Gary Reum lost more than a million dollars during this bumpy time and TPA went into receivership. When the Jim Pattison Group made an offer in 1979, the shareholders sold. Under Pattison's Air BC, TPA flourished at Prince Rupert and Port Hardy. Gene Storey, a knowledgeable and well-liked coast pilot, managed the operation. He had started flying with the Chilliwack Flying Club in 1964. He got his first job with Powell River Air and later flew in BC, the Yukon and the NWT. In 1986 he headed a group that purchased TPA (except for the Port Hardy operation, which stayed with Pattison as Pacific Coastal Airlines).

The Bristol Freighter and TPA

TPA became interested in the Bristol B.170 Freighter for heavy work in mining. This type had a solid reputation in Canada, starting in 1945, when A.G. "Tim" Sims was hired by Bristol to take it on tour in North and South America. Sims, who had grown up in Montreal in the 1920s, worked before the war in Alberta and Quebec. Along the way he took up flying and was with Ferry Command. As the war tailed off, he became a sales rep at British Aero Engine Ltd. of Montreal, where J.N. Baird was manager. As a rep for Bristol products, BAEL was interested in all postwar possibilities. In an April 9, 1945 memo to Sims, Baird mentioned an astounding bit of information—floats were being seriously considered for the Bristol

A typical Trans Provincial scene—Goose BXR servicing a sport fishing operation on BC's north coast. BXR was exported to the US in the early 1990s. (Henry Tenby)

Freighter, which was soon to be demonstrated overseas:

I left a questionnaire with Bristol, which sets out in detail all the information Edo need to have on the type of machine Bristol are building, so that they will be in a position to see whether the present DC-3 floats are suitable, or whether something else will have to be designed. Prior to leaving England, Bristol agreed that the floats will be made over here (rather than Saunders Roe, as at one time contemplated by them), and I cabled Crowles on March 29th, to ask him if the Edo float data had been mailed. I think it is quite important to get this information quickly, so that when the machine comes out, the floats will be available. I am quite sure, also, that they should be made here by MacDonald Brothers, and the sooner we get this under way the better.

Baird sent Sims on a familiarization tour of the British aircraft industry. He visited Bristol and reported that it would equip the B.170 with Hercules, not Perseus engines; that on the Canadian tour, "Bristol would be responsible for the cost of floats and skis" for the aircraft; and that a list of technical requirements needed

to suit the aircraft to use in Canada be submitted to the design office by Mr. Sims." The matter of Sims taking the plane to Latin America was also discussed. Discussions about floats and skis seemed to be of great importance at this time. In concluding his memo, Sims noted: "I believe the Hercules Freighter to be an extremely well-deigned and robust aircraft with comparative performance figures exceeding anything so far in sight... it may well be responsible for promoting a completely new low rate airborne market." In these predictions Sims could not have been more correct.

On August 2, 1946 Sims landed the first Bristol Freighter (G-AGVC) in the New World, touching down at Goose Bay on August 2, 1946 after a flight from Greenland. The plane brought the first commercial air freight across the Atlantic—a load of ladies' autumn fashion wear. No sooner had it landed at Dorval from Goose Bay than the clothing was whisked to Eaton's downtown store. In Montreal the Bristol was flown by Grant McConachie of CPA, who did stalls and single engine flying, and was favourably impressed. Sims also demonstrated a single-engine takeoff—this was a serious tour! G-AGVC next flew to Avro at Malton, where long-range tanks were removed

TPA Bristol Freighter C-FDFC over mountain terrain in the spring of 1989. (Grant Webb)

and a new paint scheme applied. Within a few days it was on display at the Toronto airshow, and impressing the crowds by flying in a new Studebaker Champion each day. Next it flew west with 25 troops for an army exercise at Carberry, Manitoba, then did demonstrations for the military in Ottawa. Three RCAF pilots from Test and Development evaluated G-AGVC at Rockcliffe. In time all this good work paid off, for the RCAF ordered several examples. Later, Sims demonstrated the Bristol to TCA in Winnipeg, but felt the prospects of a sale were poor: "At present they do not appear to be too interested in an aircraft of this type for they are heavily committed to DC-3s and the Douglas DC-4M." A freight run to Yellowknife followed, with 7,400 pounds of supplies, the biggest load ever brought there by air. The return to Edmonton was a medevac carrying an Eskimo boy, the sole survivor of an explosion at

Bristol AVD while being demonstrated in the west by Tim Sims. It was shot at Edmonton by William Kensit.

distant Cambridge Bay. The visit to Canada was brief, but the Bristol made a lasting impression. After a stop in Vancouver for a 100-hour check, it was off down the US coast and through Latin America. Sims later recalled, "Loading cars in and out of the machine was a task at which we soon became adept and one which, by popular demand, we carried out at every stop." The Bristol's clamshell nose was its main feature, once one got used to its homeliness. Bristol described the Freighter as "bulky" and "bull-nosed". Through the 1950s various Freighters worked on the DEW Line, carried freight for TCA and supported the RCAF in Europe. In the 1970s Wardair used several in the specialized Arctic market

TPA became interested in the Freighter through mine developer Reg Davis (Skyline Gold Corp), who was opening the Johnny Mountain gold mine at Bronson Creek. Years earlier he had been impressed by a Bristol on another project. In 1987 he and Jim Storey formed a joint venture, with Davis financing two Bristols to fly concentrate the 52 miles between Bronson Creek and Wrangell, then return with cargo. Two ex-RNZAF Freighters were located with Hercules Airlines in New Zealand. The first arrived in BC in September 1987. This was enthusiastically reported in *Propliner,* the journal of the world's prop-

driven airliner fans. In its winter 1987 edition *Propliner* ran a photo by John Kimberley with the caption: "Canada once again proudly boasts an airline flying a Bristol Freighter. This amazing news follows the delivery of Hercules Airlines' Mk.32M ZK-EPD to Trans Provincial Airlines of British Columbia as C-GYQS. This same aircraft once flew for Instone Air Line from Stansted and Lydd as G-AMLK. Departing Auckland on Sept. 19, 1987, the Freighter crossed the Pacific Ocean staging via Honolulu and Oakland prior to arriving at Prince George, BC, in a flight time of 52 hours... there is even speculation that further aircraft may be on the way." YQS made its first revenue flight on October 19, operating Terrace-Johnny Mountain. On December 10 it was joined by YQY, allowing YQS to fly to Abbotsford for a detailed corrosion inspection.

In 1988 Cominco, whose Snip Mine lay below Skyline's operation, also began using Freighters from the 3,000-foot Bronson Creek site. Both companies found them ideal. They churned back and forth on their 20-minute runs, doing six or seven trips a day at a cost of about $2,500 per round trip. This was no problem for either a DC-3 or a Bristol, but the latter could carry 10,800 pounds of cargo (usually four bags of ore concentrate) compared to 6,500 for a DC-3. As a tanker, the Freighter carried 1,658 imperial gallons in a single aluminum tank. Fuel was what the Bristols usually carried back from Wrangell.

On June 21, 1988 Capt Wayne Lebeau was landing YQY at Bronson with jet fuel. The right main gear ran off the runway into soft gravel; YQY ground-looped and broke up. Later it was cut up, some recyclable parts to be flown out, others to be bulldozed. The main item salvaged was the fuel ferrying tank. YQY was replaced by C-FDFC, which arrived from the UK in January 1989. Tail wheel damage to YQS and DFC led to the purchase of C-GTPA. Another old Kiwi machine, TPA, which reached Prince Rupert in October 1989, most recently had been on display in the RNZAF Museum! It was not able to work for several months pending installation of a reinforced cargo door. TPA was damaged one day, when the nose doors blew open on takeoff from

Johnny Mountain. The propellers were struck and TPA was grounded until parts arrived from New Zealand.

The Bristols found other work. There were contracts at Windy Craggy and Atlin in northern BC. In the spring and summer of 1990 DFC was far from home, resupplying DEW Line sites and carrying mine equipment on Baffin Island. It returned in August in the midst of a fuel haul to Johnny Mountain, but the mine was abruptly closed in October, depriving TPA of needed revenue. Meanwhile, Cominco tried an alternate means of freighting. It purchased a 96,000-pound API 88-100 hovercraft powered by four 525-hp Deutz diesels. By hovercraft the trip up the Stikine and braided Iskut rivers was 72 miles. It took 2 1/2 hours, but operated only by day. The hovercraft worked reasonably well and carried 24,000 pounds, but often was delayed by technical snags and weather. In 1996 the hovercraft was pulled from the job after charges from environmental campaigners that it was disrupting salmon spawning in the Stikine and Iskut rivers. Aircraft, which operated more cheaply than the hovercraft, took up the slack.

In 1991 TPA operated 10 Convair 580s (most leased from Kelowna Flightcraft for FedEx overnight runs, but also used on sport fishing charters from Vancouver and on fire fighting support.), three Twin Otters, three Goose, six Beavers, three Otters and three Cessnas. In 1992 TPA entered a new era—it established a Boeing 727 charter operation for FedEx. Mismanaged, this gobbled up money. Suddenly TPA lost its FedEx contract and on March 19, 1993 entered bankruptcy. About 100 employees lost their jobs. Rivals like Harbour Air took over TPA's float bases. Gene Storey joined Harbour Air at Prince Rupert and CanAir Cargo became Canada's primary Convair courier operator.

The Bristol fleet was inactive at Terrace for more than a year after TPA folded. It then was taken over by Hawkair Aviation Services (chief pilot David Menzies, chief engineer Paul Hawkins), which put C-GTPA back to work for Cominco. The first contract was lifting 50 tons of explosives to Bronson Creek. Hawkair also bid on an environmental clean-up job at Windy Craggy. C-FDFC was sold to John Duncan in the UK and on September 13, 1994 departed Terrace on the North Atlantic route. Nearing Scotland it had engine trouble, so was escorted to Glasgow by a SAR Sea King. Later in 1996 it was written off at a UK airshow. C-GTPA was working the Bronson Creek-Wrangell route in 1997, making as many as four return trips daily. C-GYQS was in long-term storage at Terrace.

Bristol 170 Freighter Mk.31

Span: 108'
Length: 68' 4"
Height: 21' 8"
Empty weight: 29,550 pounds
Max. takeoff weight: 44,000 pounds
Max. speed: 225 mph
Cruise speed: 160 mph
Range: 820 miles
Engines: Two 1980-hp Bristol Hercules 734

Bristol Freighter scenes by Grant Webb. He shot C-FTPA (left) after it was sidelined by wind and fog at Bronson Creek over the winter of 1992. Then, YQS settling onto the gravel at Bronson Creek in a scene from the summer of 1988. Next, YQY being gutted at Bronson Creek, where it crashed June 21, 1988. Many valuable spares were salvaged. The aerial of the Skyline mine at Johnny Mountain over the winter of 1988 shows a Bristol and Twin Otter in with supplies. (Grant Webb)

Central Mountain Airlines

Central Northern Air Service was formed in Smithers in 1982 by partners Tom Britton and Mel Melisson. It served the mining, forestry and fishing/hunting markets in northern BC with light planes— Beaver, Beech 18, Ce.206, Islander. Melisson disappeared on a flight in 1983. Under airline deregulation Neil Blackwell and Doug McCrae took over the company in April 1987 and re-launched it as Central Mountain Airlines. Its market would focus on business travellers, especially in mining. CMA viewed the Beech 1900C as the best airplane for this market. Its first (C-FCMA, from Australia's New Hazelton Air) was delivered July 20, 1990.

By 1992 three 1900s were in service on skeds like Terrace-Vancouver and Terrace-Smithers. Each was logging about 160 hours a month. As well, there were three Beech King Air 200s, a Beech 18, two Ce.185s, two Ce.206s, a Navajo and three Turbo Otters. CMA's hangar at Smithers, formerly a provincial water bomber base, was modernized in 1990. Of 110 employees about 35 were pilots; 14 were in maintenance. By mid-1994 CMA was serving communities, often in direct competition with Air BC and Canadian. CMA had done well in establishing and streamlining operations, but was struggling financially. It dis-

(Left) Central Mountain's main types on the tarmac at Smithers over the summer of 1989: Turbo Otter, Navajo, King Air 200, Beech 18, Cessna 206 and DC-3. (Grant Webb)

(Below) The Beech 18 got many a small local carrier going in the 1960s-70s. In the late 1990s it still was playing this role and giving reliable service. Note the retrofitted cargo door on CMA's Beech. (NA/McNulty Col. PA191789)

(Below) CMA's renovated wartime hangar at Smithers in April 1992. The company's DC-3 C-FAAM was waiting for work. (Larry Milberry)

(Above) Off-loading the King Air at Bronson Creek as passengers wait to board. (Larry Milberry)

A pilot's view about a mile back from landing at Bronson Creek. Capt Gary Meier and FO John Hartman were on final with Central Mountain's King Air 200 C-FCGX. Their approach was for an uphill landing; they departed downhill—standard procedure at this strip. Then, a wider view of the approach with the frozen Iskut River in the foreground. (Larry Milberry, Jim Smith)

While Bristol Freighters usually hauled the gold concentrate from Bronson Creek, sometimes they were down for maintenance. In this case the scene is aboard a Buffalo Airways DC-4, which was filling in. The nearest sack is marked with its weight—2,817 pounds. Snip Operations was a branch of Cominco Metals, a company involved in aviation since the early 1930s. (Jim Smith)

posed of a 1900 and two Turbo Otters and in June 1994 filed for protection under the Bankruptcy and Insolvency Act.

Many small carriers like CMA appeared since WWII. Usually run by local businessmen, they count on the loyalty of the locals, but that only goes so far. While CMA had modern Beech 1900s, those coming or going in Northern BC had the option of taking a larger, more comfortable, faster Dash 8 or BAe 146. If they could, that is how they travelled. The smaller carrier bent over backwards to attract customers

with incentives like more convenient schedules. However, at any moment a rival regional could make life miserable for the likes of CMA by slashing fares. The commuter lines faced other problems. Often they were undercapitalized, and always were losing essential staff—just when a pilot was trained and upgraded to captain, he would be offered a job by a bigger airline.

A typical CMA sked was F591 from Smithers to Bronson Creek on April 3, 1992. Pilots Gary Meier and John Hartman took off

in King Air C-FCGX with three passengers and freight. They covered the 170 nm at 195 KIAS at FL160 and landed from the up-slope end of Bronson's 4,500-foot gravel strip. Four passengers and a pick-up piled with freight were waiting. In only 15 minutes CGX fired up, taxied to point downhill towards the Iskut River, and roared off. On another sked (F403), pilots Rob Johnson and Kent McRae flew Beech 1900 CMV the 57 nm from Terrace to Smithers on April 2, 1992. Eleven passengers were aboard for the 18-minute hop. Johnson was typical of young pilots entering the business. He had been flying just 4 1/2 years, his first two gaining experience on the coast, and in the Yukon and Manitoba.

As its financial position declined, CMA had to reduce service on its prime Smithers-Vancouver run to once-a-day on February 28, 1994. The reorganized flight stopped at Prince George, so for 1994 CMA was out of the non-stop Smithers-Vancouver market. The regionals had that market in their pocket and the "little guy" was on the ropes. CMA was able to hold on and tried other markets, including Vancouver/Victoria-Campbell River. In 1995 it was serving 11 centres and had a fleet of 10 aircraft, but it was a struggle, especially in view of competition with Air BC and CRA. In 1995 CMA won a three-year medevac contract with the BC government, which recently had disbanded its own air ambulance service. For 1996 CMA was serving Smithers, Terrace, Prince George, Watson Lake, Dease Lake, Kamloops, Vancouver and Victoria. By this time it was operating the Beech 1900D.

The Photographer Pilot

Grant Webb flew for CMA in the early 1990s. Born in Kamsack, Saskatchewan in 1958, he came by his interest naturally—his father had spent 20 years as an RCAF photographer. Grant got his licence at the Calgary Flying Club in 1973; his first job was instructing at Castlegar. From 1977-80 he flew the 185, Beaver, Beech 18 and Norseman for Walsten Air Service in Kenora. In December 1980 he joined North Coast Air Services, where he flew the last working Fairchild Husky, CF-EIM. Like most professional Husky pilots, Webb found the plane a good performer, recalling: "With a load the Husky climbed initially at 2,000 feet a minute", but he criticized its small fin and rudder: "You could get washout in turns, so you had to keep sharp on the rudder pedals."

In 1986 Webb moved to La Ronge to fly the Beaver, Otter and Twin Otter on mining work. Next season he flew Twin Otters for TPA, then

A CMA Beech 1900C and an Air BC BAe146 at Terrace on April 2, 1992. Then, Beech 1900 pilots Rob Johnson and Kent McRae at Smithers after flying in from Terrace, a 20-minute hop. (Larry Milberry)

was promoted to the Bristol Freighter, flying tough 20-day shifts on the Bronson Creek-Wrangell run. In January 1989 Webb was hired by CMA and settled in Smithers. His main job was flying DC-3s, mostly C-FAAM, a 27,500-hour ex-Austin Airways freighter. At 16,900 pounds empty and 26,900 all-up, it could haul 7,600 into Bronson Creek. Webb loved the DC-3, which he found nicer to fly than the Bristol. He also flew the Beech 18 and Turbo Otter. The latter became a favourite, except for being slow on takeoff. After it reached 80 knots, Webb found it a joy—fast and able to get high enough to clear terrain and the rough mountain air. It carried a hefty payload—as much as 2,900 lbs. "This airplane walks on a Twin Otter," is how he summed up the Turbo Otter.

Besides his love of flying, Grant Webb became an avid photographer, airplanes being his favourite subject. He started with 35mm cameras, but branched into larger format. Day or night, in clear weather or bad, he was out with his cameras. He lay the groundwork for a photography business, and his framed enlargements became popular in offices and airport terminals. In January 1993 Webb was busy hauling gold concentrate to

Grant Webb at Smithers, BC in 1992 during his Central Mountain days. In the second view Grant is busy aboard DC-3 AAM strapping down some heavy mining gear. (Larry Milberry)

Wrangell, a trip he had made hundreds of times. On January 13 he had two pallets of concentrate and some empty fuel drums on AAM for the day's first flight. The weather was good, but moderate to severe turbulence and wind shear below 4,000 feet was forecast for the Iskut River valley. After takeoff AAM climbed on course, turning steeply through 270 degrees. As it headed back across the airstrip at about 800 feet above ground AAM rolled suddenly to port, and went straight into the frozen Iskut River. Webb and co-pilot Manfred Ernest Harrichhausen were killed. No pre-impact failure or system malfunction could be determined, but the plane was so badly damaged that nothing could be certainly concluded. AAM was 1,700 pounds under maximum permitted weight. The Transportation Safety Board determined that it would have had a stall speed of 78 knots while in its steep (45°) turn. With climb power AAM would have stalled as low as 68 knots in a tight turn due to increased propeller thrust. If the aircraft had a slightly forward C of G, it would have stalled gently and with adequate warning. With C of G at limits (as was the case on this flight) the stall would have been harsh, with an ensuing rapid roll. For recovery this would have required an immediate dive to gain speed. Otherwise the plane would fall uncontrollably and recovery would require more altitude than was available.

Campbell River

Long a float plane haven, Campbell River (1997 pop. 21,500) is about half way up Vancouver Island on the Island Highway. The busy fishing, forestry and tourist centre has seen a great deal of flying from HS-2Ls to Wacos to Huskys, Beech 18s and Twin Otters. In 1959 the town opened an airport and PWA began DC-3 service. Today, flights to Campbell River arrive at a modern airport with a 5,000-foot runway. There is a modern terminal and flight services station, daily skeds, corporate visitors (many US fishermen), a flying club, VIH helicopter base, private planes and Air Cadet squadron. Any visitor in 1995 would have noticed the local operators' signs displayed around the terminal— Armstrong Air, Timberline Air, Western Straits Air, VIA. Another sign (1993) listed fees for airport services— turbo fuel $.68L, landing fees e.g. $2.01 for jets up to 21,000 kg, and parking fees e.g. $9.84 per day for a 10,000-30,000-pound plane. Airport manager Dave Burns noted how Campbell River passenger and aircraft move-

The terminal was quiet when this photo of Campbell River airport was taken on August 10, 1993. Commercial, corporate and private aircraft are based here, and many sked and charter flights come and go daily, although the spit records far more aircraft movements. (Larry Milberry)

about the Short, they winced as only pilots can do who long for promotion, or have been bumped from larger equipment.

For the summer of 1993 Air BC and CRA schedules indicated two dozen weekly Dash 8s and Short 360s from Campbell River. Another regular was CRA's 48-seat CV580 C-FKFW, dedicated to fly-in fishing. August 12 it landed at Campbell River as F5212 with 40 fishermen for Painter's Lodge. As they drove off in Painter's big highway coach, another 40 boarded for Vancouver to connect to cities across the US. The baggage handlers were busy for 30 minutes loading luggage, gear and boxes of frozen salmon. Pilots Randy Harris and Lorne Schults were praiseworthy of their old Convair, the last in the fleet, and regretted that CRA soon would terminate it. Meanwhile, Air BC F1589 landed and let off 18 passengers. Another 18 boarded, mostly Japanese tourists, and the Dash 8 flew to Comox and Vancouver. Later in the day KFW was seen at Vancouver loading fishermen for Sandspit.

Besides airport activities most local Campbell River traffic originated on "The Spit", a sand bar at the mouth of the Campbell River owned by forest industry giant Fletcher Challenge. Bob Langdon, who founded Island Airlines in 1959, first promoted the spit for seaplanes. BC Airlines and others followed. Within a decade, the spit was reputedly the busiest float base in the world. It also was home to Westmin Resources Discovery Terminals from where concentrates from its Buttle Lake base metals mine were shipped. While the float operators and some tug and salvage outfits lined the estuary side of the spit, a trailer park and camp ground occupied the Strait of Georgia shore.

Vancouver Island Air

Vancouver Island Air ("The little airline that's big on service") was the main spit operator in the late 1990s. Larry Langford founded VIA in 1985 after flying the coast for Gulf Air, Trans Mountain and Island Air. He began with a Cessna 185 and a Beaver, but by 1993 had two Beech 18s on floats, two Beavers and two Ce.180s. At peak season he employed nine pilots.

While the Twin Otter naturally held much of the multi-engine market on the coast, other types held their own. On wheels, the Aztec, Navajo and Islander seemed irreplaceable. For a good float performer, the Beech 18 remained a solid bet as far as Langford was concerned.

ments reflected the region's ups and downs: 1985—34,542/36,401; 1987—71,468/46,199; 1989—85,824/50,519; 1992—64,153/38,026. Influences were many, one that always was monitored by US fishermen being the level of the Canadian dollar.

A typical flight to Campbell River was Air BC F1589 of April 4, 1992—a 37-seat Dash 8-102 from Vancouver. The pleasant half-hour trip with 27 passengers was conducted by Capt Curt Horning and FO Garrett Smith. Ca-

nadian Regional Airlines also served Campbell River in the 1990s, one of its skeds being F1260 (Short 360 C-GTAX) of August 9, 1993. The 36-seater operated with an 75% load factor, not unusual for the busy tourist season. As the flight attendant passed around a tray of packaged cookies and chocolate bars, one passenger reminisced "how they used to offer us great big huge muffins". On a return to YVR on the Short the pilots had their chance to moan. When someone remarked favourably

VIA's dock on August 10, 1993 with its three main types present--Cessna 180, Beaver, Beech 18. Maintenance operator Sealand, VIA's next door neighbour, also is shown. (Larry Milberry)

C-FCSN, bought from Kyro's in Geraldton, Ontario in 1989, was his first. It came with more than 8,000 hours and went straight into the shop for a 7-month refit. This included improvements like flush riveting, a two-piece windscreen and a 400-pound saving in weight. Once at work CSN was good to carry 2,200 pounds at 140 knots.

On April 5, 1992 Peter Killin took off from the spit with CSN at 10:53 for Kingcome Inlet, an Indian reserve 70 nm northeast. Killin was born in 1953, the same year Beech rolled out his plane as RCAF Expeditor No. 1441. He started flying in 1973, did maintenance in Kelowna, got his first flying job with Air

Dogrib in the NWT, then moved to Tofino, flying for Pacific Rim. A dyed-in-the-wool aviation buff with the motto "Buy ugly, sell pretty", Killin had owned 20 planes over the years. At the time he had a Cessna 140 in his car-port as a winter works restoration project. This day Killin took CSN low over the coast mountains in ideal VFR conditions, pointing out various sights including logged-out patches which he had seen start back as healthy forest over his 18 years in the area. After 38 minutes Killin set down in the Kingcome River and taxied to the dock to let off two loggers and an artist wanting to record some local scenes.

Next came a 12-minute hop to the Interna-

tional Forest Products' Eclipse camp (on Simoom Inlet), where Helifor was logging with Vertol 107 C-GHFI. This was a 25-man floating operation with everything based on a 340-foot barge, which moved from inlet to inlet. There were seven pilots for the Vertol and a Hughes 500D. Jeff Briggs was chief pilot, while Gordon Ashcroft headed maintenance. Heli logging is intensive business. In 9 1/2 months in 1991 HFI flew 2,400 hours, 338.3 in one month. On a regular day it did 225 lifts. (In 1993 a new world record in heli logging was set when a Rocky Mountain Helicopters crew in Alaska—four pilots, four mechanics— flew an S-61 2,649.9 hours from April 1992 to April 1993.) Although it mainly logged, HFI was available for odd jobs, and recently had fought fires, set towers, and slung the hulk of an F.27 onto a 10,000-foot glacier near Invermere for filming the movie "Alive". Following lunch in the Eclipse cook house, Pete Killin corralled his passengers and made the four-minute hop to another camp—Scott Cove. Here the logs from the Helifor operation awaited barging out. CSN now headed home, but on the way, the VIA dispatcher called to have it backtrack to pick up two Fletcher Challenge loggers at Chatham Point on the north tip of Quadra Island. By the time CSN finally plunked down inside the spit at Campbell River, it had logged 2.2 hours.

In 1991 Larry Langford bought Beech C-FVIB in California, where it had been with a parachute club. He put it into Dave Nilson's overhaul shop in Campbell River. Nilson had gotten into aviation through Air Cadets, where he earned his pilot's licence. His first job was with the Regina Flying Club, and in 1953 he joined Max Ward, who was operating a Beaver, Otter and Husky. After two years in Yellowknife, Nilson spent two years with de Havilland at Hatfield, then came home to work for Athabaska Airways and Sherritt Gordon Mines, where he got to know Husky CF-EIR. It was an excellent machine. According to Nilson, however, for moving camps the Beaver was better for the heavier loads. EIR was ideal for bulkier items like canoes and tents. It also towed a magnetometer. Nilson later worked with Okanagan and Island Airlines.

In Nilson's shop VIB was given a Hamilton extended nose with a 67-gallon auxiliary gas tank and room for 330 pounds of freight. For floats a pair of Bristol 7850s was lengthened 33 1/2" to become 8500s, and modified with freight lockers. Weight-saving included replacing 150 pounds of plywood floor with 38 pounds of honeycomb composite. The goal was to reach a basic weight of 6,200 pounds and all-up weight of 9,300 (a standard Beech 18 on wheels weighs 6,400 empty and 8,725 all-up). After two years and about $500,000 VIA was ready to fly in September 1993. With all the improvements it seemed like a new plane, so Langford christened it with a new name—Sea Wind. It would cruise at 170 knots, and carry 2,500 pounds or 12 passengers. Langford had done a lot of research into his Sea Wind. He looked beyond VIB to the concept of an amphibious, 12-passenger Beech

Aboard the Helifor barge at Eclipse: the big Vertol logging machine, then the Hughes 500D "runabout" with Beech pilot Peter Killin. In early 1997 Helifor had three Vertols on the BC coast. (Larry Milberry)

Scenes at Vancouver Island Air

(Right) A closer view of VIA's dock, this time on April 5, 1992. (Larry Milberry)

(Below) Beech C-FCSN leaving Campbell River for a trip up the coast, then getting scrubbed down to get off the day's accumulation of salt. (VIA, Larry Milberry)

(Right) The crew that completed a seven-month rebuild of C-FCSN at Nilson Aircraft of Campbell River in 1990: David L. Nilson, Dan York, J. David Nilson, Keith Hamilton (VIA), Roy Sanders (VIA) and Bill Ripley. Nilson Aircraft was typical of small, innovative Canadian aircraft firms. It sought business in any corner. In the early 1990s it did the engineering for VIA's improved Beech 18—the Seawind. In a letter of April 1995 David Nilson noted: "We are busy rebuilding floats for the Beech 18 and marketing our shoulder harness kits for Beech 18s, Beavers and Otters. We are trying to get our Otter seaplane main legs into production. My son just finished some compound skins for the air scoop on a P-51 and I am making an instrument panel for a 1943 Fairchild 24." From December 1994 to August 1995 CSN went through another rebuild at Nilson's. (Larry Langford)

(Above and right) VIA's Beech 18 Seawind was built in Dave Nilson's Campbell River shop. (VIA)

Beech aficionados Larry Langford of VIA (left) and Dave Cummings, who was visiting Campbell River from Portland, Oregon on August 13, 1995 with Beech 18 N42C. (Larry Milberry)

Totem poles in an Indian burial ground near The Spit at Campbell River. Then, contrasting ancient native culture, the huge Elk Falls forest products mill not far away. From its earliest days aviation was intertwined with BC's fundamentals—the sea, the forests and the mountains. (Larry Milberry)

these for now. These darned old crates are always falling apart."

On a trip of April 10, 1992 Frankham took off from Campbell River airport in Turbo Otter C-FEBX for a rough gravel strip on the side of a mountain at Moh Bay, 35 nm away. Mike Farrell, flying Islander C-GYMW, followed with four passengers. EBX, the 38th Otter, was delivered to the RCAF in 1954 as No. 3680. Its first assignment was to No. 121 Communications and Rescue Flight at RCAF Station Sea Island, Vancouver. In 1964 it joined the RCMP as CF-MPK, and in 1972 was sold to EB Exploration of Ottawa as CF-EBX. Two years later it returned to the RCMP, but crashed at Goose Bay in 1981. As bent planes often do, it reappeared and became the second of the Vasar conversion Turbo Otters. In April 1993 it was on Edo 7940 amphibious floats, making it ideal for I and J Logging's many land or water locations.

Sitting in the cockpit of EBX, it was hard to believe that this was an Otter. When Frankham started up, there was no coughing, banging and smoking as with any "self respecting" Otter. Instead came the well-mannered swoosh of a PT6. Being light, EBX took off in moments from R11, climbed quickly and cruised the 20 minutes to Moh Bay at 130 KIAS, 25-30 knots faster than a "steam-powered" Otter. Awaiting

with PT6A-27s, the same as a Twin Otter on Wipline 13000s. The Beech would fly faster, yet burn less fuel than a Twin Otter. Potential buyers were calling. There was the Californian millionaire who liked the idea of an amphibious Turbo 18 for fishing trips. An Alaskan sightseeing operator also called for details. By 1993 the Sea Wind was in the air, but there were delays—after a series of test flights it went back to the shop for a new single-fin tail, similar to that on some Westwinds. It was test flown in this configuration in 1994 and finally entered service.

Western Straits

With the coming of deregulation corporate flight departments could charter their aircraft to earn additional revenue. One such operator in Campbell River was Western Straits Airlines, part of I and J Logging, a company with operations in the Charlottes, Terrace, Bella Bella, etc. In 1993 Lee Frankham was its operations manager, running a Beaver, four single Cessnas, an Islander and a Turbo Otter. A veteran bush and coast pilot, he was steeped in aviation from his childhood—his father, H.D.L. Frankham, had been with Canadian Airways and CPA. Lee started flying in Winnipeg in 1947. He worked for Central Northern Airlines, Echo Exploration, the Hudson's Bay Co. and Riverton Airways in Manitoba, and did a stint in Newfoundland. A fellow pilot remembered Frankham as a practical joker. One trick was asking his passenger as they flew along, "Hey, did you hear that!", as if something was wrong with the plane. Then he'd grope around on the floor and come up with a handful of bolts, springs and odd parts, hand them to the passenger and say, "Here, hold

(Below) A view of Turbo Otter EBX at home base. Western Straits folded late in 1995. Waltair moved into its hangar at Campbell River. (Larry Milberry)

The strip at Moh Bay (left) was carved into the side of a mountain. Then, Frankham's passengers waiting to board with their hefty kitbags. Western Straits' Islander (below), flown this day by Mike Farrell, off-loads fuel and freight at Moh Bay. (Larry Milberry)

CHL's Campbell River base (above) on August 10, 1993, and (below) AStar C-GIWH and Jet Ranger C-GIWC there on August 12. (Right) Base engineer and helicopter history buff Brent Newberry, and base pilot Duncan Handley with Jet Ranger C-FAHU on a local photo flight. (Larry Milberry)

the two planes was a crowd of burly loggers and their hefty kit. The loggers crammed aboard EBX, a few others squeezed into YMW and the aircraft returned to Campbell River. The logs were written up for the day, and showed 14,849.8 flying hours for EBX and 4,928.1 for YMW. A year later, Western Straits had gone the way of many small operators. Due to the economy it had trimmed fleet and staff and was busy sniffing out any business. In July it had EBX on lease to Wagair at Seal Cove to earn a few extra dollars.

Canadian Helicopters Ltd.

In 1993 Campbell River had seven helicopter operators. "There are seven," said one pilot, "but work for three." He probably was right, but the Blackfish, Canadian, E&B, Forest Industries Flying Tankers, Highland, Long Beach, and Vancouver Island helicopter companies all were hanging tough. Canadian and VIH were the veterans. Canadian had opened its base in 1966, when it was Okanagan. Its hangar sat on what old-time Okanagan engineer George Crawshaw had determined was an Indian longhouse. He had been working in Campbell River since day one of the Okanagan base. Also on staff in August 1993 were chief pilot Mark Johnson, pilot Duncan Handley, engineer Brent Newberry and office manager Hope Benavidez. Handley and Newberry were newcomers. Handley, who had started flying in 1978 and was a graduate of the Selkirk College

aviation program, recently had left the air force. He had two tours on Sea Kings with 443 Squadron and one instructing on Jet Rangers at Portage la Prairie. Newberry, a BCIT graduate and diehard helicopter history fan, had been with the company since 1980.

On August 10, 1993 Canadian's base had C-FAHU, a 12,375-hour 1971 Bell 206B and C-FFHS, a 7,361-hour 1980 AS350B. FHS and FHP were the first two Ariel-powered "350s" in Canada (FHP crashed on Christmas day 1992 at Canmore, killing a load of sightseers). The machines worked on a great many jobs. They stocked lakes with rainbow trout fingerlings, supported logging, fought fires, set fires (with the drip-torch system), set poles, harvested seed cones from atop trees, poured cement, flew communications techs around to their sites, did medevacs, and took tourists fishing, hiking and sightseeing. Their gear included Bambi Buckets for fire fighting, buckets for seeding fish, buckets for pouring cement, cables for slinging 150 feet down through trees, air ambulance and marine rescue kits, and overnight spare and tool kits for when a machine was away.

Maintenance

In 1993 Sealand Aviation occupied the old Island Airlines hangar on the spit. Started in 1980 by SAIT graduate Bill Alder, it did inspections, mods and overhauls on a variety of types. Through the 1980s it had years of putting

through an airplane a day, but in the early 1990s business had dropped to one plane a week. With about five employees, Sealand was like many small Canadian maintenance operations. It served commercial operators e.g. Coval Air, but also worked on forest, fish company and private planes. Typical was Beaver C-FPMP owned by C.H. McLean and Sons, a logging firm. PMP was hauled into the hangar the morning of August 10, 1993. Within an hour its wings were off and the mechanics were busy with a detailed periodic inspection. In 1990 Alder and partner Ed Wilcock formed E&B Helicopters to sell, service, operate and train with Robinson R22 helicopters. The tiny R22 was ideal for jobs like moving fallers around balloon logging sites and flying buyers among small log parcels. At $295 an hour it rented for about half the Bell 206B. In late August 1993 E&B brought the first R44 four-seater to BC.

Sealand was known for its Beaver cockpit extension mod. To date it had installed 37 kits, which involved pushing the cockpit back 28". The heavy battery behind the rear bulkhead was replaced by a lightweight unit in the engine compartment. Small maintenance companies like to keep a rebuild or two in the works. On a quiet day someone usually devotes an hour to the current project. In 1993 Sealand had two basket case Beavers, including CF-JBP. Also piled in the back of the hangar was Widgeon N2PS with Lycoming GO-480s. Alder had located it in Jacksonville, where it had been dormant for 12 years. A 1943 model, it had been used last by the State of Florida on park duties.

Across the river from Sealand at Dave Nilson's, when the lads were not working on the

Trans-Provincial's Otter RNO in for a 100-hour check at Sealand; then, a pair of E&B R22s. (Larry Milberry)

Seawind, they were picking away at a Beaver, and an Air Cadet Cessna L-19. At Coval's hangar two Cessna 185 wrecks were rising from the rubble, while derelict Otter QEI waited attention. Eventually, all such projects turn to profit, when the finished product takes to the air.

Harbour Air

In the mid-1990s Harbour Air, owned by Greg McDougall, was Vancouver's busiest float operator. Based on the middle arm of the Fraser River along Vancouver International Airport's historic South Side, in the summer of 1992 it had eight Beavers, seven Ce.185s, a Turbo-Otter and six Twin Otters. It also had subsidiary Jet West with two King Air 90s; 16 pilots were on the payroll. The Cessnas were mainly for forestry, the larger planes for tourism. The Twin Otters were chartered each summer from Kenn Borek, who took them back in the fall for Arctic, Antarctic and other offshore charters. The usual pattern with the Twin Otters was that two would show up in April, build to six, then peter out in late September until all were gone a month later.

Operations manager in 1992 was Peter Evans. He had started in the bush with Barney Lamm's OCA in 1974, and moved to the coast in 1979. Chief pilot Bill Pennings had flown Meteor fighters in the Dutch Air Force in his formative years. He learned about Canada from his mates, who had trained there in the postwar NATO scheme. He emigrated in 1959, finding work in Vancouver as an electronics tech. In 1965 he got back into flying and spent seven years in Whitehorse. In 1972 he joined Air West and stayed through the Air BC transformation.

For 1992 Harbour Air carried about 25,000 passengers to sport fishing camps with names like Hoeya Hilton, Sonora Island Lodge, Joe's Salmon Lodge, and King Salmon—17 camps in all. The pace was hectic with flights from sunrise and to sunset. Sixty to 70 trips a day were standard, and a one-day high of 84 was recorded. Throughout any summer's day the Harbour Air radio room buzzed with 70-80 transmissions an hour. A typical trip was on April 13, 1992 when 22 fishermen, mainly from the US, flew 160 sm up the coast for thousand-dollar-a-day fishing at the Pacific Springs Lodge, a floating operation on Knight Inlet. The men and their gear were loaded aboard Twin Otter C-GGEN and Turbo Otter C-GUTW. In earlier years fishermen went from Vancouver aboard Navajos to Port Hardy, thence to the camps in float planes. Any party comprised a cross section from teens to seniors. One fellow this day hailed from Vacaville, California. He was an ex-USAF pilot who had flown everything from P-40s to B-29s, B-52s, F-101s and FB-111s.

The sport fishing crowd appreciated Harbour Air's direct service, but not the added expense of the pricey Twin Otter. This led the company to evaluate the PT6A-135 turbine Otter offered by Aero Flite of Vancouver. Simple math showed that the Turbo Otter had potential: the Twin Otter burned 600 gallons hourly; the Turbo Otter, 370. The Twin Otter required two pilots, the Turbo Otter, one. The planes carried about the same load, but on a long haul the Twin Otter had to stop for fuel at Port Neville, 155 sm en route. Although slower, the Turbo Otter could go nonstop. First

(Above) Harbour Air's Vancouver base on April 18, 1991. (Larry Milberry)

(Left) Passengers mill around the terminal awaiting a flight to the North Pacific Springs fishing lodge. (Larry Milberry)

(Right) An Otter of Harbour Air poses along the coast for photographer Paul Duffy.

A Harbour Air Twin Otter seen at Vancouver by Christopher Buckley on August 15, 1991.

A Kenn Borek Twin Otter 100 passing through Prestwick in May 1996. Such aircraft fly for Harbour Air in summer and Kenn Borek in winter, but also do overseas leases. In 1996, for example, several Kenn Borek Twin Otters were listed to Maldivian Air Taxi in the Indian Ocean. This example was originally Jamaican Defence Force JDFT-1 in 1967, then C-GIAW of Air West. It later was alternately Kenn Borek and Harbour Air. Here it was returning from a Maldive lease, where it had been 8Q-IWW. (Wilf White)

developed by Vasar of Bellingham, Washington, the Turbo Otter enjoyed good initial sales. By 1992 about a dozen had been sold to customers from Alaska to Quebec and Sweden. The first Aero Flite conversion was C-FEBX.

When TPA folded in 1993, Harbour Air took over its Seal Cove float base so, in mid-1993, it had about 40 each of airplanes and pilots and was straining with the sudden pressures of bigness. Like any coast operator it was earning most of its revenue from tourism, unlike former days when fishing, forestry and mining were big. BC government policies in the early 1990s had killed much of mining and greatly reduced forestry. Meanwhile, commercial air carriers lost business as fishing and logging firms, trying to trim overhead, were flying their own Cessnas and Beavers.

On August 9, 1993 Bill Pennings and Mike O'Hanlon of Harbour Air were in Campbell River, weathered out and awaiting the green light from radio operators up the coast that the clouds were lifting. O'Hanlon had a unique background for a coast pilot. Today he had a Cessna 185, but here was a pilot who had

started in the RCAF in 1961, flown the Neptune, Argus, and CF-104, and finished in 1971 instructing on Tutors. That year he started his civil career and worked his way along like anyone else, flying the Norseman and Navajo before joining Harbour Air in 1987. At the dock this day were four orange Beavers from NW Aviation of Renton. They too had been coming up the coast, heading for River's Inlet Resort with a crowd of Shriners eager to wet their lines. Pennings and some other old timers were talking about hiring young pilots. There were few jobs, but Pennings had hired a half dozen dock hands—young fellows with new commercial tickets. He was following the tried-and-true method of putting them to work loading planes, gassing them, tending to passengers, etc. The sprogs would get a bit of flying, some-

times taking a 180 on a local trip. Mainly, however, they slugged it out on the dock for low wages. Pennings would see who had the right stuff—who would work the hardest and keep a smile on his face. The following year, that's the one who would have a flying job.

Tofino Air

The west coast of Vancouver Island is a natural treasure. Each summer thousands of tourists travel there to enjoy the quaintness of Tofino and Ucluelet, the incomparable beauty of Pacific Rim National Park with its miles of beaches, and the hulking grey whales frolicking offshore. Most visitors reach Tofino along tortuous Highway 4 from Port Alberni, alternately traversing stunning mountain vistas and clearcuts that resemble moonscapes. Tofino, named for an 18th century Spanish hydrographer, and Ucluelet, meaning "people with a safe harbour", had a combined 1997 population of 3,000. Up and down the coast are native fishing communities and logging camps, from San Mateo to Nootka to Kyuquot and remote Cape Scott. Aviation came late to the area, although there were spurts of activity like the gold rush at Zeballos. What really brought the airplane was WWII, when the RCAF established coastal patrol and fighter squadron bases on Vancouver Island. A squadron of P-40 Kittyhawks moved to Tofino, and seaplane bases appeared at places like Ucluelet and

Coal Harbour. Sharks, Stranraers and PBYs became common as they watched for any Japanese incursion. The old Tofino airport remained active in the 1990s. Tourists would arrive in their own planes, or aboard Convairs, to be ferried on in floatplanes to fish for prize salmon. In town, floatplanes bobbed at the dock, and the occasional chopper popped in at the Coastguard heliport. The local operator was Tofino Air, a small company run by partners Doug Banks and Gary Richards. In 1993 it had Ce.180s C-GHBX, HZR and IDX, and

Signs of clearcut logging assault the eye along the road from Port Alberni to Tofino; but humanity demands that nature give up its resources. The trees will come back—the old-time pilots have watched regeneration progress over the decades that they have flown the coast. (Larry Milberry)

Gary Richards heading up to Mooyah Bay in Tofino Air Ce.180 C-FIDX on April 9, 1992. IDX was no CF-100, but a diehard aviator is as happy in one type as the next. Then, the float base on a lazy day at Tofino. (Larry Milberry)

The leftovers from Gary Richards' "oyster special". His landing drew a good crowd. (Bruce McKim/Seattle Times)

Beaver C-FGCY. It mainly served Indian villages, logging/fishing camps and lodges on the coast. In peak season it employed five pilots, an engineer, two dockboys and four office staff. Richards was chief pilot. He had started flying in the RCAF in 1955 and was posted to CF-100s with 433 Squadron in North Bay. The "Clunk", as they called the CF-100, was the finest all-weather interceptor of the era; for a young lad it was a plum posting. Unlike those flying Sabres on day VFR operations, the Clunks went out day and night in fog, rain, snow or sleet. Sometimes mechanical snags

added to the challenge; that is how Gary Richards had his first big flying adventure. While patrolling at 43,000 feet, his oxygen system failed. He passed out from hypoxia, but was lucky. His navigator, F/O C.M. Alexander, realized that something was wrong. He began talking to Richards to rouse him, then kept him half awake. Groggy, Richards almost flew them into the ground, but was able to respond to his nav's commands, pull out and land.

As a hot-shot fighter pilot Richards learned to squeeze the most out of flying. That included trying a few nutty things—one day he

rolled a CF-100 at 52,000 feet! In 1964, as Richards put it, "I burned all my suits and ties and took a job with Island Airlines." For the next three decades he flew on a variety of jobs. There was a summer on a Beaver in Ketchikan, another on a tri-gear Beech 18 dropping smoke jumpers in the Yukon. He flew Twin Otters for Kenn Borek, water-bombed with Avengers for Conair, sprayed budworms in New Brunswick and crop-dusted in PEI. This was not work for a kid gloves pilot, and there were hairy moments. One day in 1974 Richards had to jump from his water bombing Avenger after ploughing through a tree south of Cranbrook. "I've never done this before," he thought at the time, "but I've seen it in the movies." Below he spotted a rescue helicopter waiting, something that prompted another fleeting thought: "I'm going to be the first guy in history to get chewed up in the rotor blades of the helicopter coming to rescue him!"

Twice Richards had some excitement in the Husky. While taking a load of Indian kids up the coast from Tofino, a rear door came open, impeding flap operations. He trimmed the aircraft, hurried back to close the door, skidded on a floor plastered with puke (the school kids always threw up) and scurried back to his seat. Another time he had a load of mourners on a wreath-laying flight. At some point a passenger shouted, "This is the spot!" All 10 mourners shifted to the back to have a look. The Husky pitched up. Terrified, Richards poured on the coal to keep the wreath-laying from turning into a mass funeral at sea!

In the 1960s Richards started flying fresh oysters from Vancouver Island to Seattle. He got hold of a clapped out Beech 18 for $10,000 (CF-RQK, ex-RCAF HB135). On a December 5, 1968 flight from Victoria he ran out of gas on approach to Boeing Field, crashed onto a road and tore through a fence. Rescue vehicles rushed to the scene. The firemen were appalled at the gory sight— a burning plane, smashed cars, oysters splattered everywhere, the pilot probably somewhere underneath them. But Richards was safe. He was working the crowd and enjoying the excitement. He chuckled when he overheard an ambulance attendants lament, "Well, boys, we may as well call the coroner. Nobody could have survived this one."

Founds on the Rebound

During his career Gary Richards put 1,500 hours on the FBA-2C. He found it an excellent bush plane, using the same floats as a Ce.180, but performing better. He appreciated the ease of loading stretchers and fuel drums, and liked the handy flap lever (which other pilots moaned about). To him the Found was "a good little four-door sedan... a mini-Beaver." In 1997 one small BC operator owned several FBA-2Cs and was working them hard in the interior. This was John Blackwell's Fawnie Mountain Outfitters of Nimpo Lake in the Anahim Lake region. CF-RXI and SOO were busy in 1997 and OZU and RXD were available for rebuild or spares. John Vandene, one of Blackwell's pilots, wrote of the Founds:

"They are used hard every day of the six-month season moving people and equipment. In late fall one is stored for the winter as the second is readied for wheel-ski operations. Due to the innovative main gear design, the Found can be changed from floats to wheels in less than two hours." Blackwell discovered that the Found required little maintenance and most repairs were made easily. In 1992 N.K. "Bud" Found visited Blackwell and flew one of his old machines. His post-flight notes included: "Although supercharging the engine would be beneficial for high elevation take-offs on hot days, CF-SOO demonstrated good float performance with flat, calm conditions at 3,600 feet ASL. At gross weight with five people and baggage it took off in about 30 seconds, then climbed to 6,000 feet at 500 fpm using cruise power. Early morning cool air helped this performance." This nostalgic flight may have provided the spark that led Bud Found to reintroduce the FBA-2 in 1997.

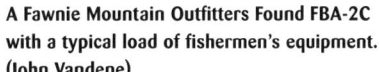

A Fawnie Mountain Outfitters Found FBA-2C with a typical load of fishermen's equipment. (John Vandene)

Prince Rupert: Of Ships and Planes

An hour and 35 minutes north of Vancouver on Air BC's BAe146 is Prince Rupert. Incorporated in 1910 as terminus of the Grand Trunk Pacific Railroad, the city of about 16,000 (double that with its hinterland) is important for fish processing and forestry. Like most such coastal centres it hugs a narrow ledge from where mountains rise steeply behind. Nearby Ridley Island and Port Edward ship prairie wheat and BC coal overseas. Prince Rupert has a busy downtown with hotels, office buildings, malls, wharves, a museum and canneries shoulder to shoulder. Out by the float base at Seal Cove, Sullie's pub was a favourite hangout for the aviation set, but so was one of the better downtown spots, great for after-work chit-chat and a few beers. One evening a local pilot, smitten by one of the waitresses, called her on his cell phone to say he'd soon be in. He asked that she go onto the patio. She did, just in time to have the pilot roar by so close that he could have reached out for a cool one! It was all the more fun when he showed up to be greeted by four civil aviation inspectors from Transport Canada. They had been on the patio as the Beaver skimmed by. Some negotiating followed; TC decided to let the infraction pass in exchange for a round of beers. In a not-so-exclusive hotel down the street one didn't find any aviators. Its clientele was announced by the stickers slapped on the elevator: "Heat and Frost Insulators, Local 118", "Brotherhood of Plumbers and Pipefitters", "Association of Injured Motorcyclists", "International Union of Elevator Constructors". This also was where aviation writers on a budget hung out.

Aviation did not come early to remote Prince Rupert. One episode began in August

(Left) An August 1993 overview of Prince Rupert's famous Seal Cove, with the seaplane base in the centre and hangars beyond. Then, a closer view of the seaplane facilities. (Larry Milberry)

(Below) The Panamanian freighter *Rubin Iris* ready to load unfinished logs at Prince Rupert August 2, 1993; then the city's forest product, grain and coal shipping facilities at Ridley Island. As these prosper, so does the local aviation scene. (Larry Milberry)

1921 when Americans Clarence O. Prest and Morton Bach attempted a Mexico-to-Siberia flight in their JN-4 "Polar Bear". They reached Hazelton, but Prest did not like the fields around there. He shipped his plane by rail to Prince Rupert for his next take off. The city fathers convinced Prest to put on a flying display. He used a sport field for takeoff, but it was too short for a safe landing. Some local Rube Goldberg types solved this problem by erecting a seine net at one end of the field to catch the Polar Bear when it landed. The plan worked, but on one landing a propeller was broken. A replacement had to come from Seattle and Prest was ready to depart on September 22. That day, sad to say, a sudden wind arose. It snatched the Polar Bear and smashed it to pieces.

In April 6, 1924 six US Army Douglas seaplanes reached Prince Rupert on their way to Alaska and around the world in five months. A few weeks later the rescued crew of a Vickers, downed off Alaska on its round-the-word attempt, arrived by steamer. The German aviator von Gronau, who seemed to be everywhere in the Thirties, landed his Dornier Wal in August 1932, while working his way around the planet. Between the wars Prince Rupert was visited by many government flights concerning fisheries, Indian settlements, etc. During WWII it thrived as a resupply base for the North Pacific theatre. The RCAF built a seaplane base at adjacent Seal Cove from where Sharks, Stranraers and Cansos patrolled.

With the 1950s aviation began thriving up the coast. QCA and PWA ran bases from Bella Coola north. The industry never looked back, although it was not always clear sailing. In mid-1993 Prince Rupert enjoyed 32 weekly Air BC BAe146 and CAIL 737 flights. Central Mountain offered daily Beech 1900s from Vancouver. All these flights operated into Digby Island, a short distance offshore. From the airport passengers travelled by bus and ferry to downtown Prince Rupert. A typical Prince Rupert flight was Air BC F1599 of August 1, 1993. The aircraft was C-FBAB with Capt Kurt Miller and FO Tim Gale. Aboard were 49 passengers. At Vancouver BAB taxied behind KLM's 747 PH-BUW and American's MD80 N9081, then waited for a Horizon Dash 8 and CAIL 737 to land before getting off R26 at 132 knots. It soon was parallelling the Coast Range at FL310 and ticking off the 423 nm along track.

As he had a year earlier, Capt Miller praised the pilot-friendly BAe146 as all aboard enjoyed a perfect evening. FO Gale chatted about his flying career—another story of grinding up the unpredictable ladder from PPL in 1980 through a string of bush and coast carriers to that big break in 1988 when Air BC hired him from Harbour Air. The fellows also discussed that ever present topic of aviation safety. The recent crashes of Wilderness Air and Wagair Beavers was big news—12 lives lost in what looked like two avoidable crashes. Eighty miles back F1599 started down and landed on R31 at 1754. Those for Prince Rupert deplaned to wait in the terminal till called for the bus. Soon Miller and Gale had their passengers

Prince Rupert airport on Digby Island in a familiar scene—R31-13 socked in, waiting for the morning fog to burn off or blow away. Then, boarding the bus at the airport for the ferry to town. (Larry Milberry)

away for the hop to Terrace from where they turned south for Vancouver. A nice evening's work in about three hours of flying.

The Terrace-Vancouver route is entirely over mountains. For the Air BC run of August 9 the pilots were Capt Jens Thomassen and FO Graham Goode. They arrived at Terrace from the 89-nm hop from Prince Rupert with 34 passengers and quickly got away with a light load of 20. Thomassen was another former Air Cadet who went into commercial aviation flying the Otter and Twin Otter at Northern Thunderbird. Goode had trained in 1972 at Skyways in Langley. His first job was as a dock hand for Airspan. Here he met a number of furloughed CP Air pilots sweating through another recession. In 1974 O.J. Wieben hired Goode as a helper, then let him fly the 180 at Pickle Lake. Goode moved to Big Trout Air, then to Bearskin for six years. There he graduated to the Beech 99 and ST-27. Finally, he spent 2 1/2 years at City Express before signing on with Air BC in 1987.

At 1941 hours, as F1599 headed down to Vancouver, an interesting call came over the radio: "Russian Air Force 788", was the fellow's call sign (spoken without accent). "Estimating Vancouver at three five", the voice added. 788 now identified himself as an IL-76 transport shepherding six SU-27 fighters of the Russian Knights demonstration team. He wanted permission to overfly Stanley Park before landing in Abbotsford for the annual airshow. F1599 was soon on a right hand visual approach for Vancouver's R26. It landed at 2010. Just as the passengers deplaned, the IL-76 with its gaggle of fighters passed low towards downtown. Next day it was learned that the voice of "788" was that of Ian Struthers, ex-CF-18 pilot, Canada 3000 captain and "Flying Events" man with the Abbotsford Airshow. He was aboard 788 assisting the Russians, who were unfamiliar with Canadian ATC.

North Coast Air Services

Although Digby Island is a vital link for this region, the real action at Prince Rupert is at the float base across the channel at Seal Cove. The oldest operator there was North Coast Air

Services, founded after the war by Jack Anderson. Born in 1921, he grew up on a farm in Virden, Manitoba. He joined the RCAF in 1940 and trained at Virden and Gimli on Tiger Moths and Anson IIs before going overseas. Following a bomber tour he was posted to the East Coast on PBYs. After the war he worked for Central Northern Airlines, did some survey flying in BC, and budworm spraying in New Brunswick. He returned to BC in 1953, where QCA hired him to fly Stranraers. Much of his work was supporting mining from Prince Rupert. Anderson liked the Stranraer. "It was all lift and cruised at 135. It was ideal in tight spots and carried a good load," he recalled.

On his first day in Prince Rupert Anderson decided that he liked the place. He stayed with

North Coast's Seal Cove office; then, company owner Jack Anderson in August 1993. He died in late 1996. (Larry Milberry)

The rare Twin Bee occasionally appeared in BC, including at North Coast. This example was visiting Vancouver in June 1982. (Larry Milberry)

Jack Anderson's dormant hovercraft CH-NCH at Seal Cove on August 2, 1993. In 1997 it was in a Prince Rupert scrapyard. (Larry Milberry)

North Coast C-46 FNC at Terrace in July 1979, then its Bristol. These big freighters were unsuitable to North Coast's daily needs, but flew on special contracts. (Les Corness, NA/McNulty Col. PA191767)

buy a couple of C-46s. "These were a nuisance", he said, "since you had to chase all over the country finding work for them."

In early 1970 Anderson purchased a DC-3 in the UK for £7,500 and ferried it home as CF-CQT. It had served in the South Pacific with the US military, then Qantas and Trans-Australia Airlines as VH-AFA. In 1957 it was sold to West African Airways as VR-NCO. In 1964 it went to Air Turas of Dublin, then worked in the UK for Handley Page supporting Jetstream certification. When Handley Page disappeared in 1970, Anderson bought the DC-3. It worked from Prince Rupert, but later supported drilling in the Arctic and hauled fish in Manitoba. In the mid-1980s it was freighting with Air Manitoba. With more than 50,000 air-frame hours, it was retired in 1990.

Along the way Jack Anderson also operated the Twin Bee, Widgeon and Mallard. He bought the Twin Bee for Masset in the Queen Charlottes, but found it difficult keeping staff out there and gave up the idea. On March 5, 1974 North Coast's Mallard HPA crashed near Prince Rupert. A DOT summary of the event noted: "While in a turn shortly after takeoff, the aircraft struck the side of a hill. The engines were producing power and no mechanical fault was found." Six of nine aboard were killed, including Jack Anderson's son Bob, the pilot.

In 1986 Anderson gambled on another idea. He purchased an SR6N hovercraft. Powered by a 1,500-shp Rolls-Royce Gnome turbine, it could carry 38 passengers or 10 tons of freight, but the concept was not right. After all, it was hard enough finding 4-5 passengers for a Beaver, and the SR6N was expensive to run. After two seasons Anderson laid it up and started looking for a buyer. In 1993 he sold out to Clarence Hogan of Tsayta Air in Fort St. James. He was ready to retire and not at all regretful about it. He recalled how great it had been in years gone by, when the fishing fleets bought every flying hour they could scrounge. Fishermen, spare parts and supplies were flown to and from the boats. Planes were used to spot herring. But now the fishing seasons were shorter. Coast fishermen no longer worked the Bering Sea or the waters far west of the Queen Charlottes. Boats rarely broke down any more. If they did, companies delivered the parts with their own Cessnas. As to forestry, Anderson recalled that in 1972, under Social Credit premier Dave Barrett's government, North Coast served 18 logging camps. Under the NDP a few years later that figure fell to zero, when the NPD increased stumpage on spruce to a higher price than the loggers could get on the market. Companies shut down, some seeking better opportunities in countries like Malaysia.

Anderson also commented about the absence of cruise ships in Prince Rupert. He felt that the city had driven them away by over-

QCA until PWA bought it, then set up his own company in 1957—he put $500 down on a new $15,000 Cessna 180, financing the rest. Business was good. When Ottawa made it easier for small operators to get licences for heavier equipment, Anderson bought Husky CF-EIM from Husky Aircraft in Vancouver. He claimed: "There wasn't anything you could dislike about that airplane. It was great with long loads and ideal for offloading at a beach. I preferred the Husky to the Otter. You could do almost the same amount of work, but use only 18 gallons of fuel an hour. The Otter guzzled 30." EIM sank at Prince Rupert in February 1985. Pilot Grant Webb went down in scuba gear and

got a cable on the wreck. It was raised and the leftovers were donated to the Canadian Museum of Flight and Transportation.

When mining was booming in BC's Stikine region and the Yukon in the mid-1960s, Jack Anderson established a multi-engine operation. His first type was Scottish Aviation Twin Pioneer CF-STX. Next came Bristol Freighter CF-UME. Anderson ferried these from the UK. His policy was to shop for the ideal plane for a job. The trouble was that once the job ended he had to peddle the airplane. This wasn't so bad with his Bristol Freighter—it went through the ice to the bottom of Baker Lake north of Whitehorse. In the 1970s Anderson teamed with Pete Lazarenko to

699

charging for docking. They went instead at Ketchikan. There the local airlines ran a booming sightseeing business with cruise ship passengers. Another big complaint from Anderson was the way Transport Canada demanded that small companies conform to the same standards as those demanded of Air Canada, e.g. just like Air Canada, the small outfits were expected to have bilingual safety information cards in their little Cessnas. No matter that none of their customers spoke French. (Of course, this really wasn't TC's fault—it was under orders from those in Ottawa enforcing things like bilingualism to the future detriment of Canada.) As far as Jack Anderson was concerned, he had had his best years in aviation.

Business wasn't much fun any more, and Transport Canada had lifted his operating certificate. It was a good time to retire.

Prince Rupert in the Nineties

There's a predictable answer when asking a small air operator, "So, how's business?" "You'd never want to be in aviation," is the usual reply. The questioner might (but probably wouldn't) then ask, "Tell me, why have you stayed in such a lousy business for 30 or 40 years?" The answer is simple—the operator loves aviation and has made a very good living in it. Of course, times can be bad and operators can hurt, but most survive and the economy moves to the next upswing. In the 1980s and

early 1990s Canada was suffering. Even so, on August 2, 1993 there were 16 planes at the Seal Cove dock at 7:30 in the morning. The sun was rising and starting to beat back a light fog. Eight Beavers, seven Cessnas and a Twin Otter soon were dispersing to coastal and island centres. Adjacent to the docks is a small peninsula with a hangar for the Coast Guard's S-61, one for the RCMP, whose famous Goose CF-MPG still was in use, one for VIH's helicopters, and the former TPA hangar, built by the RCAF in WWII. The S-61 always was busy in lighthouse support and maintenance of navigation aids. The Goose flew on law enforcement. VIH catered to any need, and kept Bell 222 ambulance C-GVUT on 24-hour standby. Jet Rangers C-GTPH, TWP and VIT did charters. On August 2 VIH's Hughes 500 C-FQHB was also on the ramp. Rick Woodward of Tsayta Aviation was one of the pilots warming up his Beaver as the sun rose. He had started at Pitt Meadows in the mid-1970s. He went to Ontario for his first job—flying Cessnas for Bearskin—then moved to Kyro's, La Ronge and Tyee before joining TPA in 1988. On March 19, 1993 he flew TPA's last sked—Douglas Channel return in Otter C-FKLC. When he landed, he was out of work, but Harbour Air took over, Wagair moved in to challenge, and Tsayta was in the midst of acquiring North Coast. There were jobs around. Also on the dock early August 2 was sprog pilot Mark O'Brien. He had started in Nakina with Lunenberger Air Service. This day he had

(Below) Harbour Air Twin Otter C-FQHC at Seal Cove on August 2, 1993. By 1997 the Twin Otter, Turbo Otter and Beaver were the main commercial types here. (Larry Milberry)

(Above) Early morning at Seal Cove. One by one these floatplanes would head out to start the day's work. One of the first tasks (below) is pumping out the floats. (Larry Milberry)

(Right) This VIH Bell 222 air ambulance was based at Seal Cove in the early 1990s. (Larry Milberry)

an early charter to Masset on Graham Island in Ce.180 TWP. He followed the 220 radial from Seal Cove, cruising at 125 mph 1,200 feet over the water. It was a quick turn-around—a rancher from Burns Lake boarded after a fare-well embrace with his forest-worker girlfriend. After lunch O'Brien flew a charter to Queen Charlotte City to pick up a Salish lady visiting from Vancouver. She talked about the Haida of the Queen Charlottes and how they were pros-pering by hard work in the local fishery. Their well-kept draggers crisscrossed the waters as TWP buzzed back to Seal Cove. Later in the day O'Brien flew what he called "the Port Simpson Express", a 10-minute, 21-nm sked from Seal Cove to the fishing centre of Port Simpson. Two passengers and a load of freight went out, one passenger came back.

Wagair was a busy coast airline in the 1990s. Formed in 1982 by pilot Al Beaulieu, mechanic Chris Kent and fisherman Vivian Wilson, in 1993 it had two Beech 99s and four Navajos, a Goose, two Otters, five Beavers and a Cessna 185. The twins generally worked from Vancouver to points like Bella Bella. The Otters at Prince Rupert and Bella Bella often were preferable to Beavers when villagers turned up with big loads of groceries, which was often enough. They also suited logging camp moves, and heavily-equipped hunting parties. Wagair's associate company Pacific Aircraft and Salvage of Vancouver specialized in Beaver mods and rebuilds.

Sandy Parker was with Wagair in 1993. He had started at Nanaimo in 1969, then took time for some world travels. He later studied at Simon Fraser University, went to North Coast for seven years, then flew for Powell Air. In 1988 he moved to TPA on the Goose and Twin Otter. On August 2 he made several trips from Prince Rupert with ex-US Army L-20A Beaver C-FWAC. It had the rear bulkhead extension and two comfortable three-seat benches behind the cockpit. The prop had blades shorter than a regular Beaver's, producing less saltwater prop erosion, and less noise. Parker's first trip was with a passenger 40 nm south to Kitkatla, a Haida village thought to be the oldest continu-ously inhabited site on the coast (5,000 years).

(Below) Ketchikan scenes from August 3, 1993: one of the many cruise ships that dock each summer, and some of Ketchikan's dozens of float planes. A string of summer cruise ships boosts the local economy. Many passengers take sight-seeing flights before their boats push off. (Larry Milberry)

Next, Parker took three passengers 79 nm north to Ketchikan on the Alaska Panhandle. Along the way the conversation was full of lively topics. Like most earning a livelihood on the coast, Parker had his views about logging. He disapproved of harsh techniques, where logs are dragged out, ruining topsoil structure. He touted selective logging with balloons or helicopters. Letting forests sit didn't make much sense to anyone on the plane —left alone, trees were destined to fall or burn. In the long run, any logged or burned areas would sport mature growth. This knowledgeable coast pilot had a straightforward view of things, un-like many enviroloonies.

Fishing and its importance to aviation was a natural topic. Someone groused about the po-litically correct weirdos in Ottawa using public funds in their drive to bend and twist the Eng-

Coast pilot Rick Woodward ready for work on an August 1993 morning at Prince Rupert. (Larry Milberry).

A Waglisla (i.e. Wagair) Beech 99 near Vancouver on September 1, 1991. (Henry Tenby)

Sandy Parker piloting his Wagair Beaver to Ketchikan on August 2, 1993; then, coming up to Ketchikan. Sandy was taking some film makers in from Prince Rupert. Note the airport across the channel from town. In 1996 Wagair went out of business after two bad accidents. (Larry Milberry)

lish language out of shape to suit their far-out social agenda. The latest was a scheme to outlaw the word "fisherman". Henceforth, the kooks decreed, the world should have no more fishermen. All would now be "fishers". Coast fishermen of both sexes only chuckled. As WAC approached Ketchikan, low, flat Annette Island appeared on the nose. Fifty years earlier RCAF airmen defending against Japanese invasion had slogged in its muskeg to make a serviceable base for P-40s and Bolingbrokes. Parker now concentrated on his approach and the radio crackled with calls. He landed in the rough channel, the strung-out town to his right. Cruise ships, fishing boats and a couple of dozen bushplanes lined the docks. This was a strong Otter base, indicating a larger market than Seal Cove's. A Scenic Airlines Twin Otter on $300,000 Wipline floats was working the tour market here. Beavers kept coming and going, along with 185s with belly panniers not usually seen in BC. At the airport (cross-channel from downtown) sat a civil Hercules; as WAC took off for home, a 737 was on final. From Ketchikan northward, aviation was king, with float bases to make Seal Cove look tiny.

On a December 4, 1993 take-off from Prince Rupert in Wagair Goose C-FUMG Sandy Parker lost an engine. The plane apparently stalled and crashed. He and one of his four passengers died. UMG, delivered to the US Navy in August 1945 as No. 87751, had been the last Goose built.

Prince George

Located at the junction of the upper Fraser and the Nechako rivers, Prince George is in BC's Fraser-Fort George Regional District. The Hudson's Bay Co. started trading here in 1805 at a post called Fort George. Things remained quiet till the Grand Trunk Pacific Railway came through before WWI and the town became a rail hub. A lumbering boom after WWII brought prosperity. By 1960 Prince George was northeast BC's main urban centre. By 1997, in spite of the downturn in forestry, it had a regional population of 75,000 and still had more than a dozen lumber and pulp mills. Its busy rail yards served forestry and coal mining. Its main roads were Hwy 16 running east-west from Edmonton, and across to Prince Rupert; and Hwy 97 leading north to Dawson Creek and south to places like Quesnel and Williams Lake.

Isolated and deep in the mountains, Prince George saw few airplanes in the early days, but August 2, 1920 was a red letter day. The US Army's First Alaska Air Expedition came to

With its humble beginnings as an isolated trading post, Prince George made its mark in the forest industry after WWII. Here its mills are seen on August 11, 1995. The view is from an Air BC flight on approach to Runway 15. The city lies beyond. (Larry Milberry)

town —eight men and four D.H.4Bs led by Capt St. Clair Streett. The expedition had begun at Mitchell Field, New York on July 15. Its arrival in Prince George had a tense moment when one machine went on its nose and another crushed a wing. Repairs took until August 13, when the team left for Hazelton. There a farmer had prepared a good strip in one of his fields. The expedition was a success and was back at Mitchell Field on October 20. By the late 1920s Prince George was seeing more of aviation. In May 1929 Western Canada Airways sent Paul Calder and his cameraman there on a government survey. Using Junkers CF-ABK, their work was done by mid-October. Meanwhile, a WCA inspection tour arrived in July with a Fokker VII full of VIPs (including the premier of Manitoba). Later in 1929 Andy Cruickshank was flying Super Universal G-CASQ from Prince George. Soon he was called to the Arctic coast to help search for the MacAlpine expedition. Through the 1930s Canadian Airways and various small carriers

operated from Prince George and Fort St. James, about 65 nm northwest. Pilots like Russ Baker, Sheldon Luck and Grant McConachie tapped the area for every nickel's worth of business. After WWII Baker and Walter Gilbert formed Central BC Airlines and won forestry contracts around Prince George and Fort St. James. Baker later founded PWA and maintained a Prince George VFR base until 1967. Thereafter PWA operated Prince George skeds with Convairs, competing with CP Air DC-6Bs and 737s to Vancouver. In the mid-1990s Prince George enjoyed numerous daily skeds, the main carriers being CAIL, Air BC, CRA and Central Mountain. Local operator Northern Thunderbird Air (NTAir) had its head office here.

A typical sked was Air BC F1813 of August 11, 1995, an 84-seat BAe 146-200 C-FBAV piloted by Capt Bill Walters and FO Doug Williams. It started at Gate 9 in Vancouver with 26 passengers. It was pushed back at 78,000 pounds (15,000 under AUW). Since

Early aviation scenes at Prince George. (Left) The July 1934 visit of US Army Martin B-10s. Under Col. Henry H. Arnold they were on a training exercise from Washington, DC to Fairbanks with stops in Winnipeg, Regina, Edmonton, Prince George and Whitehorse. Then, a 1938 riverfront scene with United Air Transport's Fleet Freighter CF-BDX. Soon afterwards it was wrecked stalling into the trees along the Liard River. In the foreground is a UAT Norseman. It burned off Nanaimo on October 30, 1940. (Fraser-Fort George Regional Museum P981.35.107, P984.17.3)

Air BC's BAe 146 C-FBAV at Prince George on August 11, 1995; then its crew of Capt Bill Walters and FO Doug Williams. (Larry Milberry)

delivery in February 1989 BAV had flown 15,600 hours, an average of about 2,600 yearly. The flight was airborne at 0911 at 119 knots and climbed at 250 KIAS for 9,000, then 31,000 feet. It would take 57 minutes and cover 311 nm. Capt Walters had started in 1966 at the Victoria Flying Club. He got his first job two years later as a Canso co-pilot for Flying Fireman. Through the 1970s he flew with a number of operators the Mallard, Turbo Mallard and Twin Otter. The Turbo Mallard was his favourite. It was pleasant to fly and, at 170 knots, was the fastest plane on the coast. Walters also liked the BAe 146, noting: "The 146 is old technology designed to a 'T'. It's built like a tank." FO Williams had started at the Langley Flight Centre in 1978. He earned a commercial ticket with Juan Air at Victoria and in 1984 was flying Cessnas and Otter ODK

Prince George airport in the early 1990s. Nearest is the RCMP hangar, centre is NTAir's. The passenger terminal and tower are beyond with a CAIL 737, Air BC BAe146s, air force Buffalo and Labrador, and various light planes. (NTAir)

with Huronair in Pickle Lake. He went to Australia for Air Whitsunday, flying Great Barrier Reef tourist trips with float Beaver VH-HAQ. He also did a stint on a Beaver with Aquatic Airways of Sydney. Back home he flew for Kenn Borek and Harbour Air before joining Air BC in 1988. At 0947 F1813 was cleared to FL200, then (0951) FL120. It landed on Prince George's R15 at 1006, discharged its passengers and cargo, and in about 20 minutes boarded for the return to Vancouver. From there Walters and Williams carried on to Edmonton to overnight.

NTAir

The story of NTAir dates to May 1971 when Thunderbird Airlines and Northern Mountain Airlines combined. Northern Mountain recently had taken over PWA's VFR base at Fort St. James. Several local aviators and businessmen formed the new company—Keith Carr, Ben Lloyd, Harold Mallory, Ed McPherson, Milton Richie, Dean Shaw. Richie was president. In the early years NTAir operated the Beaver, Ce.185, Ce.337, Ce.402, Goose and Otter. The Beech 18 quickly found a place— CF-BAS (on floats) and CF-WPO (wheels)

were acquired. In 1976 DC-3 CF-JUV was added for sked work. It operated on contract to PWA serving communities in the Okanagan. In 1974 the Northern Mountain operation split off as Northern Mountain Helicopters. The fixed wing side became Northern Thunderbird Air headed by Jack Stelfox. The fleet was mainly single-engine bush planes, but the Beech 18 remained important. Pilot Les Bower, with 14,600 hours on the Beech 18 by the time he retired in 1994, recalled one freighting job for a logging company that involved 1,500 trips, mostly with a Beech on skis. In November 1975 Twin Otter C-GNTB was added for skeds and charters, but its days were numbered —on January 14, 1977, while approaching Terrace, it crashed on Little Herman Mountain. Twelve died. Earlier (August 4, 1976) NTA lost Beaver CF-JOM in a mishap that killed two, so things were not always easy with the company. In the late 1970s NTAir had as many as 18 aircraft and 80 employees. There were various skeds, including Prince George-Smithers-Terrace-Prince Rupert; and Prince George-Mackenzie-Dawson Creek-Grand Prairie-Edmonton. The PWA sked operated Prince George-Quesnel-Williams Lake-Kamloops-Kelowna- Penticton-Cranbrook. There was also a sked to Dease Lake to serve coal miners. Through the 1970s-80s VFR business was eroded by the Bell Jet Ranger, which was being aggressively marketed in BC.

In 1991 Stelfox sold to Prince George logging company Joe Martin and Sons. In 1995 its fleet comprised: Beech 18 C-FBCE, Cessna 337s C-FVAS and YOC, Navajos C-GIPP and ZEB, a King Air 100, Twin Otter 300 C-GNTH and Aero Commander 500 C-FVNF. Besides its main base in a WWII hangar at Prince

NTAir's Beech C-FBCE (above) after a fender-bender at the Swanell logging camp, 222 nm northwest of Prince George in the early 1990s. Pilot Les Bower and AME Gary Cartwright ponder repairs. Once props were flown in and the plane was on blocks, it didn't take long—BCE was home in 24 hours. Then, CF-BCD after pranging at Germansen, site of a placer mine 160 nm northwest of Prince George. Here Gary Cartwright, formerly of Norcanair and Alert Bay, gets down to the slicing and dicing. (NTAir)

The Piper PA-23 Apache and Aztec gave much to aviation in the Canadian back country before the years of the Beech 99, Metro and Twin Otter. By the 1990s, however, many of these little toughies were languishing against the fence. Here NTAir's Aztec C-GNTE flies near Prince George in the 1980s with one of the company's Beech 18s. NTE was noted at Yellowknife in 1995, mired and partly cannibalized. (NTAir)

NTAir AMEs Bill Aleekuk and Gary Cartwright at work in NTAir's hangar. (Larry Milberry)

Convair 580 compared to Convair 5800		
	CV580	*CV5800*
Length	81' 6"	95' 9"
Span	105' 4"	105' 4"
Height	29' 2"	28' 1.6"
Wing area	920 sq.ft.	920 sq.ft.
Empty weight	28,380 lbs.	33,166 lbs.
Max. weight	58,140 lbs.	63,000 lbs.
Passengers	52	77
Payload		21,800 lbs.

The Convair 5800

A newsworthy BC aircraft industry story from the 1990s involved the Convair CV580 powered by 3,025-shp Allison 501 turbines. The first Allison conversion, a USAF C-131, had flown in June 1954. By 1960 Pacific Airmotive in California was leading the way in marketing the concept. The first Canadian CV580 was Shell Oil's Toronto-based CF-KQI. Next, Eastern Canada Stevedoring converted CF-ECS. The RCAF followed with a fleet conversion of its CV540s. The CV580 eventually served many Canadian operators—CanAir Cargo, Inter-Canadien, Kelowna Flightcraft, SEBJ, Soundair, Time Air, TPA, etc.

Through the 1980s Kelowna Flightcraft developed a specialty in Convair 580 repair, overhaul and modification. It converted several ex-US military C-131s to CV580s and certified a CV580 cargo door. Jim Rogers and Barry Lapointe spearheaded this work, while developing a grander plan. They didn't need any lessons about the soundness of the CV580 and knew that it was in resurgence. While a CV580 in the early 1970s might fetch $100,000, this quintupled in a few years. There were lots of airframes, since the USAF and USN had retired the C-131s. It was no surprise in the late 1980s when Kelowna Flightcraft engineered a stretched CV580. The prototype, an ex-USAF C-131F, was designated "Convair 5800". It had a 120 in. x 72 in. rear cargo door and accommodated six 88 x 108 inch standard airline pallets. It cruised at 280 knots over 1,000 nm with reserves. The cockpit was a full EFIS installation designed by KFC. In spite of KFC's efforts, by 1997 only two CV5800s had been converted.

Regional Competition

For competing regionals, keeping on top of industry changes is a full-time job that almost qualifies as intelligence/counterintelligence. Each operator watches the next for the least change in schedules, fares or quality of service. The job for each regional is to innovate where competitors can't. Advertising is the main tool in this battle. Otherwise, professional employees, keeping the customer informed, and polite and friendly service remained key in attracting and retaining business.

In January 1993 Time Air reported a list of changes. Effective December 17 there was a general fare increase of 5% on long-haul routes and 3% on short-haul (under 250 miles) Starting on January 6, it matched a seat sale being run by Air BC. The advance sale period for it was January 6-25, seven days prior to depar-

George, NTAir ran a base at Mackenzie, a mill town about 75 nm north on Williston Lake. Business was generally split 50-50 between skeds and charters. At this time the company employed about 20, half of whom were pilots. In 1996 it acquired CatPass 200 C-FCGM and inaugurated skeds from Prince George and Mackenzie to Vancouver. Meanwhile, Northern Mountain Helicopters had its head office in Prince George. It began with a fleet of Jet Rangers working throughout northern BC, the Yukon and NWT. By 1993 it had 12 permanent and seven seasonal bases and some 50 helicopters from the Jet Ranger to the Bell 205, Hughes 500 and Soloy Bell 47.

ture with stipulations for the traveller being a Saturday night, travel to be complete by April 30 and downgrades not allowed.

Meanwhile, Central Mountain raised its fare 5% on Smithers-Vancouver. Effective January 15 Time Air did the same. The fare Vancouver-Port Hardy was altered to $145 for the period January 11-April 30 to remain competitive with Pacific Coastal Airlines. For the same period a certain fare ($369) was reintroduced Watson Lake-Vancouver to remain competitive with CMA. As of January 7, other fares were reduced to remain competitive with Air BC rates.

Starting January 4, Time Air reduced capacity to a number of city pairs: Calgary-Saskatoon, Calgary-Edmonton, Kamloops-Vancouver and Seattle-Vancouver. The company, watching its competitors with an eagle eye (as they were watching it) noted that Air BC recently had reduced Calgary-Saskatoon and Seattle-Vancouver service, and that Horizon Air had dropped Seattle-Calgary on January 31. Later in the spring Time Air was happy to see a report from Air BC's Iain Harris that on 1992 sales of $165 million the airline had lost a million dollars. In response to slowing business it had grounded two Dash 8s and a Jetstream.

Air BC dealt CAI a jolt in April 1993 by opening Edmonton Municipal-Vancouver service with 85-seat BAe 146s. Meanwhile CRA introduced a choice of three fares on its Calgary-Edmonton "airbus". As announced in May, a regular fare of $226, four-day fare of $190 or seven-day fare of $126 could be selected. This was in response to a 20% drop in traffic through 1992, which CAIL attributed to more people opting for a three-hour drive rather than paying fares they felt were too steep. Statistics Canada reported that 1989, '90 and '91 figures for passengers flying between Calgary and Edmonton were 331,989; 288,360 and 259,750. This was linked to Canada's ongoing recession. To launch its new fares, CAIL adopted the slogan, "If time is money, then driving between Calgary and Edmonton is highway robbery."

In June 1994 CRA noted some other competition-related details. Air BC had announced three extra Vancouver-Seattle dailies and one on Vancouver-Portland; North American Aviation had dropped Edmonton-Fort McMurray; Kelowna Flightcraft was increasing summer Vancouver-Sandspit fishing charters. CRA announced summer package tours in Alberta and the NWT and some special family and student rates. There was a special June Timmins-

Toronto rate with a $62 saving, and a discount fare to introduce new Sault Ste. Marie-Thunder Bay service. In Quebec CRA was busy matching fare changes introduced by Air Alliance. It announced a Western Canada load factor for March 1994 of 49.98% compared to 49.15% for the previous March. In mid-1994 CRA noted Thunderbird Air operating a BAe 748 Vancouver-Masset. By this time CRA had abandoned its CV580 service to the Queen Charlottes from Vancouver, although CanAir Cargo had stepped in with Convair service in that market.

On the promotional side in mid-1994, CRA sponsored such activities as hockey awards in Kamloops, golf tournaments in Vernon and Regina, a relay race in BC, an airshow at CFB Moose Jaw, a cycling event in Sault Ste. Marie, a symphony orchestra concert in Timmins, a Science North promotion in Sudbury, Secretaries' Week along the North Shore, and a lobster fishing trip to Îles de la Madeleine. Pilot strikes hit both CRA and Air Ontario in 1994. These hurt each company, but they recovered quickly once staff returned to work. In the case of CRA, it recovered lost market share by immediately introducing short-lived bargain fares, e.g. Calgary-Winnipeg $159.

BC's early airlines clung mainly to the coast, capitalizing on lucrative skeds to Vancouver, Victoria and Seattle, running south to San Fransisco and connecting to US transcontinental lines. The exception was TCA. It crossed the Rockies to Lethbridge, thence to Regina, Winnipeg and, in the early 1940s, to Fort William, Chicago, Toronto, etc. This was a typical scene at Vancouver in the late 1930s with a Canadian Airways Rapide, and a United Airlines Lockheed 10 and Boeing 247. (Gordon S. Williams)

(Left) Even after WWII some of the vintage pre-war types inhabited BC, as illustrated by Fairchild 82A CF-AXE. In 1951 it was in service with Associated Air Taxi of Vancouver. (Finlayson Col.)

Modern Beavers and Cessnas became BC's main coast workhorses in the 1950s, replacing types like this Seabee at Prince George. Here are Cessnas of Trans Mountain, BC Air Lines and Island Air. (Fraser-Fort George Regional Museum P982.53.66, Wilf White)

As everywhere in Canada, the Beech 18, DC-3 and Convair grew in importance in BC from the 1950s. Here are Pacific Coastal and Air Rainbow Beeches; and a Harrison Air DC-3 and Convair 440. (Wilf White, Graham Wragg, Larry Milberry).

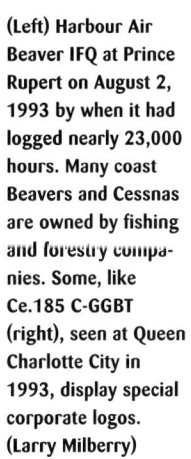

(Left) Harbour Air Beaver IFQ at Prince Rupert on August 2, 1993 by when it had logged nearly 23,000 hours. Many coast Beavers and Cessnas are owned by fishing and forestry companies. Some, like Ce.185 C-GGBT (right), seen at Queen Charlotte City in 1993, display special corporate logos. (Larry Milberry)

What supports the air carriers on the coast—fishing and forestry. Here are scenes from Queen Charlotte City in August 1993. (Larry Milberry)

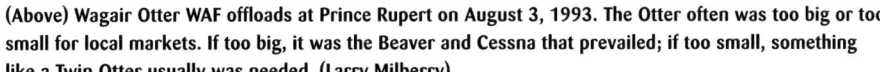

(Above) Wagair Otter WAF offloads at Prince Rupert on August 3, 1993. The Otter often was too big or too small for local markets. If too big, it was the Beaver and Cessna that prevailed; if too small, something like a Twin Otter usually was needed. (Larry Milberry)

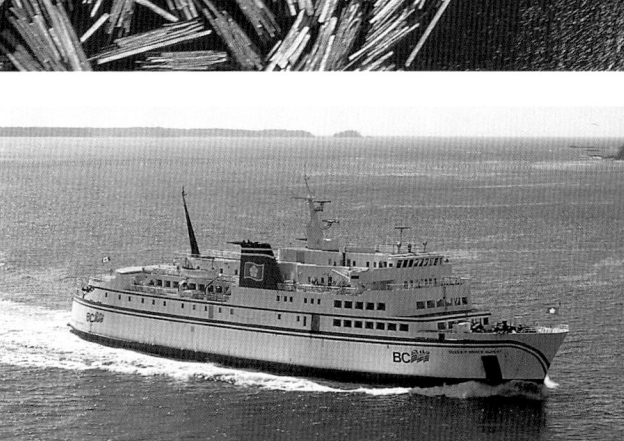

Masset is a typical community in the Queen Charlotte Islands. The links to the mainland for such towns are the airplane and the ferry. There are few airports, although the one at Sandspit is an exception. (Larry Milberry)

707

(Above) Through 1985 Canso JCV served a floating sport fishing base on the BC north coast. JCV later found a career flying exclusive tours up the Nile from Cairo to Victoria Falls. (Kenneth I. Swartz)

Flying in BC has produced many leading figures over the decades. In 1960 Al Campbell (left) founded Tyee Airways of Sechelt, specializing in moving loggers among camps. In 1989 he sold to a local native group. Herman Peterson (above) was well-known in the interior around Atlin. He started flying in Quebec in 1935, then moved to the Yukon in 1942 for a lifelong career serving mining. (Kenneth I. Swartz)

(Above) Action on the coast. A Wilderness Ce.185 cruises along for the camera, while Randy Hanna's Helio Courier gets away from the Fraser River at Vancouver International. (Henry Tenby, Larry Milberry)

(Right) Air BC found the Twin Otter suitable for float and wheel operations. Here C-FGQE waits at Vancouver on April 18, 1991. (Larry Milberry)

(Above) An ideal view of BAe146-200 C-FBAV over Vancouver. It was delivered to Air BC in early 1989. Able to use runways 1,000 feet shorter than those required by a 737, the 146 was ideal for regional airports. In the 1990s this type operated daily skeds as far east as Winnipeg; and night and weekend charters for sports teams, cruise boat passengers, Las Vegas gamblers, even shoppers eager to add West Edmonton Mall to their list of conquests. While most of its routes were profitable by this time, on April 6, 1997, following a two-month strike, Air BC cut skeds to Castlegar, Cranbrook, Dawson Creek, Nanaimo and Penticton. Central Mountain Airlines took over these runs using the B.1900D. (Air BC)

(Right) Vancouver was home to all the CPA jets. Here four DC-8s await trips from there in a scene from February 1967. (Les Corness)

(Below) Close in with a stretched CPAir DC-8. (CP Air)

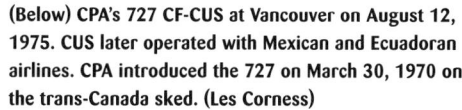

(Below) CPA's 727 CF-CUS at Vancouver on August 12, 1975. CUS later operated with Mexican and Ecuadoran airlines. CPA introduced the 727 on March 30, 1970 on the trans-Canada sked. (Les Corness)

(Right) CPAir briefly operated 737-300s. This example was at Vancouver in May 1985. Soon it moved to VASP in Brazil, thence to Markair in Ireland. (Gary Vincent)

(Above) C-FCRA "Empress of Japan" flies by at the August 1985 Abbotsford International Airshow. CRA was a 1973 model and remained in service till swapped to Pakistan International Airlines for DC-10s in 1986. Then, another great Abbotsford scene as CAIL DC-10 C-GCPF comes down the show line with its gaggle of smoking Pitts as a backdrop. (Kenneth Swartz, Bob Arnold)

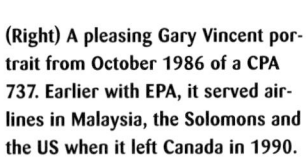

(Right) A pleasing Gary Vincent portrait from October 1986 of a CPA 737. Earlier with EPA, it served airlines in Malaysia, the Solomons and the US when it left Canada in 1990.

(Below) This CAIL DC-10 was in the midst of a six-month tear-down ("D" Check) at Vancouver in October 1994. (Larry Milberry)

July 31, 1993—an everyday scene at Vancouver International. A CRA Dash 8 and Short 360 await, as a CAIL DC-10 arrives. Beyond are 747s of Cathay, Japan Airlines and KLM. (Larry Milberry)

(Below) Air Canada DC-8s at Vancouver on June 6, 1968. CF-TJP was sold in 1984, but remained active as freighter N802CK in the mid-1990s. (Les Corness)

(Left) For the photographer there's always a new way to look at a subject. Here Air Canada DC-9 C-FTMA appears in a panel of windows at Vancouver's terminal on August 1, 1993. (Larry Milberry)

(Below) Bearing the latest colours, Air Canada's B.747-400 C-GAGN, delivered in August 1991, arrives at Vancouver. (Jeff Wilson)

Martin JRM-1 Mars CF-LYJ during an early practice drop. This water bomber was conceived by Dan McIvor, a veteran coast pilot who was involved in aerial fire fighting after WWII using types like the Beaver and Goose. He heard that the US Navy was selling its fleet of Mars. He was too late to bid, but in 1959 visited Hugo Forrester, the scrap dealer in California who had purchased the lot for $26,000. The scrap man flipped the package to McIvor's group. With the help of two US Navy pilots, Dan McIvor, Harold Rogers and Jack Edwards ferried the planes to Victoria, where Fairey Aviation turned the first into a 7,200-gallon tanker. Meanwhile, McIvor convinced several forestry companies to form a consortium, Forest Industries Flying Tankers, to operate the Mars. The first trial made the Mars an instant celebrity—on 22 drops, the prototype tanker (Marianas Mars) quelled a difficult fire on Ramsey Arm. (Canadair Ltd. 40831)

CF-LYM Caroline Mars was badly damaged at Victoria by Hurricane Frieda. The Mars first flew in December 1941 at Little River, Maryland. Only seven were built: the prototype PB2M-1 patrol bomber, soon converted to a transport; five JRM-1 transports (the first was soon wrecked, the others were dubbed Marshall Mars, Marianas Mars, Philippine Mars and Hawaii Mars); and one JRM-2, Caroline Mars. They set many distance and payload records for the US Navy. They were its air transport mainstay on the San Francisco-Honolulu supply run, serving to 1956. Marshall Mars burned in Hawaii in 1950, Marianas Mars crashed June 23, 1961 during fire fighting ops (the four-man crew was lost) and Caroline Mars was wrecked by Hurricane Frieda on October 12, 1962. (Ryan Bros. Photo Centre)

(Above) The FIFT base at Sproat Lake on April 9, 1992 with Mars, G21s and an S-64 present; then a closer look at Hawaii Mars (CF-LYL). These views were from FIFT Jet Ranger C-FIFT flown by Paul Greenwood. (Larry Milberry)

(Right) Philippine Mars (CF-LYK), roars down Sproat Lake during training in April 1988. Pilots John de Bourcier and Jack Waddington and FEs Roy Copeland and Mike Johnson were aboard. (Michael O'Leary/FIFT)

Philippine Mars received a lot of heavy attention in 1992. The plastic tent over the starboard wing enabled the overhaul crew to work on rainy days. Then, views of the engine maintenance section and the Mars cockpit and engineer's panels. Keeping the Mars in service so long has a lot to do with the availability of engines and props, and good shops for overhauling them. In the 1990s there was a shortage of props. FIFT arranged to buy those on a Constellation stored at Mont Joli. (Larry Milberry)

Hawaii Mars showing off at Abbotsford on August 11, 1995. The Mars has a wingspan of 200 feet, is 120 feet long and weighs 162,000 pounds all-up. It uses four Wright R3350s and has a cruise speed of 215 mph. (Larry Milberry)

Next to the Mars the DC-6 was Canada's biggest water bomber from the 1960s onward. Conair had a large fleet, including C-GIOY, seen at Abbotsford in August 1987. By 1997 Conair 's DC-6 fleet was waning. Electras were infiltrating the Canadian market and C-130s were on the horizon. (R.W. Arnold)

On August 19, 1983 a 407 Squadron Aurora was training over the BC interior. Passing the Kamloops area it overflew Conair DC-6 "Tanker 50" working a fire in typical desert-like country in the area. (Larry Milberry)

(Above) Other Conair fire bomber conversions. Fokker F.27 C-GSFS began with the Nigerian Air Force in 1972. Conair converted it in 1986 and leased it to the Sécurité Civile in France. Ralph Bolton, the Canadian captain, was one of those killed when it crashed near Aires, France on September 4, 1989. Then, Turbo Firecat C-FKUF landing at Abbotsford on August 6, 1993. (Kenneth I. Swartz, Larry Milberry)

From the old days... Conair Avengers at Kamloops in the 1960s (below), then Conair A-26s there on August 2, 1974. The A-26 soldiered on with Conair to the late 1980s. In the inset view, helpers mix fire-bombing slurry for the A-26s. (R.W. Gleasure)

Vancouver views by avid photographer John Kimberley... Harbour Air's terminal as a Beaver is launched. Spotters can watch all the proceedings from the pub while quaffing a favourite beer. Then, a West Coast Air Twin Otter lands on the Fraser River at Vancouver. This company was formed in 1996 by Allan Baydala after he took over Air BC's historic Vancouver-Victoria float operation. Note the sightseeing windows on Twin Otter C-FWCA.

(Left) In 1996 Central Mountain of Smithers added the Beech 1900D. Developed from the B.1900C (first flight September 1982) , this model first flew in March 1990. It uses 1,279-shp PT6A-67Ds, while the "C" has 1,100-shp -65Bs. Both models are 19-seaters, but the "D" offers more headroom and is about 30 knots faster—max. cruise 280 knots.

(Right) This gaudily-painted Air Transat B.757 was at Vancouver in early 1997. Canada's main charter carriers—Air Club, Air Transat, Canada 3000 and Royal Air— all made the most of Vancouver, using it as a terminus for domestic charters from the east, and as a jumping-off point for sun spots like Hawaii.

(Right) A Canada 3000 A320 gets away from Vancouver.

(Below) CAIL DC-10 C-FCRE carried a special paint job into 1997 that included the autographs of hundreds of company employees.

One of Canada's great airshows is held annually at Abbotsford, up the Fraser Valley from Vancouver. Here the public can see the latest in aerospace technology, like these Brazilian air force Tucano trainers that visited in August 1995. The Tucano is one of dozens of important types using Canada's P&WC PT6 turbine engine. (Brent Wallace, Larry Milberry)

Although helicopters serve Canada from coast to coast to coast, British Columbia is the heartland of Canadian helicopter operations. Here, Canadian Helicopters Ltd. Aérospatiale AS350B AStars C-GRGJ and C-GFHS fly over Vancouver on January 21, 1990. The five- or six-seat AS350B found favour in the market a notch higher than the Bell 206B. With a cruise speed around 130 knots it is one of the fastest light helicopters. (Brent Wallace)

(Left) Okanagan Helicopter workhorses—the Bell 205 and Bell 206. These were at Vancouver in September 1977 in the days before CHL took over Okanagan and changed to the now-familiar red, white and blue colours. (Brent Wallace)

(Above) CHL AStar C-GRGU departs Terrace on April 2, 1992, while a B.206B waits for its next charter. (Larry Milberry)

(Above) An S-58T departs Langley, BC on August 13, 1995 for a site near Hope, where it was slinging for a selective logging operation. (Larry Milberry)

(Below) A 442 Squadron Labrador near CFB Comox on April 8, 1992. The "Lab" cruised at about 120 knots. Its crew normally included two pilots, flight engineer and two SARTechs. It had been Canada's frontline search and rescue helicopter since 1963. Empty weight was 13,500 pounds; but with fuel (3,000 pounds), SAR kit, tools and spares (1,300) and crew (1,200) a Lab grossed out at 19,000 pounds. In 1996 Ottawa deferred selecting a Lab replacement for at least another year. It was likely that the Lab, refurbished time and again, would see the year 2000 in CAF service. For early 1997 ATG still operated 13 Labradors. (Larry Milberry)

(Above) John Kimberley captured this scene on Kodachrome in September 1988—one of Helijet's S-76 commuter helicopters waiting for a Wardair 747 to land at Vancouver International before departing for downtown Victoria.

(Left) This mighty Kamov KA-32 from Russia began heli logging in BC in 1992 with Vancouver Island Helicopters. It proved a safer (better power reserves), heavier lifting machine than traditional types like the S-61. The Kamov was shot at Abbotsford on August 6, 1993. (Larry Milberry)

Al Martin – Aviation Fan

I n the 1920s-30s many young Canadians were taking an interest in aviation. They read everything they could, haunted airports, took flying lessons, exchanged stories, etc. The camera was the most popular means of recording the excitement—new planes kept appearing, there were airshows all summer, it was every youngster's dream come true. Fans kept busy with stories like the Bremen, R-100 or Short "Cambria", the Trans-Canada Air Pageant, daily news from the gold fields, and visits by aviators like General Balbo, Frank Hawks, Charles Lindbergh or Wiley Post. The enthusiasts were everywhere to capture events on film, fellows like John Davids, Fred Hitchins, Fed Hotson, Al Martin, Jack McNulty and Ken Molson. Elmore Owen Martin, usually known as Al, was born in Winona, Ontario on February 2, 1923. When he finished high school in 1939, he went to work at McKinnon Industries in St. Catharines, then joined the RCAF in 1942, training as an air gunner. He was overseas from October

Al Martin at Rockcliffe for the Centennial year airshow on June 10, 1967, then with his longtime aviation associate Jack McNulty at Toronto Island Airport on August 30, 1969. (Larry Milberry)

1943 to February 1945, completing a tour on bombers.

Postwar, Al learned to fly in Hamilton. He instructed, ferried several aircraft around Canada and worked for the City of Hamilton. In 1951 he started with TCA and moved up in customer service. During his TCA years at Malton, Al met a few youngsters learning the basics of photography and the ins-and-outs of local aviation. He always had time to chat with them at his counter. If there was something interesting on the ramp, he would escort the fellows for some shooting; and often alerted them to upcoming visits—the first Vanguard in Canada, an American Airlines Electra press flight, a Ford Trimotor, etc. Al also hosted evenings where his friends got a chance to show their slides and exchange ideas. Late in 1962 he told them about plans to establish a

society dedicated to Canadian aviation history. He invited them to the first meeting where for $2.00 they became members of The Early Birds Association of Canada. This group evolved into the Canadian Aviation Historical Society, Canada's pre-eminent aviation history society. Al and his aviation buddies also spent hours on the phone, where subjects ranged from aviation to photography, classical music, books of all kinds, the BBC shortwave news, politics and a host of other topics

In the late 1960s Al moved into public relations with Air Canada. One of his biggest assignments was also his toughest—handling press matters after the crash of an Air Canada DC-8 on July 5, 1970. When he retired in November 1985, Al and his wife, Gay, moved to the West Coast. He passed away in White Rock, BC on May 9, 1993. Shortly before *Air Transport in Canada* went to press, Al's photos became available, but too late to be integrated throughout the text. This special chapter is presented as a tribute to Al Martin—one of Canada's pioneers in aviation photography.

(Right) Al Martin enjoyed the company of fellow aviation enthusiasts. These are the fellows at the first CAHS meeting (Toronto, 1962): Jock Forteith, Al, George Morley, Bill Wheeler, Herman Karbe, Jeff Burch, Charlie Catalano, Harry Cregen and Roger Juniper. (Ernie Harrison)

(Below) The post-WWII aviation buffs were interested in a range of photographic concepts. A good angle was always important, and clutter of any kind was avoided. Sometimes the fellows would wait an hour for a vehicle to move away from a plane or for someone to close a cabin door, just to get that "ideal" shot. In time most photographers realized that the "ideal" shot could be rather dull . Capturing a plane in different environments also was of interest, as with these Seabee views. CF-GAF was at Toronto Island on a charter from Orillia on September 8, 1953. The light was perfect and the setting natural. Al shot CF-ECX on wheels at Malton on September 15, 1951.

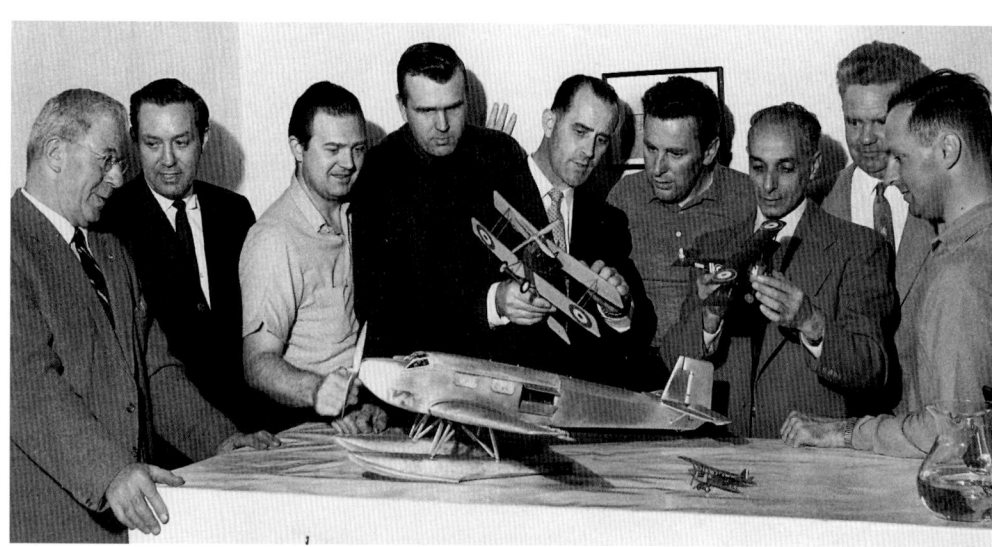

Amphibians always held special interest, whether a humble Seabee, "sexy" Royal Gull or lumbering Canso. Al Martin shot this Royal Gull (right) at Malton on April 8, 1956. Built in Genoa, Italy and distributed in Canada by Timmins Aviation, the Royal Gull made a brief hit in this country, but turned out to be of marginal use commercially. CF-IZA crashed in Belwood Lake north of Guelph, Ontario on June 17, 1957. Owner J. Thomas and his five-year-old son were killed. Pilot Bob McLaren and passengers Guy Mills, Jerry Overman and Claude Walker were injured. Then (below), Wheeler's Canso CF-EMW in a typical winter scene at Dorval in the mid-1950s.

(Below) Aviation enthusiasts worth their salt are intrigued by anything that flies, although in recent times the so-called "specialist" has appeared—the fellows who only shoot airliners, jet fighters, etc. Al Martin could discuss any aspect from his favourite homebuilt (the Bowers Flybaby) to bushplanes, heavy transports and helicopters. The specialist would be at a loss in a conversation with him. Here is a selection of helicopter photos from Al's files. The first is a rare Sikorsky Hoverfly II (R-5), one of the earliest helicopters to visit Canada. No date is given, but the setting was the old Hamilton municipal airport. The RAF pilot apparently mistook it for the nearby RCAF station at Mount Hope. This machine was last heard of in the UK with the Aeroplane and Armament Experimental Establishment mid-1945.

(Below) RCAF Sikorsky H-5 (S-51) at Toronto Island on August 24, 1951 for the Canadian National Exhibition airshow. This was the sort of "set-up" shot a photographer dreamed of. There's action, ideal light, a good angle to show the H-5 to full advantage, and interesting background. Then, a similarly well-taken view of "Navy 963", an HO-4S at the TIA on September 19, 1953, also for the CNE. Later in the day it was called upon when an RCAF Sabre crashed off the CNE, killing S/L Ray Greene.

(Right) Toronto Island Airport was a favourite hang-out for Al Martin. In his files are hundreds of photos taken there, including this S-55 from June 26, 1958. Most of Al's TIA shots are of static subjects, since it was not always possible to catch the action. In this case Al was lucky that HNG was running up. TIA always was a great place for the photographer, since there was nobody to yell, "Hey you, what do you think you're doing? Get outta here!" Anyone interested in aviation was welcomed. With today's security most airports now are off limits to the photographer.

Further views of the rare Doman LZ-5 (at right, Malton, October 8, 1955) and Bristol Sycamore (Oshawa, November 14, 1956). For the latter Al got some height to gain a more interesting angle.

(Right) The Bell 47 was the most popular light helicopter in Canada from the late 1940s into the early 1970s, when the Bell 206 Jet Ranger took a solid foothold. Al Martin recorded this Bell 47B3 at Vancouver on February 8, 1953.

(Below) This early 1940s photo of RCAF Northrop Deltas likely was among the many negatives in Al Martin's collection that he traded with fellow hobbyists like Pete Bowers, Jack McNulty, Harvey Stone and Gordon S. Williams. Although such collectors carefully filed each negative in an envelope labelled with all the basic "gen", the name of the photographer was not always included. This looks like many a Gordon S. Williams Delta photo. Gordon, who lived in Seattle and worked at Boeing, covered the Canadian West Coast, then traded spare negs. He was more likely to photograph planes without worrying that people were wandering around.

(Above) RCAF Expeditor 1547 was at London, Ontario on December 7, 1955. Rarely were RCAF aircraft spotted with station markings like these. When they were, the photographer had a real "find" and something to crow about to his buddies. Besides the North Bay lettering, 1547 bears code letters unique to its station. Enthusiasts knew the dozens of letter combinations, so could tell where any RCAF plane was based.

(Below) The RCAF's North Star "Rockcliffe Icewagon" at Malton on October 27, 1952. Here was the photographer's ideal situation—a beautiful airplane running up, good lighting, and the opportunity to choose that favourite angle—three-quarters front, sunny side.

(Left) Being at Malton in the 1950s was perfect for the aviation fan. Among other things there was all the action at Avro Canada. Here the Jetliner has drawn a crowd on the Toronto Flying Club ramp on August 14, 1953.

Avro's flying test bed B-47 being towed into the TCA hangar for weighing on May 27, 1958. Once again, the photographer would have been pleased with his subject matter and the conditions.

Like any photo buff Al Martin was keen to shoot the wide variety of odd-ball aero survey planes visiting Malton. The Spartan P-38 at left was there on May 14, 1954, parked in front of Jack Sanderson's hangar. A veteran of the 1947 Bendix Trophy Race, this P-38 ended in a crash outside Ottawa on November 29, 1955. Kenting's Mosquito (below left) was shot on August 30, 1953, while its Hudson, bearing special Colombo Plan titles, was on the Genaire ramp on December 8, 1955.

Famous bush types of the early 1950s that interested the photographer. The Saskatchewan government's Husky SAQ (right) at Toronto Island on July 29, 1951; Norseman GUN (below left) at Vancouver on February 8, 1953; and Cub JXV (below right). The uniqueness of a type like the Husky, different colour schemes, or an unusual undercarriage arrangement like the Cub's Whittaker tandem wheels for off-strip operations, the weather and general setting—such details affected whether or not an enthusiast would take a picture. One might wonder what a Norseman was doing at Vancouver on wheel-skis. Research done by CAHS member Bruce Gowans reveals that in February 1953 GUN left Queen Charlotte Airlines for Hall's Air Service of Val d'Or, Quebec. It's an easy guess that QCA sold the plane on wheel-skis by arrangement with Hall's, and that the photo was taken just before GUN left the mild west coast for the ice box conditions of northern Quebec.

A nice trio of corporate planes from Al Martin's collection: Shell Oil's Dove (above, June 1953), a Lodestar (right, Malton, October 12, 1951), and Finning Tractor's Cessna 195B (below, Vancouver, September 25, 1956). The tiny script on the Cessna reads "Pacific Ventures (Aircraft Lease)". Enthusiasts usually were delighted to copy such details in their pads. Sometimes, however, they failed to list dates, places and other details, leaving holes in the record.

(Below) Al Martin shot MCA's DC-4 at Malton on July 8, 1956. He noted on the negative envelope: "Depicted is the aircraft just leaving the south end of the new TCA hangar. I had worked this charter flight to London, England. On board was the Canadian Legion, Kitchener Branch." MCB earlier had been United Airlines "Mainliner Colorado River" and "Mainliner Paul Bunyan". Later it was CF-NAB with Nordair, then PP-BTT in Brazil, where it was scrapped in 1969. Note the many folks who showed up to see off their friends and relatives. About a year later a similar crowd waited at Malton to greet DC-4 CF-MCF on a Legion trip home from England. Imagine their dismay when the public address system announced that the plane would not be arriving—it had crashed en route, killing all 79 aboard. Aviation could dishearten as quickly as it could delight.

(Below) CF-TGB at Malton for the inaugural TCA Super Connie flight to Prestwick and London on May 14, 1954. Al noted that the trip was made two hours faster than by North Star. Note the crowd on the observation gallery of Malton's old terminal. Even on the most blustery winter day an aviation enthusiast might be found there, waiting for any bit of action.

(Right) A simple snap of a TCA Viscount at Malton on April 28, 1955, but Al Martin's notes give it special interest: "While about to operate as Flight 325 from Idlewild, New York to Dorval, CF-TGL was demolished by a Seaboard and Western Airlines Super Constellation [N6503C]. The latter went out of control on take-off, caught fire and careened into TGL. TCA Stewardesses Jeannine Bedard and Karin Foch scrambled to safety when warned from the ramp of the impending collision."

(Left) Working for TCA meant that Al Martin easily could get to places like Vancouver when he had spare time. This made him the envy of his buddies, most of whom were stuck in and around Toronto. Al shot BCPA's classy-looking DC-6 VH-BPF "Endeavour" arriving in Vancouver from San Francisco on February 8, 1953. BPF later served Tasman Empire Airways and the RNZAF before going to Laos. It was lost in a crash in Java on February 1, 1972.

(Right) Whether a grand DC-6 or humble Beechcraft, Al Martin kept track of them all. This attitude, as well as how he followed music, world events, ham radio and other topics, made him an especially interesting friend. Al shot Chinook Flying Service's modified Beech AT-11 at Calgary in September 1956, noting of it: "I was told that this aircraft did much DEW Line work."

(Below) On May 10, 1961 Al Martin visited Dorval for Vanguard familiarization, but noted: "Both the scheduled Vanguard flights were cancelled and replaced by Super Connies (which someone nicknamed 'Conguards'). At Dorval there were North Stars lying around dormant, awaiting buyers." In this historic line-up are the North Stars with a Vanguard and DC-8.

(Above) C-46 CF-IHX at Dorval. The absence in Al's files of any gen about this photo suggest that it was a "trader" taken by someone else. IHX was with Dorval Air Transport 1955-59, and ended in a crash in Venezuela in 1962.

Air Transport in Art

Tom Bjarnason's representation of Canadian air transport over the decades: a Vedette, North Star, Beaver, Global Express, and the Space Shuttle, in which Canada is deeply involved.

With its first book in 1981 CANAV recognized the role of art in aviation history. This was amidst an atmosphere where few publishers would spend a penny on an original painting. Their aviation book jackets usually were cooked up by recycling paintings from the Canadian War Museum, etc., and using them gratis.

In books like *The Avro CF-100* and *Sixty Years*, CANAV showed various airplane types using technical side views to illustrate colour schemes. The jacket art for *The Avro CF-100*, *The Canadair North Star* and *The De Havilland Canada Story*, all rendered by Peter Mossman, raised standards in the Canadian aviation book trade to a new high. With *The Canadair North Star*, CANAV introduced a modest gallery of aviation scenes—artists and illustrators created seven interpretations in various styles, using acrylics and air brush. All these paintings later became the foundation of Canada's Air Force Art Collection, and went on permanent display in the Billy Bishop Building at CFB Winnipeg.

Other art followed in successive CANAV titles. Sometimes paintings were controversial, riling those without much artistic sense (and who are proud of the fact). Especially critical were those who accept a painting only if it is rigorously photographic in execution. Such viewers wouldn't give you a penny for a Wooten or a Bradford, if they happened to miss a rivet. For them, the only satisfactory solution is a photograph. Another category of critic turned out to be other artists, some of whom proved to be ruthless (regarding anything not painted by themselves or their personal favourites), but these rarely included seasoned professionals.

For *Air Transport in Canada* CANAV presents the largest gallery of original Canadian aviation art yet published. Ten artists and illustrators have contributed. Brush and air-brush styles are shown, depending on the artist's approach. This time we have included examples of computer art, the first displayed in a Canadian aviation book. Had the Masters had this technology, you can be sure that it is something in which they would have revelled. In years ahead people will be used to "painting" with computer. Let these examples show where it started in Canadian aviation art.

The Artists

Geoff Bennett, Atlantic Canada's best-known aviation artist, was born in England in 1930. After studying architecture, he was conscripted into the RAF. He learned to fly in Canada in 1953-54 under the famous NATO training plan. In 1957 he joined the RCAF, where he flew everything from the Harvard to the Argus. He left the air force in 1977 and joined the DOT. Over the years he qualified on 15 types and logged 11,000 hours.

Geoff had been experimenting with art media over the years, but got serious only after he was posted to RCAF Station Greenwood in 1967. There he took up painting with acrylics which, as he put it, "suited me beautifully." His first serious work was an Argus commission. In the years that followed, 400 of his 1,500 works would by of the Argus. His first commission for a book was done for CANAV in 1986—his now famous 413 Squadron F-86 jacket art for *The Canadair Sabre*. In 1994 Geoff's painting of an Antonov AN-25 won first prize in the National Aviation Museum's prestigious Artflight exhibition. An unpretentious fellow, Geoff used to refer to himself only as "one of Canada's famous brushpilots".

Tom Bjarnason was born in Winnipeg in 1925. As a young man he was an avid hockey player, and served in the Canadian Army in WWII. Later he attended the Winnipeg School of Art, and Meinzinger's Art School in Detroit; then worked as an illustrator in Windsor (Ontario), Toronto, London (UK), Stockholm and Dusseldorf. In 1970 he returned to Canada, working mainly from his studio in Toronto's Yorkville district. Tom's work became well-known in such publications as *Star Weekly*, *Weekend*, *Reader's Digest* and the Canadian Aviation Historical Society *Journal*. It also appeared on Canadian postage stamps (e.g. a series of 16 sailing ships and four city street scenes). When he had the opportunity, Tom painted under the Canadian Forces Civilian Artist Program, covering such types as the CF-104 and CF-18. Several of his CFCAP paintings are with the Canadian War Museum. Two of his best-known aviation works are those on the covers of CANAV's *Sixty Years* and *Power: The Pratt & Whitney Canada Story*.

Tom lectured at the Ontario College of Art, and the Icelandic College of Arts and Crafts. He took an active part over the years in such organizations as the Canadian Association of Photographers and Illustrators (which awarded him its Lifetime Achievement Award in 1992), The Arts and Letters Club of Toronto, and the CAHS. In the 1990s Tom was living in Port Hope, Ontario, where he developed a new style of art incorporating recycled hi-tech components. One of his first clients for this work was NavCanada, the air traffic control and air navigation institution headquartered in Ottawa.

Dugald Cameron is one of the leading British aviation artists. In 1961 he graduated from the Glasgow School of Art, then worked on many industrial design projects for companies like Honeywell Controls, Rolls-Royce and Yarrow Shipbuilders. He lectured at various institutions. In 1976 he was appointed Governor at his alma mater; became its Head of Design in 1982; and received from it an honorary professorship in 1991. Dugald is well-known for the hundreds of commemorative colour profiles he created under the "Squadron Prints" name. He authored books, such as *Glasgow's Own* (history of 602 Squadron) and *Eagles in the Sky* (RAF 75th anniversary history). He served in the RAFVR, earned a pilot's licence, pursued a hobby in railway history, and (1996) was named a Companion in the Royal Aeronautical Society.

Stéphane Cochin is an innovative Montreal artist whose interests range from air brush paintings to industrial design. He was born in 1964 in Grande-Rivière in Quebec's Gaspé region. "I am the son of a merchant marine officer from Belgium and a nurse from Caraquet, New Brunswick," he noted in 1995. "I've been drawing as long as I can remember." Stéphane studied technical illustrating and industrial design in college and at the University of Montreal. In 1992 he won first prize in an automotive design competition sponsored by Ford of Canada. His interests focused on air brush painting (including box art for plastic model kits), and creating original vacu-form plastic models like the Martin Seamaster, and solid resin kits like the Lockheed X-31. In the 1990s Stéphane was operating as Stratosphere Models.

Robert Finlayson was born in Hamilton in 1930. As a boy he followed every aspect of aviation and took to the air for the first time at age seven. His wartime hero was his older brother Ross, who flew Mosquitos with 409 Squadron. After the war Bob pursued a career in photography. Work he did for the Foundation Co. of Canada included photographing famous bridges like the Burlington Skyway. Bob also was an avid birder, a hobby he got into in the 1940s. He sketched with pencil for years. Then, in 1960 he started to get serious about painting in oils, concentrating mainly on airplanes and the native birds of Southern Ontario. His first published airplane oils appeared in CANAV's *Sixty Years* in 1984.

Garfield Ingram was born in Oakville, Ontario in 1954. He studied in Toronto at the Ryerson Polytechnical School of Interior Design, graduating in 1983. Since then he worked in architectural rendering. Through the years Garfield also pursued his hobby of plastic modelling, becoming one of the top people in Canada at this craft and winning various awards for his creations. In time he turned to painting. The cover on Vanwell's *Canadian Squadrons in Coastal Command* and the UN Buffalo shown in this book are his first published works.

Ron E. Lowry, born in Toronto in 1934, has been involved in aviation since boyhood days. As a young man he worked at Avro Canada at the height of the CF-105 era. Then he joined Bell Canada, where he spent 31 years, retiring in 1986. Along the way Ron became known as a top modeller, winning acclaim in competitions all over North America. Several of his models were commissions for such collections as the Smithsonian Institution, the National Aviation Museum and the Sikorsky Museum. The first of Ron's paintings to be published appeared in CANAV's *De Havilland Canada Story* (1983). His reputation soared thereafter and he barely was able to keep up with the flood of commissions. When he retired, Ron took his aviation interest a step further—he learned to fly at the Brampton Flying Club, then bought a Cessna 150.

David Nilson, who was born in Regina in 1932, is one of the West Coast's best-known aviation figures. After an air engineering career in Alberta and the NWT, he went to the Okanagan in 1960, then to Campbell River. There he established Nilson Aircraft Ltd., which gained a top reputation, turning out such fine products as VIA's Beech 18 Seawind conversion. One can argue that something like the Seawind is a work of art unto itself. Although he took some art courses, David was mainly self-taught. Since retiring from aviation in 1996 he had more time to paint, but also turned his hand to sculpting, writing and other creative activities.

Jan Stroomenbergh was born in Driebergen, The Netherlands in 1946. He came to Canada in the early 1950s aboard a KLM DC-7C. This was his first airplane ride and remained a memorable event in his life. After high school Jan studied technical graphics at Medway College of Design in England. After graduation he worked as an illustrator at places like Northwest Industries in Edmonton and Bristol Aerospace in Winnipeg. On the side he turned out dozens of aviation paintings, often commissions from people at work. In 1997 Jan was living in Vancouver, where the ever-busy aviation scene was to his liking.

Alfred Wong was born in Hong Kong in 1961. He studied fine art at York University in Toronto, earning a bachelor's degree in 1985. He established himself as a freelance illustrator, finding work in a broad sector—the *Toronto Star*, the *Globe and Mail*, the *Medical Post*, etc. Alfred also did some prop design for television productions. Meanwhile, he pursued his hobby as a model builder and was an active member of the International Plastic Modellers Society.

Robert Finlayson. Oils. RCAF Canadian Vickers Vedette "ZN" over the air station at Cormorant Lake, which was northeast of The Pas, Manitoba. The Vedette was the first indigenous Canadian design to reach production. In RCAF service its duties included forestry and smuggling patrols. The last ones retired early in WWII, when they still were being used as flying boat trainers on the West Coast. Of 60 built 1924-30, not one Vedette survived, although a superb replica was built by the Western Canada Aviation Museum.

Robert Finlayson. Oils. Grumman Goose 917 on British Columbia's north coast. The Goose was important in BC during the war. It did air-sea rescue, carried passengers, mail and freight to remote outposts like Tofino, and supported RCAF fighter squadrons in Alaska. This example was with No. 13 (Operational Training) Squadron and No. 122 Squadron at RCAF Station Patricia Bay, near Victoria. In July 1942 it was involved in the tragic incident related in Chapter 12.

Robert Finlayson. Oils. Boeing B-17F Flying Fortress of 168 (Heavy Transport) Squadron
near Gibraltar. Formed in October 1943, No. 168 carried the mail for Canada's troops
overseas, operating from Rockcliffe to the UK, Gibraltar, North Africa, Italy and Egypt
with B-17s, B-24s and Dakotas. B-17 No. 9202 made the squadron's first return trip,
leaving Rockcliffe on December 22, 1943 under F/L W.R. "Bill" Lavery; returning from
Prestwick on January 11, 1944 under F/L W.H. "Wess" McIntosh. On November 4, 1946
No. 9202 crashed near Muenster, Germany, while on a mercy mission carrying
penicillin to Poland. All the crew perished.

Robert Finlayson. Oils. Douglas C-47 Dakotas of 437 (Husky) Squadron at Arnhem in September 1944. Arnhem was a crucible for the Allied air forces, which lost hundreds of aircraft in the brutal action there beginning on September 17, 1944. Dakotas and Stirlings bore the brunt of the transport duties. Flying low and slow, many were brought down by fierce enemy fire. These two Dakotas are over farm fields, where Horsa gliders have landed with troops and equipment to take on the Germans. A belly-landed Stirling adds to the scene.

Ron E. Lowry. Airbrushed. TCA Lockheed 18 Lodestar CF-TCY at night over a prairie cityscape. The Lodestar brought the travelling public new standards in speed and comfort, and served TCA into the late 1940s. Once sold, many of the fleet were converted to luxurious corporate planes for companies like BA Oil, T. Eaton Co. and Massey Harris. This example moved to the Department of Transport and ended in the Canadian Museum of Flight and Transportation in Langley, BC.

David Nilson. Acrylics. Queen Charlotte Airlines Supermarine Stranraer CF-BXO in a typical BC coastal setting. QCA made good use of the rugged, reliable "Stranny", especially from Vancouver, serving Alcan's burgeoning aluminum operation at Kitimat. Built under licence by Canadian Vickers in Montreal, CF-BXO served into the 1960s. At the end of its days it was acquired by the RAF Museum in Hendon, England, where it became a premier display—the last complete Stranraer in existence.

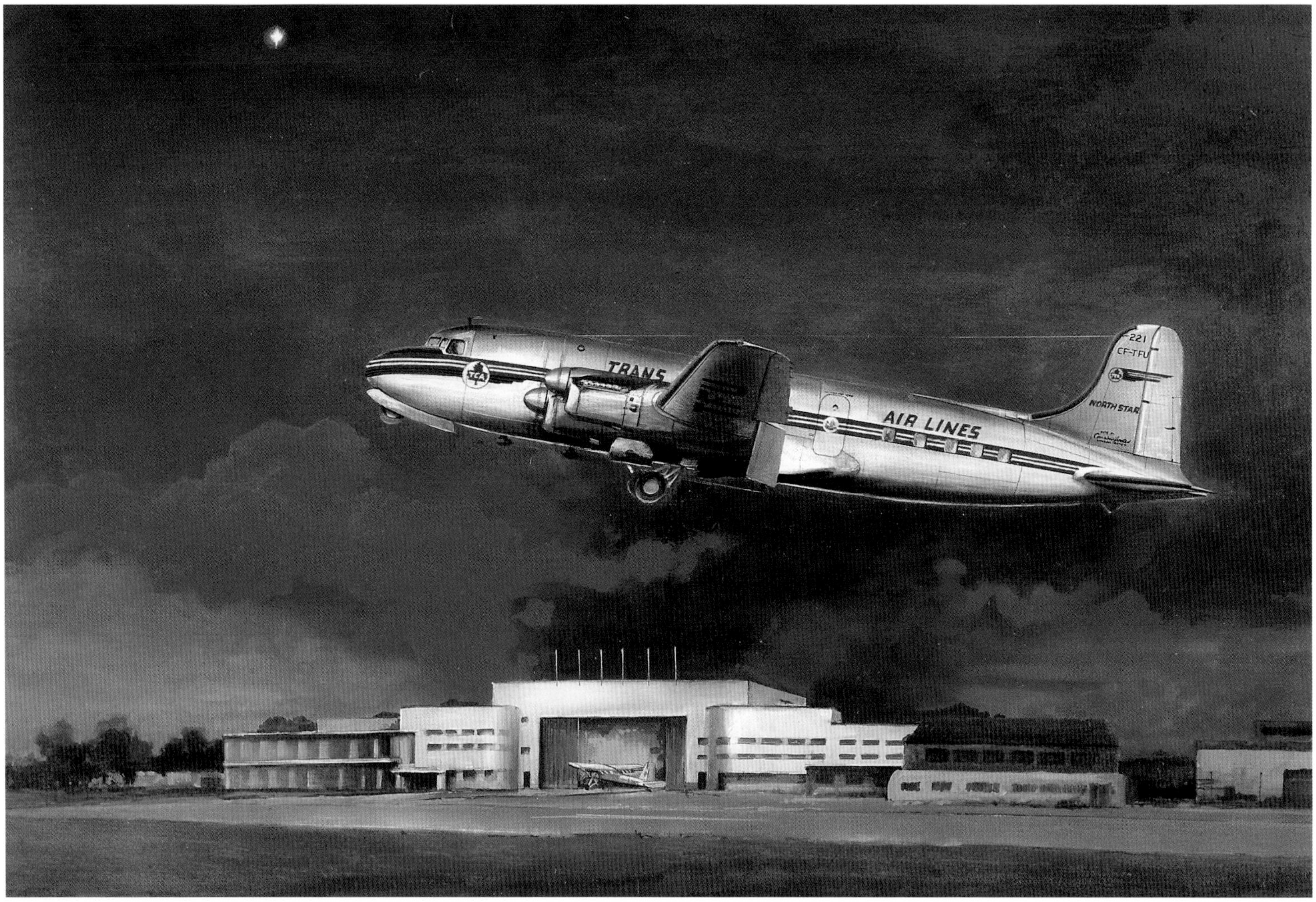
TRANS
AIR LINES
221
CF-TFU
NORTH STAR

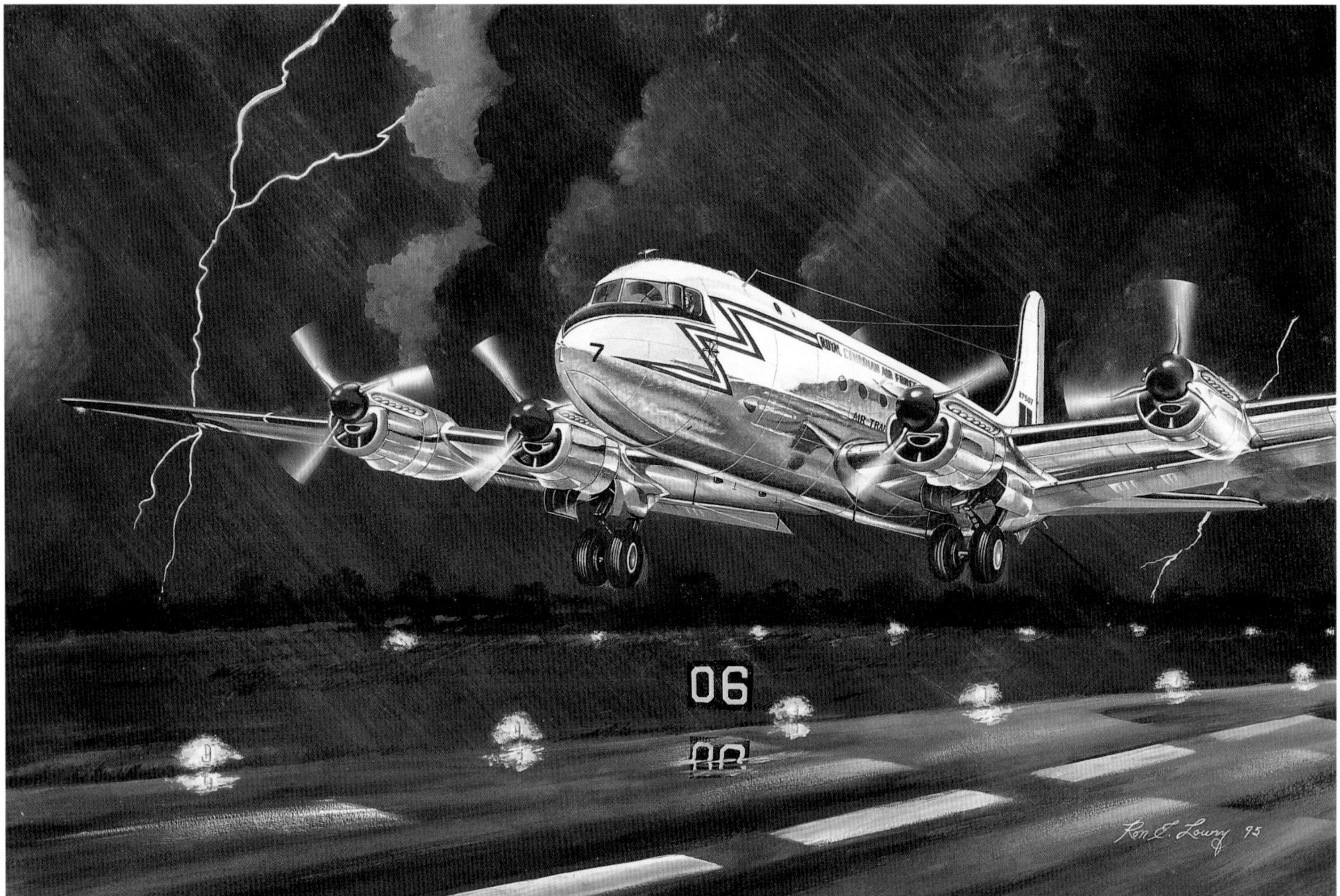
06

Facing page

Dugald Cameron. Acrylic. Canadair North Star CF-TFU of Trans-Canada Air Lines departing Prestwick for Canada in the early 1950s. The North Star served well on the Atlantic in the period before TCA replaced it in 1954 with the bigger, faster Super Constellation. The backdrop is the classic Palace of Engineering, erected for the 1938 Empire Exhibition in Glasgow, and still used in the 1990s by British Aerospace.

Ron E. Lowry. Air brushed. North Star 17507 of 426 Squadron departs Trenton on operations as the weather boils and flashes behind. The North Star put in nearly 20 years with the RCAF, finishing a grand career in 1965 without having lost a single passenger or crew member. In spite of a lot of ill-deserved bad press, it holds a place of honour in aviation history. The only surviving North Star, an ex-RCAF example, may be seen in Canada's National Aviation Museum.

Dugald Cameron. Acrylic. De Havilland D.H.106 Comet 1A of 412 Squadron crossing the English Channel from France to the UK. The RCAF received two Comets in April 1953. With these it became the first operator in the world with scheduled trans-Atlantic passenger jet service. Canada's two Comets served admirably into the 1960s, then were sold to an entrepreneur, who hoped for a sale in Latin America. He combined the best parts from each to create CF-SVR, but his sale was blocked by Ottawa. SVR met its end in a Miami scrapyard.

Alfred Wong. Acrylic. Beech 18 Expeditor No. 1597 banks away from the setting sun. Beginning in 1940 and lasting till 1968, the Expeditor was a jack-of-all-trades in the RCAF, working in such roles as VIP transport, pilot and nav trainer, and search and rescue. Many ex-RCAF Expeditors still were serving Canadian bush operators in the late 1990s.

Facing page

Alfred Wong. Computer art. An Air Transport Command Fairchild C-119 Flying Boxcar and Canadair CC-106 Yukon. From Arctic icebox to torrid tropics, these great types served the RCAF in the 1950s-60s.

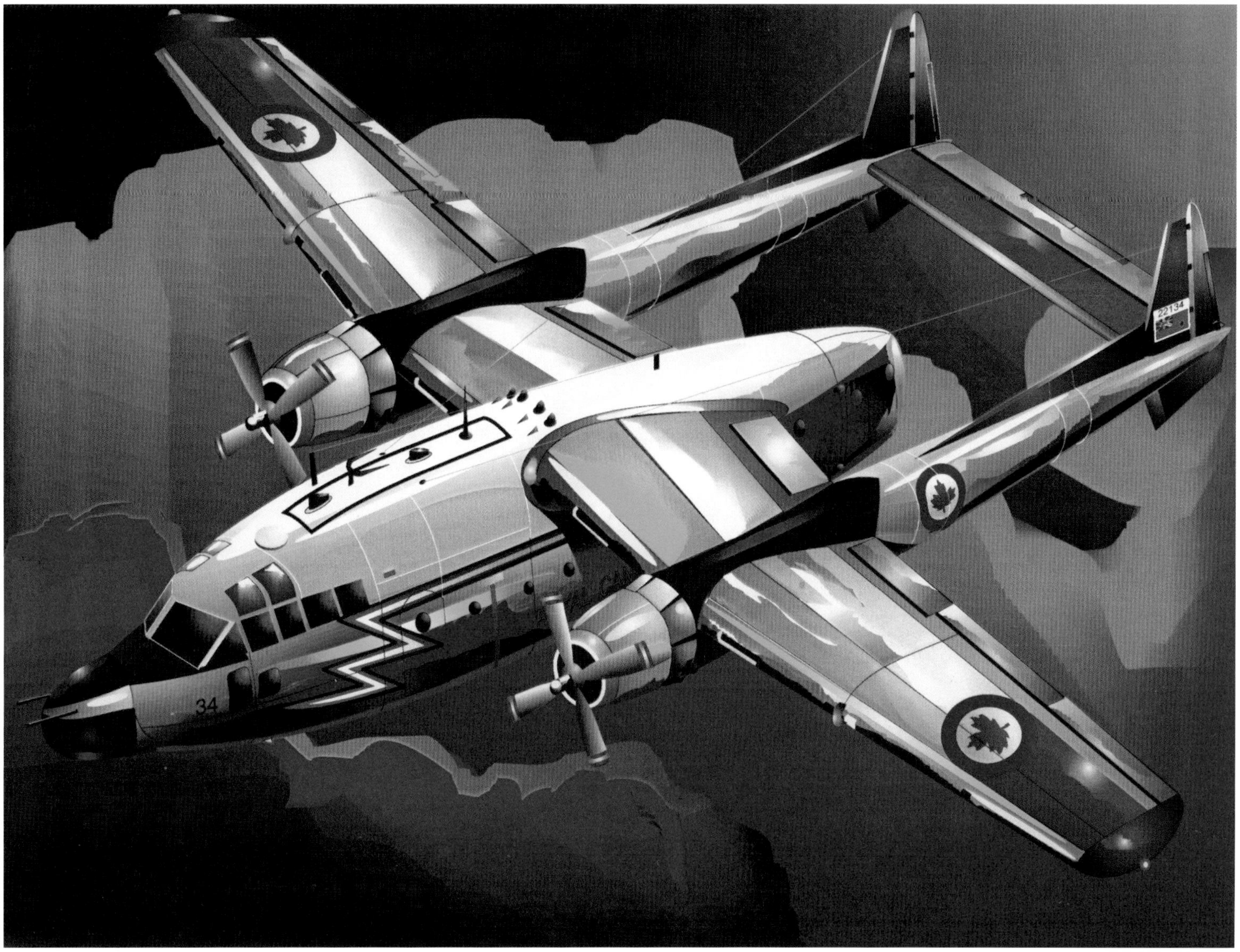
22134
34

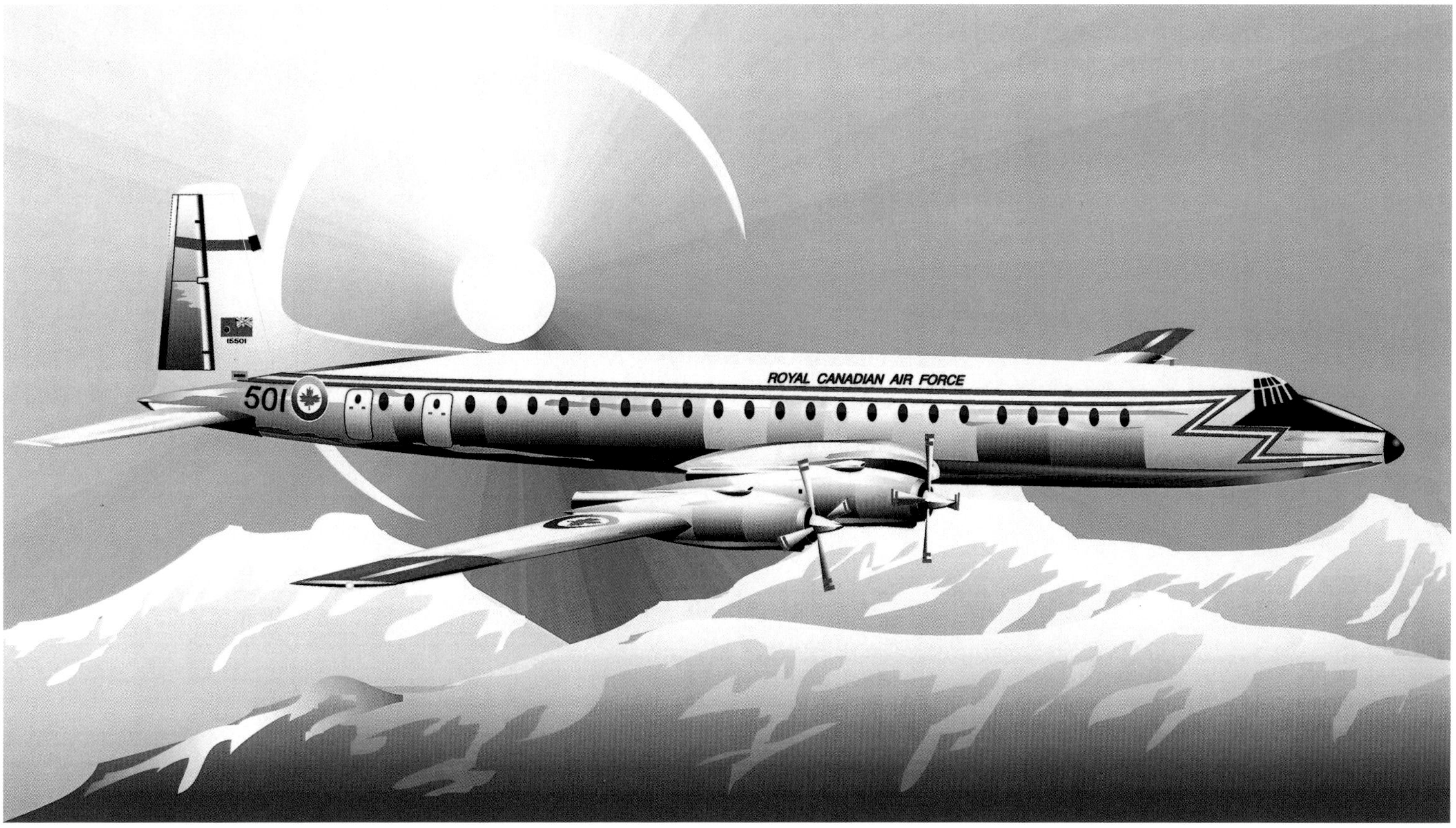
15501
ROYAL CANADIAN AIR FORCE
501

Alfred Wong. Computer art. An RCAF Vertol H-21B, the type used first during the construction phase along the Mid Canada Line. H-21s also specialized in search and rescue, until replaced by the Vertol CH-113 Labrador. This art form is new. Alfred explains: "These drawings were made freehand on an IBM486 PC using the CorelDraw program and a mouse. No scanned images were used. All the individual shapes which make up the aircraft and their details were drawn by manipulating basic shapes with the bezier and node tools. The colours, shades and fills come from the stock palette within the program. Unlike traditional painting, the fills and shadings are restricted by the forms and kinds of shading presets allowed by the program. However, the computer is able to render crisp, sharp images with excellent colour saturation."

Robert Finlayson. Oils. A Spartan Air Service aero survey Lockheed P-38L Lightning over Whitehorse. The modern maps of Canada's north owe much to the work done by Spartan's P-38s and Mosquitos in the mid-1950s. CF-HDI, modified from one of the rare two-seat P-38 night fighters, was a favourite of Spartan's famed chief pilot, Weldy Phipps. It later ended in a tragic crash.

Ron E. Lowry. Airbrushed. The PBY-5 Canso replaced the Stranraer on the BC Coast. This painting represents PWA's CF-GLX taking off from a remote BC inlet. Like DC-3s and Beech 18s, commercial Cansos were stalwarts, backing the HUT airlift, the Mid Canada Line and the DEW Line. They hauled fish in Manitoba, fuel in the high Arctic, sportsmen on BC's North Coast, and school children in James Bay. They bombed forest fires from Vancouver Island to northern Saskatchewan to Newfoundland. Seemingly irrepressible, several Cansos remained at work in the late 1990s, fire bombing with Buffalo Airways in the NWT.

Robert Finlayson. Oils. Curtiss C-46 Commando and Douglas DC-3 of World Wide
Airways in a typical DEW Line setting of the mid-1950s. World Wide, a legendary air
carrier in this era, was headed by the flamboyant Don McVicar. The classic C-46 and
DC-3 were the backbone of Arctic resupply into the 1970s. A few still served at the turn
of the century, notably with Buffalo Airways in the NWT.

Robert Finlayson. Oils. TCA Lockheed L.1049G Super Constellation CF-TEU over Scottish countryside around Prestwick in the mid-1950s. This type was one of the grand postwar trans-Atlantic airliners. But its days were cut short by the appearance of the first big jets. CF-TEU ended its days in 1966 in a Lancaster, California scrap yard.

Robert Finlayson. Oils. Bristol Britannia CF-CZA "Empress of Buenos Aires" of Canadian Pacific Airlines over the Andes. The Britannia, like the Lockheed Electra, filled the gap between the great piston-engine propliner era and the new passenger jets. Passengers were impressed by the speedy "Whispering Giant", as the Britannia was dubbed.

Stéphane Cochin. Airbrush. Noorduyn Norseman Mk.IV CF-FQI over the Chubb Crater
(aka New Quebec Crater and Pingualuit Crater) in Quebec's Ungava region. It was
serving Sept-Îles Air Service at this time (1963). While this book has the most complete
Norseman history yet published, the artist was thoughtful enough to tell a little about
Chubb Crater, where the Norseman was perhaps the first airplane to visit: "The crater
was formed about 1.3 million years ago when struck by a meteorite 170 meters across.
It is one of the best preserved meteor craters in the world. Historically, it was a
meeting place for the Inuit during hunting and fishing expeditions." Norseman FQI was
active in 1997 with Viking Outpost Camps of Red Lake, Ontario.

Robert Finlayson. Oils. De Havilland Canada DHC-3 Otter CF-ODJ of Ontario's
Department of Lands and Forests in a fire fighting scene. The Otter first appeared as a
"flying fire truck" in 1957, equipped with two float-mounted, 97.5-Imp. gallon tanks.
Next came a single 180-Imp. gallon tank under the belly. These tanks were filled in
seconds using small probes as the Otter skimmed along a lake at high speed. Later
equipped with floats having built-in water compartments (220 Imp. gallons total), the
Otter proved an excellent fire fighter till replaced by Twin Otters and CL-215s
in the 1980s.

Geoff Bennett. Acrylics. Grumman Avengers on budworm spraying in northern New Brunswick. From the 1960s into the 1990s these sturdy WWII torpedo bombers led the battle against the equally sturdy budworm, that scourge of Canada's spruce forests. When not fighting its pesky arch-enemy, the Avenger often was busy with forest fires in BC or New Brunswick. By the late 1990s Avengers had disappeared from forestry work and moved into the ranks of the classic warbirds.

Jan Stroomenbergh. Acrylics. Transair DC-3 CF-TES cruises over the prairies in a typical springtime setting. CF-TES had been in the RAF in WWII as FL547, then joined TCA in 1947. Later careers were with Transair and Lambair. TES found its final home with the Western Canada Aviation Museum in Winnipeg.

Geoff Bennett. Acrylics. The Handley Page Dart Herald was one of many DC-3
"replacements" ordered by the airlines after WWII. It was an excellent design—faster,
quieter and roomier than a DC-3. So far the DC-3 has outlived most of its replacements,
some of which themselves have become extinct. This EPA Dart Herald is depicted in
typical Atlantic Canada flying weather.

Robert Finlayson. Oils. Modern turboprops like the Beech 99, Beech 1900, Metro, ATR, Brasilia and Dash 8 eventually pushed the last of Canada's DC-3s onto the northern frontiers. Here Canadian Partner Embraer EMB-120 Brasilia is represented over Timmins airport in Northern Ontario. This speedy 30-seater was used by Canadian Partner between Toronto and Winnipeg, but its days were brief. By June 1993, three and a half years after the first of six had been delivered, the Brasilia was phased out.

Garfield Ingram. Acrylics. The De Havilland Canada DHC-5 Buffalo served the Canadian air force from 1968 to the turn of the century. Here Buffalo 115452 of Canada's 116 Air Transport Unit is over Port Said, Egypt in 1978. Two 116 ATU Buffalos were based at Ismailia in the Sinai in this period, supporting UN peace keeping.

Ron E. Lowry. Air brushed. Air Transport Group's Lockheed C-130 Hercules 130326 (call sign "Canuck 36") at Jijiga, Ethiopia on a muddy day in November 1991. It was delivering UN food relief from Djibouti to famine victims in the Ethiopian desert. Such operations were a trademark of ATG in the 1990s. The Canadian Hercules, the perfect plane for the job, turned up wherever tragedy struck, from Yugoslavia to Kurdistan, Ethiopia, Somalia and Rwanda.

Other Western Carriers

A book such as *Air Transport in Canada* never would be published if the point was to squeeze in everything. Aviation history researchers and writers eventually will fill the gaps. Meanwhile, this chapter takes a look at some more case studies from Western Canada.

Max Ward, whose early days are covered in Chapter 19, had been an RCAF instructor pilot during WWII. His military years at an end, he took a job in 1945 flying with Northern Flights, a Yellowknife company owned by Jack Moar, but Ward was keen to be his own boss. In 1946 he bought a Fox Moth and began operations from Yellowknife as Polaris Charter Co. In 1947 Ottawa required that air carriers hold an operating certificate. Ward did not have one, so formed a partnership with George Pigeon, who did. With Fox Moth CF-DJC and a Stinson 108 they operated as Yellowknife Airways.

In 1949 Ward left to fly with Associated Airways. As prospects improved in mining, he gambled and ordered one of the first Otters. DHC delivered it to him in Yellowknife in 1952—Wardair was in business. As business increased, aircraft were added—by 1962 the fleet numbered three Otters, a Beaver, a Bristol 170, a Beech 18 and a DC-6B. The latter, bought for overseas passenger charters, came from KLM. Business centred on Europe and the Caribbean and the first four years kept the DC-6B busy. Ward ordered his first Twin Otter in 1966 and had six by 1973. Now he foresaw a charter boom—the time seemed right to enter the jet age. Wardair ordered Canada's first Boeing jet liner—B.727 CF-FUN. It entered service in April 1966. Two years later, with business soaring, Wardair accepted its first 707, another a year later. In 1972 a 747 was ordered. An ex-Braniff machine, it was delivered April 23, 1973. In late 1974 a second 747 (ex-Continental) arrived.

In 1975 Wardair carried 161,000 round-trip ABC passengers, 125,000 from Canada (56,000 of whom used Toronto) and 36,000 from the UK. Two 707s and two 747s did all the work. At a peak there were 15 weekly flights to Gatwick, four to Manchester and two to Prestwick. Vancouver was the company's second busiest location, but Calgary, Edmonton and Winnipeg all expanded. Winter flights were on the upswing. By this time Wardair had an international reputation as one of the finest global air carriers. So good was the service that after one trip a passenger usually was hooked on Wardair. The company prospered through the 1970s-80s, then Max Ward decided to build a fleet of new Airbus A310s for long-haul, and MD-80s for something new that he wanted to try—competing with Air Canada on domestic routes. Soon, however, Wardair ran into expansion trouble. When the opportunity arose in

(Above) Three ex-TCA Bristols in Wardair service at Yellowknife. They didn't fit in with TCA but were ideal for Wardair, especially serving the mining industry. (CANAV Col.)

(Right) The original Wardair CF-FUN "Cy Becker" was this Boeing 727 seen on delivery from Boeing in April 1966. In May 1973 it was sold to Cruzerio do Sul of Brasil. In 1997 it was with Aerocar Colombia as HK-3770X. (Boeing)

(Right) Boeing 707 CF-ZYP "W.R. 'Wop' May" on a charter at Toronto about 1970. It served Wardair 1969-78, then Austrian and Sudanese carriers. In May 1981 it was seized by the US government, was stored at Davis-Monthan AFB, then joined the USAF as 85-6973. (Larry Milberry)

(Above) The tarmac was covered with Wardair planes on the day in April 1973 the company accepted its first 747–CF-DJC "Phil Garratt". Also present were Fox Moth DJB, 727 FUN and 707 FAN. DJC served till the 1989 CAIL takeover. It later served Nationair, Garuda of Indonesia and Saudia. In the early 1990s it was in storage in the Arizona desert. (Boeing)

(Right) Wardair was an early user of the Dash 7, this example (c/n 7, C-GXVF) being named for Don Braun, one of Wardair's early pilots. XVF spent only a year with Wardair before being sold in the US as N27AP. In the early 1990s it was in Alaska with Markair as N677MA. (DHC 46506)

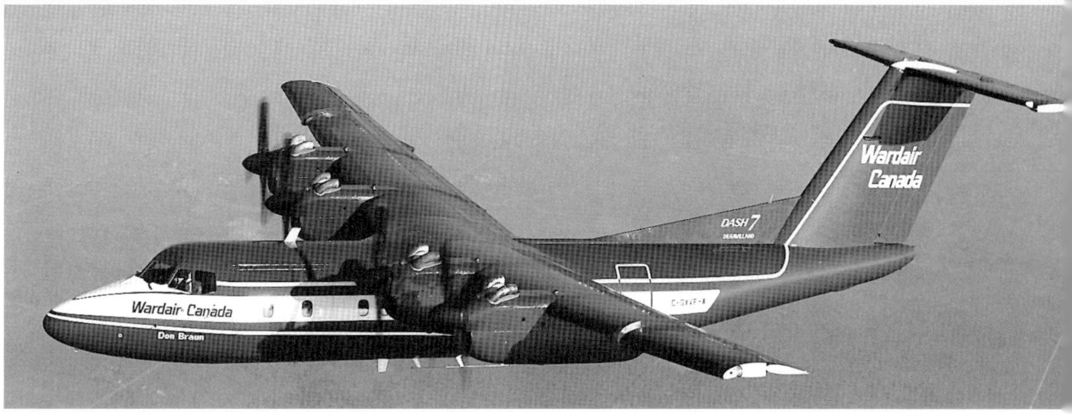

1989, Max Ward sold to PWA Corp. (Canadian's parent company). He chronicled the whole story as he saw it in his 1991 autobiography *The Max Ward Story*.

Eldorado Aviation
In 1944 the federal government formed Eldorado Mining and Refining to mine and process radioactive ores at Port Radium, at the east end of Great Bear Lake. The only way to transport ore from the site was by air. A division called Eldorado Aviation was formed in Edmonton; for 1944 two Norsemen were used (earlier, Mackenzie Air Service's Bellanca Air Cruiser had become famous carrying Eldorado ore). In 1946 Eldorado added a DC-3. It was

(Right) In the 1930s-40s Bellanca Air Cruiser CF-AWR, flying first for Mackenzie Air Service, then Eldorado, and CPA, supplied the mine at Port Radium, and carried out sacks of ore concentrate. Various types followed over the years. (CANAV Col.)

joined in 1947 by a DC-4 and a C-46. The operation was ideal since there were good loads in both directions. Summer flying was from a 4,500-foot strip 40 miles away at Sawmill Bay; winter operations were from the ice at Port Radium. The mine closed in 1960, but air ops continued to Eldorado's site at Uranium City on Lake Athabasca. When it closed in 1983, Eldorado sold its fleet which, by that time, included a 737. In 1975 Echo Bay Mines formed as an offshoot of Eldorado, using DC-3s, then a Convair 640, Hercules and 727. In 1989 Echo Bay became part of Wescan Transportation Services of Edmonton, by when it had weekly skeds to sites in the NWT.

International Jet Air

In the late 1960s Roy Moores set up Moores Aviation in Calgary. Using Turbo Commanders, etc., he looked for charter work. Business grew when he became a partner with surgeon and entrepreneur Dr. Charles Allard in a new company, International Jet Air. It carved out a moneymaking niche in Alberta's booming oil and gas industry. In 1970 IJA purchased the licences of Great Northern Airways, which had been Dawn and Gordon Bartch's company. The latter had begun as a small Beaver and Cessna operation, Connelly-Dawson Air of Dawson City. Ron Connelly eventually left and started Conair. GNA grew and moved to Calgary in the late 1960s, by when it had types like the DC-3, DC-4, F.27, Otter and Turbo Beaver. When

Panarctic Oil asked GNA to set up an Electra operation to rotate crews at Arctic drill sites and keep the sites supplied, GNA recommended Britannias instead (the concept was stillborn). In 1968-69 GNA wrecked F.27s at Inuvik and Resolute Bay. This helped put it out of business.

IJA's fleet eventually included the Electra, F.27, Jet Commander and Jetstar (Allard's private plane). The Electras were Canada's first. Excellent examples were bought for $250,000 (passengers) and $350,000 (combi with cargo door). Several were placed on the line, others stored at Calgary. CF-PAB inaugurated service in 1969, followed by IJM the same year, IJV (a 13-ton freighter) in 1971; and IJW, IJY and PAK in 1973. IJC and IJJ were bought in 1971, but were flipped within months at a good profit.

Using IJY, a daily 1½-hour Whitehorse-Inuvik sked was begun supporting exploration in the Beaufort. Roy Moores then set up Truck-Air, whereby tons of cargo moved overland to Fort Nelson from where it went to Inuvik by Electra two or three times a week. Eureka, Johnson's Point, King Christian, Rae Point and Resolute Bay were some more distant destinations. The Electras were at home on either gravel or ice. Calgary-Edmonton-Tuktoyaktuk became a key Electra passenger service operated for Dome Petroleum. It involved IJW (passengers) and IJV (combi). This continued to 1980, when Dome bought 737 C-FDPA. It gave Dome some independence in re-supply

and crew rotations, but cost a great deal more per seat mile than an Electra.

Dome became the biggest of the players in the Arctic oilfields. It built a large fleet including the 737, three Citation IIs, a Gulfstream, HS125, Twin Otter, King Air 90, S-61, BO 105, and two S-76s. The flight department under Bud McMurchie grew to about 150 people. Herb Spears worked for Dome in operations. Boeing was incredulous when he asked how much sea ice it would take to support a 737. Boeing had no idea, but Dome soon determined that at a 50,000-pound AUW 5½ to 6 feet were needed. Over the years DPA operated regularly from 7,000-foot ice strips established beside a number of frozen-in drill ships. Support for one ship cost a half million dollars a day. In 1988 Dome was taken over by Amoco, which sold most of the aircraft.

IJA had other interests. Beginning in 1971 it operated Electra IJR for Imperial Oil. It also ran a host of specialized charters. It carried the Edmonton Oilers hockey team from game to game. Basketball's famous Harlem Globe Trotters used IJA, as did singer Ann Murray on a northern tour. CF-IJW worked gambling centres like Reno and Lake Tahoe. Electra pilots Bob Bowman, Harry Brown, Glen Long, Ken Peters, Joe Redmond and others kept busy as IJA flourished. A top staff of engineers like Fred Woodhall and Stafford Wagstaff kept the fleet in shape. IJA suffered one disaster, however—on October 29, 1974 PAB crashed through the ice at Rae Point with serious loss of life.

IJA formed a partnership with the Russians to market the Yak-40 commuter tri-jet in North America. A Yak-40 toured North America, but no interest developed. It was a decent plane, but primitively equipped, especially in avionics. The Soviets were not prepared for the sophisticated North American market. Dr. Allard's communications interests gradually topped his list of activities. In 1975 he sold IJA except for the Jetstar. Bob Engles of NWT Airways quickly realized an opportunity and bought the Electra operation. Along with it came IJA's customers and, under NWT Air, the operation prospered. Roy Moores left Calgary for Toronto, where he established a bizjet charter service and an important new airline—Worldways Canada.

Saskatchewan Government Airways

Saskatchewan got into aviation with four ex-RCAF Vedettes acquired between 1933-36. These were mainly for forestry, much as were Manitoba's Vedettes, but all were retired by 1937. From then through WWII Saskatchewan had little to do with aviation. In 1946, however, M&C Aviation of Prince Albert sold to the province and Saskatchewan Government Airways came about, offering air transportation to forestry, mining, commercial fishing, trapping and tourism. It also did air ambulance, photo and postal work. Revenue in 1948, the first full year of operations, was $268,271. The SGA annual report for 1949-50, submitted by chairman J.H. Brocklebank, summarized the flying done:

Waco YKS-6 CF-AZQ was the Saskatchewan government's first postwar aircraft. It came to Canada in 1936 for the New Brunswick Department of Lands and Mines. It went to the DOT in 1940, was an instructional airframe with the RCAF, served Curtiss-Reid Flying Service at Cartierville, then went to M&C Aviation in PA. M&C sold AZQ to Saskatchewan in January 1945. Aklavik Flying Services bought it in July 1948. On March 9, 1950 it was damaged beyond repair landing north of Aklavik. Then, a view of SGA Waco ZQC-6 CF-BDM at La Ronge over the summer of 1947. (Glass Col., Archives of Saskatchewan)

Saskatchewan Government Airways

(Above) CF-SAF was one of the Tiger Moths bought by Saskatchewan for a few dollars each from War Assets. (Glass Col.)

(Left) A group of forestry men on a trip to Stony Rapids. Pilot Floyd Glass is at the rear. (Glass Col.)

(Below) Norseman SAN awaiting frozen fish in front of the La Ronge fish filleting plant, then at Fort Smith, both in early postwar days. The Norseman was the backbone of SGA operations into the 1960s. (Saskatchewan Archives Board)

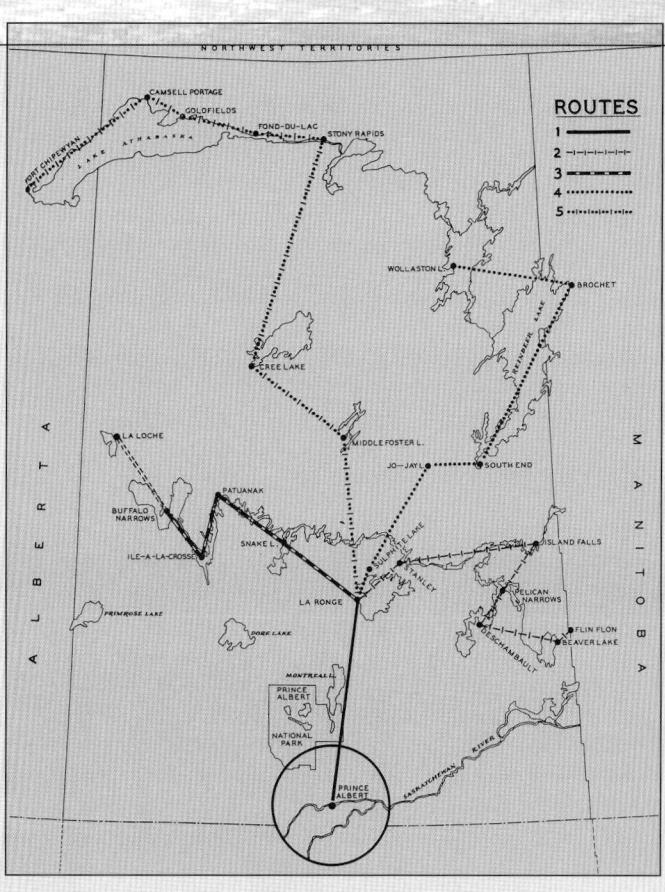

(Above) SGA routes in 1949. Then (left) an SGA Beech 18 in the late 1950s, shortly before it joined Southern Provincial Airlines in Toronto. (Saskatchewan Archives Board R-B5621)

753

(Top) The SGA dock at La Ronge on July 12, 1957—a typical scene with Cessnas, Norsemen, and a Beaver departing. Then (above) SAG Otter CF-JFJ on Lac la Ronge in July 1958. It later flew for Norcanair. (Saskatchewan Archives Board R-B6765)

(Right) An aerial of Uranium City, Saskatchewan in August 1957 as a SGA Ce.180 passes over. Thirty years later, "U-City" was almost a ghost town. (Saskatchewan Archives Board R-A13493)

(Below) DC-3 CF-SAW loads at UC for the sked to PA. It had careers in the UK, Portugal, and Greece before coming to Canada, where it served SGA 1954-61. It later was CF-NAO with Nordair, Pem-Air and Austin Airways, then migrated to Florida in 1977. (Saskatchewan Archives Board R-B6782-1)

(Right) SGA water-bombing Cansos do a practice drop at Moose Jaw on July 11, 1992. (Larry Milberry)

(Below) A Saskatchewan government Navajo at Swift Current on August 10, 1976. (Larry Milberry)

The third complete year of operations of the Saskatchewan Government Airways ended on October 31, 1950. The year has seen a marked increase in the corporation's volume of business, indicating that Saskatchewan Government Airways is becoming firmly established as an air carrier which can be depended upon to provide safe and practical service. The contribution that the airways is making in assisting the provincial government in its mineral development program in northern Saskatchewan is now beginning to show concrete results. The service provided by the corporation to the newly formed mining companies, prospectors and other residents of northern Saskatchewan has become an essential part of the daily life of this area of the province.

The year's operations showed that the corporation's aircraft carried a total of 7,096 passengers, 1,877,486 pounds of cargo, and 72,999 pounds of mail. Compared with the operations from the previous year, these figures represent increases in passenger traffic of 86%, 29% in volume of cargo, and 89% in pounds of mail carried. Because of the increased business with mining companies in the Lake Athabasca area, a base at Stony Rapids was applied for and obtained from the Air Transport Board. Aircraft were maintained at Stony Rapids and Goldfields during the summer months, and one machine will be retained in the area during the winter months.

The base at Lac la Ronge continues to develop and the majority of the cargo transported originated there. Because of Lac la Ronge's strategic location in the central part of the north half of the province, the competitive rate for flying freight from this base compares favorably with the rates from points similarly located in Alberta and Manitoba. Resident employees are now being maintained at Lac la Ronge the year round to handle the business there.

The construction of airstrips by the Department of Natural Resources in various parts of the north is an added advantage, not only for Saskatchewan Government Airways, but other air operators as well. Passengers and cargo can now be transported between airstrips faster and more economically than by float-equipped machines.

During the fiscal year under review, the fleet of aircraft owned by Saskatchewan Government Airways consisted of one Canso, three Ansons, four Norseman, two Beavers, three Stinsons and four Tiger Moths. The amount of business brought in to the corporation's overhaul shop continued to be about the same as the previous year. The shop not only provides a valuable service to many northern operators, including those based in Manitoba and Alberta, but also allows the corporation to reach a very high standard in servicing and maintaining its own aircraft. This is a major factor in the excellent safety record enjoyed by Saskatchewan Government Airways. There were no increases in fares or tariffs during the fiscal year under review. Wages were increased and the staff slightly enlarged, but due to the increased volume of business and the gain on the sale of capital equipment, the corporation was able to show a net operating profit of $23,383 on expenses of $271,985.

The building of Mid Canada Line stations across Saskatchewan in the 1950s brought a welcomed flood of charter business. Revenue hit $930,828 for 1960. A specialty of the SGA was using smoke jumpers in fighting forest fires. Norsemen were dedicated to this. When a fire erupted, they were ready to fly from the jumpers' base at Lac la Ronge. If required, jumpers then would go in to work the fire.

SGA became a full-fledged regional airline when it bought a DC-3 for the Prince Albert-Uranium City sked (replacing Ansons). A second DC-3 was added in 1957. By this time the mining boom town of Uranium City had a population of 5,000. In discussing the likelihood of a DC-3 replacement, SGA manager I. Macleod noted in August 1958 (to Edwards): "It would appear they will be a desirable type

cenced in 1938, then bought CEZ and Avian CFA for $450. He ran a little business from the local field. When a passenger accidentally hit the throttle one day and ran CFA into a fence, Glass combined the best of parts from his two Avians and carried on. He also earned an air engineer's licence early in his career.

In WWII Glass was a flying instructor at No. 6 EFTS, which had been set up at PA to train pilots on Tiger Moths. After a year and a half he trained on Harvards at No. 6 SFTS at Dunnville, then was posted to Arnprior to instruct at No. 3 FIS. From there he hoped for an overseas posting, but instead went to No. 24 EFTS at Abbotsford. He took his release in December 1944 and returned to PA. In January 1945 the CCF (Co-operative Commonwealth Federation) provincial government purchased Waco CF-AZQ for natural resources work in the north and hired Glass to fly it. He recommended that Saskatchewan rebuild its dormant air service and went to Ottawa to arrange licences with C.D. Howe. Still a powerful figure and as opinionated as ever, Howe told Glass, "Young man, you have a good idea and can have your licence, but I still have no use for that CCF government of yours."

In 1947 Saskatchewan, on Glass' recommendation, purchased a Canso, five Norsemen,

January 7, 1993 at Prince Albert. Minus 30 degrees. Passengers shuffle out to their Athabaska Airways Beech 1900C sked heading to points north. (Larry Milberry)

for some years to come, particularly for routes of this nature which cannot economically stand large capital outlays for newer equipment, such as the Friendship and the Caribou... What we actually require is a smaller twin-engine aircraft as fast or faster than the DC-3, with a load capacity of 10-12 passengers to be operated as a strictly passenger type, supplemented by the DC-3 as a freighter." For 1958 the SGA fleet included two DC-3s, an Otter, four Norsemen, five Beavers, nine Ce.180s, a Ce.170, a Ce.140, a PA-18. A separate air ambulance fleet included four Ce.195s and two Beech 18s.

Athabaska Airways
In 1955 Floyd Glass of Prince Albert founded Athabaska Airways. In the 1930s his family had moved to PA to farm and operate a creamery. Floyd got "the flying bug" at about age 12, when a barnstormer operated from his father's farm. He learned to fly in Avro Avian CF-CEZ owned by the Saskatoon Aero Club. He was li-

and three Ansons from War Assets. The Ansons, only $800 each, were "like new". Glass also picked up five surplus Tiger Moths for forestry patrol. Bases were established at PA and La Ronge. Weekly skeds were organized into Cree and Chipewyan Indian settlements as far north as Stony Rapids. Flying into this isolated region, Glass saw the misery in which the Indians lived. In one settlement he saw only adults, so asked where the children were—they were all dead from chicken pox, which recently had swept through. It was clear how badly the Indians needed air transport for medical and other needs.

Glass was a proponent of airstrips, where landplanes like the Anson could be used instead of more costly float operations. The government put in its first northern strip at La Ronge. The DC-3 was added for U-City. Glass stayed with the SGA through its formative years, then left to fly for QCA at Prince Rupert. He returned to found Athabaska Air-

Prince Albert and La Ronge

In and outside Athabaska Airways' hangar at PA in January 1993 and (inset) the company's modern electronics shop. FCH was one of the company's sizeable fleet of reliable Cessna twins. Then, Floyd Glass and son Jim. (Glass Col., Larry Milberry)

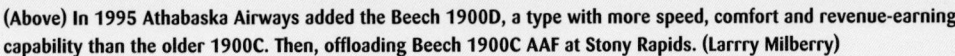

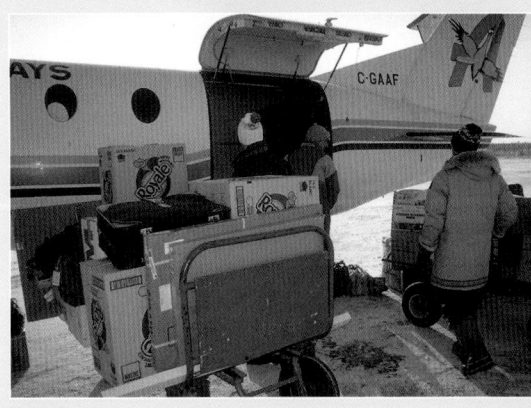

(Above) In 1995 Athabaska Airways added the Beech 1900D, a type with more speed, comfort and revenue-earning capability than the older 1900C. Then, offloading Beech 1900C AAF at Stony Rapids. (Larrry Milberry)

(Above) Mrs. Mamie Glass with son Jim (standing), daughter Carol and Athabaska Airways' first helicopter, Bell 47G2 CF-IKS. Floyd Glass purchased it from Western Helicopters of Port Credit, Ontario. Jim became a pilot in 1979 and later was general manager of the airline. (Glass Col.)

(Right) S-55T C-FUNT ready for a day's work at PA with engineers Dale Janeczko and Don Stephen and foreman Peter Vis. Their gear includes a 200-gal. firefighting bucket and some cargo netting. (Glass Col.)

ways with a new Ce.180. He soon had enough business for another, but this was small potatoes compared to what was to come with Mid Canada Line and uranium exploration. Of course, there was always summer tourist business—before long Athabaska Airways had 18 aircraft. Its first helicopter was a Bell 47G2

purchased in 1961 and soon it was busy on mining and natural resources contracts. Eventually the company had a big rotary wing operation, and added the first of several Sikorsky S-55s in 1970. An S-58T came later. The first twin was a Cessna 310 salvaged in 1959 from a lake 150 miles northwest of La Ronge. Glass

bought it from the insurance company for $3,200 then dismantled and overhauled it on the lakeshore. After freeze-up it was reassembled and flown off the ice. This was the first small twin in the area and was popular for business and tourist charters.

In the 1970s Athabaska Airways had as

(Above) A line of Athabaska Airways S-55Ts at the Glass farm in PA. (Glass Col.)

(Left) Sgt Brian C. Blacklock readies RCMP Twin Otter MPJ for a patrol from Prince Albert on January 8, 1993. MPJ served about 15 detachments as far north as U-City. (Larry Milberry)

Les Martin was working for Transport Canada in the PA flight service station in January 1993. The FSS provided such information to flights as weather, traffic and radio frequencies. (Larry Milberry)

(Above) An AirSask Navajo at La Ronge on January 7, 1993. AirSask (formerly Pinehouse Airways) was run by Pat Campling, whose main operation at this time was La Ronge Aviation. Years earlier he had run Mackenzie Air of Edmonton. Then (right), a La Ronge Beech 99 the same day with the crew of Landon Zenuk and Robin Wohl. (Larry Milberry)

(Left) Over-wintering water bombers at La Ronge in January 1993. (Larry Milberry)

many as 35 aircraft including Beavers, Otters and Beech 18s. It remained a solid operation through the decades, but in the 1980s there was a slowdown in charter work. This was partly the result of roads being built e.g. to Buffalo Narrows. Athabaska countered by opening some skeds with Cessna 404s and developing these into a feeder arrangement with Time Air. In February 1991 Glass added Beech 1900C C-GAAF and began commuter skeds to the south. The 5-days-a-week service flew Regina-Saskatoon-PA-La Ronge-Stony Rapids. A second 1900C (C-FCMV, ex-Central Mountain Airlines) arrived in May 1992. Other skeds/ charters were being flown to places like Beauval, Buffalo Narrows, Fond du Lac, Pine House, Points North, U-City and Wollaston Post. By 1993 Athabaska Airways also was serving Northern Saskatchewan mines like Cluff Lake, Key Lake and Collins Bay, supplementing Time Air, which ran F.28s to and from the mines.

Support to mining traditionally has been boom-or-bust. During exploration and staking, mining always needs air transport. When a mine is developing, the need is also great; sometimes all construction material must be airlifted. Once a mine is running, air transport may fade, if roads develop. If not, scheduled air service is required. The downside comes when a mine closes, and in Saskatchewan there could be no better example than U-City. It boomed in the 1950s-60s as a jewel in the mineral-rich Canadian Shield. There were daily flights by SGA and Eldorado with DC-3s and DC-4s. Then the ore dwindled and the world market for uranium softened. In 1982 mine and mill operations ceased. A decade later the once booming community of several thousand had but 150 inhabitants.

In 1993 Athabaska Airways had two 19-seat Beech 1990Cs, two Twin Otters, a Cessna Conquest, a Ce.401, two Ce.402s, two Navajos, four Ce.310s, a Turbo Otter, two Beavers, a Turbo Beaver, two Ce.172s, five Ce.185s, four Bell 206Bs and five Sikorsky S-55Ts. The latter saw years fighting fires with 200-gallon buckets, and transporting firefighters and equipment. On a 1992 charter one of them slung a stranded Ce.206 from the bush near Flin Flon.

Typical flying with the Beech 1900C took place January 7, 1993 with pilots Corey Siemens and Ross Stewart. Siemens had started flying with his family's tourist business, while Stewart's experience included flying Cansos for Saskatchewan, and Twin Otters for Kenn Borek in Antarctica. Mid-morning on the 7th they flew AAF (Flight 202) PA- Regina- Saskatoon and back to PA, where they parked at the modern terminal. It was a clear, frigid morning with ice fog covering the area (the result of city smog, the plume from the pulp mill, and crystals produced by the snow-making operation at a nearby ski hill; there was no wind to dissipate this airborne matter, so visibility was restricted). In the terminal 14 passengers waited. A display on one wall, including a Tiger Moth propeller, noted the history of wartime aviation at PA, e.g. the EFTS had trained 2,647 pilots. Another plaque recorded June 24, 1985 as the opening date for the new terminal.

Soon the passengers boarded and AAF taxied and took off on the 117-nm leg straight north to La Ronge. It climbed at 2,000 fpm, levelling at FL200 at 220 KIAS. In 35 minutes F202 touched down at La Ronge. A regional hub, it logged about 23,000 aircraft movements in 1993. It was named for Jimmy Barber, DFC and was the main base for the provincial government's fire fighting fleet. At this time of year the CL-215s, Firecats and Cansos sat in snowdrifts or were hangared for overhaul. Both Athabaska Airways and La Ronge Aviation/AirSask used the airport for maintenance.

Next destination for AAF was Stony Rapids, a native community 248 nm and 1:10 hours north. It was the farthest north point on Athabaska's sked run. Departure was delayed 30 minutes because of ice fog at Stony Rapids,

but AAF finally left with 14 passengers. On arrival the freight quickly was offloaded and moved into Athabaska's hangar. The passenger waiting room was abuzz for several minutes as passengers met family and friends, and seven downbound travellers waited to board. In the hangar were a Ce.310 and a Navajo, while in the crisp -40° air sat an Athabaska Turbo Otter, Ce.185 and Jet Ranger. The company had four pilots here. The Turbo Otter, in its third year of work, was steadily busy. Floyd Glass noted that if the conversion wasn't so expensive he would like another. AAF was 30 minutes on the ground before leaving for La Ronge and PA as F203. It overnighted at PA, and next morning flew the sked to Regina to start the routine over.

In 1997 Athabaska Airways was typical of small, well-established Canadian air carriers. It was a family business with Floyd Glass at the head, and sons Jim (general manager) and Daniel (operations manager) involved. The fleet included two Beavers, a Beech 1900C, two 1900Ds, two Bell 206Bs, four Ce.185s, four Ce.310s, two Ce.402s, a Ce.404, a Ce.441, two PA-31 Chieftains, three Twin Otters, two turbine Otters and four S-55Ts. The family farm was still busy—3,000 acres that included a 3,500-foot airstrip, where the S-55T fleet over-wintered. There was also a family-owned island in the Bahamas to which Floyd Glass commuted in a Ce.310.

The Rise and Fall of Norcanair

Saskatchewan Government Airways prospered through the 1950s-60s, providing service throughout the north. It invested more in its operation than the typical commercial operator would have, considering the limits of the market. From 1948-61 SGA earned a profit in all but two years. In 1962 it was renamed SaskAir. The operation had been formed and run for years by the CCF government. In 1964, how-

ever, the free-enterprise Liberal party took over Saskatchewan. It had no use for government "interfering" in business—it sold SaskAir to a PA-based company which renamed it Norcanair. Included in the deal was a guarantee of $275,000 in annual government business, and 75% of government expenditures on air travel in areas served.

Norcanair expanded prudently, adding some DC-3s and a Bristol Freighter. The Bristol found work on northern jobs, including with Hydro Quebec. In 1972 Norcanair took over PA-Saskatoon-Regina from Transair, first using a Twin Otter leased from Midwest Airlines, then buying F.27Js from Hughes Air West of California. By this time the socialist NDP government had power in Saskatchewan, but it steered clear of the airline business. From 1976-81 Norcanair operated Regina-Minot, where passengers connected with US carriers. Meanwhile, it gradually focused operations in Saskatoon instead of PA. Several Canso waterbombers were acquired and leased each summer to the province. These later were sold to Saskatchewan, when it re-formed a natural resources aviation branch in the early 1980s.

By 1979 Norcanair was facing problems, including being cited for unfairly dismissing 20 employees (whom it was ordered to reinstate). In 1981 the owners had Norcanair for sale. Saskatchewan, anxious that Norcanair remain in its hands, offered $5.3 million for its southern routes, plus equipment and facilities. Saskatchewan would take over the main sked using 737s. This plan went out the window when the 1982 provincial election installed a Conservative government. The pro-business Conservatives did not want Norcanair, and High Line Airways of Saskatoon became the new owner. Formed in 1975 by Albert Ethier as an offshoot of his electrical contracting firm, High Line had some King Airs and a Convair 640. It continued operating as Norcanair, a

Norcanair took over from SGA. Here its DC-3 KBU was at Resolute in the 1970s. KBU came from the US in 1957, first to SGA, then Norcanair (1965), Aero Trades Western (1976), High Line Construction (1978), Slate Falls Airways (1981) and Bearskin (1983). It returned to the US in 1987. (via Gord McNulty)

Norcanair used a variety of types over the years, from F.27 to F.28 and this Brazilian-made Bandeirante. (NA/McNulty Col. PA191783)

name by then well-known among travellers. Frequent skeds were operated on southern and northern routes. A Saskatoon-North Battleford-Lloydminster-Edmonton Bandeirante sked was opened in 1984. The fleet was modernized. Two old planes (prototype Beaver CF-FHB and the Bristol Freighter CF-WAE) were donated to museums. In 1985 Saskatoon Regina Minneapolis was opened with a Convair 580. The next year the first of two F.28s appeared. In 1986 Ethier formed an alliance with CP Air, which needed Norcanair as a Winnipeg, Edmonton and Calgary feeder. Suddenly, in March 1987, PWA took over CP Air, and within a few weeks Ethier decided to sell to Time Air, itself 45% owned by PWA. Thus did Norcanair disappear. Ethier agreed to stay out of the airline business for three years after the sale. When he learned that Time Air might pull out of northern Saskatchewan, he decided to re-enter that market in 1990 with a Convair 640. Using the Norcanair name, he foresaw business in fishing, ski and sun charters; and was awarded sked licences to serve Baker Lake, plus PA-Saskatoon-Regina. For the 1990 tourist season the Convair, based in Minneapolis, logged 400 hours on fishing charters. In June 1990 Ethier leased 748 C-GQTG from Air Creebec for a daily Saskatoon-PA-La Ronge-Stony Rapids-Fond du Lac-Uranium City sked. In the fall QTG was off lease and the Convair took over. Ethier foresaw two F.28s for expanded skeds to the NWT; and mining and sun charters. In May 1992, however, Transport Canada pounced on Norcanair for safety violations, pulling its operating certificate. Complaints focused on unserviceable warning lights and a faulty de-icing system. Norcanair remained grounded into 1997.

Air Manitoba

Manitoba's main regional airline in the 1980s was Winnipeg-based Air Manitoba. Its history dated to the 1950s to a family operation started by Peter Lazarenko. It was involved in a variety of enterprises like the fresh water fishery, running steamboats, selling fuel, hauling on winter roads and operating tourist camps. Lazarenko started commercial fishing in 1935 and learned to fly on a Taylorcraft with Charlie Graffo in 1937. About this time the family

business started using Wings Ltd. and CPA to fly fish from the north to Lac du Bonnet from where it was trucked to Winnipeg. In 1949 Lazarenko formed Northland Airlines using the Anson, Crane, Fleet Canuck, Norseman, Reliant, Tiger Moth and Waco. Commercial fishing boomed after 1950, providing hundreds of jobs. There were processing operations at God's Lake, Island Lake, Warren's Landing and Wheatcroft. A fishery later was opened in Sandy Lake, Ontario. Bigger types like the Barkley-Grow, Lockheed 10 and Lodestar were added. In 1967 Lazarenko bought Transair's northern operations. He had a maintenance base at Netly, north of Winnipeg; and later used Gimli as his main Canso base. In 1970 Lazarenko merged with Midwest Airlines and a few months later the new company merged with Transair.

Ilford Airways (formed in 1954) and Riverton Airways (1952) were two contemporary Manitoba companies. In 1960 they merged under businessmen like Jack Duncanson and Connie Geff. Ilford-Riverton operated Beech 18s, DC-3s and C-46s, but went into receivership in 1972. A group including Pete Lazarenko, Joe Mazur and Barney Lamm then formed Northland Outdoors Canada to assume Ilford-Riverton's debts and keep it going. After several years Ilford-Riverton was again solvent. Through the 1970s and early 1980s it flew the C-46, Canso and DC-3, mainly serving Indian reserves. By this time Lazarenko was out of fish hauling. The business had come under a marketing board, the policies of which Lazarenko disliked; and independent native co-ops gradually took over the trade. Lazarenko assisted the natives in getting their co-ops running.

The Ilford-Riverton partnership changed as the years passed e.g. Barney Lamm left and Ron Williams came in. The airline wanted to change its name to Air Manitoba, but the provincial government would not allow this. Thus, the fleet carried the titles "Northland Air Manitoba"; gradually the "Northland" appeared in small letters, then disappeared. By the late 1980s the BAe748 was added to the Air Manitoba fleet. In 1983 Stan Deluce, on account of his expertise with the 748, was brought in as a partner. For 1989 Air Manitoba carried 57,000

Some of the famous Lake Manitoba freighters, including *Keenora* (another photo of it, taken by Arvid Dancyt some 60 years earlier, appears in Chapter 27). They were seen in 1992 in Selkirk's marine park. Northland's land, water and air transportation interests proved one of Manitoba's most successful enterprises. Ships and planes remained busy in the Lake Winnipeg fishery in the 1990s. Annual quotas included figures such as two million pounds of Walleye and 1.5 million pounds of perch. New York, Boston and Chicago were still the big markets for this multimillion dollar annual bonanza. (Larry Milberry)

passengers and 18,000 tons of freight. Its director of maintenance in the early 1990s was Tom Phinney. He had begun his aviation career in 1973 with Ontario Central Airlines, then went to Ilford-Riverton and stayed into the Air Manitoba era. When he started, Air Manitoba was mainly a DC-3 operation with C-FADD, C-FETE , C-FIKD, C-FCQT and C-GWYX. It also had two Ce.185s, two Aztecs, a Geronimo (hopped up Aztec), a Norseman and an Otter. It later added the Ce.401, Ce.402 and Canso C-FOWE.

The DC-3s carried much history. ADD was a 1944 model delivered to the US Navy. Later it flew with a variety of US operators, including the Broken Box Ranch in Oregon. In October 1973 it came to Canada for Ilford-Riverton. Undetected metal fatigue led to a disastrous end in 1986—pilots Keith Peddie and Russ Clarke died when a wing came off. CQT began in 1943 with the USAAF in Australia. It then served Australian civil companies, was in Nigeria in 1958 and in Ireland with Aer Turas in 1964. It next flew in the UK, and in 1970 was with North Coast of Prince Rupert. Ilford-Riverton took it over in 1975.

ETE joined the RAF new in October 1943. It served overseas, but in September 1946 was on Canadair's line being turned into a VIP plane for the T. Eaton Co., Canada's premier department store. T. Eaton, which had had R2000s installed in ETE, sold it to the Hudson's Bay Co. in 1963.

One of Air Manitoba's C-46s loads groceries at Winnipeg for Indian reserves to the north. (Larry Milberry)

Air Manitoba C-46s ready to work at Winnipeg in a scene from the early 1990s. Then (inset), a close-up of the exotic art worn by "The Ancient Lady". (Larry Milberry)

Ilford-Riverton took it over in 1975. In 1979 it was exported to the US as N62WS. There it fell from honour—the new owners used it on smuggling operations to Mexico and Central America. In 1980 it was impounded in Indiana as a drug smuggler. Some time later it was being used by the CIA to run guns to the Contra rebels in Nicaragua. The Sandanista military in Nicaragua was especially vigilant about CIA air ops. One day when ETE (against ICAO regulations, the CIA continued using this registration) came through on a paradrop, they blew it out of the sky.

IKD was another ex-RAF WWII machine. Postwar it was N42F, and in 1955 entered Canada as an executive conversion for the Ontario Paper Co. It served for a decade before moving to a series of operators, the last being Ilford-Riverton. In the 1990s it was sold in France. The Air Manitoba DC-3 with the longest Canadian history was WYX. It had been delivered to the RCAF new from the factory in May 1944 and it stayed with the air force for nearly 31 years. It was bought by Ilford-Riverton in 1975. By 1990 the Air Manitoba fleet had changed—four C-46s, six DC-3s and five 748s. Of the DC-3s, C-FIKD and C-FSCC were passenger planes, C-FAOH and C-FGHL were combis, and C-FCZG and C-FCQT were freighters.

While the 748 was Air Manitoba's passenger plane, the C-46 was the main freighter—for 1989 it carried the majority of Air Manitoba's 18,000 tons of cargo. C-46 TXW was a full-time tanker equipped with two 1,700-Imp. gal. aluminum tanks bolted to the floor. Gone were the days of either 45-gallon drums or rubber bladders. The "threes" were busy from May to October on fishing and hunting charters. Spring saw them hauling fuel, usually on skis. Air Manitoba also had Dash 8 C-FDOJ, leased to Air Alliance. The company had 140 employees in November 1990 of whom 21 were pilots (some being dual qualified). The maintenance department numbered 45-50. Air Manitoba kept the C-46 alive in Canada into the 1990s. Its first was C-FFNC, acquired in 1983 from the Lambair bankruptcy. It suffered

fire damage at Winnipeg in the winter of 1989, was sold for salvage to Everett Air Fuel, repaired and ferried to Alaska.

Pilots liked the C-46. It handled nicely, was smooth, fast and got in and out of short strips with ease. As far as management was concerned, the C-46 was a good choice. It didn't cost a million dollars to buy—a couple of hundred thousand was plenty. In a 1990 interview Tom Phinney pointed out that a C-46 required only four maintenance hours per hour flown compared to six for the 748 and three for the DC-3. It burned 1,000 pounds of fuel per hour compared to the 748's 2,000.

By the early 1990s C-46 parts still were available—when Air Manitoba damaged a C-46 at Winnipeg and needed a new wing, one was trucked from Texas. Spare R2800 engines were plentiful, with overhauls available at Precision Air Motive in Seattle and Mill Air Engine and Cylinder in Miami at about US$28,000. A comparable overhaul on a 748's Dart cost $250,000. The highly touted fuel-saving "18-60" mod on a Dart cost US$151,000, taking 3-4 years to pay for itself.

Such figures convinced Air Manitoba that the C-46 made sense in their kind of operation.

The C-46 carried a hefty load—14,000 pounds (11,000 for the 748). Although it had double freight doors, loading a C-46 was awkward because of a steeply-sloping floor. Forklifts and a motorized winch eased the burden, but a lot of elbow grease was essential. Offloading was easier, manpower being provided by the pilots and a helper hand-bombing the load onto trucks. Cruising at 170 knots, a C-46 easily could make two return trips daily from Winnipeg up through the northern lake region. Depending on demand, Air Manitoba logged 600 to 1,000 hours per C-46 yearly.

Whistle Stopping with Air Manitoba

A typical C-46 trip was operated with C-GIBX on November 16, 1990. Capt Mark Mesdag, FO David G. Jackson and helper Dwayne Long comprised the crew. Mesdag was a graduate of Confederation College in Thunder Bay. He found work in the bush, building hours on the Beech 18 at Nestor Falls. He joined Air Manitoba in 1980 and by late 1990 had 1,100 hours on the C-46. Jackson had started his career in 1985, flew from Kenora, on the coast from Prince Rupert, then joined Air Manitoba in 1989. Long had started at Air Manitoba as a loader in 1988, was working on his pilot's licence at the Winnipeg Flying Club, and dreaming of flying the C-46.

IBX had been loaded the night before—it carried 13,050 pounds of groceries and other freight. At 0715 it taxied at 47,997 pounds. The pilots did their run ups and at 0728 IBX was rolling on R31, sending the 4,000 horsepower roar of twin R2800s across the field and into slumbering neighbourhoods. Rarely was such a tumult heard in Canada in the age of whispering turbines. Takeoff in a heavily laden C-46 is an event. The pilot focuses on the business at hand—throttles, instruments, control column, rudder pedals. As power is set, a clamour fills the cockpit. The instrument panel rattles and the dials blur—the plane trundles away. Things relax a bit once the tail rises. The

Pilots Marc Mesdag and Dave Jackson in their office aboard C-46 IBX on November 16, 1990. (Larry Milberry)

amateur observer looking over the pilots' shoulders secretly is happy about R31—it's 8,700 feet long. At long last IBX goes flying. The sense is that all this is barely happening—the initial rate of climb seems imperceptible, but everything is working. The sun edges over the horizon back under the tail as IBX heads for Island Lake, 296 miles northeast.

Capt Mesdag levels at 3,500 feet and sets power at 960 hp a side. His plane slices easily through the cold, clear sky at 150 knots. In time the power is eased; we are descending. Ahead in the flat, ice-bound country that is Island Lake someone picks out a snow covered runway. Bank left a bit and line up on R30—4,000 feet of crushed gravel. An hour and 23 minutes after groaning free of the runway at Winnipeg comes touchdown at Island Lake. After rolling almost to the end of R30, IBX backtracks to the tarmac. Island Lake is typical of the strips across northern Manitoba that the provincial government runs. They are good ones, but gravel is hard on tires and brakes. While a C-46's props, well clear of the runway, were not susceptible to gravel damage, the 748 suffered expensively—a good used propeller cost at least $60,000. Another argument for C-46s.

As IBX shut down, a truck rolled out. Capt Mesdag nonchalantly opened a side window, squeezed his way through, hoisted himself to the roof, shuffled down to the tail, slid onto the stabilizer and hopped to the ground. The others followed—there was no way back through the cabin, chock-a-block with cargo. The crew swung open the cargo doors, clambered aboard and started chucking boxes onto the truck. It was cold, but the faster one worked the warmer he became and the sooner it was time to clear out. Meanwhile, nature's icebox kept away the gawkers, who usually skulked around as the C-46 offloaded. Today they kept warm in the airport office, garage and café.

IBX had landed at Island Lake at 0851. An hour and three minutes later Dave Jackson was taxiing out for his leg—a 19-minute hop (74 nm) to God's River with the rest of the freight. Takeoff weight was 36,696 pounds —lighter by five tons of fuel and freight. IBX slid onto 3,400-foot R27. Things were quiet except for the local truck. An even quicker turnaround, and IBX was leaping into the air homeward bound. It covered the next 372 miles in 1:35 hours, cruising at 170KIAS. The crew chatted along the way, covering all kinds of topics (mostly aviation). Dave Jackson spoke keenly about the 748 course he was soon to take. Before long he moved up to the big turboprop. He loved his flying, so it was a gloomy day on November 10, 1993 when Dave, Capt Abe Hiebert and crewman Greg Pierce lost their lives in 748 QTH, which crashed at Sandy Lake, Ontario. A remembrance in the newspapers noted: "Dave was a selfless, caring person who gave freely of himself to his family, friends and community. He will be remembered for his easy-going approach to life, quick wit and his ability to accept people just as they are. He craved the simple and the uncomplicated, and lived with an integrity and a loyalty

Capt Peter Hildebrand at work in 748 AGI en route Winnipeg-Gods Lake Narrows on February 7, 1992. Then, helping to off-load there. (Larry Milberry)

(Below) The terminal at Gods Lake Narrows. (Larry Milberry)

to his vision of family and the world." Capt Hiebert also was well remembered.

Since joining Air Manitoba, the 748 gained in importance; by 1994 it was the backbone of the fleet. The C-46s and DC-3s had been retired at last. This was the goal of the Deluce family, whose control in the company strengthened in the early 1990s. To the Deluces, who had done more to establish the 748 in Canada than any others, the days of the piston airliner were over. "Round engines" were no longer cost-effective, high-octane fuel was getting less and less common, and in winter there was the hassle with cold weather starts. It was time to standardize to turbines.

A typical day for an Air Manitoba 748 was February 7, 1992. Aircraft AGI (Flight 403), by then a 23,700-hour veteran, was ready on the ramp for early departure. Configured as a 28 pax combi, it took on six passengers and 4,300 pounds of cargo for its run. The crew was Capt Peter Hildebrand, FO Ed Moon and crewman Greg Pierce. Hildebrand was a Winnipeg lawyer, who found flying more to his liking. He had five years on the 748. Moon, who started flying at the Winnipeg Flying Club, had joined Air Manitoba as a crewman on C-46s.

Hildebrand fired up at 0803 hours and taxied to R31 behind two CAIL 737s. At 41,319 pounds, AGI rotated at 110 knots. It climbed out at 1,200 fpm, heading for FL130. Once level it cruised at 200 KIAS. Destination was Gods Lake Narrows, where it landed at 0940 to offload 3,000 pounds. As usual the airport was quiet, although a Ministic Air Aero Com-

mander was just departing on a feeder sked. It was a quick turnaround for AGI—on these milkruns there's no time to dilly-dally. Eleven passengers boarded and at 1002 Ed Moon was hauling back on the wheel, heading to nearby God's River. At 1009 the gear was down again and at 1012 AGI landed at 80 knots. Another 1,000 pounds went off, and at 1027 it was on to Oxford House with 16 passengers. More of the same, then AGI departed with 13 passengers for Thompson. A hundred miles and 35 minutes later AGI touched down on R05—5,800 feet of smooth asphalt. It backtracked and taxied to the terminal.

Thompson's population in 1992 was some 14,000. Development of its mine site began in 1957, shortly after vast nickel deposits were located by Inco using electromagnetic aerial surveying. Construction of a modern townsite followed and nickel production began in 1961. Although the site was served by the CNR, aviation became key to the city's growth. At first Thompson was served by Transair skeds—DC-3s, then YS-11s, F-28s and 737s. It was a main base in the 1960s-70s for Lambair, but Calm Air was its major operator in the 1990s. It had been started in May 1962 by Carl Arnold Lawrence Morberg in Lynn Lake. His first plane was a Ce.180. The company expanded —in October 1970 it absorbed Chiupka Airways.

Like all operators Calm Air regularly applied for licences to the ATB. An example from 1971 shows how the process worked. Calm Air applied for a licence to operate a "Class 3

Irregular Specific Point" service from Lynn Lake to Wollaston Lake and Stony Rapid. This was described by the ATB in Decision No. 3188 of July 16, 1971:

Calm Air Ltd. stated that it has substantial facilities at its Stony Rapids and Lynn Lake bases; that it operates a fleet of 12 aircraft and is contemplating the acquisition of a Twin Otter; that the uranium development on Wollaston Lake by Gulf Minerals has a requirement for air transportation; that there is an annual requirement to haul 500,000 pounds of fish from Wollaston Lake to Lynn Lake with some 400,000 pounds of goods on the back haul; that the applicant expects additional revenue traffic to be generated by the Hudson Bay and other stores, a food catering organization and residents of the Lynn Lake area, and that one of the principals of Calm Air Ltd. operates a sports fishing lodge and outcamps in the Stony Rapids area which requires air transportation both for guests and supplies.

The ATB felt that the latter was really Calm Air's main interest. In cross-examination it learned that the mine had good air service from Prince Albert and La Ronge. Norcanair objected to the Calm Air proposal, stating that it provided ample service throughout the region, and that it recently had terminated a Lynn Lake-Wollaston Lake run "due to lack of traffic demand"—it noted that for 1969 it carried but 10 passengers and 1,090 pounds of freight on the route. After listening to arguments from both sides, the ATB ruled against Calm Air. In many earlier and subsequent cases it ruled in Calm Air's favour. In Decision No. 3687 of August 17, 1973 it approved a Calm Air licence for Pukatawagan, Flin Flon, Lynn Lake, Wollaston Lake and Co-op Point (with no local traffic rights between the latter two towns, and no non-stops between Lynn Lake and Flin Flon). Calm Air convinced the ATB that there was adequate demand for additional air service to those centres, in spite of Norcanair's objections.

In 1971 Calm Air added a Twin Otter. Two others, then 748 C-FMAK were acquired from Gateway. DC-4s PFG and PSH joined the fleet

in 1982 for a resupply contract to the mine at Hidden Bay, Saskatchewan. On June 16, 1987 PFG was landing there after taking cargo to Kasba Lake in the NWT. As it landed, its right gear struck the lip of the runway. The landing was completed but PSG was so severely damaged that it later was scrapped. Other types used by Calm Air over the years were the Beaver, Ce.180, DC-3, Found FBA-2C, King Air, Navajo and Skyvan. Skeds gradually expanded to include Churchill, Flin Flon, Lynn Lake, Rankin Inlet, The Pas, Thompson and Winnipeg. Calm Air became a 45% PWA-owned partner in 1987. In the 1990s its fleet centred on the 748, Saab 340 (first Canadian operator) and Navajo.

During AGI's Thompson stop there was little Calm Air activity at the terminal, except for a Navajo. A Northwinds Navajo awaited passengers—this was a small airline run by the Keewatin Tribal Council. Across the field was one of Air Manitoba's C-46. Pilots Mark Mesdag and Jeff Schroeder were in town hauling for local reserves. On a wall in the passenger terminal a plaque honours the crew of Lambair's Bristol which crashed in 1974: "Given in memory of pilot Don Boone and co-pilot Bob Hildreth who lost their lives at Rankin Inlet, May 31, 1974 in helping to make our Canadian North a better place to live."

For the downbound half of the day AGI carried 13 passengers from Thompson, 27 from Oxford House, 25 from God's River and 13 from Gods Lake Narrows. The day's flying

Flight 403 Up-bound—February 7, 1992

Leg	Distance	Passengers	Freight
Winnipeg–Gods Lake Narrows	343 sm	6	4,288
Gods Lake Narrows–Gods River	26	11	1,340
Gods River–Oxford House	50	16	477
Oxford House–Thompson	119	13	320

was eased by a GPS, and on the last leg, when everyone was keen to get home, it told the crew that Winnipeg was exactly 104 minutes away, once the wheels came up at Gods Lake Narrows. The final leg was made at 275 KIAS compared to 215 upbound. AGI touched down in Winnipeg at 1545 local. The log book showed 4.9 hours of flying since leaving Winnipeg that morning.

In late 1993 Air Manitoba was a small operation compared to 1990. The fleet was 748s AGI, FFS, GGE and TTW and a Caravan leased from Kelner of Pickle Lake. C-46s IBX, IXZ and TXW had gone to Kenya for humanitarian operations in Sudan. Operating as Relief Air Transport, they were managed by Bill and Bruce Deluce. C-46 AVO was being refurbished, and TPO had been sold to Buffalo Airways, along with Air Manitoba's last DC-3s. The company was down to a staff of 85. In the early 1990s the Deluces forced the Lazarenkos from Air Manitoba. In January 1994 the company was audited by Transport Canada. Its inspectors didn't like all they saw, and revoked Air Manitoba's operating certificate. It was grounded for some time, but gradually resumed business hauling freight.

Relief Air Transport

By mid-1993 Air Manitoba had made up its mind—the C-46s must go. There would be no more discussing the pros and cons of the C-46 and 748. The demand for humanitarian airlift in East Africa was high at this time. The Deluces studied the situation and determined that the C-46 had prospects as a desert workhorse, now that its days over prairies and boreal forest seemed over. Contracts were found with non-government operations (NGOs) in Kenya. In August 1992 the first C-46 ("Ancient Lady") ferried Winnipeg-St. John's-Lisbon on its way to Nairobi. Touchdown in the "old world" was on August 19, and the next day the plane pressed on to Athens, a flight of 8:37 hours. On the 23rd the trip continued to Nairobi via Jeddah.

September 20-23 aircraft IBX followed along to Nairobi. In November IBZ arrived. The C-46s were soon at work on Operation Lifeline Sudan, based at Lokichoggio, Kenya. The work lasted about two years. One C-46 was wrecked on takeoff at "Loki", the others returned to Canada in 1995. A fourth machine, AVO, was being rebuilt for East Africa—it had been used by Air Manitoba for parts over the years. It remained in Canada and was bought by Joe McBryan of Buffalo Airways, who also organized repairs on TPO. It had been unserviceable at Pickle Lake since 1989. On September 1, 1994 TPO was ready to ferry to Winnipeg. McBryan had its nose emblazoned with graffiti for the occasion: "Down and locked and laughing"! Jim Smith ferried TPO to Winnipeg, from where it carried on for major repairs in Seattle.

Perimeter Airlines

In 1960 W.J. "Bill" Wehrle was part of a group of investors that set up an executive air transport operation in Winnipeg. In 1964 he bought out his partners and founded a small IFR flight training school, Perimeter Airlines. In the early days it used a Debonair, a Musketeer and a Travel Air. Wehrle had begun in aviation in 1953, learning at the Winnipeg Flying Club. He instructed there from 1955-59, specializing in IFR ratings. It was a busy time for this work, as pilots were needed to fly on the DEW Line. Meanwhile, Wehrle also did some corpo-

AGI stopping at Thompson. It first belonged to Air Gaspé, where it began in 1981. Regionair, Calm Air and CAIL all used it thereafter. It joined Air Manitoba in 1988. (Larry Milberry)

Bill Wehrle in Perimeter's engine shop on July 19, 1992. (Larry Milberry)

C-FIHE, one of Perimeter's colourful Metros, arrives in snowy Winnipeg on January 11, 1993. Straightforward, fast and reliable, the Metro was the backbone of many small carriers in the 1980s-90s. Its lineage began in the early 1960s with the Merlin I, a modified Queen Air with Lycoming piston engines. The SA26T Merlin IIA (MTOW 9,800 pounds) followed with PT6A-20s. The SA26AT Merlin IIB (10,000 pounds) was introduced in mid-1968 with Garrett TPE 331s. The SA226T Merlin III (12,500 pounds) had a two-foot fuselage extension, larger tail with the elevator mid-way up the fin, and larger Garretts. The SA226TC Metroliner and 12-passenger Merlin IV/IVA corporate plane appeared in 1969. In 1974 the Metro II (square windows) replaced the SA226TC. The Metro IIA of 1980 bumped up gross weight, so had bigger engines. Still heavier Metro IIIs followed. Worldwide some 750 Merlins and Metros were in service in 1997. (Larry Milberry)

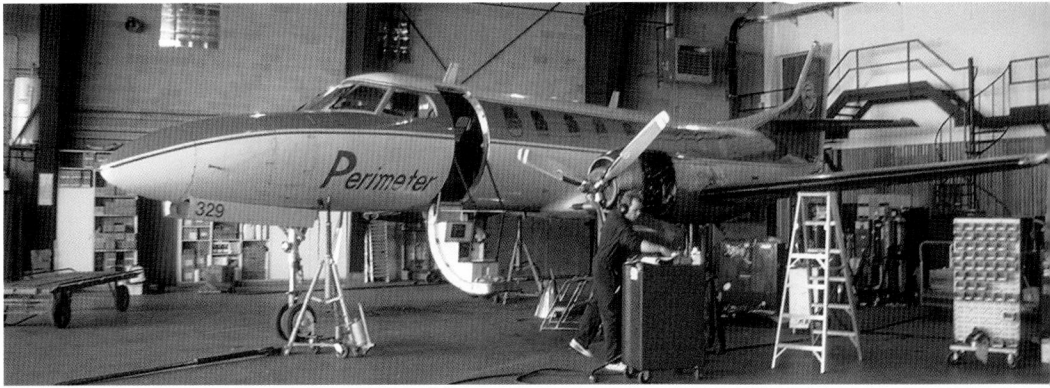

A Perimeter Metro is serviced at Winnipeg on May 6, 1995. At this time Perimeter's fleet numbered 14 Metros. Besides its airline fame the Metro was important in Canada with overnight courier operators, its attractive features for them including being readily available on the used aircraft market, affordability, low operating costs, and speed. In the early 1990s used Metro IIs sold as low as US$400,000 for a 1970 model to US$850,000 for a 1981. (Larry Milberry)

rate flying on a Cessna 310. By the time he opened Perimeter, he was solidly grounded in general aviation.

Although business was slow, Perimeter gambled and added Beech E18S CF-ANT. It was soon in demand—the " Super 18 " was, for the day, a classy way for executives to travel. Perimeter also had a Beech distributorship, so sales and service became important— Barons, Bonanzas, Debonairs, etc., were sold to many operators in the Perimeter region. The company began flying a Queen Air, added two flight simulators and had as many as 10 instructors. By this time Perimeter was a full-service FBO with avionics, sheet metal, wheel/brake and upholstery shops. In 1976 it took over some northern routes from Transair. These came with two Twin Otters, and the first of several Metro IIs was added. Such communities as Norway House, Cross Lake, Oxford House and Gods Lake Narrows were served. Demand was such that three DC-3s were put to work, while the Metros were leased in California. Specialized contracts were negotiated that brought Lear 35A C-GZVV and King Air A100 C-GNEX, the Lear for five years, the Beech for 12. Meanwhile, DC-3 business was steady but, unlike big operators such as Austin Airways or NWT Air, Perimeter did not have the infrastructure to maintain its fleet economically—the DC-3s cost too much. To ensure that passengers were properly served and that the ever-growing air cargo traffic was accommodated, they were retired after five years. The Metros returned from California. Cruising at 255 knots and being pressurized, they were a step ahead of the Twin Otters, which soon departed. A UPS courier contract brought an F.27, but it proved uneconomic. Smaller Beeches later proved their worth on most Perimeter courier legs.

Wehrle was the first Canadian operator to appreciate the strengths of the Swearingen/Fairchild Metro II. The 19-seat commuter had a rugged airframe, large rear cargo door and straightforward Garrett TPE-331 fixed-shaft turbines. It could carry a 2,500-pound load as a cargo plane. Wehrle found that the "331" was

more intricate than the PT-6. It needed more spares on hand, and greater training, but was fuel efficient. For his routes it was the best engine in the long run. Besides, it was priced right, and Wehrle found that Fairchild and Garrett were reliable when it came to support. Perimeter established Canada's only certified Garrett overhaul shop, taking in engines from operators across the country (it also established an overhaul shop for the Lycoming IGSO 480 piston engine).

By 1993 Perimeter Airlines had some 120 staff. It was a complete FBO and was operating nine Metro IIs, and several Beech Barons, Queen Airs and Travelairs. It focused on its traditional northern Manitoba routes, parts of the old "Prairie Milkrun" (including Brandon and Dauphin), courier work, corporate charters, and IFR flight training. The milkrun skeds were still money losers, but Perimeter was not receiving a government subsidy to fly them. They were primarily operated for the chartered banks as courier runs. The small Travelair normally would have operated such runs, but Perimeter put on a Metro to carry courier sacks and what few passengers used the routes.

In the mid-1990s Bill Wehrle looked closely

at some new equipment. What impressed him was the Do.228 with TPE331s. He liked its high wing, which was ideal for gravel strips—the props were less prone to damage, and the engines to FOD. One 228 was ordered, but this later was cancelled, based on the Dornier's lack of pressurization and of a rear cargo door, two features offered by the Metro. In 1995 Perimeter was operating 14 Metros and keeping them busy on newly added passenger runs—ones lost by Air Manitoba, when Transport Canada suspended some of its privileges.

The Ubiquitous Metroliner

The Merlin/Metro family originated with Ed Swearingen, founder of Swearingen Aircraft in San Antonio. He brought great expertise to the industry, particularly in aircraft modification—he had worked for Bill Lear and Dee Howard, both renowned for conversions. He also designed for Piper, where he brought about the Twin Comanche. His earliest conversions at San Antonio were installing more powerful engines in the Queen Air and Twin Bonanza. Swearingen next designed the Merlin I. It mated Queen Air wings to a new pressurized fuselage. At the same time he built the

Merlin II with PT6A-6 turboprops; it immediately superseded the Merlin I (only the prototype of which was built). The eight-seat Merlin II, aimed at the corporate market, flew April 13, 1965 and was one of the first PT6 applications. It was certified in July 1966 and 36 were built with 550-shp PT6A-20s, before Swearingen switched to the 665-shp Garrett TPE-331 (he was able to get easier terms from Garrett than from P&WC). The Garrett-powered Merlin was certified in June 1968 and Garrett assumed the national marketing task for Swearingen.

The Merlin II soon was stretched (19 seats) and given a new wing to become the Metro. It had a gross weight of 12,500-lbs, the limit then allowed for commuter operators by FAA regulation FAR Part 23. Swearingen now allied himself with Fairchild in a manufacturing arrangement. The Metro prototype flew on August 26, 1969 with Garrett TPE 331s. Production began, but overhead drove Swearingen out of business in 1971. Fairchild owned several sets of Metro wings for which it had not been paid. In early 1972 it arranged to take over Swearingen's San Antonio plant. Swearingen remained with the new company, which focused on delivering a few aircraft that year. Garrett continued in national sales for Swearingen (but gave this up to Fairchild in 1974).

In 1973, 36 Merlins and Metros were delivered. The Metro entered service with Commuter Airlines of New York in March 1973. By late 1975 Swearingen made a small profit for the first time. It reduced the work force to below 400. In 1975 an improved product, the Metro II, was introduced, and it changed the company's fortunes forever. Commuter airlines began lining up to place orders for the attractive, economical mini-airliner. Since the early 1980s many Canadian operators used Metro IIs. Air Alma, Air Dorval, Air Toronto, Alexandair, Intair, Inter-Canadian and Perimeter Airlines operated them on skeds. Its drawback as a passenger plane was its noisy engines compared to the quieter PT6s.

The Metro II was designed from the outset to be readily convertible to cargo configuration. Cargo Metro IIs have several attractive features: they are cheap, usually readily available, carry a good load, and are fast and reliable. As such they became popular with Canada's overnight courier services. The largest fleet was operated by Soundair Express, Air Toronto's cargo branch. When the Soundair empire collapsed in 1991, Jet All became Canada's biggest Metro II operator, but it followed Soundair into bankruptcy.

In December 1981 the prototype Metro III took to the air with a longer wing and bigger engines (1,000-shp Garretts). In 1992 Bearskin Air of Thunder Bay became the first Canadian operator. The newest version, the Metro 23C, appeared in 1991 with 1,100-shp Garretts and an AUW of 16,500. It has a 5,000-lb payload and one of the best ratings for fuel economy in its class—according to Fairchild, it could fly four 150-mile sectors without refuelling.

(Above) Selkirk Air's Beech 18 comes across the highway from the airstrip and down to the Red River to be launched for the 1995 season. It was RCAF HB112, serving 1943-65. OCA was its first civilian owner. (Larry Milberry)

Lac du Bonnet and Selkirk

Two busy float bases north of Winnipeg are Lac du Bonnet at the mouth of the Winnipeg River, and Selkirk on the Red River. Lac du Bonnet was first visited by airplanes in early post-WWI days. It became the key air head for operations into the Red Lake mining area after 1926, and later was main base for the Manitoba Government Air Service. Since the 1960s Selkirk developed a strong tourist focus, mainly for fly-in fishing. One of the pioneer aviators in Lac du Bonnet was Merrick George "Shorty" Holden. He was born May 23, 1911 on a farm near Windsor, Ontario. When an itinerant instructor appeared in the area with a 40-hp Taylorcraft, the youthful Holden scrounged enough money to start flying. For $100 the instructor offered eight hours of dual plus some solo time, but the deal fell through. Holden pursued his licence on a Gipsy Moth. He earned a commercial and in 1936 was hired by the Starratts of Hudson. The first winter he flew Gipsy Moth AGX chasing tractor trains. The Starratts kept several trains, or "swings", on the move all winter, supplying places like Pickle Lake, Uchi Lake and Gold Pines. There usually were four sleighs per tractor. As there were no mobile radios, the Moth kept tabs on swings moving in various directions. A typical day's flying on AGX was February 10, 1937. Starting at Hudson, Holden flew to Manitoba Point (40 minutes), Steamboat Narrows (15), Bluffy Lake (40), Manitoba point (30) and Hudson (25). This was all great stuff for a young pilot on his first job.

On March 23, 1937 Holden checked out on the Super Universal at Sioux Lookout. Starratt had two—CF-AJA and CF-AJB. He took naturally to the Fokker. In a 1995 interview he noted: "It was the most stable airplane I ever flew. The Fokker was an excellent workhorse, and easily would carry an 1,800-pound load, plus gas. You could reduce power gradually, while pulling back and the Fokker wouldn't fall from under you. I tried to spin it on different occasions, but it wouldn't co-operate." For June 6, 1937 Holden flew AJA from Hudson: Gold Pines (45 minutes), Red Lake (30), Uchi Lake (45), Sioux Lookout (50), Hudson (10). Three days later Holden made 11 round trips between Pickle Lake and Doghole, routing twice through Osnaburg (6:15 hours).

For 1938 Holden added Norseman BDE and Fairchild 82 AXL, but for most of the year was

on AJB. His favourite plane became Travelair SA-6000 CF-AEJ, which he described as "one of the nicest airplanes you could get your hands on." The F.82 was another favourite. While some pilots complained about its tight cockpit, the diminutive Holden could find nothing lacking. His log showed 846:45 hours total time for 1937. Hours dropped in 1938 to 753:25, but rose in 1939 to 1,036:35. Other types Holden flew included Pacemakers AEC and ANW, Beech 17 BIF and Beech 18 BQG. He always was interested in the characteristics of a new type. To see what would happen with the Travelair on floats, he stalled it one day. It spun and recovered readily. But life wasn't always a breeze. In one logbook note Holden mentioned of Fairchild CF-BVI: "Very drafty machine, no heaters, 35° below, very chilly flying".

In the early 1940s the CPR gobbled up a host of companies under the CPA banner. As Shorty Holden put it, "We were now all flying for the same outfit". Opportunities abounded for pilots and mechanics. For 1942 Holden flew 492:15 hours, this reflecting wartime restrictions. A consolation was flying the superb Junkers bushplanes.

Holden also flew the ski-equipped Beech 18 on the Winnipeg-Red Lake-Kenora-Winnipeg run on a six-days-a-week sked. He then took a Beech to Whitehorse, but soon moved to the Boeing 247 there, a machine he found as placid as the Fokker. From one distant outpost to another—Holden was transferred in 1942 to Sandgirt Lake in Labrador. There he replaced Don Murray flying Fairchild 71 CF-BJE. He had 10 field crews to supply and move. Each team comprised five geologists and five prospectors. On his first trip in the 71, just as the plane was loaded and ready to go, Holden faced some embarrassment. No two 71s were alike. In this one the new pilot in camp couldn't find the starter. He had to eat crow and call over the mechanic to save the day.

From Sandgirt Lake Holden moved to Havre St-Pierre, where he and Allan Delamire each had a Beech 18 on floats. In September they moved the exploration parties from Labrador, then Holden ferried his Beech to Montreal for servicing. He next went to Roberval for photo work with a Rapide, but this was stymied by weather. Soon Holden was back in CPA's Central Division in Winnipeg. Starting in 1944 he added Bellancas AWR, BKV and BTW. In August 1944 he flew BKV as far as

Shorty Holden at Hudson in 1937 atop one of his favourite planes, Starratt's Fokker CF-AJA. Then, with an early Beech. In 1995 he noted: "The aircraft is Beechcraft 18 CF-BKO, formerly of Prairie Airways in Moose Jaw. This was about March 1942 at McKenzie Island about three months after the CPR amalgamated the various operators they purchased to form CPA. BKO was flown from late February to mid-April on a run from Winnipeg to Red Lake and Kenora (return) six days a week until the ice at McKenzie Island became unsafe." BKO was sold to TACA of Venezuela in 1944. (Holden Col.)

Chesterfield Inlet with a missionary priest and four sisters for the missionary school. He flew out four other sisters on their annual rotation. On the same run he stopped outbound at Eskimo Point to do some trips inland to Padley for the HBC. Fifty tons were involved. On August 30 he did several trips for 10:35 hours. While he was in Winnipeg he had time for a Lodestar check-out and to ferry a Rapide Vancouver-Winnipeg. Holden spent the winter of 1944-45 at Lac du Bonnet. While there he decided to experiment in wild rice, a crop that had interested him for some time. Over the winter he designed and built a wild rice harvester—a freight canoe with the bow removed and replaced with revolving paddles. These would remove rice from their stalks and deposit it in the canoe. Holden now began devoting more time to wild rice. The 1946-47 seasons were poor, but he harvested good quantities from outlying lakes in the next few years. In 1948 he left CPA to fly for CNA till 1950. In January 1952 he purchased Waco BDJ from Elk Fish Ltd. of Red Lake. In June 1954 BDJ was sunk by winds at Lac du Bonnet. With his insurance settlement Holden bought a Seabee from CNA. By this time he was also in mining, so got out of bush flying to concentrate on these affairs. He developed an expertise in geophysical survey, establishing a company called Central Geophysics. Art and Don Gaffray of Pine Falls also were involved in wild rice at this time. About 1960 they proposed to Holden that they could raise the productivity of his lakes in exchange for a half share of the increase. In 1969 the Gaffrays bought into Holden Wild Rice. By this time the wholesale price for their product soared as high as $5.00 a pound. The market, primarily in the US, was willing to pay almost anything for the delicacy. A co-op in Grand Rapids, Minnesota controlled most sales

The Gaffrays established Silver Pine Air Services, which eventually included two PZL Otters for wild rice and general charter work.

Holden and his son Pat kept up an involvement in aviation. In 1970 they attended an auction of surplus military aircraft in Tucson, where about 100 L-20 Beavers and 30 helicopters were sold. The bidding for L-20s started at $17,000, but soon began to rise as planes were snapped up. The Holdens bought No. 50-32810 for about double what they would have paid earlier in the day, had they been quicker into the game. They ferried their L-20 to Lac du Bonnet and converted it for radiometric surveying. With its price still high, uranium was their object. They operated mainly in Manitoba and Ontario. When the bottom fell out of uranium, they sold the Beaver. Shorty Holden flew into the 1990s. He had a Cessna L-19 on floats, which was useful in the wild rice business. In 1994 he finally quit flying—he was 84. In 1995 he was working every day at his geophysical and wild rice office on Oak Street in Lac du Bonnet. Shorty passed away on April 20, 1996 after a brief illness.

Shorty Holden with son Pat at Lac du Bonnet May 7, 1995. (Larry Milberry)

A snap taken by Shorty of Starratt's ill-fated Beech BGY fitted with skis. It was lost in January 1941.

Manitoba Scrapbook

Air Command maintains a historic collection of many of its famous aircraft at Winnipeg. The original collection from the 1980s comprised this Dakota and B-25, along with an Expeditor. (Larry Milberry)

(Right) Selkirk Air's hangar on a busy May 7, 1995. Aircraft that had been stored through the winter were ready to work again. The staff was busy as one plane after another had its turn to be checked over, washed, launched and test flown. Selkirk Air was started by Bob Polinuk in the mid-1960s. (Larry Milberry)

Nick Zahorodny pumps some oil to replenish his hungry PZL. The veteran Manitoba bush pilot earlier had run his own operation at South Indian Lake. (Larry Milberry)

Selkirk Air's Otter GSL home on May 7, 1995 after a job at Cherrington Lake, Ontario. Nick Zahorodny was at the controls. GSL was Otter No. 166, originally delivered to the US Navy in October 1956. After later tours with the Missouri National Guard and the Civil Air Patrol it joined Parsons Airways of Flin Flon in 1979. Later it was converted to the 1,000-hp PZL engine. (Larry Milberry)

Otter GSL in storage at Selkirk over the winter of 1991-92. (Larry Milberry)

(Below) Expeditor KAK (ex-RCAF 1418) derelict at Selkirk in February 1992. It had gone to Kier Air of Edmonton when released from the RCAF in 1965 and later flew with Theriault Air Services of Chapleau and Air Windsor. Such old warhorses often were revived for new careers, but this always depended on economic times. (Larry Milberry)

(Below) Another old wreck on Bob Polinuk's back-40 in 1995. (Larry Milberry)

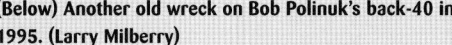

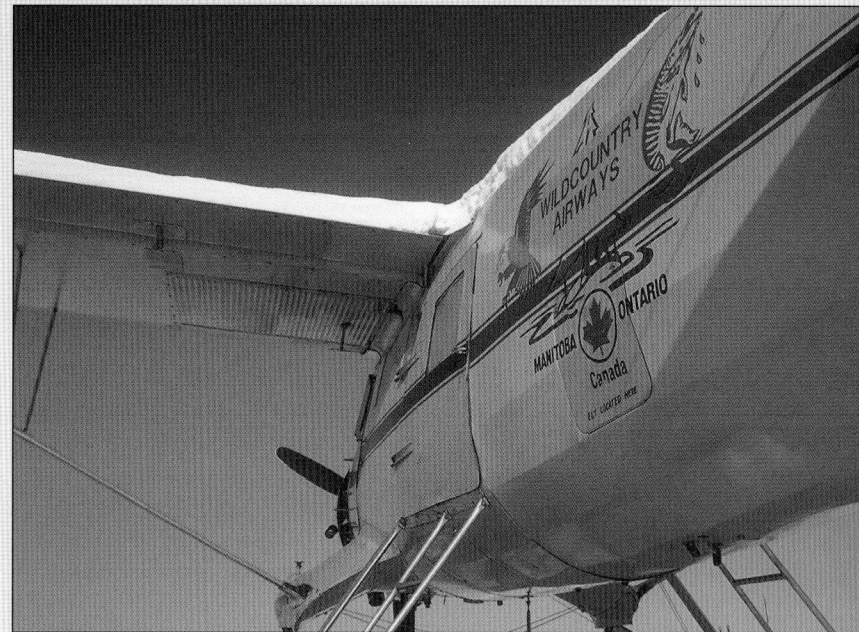

(Right) Norseman OBR wintering at Selkirk in January 1993. It had been on lease to Wildcountry of Red Lake. (Larry Milberry)

(Below) Don and Art Gaffray in their hangar in May 1995. They had been in aviation since 1961, specializing in wild rice and doing charters. In the 1970s-90s they operated as Silver Pine Air Service. It changed hands in 1994, becoming Eagle Aviation. The Gaffrays had the first PZL Otter, C-FIFP. It was the test bed for a 600-hp PZL, but this engine gave way to the 1,000-hp version. Shown is a detail of IFP's PZL as it was being inspected before the 1995 season. (Larry Milberry)

The back lot at the Gaffrays' strip on Highway 11 south of Pine Falls. All sorts of interesting early technology was scattered around, including the hulk of Beechcraft HVF. (Larry Milberry)

(Right) Eagle's Beech (formerly CF-PCL) at the dock at Pine Falls. Over the winter of 1992-93 it was restored by Rollie Hammersted of Reddit. In late 1996 Eagle became Blue Water Aviation Services (Larry Milberry)

(Below) Having an agricultural economy, Manitoba supports many small aerial application operators. These Cessna Agwagons, owned by Arty's and just starting the season, were shot at Winkler on May 10, 1995. (Larry Milberry)

Manitoba government Cessna Citation C-FEMA air ambulance framed by Otter ODY during a medevac of March 24, 1992. Then, a closer view of EMA, and of ODY's and EMA's tail markings. (Larry Milberry)

(Above) Night courier action at Winnipeg. FedEx's 727 N180FE was shot in October 1991. It originally was United's N7060U. It later was C-GBWS of Calgary-based Brooker-Wheaton, then Max Ward's Edmonton-based Morningstar Express. (R.W. Arnold)

(Right) The Falcon 20 was the first bizjet dedicated to courier flying. Soundair's C-FONX was seen on October 30, 1989. It was operating Winnipeg-Wilmington (Ohio)-Dorval-Ottawa for Purolator with pilots Tom Hamilton and Steve McLaren. (Larry Milberry)

Heavy metal at Winnipeg. Like any major airport, Winnipeg offers endless hours of enjoyment for the spotter and photographer. Each day brings its surprises—some new type or operator. John Morrison shot the brightly-coloured CPAir DC-10 (on lease from United) in September 1985, and the Canada 3000 757 in February 1991. Bob Arnold caught An-124 CCCP-82045 in September 1992. It was making off with a hefty load of Canadian-made farm equipment, including tons of prefabricated grain storage bins.

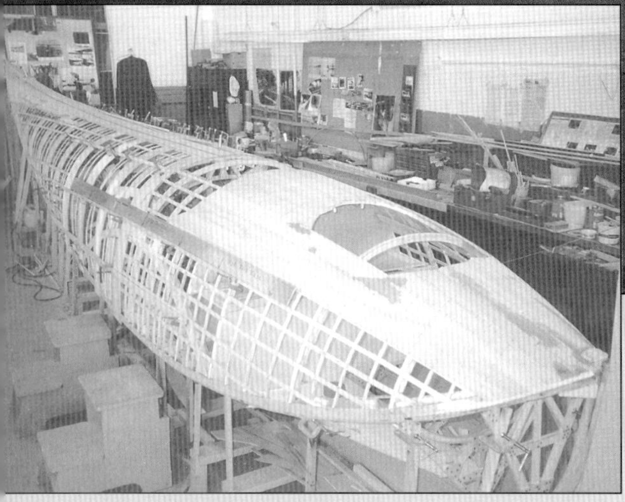

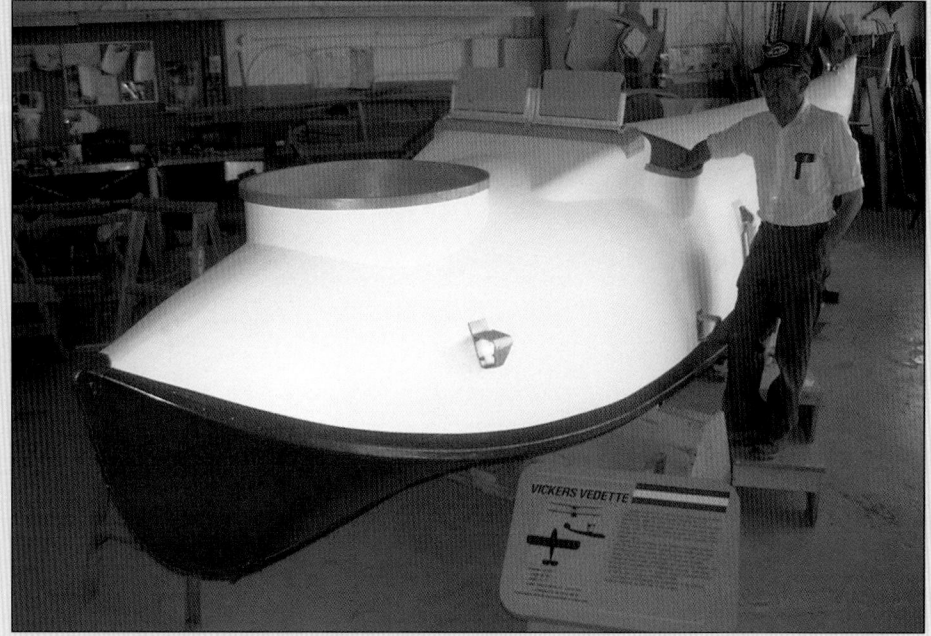

(Above) A wide view of Winnipeg International Airport on July 18, 1973. Air Canada recently had retired its large Viscount fleet and many were awaiting disposal here. (Larry Milberry)

This Canadian Vickers Vedette was under restoration at the Western Canada Aviation Museum in Winnipeg for more than 20 years. It's shown above in 1995, then a (right) year later, by when it was nearing completion. Doug Newey and Paul Latocki, pioneers from Canada's aviation industry, worked on this Vedette. They had been Canadian Vickers men in the 1920s, so brought (and passed on) rare skills to the project. In the second view WCAM volunteer Tony Morien inspects the cockpit. (Larry Milberry)

(Below) This Ju.52, shot in 1985, was converted by Bristol Aerospace from trimotor to single-engine configuration for the WCAM. It is a replica of Canada's original CF-ARM, which was scrapped after WWII. The WCAM got its start in 1972 through the efforts of dedicated volunteers like Gordon Emberley. His original cadre was the Manitoba Aircraft Restoration Group. Its first project was to look for Vedette remains around Cormorant Lake. The group became the WCAM in January 1974. (NA/McNulty Col. PA191785)

This vintage Manitoba Government Air Service Beaver (above) is another WCAM treasure. Bellanca CF-AWR (right) is a long-term restoration project. (Larry Milberry)

Formerly CF-EIL, this Fairchild Husky was re-manufactured in the 1970s by Saunders Aircraft of Gimli. Hopes of a Husky revival didn't materialize and the plane ended its days in the WCAM. (Larry Milberry)

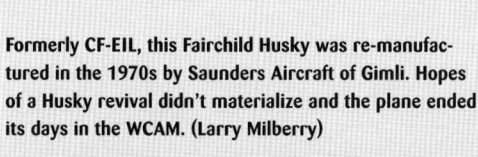

(Above) When Norcanair retired Bristol Freighter CF-WAE, it was donated to the WCAM in 1983. Occasionally the museum had enquiries from operators wanting to buy WAE, but it was in the museum for keeps. Here WAE cruises over the prairies during its working days. (CANAV Col.)

(Left) An ex-Air Canada Viscount in retirement at Gimli in February 1992. (Larry Milberry)

Jets in Colour

(Above) The Avro C.102, North America's first jet transport, represents a historic starting point for commercial jet travel in Canada. This model was in the Reynolds Museum in Wetaskiwin, Alberta in 1993. (Larry Milberry)

(Left) North America's first jet transport in service was the de Havlland Comet. The RCAF accepted two in 1953. They operated the first scheduled trans-Atlantic jet flights. Comet No. 5301 of 412 Squadron was shot at Uplands by Turbo Tarling in June 1958.

(Right) The Boeing 707 was a valuable part of Wardair's fleet. Here CF-ZYP finishes a charter at Toronto on April 29, 1972. It served Wardair 1969-78, then went to Austrian and Sudanese carriers. In May 1981 it was seized by the US government, was stored at Davis-Monthan AFB, then joined the USAF as 85-6973. (Larry Milberry)

(Left) Other Canadian companies favoured the 707. Here a sparkling Quebecair 707, C-GQBH, sits at Toronto in October 1978. Originally American Airlines' N7523A, it was a 1959 model. It was severely damaged landing at St. Lucia on February 19, 1979 and was sold for scrap. (Gary Vincent)

(Above) One of several 707s used by PWA lands at Vancouver. Bought from Northwest Airlines in 1973, it later was C-GRYO of Ontario Worldair (left). While 9G-RBO of Gas Air Nigeria, it was wrecked while landing at Ilorin, Nigeria on April 29, 1992. (John Kimberley, Gary Vincent)

(Left) British Caledonian 707 G-AXRS at Malton on July 22, 1972. In 1984 it became 5N-AOQ in Nigeria. (Larry Milberry)

(Right) G-AXGW, landing at Malton on July 2, 1971, was one of the last 707s delivered to BOAC. In 1981 it was sold in Yemen as 7O-ACO, where it survived into the 1990s. (Larry Milberry)

(Below) One of the five 707 combis (CF designation CC-137) which joined the CanForces in the early 1970s. Operated by 437 "Yukon" Squadron at CFB Trenton, they served well for a quarter of a century, two as aerial tankers for CF-5s and CF-18s. Tanker 13704 is seen at Keflavik on January 22, 1993, where it was on Operation Rhine Prosit—refuelling CF-18s between Germany and Canada. (Larry Milberry)

(Above) Flight engineer WO Red McLean of 437 Squadron monitors his panel during a 707 flight of July 20, 1988. Then, a pair of CF-18s practices aerial refuelling routines with 13703 on June 26, 1994. (Larry Milberry, Tony Cassanova)

(Left) CC-137 No. 13705 at Krasnoyarsk, Siberia during a February 14, 1993 Red Cross relief flight from Trenton and Helsinki. (Larry Milberry)

(Left) Wardair's famous "FUN" at Prestwick in 1966. Delivered that year, FUN was the first Boeing jetliner for a Canadian airline. It served till 1973, then was sold to a Brazilian airline. (Wilf White)

(Right) CPAir 727 CF-CPN landing at Malton on March 28, 1972. CPAir operated four 727s mainly on domestic skeds. (Larry Milberry)

(Left) C-GYNA, seen at Toronto on September 3, 1989, served Air Canada 1980-91, then joined Federal Express. Air Canada operated 42 727s beginning in 1974. (Larry Milberry)

(Right) First Air 727 combis at Iqaluit on August 15, 1992. C-GVFA, seen taxiing in, started with South African Airways in 1971. It later served FedEx and Purolator, and migrated north in 1988. Even though old-technology engines made for gas guzzling in the 727, there were smart reasons to keep it going. In the 1980s a good 727, available for as little as $1 million, offered fast, safe, comfortable and profitable transportation. Few planes in its class were as versatile and rugged. By day First Air's 727s flew the Arctic, where they were at home on gravel strips; by night they carried freight in the south; on weekends they visited holiday spots in the US, Mexico and Caribbean. (Larry Milberry)

(Left) Front end crew of First Air 727 C-FRST (Flight 806) on August 12, 1992 while en route from Ottawa to Iqaluit: Capt Gord Wallace, FO Stephan Ekiert and SO Lowell Teasdale. Having a cockpit crew of three added to the cost of running a 727, making two-pilot equivalent types like the A320 tempting for operators. (Larry Milberry)

(Left) A Greyhound 727 departs Vancouver for points east. This no frills carrier, a union between Greyhound in the US and Kelowna Flightcraft of BC, started flying in 1996. Kelowna provided the licences and aircraft. Dick Huisman of Greyhound and Barry Lapointe of Kelowna Flightcraft were the key men in launching the operation, which served points from Vancouver to Toronto. While people liked the cut-rate fares, Greyhound wasn't able to offer the frequencies demanded by the business traveller. It suffered a $31-million loss in its first year. (John Kimberley)

(Right) C-FBWY of Brooker-Wheaton Aviation working for FedEx at Calgary on July 5, 1993. BWY was an old United Airlines 727, which went to FedEx in the US in 1990. The 727 dominated small-freighters (30-ton payload) in the 1990s by when there were some 1,300 freighters globally. For 1997 about 550 were categorized "small", 415 being 727s. The 727's popularity resulted from its availability at good prices, having good conversion programs in place, and Stage 3 hushkits on the market. (Larry Milberry)

(Left) Montreal-based Royal Air was another Canadian charter carrier with 727s. This example was landing at Toronto in April 1994. (Andy Graham)

(Right) The world's most popular jet airliner is the Boeing 737, used by most major Canadian carriers. First flown in April 1967, more than 3,000 had been delivered by 1997. By late-1997 Boeing had pushed deliveries to 17 per month and was offering three new versions. Here CPAir's C-GKCP lands at Toronto on September 4, 1983. In 1987 CAIL's 737 fleet numbered 66. These usually carried 113-115 passengers; but in 1993 CAIL reconfigured two as 127-seaters for Canadian Holidays southern runs. This was significant enough to generate an extra $4 million per year per aircraft. As open skies clicked in in the early 1990s, and CAIL began adding routes such as Chicago and New York from Toronto, it needed every 737 it could muster. Thus did it hand over its Halifax-St. John's service to Air Atlantic to gain 737s. By mid-summer 1995 CAIL had 64 daily flights to 16 US centres, 42 daily Calgary-Edmonton shuttles, 31 Calgary-Vancouver, 14 Edmonton-Vancouver, 41 Toronto-Montreal and 33 Toronto-Ottawa, mostly operated with 737s. CAIL's 737 fleet still numbered 43 in 1997. Then, C-FPWC departing Edmonton Muni in May 1977. On February 11, 1978 this aircraft crashed while attempting a go-around at Cranbrook in snowy conditions. (Larry Milberry, John Kimberley)

(Left) Boeing 737 "Fort York" as seen on January 4, 1974 from the ever-popular photographers' vantage point atop Toronto's Terminal 1. This 737 came new to Transair in March 1970, but later joined the CPAir fleet. In 1982 it went to PanAm, thence to Euralair of France. (Larry Milberry)

(Left) CF-EPO at Toronto's Aeroquay (Terminal 1) in August 1978. These were the heady days when Canada's regionals were flexing their muscles and winning skeds far outside their traditional domains. This 737 served EPA/CPAir/CAIL 1970-87. (Larry Milberry)

(Below) Nordair's Boeing 737-242C CF-NAP at Malton on May 12, 1972, 10 months after it came new from Boeing in Seattle. Then, 737 C-GNDG with Canadian North at Great Whale on January 23, 1992. (Larry Milberry)

C-FFPW serving Plummer's Lodge on Great Slave Lake on August 3, 1996. Each summer Canadian North 737s operated here and at Plummer's camp on Great Bear Lake, coming and going with ease on 5,000-foot gravel strips. Then, FPW in earlier markings at Toronto. For 1997 the 737 remained the backbone of CAIL's domestic fleet, but had a more meaningful role. With Open Skies, many were pulled from the money-losing Vancouver-Edmonton-Calgary and Toronto-Ottawa-Montreal triangles and placed on profitable trans-border runs. Smaller CRA F.28s were given a bigger role on the highly-competitive triangles. (Larry Milberry, Richard Beaudet)

The 737 was a solid part of NWT Air's operation since 1989, when ex-Wein Air Alaska machines NWN and NWI were acquired. Here NWI awaits DEW Line freight at Winnipeg on November 14, 1990. (Larry Milberry)

(Right) One NWT Air 737 awaits passengers at Yellowknife on August 18, 1995, while Capt Ray Fisher taxis away in another on the trans-Territorial sked to Rankin and Iqaluit. (Larry Milberry)

(Left) Eldorado Mines' new 737, which replaced the company's two DC-4s, was at Edmonton on June 20, 1980. Its registration indicated Eldorado Nuclear Ltd. It served to March 1983, when it went to Esso Resources. In December 1986 it was sold to Aloha Airlines. (Les Corness)

(Right) Nordair 737 C-GNDM in March 1986. The following year it joined CAIL after a corporate merger. In the mid-1990s it was on lease to Cayman Airways in the West Indies. (Henry Tenby)

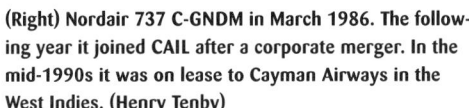

(Left) Boeing 737 QBH, at Dorval with a BAC 1-11 (nearest), joined Quebecair in 1981. The 737 proved uneconomic for Quebecair routes. The company sold its last in 1987, reverting to the smaller BAC 1-11. In 1997 QBH was serving CAIL. (Richard Beaudet)

(Left) Toronto-based courier operator Jetall made a big step from Metro and Convair 580 to the 737, one of which is seen at Toronto on September 5, 1996. In 1996 Jetall folded and its 737s sought other homes. (Larry Milberry)

(Right) Astoria was one of many short-lived Canadian airlines of the 1980s-90s. Using two 737s, it offered budget fares between Toronto and Montreal. It folded in September 1995, four months after start-up. A similar failed upstart was Newfoundland-based Triton Air. It appeared in 1992 with a 737 and just as quickly vanished. (Larry Milberry)

(Left) WestJet of Calgary was another Canadian upstart of the mid-1990s. Into 1997 it was operating five B.737s between Victoria and Winnipeg, and was earning profits. WJK was at Edmonton International on July 2, 1996. (Henry Tenby)

(Right) While the DC-3 may be the most renowned transport, the Boeing 747 holds a lofty place of its own. Many in the business agree that it brought aviation further and made the greatest impact for social good of any plane. The 747 first flew in February 1969 and remained in production since. Air Canada was an early 747 operator, CF-TOA being delivered in February 1971. Evening light produced this view of CF-TOC at Toronto in June 1980. Delivered in June 1971, TOC was Air Canada's third 747. It was first in the fleet to carry the company's new colour scheme, dark geen tail and red maple leaf, revealed on December 1, 1993. Air Canada, which had nine 747s in 1996, planned to dispose of its model 100s by the end of that year, its 200s in 1997, but this was deferred due to growing business. (Dave Thompson)

(Left) Ruler of the skies! Air Canada's 400-ton, 233-foot Boeing 747-433 C-GAGL gets away from Toronto on August 28, 1993. The -400 shunted older jumbos aside. While providing the comforts expected by an ever-more demanding traveller, it guaranteed efficiencies to strengthen operator profits. By the late 1990s older 747s were finding work as freighters, or were being "parted out" in scrapyards like Mojave. (Larry Milberry)

(Right) B.747-233B Combi C-GAGB is towed to its gate at LBPIA's Terminal 2 on September 5, 1996 as a 767 departs on R33. For 1996 Air Canada, the world's 22nd leading airline, averaged 550 sked flights daily. It carried 12.6 million passengers to 525 destinations in 110 nations, reaping $3.98 billion in passenger revenue. (Larry Milberry)

(Left) C-FCRA "Empress of Japan" was CPAir's first 747. Delivered in November 1973 it served till 1986, then went to Pakistan International in a deal that brought a number of PIA DC-10s to Canadian. This day in September 1980 CRA was at Toronto's Terminal 1, while a CPAir DC-8 taxied out. These were the days when one could see the tarmac plastered in CPAir orange—727s, 737s, 747s and DC-8s. In 1997 CAIL was operating four 747-400s on such runs as Vancouver to Beijing, Hong Kong and Taipei. Aircraft types in a fleet like CAIL's often switch routes, depending on seasonal demands, what the competition is up to, unexpected pressure on a route, etc. In September 1991, for example, CAIL replaced the DC-10 with the B.767ER on service from Canada to Buenos Aires, Rio de Janeiro, Santiago and Sao Paulo. (Larry Milberry)

(Above) CPAir's 747 C-FCRD touching down at Toronto in June 1978. It became AP-BCO of PIA in 1986. Then (above right), CAIL dedicated this B.747-475 to Russ Baker, founder of PWA. It was delivered in February 1991 and was shot at Vancouver. CAIL first ordered the -475 in the summer of 1988 (three plus options for four). US airliner manufacturers continued to lead the world in the late 1990s, but were being challenged by Europe's Airbus Industrie consortium. Of 432 jetliners built worldwide in 1994, US companies accounted for 309. Boeing Commercial Airplane Co. noted orders in 1995 for 346 jetliners (69.7% of the world's orders) worth $31.2 billion. That year Airbus sold 106 planes, or 14.8% market share;. McDonnell Douglas had 9.9%. In late 1996 it merged with Boeing, a move that greatly alarmed the Europeans. (Gary Vincent, Jeff Wilson)

(Above) Wardair 747 C-FXRA "Herbert Hollick-Kenyon" buried in snow at Vancouver in December 1980. It came to Wardair in June 1978. Besides its Wardair routes, XRA was on two leases to Libyan Arab Airlines, then went to British Caledonian in 1986, and to Philippine Airlines in 1991. (Maxwell Col.)

(Left) Jumbos hold for taxi instructions at LBPIA in this September 1987 view from atop Terminal 1. Closest is Wardair's C-FDJC "Phil Garratt", bought new in 1973. Like FUN it went to Nationair when Wardair ceased operations. Air Canada's combi C-GAGA, a 1974 model, remained in service in 1997. (Richard Beaudet)

(Right) Alitalia began flights to Canada using the DC-7C between Rome and Montreal in the 1950s. With much work it gained access to Toronto with its huge population of Italians. Alitalia inaugurated Rome-Toronto in November 1972. Here its I-DEME, the 42nd 747, arrives in Toronto on July 6, 1973. ME later had a gypsy's career, serving carriers like Aer Lingus, SAS, Icelandair, Air Algerie, ONA, PIA, People Express and Continental, mostly on lease from Boeing Equipment Holding Corp. In 1997 it was in storage in the Arizona desert. (Larry Milberry)

(Above) B.747-467 VR-HUD taxies at Vancouver, heading to Hong Kong on July 31, 1993. Cathay Pacific burst onto the Canadian scene in the late 1980s, capturing a great deal of trade. Traditionally, Cathay used British, Australian and Canadian pilots, but in 1995 began hiring Americans to crew freighters. By basing them in the US, Cathay saved millions in cost-of-living overhead. For the first half of 1996 Cathay earned US$213 million on sales of US$1.97 billion. Meanwhile, Canada's two major airlines consistently lost billions. As to the venerable 747 airframe, in 1997 Boeing was looking at new stretches. The proposed 747-600, with a new wing and engine, would seat 546 (420 in the -400) and fly 7,500 nm (300 miles further). In the next photo, Argentine 747 PP-VNA, a 1981 model, arrives at Mirabel from Buenos Aires on July 23, 1994. (Larry Milberry)

(Below) B.757 C-FOOA of Canada 3000 at LBPIA. In November-December 1993 Canada 3000 operated its first round-the-world charter using a 205-seat 757. This exotic trip was for well-to-do holidayers. Stops included Nadi, Alice Springs, Katmandu, Mombasa, Cairo and Bermuda. This gave Canada 3000 valuable experience in the Pacific, where it hoped to inaugurate service with the A330 in 1998. In 1996 Canada 3000 had 168-seat A320s C-GVXA, VXB, VXC, VXD and VNY; and 228-seat 757s OOB, OOE, OOG, OOH, XOC, XOF, XOK and XOO; and had plans for two 340-seat A330-200s. (Tony Cassanova)

(Above) Air India B.747-200 VT-EFJ at LBPIA in September 1995. Considering India's huge population, Air India was a modest-size carrier at this time (fifteen 747s, eight A310s, three A300s). Its 747 VT-EFO was destroyed by a bomb off Ireland on June 23, 1985 en route Toronto-Montreal-Heathrow. Two days later Air India suspended Toronto service and did not resume till May 1991. Indian separatists based in Canada long were suspected of destroying VT-EFO, and of attempting to bomb a CPAir DC-8 in Japan the same day. (Larry Milberry)

(Bottom) Entrepreneur Brian Child's Toronto operations included Soundair, Air Toronto and Odyssey International. The multi-tiered concept included Soundair with Convairs and Metros doing courier and passenger charters, sked carrier Air Toronto flying commuter routes to the US mainly with Jetstreams, and holiday carrier Odyssey International with 757s. Bit by bit the operation slid into a financial hole, then failed in 1991. Nationair took over most of Odyssey's business, and the leases on two of its 757s. In this view 757 C-FNBC was at Toronto with a Convair and a Jetstream. In the mid-1990s NBC was N757GA with Guyana Airways. (Kenneth I. Swartz)

(Above) Boeing 757 C-GTSE of Air Transat at Toronto on September 5, 1996. (Larry Milberry)

(Right) The Boeing 767, first flown in September 1981, entered service a year later with Delta Airlines. It quickly became a favourite around the world. Stretched and extended-range versions soon were in use. Air Canada accepted its first 767 in October 1982 and had 29 in service by 1997. Here C-GAVA departs Toronto on September 3, 1989; while C-FUCL arrives at Vancouver on August 13, 1995. UCL originally served China Airlines, then Air New Zealand. It reached Air Canada in May 1985. (Larry Milberry)

(Left) PWA briefly operated two 767s. Here C-GPWB sits at the gate at Winnipeg in August 1983 a few months after delivery. Both aircraft were leased to Air Canada in 1985, when they proved unsuitable for PWA's operation. (John Morrison)

(Right) DC-8-63 C-FTIO lands at Winnipeg in April 1991. It started with Air Canada in 1969, and became a freighter (-63AF) in 1980. In 1984 it was upgraded to a -73AF and served into the 1990s as N802DH with DHL Worldwide Courier. (John Morrison)

(Below) DC-8-43 TJK at Malton March 16, 1973 alongside DC-9-32 TLT. TJK served 1961-76, then went to Cubana as CU-T879. On March 18, 1976 it collided in flight with a Cuban Air Force An-24. The DC-8 survived; it was scrapped in Havana in 1978. (Larry Milberry)

(Above) CF-TIQ followed a career similar to TIO's. Here it was departing Toronto on March 26, 1972. In 1980 it was upgraded with CFM-56 engines (Stage 3). In the 1990s it was with DHL Worldwide Courier. (Larry Milberry)

(Right) Airport security became crucial in the 1960s, when crazies and losers started hijacking planes to Cuba, and other loonies started bombing them. Here the RCMP conducts security on the ramp at Malton in 1975. DC-8 TJR/818 later flew with companies like Arrow Air and Quebecair. On February 15, 1992, while operating as 9G-MKB of MK Air Cargo, it crashed approaching Kano. (Larry Milberry)

(Below) DC-8-43 "Empress of Rome" at Vancouver on June 6, 1968. Delivered February 22, 1961 it was CPA's first DC-8. It left for ARCA of Colombia and reportedly was scrapped in Miami in the 1980s. (Les Corness)

(Above) CPAir DC-8 CF-CPL arriving at Toronto on July 6, 1973. After a long career in the passenger business, CPL became a freighter, serving with Emery Worldwide in the 1990s. (Larry Milberry)

(Left) This DC-8-61 began in 1967 with Trans International, but served with a host of carriers afterward. Its Nordair period as C-GNDA was 1974-78. It last was heard of as N702UP in the 1990s with UPS. Here it was at Prestwick in August 1977. (Wilf White)

(Right) C-FCPP launches forth from Toronto on September 3, 1989. A DC-8-63, it was delivered to CPA in 1968. It joined Worldways in 1983, serving nearly four years. In 1997 it was N819AX of Airborne Express. (Larry Milberry)

(Bottom) C-GMXR was an old Swissair DC-8-62 that Montreal's Nationair operated 1985-91. When Nationair folded, MXR carried on, being with Cargo Lion of Luxembourg as late as 1996. The loss of Nationair DC-8 MXQ is Canada's most infamous air disaster. At 0124 on July 10 it landed at Accra under Capt Allan, staging to Jeddah. The flight engineer had undercarriage repairs in mind, but could not get to his spares for lack of a key. Soon Capt Allan's dispatcher was urging him to press on in order to keep the contract on schedule. He concurred and at 0447 MXQ left on a six-hour leg, repairs incomplete. The tired crew checked into a hotel in Jeddah, but two mechanics worked on MXQ's snags—it was due to take off for Nigeria at 2000. Stories differ as to why, but they did not change any tires. One mechanic later was quoted as

saying the tires were "fully serviceable, but coming close to our limits."

MXQ did not depart at 2000 hours, but sat on the tarmac overnight. According to Nationair president Robert Obadia, it was at his broker's suggestion to hold for a cash advance from the client. In 30°C weather MXQ boarded its passengers next morning. At 0829 it started its takeoff. After only 500 feet a port tire exploded, but there still were 11,000 feet of runway. For reasons unknown the pilots did not abort. With DC-8s, which did

not have heat sensors in the landing gear wells, pilots would let the gear hang for a few minutes, especially after a long roll on a hot day; but the gear was raised immediately. Another tire blew, damaging electrics and hydraulics. Fire spread into the wing and cabin. The hydraulics for the spoilers are in the port wheel well. If the spoilers were disabled, control would be difficult. The crew managed to turn back, but 12 minutes after takeoff MXQ dove into the desert. All 261 aboard perished, including the 14 Canadian crew. Nationair denied responsibility for the accident. It diverted attacks on its operating procedures by saying that the tires might have blown if they hit debris on the runway. Nationair spoke of such "evidence" being swept away soon after the crash. It also suggested the tire could have blown if a wheel rim disintegrated. (Henry Tenby)

(Right) One of Air Canada's early DC-9-10s at Malton in September 1967. A year later it was sold to Texas International as N5726. It next was N626TX of Continental. On September 15, 1987 it was lifting off in snow at Denver when it rolled into the ground, killing 28 of the 82 aboard. (T.R. Waddington)

(Left) DC-9-32 CF-TMD departing North Bay on April 2, 1980. Delivered to Air Canada in September 1968, it put in more than a quarter century of good service, paying for itself many times. Over the years seating in Air Canada's DC-9-32s ranged from 92 to 108. (Dave Thompson)

(Right) DC-9 CF-TMJ in Air Canada's 1990s colours at Vancouver on August 13, 1995. An A340 and 767 are beyond. While Air Canada continued selling DC-9s in the Philippines in 1997, it planned to keep at least 15 for the foreseeable future. (Larry Milberry)

(Below) CPAir's DC-10 C-FCPE "Empress of Alberta" at Vancouver on July 14, 1987. The DC-10 first flew in August 1970 and entered service with American Airlines a year later. Then, an oddball view of a CPAir DC-10 touching down at Calgary on a March 1981 flight from Toronto. (Henry Tenby, Larry Milberry)

(Right) CAIL DC-10 CPF rotates off R33 at Toronto on September 3, 1989. It had been with Canadian since new in 1980. In later years the DC-10 proved ideal for CAIL's Pacific services. (Larry Milberry)

(Left) Wardair's DC-10 C-GXRC "W.R. 'Wop' May" crew training at North Bay in November 1978, the month it entered the fleet. When Wardair folded 10 years later, XRC went to the GPA Group, thence to leases with Finnair, World Airways, Garuda and Aero USA. In 1997 CAIL had 80 aircraft with a young average age of 13 years, although its nine DC-10s were much older at 17. (Dave Thompson)

(Right) An American Airlines DC-10 at Toronto in December 1974. Having launched DC-10 service in 1971, American still had nearly 50 25 years later. It lost only one over the years—N110AA crashed at Chicago in May 1979, when an engine separated on takeoff. (Larry Milberry)

(Below) The MD-11 followed the DC-10. Launched in December 1986 as a stretched and modernized DC-10, it flew in March 1989. With orders from FedEx, Finnair, Korean Air, Swissair, etc. the MD-11 looked promising. But McDonnell-Douglas was unable to compete and succumbed to Airbus and Boeing. In December 1996 it agreed to merge with Boeing and phase out MD-11 production. No Canadian carrier ordered an MD-11. This Swissair example was departing Toronto on July 22, 1989. (Larry Milberry)

(Right) The L.1011 Tristar wide body served Air Canada 1973-96. The prototype flew in November 1970. Production ended at 250 in 1984 after a recession led companies like Air Canada, Delta, Pan Am and Trans World to cancel options for some 40 aircraft. Several Canadian firms had profited from the Tristar. Fleet of Fort Erie made dorsal fairings, main landing gear doors and cowlings for No. 2 engine; NWI of Edmonton made nose landing gear doors and floor and underfloor components; Bristol of Winnipeg made the intake and exhaust ducts for No. 2 engine. After serving Air Canada 1973-90, L.1011 C-FTNH, shown departing Mirabel, went into storage at Marana, Arizona. Another Air Canada Tristar (TNJ) was sold in 1992 to Orbital Sciences Corp. and converted to launch space vehicles. (Richard Beaudet)

Eastern's Tristar N315EA in Air Canada colours at Sea-Tac, Washington in April 1980. It frequently was seen as C-FTNC and went to Air Transat in October 1987. About 200 Tristars remained in use in 1996. (John Kimberley)

Tristars stored at Marana, Arizona in May 1991. Before long, Air Canada missed the steadiness and profitability of its long-range Tristars (although this had not been the case in their rocky early years). It brought several back till the winter of 1995-96, but by then delivery of new aircraft had caught up with expanding demand. The old-technology Tristars, much loved by passengers and crew, came to the end of the line. TND, TNG and TNL were the last in Air Canada service. They soon started going out to the charter carriers, e.g. TNL left Marana for Air Transat on April 17, 1996. (Christopher Buckley)

C-FCXJ, on lease from British Caledonian Airways, at Toronto in February 1989 with the blast of smoke associated with a Ten-eleven start. CXJ began in 1975 with British Airways and remained active with Caledonian in the 1990s. (Tony Cassanova)

(Below) British Airways VC-10 G-ASGA waits for an evening departure from Toronto on December 29, 1979. The VC-10, first flown in 1962, was one of the glamorous airliners of the 1960s, although only 54 were built. GA served BOAC/BA 1965-81, then became RAF ZD230. A number of transport and tanker VC-10s remained in RAF service in the 1990s. (Dave Thompson)

(Above) Air Canada's historic monopoly restricted meaningful airline competition in Canada for decades. Into the 1960s competition amounted to nothing but a few carriers running the clapped-out likes of DC-4s and Super Constellations. In time Ottawa forced Air Canada into a free market mode; airlines like Air Transat, Canada 3000 and Royal Air appeared and prospered. Air Transat Tristar TNA had begun in 1972, alternately working for Eastern Air Lines and Air Canada, depending on which was busy enough to need it. While on duty with EAL it was N312EA. It finally went to Air Transat in 1988. Here it was at Toronto on August 27, 1991 with Nationair 747 FUN (ex-Wardair) at the farther gate. (Christopher Buckley)

(Left) Wardair introduced the Airbus family in Canada, but soon sold to PWA Corp. Here two Wardair Airbus A310s wait at Toronto in June 1989, C-FNWD "Jack Moar" being nearest, C-FGWD "Z. Lewis Leigh" farthest. Then (below), a candid photo of Z. Lewie Leigh at the controls of Wardair's A310 simulator in Toronto in June 1988. The new technology "glass cockpit" of planes like the A310 astounded old-time bush pilots like Lewie, but they invariably praised this progress. While such instrumentation, symbolized by CRTs (cathode ray tubes), improved operations, it had its critics. In January 30, 1995, for example, *Aviation Week* noted "a growing body of evidence that pilots are encountering difficulties interfacing with highly automated cockpit systems. A series of incidents and several fatal accidents are evidence that something is amiss. In more than a few cases the pilots involved can be heard on cockpit voice recorders questioning what the automatic flight system is doing. Some automated cockpit systems obviously are too complex, because they continue to surprise even experienced pilots." As with anything new, the glass cockpit had to earn pilots' respect and outgrow the bugs. (Larry Milberry)

(Above) "Z. Lewis Leigh" departs Toronto on September 3, 1989. In 1990 it became HS-TID of Thai International. On July 31, 1992 it was approaching Kathmandu in poor weather and without radar control. Thai procedures required full flap and slats by 13 nm DME (distance measuring equipment on the ground) for this approach, but a snag obliged the crew to cycle the flaps. The captain, a 14,000-hour man who had been into Kathmandu many times, decided to re-fly the approach. Seven miles southwest of the Kathmandu VOR he turned right and climbed from FL105. On a heading of 025° he reported ready to proceed to the Romeo fix (navigating point) 41 nm southwest of the VOR to begin descent.

The plane passed the Romeo fix, apparently because the pilots were busy loading data into the FMS (flight management system). It continued 16 nm northwest of the VOR and altered heading to 005°. A minute later the GPWS (ground proximity warning system) sounded, but the captain disregarded it as false. Seconds later the plane crunched into a mountain at 16,000 feet ASL at 300 knots. All 113 aboard died. No one had realized that the A310's progress was unorthodox. No headings were forwarded from either end; only DME was discussed. Apparently there was a cockpit-tower language problem. One publication noted that communications difficulties, a climbing and then a descending turn, and reprogramming the FMS may have reduced crew effectiveness in interpreting navigation data. (Larry Milberry)

(Above) Five ex-Wardair, ex-CAIL A310s went to the CanForces designated CC-150 Polaris. Here 15004 waits at Trenton on January 12, 1994. The old CAIL fleet number (205) on the nose gear door identifies this Polaris as former C-FNWD "Jack Moar". (Tony Cassanova)

(Above) Wardair A310 C-GLWD "C.H. 'Punch' Dickins" departs Toronto on September 3, 1989. (Larry Milberry)

(Right) In 1997 Montreal-based Air Club operated four A310s. This example, at Mirabel on July 29, 1994, originally was PanAm N819PA "Clipper Morning Star". It came to Canada in June 1994. (Larry Milberry)

(Left) Air Canada A320 C-FDSN at Toronto on September 6, 1992. The night shift was doing an engine run. Air Canada ordered 34 A320s in the summer of 1988, the first entering service in March 1990. Aircraft DSN, the 126th A320, joined the fleet in November 1990. The A320 was the first civil airliner with fly-by-wire controls. With other innovations like automatic fuel management, such types quickly replaced the jetliners of the 1960s-70s. At Air Canada that meant an end for the long-serving 727. (Dave Thompson)

(Right) The latest generation of jetliners is all about economy. In Air Canada's experience, for example, a 112-seat A319 burns 16% less fuel than a 92-seat DC-9. Here Air Canada A320 C-FPDN is at Vancouver on August 13, 1995. The A320 first flew in February 1987. Sales caught on quickly and the 500th was delivered (to United) in January 1995. Of 916 new airliners sold in 1996, Airbus accounted for 301 compared to 559 for Boeing. The world's airlines were booming in the mid-1990s. Forecasts were for a tripling in business over the next 20 years, a third of the growth in Asian-Pacific, the fastest growing sector. By 2015 80% of the 1995 fleet of 7,300 airliners would be replaced and the fleet would have doubled in size. (Larry Milberry)

CAIL A320-211 C-GQCA at Vancouver on September 27, 1995 after a four-hour flight from Toronto. The crew included Sharon Rush, Robert Leblanc, Regina Inciura, FO Jim Rawlings, Ellen Racette, Capt Bruce Laxon and Susan Iwaskiw. This A320 was not owned by CAIL, but by GPA Airbus A320, an Irish-based leasing company. Today's airliners typically belong to the likes of GPA, banks, insurance companies and other powerful interests. A lessor, however, may own only the airframe, while engines, avionics, etc. are the property of other investors. Such ownership is necessary, considering the cost of a new A320—perhaps US$50 million— double or triple that for larger planes. By the 1990s few major airlines owned even one of their planes. At the 1997 Paris airshow Airbus announced the go-head for its 8,500-nm, 900-passenger A3XX, the first commercial jet with an MTO above 1 million pounds. Airport authorities already were gearing for this class. While Airbus was bullish on super jumbos, Boeing remained reserved, even though about two dozen leading airlines were working with Airbus on design details. (Larry Milberry)

(Above) Canada 3000 (C-GVXB) and CAIL (C-FLSS) A320s at Toronto on June 23, 1993. Canada 3000 introduced the A320 on June 7, 1993. In Airbus Industrie's 1995 awards, praise was heaped on Canada 3000, which had average utilization for the year of 4,680 hours for each of its five A320s—12.8 hours daily per machine. (Larry Milberry)

(Right) Andy Graham photographed this Airbus pair at Toronto on June 31, 1995. Nearest is A320 C-GTDC of Toronto-based Skyservice, then Air Canada's A340 C-FTNP. Airbus cockpit commonality helps pilots switch from type to type.

(Below) The 295-seat Airbus Industrie A340, launched in 1987, flew in October 1991. Air Canada placed an early order, introducing the A340 on Vancouver-Osaka in mid-1995. To augment business, it entered a code-sharing agreement with All Nippon Airways on this route. John Kimberley caught this Air Canada A340 on departure from Vancouver. Henry Tenby shot one of Cathay Pacific's landing at Hong Kong on January 1, 1997. Cathay operated the A340 daily between there and Vancouver. By this time the Pacific was the world's fastest-growing travel market. Encouraged by the success of the A340 there, Airbus was studying a three-class, 375-seat, 7,000 nm-range stretched version. (Larry Milberry)

(Above) The first A319 flew August 25, 1995 at Hamburg, then commenced a 650-hour test program aimed at certification in March 1996. The 124-seat inter-urban plane is seven frames, or 147", shorter than the 150-seat A320. Air Canada, Air Inter, Lufthansa and Swissair were the initial customers. Like all modern Western airliners it was offered with a choice of engines: CFM56-5A, '5B and IAE V2500. The A319, A320 and A340 cockpits have nearly complete commonality, a strong marketing feature in an arena that saw stiff competition from the 737 and MD-87. By late 1997 A319/320/321 production had reached 14 per month. (Airbus Industrie)

(Above) Capt Pat Dayman aboard Air Canada A340 C-FTNP en route Toronto-Vancouver (1,842 nm) on August 8, 1995. Co-captain on the flight was Gilles Larue. Kevin McElrae was the FO. Normally, the A340 has two pilots, but this day some extra training was in progress. (Larry Milberry)

(Left) The Russians are coming! Transports from the USSR have visited Canada since the 1950s, but never with the openness of the 1990s. The Antonov An-225 Mriya (Dream) was the world's largest aircraft in the 1990s. One was built, primarily to carry the USSR's Buran space orbiter. When the Buran program was scrapped, the AN-225 made a few commercial trips, but was grounded soon after this visit to CFB Namao (Edmonton) in May 1992. The Mriya was the first plane to exceed 1 million pounds (MTOW 1,323,000 pounds). Wing span was 290', length 275' 7", height 59' 4.5". Mriya used six Lotarev turbofans each of 51,590 pounds thrust. (Larry Milberry)

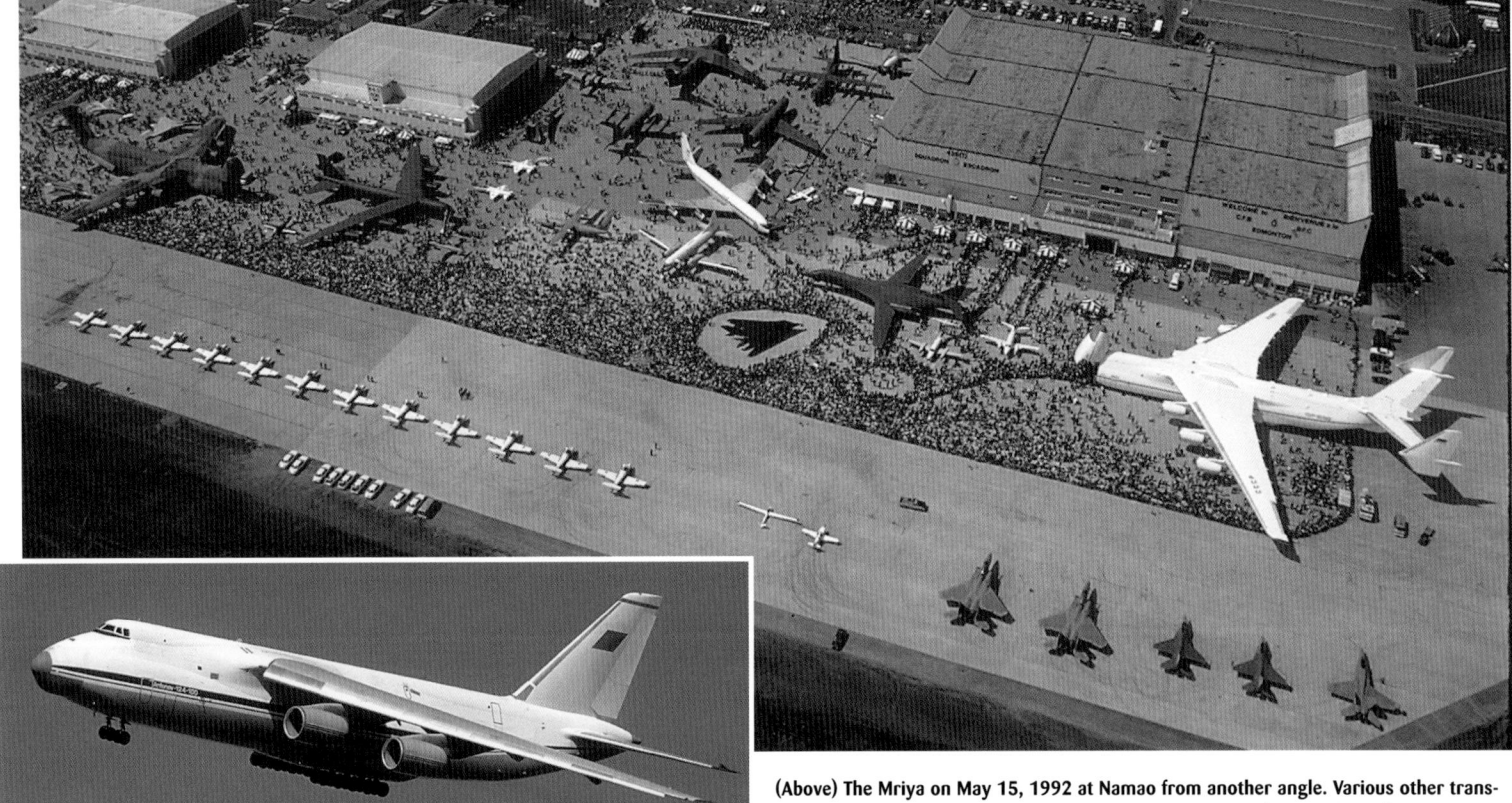

(Above) The Mriya on May 15, 1992 at Namao from another angle. Various other transports from a C-5 to an A-26 waterbomber are on display . (Larry Milberry)

(Above) An-124-100 UR-82008 over London, Ontario on June 4, 1994. Built in 1985, this was one of the earliest examples. Recently it had hauled space equipment to Brazil for NASA, airport jetways from Utah to Bogota, and 216 ostriches from Namibia to Gatwick. It was in London to collect a GM locomotive for Dublin. By this time the AN-124 was gaining popularity with shippers. Air freight had begun soaring after the Arab oil embargo of the 1970s. Shippers became convinced of the benefits of fast delivery, reduced warehouse and insurance charges, cheaper crating and quicker settling of accounts. From 1989-95 the market for outsized air cargo grew 500% annually. Prospects for the An-124 improved with agreements among Antonov and Western manufacturers to replace inefficient CIS equipment, e.g. Antonov hoped to replace the An-124's unreliable Progress D-18T engines with GE CF6-80C2s. The impact of the An-124 forced the West to review air freight commitments. In the mid-1990s, for example, McDonnell Douglas was studying commercial markets for its military C-17 Globemaster III. (Tony Cassanova)

(Above) One-two-fours at Mirabel on July 29, 1994. Volga-Dnepr/Heavy Lift was en route to Prince George, BC. Ruslan/Air Foyle UR-82008 (subcontractor, Toronto-based Skylink) was heading to Entebbe and Mombasa with relief supplies for Rwanda. (Larry Milberry)

(Above)The Il-62 airliner (186 seats) first flew in 1963. It often visited Canada with airlines like Aeroflot (Russia), CSA (Czechoslovakia), Cubana and LOT (Poland). This Aeroflot Il-62 was at Mirabel on July 29, 1994. The Aeroflot Il-86 (right) was at Gander on February 28, 1992, likely on a Cuban flight. First flown in 1976, the 350-seat Il-86 was the USSR's first wide-body airliner. (Larry Milberry)

(Left) The 85-seat Fokker F.28, first flown in May 1967, was a multinational venture. The Netherlands funded development; MBB and VFW of Germany, and Shorts did subcontract manufacturing. Production ended in 1970 at 245 aircraft. Air Ontario had two F.28s and inaugurated service June 1, 1988 with a sked from Toronto to Sault Ste. Marie. C-FONG was shot at Toronto. A 1973 model, it served less than a year, going on to Papua New Guinea and Iran. On October 12, 1994 it ended its days crashing into an Iranian mountain. In 1996 Air Canada connectors like Air Ontario, operating some 600 daily flights to 92 destinations, carried 3.9 million passengers. (Larry Milberry)

(Left) Canadian Regional F.28 C-FTAV, nicknamed "Silver Bullet", at Hay River en route Yellowknife as F1142 on June 25, 1993. Capt Pat Meek and FO Randy Hulkenberg were on the flight deck. Built in 1971, TAV originally flew for Transair (1971-79), went to Papua New Guinea for Air Niugini, then returned to Canada for CRA. (Larry Milberry)

(Below) CRA F.28 captains Mark Whitehead and Earl Dyck ready to operate F1991 from Toronto to Thunder Bay on May 1, 1996. (Larry Milberry

Photographed at Dorval in March 1987, this Quebecair F.28 started in 1969 with the Norwegian operator Braathens SAFE. It served briefly—along with several pilots it went to the French airline TAT in 1989. The F.28 reappeared in Quebec in 1994, when Inter-Canadien began Quebec City-Toronto skeds, then added runs like Sept-Îles to Goose Bay. By 1997 F.28s use still was expanding; CRA had 18 in the fleet. One of its busiest routes was a five-times-a-day Toronto-Thunder Bay service. (Kenneth I. Swartz)

(Below) F.28 C-FAIF at Toronto in November 1994. It served airlines in Ghana, Norway, Peru, Turkey, Indonesia and the US before putting in a stint in 1994 with short-lived Atlantic Island Airlines of Charlottetown. AIA had planned a fleet of six F.28s. This F.28 later joined CRA as C-FCRU. (Andy Graham)

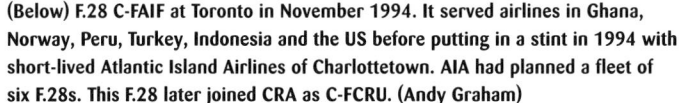

(Above) Norcanair F.28 C-GTEO at Minneapolis in July 1986. Delivered to Itavia of Italy in 1970, it was one of the last F.28s. It went on to Time Air and CRA. Norcanair had taken over SGAS in 1965 and thrived for many years. It later failed, then re-appeared in 1991 using a BAe 748, Convair 640 and F-27J. This time its existence was brief—Transport Canada grounded it within a year over maintenance matters, and the Saskatchewan government finished it off by calling a loan of more than $2 million. (John Morrison)

(Right) Fokker prolonged its jetliner concept with the F.100, a stretched F.28 with bigger engines and as many as 119 seats. The F.100 first flew in November 1986; Swissair was the launch customer. Inter-Canadien was the first in North America with F.100s, launching service in January 1989 in the busy Quebec-Ontario commuter market. The operation folded in less than two years. Here Bagotville passengers board an F.100 for Dorval in May 1989. Another short-lived deal was with Wardair. In the summer of 1988 it ordered 16 F.100s, but this plan died when Wardair sold to CAIL. In 1995 Fokker, with 10,000 employees, went under. Bombardier, whose subsidiary Short Brothers made F.50 and F.100 wings, considered a takeover but did not follow through. (Larry Milberry)

(Left) The rugged, long-lived BAC 1-11 and BAe146 also served Canada's short-haul markets. Here one of Quebecair's 79-passenger BAC 1-11 lands on R24L at LBPIA September 5, 1987. QBO was built in 1967 for British Eagle and came to Canada in 1969. It returned to the UK in 1984 for service with various carriers and in 1991 became 5N-IVE of Nigeria's Okada Air. (Larry Milberry)

(Right) Air Atlantic BAe146 C-FHAV boarding at Gander on February 28, 1992. Originally Air Wisconsin N601AW, it came to Canada in May 1990. By 1997 there had been 18 Canadian-registered BAe146s. (Larry Milberry)

A short-haul passenger jet interested Canadair from earliest Challenger days. This led to the 50-passenger CRJ, which first flew in May 1991. It was a bit early for the airlines to appreciate a pure-jet commuter plane, so sales started slowly. Forward-thinking DLT of Germany was the launch customer; other orders followed, especially one in August 1993 from Air Canada for 24. Before long, commuter lines with high-density routes into major hubs realized that the CRJ was the plane to replace types like the ATR, Saab 340 and Dash 8. Cincinnati-based Comair had some 50 CRJs by mid-1997. Canadair and its marketing arm, Bombardier Regional Aircraft Division, now could see that its CRJ go-ahead decision had been well-timed. Here Air Canada CRJ-200 C-FRIL lands at Ottawa in March 1996, while C-FSJJ is seen being de-iced at Toronto the same month. (Andy Graham)

A CRJ-200 in Lufthansa colours at Dorval on June 10, 1994. It carries a temporary, pre-delivery Canadian civil registration. Then, Comair's CRJ N916CA delivers passengers to Terminal 1 in Toronto on July 13, 1996. In January 1997 Bombardier announced the go-ahead for the 70-78-seat CRJ-700. In June 1997 Comair placed the first big CRJ-700 order—25 aircraft. (Larry Milberry)

One of the Arctic's great aviation symbols—NWT Air's Lockheed L.100-30 Hercules C-GHPW gets airborne on June 29, 1993 from the bumpy strip at Talston River 28 nm from Fort Smith. It had just dropped off a "Cat" for a heavy construction job. HPW began in December 1978 with PWA, then went to NWT Air in November 1983. It worked all around the Arctic on drilling, mining and general support, and globally in Angola, Kuwait and Rwanda for the UN and Red Cross. There were many other contracts such as freighting auto parts from Ontario to Mexico, and spare engines from place to place for the airlines. (Larry Milberry)

Aviation in the Northwest Territories

Getting Around

The Northwest Territories are vast, comprising 34.4% of Canada's landmass. But the region is sparsely-populated—for 1997 it had but 66,000 inhabitants. Nonetheless, the NWT is a treasure house of minerals and other resources. Over the centuries these brought human involvement and settlement, one result being a complex transportation infrastructure. It was the region's waterways that were the original highways. These remain important, especially the great Mackenzie with its summer barge trade. Other parts of the Arctic have excellent salt water connections, and trans-Arctic shipping via the historic Northwest Passage is now common. Each summer ships and barges sail from ports like Halifax and Montreal through Hudson and Davis Straits, reaching remote locations like the Polaris Mine on Little Cornwallis Island. They deliver supplies and return with ore concentrate for southern refineries. Other ships reach the western Arctic through Bering Strait. The main problem is a short season and the cost of building and operating ice breakers.

Roads are few in the NWT. The development of new ones is rare. Over what roads there are, like Edmonton-Hay River-Yellowknife, or Whitehorse-Dawson-Inuvik, distances are great and weather daunting. As to railroads, the North rarely has enticed them, and none has reached the NWT. (Elsewhere, railroads have edged north, but only reluctantly and at inordinate expense. One went in from Skagway to Whitehorse with the gold rush of 1898. The Hudson Bay Railway was built in the early 1930s from The Pas to move grain to tidewater at Churchill. In the post-WWII years the Quebec North Shore and Labrador Railroad pushed into Labrador from Sept-Îles to exploit iron ore.) There is little likelihood of new northern railroads. The only practical means of transportation in the Arctic, after centuries of effort, turns out to be the airplane. It is the lifeline for those 66,000 people—reliable, versatile, fast, safe, relatively cheap and everywhere.

What are the roots of Arctic aviation in Canada? Most histories consider the 1926 Hudson Strait Expedi-

tion as *the* pioneering event. Headed by S/L T.A. Lawrence, it put an aviation detachment into the Ungava and was a huge success. The expedition spun off from a 1922 venture, when Ottawa sent S/L R.A. Logan north aboard the steamer *Arctic* to investigate climate and other natural conditions, for the airplane was being considered as a link for scattered RCMP posts. After his summer aboard *Arctic*, which included visits to remote posts, Logan reported: "There is no reason to suppose... that when the need arises, with special machines and precautions, aircraft cannot be operated to advantage in the Arctic... Sufficient is known to justify despatch of a small party with machines to the far north for further investigation." Logan suggested that a detachment with two aircraft spend a year exploring operating conditions. In time the Hudson Strait Expedition was despatched. Thereafter, aircraft slowly made in-

With prospecting and mining in the NWT came settlement. The rare view below shows Yellowknife in the late 1930s. Gold gave the NWT a big start in mining. Carpenter foreman T.H. Maranda took the snap of the Consolidated Mining and Smelting operation at Yellowknife about 1938. (R.C. Davis Col./NWT Archives, Maranda Col. N91-032:0034)

Aviation supported the early prospectors, then mine development. G.W. "Gil" McLaren was one of the early pilots in the area, flying for Mackenzie Air Service of Edmonton. Here he is in typical winter apparel. (G.W. McLaren Col.)

MACKENZIE AIR SERVICE

Schedules

Air Mail - Passenger - Express

Passenger Fares

and

Express Rates

Effective June 1, 1938
Cancelling Previous Fares, Rates and Schedules.

SERVING ALL
MINING AREAS
AND
TRADING POSTS
IN THE
CANADIAN NORTHWEST
FROM EDMONTON TO THE ARCTIC

MACKENZIE AIR SERVICE LIMITED
W. Leigh Brintnell, President
MACDONALD HOTEL
EDMONTON, ALBERTA, CANADA
Telephones: 21261, 21361, 25428

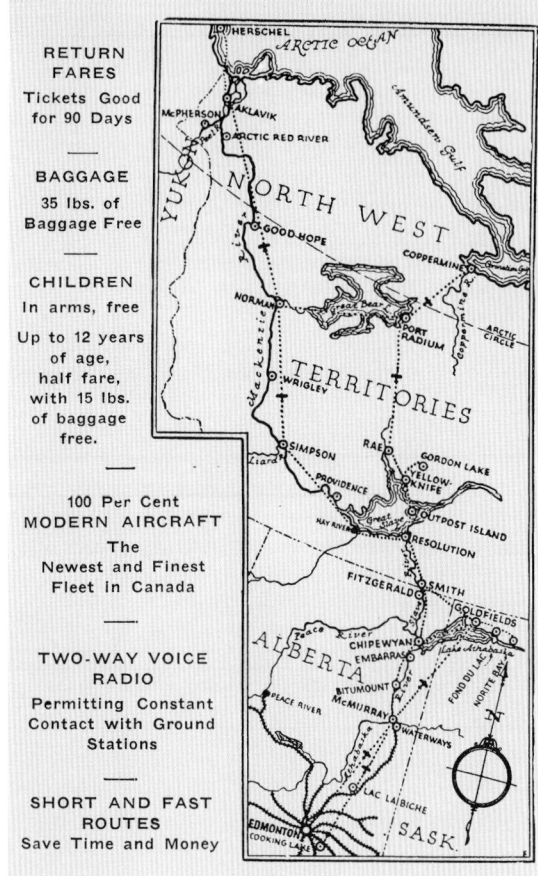

RETURN FARES
Tickets Good for 90 Days

BAGGAGE
35 lbs. of Baggage Free

CHILDREN
In arms, free
Up to 12 years of age, half fare, with 15 lbs. of baggage free.

100 Per Cent
MODERN AIRCRAFT
The Newest and Finest Fleet in Canada

TWO-WAY VOICE RADIO
Permitting Constant Contact with Ground Stations

SHORT AND FAST ROUTES
Save Time and Money

(Left and above) Excerpts from a pre-WWII MAS schedule for its Mackenzie River service. (Rutledge Col.)

roads, especially with WWII, when airstrips were built at places like Fort Chimo and Frobisher Bay, to serve warplanes ferrying to the UK.

The Postwar Evolution

An oil boom on the Mackenzie in the 1920s, the staking rush around Yellowknife of the 1930s, wartime projects in the 1940s—all focused attention on the NWT. After the war Edmonton, Winnipeg and Montreal became DEW Line hubs from where materiel went north. They became headquarters for burgeoning airlines—Associated and PWA in Edmonton; Transair in Winnipeg; Nordair, Wheeler and Worldwide in Montreal. Collectively these carriers brought many benefits to the Arctic. On a lesser scale a number of small entrepreneurs appeared after WWII. Using types like the Anson, Bellanca, Fox Moth and Stinson, they

catered to mining, trapping and trading. Typical were Jim and Chuck McAvoy —mainstays in Yellowknife aviation since 1949. They operated McAvoy Air Services. On June 9, 1964 Chuck took off from Turner Lake, north of Yellowknife, bound for another lake with two young mining men. The plane, Fairchild 82 CF-MAK, never was seen again. A search with RCAF, USAF and civil aircraft ran into July, but nothing was found. Jim McAvoy remained in Yellowknife aviation for many years, not retiring till the early 1990s. Such were the hearty Arctic personalities. Many came and went— they put modern NWT aviation on a solid footing.

By the 1960s Ottawa was taking a keener

Z.L. "Lewie" Leigh (left) and John D. "Jack" Hunter also were NWT aviation pioneers. Lewie flew Fokkers and Fairchilds there from the early 1930s; Jack often toured the region in a DOT Lockheed 12 from his base in Edmonton. Here they were in 1994 at Lewie's home in Grimsby, Ontario. It had been more than 60 years since they first flew together. (Larry Milberry)

(Right) Senior Skyrocket CF-DOH near Yellowknife in 1949. This was Northwest Industries' improved 1930s Bellanca 31-55. Stan McMillan flew the protoype on February 28, 1946. Only 13 were built, due to a supply of cheap, war surplus Norsemen and to the appearance of the Beaver. Registered on August 23, 1946, DOH was sold on June 27, 1947 to McDonald Aviation of Edmonton, in whose markings it's seen near Yellowknife. On September 2 the following year, while on lease to QCA, it crashed at Growler Cove, north of Campbell River. Pilot Wally Britland and fisheries inspector Ken Weaver died. There was a cursory investigation—people presumed that DOH ploughed in, maybe while patrolling low to surprise illegal fishermen. In 1983 the CMFT salvaged DOH. Museum founder Ed Zalesky traded its engine to Red Deer for that in the city's display Harvard. He tore down, sandblasted and painted DOH's engine. In the process he found that it had serious internal damage consistent with a failure in flight, not impact. (NWT Archives, Busse Col. N79-052:3502)

(Left) While aviation developed cautiously down the Mackenzie and around Great Slave and Great Bear lakes, it was slower to gain a foothold beyond. This view shows the RCMP patrol vessel *St. Roch* at Cambridge Bay over the winter of 1944-45. An RCAF Norseman VI was on hand, but few were such visits until DEW Line times. (CANAV Col.)

(Right) Air transport expanded after WWII as exploration boomed. These views show bushplanes on the frozen bay in front of Yellowknife about 1950. Anson CF-FFC (ex-RCAF 12543) is seen with a dog team and horse-drawn sledge. Ansons were valuable in the Arctic. (NWT Archives, Busse Col. N79-052:4712)

(Left) As everywhere, accidents have plagued NWT aviation. Here the HBC's Canso CF-BSK burns near the Yellowknife Hotel on February 9, 1947. A freighter or 17-passenger plane flown by Harry Winny and Fred Bradford, it was purchased to serve remote posts. Fuelling for its first trip, it exploded, severely injuring engineer Les Templeton and crewman Will Kennedy. A local newpaper noted: "The Yellowknife Volunteer Fire Brigade responded to the alarm and the truck and bus were on hand in record time. Assisted by the Hudson's Bay Company's stationary fire unit, excellent work was done in the saving of nearby aircraft and eliminating the possibility of the fire spreading to buildings in Yellowknife Bay." The HBC replaced BSK with Canso FOQ. (NWT Archives, Busse Col. N79-052:3490)

(Right) While today we see Twin Otters landing at Yellowknife, this scene was typical of the early-1950s—a PWA Stranraer arriving as a Norseman awaits business. (NWT Archives, Busse Col. N79-052:3743)

(Above) A new Mack arrives for the Yellowknife Fire Department aboard a Wardair Bristol 170. (NWT Archives, Edmonton Air Museum Col. N79-003:510)

(Right) Two views of Yellowknife with its seaplane docks about 1960. Many of these buildings still stood in 1997. In frontier days operators here scurried to consolidate business, squeeze out competitors and stave off new ones. In time a handful would control air transport in such a market. As years passed, one operator with his family often dominated much of a town's affairs, running not only aviation, but fuel concessions, trucking and water transport, auto sales and rental, food and dry goods sales, travel and agencies, hotels, restaurants, outfitters, tourist camps, etc. If a competitor appeared, the owner might turn out to be related by blood, marriage or previous business affiliation. Family members often entered politics, further consolidating their influence. Thus could a dynasty arise from a little air operation established decades earlier. (Provincial Archives of Alberta, John Davids Col. D695, NWT Archives, Busse Col. N79-052:4855)

interest in the NWT. Yellowknife became capital; centres like Rankin Inlet became seats of regional government. Air services were needed more than ever. But there still was no regional airline. Instead, the NWT was served by numerous small carriers. Wardair had some Otters and Bristol Freighters in Yellowknife; Transair and Lambair operated north from Churchill with Otters, DC-3s and Bristols; Bradley had a base in Frobisher with everything from Super Cubs to DC-3s, Atlas Aviation was in Resolute. Bob Engles of Yellowknife mused about one big NWT carrier, but there only were 50,000 people in those days. How could they support a major airline? Somehow, difficulties were overcome and an excellent air transport system was established.

In the 1990s aviation in the Arctic and sub-Arctic involved three main carriers: Bradley/First Air of Ottawa; Canadian North of Edmonton; and NWT Air of Yellowknife. As well, there were several mid-size carriers like Buffalo Airways of Yellowknife and Air North of Whitehorse with aircraft as big as the DC-4; and a host of smaller local carriers with singles and light twins. Edmonton, Winnipeg, Ottawa and Montreal were Canada's north-south gateways. Each had historic roots e.g. Canadian North in Edmonton was based on the old CPA/PWA operations; NWT Air from Winnipeg parallelled old Transair and CPA routes; and First Air flew along old Nordair, Wheeler and Worldwide routes.

(Right) An overall view of Yellowknife in July 1992 with its modern CBD in the foreground and Old Town beyond. Since it became territorial capital, Yellowknife has grown steadily. (Larry Milberry)

(Above) Early on NWT Air had two of the bush operator's favourites—the Beech 18 and Otter. (via Tony Jarvis)

Bob Engles at home in Yellowknife in the summer of 1993. He remained involved in aviation through the 1990s, owning three aircraft in the NWT Air fleet, as well as Citation II C-GLMK and Ce.185 C-FNWU. (Larry Milberry)

(Left) Bob Engles bought his first DC-3 in 1964. CF-NWU, seen at Edmonton on July 22, 1971, had been USAAF 41-18689 in India and China during WWII. It joined NWT Airways in July 1971. It was later with Buffalo Airways, then went to Florida in 1991 as N62BA. (Les Corness)

NWT Air

In 1961 Bob Engles, a Yale graduate in industrial administration (Class of 1950), established Northwest Territorial Airways in Yellowknife. He had begun flying at Syracuse, NY on a float-equipped J-3. He earned a commercial licence at Boeing Field, Seattle in 1954, then spent a summer flying for a family-owned lodge at Bear Lake, BC. In 1957 he bought Beaver CF-IFG. In 1958 came a call from Max Ward asking Engles if he would bring IFG to Yellowknife to work for him on a McGill University research contract. Engles spent that summer on Bellot Strait with the McGill crew. He next took an air transport rating with American Flyers in Dallas, flew the Bristol Freighter in 1959-60 with Ward's senior pilot, Don Braun, then struck out on his own. But he needed some waterfront in Yellowknife to get an operation going. He sat down in a local bar with cat skidder Frenchy Lamoreux, who had some frontage that included CPA's old building, then being used by La Ronge Aviation. Engles bought it for $10,000. He located a good Otter in Colombia—aircraft No. 54. Operations began in 1962 with Engles flying the Otter—CF-NTR. Pilot Darrell Brown soon was hired. Engles was competing head-to-head with PWA's VFR operation run by Stan McMillan, and with Wardair. They didn't think much of the uppity new competitor; but Engles was determined.

Gradually, NWTA added clients, and aircraft like the Beech 18 (1963) and DC-3 (1964). His first DC-3 was the famous CF-BZI, ordered in 1942 by the USAAC. After the war Maritime Central Airlines put it into service. In 1953 Interprovincial Pipe Lines of Edmonton bought BZI as a VIP aircraft. It finally went to NWTA and flew until damaged on Somerset Island in 1971. Joe McBryan of Hay River then took it over. He and Bob Engles eventually donated BZI to the Calgary Aerospace Museum, where it was restored with NWT Airways colours and wheel-ski undercarriage. Another NWTA DC-3 was CF-NWU. The 1941 model served in India and China, went to Turkey after the war, then became N51F, a VIP machine with General Mills of Minneapolis. NWTA acquired it in July 1971 and kept it in VIP configuration to fly judges, bishops and corporate executives around the NWT. It later was sold to a Fort Lauderdale tour operator, but Engles reserved the registration for his private Ce.185.

Beginning in 1968 Max Ward began using Twin Otters in the NWT. The attractive turbine was a hit with mining companies and other customers. However, for many jobs NWTA's DC-3 and Beech 18 were preferred. The DC-3 was as fast as a Twin Otter and had greater range, and the Beech 18 itself was peppy enough. These older types also had the advantage of lower charter rates, so having Twin Otters wasn't a guarantee that the operator would silence the competition.

With the great distances in the NWT, the need was clear for improved air service to better connect communities with the south. Engles planned to fill this need. In 1968 he applied to the ATB to operate a DC-6 or Hercules. The Herc was especially needed—companies with drill rigs on Canadian sites had to charter Hercs in Alaska, when a move was in the works. In an October 15, 1970 item *News of*

the North reflected the Canadian air transportation shortfall: "Curiously, one of the hardest hit is the Panarctic consortium in which the federal government has a substantial interest. Panarctic is faced with the problem of how to get 80,000 gallons of aviation fuel and other supplies from various points, where they were landed by ship, to its four well sites in the High Arctic." Engles' Herc application was refused, but the DC-6 was approved—in 1970 CF-CZZ joined NWT Airways.

In 1993 Engles reminisced about the DC-6, referring to it as "a wonderful bushplane that did everything and more that the DC-3 could do." With its 2,500-hp R2800s the DC-6 covered great distances at high speed. Passengers travelled in comfort, and loads up to 25,000 pounds were handled with ease, using large fore and aft cargo doors. When he sold his last DC-6 (CF-NWY, to Ethiopian Airlines in July 1978), Engles got more for it than he paid for the Electra purchased as a replacement. From Otter to Beech 18, DC-3, and DC-6 NWT Airways gained the maturity to let it move into turboprops and jets. In 1975 Engles introduced Electra CF-IJV, bought with a load of spares from defunct International Jet Air. Transfer of IJA's Class 4 service Calgary-Edmonton, and its Electras was approved in late 1976 by the Air Transport Committee (Engles operated the service under a temporary arrangement till then).

The Lockheed L.188 Electra was one of the first great turboprops. It sprung from interest shown by American Airlines and Eastern Air Lines in a four-engine design. The prototype flew in December 1957 and was certified the

(Above) NWT Air DC-6B C-FCZZ started at CPA in 1958, left for Wardair in 1962, returned to CPA five years later, then served PWA and NWT Air. It moved to the US in 1979 to bomb fires, then joined Conifair in April 1983 as C-GBZC. It went next to France's Protection Civile. It returned to Conair in 1988. Here it refuels at Edmonton Muni in May 1977. (John Kimberley)

(Left) Electra IJV on an ice strip in the Beaufort Sea at the height of oil and gas exploration. (via Tony Jarvis)

following August. It was a hit wherever it went, but its reputation was sullied by accidents, the first in September 1959, when a Braniff machine disintegrated in flight. Northwest had a similar loss in March 1960. The fleet was grounded and the cause traced to engine nacelle design that allowed vibrations that could flex the wing to destruction. A fix was engineered and the Electra went back to work, but the times were changing—airlines were moving to jets—DC-9s and 727s soon overshadowed the Electra. Production ended with the 170th example. The classy Electra was relegated to lesser carriers, where it did excellent work for many more years.

The Electra proved to be a superb northern performer and a great money maker. NWTA added CF-NWY in 1977, followed by IJR (bought from Imperial Oil in 1983), NWD in 1983 and NWC in 1984. Other Electras were leased as needed. They crisscrossed the Arctic and by 1983 were landing at Vancouver, Calgary, Edmonton, Winnipeg and Toronto on a nightly courier service on behalf of Air Canada, an operation dubbed the "Gold Label" run. While the Electra handled loads at major hubs, NWT Air DC-3s connected places like Regina and Saskatoon to the Electra stops. In 1985 the Electras began competing head-to-head with PWA 737s on the lucrative Yellowknife-Edmonton passenger route.

In 1979 Wardair had closed its Yellowknife operation. This happened when NWT Air successfully outbid it and First Air to win the first trans-NWT sked. The route was Yellowknife-Rankin Inlet-Frobisher Bay using a 59-seat Electra combi. The service met Calm Air at Rankin Inlet and Nordair at Frobisher, allowing travellers to connect southward. The loss to NWT Air irked Max Ward. He wrote in his autobiography that the Air Transport Committee "gave the route to Northwest Territorial Airlines. We didn't even come second in the contest... we came last." The premise of Ward's

book is that Ottawa had it in for him and that Wardair hadn't a fair chance, whatever the issue. But NWT Air likely won the sked fairly. For one thing it offered fast, comfortable Electras compared to Wardair's smaller, slower, shorter-range Dash 7s. NWT Air's first southern connections came in 1981 with nonstop Yellowknife-Winnipeg service. This was in conjunction with Air Canada, which handled reservations, ground handling and cargo arrangements.

The trans-NWT service faltered, but eventually succeeded. (By 1997 NWT Air was running the service with 737s, First Air with 727s—northern Canadians were being well served.) Meanwhile, NWT Air DC-3s provided feeder service, and carried vast amounts of freight. By 1985 the company had five DC-3s, five Electras and two Hercules. Two Electras were sold to Falcon Cargo of Sweden in 1986, while another was leased from Reeve Aleutian. In April 1987 NWT Air became an Air Canada Connector. With introduction of Air Canada's modernized DC-8 cargo fleet, the Electras were pulled from trans-Canada cargo service. In late 1988 NWT Air phased out the DC-3. C-GNWS and C-GWZS went to Buffalo Airways to assume work formerly done by NWT Air. While the DC-3s disappeared from NWT Air, two 737-210C Advanced Combis were added in 1988. These were select 1975-built machines with noise-compliant engines—good for operations to 1999, by when engine hush kits would be needed to meet regulations. The ex-Wein Alaska aircraft were C-GNWI; and C-GNWN. They were busy from the beginning with NWT Air—in June 1992 each was logging 9.2 hours daily. The 737s employed about 20 pilots. Many had started with Bob Engles in the 1960s. A group like this was rare—experienced northern pilots, who had learned the ropes as they moved up in the challenging environment of Arctic aviation.

Through the years since Engles began, Yel-

lowknife's population grew from 4,000 to more than 15,000. Tourism expanded as travellers discovered the lure of the Arctic. This added to business, and there were even southern holiday charters. An extra plum came when NWT Air won the North Warning System vertical resupply contract from rival PWA. In September 1988 Engles sold 90% of his interests in NWT Air to Air Canada. Of this Max Ward noted that Air Canada paid "$14 million for a bunch of old aircraft." To put it charitably, this was an oversimplification. NWT Air would have looked good to any investor in 1989, and its fleet was far from "old". Besides, had not Ward himself begun with used planes like the Bristol and DC-6? Engles remained chairman and CEO of NWT Air till January 1990, then retired. The airline by now was wholly owned by Air Canada, but Engles kept its 737s and the Hercules, leasing them to Air Canada.

In 1991 NWT Air's Electras were retired. The last three (IJR, IJV, NWY) were sold to Air Bridge in the UK. Bought for about $300,000 each, they sold for around $2 million each. Several NWT Air pilots and engineers accompanied the Electras to help Air Bridge get started. Others were laid off in a job market that was anything but encouraging. In its spring 1991 issue *Propliner* reported: "Air Bridge Carrier's Lockheed Electra fleet was further expanded on the evening of May 13, when former NWT Air Electra C-FIJV landed at Castle Donnington after a trans-Atlantic ferry flight via Keflavik. This brings Air Bridge's Electra fleet strength up to four aircraft, although C-FNWY is currently in storage at Castle Donington pending a decision on her future. Electra C-FIJV was soon pressed into service on Air Bridge's popular night freight schedules, positioning from Castle Donington to Birmingham on the evening of May 17 prior to operating the ABR 411 schedule to Brussels." Thus do good old aircraft go on and on.

Air Canada's takeover brought changes. By

early 1991 NWT Air was on thin ice, so Air Canada placed operations in the hands of Air BC, with Darrel G. Smith in charge. Once the company was back in shape, Dan Murphy, an Air Canada manager from Montreal, became company head. Under Air Canada NWT Air no longer could operate its usual charters. US destinations were out. This placated Air Canada pilots, who feared tiny NWT Air invading their domain. By 1994 the companies merged pilot seniority lists. Now NWT Air could hire only from Air Canada's list of furloughed pilots. On a day-to-day basis NWT Air seemed uncertain about what Air Canada might do next. Especially concerned were the Herc crews—they wondered how long Air Canada would want a lone Herc in its fleet. In July 1994 Air Canada decided to divest itself of NWT Air, and a deal was arranged to sell it to its employees and an Inuit group—Sakku Investments. This fell through. In 1997 NWT Air was still a lean operation with three 737 and the Herc. Headquarters in Yellowknife had about 110 staff, while some 30 were in Edmonton and a few in Winnipeg.

The 737

Flight 953 of July 12, 1992 was a typical NWT Air 737 sked. Aircraft C-GNWN was configured for takeoff at 92,900 with V1 speed

NWT Air's 737 NWI near Yellowknife on March 22, 1992. In 1994 Air Canada, which bought NWT Air in 1987, negotiated to sell the company to the Inuit company Sakku Investments and the employees of NWT Air. No deal took place but in June 1997 the company was sold to First Air. (Henry Tenby)

(critical engine failure speed) of 126 KIAS and VR (rotation into flight) of 131. Just after noon Capt Roland Brandt and FO Keith Kruger taxied at Edmonton to the end of R02. At 1211 they had NWN and its 55 passengers airborne and climbing for FL350. Destination was Yellowknife 550 nm north. The pilots were old hands. Brandt, with 19,600-hour (10,000 on DC-3s), had come to Canada from East Germany at 21. He took a private pilot's licence with the Wong brothers at Central Airways in Toronto (Central was one of those schools that, year after year, produced good young pilots for commercial aviation). At first Brandt could not find a flying job; but in December 1968 Frank Russell, head of maintenance at Austin Airways, hired him as a helper. He worked at Sudbury and Moosonee, then was sent to Nakina to fly Ce.180 LFM and Beaver GQQ under veteran Austin pilot Elmer Ruddick.

Brandt next got on as a Canso co-pilot with

Capt Bob Pettus, then moved to the DC-3. Like many, Brandt had his differences with the famous J.C. Bell, Austin's chief pilot. He would get the brush off from Bell, which he attributed to language—Brandt was still learning English. Then, just as he was to get his chance to move up with Austin, Brandt was called to Germany on a family matter. Bell easily could have given the job to another pilot. But when Brandt returned, the job awaited. This was the real Jim Bell in action. Brandt's time at Austin Airways was one not to be missed. As he put it, "Austin Airways was the tail end of a historic era in aviation". Brandt left Austin for an opportunity at Nordair. He did courses on nearly all types in the fleet—C-46, Canso, DC-3, Electra and Skyvan. He went on the line on the Electra, but was laid off a week later! From Nordair he worked for several carriers—Fecteau, PemAir, etc. In April 1975 NWT Airways hired him on the DC-3 from where he progressed to DC-6, Electra and 737.

Keith Kruger had joined the army out of high school, spending three years as a military policeman. Back on civie street he got into aviation as a dock hand with Lambair for $150 a month. He learned to fly in 1970, but didn't get a job until 1978, when Calm Air hired him to fly a Ce.185. Later he worked for Ilford-Riverton Airlines, flying Cessnas and Otters;

and went to Papua New Guinea to fly for the Lutheran Church. Later, he hauled fish at Buffalo Narrows in Norseman BHS, then worked with a series of operators including the Manitoba Government Air Service, where he flew his favourite among the bushplanes —the DHC-2T Turbo Beaver. In 1981 Kruger moved to Buffalo Airways on DC-3s, flew at Austin Airways for six months, then joined NWT Air in 1985.

As F953 cruised towards Yellowknife and the boys on the flight deck chatted about the "good old days" of aviation, a contrail appeared far ahead. The pilots knew who it was—southbound for Edmonton was Capt Rick Van Ness in Echo Bay's Boeing 727. Echo Bay ran its own airline serving the Lupin Mine, 212 nm north of Yellowknife, and presently was doing mine charters to Russia. At 1317 Brandt and Kruger started down, and landed on R33 at Yellowknife 18 minutes later.

F953 shut down beside a Buffalo Airways DC-3. The flight continued the same day to Inuvik. On that leg two days later Capt Brandt was accompanied by FO Bruce Sinclair aboard NWN. The trip covered 587 nm in about 1.5 hours. Sinclair also was widely experienced. He had earned his wings at Oshawa as an Air Cadet in 246 Squadron. He recalled his first plane ride at age 10—a 1958 trip with his family from Toronto to Nassau in a TCA North Star. Sinclair got into commercial aviation in 1976, instructing for Trentair at Peterborough. Later he worked for Alberta Northern, flew an aerial survey King Air in South America, then was hired by Dome Petroleum in its heyday. Dome had 185 people in its flight department and operated a 737, Gulfstream II, Citation, two King Airs, Aztec, Twin Otter and helicopters as big as the S-61. When laid off with the demise of Dome's oil activities in the Beaufort, Sinclair flew DC-3s for Skycraft, then did two years on Dash 8s with City Express before joining NWT Air in 1989. At City Express he had met Wess McIntosh, then well into his seventies but still flying Dash 7s. McIntosh, renowned in Canadian air transport circles, became one of Sinclair's heroes. Sinclair reminisced about the DC-3 as his favourite aircraft, although he greatly enjoyed the 737. "I always want to make sure I have two Pratt & Whitney engines out on the wings", is how he summed up his feelings about airplanes.

On July 16, NWT operated one of its regular trans-territorial flights—F951 Yellowknife-Rankin Inlet-Iqaluit return. Just a few years earlier, it was inconceivable that such a service could work. Today, the planes are often full. Capt Ray Fisher and FO John Evans crewed the flight deck as 737 NWI taxied with its 36 passengers at 10:02 to start the day's work. The leg to Rankin Inlet was 613 nm, to Iqaluit from there was 637 nm. FO Evans had worked his way up in the usual fashion over the years. He started in 1975 with O.J. Weiben in Pickle Lake, then worked in 1976 for Stan Deluce at White River. Next came what Evans called "my best summer ever"—flying for Elmer Ruddick of Austin Airways at Nakina. Austin next posted him to Pickle Lake on the DC-3 and Twin Otter. Tours with Air-Dale and Buffalo Airways followed, then Evans joined NWT Air in 1985.

F951 landed on Rankin Inlet's 6,000-foot gravel strip at 11:40. The airport, run by the NWT government, had a "sensitive bird sanctuary" noise abatement restriction in effect from May to August for aircraft departing on R31—planes were to bank as soon as possible after takeoff, lest they offend the birds! Curiously, the town dump was near the end of R31— an ideal gathering spot for the local feathered set. Also off the end of the runway was a large white cross and the wreck of a Navajo that crashed in 1991, killing six, including the NWT's Catholic archbishop. At this time Rankin Inlet (population 1,750) had about 35 daily aircraft movements, mostly Calm Air, First Air, Keewatin and NWT Air. On July 16 there were few aircraft—a Cherokee 140, a Lake, a Ce.182, a Bell 206.

An overall view in 1973 of Rankin Inlet, capital of the Keewatin District. The airport, as with most NWT communities, is conveniently located. Inconveniently located was the town dump (bottom cente), which attracted gulls right to the end of the runway. In the early 1990s Rankin became a base for NORAD fighters, but the USSR dissolved, taking with it the Soviet bomber threat to North America. Fighters almost never visited Rankin Inlet after the base opened, but the town had one of the finest Arctic airports. (Larry Milberry)

Jack Lamb, while airport manager at Rankin Inlet in 1992. He later retired to Winnipeg. (Larry Milberry)

(Right) CF-BCF, one of the original PT6 Beech 18s, was converted by the BC government. In those days "Flyin' Phil" Gagliardi ran the BC air fleet. He became infamous for roaring around in the people's Beech 18s—132 trips in 1955 alone! BCF is seen at St. Andrews airport while with Keewatin Air. In 1990 it was donated to the CMFT after being damaged by fire at Rankin on May 4, 1990. Then, a Keewatin Air medevac Merlin at Rankin on July 17, 1992. (Jan Stroomenbergh, Larry Milberry)

The big project was construction of a CF-18 FOL (forward operating location). Mountains of material had been barged in and stockpiled so the airport could be paved in 1993. Otherwise, there was a terminal building, and a hangar operated by Keewatin Air—a Merlin II air ambulance and charter carrier. Winnipeger Bob May owned Keewatin. He had flown earlier for companies like Lambair and Transair. In the mid-1970s he bought Moose Nose Lake Airways at Ilford from Dave Rondeau; and added bases at Gillam and Rankin. He later sold his Manitoba bases and developed Keewatin Air at Rankin. He built up a fleet of Beavers bought from the Sultan of Oman, but also had Cessnas and Otter C-GKYG.

About 1980 Keewatin leased Beech G18 C-FPAV from owner Jim Smith. He had bought PAV in Texas and had a cargo door and new floor installed. Bell Canada became a steady customer for PAV. Next, Keewatin added two PT6 Beech 18s: Westwind BLI and Tradewind BCF. Bob May had BLI's roof raised a foot to conform to Super 18 standards. This enabled bulkier cargo to be carried. Eventually BLI was damaged at Whale Cove, then was sold in the US. BCF was damaged in a hangar fire at Rankin. Meanwhile, Smith leased PAV to Icarus in the Magdalen Islands, where it earned revenue freighting before going to the US. Later in the 1980s Keewatin built a fleet of vintage PT6 Merlin IIs, which proved to be good air ambulances.

Just before F951 departed Rankin, Calm Air's 748 sked arrived and taxied in ahead of its own dust storm. NWI then roared down R31, rotated, and set course to overfly Southampton Island, Nottingham Island, and the Meta Incognita Peninsula. It landed on Iqaluit's paved 9,000-foot runway, being next

in line behind a BAe 146. The 146, operating with call sign "Rainbow", turned out to be from the RAF Queen's Flight. As NWI pulled in with its 30 passengers, Prince Philip was strolling across the ramp from "Rainbow". NWI made a fast turnaround and headed back to Rankin Inlet with 85 passengers, this time as F956. Another quick stop, and it was on the move. As it taxied, the dispatcher radioed sarcastically that the runway was nice and smooth— Capt Fisher would be *sure* to make a good takeoff. Fisher's disrespectful FO replied, "You'd be surprised," and away they went with a full house—111 passengers, most heading for a weekend music festival in Yellowknife.

On NWT Air's July 18 sked to Rankin Inlet and Winnipeg one of the passengers was a typical Arctic traveller, Timothy Akerolik of Whale Cove. He had been born in Coral Harbour in 1939, when it was still a traditional Eskimo community of about 100. Early in the war the US military arrived to put in a runway—part of the overseas ferry system. Timo-

thy trained as a mechanic and recalled his first flight—with Wheeler Airlines to work on a DEW Line site. He later flew with many DEW Line carriers. One day the Wheeler Otter he was in struck a hill in weather, but all survived. Timothy worked eight years as an aircraft refueller at Coral Harbour and Baker Lake. He was still a serious hunter and fisherman, and a great traveller across the Arctic. He recently had been to such far-flung settlements as Pangnirtung, Pond Inlet, Cape Dorset and Fort Smith visiting friends and relatives.

F965 was an NWT Air 737 Iqaluit-Yellowknife sked on July 2, 1993. The usual stop at Rankin Inlet was bypassed this day—the runway was closed for paving. Thus did 737 C-GNWN (28 passengers) sail past Rankin on a straight 1,222 nm-run. In charge was another senior pilot, Bob Bowman. There were three hours to chat with him and co-pilot Willie Blake. Bowman had been flying since 1962, when he started at the Calgary Flying Club. He had his commercial by 1965 and spent that summer crop dusting around North Battleford. In 1966 he got a job as navigator on Spartan's Beech 18 CF-MJY at Mould Bay on Prince Patrick Island. In 1967-68 he did some instructing and in 1969 went to Great Northern Airways to fly an Aztec. A promotion to the Edmonton-Rae Point F.27 service followed, then Bowman joined International Jet Air on the F.27 and Electra. The Electra became his favourite. It was an excellent freighter for the drill work being done—its cargo door could handle outsize loads, and for the passengers it had plush seats. It was also fast, cruising at 340 knots. Electra IJR had long range tanks (44,000 pounds of fuel). With a burn of 4,000 pounds an hour it had outstanding range. Bowman logged 11,500 Electra hours before the fleet retired. Flying became a bit mundane since he moved to the 737, so he enjoyed reminiscing about the Electra. He recalled that when Dome was chartering NWT Air Electras to support drill ships in the Beaufort Sea, an Electra cost about a third what it later cost Dome to run its own 737. When F965 shut down at the Yellowknife terminal, its airframe hours were noted as 45,527 hours. About a year earlier (July 12, 1992) it had 43,273 hours, so in a year it had logged more than 2,000 flying hours.

Red and White Hercs

A trademark in the Arctic since 1983 has been C-GHPW, NWT Air's famous L100-30 Hercules. HPW had flown with PWA 1978-83, having been bought from Lockheed to haul CF-104s between Canada and Germany. Earlier, NWT had three other Hercs, following acceptance by the ATB in 1975 of its application to operate the type. The first was CF-NWF, an ex-Safair machine leased from October 1978 to November 1983. CF-NWY was leased from Safair from May 1979 to November 1983. Meanwhile, CF-PWK came from PWA in April 1981, but was destroyed by fire while de-fuelling at Paulutuk, near Inuvik on April 11 a year later.

A 1979 model, HPW had 21,000 hours by 1993 and was logging about 900 hours in a good year. One of those with most experience on it was Anthony T. "Tony" Jarvis. He had started in aviation as an Air Cadet, making his first flight as a student pilot at the St. Catharines Flying Club on June 23, 1971 with instructor Leon Evans. In February 1973 Jarvis

was presented with his private pilot's licence. On January 29, 1974 he took his first job—Jim Bell of Austin Airways hired him as a crewman. That day old-time Austin pilot Moe Sears took Jarvis along in Otter CF-BEW from Moosonee to Fort Albany. Working for Austin was valuable experience for any young pilot—it introduced him to grass roots commercial aviation, to the school of hard knocks, to some of the legends in bush flying. Even if it meant living in a cheap boarding house, and working like a slave from dawn to dusk, this was a golden opportunity. For the young mechanic or pilot tough enough to survive, it gave experience that money couldn't buy. A year at Austin sped by, and Jarvis moved to Sudbury Aviation to instruct. In May 1977 he travelled west to fly for Ilford-Riverton. Oxford House, Red Sucker and Sachigo now entered his vocabulary. They were among his many stops as he flew Ce.180 CF-KMF, Ce.185 AUI and Norseman FOX to Indian reservations. On August 17, 1977 Jarvis happily logged his 1,000th hour as pilot in command. The following June he went north to Thompson for Ellair, became its chief pilot, and started flying multi-engine, first Dornier Do.28 AFC, then Beech 18s KAK and OWQ. Apache AQL and Skymaster PPR were also in the fleet.

In May 1980 Jarvis moved further west and north—to Fort McMurray to work for Contact Airways flying the Navajo and Ce.206. He now felt that he had enough experience to try his hand with a big carrier. In 1981 he applied to NWT Air and was hired as a co-pilot on the DC-3. His first trip was in C-GWZS from Yellowknife to Coppermine on June 5. For the next two years he added DC-3s NTF, NWS, NWU and WIR. The work was routine— passengers and freight, but there was some variety. Aircraft NWU had a VIP interior for 16 passengers and was used to fly the territorial judiciary to outlying settlements, where court was held. NWT also carried the "flying bank" (Canadian Imperial Bank of Commerce) from place to place, so people could do business without travelling out. In December 1983 Jarvis went on HPW, starting with simulator work with the air force at Namao. He made his first flight on December 20 and soon was flying in the Arctic supporting oil exploration. Week after week was spent moving rigs for Panarctic. Each rig along with its support infra-

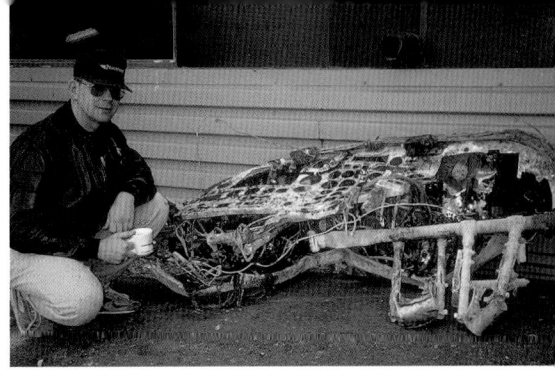

NWT Air Herc captain Tony Jarvis proves that one fellow's scrap is another's treasure. One day at Hall Beach with HPW he spotted some junk near the runway and had it excavated. The mangled scrap turned out to be the instrument panel from a DEW Line York that had crashed decades earlier. Here Tony poses with his "archaeological" find. At this time the remains of many aircraft existed in the NWT, including: Alert–Lancaster; Cape Perry–C-46; Hall Beech–Aero Commander, North Star 17520, York HFQ; Hall Lake–York HMX; Beaver Lodge Lake–Bristol 170 TFZ; Pelly Bay–C-46, DC-4; Pelly Lake–Mosquito, Ventura; Resolute Bay–Lancaster. (Larry Milberry)

The wreck of C-54 CF-PBC at Pelly Lake in the 1990s. This old freighter worked for the US Navy, PanAm, Japan Airlines and Philippine Airlines before coming to Canada in 1971 for the Koonui Co-Op of Pelly Bay. (Tony Jarvis)

Ever interested in junk, Tony Jarvis was happy to pose in 1992 with this fine piece of Iraqi armour in the desert. He was in Kuwait with HPW returning Iraqi plunder to Kuwait. (CANAV Col.)

(Below) Hercules HPW near Yellowknife on March 2, 1994. It was training with the crew of Capt Tony Jarvis, FO Keith Kruger and FE Will Brander. Henry Tenby did the photography from an Air Tindi Navajo.

(Right) The low, gravelly terrain around Igloolik in late June 1993. (Larry Milberry)

(Below) HPW delivers materiel needed at Igloolik to build a new airport terminal. (Larry Milberry)

Capt Murray Burr at Igloolik with a crowd of local lads keen to tour mighty HPW. (Larry Milberry)

structure of buildings, vehicles, etc. took about 150 trips. HPW operated 24 hours a day in almost any weather.

On August 31, 1988 Tony Jarvis left Yellowknife on his first overseas Herc job. They flew to Halifax, then Santa Maria in the Azores. On September 2 they pushed on for Dakar, Abidjan and Luanda on the coast of Angola. HPW spent the next three years there for the Red Cross. Each crew would work two-three months in Angola before getting leave in Canada. Jarvis flew many of the 45- to 50-minute trips with relief supplies between Benguela and Huambo, a route taking HPW over territory, where government and rebel troops often were fighting. The NWT crews hoped that those below would not get trigger happy.

In March 1989 Jarvis was promoted to captain. He returned from Angola on March 19 and four days later had HPW on a job between Gillam and Fort Severn, Manitoba. March 28 he was operating Tuktoyaktuk-Valdez, carrying equipment to fight to the *Exxon Valdez* oil spill. Jarvis did five other Angola tours and was in Iraq December-February 1991 on Operation Desert Recovery. This was a UN contract returning Iraqi plunder to Kuwait. A number of treasure-laden safes were among the items returned. Seven trips were logged. Jarvis described the first from Kuwait on December 15:

A souvenir of HPW's Angolan days hangs in NWT Air's offices in Yellowknife. (Larry Milberry)

"On landing we are briefed almost immediately for our first flight to Iraq. Our main contact here is an ex-Air America pilot on loan from the US State Department. On first meeting, you would believe him to be, as we say in the North, 'bushed'. It does not take long to establish that this is incorrect. A UN pilot tells us about the route and the airport. Higher officials tell us of the flight clearance and still higher ones tell us of our overseeing eye-in-the-sky, a USAF AWACS. Not a gnat moves in Iraq that they don't know about. We take the first team in. They will fly two Hawk aircraft back to Kuwait. We pass Babylon at FL200 and begin descent into Al-Habbaniyah airport. No navigation facilities. Our INS guides us into final."

The next day 17 safes were returned from Iraq. On the 29th zoo animals were flown to Kuwait City from Bahrain—the Iraqis had pillaged the zoo in Kuwait! On January 9 the NWT crew visited the "Highway of Death", where allied aircraft clobbered an Iraqi convoy fleeing Kuwait. On February 11 they left the desert, their contract cut short by Iraqi intransigence concerning the smooth-running of repatriation flights. Next day HPW stopped for fuel in St. John's. It left on the 13th for a 7 1/2-hour hop to Winnipeg, where a snow blower spare part was loaded for Baker Lake.

On the 14th HPW landed there in -28° weather.

In 1993 HPW worked the Canadian North on a series of typical jobs. One was to Igloolik, a settlement off the northeast tip of Melville Peninsula (69° 21' 9"N. lat., 81° 49' 2" W. long.) 886 nm northeast of Yellowknife. Igloolik needed a new terminal building and materiel for it was to arrive as four Hercules loads. The second load went in from Yellowknife on June 26 with the crew of Capt Bob Barrett, Capt Murray Burr, FE Peter Ormiston and LM Stirling Forrester. Also aboard was Sam Denhaan from the NWT Dept. of Public Works. HPW got airborne from Yellowknife as F701 carrying 36,000 pounds of construction material. All-up weight was 145,000, making for a required field length of 5,750 feet and a rotate speed of 116 knots. HPW was level at FL190 at 1115 local time and cruising at 210 KIAS.

Capt Barrett had been flying since 1957, when he started at the Brampton Flying Club. His first profession had been teaching, and he worked for years in "phys ed", while doing part time flying with companies like Air Muskoka. In 1974 he went full time with Bradley, flying the DC-3, Twin Otter and Citation; moved to Borek at Resolute Bay on the Twin Otter; then back to Bradley on the 748. He also spent two winters on the DC-3 at British Virgin Island Airways. He flew the Convair 640 for Worldways and DC-4 and Falcon 20 for Soundair, then helped start Inter City Airlines, an Oshawa company with two 748s. It failed, costing the partners a fortune. In 1988 Barrett joined NWT Air to fly the Electra and C-130.

Capt Burr had started at the Winnipeg Flying Club as a 16-year-old. He instructed at St. Andrews till he had a thousand hours and Bill Wehrle of Perimeter Airlines would hire him. There he started as co-pilot on the Metro. Over

Around Igloolik on June 26, 1993. As the ice slowly melted there definitely was no hurry to salvage skidoos and komotiks before they went to the bottom. The Igloolik Research Centre is an Arctic landmark. (Larry Milberry)

several years he moved to captain on various types. He ended with 1,000 hours on the Metro, 2,000 on the DC-3 and 3,500 on the Queen Air, then joined Calm Air to captain 748s for four years (4,000 hours). He joined NWT Air in 1989, flew the Electra for 1,000 hours, then moved to the Herc.

Flight engineer Ormiston had been 11 years with NWT Air. As a student he had taken business at Lethbridge College and aviation at the Southern Alberta Institute of Technology. In a good year at NWT Air he could log 1,100 hours, but 1992 was slow—350. Loadmaster Forrester had joined NWT Air at its Winnipeg DEW Line resupply operation. By 1992 he had been five years on the Herc, a job that had taken him around the world.

At 1351 HPW started down for Igloolik, passed into cloud at 5,000 feet, poked its nose into the clear about 1,000 AGL and landed on the 4,600-foot gravel strip. It was 3:03 hours after leaving Yellowknife and HPW had burned 16,000 lbs. of fuel.

Igloolik is a hamlet of about 1,000. It's noted for Thule archaeological sites dating back 2,000 years. Inuit here still subsisted on fishing and hunting (not to mention government cheques). In 1613 British explorer Thomas Button became the first white visitor here, but it wasn't till 1939 that the HBC set up a trading post. By now important meteorological, biological and anthropological centres operated at Igloolik—the Northwest Territories Science Institute. Like all Arctic settlements Igloolik depended on the airplane. In 1993 it had First Air 748 skeds five days a week to neighbouring settlements like Hall Beach and Pelly Bay, connecting to Boeings at places like Iqaluit, Cambridge Bay and Yellowknife. Occasional charters landed at Igloolik and helicopters often flew on summer science projects. The visits by HPW in 1993 were a rarity, and reportedly the first by a Hercules, although an NWT Air 737 earlier had used the Igloolik ice strip. As it turned out, June 26 was a fairly busy day at Igloolik, with the Herc, two 748s and a Navajo. As recently as 1950 places like Igloolik never had seen an airplane, and it was years later before they had air service of any kind.

Offloading HPW at Igloolik was a demanding job. The only equipment available was a multipurpose front end loader. The loadmaster carefully orchestrated the job and HPW fired up about an hour and a half after landing. It took off for the 40-nm hop south for fuel to the North Warning base at Hall Beach. HPW was on the ground at Yellowknife at 1917 hours to join a busy scene of 737s, DC-4s, 748s, DC-3s, CL-215s, Twin Otters and dozens of smaller types.

Buffalo Airways

An important NWT air carrier in the 1990s was Buffalo Airways of Hay River and Yellowknife. The company, with about 40 employees in 1997, was formed in 1968 with a lone Ce.185 by ex-RCAF pilot Bob Gauthier. His licence dated to Hay River Air Service, a company started in 1961. It had bases at Hay River, Fort Simpson and Fort Smith with a Ce.180 and a separate licence for each base. It competed directly with PWA's VFR operation. Later, Hay River Air Service was sold to Northern Mountain Airlines of Prince George. Gauthier's piece of the Hay River operation was its Fort Smith licence, which he re-christened Buffalo Airways. In 1970 Gauthier sold to Joe McBryan. He had begun flying at Hay River in 1961 at age 16, learning on a PA-12 owned by a local consortium. Once he got his commercial, McBryan flew a year with Carter Air Services at Hay River, then spent four years with Gateway, flying bush planes from Fort Smith. In 1970 he joined Great Northern Airways at Whitehorse, getting experience on the DC-3 and DC-4.

McBryan started Buffalo Airways with Norseman CF-NJV and a Ce.185. He gradually expanded with types like the Beaver, Beech 18,

(Above) One of Buffalo's DC-3s at Yellowknife on August 18, 1995. Capt Bob Burns is sauntering towards his aircraft, while groundcrew finish loading. Soon Burns was on the wing to the Colomac Mine. (Larry Milberry)

(Left) Buffalo Airways owner Joe McBryan (left) with FO Marc Vandevaghen fly the Yellowknife-Hay River DC-3 morning sked on July 13, 1992. Buffalo Airways began daily service on this route in 1987 with a Ce.402. Die-hard aviation buff McBryan started early into aviation—in 1944, when he was only 10 days old, he and his mother were flown from Cameron Bay (where his father was diamond drilling for Eldorado Mines) to Yellowknife by legendary Arctic pilot Ernie Boffa. (Larry Milberry)

C-FLFR at Boart Longyear Inc.'s exploration camp on Upper Carp Lake on May 2, 1996. The snow was just sticky enough and not too hard for ski operations. Low temperatures in the ensuing days hardened the surface, making it less desirable. The DC-3 by this time had more than a half-century of NWT history. In the second photo Capt Jim Smith (right) and FO Will Failing check their skis before returning the 80 nm to Yellowknife. Well-rigged skis cost no more than a 5-knot speed penalty. Poorly rigged, they can cost 10 knots. (Larry Milberry)

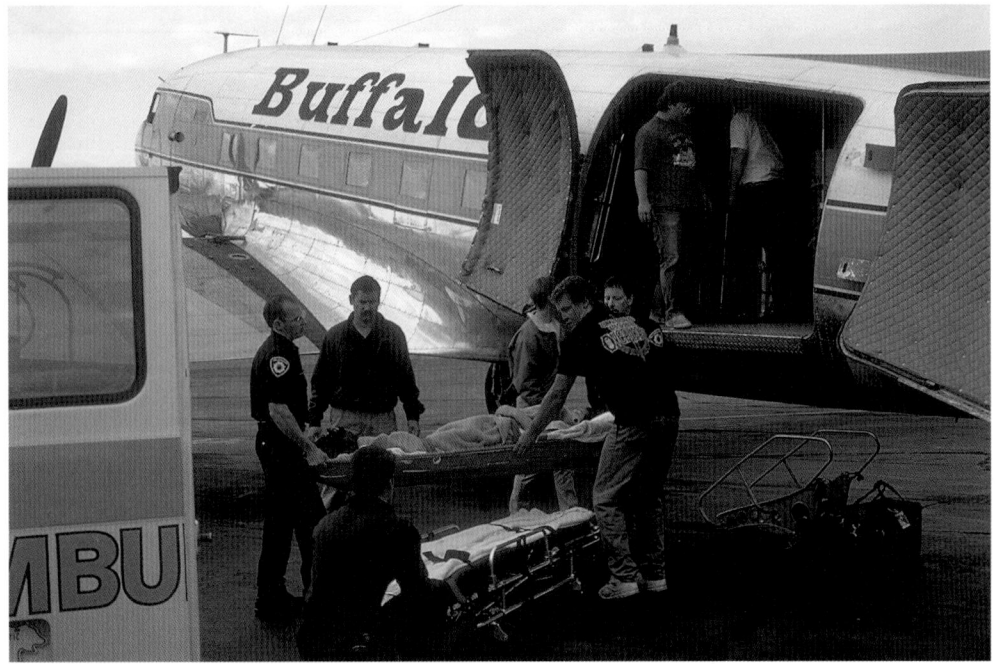

A patient arrives in Yellowknife from Hay River on a Buffalo Airways passenger-courier-medevac flight on July 14, 1992. (Larry Milberry)

The Buffalo Airways fleet was famous for its individualized aircraft artwork. DC-3 C-FROD was "Hot Rod". It was wrecked force-landing in the bush near Fort Simpson after running out of gas on June 26, 1994. Capt Chris Wells got out OK, but FO Kevin Woelk was badly injured. In 1995 Buffalo Airways acquired a new "ROD" by re-registering C-GPNW. Then, DC-3 "Summer Wages". (Larry Milberry)

Navajo and Otter. He added a helicopter operation with Alouettes, Gazelles, B.206s and Hughes 500s, building the fleet to a dozen machines. The boom years of the 1970s gave way to recession, and Ottawa's National Energy Policy, which effectively shut down exploration in the West and North. As if it wasn't hard enough to keep an airline running, McBryan was hit by a costly business-deal-gone-bad. He went broke. All he had left was DC-3 CF-RTM, acquired in 1977; but he was determined to come back. In 1981 he started Buffalo Air Express, and began riding the air courier wave. He had bases, telexes, vehicles, etc. and a keen, young staff—pilots with no immediate prospects of flying jobs who would bide their time sorting packages and driving truck. Buffalo Air Express began flying priority freight on January 1, 1982, using a Travelair. McBryan formed interline connections with national courier services. Hay River became his hub, with priority packages arriving overnight by road from Edmonton for morning distribution by air. Slowly, McBryan built his operation, centring on the DC-3, an affordable, reliable, fast (enough) type.

In 1993 Buffalo Airways operated five DC-3s and two DC-4s. As with the Beech 18,

the DC-3 suffered a period in the 1980s of waning popularity. Since northern operators were rushing to the Twin Otter, McBryan studied it. He concluded that by the end of a year, a DC-3 and Twin Otter would cost about the same to run, do about the same amount of work and have roughly equal profitability. Since a Twin Otter cost $1 million compared to $150,000 for a good DC-3, McBryan decided to stay with the latter. In the hands of a good operator, it was a solid money-maker. McBryan explained how success with a DC-3 in the 1990s depended on the dedication of the maintenance staff, and respect by pilots. The DC-3 no longer was easy to maintain. Plenty of spares were needed, and everyone had to work extra hours when a plane needed servicing.

To give an idea of the great work that a DC-3 can do nearly six decades after it first flew, Buffalo Airways spent 30 months in 1989-91 serving the 270-man Colomac Mine, 140 nm north of Yellowknife. DC-3s made 1,200 trips with no serious snags or delays. In 1992-93 DC-3s were supporting the largest claim-staking rush in Canada's history in the Lac le Gras area northeast of Yellowknife. This time the treasure was not gold, but diamonds. Over the 1992-93 winter Buffalo had three ski-

equipped DC-3s serving the diamond camps. In the summer of 1995 several were dedicated to bimonthly crew changes at Colomac.

To show his enthusiasm for the DC-3, Joe McBryan brought CF-PNR to the DC-3 rally at Expo 86 in Vancouver, and dedicated it in honour of his friend Murray Crosby. In 1992 he bought DC-3s from First Air (C-FLFR and C-FMOC) and Transport Canada (CF-CUE and CF-DTB). The plum in this batch was C-FCUE. The 1942 model served in North Africa in WWII as 42-93108. Charles Babb bought it surplus and sold it to CPA in 1947, where it became CF-CUE. It was certified as a cargo plane at 26,900 lbs. and for passengers at 25,200 (empty weight 17,186). CUE's original log shows that it arrived January 19, 1947 in Winnipeg from Las Vegas with 3,670:55 airframe hours. CPA converted it and initially sent it to work in Yellowknife serving Indin Lake. In the summer of 1993 Joe McBryan was using it on the same run to supply the Colomac Mine, only now CUE had 30,000 hours. In boasting about the DC-3, McBryan's most emphatic point was the success of the NWT Airways operation, which lasted 1964-1985. Five aircraft each logged about 1,000 hours a year. In all that time NWT Air's only problems were two gear snags and a damaged propeller. No passengers were injured. McBryan suspected that this DC-3 record probably was unsurpassed.

Buffalo Airways' DC-3 fleet began with five low-time, ex-Canadian Forces Dakotas. It stored away five more airframes and a great many spares bought from CADC. In the early 1990s each aircraft on the line was logging 700 hours yearly. A typical trip was flown July 13, 1991. C-GPNR left Yellowknife for the 50-minute sked to Hay River with pilots Joe McBryan and Marc Vandevaghen. Kevin Golinowski, a young pilot hopeful of getting into the right seat of the DC-3, was flight attendant for a handful of passengers. PNR climbed out at 110 KIAS to 4,500 feet, then cruised along at 145. It was a 22,087-hour machine by this day's flight. A look at J.M.G. Gradidge's *The Douglas DC-3 and Its Predecessors* showed PNR's history: Ex-USAAC 42-93423, delivered May 17, 1944 to the RAF at Dorval (tail number KG602). The RAF used it on 575 and 512 Squadrons. The RCAF inherited it in 1946 and used it as a navigation trainer. Its tail number changed in 1970 to 12932. No. 429 Squadron operated it thereafter at Winnipeg. It was stored in Saskatoon in 1975, then sold in October 1980 to Buffalo Airways.

Buffalo Airways and the DC-4

In 1990 Buffalo Airways began DC-4 operations. This type found a worthwhile niche in the NWT 30 years after its DEW Line heyday. Aircraft C-FIQM and C-GPSH were purchased. IQM, with 36,000 hours by July 1992, was a famous example. Delivered to the USAAF in October 1945, it came to Canada in 1956 to work on the DEW Line for Wheeler Airlines. It later flew for Nordair and in the 1980s water bombed in California with Aero Union. In 1987 it flew tourist charters from Chile to the Antarctic as N4218S. On March 29, 1990 it pranged at Gun Barrel Inlet on Great Bear Lake when the nose gear collapsed. Seven mechanics took two weeks on repairs (three prop changes, two engines), then IQM, with a sad-looking patch on the nose, and flying gear down, ferried to Calgary in May for interim repairs. On March 1, 1992 it departed for Tucson for overhaul at Hamilton Aviation, and finally returned to Canada ship shape.

With 62,000 hours PSH had started with American Airlines in January 1944, then went to Qantas as VH-EBN "New Guinea Trader". It

Buffalo Airways DC-4s load at Yellowknife on June 26, 1993 during operations to Pelly Bay. In 1980 *Wings* magazine of Calgary advised young pilots to steer clear of such old types. This prompted Carl Millard, one of the leading Douglas propliner operators, to write: "Your reference to the DC-3 and DC-4 being 'clapped out antiques' pretty much defines your knowledge of transport aeroplanes... it is obvious that you are not very much aware of the equipment that is being used today to develop Canada, particulary the North..." *Wings* had been premature with its views, but the time still comes for every plane to retire. In the late 1990s old Douglas transports were fading from the commercial scene. (Larry Milberry)

plied the South Pacific for years, and still had the oceanic cockpit to prove it—spacious, as it once housed navigator and radio operator besides pilots and flight engineer. It was imported from "down under" in 1979 by Calm Air of Thompson. It went to Basler in Oshkosh in 1985, becoming N7171H. In 1990-91 it was at Mil-Air Engines in Opa Locka, Florida, where it was overhauled, then returned to Canada as an overnight courier plane for Soundair of Toronto.

Equipped for Buffalo Airways' needs, a DC-4 was valued in 1993 at about $600,000. This included a heating system worth $50,000. With steady work the aircraft was worth the investment. Two contracts in 1992 showed what the geriatric DC-4 could do. February 5 to March 13, 1992 Joe McBryan sent PSH with two pilots to Wrangell to haul concentrate for Cominco. This usually was handled by TPA's Bristol Freighters and Cominco's hovercraft, but a 2.3 million-pound backlog had built up. The concentrate had to be at dockside in Wrangell in 35 days to meet a deadline for Japan. PSH arrived with two pilots, two mechanics and a load of spares. The job was smoothly completed in 150 trips. A later contract, hauling to Pelly Bay, also operated without delays. Buffalo Airways beat out the NWT Herc in the bidding for this job; 20 return trips were made on the 830-mile route from Yellowknife. The DC-4s took off at 74,000 pounds—39,000 pounds basic weight, a 20,000 pound load, 15,000 pounds of fuel.

On July 16, 1992 Buffalo Airways despatched PSH on a cargo run to Gjoa Haven (population 900), an Inuit settlement 590 nm to the northeast on King William Island. Aboard were Capt Jim Smith, FO Sean Loutitt, FE Greg Elliott and LM Kevin Woelk. PSH got airborne at 12:10 hours at 107 knots. It climbed at 130 KIAS at 200-300 feet per minute, levelled at 5,500 feet and cruised on at 150. Capt Smith had flown for operators like Calm Air, First Air, Keewatin Air, Lambair, and OCA. Each year was busy—in 1986 he logged 1,149 hours in 239 days of flying; 1,027 hours in 166 days for 1987. He had years on the DC-4 with Calm Air, then with Kenn Borek on the Antarctic contract November

Capt Jim Smith and FO Sean Loutitt on the way to Gjoa Haven. (Larry Milberry)

1987 to March 1988, when IQM operated from its base at Punta Arenas, Chile. There Smith flew with Bob Craig and famed Antarctic pilot Giles Kershaw, later killed in Antarctica flying a gyrocopter.

Smith was of the breed who preferred "the old airplanes", as he termed them. In 1992 he had 3,500 hours on DC-3s, 3,000-748s, 2,900-DC-4s and 900-C-46s. His log showed hundreds of hours on Calm Air's DC-4 C-GPSG. Originally delivered in 1946 to Western Airlines as NC10202, it had gone to Australia in 1947 for a long Pacific career. It did much work with Calm Air freighting to the Collaton Lake Gold Mine. When the mine closed in the mid-eighties, Calm Air was hit badly. This is the old story in northern aviation—projects are closely tied to world stock markets. It can be boom or bust for a mine (hence for aviation) as metals prices fluctuate. Eventually, PSG broke a spar landing at Hidden Bay, Saskatchewan on June 16, 1987. Conifair came up to cannibalize it. Some enterprising locals hauled away the fuselage for a storage shed. PSG finished with 66,023.9 hours. Smith remembered another DC-4 in the region—CF-QIX operated by the Pelly Bay co-op. Taking off June 1, 1979 on a fuel haul from Thompson, Manitoba, it lost an intake pipe. The tower let pilots George Oakes and Eddie Cull know that they were trailing smoke. They landed quickly and leaped from QIX as it burned. Transport Canada later noted of the accident: "No.7 induction pipe was found to be missing from the en-

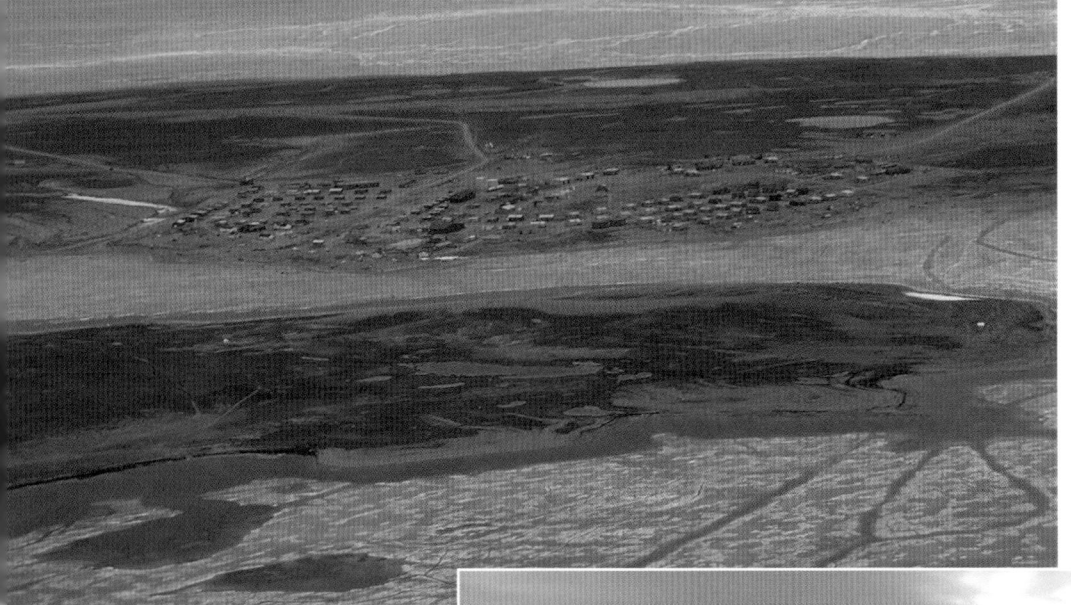

PSH was off at 93 knots at 1641. It climbed rapidly to 10,000 feet, then cruised home at 165 KIAS. For the rest of the year Buffalo's DC-4s had work here and there. In October and November the co-op at Sachs Harbour required 24 trips, hauling mainly muskox meat, 800 pounds to the box, 18 boxes at a time. The meat was destined to specialty markets in Europe.

For 1993 Buffalo's DC-4 worked as far afield as Pond Inlet. A half million pounds of cargo was hauled to Pelly Bay on 25 trips during the resupply. In May 14 loads went onto the ice strip at High Lake, a base metals development east of Coppermine. On a June 27 trip

(Above) Gjoa Haven on July 16, 1992. This day Buffalo Airways DC-4 C-GPSH brought in 18 tons of much-welcomed cargo. Then, offloading at Gjoa Haven. The day's work entailed seven hours of flying. (Below) Gjoa Haven meteorologist Rick Dwyer at work in his office. (Larry Milberry)

gine [No. 1]. This part is believed to have become dislodged after start-up, creating a torching effect, which caused the magnesium alloy crankcase to burn."

To avoid ice, much of the Gjoa Haven trip was at 2,000 feet, giving a spectacular view of the tundra. Sean Loutitt did most of the flying—this was part of the Buffalo Airways training program. Worthy young pilots, after doing their penance sweeping the hangar floor, loading planes, maybe working as FE on the DC-4, got to sit in the right seat. In time, if they proved themselves, they became FOs. Louttit had made the grade. Greg Elliott was doing his FE trips—he had worked previously with Ellair of Thompson, Canadair and NWT Air. Kevin Woelk was on his first trip north of Yellowknife, and his first ride in the DC-4 as loadmaster. Jim Smith put him in the right seat and gave him his first four-engine time.

PSH navigated surely across the Arctic using the latest nav aid—GPS (global positioning system). In time Gjoa Haven appeared in the distance, a speck on Rasmussen Basin on the southeast corner of King William Island. All around the waters still were solid, reminding one of a comment about spring made by Dan Murphy, NWT Air's general manager: "One

feature which makes the NWT different is that we normally have six seasons up here: spring, summer, fall, freeze-up, winter and break-up. You're never quite sure when any of them is going to arrive." At 1538 PSH landed on Gjoa Haven's 4,400-foot gravel strip. It was 4°C and the wind was howling. Everyone scrambled out the crew door and down the ladder. In the airport office Rick Dwyer was recording weather observations. He had come to the Arctic from Scotland 20 years earlier and liked it so much that he stayed. He had had a typical day so far—Beech King Air C-GSYN in twice from Cambridge Bay, Twin Otter C-GFYN from Shepherd Bay, First Air F843 (748) from Pelly Bay, now PSH from Yellowknife.

With the tail stand in place, to keep the DC-4 from tipping as weight shifted with offloading, Sean Loutitt got the airport forklift fired up, and the crew started dragging pallet after pallet to the back cargo door. It was a typical load of canned soft drinks by the ton, chocolate bars, cake mixes and potato chips, basically (with the exception of some Tetley's tea, baby food, Irish stew and outboard motor oil) a huge shipment of junk food for the Inuit. The only non-food cartons were three big ones with Honda ATVs. Forty-nine minutes after landing, PSH was buttoned up and its props again were turning. The GPS read 590 nm to go from YHK to YZF—Gjoa Haven to Yellowknife. At a takeoff weight of 50,000 pounds

to Bathurst Inlet PSH took in 45 drums of jet fuel for helicopters working diamond claims. It got airborne from YZF at 1455 local on the 310-nm run. A trip the day before had been the first for a DC-4 to Bathurst Inlet's 3,900-foot strip. Overgrown and soft though it was, it was useable. Capt Jim Smith was in charge this day, with FO Randy Desharnais and FE Stewart Clark. PSH, by this time with 62,658 hours, climbed slowly to 5,500 feet, then to 7,500, when the air got rough. It wasn't long before Lac le Gras passed off the starboard wingtip. Several hundred field workers were spread around this area working diamond claims. Smith recalled a bigger rush that peaked in 1977 with some 4,000 scouring the tundra for uranium with 35 helicopters supporting them from Baker Lake.

PSH approached the Bathurst Inlet strip in ideal weather. It landed at about 75 knots, heaved down the strip, turned 90° and shut down. The cargo door was opened, some heavy planks put in place, and the crew started rolling the 410-pound drums onto the gravel. There was no hurry— everyone wanted to bask a little in the tundra's 24°C. Discovered by Sir John Franklin in 1821, Bathurst Inlet is a branch of Coronation Gulf. From the airstrip could be seen the old HBC settlement—a cluster of white dots a few miles off. The HBC long ago had departed and the place was a wilderness lodge run by ex-RCMP officer Glen Warner. His Super Cub buzzed the Buffalo boys, then headed into the hills. Soon Ptarmi-

Off-loading jet fuel for Bathurst Inlet on June 27, 1993. It had been years since any large aircraft had used this 3,900-foot strip, but the versatile DC-4 made light of the job. (Larry Milberry)

gan Airways' Twin Otter IZD appeared as a speck in the sky, coming in from camp 50 nm out. Crewing it were Kevin Elke, Warren Delf and Mike Berthelet. Elke handled refuelling from one of the drums using a nifty Honda pump, while the two other men rolled drums aboard IZD. Within 30 minutes IZD was bounding down away on its tundra tires, headed to camp with a load of 45s. After 90 minutes on the ground, PSH itself was lifting off for a leisurely cruise back to Yellowknife. Next day Smith and crew flew to Pelly Bay with general cargo, then pushed south to Churchill for a load of boats. These were hauled to Pelly Bay, where the crew over-nighted, then returned to Yellowknife after putting another 16 hours on PSH.

Keeping a fleet of oil-dripping DC-3s and DC-4s in condition always was a challenge. To make sure they were ready to work each morn-ing, Joe McBryan hired Dave MacLennan to head maintenance at Hay River, and Peter Aus-tin to run the base at Yellowknife. MacLennan had begun his career with the RCAF in 1949. He had postings at Trenton and Summerside, working on Dakotas, Lancasters, etc. With such experience he was hired by MCA in 1956 to work on Yorks and other DEW Line types. MacLennan was later with Wheeler Airlines and World Wide. At Wheeler he worked on an early Canadian C-46, CF-ILJ. He recalled about it, "I think we outflew every airplane on the DEW Line with it." ILJ was a standard C-46 of the early 1950s, so carried only 8,000-9,000 pounds of cargo. Upgraded C-46s even-tually hauled nearly 15,000. He also recalled that there were often calls for ILJ to "bring an-other starter" on a trip to this site or that. That turns out to have been local code for "We need another case of whiskey". At Frobisher ILJ al-ways got refuelled in a hurry by the USAF. Once again it was the magic of a cheap bottle of

Ptarmigan Twin Otter IZD departs with a load of drums for a Cambrian Mining diamond exploration camp 50 nm away. Kevin Elke and Warren Delf flew the plane; Mike Berthelet was crewman. IZD had returned to Canada in 1990 after 15 years with Indonesia's Merpati Nusantara Airlines. (Larry Milberry)

The area around Bathurst Inlet at another time of year. This was the view from a 407 Squadron Argus on a sovereignty patrol on March 24, 1977. (Larry Milberry)

On May 2, 1996 PSH loaded 18 pallets of fresh milk and groceries at Hay River for Yellowknife, 105 nm across Great Slave Lake. It was taking part in the region's an-nual mini-Berlin Airlift, when the ice roads on Great Slave Lake become unsafe; and the mile-long ice bridge across the Mackenzie at Fort Providence breaks up and the ferry there cannot operate. In this two- to three-week period as much as 1,000 tons of cargo are airlifted. On this occasion Capt Al Fiendell, FO Ken Bews and loader Steve Kennedy crewed PSH. By this time PSH showed 63,750 flying hours in its log. (Larry Milberry)

Mechanics Bill Reid and Dave Peters of Buffalo Airways work on a DC-4's R2000s at Yellowknife. Such old radials demanded a great deal of work but, when well treated, delivered good service. (Larry Milberry)

Five Roses whiskey that kept things moving on a blustery winter's day. MacLennan did three stints with Nordair including as Frobisher base engineer. He later worked at Bradley and Kenn Borek, then went to Buffalo "for a year". That was in 1980 and in 1995 he was still there.

The C-46 Returns to Yellowknife

In 1993 Joe McBryan got interested in Air Manitoba C-46 TPO, dormant at Pickle Lake after an abortive take-off. The airport manager wanted TPO removed. McBryan had a look, but didn't decide on the spot. Later, he received a call from Pickle Lake—if he could get the C-46 off the airport, he could have it gratis. McBryan put a crew to work for two months getting old TPO into shape. The port engine was changed; a time-expired one was installed for ferrying purposes. On September 4, 1993 Jim Smith flew TPO to Winnipeg, where more work was done before it carried on two weeks later to Red Deer, then to Seattle to be readied for service. There Sorm Industries spent six weeks bringing TPO to certification standards, then it flew to Goodner Bros. in Arkansas for paint. Finally, DEW Line veteran Ron Lippert ferried it to Yellowknife.

In 1995 TPO was freighting to places like the Colomac and Echo Bay mines. A typical day's work was August 16, 1995, when it flew the Yellowknife-Fort Norman-Norman Wells-Fort Franklin grocery run. The captain was Mark Mesdag, a veteran of Air Manitoba. FO Andrew Dzal and crewman Cory Dodd completed the crew. When C-46s IBX, IBZ and TXW moved to Kenya in 1993, Mesdag was along. Flying mainly from Lokichoggio, the C-46s carried food for World Vision into war torn Sudan. The C-46 proved admirable for the hot, high, rough strips. The worst incident in

three years occurred when Jeff Schroeder belly-landed after engine failure while departing Loki. Another time TXW hit a soft spot on landing and went on its nose. For Mesdag, Sudan was routine work. His only memorable trip was a medevac with three Sudanese men to Loki—they had been torn up by a lion. Overall, Mesdag flew about 800 hours in East Africa. He also put in a few hours on Morris Catering's DC-3, flying the 2½-hour food resupply run in Somalia between Mogadishu and Hobyo. Morris Catering was building a

fish processing plant at Hobyo, but the operation ceased when Morris was gunned down there one day.

August 16 was a busy day for Buffalo Airways—11 trips to Colomac, besides other work. TPO got started at 1310. It taxied behind Canadian North 737 C-GDPA—a famous old Dome Petroleum jet; and Air Tindi's Caribou. Mesdag pulled up on the edge of the main ramp, set the parking brake and tail wheel lock, then did his run-up. Brakes squealing in C-46 fashion, TPO then moved to the taxi way to

C-46 TPO down at Pickle Lake. A heavy load and a quartering tail wind led to trouble controlling the takeoff. Then, a beautifully-restored TPO on May 8, 1994 with Joe McBryan and Calvin Bottrell doing the flying. Graham Thoburn flew the photo plane, Buffalo Airways' Ce.185. (via J.E. Smith, Henry Tenby)

Capt Mark Mesdag and FO Andrew Dzal operated TPO on its August 16 grocery run. Mark was a seasoned C-46, DC-3 and 748 captain. Andrew was on his first airline stint, having arrived in Yellowknife in 1992 from a mechanic's job with Air Canada in Toronto. (Larry Milberry)

TPO on the shuttle at Hay River on May 4, 1996. Capt Marc Mesdag and FO Scott Lippa were doing the flying. The cargo was groceries for places like Yellowknife, Deline, Coppermine, Spence Bay, Gjoa Haven and Holman. By this time TPO's log indicated 26,627 flying hours. (Larry Milberry)

GNWT/Buffalo Airways CL-215 tankers being readied at Yellowknife on August 18, 1995 to ferry to Baie Comeau, where they were urgently required to fight major fires. Then, the crews flight plan their trip. (Larry Milberry)

(Below) Ptarmigan acquired its first Gulfstream I (C-GPTA) in 1989. Aircraft C-GPTG was at Yellowknife on August 18, 1995. Beyond, a First Air 748 waits for a Buffalo DC-3 before backtracking for takeoff on R33. Ptarmigan was founded in 1961 and grew steadily. When Wardair closed its Yellowknife operation in the spring of 1979, Ptarmigan bought its Twin Otters TFX and WAB. It then added the Gateway Aviation hangar at the airport. Around 1980 owners Hettrick and Becker retired, but kept the company, employing hired managers. George Simon was the last of these, staying till Ptarmigan was sold to First Air in 1995. (Larry Milberry)

await an arriving Caravan and Twin Otter. Finally Mesdag moved out, backtracked for R15 and took off at 95 knots at 1332. TPO climbed easily at 500 fpm. First stop would be Fort Norman, 331 nm away at the confluence of the Great Bear and Mackenzie rivers. En route there was time for some magazine browsing. In the April 1989 *Flying* was a story about Vincent Burnelli and his concept of the "lifting fuselage". His ultimate product was the Canadian-built CBY-3. The article summarized Burnelli's dream project as "too much solution for not enough problem." Meanwhile TPO was humming along—264°, 135 KIAS, 10,000 feet. At 1446 it descended to 8,300. The local terrain was gently rolling with black spruce and bog. Near Fort Norman the fury of recent fires was clear—they had burned everything for miles around to the very edge of town.

TPO landed on R23 at 1535; the crew waited for the locals to get their act together. A pick-up truck stood by for some cargo, then a small stake truck showed up. There was no fork lift, so the fellows started hand-bombing the load. Most was food for Northern Stores and supplies for the Albert Wright and Mackenzie Mountain schools. Four tons in all. After 50 minutes TPO fired up and departed for Normal Wells. There it left some cargo, refuelled, and took on some freight and two engineers from Buffalo's fire bomber base. They were needed to accompany four CL-215s being despatched next day for Ontario and Quebec. TPO was again in the air at 1745, heading for Déline (formerly Fort Franklin). There it was on the ground from 1810 to 1839 before departing on the town's 2,500-foot gravel strip, and heading home in leisurely fashion. At 2030 its wheel touched at 68 knots. With its time logged, TPO was a 26,300-hour airplane. Over the summer of 1996 Buffalo Airways added more old airplanes—three PBYs for fire fighting. C-FPQM was ex-Quebec government; C-FNJE and C-FOFI were ex-Newfoundland government, now on lease from Hicks and Lawrence.

Other Yellowknife Operators

Several other carriers operate from Yellowknife, flying skeds and offering specialized services. With a staff of 45 Ptarmigan Airways was the largest of these in 1995. It had been founded in 1961 by Ken Stockall, who had flown with the RCAF and Wardair, and engineer Les Mullins. In September 1962 Mullins was on a charter in Ce.185 CF-MZZ to Virginia Falls in the Nahanni Valley. Aboard were three photo buffs, one being Henry Bussey, some of whose photos appear in this book. They failed return to Yellowknife. The initial search turned up nothing, although their wrecked plane was found the following summer. Bill Hettrick and Clem Becker then joined Ptarmigan and eventually bought the company.

In 1995 Ptarmigan had five Twin Otters for skeds to native villages and for charters. The charter market also was served with a 10-seat King Air 200 and 8-seat Citation II. Two 17-passenger Gulfstream Is were operating a daily sked to Coppermine, Holman, Fort Simpson, Hay River and Whitehorse. One complete medevac kit was stocked for use either with the King Air or Citation, and was needed several times a month. The Gulfstream I proved a good Arctic machine. Introduced by Grumman in 1958, it was the first of a new breed of turbine-powered corporate aircraft. Home Oil of Calgary bought the first Canadian example in 1959 (CF-LOO). Although popular with leading corporations, after 30 years the Gulfstream I largely had been replaced in the executive role, and had been bumped down to serve in commuter and courier roles. In 1996 Ptarmigan Airways was taken over by First Air. The G-Is were sold and replaced by 748s

Air Tindi, formed October 1, 1988 by Alex and Peter Arychuk, was another important Yellowknife carrier. It traced its lineage to Latham Island Airways, and in 1994 still employed Jim McAvoy who, along with his brother, had founded that company in the 1960s. By 1996 Air Tindi's fleet comprised the Beaver, Caribou, Ce.152, Ce.172, Ce.185, Ce.208B, Ce.310, Dash 7, King Air 200, Navajo, Otter, Turbo Beaver, Turbo Otter, Twin Comanche and Twin Otter. The Caribou was added in

1993 to serve mining. The Dash 7 (ex-N722A) appeared in March 1996. It immediately went to work on a fuel haul from Lupin. During the strike of Air Canada connector lines January-March 1997 the Air Tindi Dash 7 was seen as far south as Toronto.

Caribou Resurgence

The DHC-4 Caribou saw little Canadian service, even during its heyday. Periodically, one

Air Tindi Twin Otter ATU supplying a camp at Francis Lake 60 miles east of Yellowknife on August 18, 1995. The drill and supplies that it delivered were slung to a site about 100 meters away by a Great Slave Helicopters Hughes 500 (inset), crewed by Todd Brough and Craig Ward. This day ATU made seven trips, while the Hughes made about 20 sling loads. (Larry Milberry)

(Left) Twin Otter captain Mike Claringbull (left) and FO Dan Pagotto at Francis Lake. Mike, a veteran pilot, came to flying after a career in semi-pro hockey with the Medicine Hat Tigers. In the 1990s he spent two winters with Kenn Borek in the Antarctic. (Larry Milberry)

(Below) Air Tindi Twin Otter KBO at Mackay Lake Lodge, about an hour northeast of Yellowknife. Its 800-foot strip was "no sweat" for seasoned pilot Mike Murphy. KBO brought in fuel, drill core boxes and lumber; and took out empty drums and some caribou meat. Someone at Mackay Lake wondered if Air Tindi's Caribou could get into Mackay. One of the pilots remarked, "Sure, but then it'd be here as a permanent fixture." Mike Murphy had joined "Willy's Bandits" as a young pilot in 1979. He left aviation for drilling and blasting, but later returned to fly a Beaver for Glen Warner of Bathurst Inlet. This day he was training Warren Fix, who was upgrading to captain. (Larry Milberry)

would crop up—with Kelowna Flightcraft, Max Ward or La Sarre Aviation. In 1993 a leased Caribou (N95NC) flew briefly on DEW Line clean-up with Aklak Aviation of Inuvik. That year Air Tindi acquired Caribou C-GVYZ. It had a productive 400-hour summer working diamond camps, but mainly was dormant till the 1995 season. While carrying a DC-3 load (up to 7,200 pounds), the Caribou had an advantage of easier ops on rugged strips under 2,000 feet. This was not to denigrate the DC-3, which seasoned pilots in the NWT usually could get into 1,800 feet. On August 15, 1995 VYZ began a series of Yellowknife-Port Radium trips, where an old communications complex was being demolished. Although their

plane had been languishing about a year, pilots Tom Hanson and Peter Burgess expressed great confidence, and soon were working back and forth with loads of scrap. Their trip of the 15th began with start-up at 1610 hours. Nineteen minutes later Hanson rocked his plane back on its heels and it flew away at 83 knots. The load was eight drums of stove oil needed at the Port Radium clean-up camp, 240 nm north. VYZ climbed at 300 to 400 fpm at 120-130 KIAS. It levelled at 2,500 feet on the 311 radial. At 1645 it climbed to 8,500 feet.

En route there was a chance to leaf through various greasy Caribou tech manuals. The PSM-1-4-1A (i.e. flight manual) had first been approved January 20, 1961 and last revised February 1, 1970. Here one learned that VYZ was Caribou No. 97—a DHC-4A. The 4A differed from a straight 4 in all-up weight—28,500 compared to 26,000 pounds. Only the first 22 aircraft were 4s. Then there were the specs to be jotted down: basic weight 19,312 pounds, length 72'7", wing span 95' 7$\frac{1}{2}$", height 31' 9", horizontal tail 36', landing gear track 23' 1$\frac{1}{2}$", etc. The engines were P&WA R2000-7M2 Twin Wasp D-5s rated at 1,450 bhp/2,700 RPM on takeoff; 900 bhp/2,250 RPM for the climb. Recommended fuel was 100/130 octane. The props were 13' 1"-Hamilton Standard 43D50-353-7107A-Os. Normal airspeed was listed as 168K, never exceed speed—211, and minimum control speed—69. As to cargo hold, the dimensions were noted as: floor area 176 sq. ft., length to ramp 28' 9", ramp length 4', height 6' 3", max. width 7' 3". Time flew as the crew enjoyed the scenery, chitchatted and flipped manual pages.

At 1758 VYZ reported its position to the company radio in Yellowknife—they were 19 nm southwest of Port Radium. Soon the strip was in sight and a pass was made to take a quick look at conditions. The Caribou made a one-eighty and flew downwind. On final the crew spotted camp manager Dave Lorenzen part way down the runway, signalling to land safely passed a rough section. VYZ slid onto the gravel strip at 1806. It taxied to the camp used by Lorenzen's crew of bushmen, who were there to dismantle the hilltop radio relay "triple scatter" site. Everyone in camp worked for Arctic Remediation Services, which was on contract to NorwesTel. The ARS boys quickly offloaded the Caribou, then started shoving in lengths of heavy pipe. These had been the old pipeline up the steep hill to the radio site. Now that the station was obsolete, it and everything else had to be cut up and, for the most part, carted to Yellowknife. This was in accordance with federal environmental rules. The pilots supervised, being careful to estimate their load. When a workman suggested each piece of pipe at "about 90 pounds", Tom Hanson figured "more like a hundred", and stopped the loading at 70 pieces, not the 80 that the loader wanted. With the load secure the crew joined the camp for a hardy dinner served by cook Caron Lorenzen. At 1945 they fired up and taxied to the farthest end of available runway. They took off in a clatter that left the camp engulfed in dust. They climbed to 7,500 and followed their

Views of Caribou VYZ on August 15, 1995 at the Arctic Remediation Services camp at Port Radium. The NorwesTel site being demolished there was atop the hill beyond in the second view. VYZ languished at Yellowknife for the next two years. In May 1997 a crew from Vintage Props and Jets of Florida flew it away to a new career—freighting from Miami to the Bahamas. (Larry Milberry)

(Right) In the cook tent at Port Radium with Dave and Caron Lorenzen (left) and Air Tindi pilots Peter Burgess and Tom Hanson. (Larry Milberry)

reciprocal course. At 2107, while 74 nm back, they checked in with the company. The dispatcher asked, "How went your maiden voyage?" The reply indicated a smooth evening's work. At 2140 VYZ's tires squeaked onto R33 at 75 knots and the old stalwart taxied to Air Tindy's ramp. It was too late to worry about offloading. That was left for the morning crew.

Although the Caribou still was capable of a good day's work, it had its limitations. Its R2000 engines were costly to maintain and overall reliability meant that it might not be available when a customer needed it. The following winter Air Tindi laid up VYZ, started looking for a buyer, and acquired the Dash 7. The new plane, although expensive, offered high reliability, needed less maintenance, and carried a bigger load (although it lacked the Caribou's handy rear loading door).

Bradley Air Services and First Air

By the mid-1990s Bradley Air Services/First Air was Canada's most famous name in Arctic air transport. With some 650 employees (including 85 pilots), it operated four B.727s, nine BAe748s, five Twin Otters, two Beavers, a King Air 100 and a Dash 7. In face of persistent economic recession it was holding its own, but paring to assure survival. Bradley was formed in Ottawa in 1946 by Russ Bradley, a wartime BCATP pilot. Flight training and charter work from Uplands initially busied him. In 1950 he relocated to Carp, a few miles west of Ottawa. Here he operated the PA-18, Apache, T-50, etc., and had a Piper dealership. The DEW Line was kind even to little carriers—Bradley won some contracts beginning in

1955. Other opportunities were pursued, including budworm spraying in New Brunswick with PT-17 Stearmans. With his work on the DEW Line, flight training, aerial photography and spraying Russ Bradley was recognized as a go-getter and a survivor. One of his early deals was buying the DOT's old Beech 17 for a song, then reselling it to Spartan at a profit.

Bradley's DEW Line work involved several Ce.180s. Reg Phillips left Austin Airways in 1955 to fly one. His assignment began August 2, when he set out to ferry CF-HVG (on floats) from Ottawa to Cambridge Bay. He arrived on the 8th to begin a season carrying Western Electric staff from site to site, delivering mail, equipment, parts, groceries, etc. There also were aerial ice patrols, results of which were relayed to ships' captains. In mid-September Phillips flew to Edmonton for change-over to wheel-skis. He left on the 26th for the Arctic. Over the winter he flew on a number of searches for lost aircraft, and transported sick and injured workers. Typical were a search November 12-15 in Alaska and a mercy flight to Coppermine on the 17th. A logbook note of January 13, 1956 reads "12-8, burned man" (i.e. medevac between sites 12 and 8). Next day Phillips was on another search, but was forced down by weather. He spent two nights burrowed in a snow drift waiting out the weather. Tractor train patrols also were flown. Twice Phillips arrived over trains to find that the tractor had broken through the ice and the trains were stranded. In such cases he radioed word, or set down to lend assistance. He finished his contract on January 21. Next, he worked briefly with the Lands and Forests in

Ontario, moved to Fecteau in Quebec, then took a contract in 1957 with Bradley on Norseman CF-HAD, supporting prospectors in the Keewatin District from Eskimo Point. Phillips' helper on HAD was John Jamieson, later president of Bradley/First Air.

Welland W. Phipps went into partnership with Bradley. Phipps had been a wartime flight engineer with 405 Squadron and spent two years as a POW. He learned to fly after the war, working briefly with Atlas Aviation and Rimouski Airlines, then joining Spartan in 1949. Eventually, he became chief pilot and maintenance head. In 1956 he ran a contract for Ottawa photographer Dalton Muir, taking him as far as Alert on Ellesmere Island in Super Cub CF-HCX equipped with the tandem Whittaker undercarriage. This gear was useful but, with four wheels, was heavy, increased the turning radius, and offered little help on soft terrain. Phipps' next contract was for the Department of Mines and Technical Surveys in 1958. For geological survey it was ready to try a fixed-wing aircraft for the first time—it previously had depended on helicopters to back field crews. In the High Arctic light choppers like the Bell and Hiller were limited, especially in range.

For this job Phipps experimented with a new undercarriage for the Super Cub—he engineered and fitted a 35-inch tire. The mod included a 6-inch wheel with a heavy brake assembly. Weight (60 pounds) and speed (10 mph) penalties were the drawbacks, but great rough-terrain performance was delivered. Phipps figured that he could produce and install his kit for about $1,100. The aircraft also

AT-11 could work above 30,000. About this time Bradley acquired an S-55 for low-level electromagnetic surveying. Initially, Herbie Johnson flew it. For the 1959 season Dick Deblicquy flew it for three months, the Super Cub for three, then the S-55 for a final three, before taking the rest of the year to relax and ski. The S-55 venture led to the formation of Universal Helicopters with Bradley, Herbie Johnson and Gary Fields co-owners. Universal grew into a large operation in St. John's, specializing in offshore oil work. Bradley developed other helicopter interests about 1960, flying fire patrol contracts with the Bell 47. Phipps' innovations did not go unnoticed—in 1961 he was awarded the McKee Trans-Canada Trophy.

In 1960 Russ Bradley purchased his first Beaver and ran it on special rough-field DC-3 tires. Dick Deblicquy described the mod as "phenomenally successful on soft, rough ground." In February 1962 the first Otter joined Bradley—CF-OHD. Deblicquy put 850 hours on it by September, then flew it south with an engine on its second DOT maintenance extension. The year had been fine for weather, and the Bradley fleet piled up the revenue in the long Arctic days. For August alone Deblicquy logged 260 hours. OHD served Bradley to January 28, 1974, when it burned in a hangar fire at Carp.

The weather in 1962 may have been fine, but the Bradley-Phipps partnership was not. That year they split, with Phipps moving to Resolute Bay to run his own company, Atlas Aviation. It did a great deal of flying, especially on exploration in the Beaufort. Resolute, the most northerly Canadian port where ships could dock in the Arctic shipping season, was a vital staging base for remoter sites like Mould Bay, Isachsen, Eureka and Alert, which lie between 76° and 82° N. latitude. Atlas served these places. In 1966 Phipps bought the first commercial Twin Otter from DHC and built up a DC-3 fleet. For his Twin Otter he asked de Havilland to procure the registration CF-AAL (for Atlas Aviation Ltd.); but AAL already was an Austin Airways DC-3. Sheldon Benner of DHC proposed reserving the registration CF-WWP (Phipps' initials) and this was arranged.

In April 1969 Phipps rescued three British skiers attempting to cross the top of the world. In 51 days they covered only 90 miles of their 470-mile run to the North Pole, and they hadn't been heard from in three weeks—their radio had quit. When it was clear that the team needed help, Phipps searched for and found the adventurers at Ward Hunt Island near the top of Ellesmere Island. In another incident an F.27 crash-landed near Rez. Phipps took a Twin Otter, found the plane and landed to find all 13 aboard alive. He ferried two injured men to Rez, then returned for the others. Atlas prospered, serving places like Mould Bay, Arctic Bay, Grise Fiord and Pond Inlet. One of its pilots was an Eskimo from Rez, Markoosie. In 1972 Phipps sold to Kenting Aviation and retired to Prince Edward Island. Since then he was named to Canada's Aviation Hall of Fame,

was equipped with extra fuel, special radios, including HF and survival gear. It set off on June 15 for Rez, arriving on June 19. Phipps then flew his geologists and their equipment to camp on Melville Island, where a bedrock survey was under way. Through the summer he made about 400 landings on unprepared strips, logging 300 hours.

Beginning in 1959, Ottawa opened the Arctic to commercial exploration and staking. That season Phipps had seven Super Cubs in the region: pilots Al McNutt and Bob O'Connor on Banks and Victoria Islands, Dick Deblicquy and Ed Jensen on Bathurst Inlet, Harcourt Papps and Mike St. Arnaud on floats in the Mackenzie estuary, Weldy Phipps on Axel Heiberg, Ellesmere and Ward Hunt Islands. Of course there were upset aircraft, since landings often were on rough, soft or wet terrain with runs of only 200-300 feet at times. An old timer that season put it this way—"Our Super Cubs were sticking up like darts all across the tundra." Damage was usually minimal. The

fleet did a good job at lower rates than helicopters. The geological people were sold on the Super Cub with Phipps big tires, which someone dubbed "tundra tires". For 1959 Phipps had 35-inch tires on five Super Cubs. By 1961 he had 14. The company by now was involved in a wide range of work, flying geologists, biologists, prospectors, explorers and a host of others, and re-supplying camps. The great amount of oil exploration in the Arctic at this time benefited greatly from the Super Cub.

Tundra tires were not Weldy Phipps' only brainwave. Earlier he launched Spartan's P-38 operation—designing and building camera installations, nose extensions and perspex for P-38s and operating them in BC and the Yukon. He modified an Apache with superchargers for high altitude work. For this he developed a Super Cub testbed. One day he took the little plane to 30,000 feet, and had CF-100 interceptors up to check out the spectacle. He also supercharged the R985s in a Bradley Beech AT-11. Using superchargers from a B-17, the

and made a Member of the Order of Canada. He passed away in 1996.

In 1968 Bradley Air Services won the contract to support the Polar Continental Shelf Project, a yearly scientific undertaking involving multi-disciplinary scientific groups across the Arctic. The key to their success was air support for their dozens of small bases. Fred Alt was a young German fresh from his military service, when he visited Canada in 1963. In 1965 he began work with the PCS, eventually becoming ops manager with a chain of 50-60 camps. He watched as air operations moved from Super Cubs and Bell 47s; to Beavers, Otters and S-55s; to Bristol Freighters, DC-3s, DC-4s, DC-6s and Super Connies; and, finally, to turbines. In his 16^1/$_2$ years with the PCS Alt noted 17 flying accidents, all but one being weather related. The crash of a Bell 204 flown by a pilot from France was the only fatal incident. In 1981 Fred Alt joined Bradley as a base manager. In 1997 he was running the busy Iqaluit operation, having experienced several Arctic booms and busts. One low came in 1989, when Bradley lost its bread-and-butter PCS contract to Kenn Borek.

In 1969 Russ Bradley died in a traffic accident. In 1971 John Jamieson, Dick Deblicquy and Ian Kirkconnel bought the company. Aviation was on an upswing, and in May 1971 Bradley introduced its first Twin Otter—CF-DHT. Equipped with INS, it did outstanding work during exploration in the Beaufort. The Twin Otter would prove the finest tundra aircraft ever. Before long it replaced all but Bradley's Beavers. Also in 1971 Bradley opened the most northerly commercial air transport base at Eureka. In 1973 Twin Otters CF-ASS and ASG entered service. One was based in Halifax on a two-year contract with Mobile Oil, which was drilling near Sable Island. This aircraft had a dispatch rate of 99% and logged 1,800 hours a year.

To cope with an increase in scientific, oil and mineral exploration, Bradley purchased its first DC-3 in 1972. This type served to 1979, proving especially valuable on fuel hauls. In 1973 Bradley set up a base at Resolute; another at Frobisher Bay in December 1975; and one at Hall Beach in June 1978. Resolute was busy. Andy Campbell managed it for Bradley, and recalled 1975-76, when the company had as many as 14 aircraft and 60 people there. One Bradley machine at Rez 1966-78 was Beech G18S CF-TAE, used on ice patrol in summer. As Bradley's reputation spread, so did its reach. In 1974 one of its Twin Otters supported the US Navy's Ross Ice Shelf Project in Antarctica; this resulted in ongoing contracts. The USAF signed Bradley to serve two remote Greenland sites with a Twin Otter. A coup came in 1987, when it took over DEW Line lateral resupply from Canadian. This kept two 748s and two Twin Otters busy for years, serving points between Alaska and Greenland. Bradley, operating as First Air, introduced passenger skeds in 1973 with Twin Otter service between Ottawa and North Bay. The company fleet in 1976 included nine Twin Otters, five Otters, four Beavers, four DC-3s, two

With Capt Jim Merritt at the helm a First Air 748 sked taxis at Yellowknife for points north on July 3, 1993. C-GBFA began with Cascade Airways of Spokane in 1981. Later it was seized by the FAA and stored at Phoenix till bought by Bradley in 1988. By the 1990s a number of 748s were flying with the "18-60" mod on their Darts. This made the gas-guzzling turbines more fuel-efficient. Even though the 18-60 cost about $100,000, considering the high cost of fuel in the North, some operators felt that it would pay. First Air was owned by the Inuit corporation Makivik. (Larry Milberry)

Beech 18s, a Cessna 421 and a number of smaller types for training. On January 17, 1977 Bradley began an Ottawa-Mirabel Twin Otter shuttle under the First Air banner. Three Twin Otter (later BAe 748) daily flights were scheduled on the 30-minute run, while Air Canada operated one DC-9 flight. First Air now became the usual name for the airline operation; Bradley ran the smaller planes. Meanwhile, Bradley flying schools at Carp, Arnprior and Peterborough provided a reliable source of young pilots.

In 1979 First Air introduced the 748, its first being C-FTLD. In 1997 it had eight. The company's first jet appeared in 1986 with C-FRST, a Boeing 727-100C. By October 1988 it had four 727s—three combis, and a 125-seater (in Toronto for charters). Some overnight 727 courier work was done for Emery on an Ottawa-Montreal-Dayton run. In spite of old age the 727 did excellent work, and was well-liked by all. On a flight from Kuujjuaq to Dorval on August 14, 1992 the service aboard a First Air 727 proved as classy as on any Airbus. FAs Jimmy, Leslie and Kevin served a delicious meal, topped with a trip down the aisle with a pastry-laden trolley and liberally-poured liqueurs.

Capt Jeff Kitely commented that Bradley's 727 crews may not be raking in the "big bucks" of Air Canada and CAIL pilots, but they loved their work, especially the variety. A trip one day could be to Resolute at minus 40 degrees, and the next day to the plus 40 Caribbean! It caused a chuckle at Dorval one day when someone asked if all the passengers in the waiting room were for the First Air 727 to Ottawa. Kitely replied, "I don't think so. They aren't our crowd... they're too well dressed." They looked more like business travellers for the next Rapidair flight to Toronto than the casual folks heading to Baffin Island on First Air.

By 1993 First Air had skeds to centres in the NWT, Ontario and Quebec—from Ottawa and Montreal north to Iqaluit, then trans-territorial to Resolute, and Yellowknife, with points between. To maintain its fleet, it had engineering bases at Carp and Ottawa International. Ottawa included cargo facilities and cold storage to handle perishables for the Arctic. A standard 727 trip operated August 12, 1992 with the crew of Capt Gord Wallace, Capt Stephan Ekiert and second officer (i.e. flight engineer) Lowell Teasdale. The aircraft was C-FRST, originally of Alaska Airlines, and a veteran of 64,026 hours. It was airborne from Ottawa for Iqaluit at 1006 hours, carrying 54 passengers in the rear and 24,000 pounds of cargo up front. Capt Wallace had been with Bradley since 1976, when he started taking flying lessons at its training school. He got right-seat experience on the Twin Otter and Beech 18, made captain on them, and advanced to the 748 in 1987. Ekiert had started flying gliders at Hawkesbury and was hired by Bradley as a dispatcher at Iqaluit. He advanced to 748 captain, then moved to the 727. Teasdale had trained at the Moncton Flying Club, joined Bradley as a dispatcher, then made SO.

F860 was 2:50 hours on the 1,142-nm route to Iqaluit, cruising at FL330, the top permitted by ATC—the airways above were reserved for overseas traffic, vast amounts of which crossed the area daily. The flight was run on the company's "opti-Mach" fuel management system, tailor made for its particular aircraft and routes, and designed to save 0.5% to 3% on the fuel burn per leg, compared to standard 727 procedures. With fuel at least 15¢ more per liter in the Arctic, the system was vital to a company using millions of liters yearly. F860 started its descent to Iqaluit over Hudson Strait. It crossed the Meta Incognita Peninsula and banked towards the head of Frobisher Bay. Soon Iqaluit, with its bright yellow terminal, appeared. The 727 passed low over the head of the bay, where two Coast Guard vessels lay at anchor, along with a few fishing boats. On the tidal flats some flat-bottom barges were offloading—the annual sealift was underway. Passengers on the right got a bird's eye view of downtown Iqaluit, then the threshold flashed by. The pilot flared and touched down softly. RST backtracked to the terminal, shut down, and the passengers moved inside, where some checked in for ongoing flights. Meanwhile, ground staff went to work. Palletized cargo was offloaded and moved into Bradley's cargo centre; the cabin crew straightened out the passenger compartment for the return trip; the pilots checked in with dispatch; the Shell truck pulled up for refuelling. While RST was causing so much activity, things were bustling all around. First Air's 748 C-GYMX was waiting to load. Canadian North was on the ramp with a 737. Echo Bay's 727 was about to taxi with passengers for Russia. A CanAir Convair, in from Halifax, waited for en route weather to clear, so it could get into Rez on a crew rotation for the Coast Guard (26 passengers up, 29 back). A Coast Guard BO105 was buzzing back and forth, ferrying crew to a ship in the bay, and a Canadian Helicopters Bell 212 was fuelling for a trip to a North Warning System site.

A different First Air 727 (F863, C-GFRB) sked operated Yellowknife-Resolute-Nanasivik-Iqaluit on June 6, 1993 with the cockpit crew Capt Jeff Kitely, FO Mike Meddings and SO Richard Mugford. FRB was in its 54-seat combi configuration. The day proved easy-going in CAVU weather. The first leg (2:02 hours) carried 39 passengers and 15,874 pounds of cargo. At Rez there was little activity as F863 turned around—some CHL Jet Rangers and Kenn Borek Twin Otters were awaiting work, another Twin Otter was away on Somerset Island with some diamond hunters. F863 departed with 51 passengers and 8,811 pounds of cargo. It landed on Nanasivik's super-smooth 6,400 x 150-foot gravel runway 37 minutes later. Nanasivik was a mining community of about 350 people (the mine was about 20 miles away by road at Arctic Bay). It had six weekly skeds—two each by Canadian North, First Air and Kenn Borek. Soon Kitely and company were off for Iqaluit (1:33 hours) with 51 passengers and 5,700 pounds of cargo. At 139,000 pounds the full 6,400 feet was required for takeoff, FRB rotating at the end at 113 knots. Each day First Air operated such trips with its 727 "bush planes". Its fleet numbered six in 1997 (four combis, two freighters), each averaging about six hours

aloft per day. Although a replacement eventually would be needed, they were sure to see service into the new century.

Baffin and the 748

At 1600 hours on August 12, 1992 First Air 748 C-GDUN (F800) departed Iqaluit on its four-times-a-week sked up Baffin Island. In charge was Capt Bob Chester, with FO Yves Jolicoeur and FA Karen Duffy. F800 operated to Broughton Island (254 nm), Clyde River (201 nm) and Pond Inlet (222 nm), connecting them with the rest of the world. DUN was one of three First Air 748s at Iqaluit. Three others were in Resolute and two in Yellowknife. DUN had been handed over new on January 7, 1966 to LAV of Venezuela. It came to Canada in 1981 to fly in Quebec for Regionair, and joined Bradley in October 1984. Over the years it had piled up 41,585 hours.

The 748 originated at Avro in the UK in 1958, when Britain's aviation industry was despairing about the future of military aircraft. The prevailing mentality was that guided missiles would rule the military world—new fighters and bombers would not be needed. Avro had been committed heavily to military contracts like the Vulcan bomber, but now looked into commercial concepts—first a 20-seat feeder liner, then a DC-3 replacement. The Avro 748 resulted; the prototype flew on June 14, 1960. The 748 won favour—it was well-suited for inter-urban or inter-island services. Skyways Coach Air took the first delivery, followed by the Brazilian Air Force, Thai Airways, the RAF, etc. Production continued to 1986 with 377 deliveries. By then older 748s were down in the pecking order among operators. They had lost passenger appeal to the Dash 8, ATR42 and Fokker 50 on key routes. Pushed aside, they declined in value. By the early 1970s it was a buyer's market— Austin Airways paid as little as $100,000 for 748s in Latin America. Other Canadian airlines used 748s. EPA and Air St. Pierre were pleased with them on their Atlantic routes. They carried 44-50 passengers in comfort, and were fast and quiet enough for the passengers. Austin Airways found them excellent on short gravel

strips or ice. Their Darts had ample power (2,280 ehp for the 748-2B), the payload was a decent 11,000 pounds (double a DC-3's), and serviceability was high. Air Manitoba, Bradley, Calm Air, Inter City and Regionair added 748s in the 1980s.

At 44,700 pounds DUN lifted off from Iqaluit at 109 KIAS for its run. With 24 passengers and 4,824 pounds of cargo it climbed in gruelling 748 fashion—200 to 300 feet per minute at 155 KIAS, eventually levelling at FL150 to cruise at 180 KIAS. The pilots handflew all the way, as their auto pilot was disconnected. However, the fleet now had GPS, and new colour radars that were excellent for ground mapping. Miles from Pangnirtung the radar screen showed all the terrain features there. The flight passed across Cumberland Sound, then overhead "Pang", just as First Air F805 was taking off from there for Iqaluit—a 748 with nine passengers.

With a few hundred inhabitants Pangnirtung is perhaps the most visited community north of Iqaluit. The setting is spectacular— at the head of the fjord is Auyuittuq National Park with ice fields and glaciers. Pangnirtung is a haven for nature lovers, kayakers and hikers in the short summer. A busy spring fishery nets turbot and char through the ice for markets in the US northeast. The fish are prepared in town, and flown to Iqaluit as backhaul on the 748s. Pangnirtung has an infamous approach. On a gusty day arriving aircraft can be severely buffeted. Flight attendants are used to outbursts of panic from their passengers on such days— the "Pang screamers", as they have become known! Transport Canada suggests that inexperienced pilots steer clear of Pang. Its gravel strip is 2,900 feet long, so the 748 is restricted to 38,000 pounds going in—5,000 pounds less than normal.

F800 made its stops at Broughton Island and Clyde River, letting off passengers and unloading most of the freight—fresh bread, bananas, dozens of eggs, mail, etc. Fuel was added at Clyde and takeoff was followed by a speedy climb-out at 1,000 fpm at 37,000 pounds. DUN reached Pond Inlet at 2015, having approached down the scenic inlet between

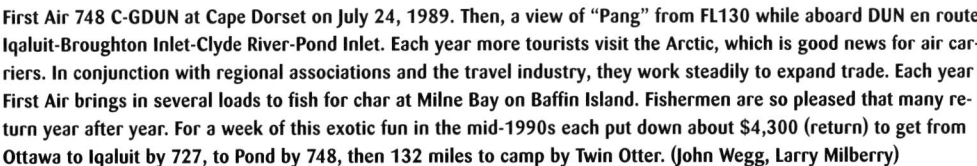

First Air 748 C-GDUN at Cape Dorset on July 24, 1989. Then, a view of "Pang" from FL130 while aboard DUN en route Iqaluit-Broughton Inlet-Clyde River-Pond Inlet. Each year more tourists visit the Arctic, which is good news for air carriers. In conjunction with regional associations and the travel industry, they work steadily to expand trade. Each year First Air brings in several loads to fish for char at Milne Bay on Baffin Island. Fishermen are so pleased that many return year after year. For a week of this exotic fun in the mid-1990s each put down about $4,300 (return) to get from Ottawa to Iqaluit by 727, to Pond by 748, then 132 miles to camp by Twin Otter. (John Wegg, Larry Milberry)

Scenes from First Air F800/801 of August 12-13, 1992. Passengers walk to the 748 at Clyde River. In the shore scene at Pond some tourists are kayaking among the floes around midnight. Capt Bob Chester is seen at work in 748 DUN, while FO Yves Jolicoeur completes paper work at Pond. Pilots hate paperwork, but it's a necessary evil. One pilot was heard telling his passengers, "Sorry folks, but we don't take off till the weight of the paperwork equals the weight of this plane." (Larry Milberry)

Baffin Island and Bylot Island. Pond was a chilly 4°C. Great slabs of ice were strewn along the shore. Four keen British tourists were kayaking among the floes. Even crazier were the local kids, whose evening sport was leaping from pan to pan. The beach was scattered with the leftovers from butchered seals. Not an idyllic scene when viewed close-up. The air was full of the barking of smelly, mangy-looking sled dogs.

Baffin is a place for the adventure tourist. Visitors in the 1990s were mainly Americans, Europeans and Japanese with a desire for the unusual, and the cash to afford steep prices for air fares and accommodations. For those who come, however, the rewards are many, although sometimes people take the Arctic for granted. On September 2, 1995 four Americans were lost when a whale capsized their boat. 424 Squadron from Trenton was scrambled to help, but too late. Only an Inuit guide survived. Even sled dogs can be dangerous. In 1992 several set upon and killed a child at Pond Inlet.

First Air's 748s found work from Yellowknife in the early 1990s. In 1995 three were there on sked and charter work. Typical of their work was F842/843, the weekend Coppermine-Cambridge Bay-Gjoa Haven-Spence Bay sked. For August 19/20, 1995 it was flown by C-GDUN configured as a combi—16 seats, the rest cargo. It departed Yellowknife at 0917 with 12 passengers and 7,760 pounds of freight. Capt Mike Cahill and FO Graham Hardy climbed out slowly to FL200, where DUN cruised along at 170 KIAS on the 318-nm route to Coppermine. Most weekends the sked was sold out. This day the passengers included Pat Lyle, an Inuit leader from Spence Bay. He had been in Toronto visiting the NWT pavillion at the CNE. He found Toronto in one of its infamous heat waves, so was glad to be going home.

At 1030 hours the crew called Coppermine for the weather, then landed at 1101 on R12.

The airport was quiet—the only other plane was a visiting Baron. Some passengers and 2,500 pounds of cargo were offloaded. DUN again was airborne at 1155, heading for FL130 and Cambridge Bay, 233 nm eastward. Someone asked the pilots to watch for the annual barge train, which reportedly was nearing Coppermine. A few minutes out the crew spotted the barge and radioed the news. Coppermine replied, "OK. Hopefully I'll have my new truck this weekend."

Along this route a great deal of smoke had wafted down the Mackenzie from forest fires to the south. As usual the pilots monitored Baffin Radio, so there were interesting flights on which to eavesdrop—British Airways, Lufthansa, United. At 1240, while 30 miles back, the 748 started down. As they got closer, they watched "Territorial 701" (NWT Air's Herc) raise a dusty rooster tail as it landed on R13. DUN followed at 1306. From here came the usual routines—passengers and freight came off, the pilots did a walk-around and checked the weather, then pressed on, overnighting at day's end before backtracking next day to Yellowknife.

Cambridge Bay was an old HBC post, RCMP outpost and mission station until a LORAN station was installed in 1947. This brought a few southerners and the number of Eskimo families soared from four to over 100. A greater influx came in 1955, when construction started on a DEW Line site. The US military built a runway suitable for C-124s, and a modern hangar. By 1996 this was one of the last DEW Line sites; it employed about 50 people. The local population was about 1,100. The airport at Cambridge Bay was busy on August 19. The local operator, Adlair, had its Twin Otter coming and going on mining work. Its King Air came in from a charter and its Lear 25B was on stand-by for a medevac. Canadian North arrived with a 737, while

NWT Air's Herc offloaded, so the place was booming for a while. In this period air travel in the NWT was becoming highly competitive. Carriers were ogling one another with takeover on their minds, and jumping on one another's routes. In 1996, for example, Canadian North increased its trans-territorial skeds from two to three, hurting NWT Air. First Air inaugurated nonstop Iqaluit-Cambridge Bay service.

Twin Otters

Bradley serves other Baffin Island centres. There are daily skeds to Pangnirtung, Lake Harbour and Cape Dorset. Lake Harbour is a half-hour run from Iqaluit by Twin Otter. The August 14, 1992 trip (F810) was flown by Dick Deblicquy and Leslie Watton. Deblicquy had joined the RCAF in 1948 and served as a navigator with 414 Squadron at Rockcliffe. This got him his first visit to the Arctic. Meanwhile, he took his private pilot's licence and purchased a Tiger Moth. He left the RCAF in 1954 and joined Bradley to instruct at the rate of $100 a month plus $1.00 an hour for flying. He began flying the Twin Otter in 1966 and piled up 16,000 hours on type (along with 2,000-3,000 hours each on the Beaver, Otter, Beech 18, DC-3 and Skyvan, and 5,000 on helicopters). Watton was a new pilot—an aviation graduate of Mount Royal College in Calgary. She had spent some time as an SO on the 727 and recently had been promoted to the Twin Otter.

F810 (C-FASG) was loaded to the gunwales with freight and passengers for Lake Harbour. Included were five sled dogs, which howled madly when Deblicquy started his PT6s. There were some local teachers on board, returnees from a summer in the south; and several Inuits. FO Watton made an attentive approach into Lake Harbour—its 1,700-foot gravel strip leaves something to be desired. After a quick stop, ASG started up to carry a few passengers to Iqaluit, including

(Above) Bradley/First Air Twin Otter ASG at Lake Harbour on August 14, 1992.

(Left) FO Leslie Watton slings the baggage for Bradley at Lake Harbour during ASG's quick turn-around for Iqaluit. She later progressed to the 748. Then, two of First Air's key men in Iqaluit on August 14, 1992. Dick Deblicquy, who retired in 1994, was one of the Arctic's most experienced pilots. Fred Alt was First Air's base manager. One of his responsibilities was running a smooth dispatch operation. Dispatch entails things like flight planning—checking notams and weather, doing take-off analyses, load planning (checking passenger weights, which vary by the season, etc.), and flight watch (ongoing communications with a flight). (Larry Milberry)

In post-WWII days Frobisher Bay (Iqaluit) was one of the busiest northern airports. Hundreds of military and civil flights refuelled here weekly. Today there are few such visits by comparison. Here the RAF Queen's Flight BAe146 stops for fuel on July 17, 1992. (Larry Milberry)

Sandra White of Toronto's Eskimo Art Gallery, who was on an annual soapstone carving buying trip to Baffin.

The Declining Trans-Atlantic Gas Stations

In the 1950s Frobisher Bay was one of the busiest northern airports. Along with Goose Bay and Gander it was a vital "gas station" for aircraft crossing the Atlantic. In the peak Cold War years much of the traffic was military, but the major civil carriers also stopped by— BOAC, KLM, Pan Am, SAS, TWA, etc. Those were the days of B-36s, C-124s, DC-7Cs, Super Connies and the early commercial jets. With the easing of East-West tensions, and the advent of true intercontinental airliners like the 747 and L.1011, few trans-Atlantic aircraft stop at these airports any more. Of course the turbine is king today, and places like Iqaluit stock little avgas. In 1991 when an ancient B-24 stopped to refuel at Iqaluit, it had to be sent on the Kuujjuaq to look for fuel. Throughout the Arctic, it will be difficult in future to find fuel for piston-engined aircraft. Although the classic propliners have gone from most of the airways, Iqaluit is still a fascinating place to watch airplanes. Its flight service station keeps a pilot sign-in book, and for 1991-92 entries included:

March 21	2 x CF-18: Capts St-Amand and Girard
May 5	Cessna 177: Ralph Machon, Zurich to Dayton
June 1	Dash 8: H. Schider, Toronto-Hamburg
June 7	Comet 4: Peter Smith, Söndre Ström-Goose Bay
June 20	C-9: Capt Steve Tanaka, Keflavik-Scott AFB
July 12	Soloy 206: Strackeryan and Polzer, YHZ-Pangnirtung-Nuuk
July 13	Pilatus PC-6: McKay & Ryan, Söndre Ström-Schefferville
July 18	Stearman: Klaus Plasa, Galesburg, Ill.-Oldenburg
July 24	Grumman Goose: Juergen Puetter, Montreal-Clearwater Fjord
July 26	MU-2B: Bignon & Castony, Paris-Anchorage
October 21	T-33: Capt Lou Glussich, North Bay-Ottawa
January 1	Cessna 441: Edmonton-Germany
January 8	C-130: Capt Dave Ross, Trenton-BGTL
January 15	Cessna 500: Washington, TN-London
January 30	C-5A: Söndre Ström-Texas
February 22	Saab SF-340: Luköping-Los Angeles
March 14	C-130: Rob Mullin, CFB Trenton
March 21	Aurora: YYR-Thule
March 28	Cessna 182RG: Klaus Plasa, Fort Lauderdale-Hamburg
March 28	Bell 212: Mike McKenzie, Kelowna-Hall Beach
May 22	Piper PA-31T: Dimitr Papazoglou, Calgary-Greece
May 30	Cosmopolitan: Capt Brian Wood, Keflavik-Ottawa
June 4	Piper PA-28: Eric Fanton, Grenoble-Montreal
July 24	Beech C-45H: Hamburg-Oshkosh

The list is much slimmer than in the old days, and gone are the famous types—not even a DC-3 turned up in the sign-in book in two years! However, it shows that there still is traffic on the North Atlantic needing a fuel stop en route, and there still are pilots game to ferry a

single engine plane across an ocean, where there would be little chance of rescue in case of ditching. One day Hollywood actor Tom Cruise passed through Iqaluit in his private jet. While he was wandering around the airport, one of the Bradley fellows said to him, "Why don't you come flying with us? We're the real top guns. We'll take you to Pangnirtung."

Other Arctic Operators

Besides Bradley/First Air the Eastern Arctic has other regular operators. Air Inuit was formed in 1979. Owned by the Makivik Corporation in Kuujjuaq, it began with a DC-3 and two Twin Otters. In 1993 the fleet numbered five Twin Otters and three 748s serving communities around Ungava Bay, Hudson Strait, Hudson Bay and in the Quebec interior. It also ran a cross-strait sked linking Salluit (formerly Sugluk) with Cape Dorset. A separate Air Inuit bush operation catering to fishermen and hunters operated two Beavers and two Otters from Kuujjuaq, and there was a contract with Hydro Québec to operate its Dorval-based Convairs.

Air Baffin was a busy operator in 1993. Formed in 1988 by Jeff Mahoney, it filled a niche parallelling Bradley routes. The company built a hangar at Iqaluit and started up in May 1989 with a Ce.421 and a Ce.206. Mahoney got into flying with the Air Cadets at St-Jean, Quebec in 1976. His concept for Air Baffin was bold —First Air was long-established and offered excellent service with large, fast aircraft. Air Baffin had only two Navajos, a Ce.337 and a Ce.206 on amphibious floats (the only floatplane based on Baffin Island in 1993). Its approach was based on frequency, making it more convenient for travellers to get around, even if the planes were smaller or slower. Four pilots were on staff in 1993.

Air Baffin's skeds covered Cape Dorset (population 1,000), Pang (1,000) and Broughton Island (350). In July 1993 Air Baffin had five weekly skeds to Cape Dorset and nine to Pang, one being on Canada Day with pilot Kevin Lamport. F102 (Chieftain C-GYRS) taxied from Iqaluit for R18 at 0930. Lamport had been slowly building experience and hoping for advancement. He had flown with Gods Narrows Air, Ignace Air and Bearskin, where he had spent three years on the Beech 99. YRS cruised north at 180 KIAS at 7,500 feet on the 160-nm route. The weather was ideal, so there was no trouble reaching Pangnirtung, which lies seven miles up a long fjord. After a quick turnaround, the Chieftain headed home with two passengers. It chased a First Air 748 back, and wasn't too far behind

Air Baffin's Jeff and Judy Mahoney with their children Jay and Owen at Iqaluit on July 2, 1993. In a December 11, 1995 plebicite 9,800 voters in the Eastern Arctic chose Iqaluit (over Rankin) as capital of the new Canadian territory of Nunavut. This prompted Jeff Mahoney to change his company name to Air Nunavut. (Larry Milberry)

(Right) Passengers aboard Kevin Lamport's Air Baffin flight head to Iqaluit for the 1993 Canada Day parade. (Larry Milberry)

when it reached Iqaluit at 1150. YRS soon was gassed and ready for its next trip—a sked to Cape Dorset, 212 nm southwest. It carried one passenger down (Jack Anawak, the MP for Nunatsiaq), and seven back.

Willy the Bandit

Since the 1980s Adlair was run by one of the Arctic's modern-day aviation celebrities, Willy Laserich. As a lad he emigrated from Germany, landing in Halifax in 1952 aboard SS *Columbia*. Within a day he had a job in a foundry in nearby Kentville, Nova Scotia. Having journeyman's diesel engine papers, he moved to a better job in a Halifax shipyard. Next came a few months in the engine room of MV *Theron*, part of an Antarctic expedition; then Laserich

Close-up of the Baffin Air crest. (Larry Milberry)

heard about tugboat prospects at Hay River. Soon far away in the NWT, Willy Laserich was thinking of aviation—the airplane seemed to offer possibilities. In 1956 he earned a private pilot's licence at the Edmonton Flying Club and got on with PWA. There were many adventures there, including a trip from Coppermine to Yellowknife, when Iris Adjukak gave birth on Willy's plane.

In time Laserich formed a charter company, Altair. He became known for his medevac efforts and in 1978 Ottawa accorded him special recognition. Otherwise, dozens of Arctic citi-

Willy Laserich started in Hay River with Stinson 108 CF-KEQ in 1959. The following year he bought Norseman CF-BTC from Northland Fish of Winnipeg. Those were tough times, but he stuck it out. Here his Lear scrambles at Cambridge Bay on a medevac to Pelly Bay on August 19, 1995. The pilots (left) were Mitch Dumont and Scott Robertson. Willy's Twin Otter was at Cambridge Bay the same day. Piloted by Ian Blewett and Dave Turner, it was hauling diamond samples from a camp 180 miles out. Bagged samples are shown waiting to be shipped south. (Larry Milberry)

Cambridge Bay's best known citizens in a photo from August 1995: NWT Commissioner Helen Magsagak and aviator Willy Laserich. Helen had just gotten in from the caribou hunt with a party of family and friends. (Larry Milberry)

Door markings on an Adlair truck. (Larry Milberry)

zens wrote letters of support to newspapers and politicians. In 1977, when he needed an extra plane, Laserich leased an Otter from Air-Dale. When he returned it, owner Bob Dale wrote: "Thank you for the care you took of our Otter BEP, while on lease to yourself. It certainly is a pleasure to get an aircraft back from a lease in the shape BEP was returned... we would be happy to lease to you any time." In 1971 Laserich had joined the Pelly Bay Cooperative. It had two DC-4s hauling fish, furs, etc. to Winnipeg, returning with groceries and other essentials. In 1974 he bought DC-4 CF-IQM from Nordair to haul fuel and supplies. Soon, however, operators like NWT Airways were raising the roof, irate that Laserich was freighting without the usual licences; but his argument was that he was entitled to do business. As to hauling fuel, he argued "purchase of fuel in place". This meant that, as a private citizen, he could purchase fuel at point "A", take it to point "B" in his own means of conveyance, then sell it to any willing buyer. On one job Laserich bought fuel in Resolute Bay for $.64 a gallon and sold it in Pelly Bay for $1.78, adding the cost of transportation. His customers were behind him on such dealings, but other air carriers fumed. In 1977 Laserich, operating as Altair Leasing, applied for a licence for a base at Cambridge Bay, where the only service was from Northward Airlines. In January 1980 the licence was awarded. But, legal

battles drained Altair's coffers and it entered receivership. The sheriff seized the DC-4; in December 1980 it was auctioned for $76,000. Within a few weeks Laserich was facing 226 charges laid by the DOT. Proceedings lasted years. In January 1982, for instance, he was fined $250 on 45 charges, 44 of which were absolutely discharged.

The Air Transport Committee in Ottawa kept Laserich grounded into 1983. During this time a DC-6 (C-GPEG) he had acquired in 1978 to fly for Dome Petroleum on Beaufort contracts, sat at Yellowknife until sold in 1987 to Northern Air Cargo of Alaska. Laserich, through the efforts of Manitoba MPs Lloyd Axworthy and Don Mazankowski, won his long-sought licence and formed Adlair Aviation (1983) Ltd, which flourished. While Laserich remained Adlair's chief pilot in 1996, son Paul was manager; son René was a company pilot. There were about 15 employees, including several who had started as greenhorns—members of the famous band known as "Willy's Bandits", who called remote Cambridge Bay home.

At this time Adlair had the only jet in the high Arctic, a Lear 25B air ambulance acquired in 1992 for the Kitikmeot region. To support this operation, Adlair hired full-time flight nurses. In 1992 the CBC aired a supposed "exposé" of the Lear, claiming that it was a costly boondoggle. As the CBC (itself often a squanderer of tax payers' dollars) tends to do, it fell on its face with the program. This reminded viewers of other CBC efforts to "expose" Canadian enterprise, e.g. the Canadair Challenger. The Challenger survived the CBC hatchet job and went on to win acclaim as one of aviation's great products.

Sights around Cambridge Bay over the summer of 1995. The wreck of Arctic explorer Roald Amundsen's ship *Bay Maud*, where it sank in HBC service in 1930 (word that the 1929 McAlpine Expedition was safe was radioed to the outside world from the *Bay Maud*); then, the local golf course. The airlines offer summer Arctic visitor packages. There are boat tours, museum and gallery visits, and Arctic cuisine at places like Cambridge Bay and Gjoa Haven. Larger centres like Iqaluit and Yellowknife offer variety; smaller places focus on local history, crafts or outdoor activities. In 1992 Cambridge Bay, Spence Bay and Gjoa Haven celebrated the 50th anniversary of the first west-east sailing of the Northwest Passage (by the RCMP vessel *St. Roch*). (Larry Milberry, Mitch Dumont)

Supply barges after their 1995 journey from Hay River. They carry everything from personal supplies to construction vehicles and prefab buildings. The Coppermine barge train, seen from a First Air 748, nears its destination on August 19. Then, Cambridge Bay's offloads a day later. The 1995 barges were slower due to shallow waters on the Mackenzie. In years when the ice doesn't go out, there are no barges; expensive airlifts are needed. (Larry Milberry)

Plummer's Lodges

About 1905 Chalmer C. "Chummy" Plummer and his wife Elzina emigrated to Canada from the US. They opened a hotel in Elfros, Saskatchewan, then moved east to Swan River, Manitoba to run a hotel. In 1929 they made a final move—to Flin Flon to run the Royal Hotel in that mining boom town. Eventually, the Plummers operated a muskrat ranch on 27,000 acres in the area and had interests in such local businesses as bowling alleys, movie houses and pharmacies. In September 1944 Chummy and his son Warren bought their first airplane, Stinson SR-8CM Reliant CF-AZV, from CPA at Flin Flon for $5,000. Warren, born March 30, 1917, got his commercial ticket on the Reliant with instruction from Jimmy Symes and Konnie Johannesson. AZV was sold to Eco Exploration of Winnipeg in December the following year. It flew into the 1960s, then ended in the Western Canada Aviation Museum, displayed in Canadian Airways colours.

In 1946 Warren Plummer bought Seabee CF-DYI. Soon after the Ce.195 appeared in 1948, he bought one (CF-FLN) for float operations. The planes were used mainly for pleasure flying—the Plummers were avid sport fishermen and hunters. In 1948 alone Chummy

and Warren made three trips into the NWT and Yukon to fish. The Seabee later was sold to the Thompson family of Cranberry Portage for shares in their Cuprus Mine. Warren now got into commercial flying. For his first season he flew the Ce.195 on lease to Barney Lamm of Kenora. The following season he bought a defunct air operation at Nestor Falls on the Lake-of-the-Woods. This became Sioux Narrows Airways. About this time Warren became part-owner with Vern Jones of Rainy Lake Airways. By this time the Plummers were interested in establishing a sport fishing lodge on remote Great Slave Lake in the NWT. For the time this was a daring concept. Nonetheless, they explored by air for a good spot, settling on one at the east end of the lake at Talthelia Narrows. This was some 1,200 miles northwest of Sioux Narrows. Work began with two Swedes building the first cabin. The family's plans, however, suffered a grave setback. Chummy and his wife were lost, when their Northwest Airlines Martin 202 lost a wing and crashed across the Mississippi River from Winona, Minnesota on August 29, 1948.

Warren pursued the Great Slave Lake project. In 1956 he flew in his first small parties of fishermen. When he returned the first time there were two fishermen from St. Louis

on the dock. When they saw the size of the trout being offloaded, each put down $500 for the first trip the following year. These early trips were gruelling. Warren, or his chief pilot Raymond "Pooch" Liesenfeld, would leave Sioux Narrows in a Beaver with two or three fishermen. If the weather was clear, they could make Great Slave by day's end, having stopped at Flin Flon and U-City. The trip was scheduled for seven days—two for travel, five for fishing. For year one there were eight trips, four each for Warren and Pooch (who passed away in 1996). Plummer's Lodge grew. More fishermen came, so more and bigger planes were needed. Norsemen BFT and BFU were purchased in 1958 from Hudson Bay Air Transport. Within two years these were sold to Parsons Airways of Flin Flon. Norseman GTP and Husky SAQ also served for a time.

The Grumman Goose seemed a good bet—it could carry six-eight passengers with gear. Warren bought one from Carl Millard of Toronto. Carl ferried it to Sioux Narrows. Warren asked Carl to fly it over to his strip, where he had a hangar to work on the plane. They drove over so that Carl could inspect the runway. It was short—maybe 1,800 feet, but it looked feasible. Carl took off from the lake and landed OK. Warren then spent the winter working on

The welcome sign at Great Bear. This camp accommodates about 50 guests. (Larry Milberry)

To serve their camps in the 1990s, the Plummers used the DC-3, PZL Otter, Beaver and Ce.206. Here their DC-3 C-FQHY (ex-RCAF 977, ex-CanForces 12958) lands at Great Bear on August 7, 1996 with passengers from Trophy camp connecting to Yellowknife on the 737. Each season QHY logged 240-280 hours on camp duties. (Larry Milberry)

QHY's 1996 crew: Capt Ed McIvor, FO Dan Crossley and crewman Doug Mair. Tom Phinney (right), an old hand from Air Manitoba, was engineer. Crossley typified up-and-coming young pilots. He had started six years earlier on QHY, serving three years as crewman. In 1990 he earned his private pilot's licence, later added his multi-engine commercial IFR ticket and moved to QHY's right seat. Each fall he returned to school in Edmonton. By the end of the 1996 season he had 1,000 DC-3 pilot hours and had passed his captain's check ride. (Larry Milberry)

(Right) Doug Mair serves his passengers on a run to Coppermine and Tree River camp on August 6, 1996. Bush planes like the DC-3, Beech 18 and Otter have one thing in common—a smell that greets all who board. It's a mixture of gas, oil and hydraulic fluid; old and rotting leather; spillage of every kind—pop, beer, jam and ice cream. And what about the planes that haul fish and quartered moose, to say nothing of passengers who get air sick? All these delights ooze into the deepest recesses under the floor boards and are sponged up by upholstery and carpeting. They fester for years, producing a unique combination that clobbers the olfactories. This may help get the die-hard crewman's sperm count going; but others may have to suppress a gag! (Larry Milberry)

(Above) Ed McIvor started flying in 1953 at Charlie Graffo's school in Winnipeg. His first job was with Taylor Airways, owned by Bob Eastman of Wabowden, Manitoba. Later, Ed was part owner of Cross Lake Air Service, which had Waco CF-BBQ. He also flew for Chiupka Air Service, La Ronge Aviation, Sherritt Gordon Mines and the MGAS. He had logged about 25,000 hours by 1997, including 8,000 on the DC-3. His favourite plane was a Twin Otter on floats. (Larry Milberry)

(Right) One of the parties that fished at Plummer's in 1996 as it was about to leave for Yellowknife: Ron Barbaro, Otto Jelinek, Bob Lloyd, Jack Finlater, Don Matthews, Ken Taylor, Peter Pocklington, Ralph Lean, Mike Harris, Gordon Cameron, Larry McNamara, Paul Wacko, Bill Wilson, Peter Cole, Glen Sather, Craig Allen and Harry Sinden. (Larry Milberry)

(Left) The view enjoyed by Plummer's guests around Great Bear Lake. Then, who knows what lies below. Near the mouth of the Sloan River, not far from Arctic Circle lodge, were the remains of an ancient bushplane. Ross Lennox inspects a float in August 1996. (Larry Milberry)

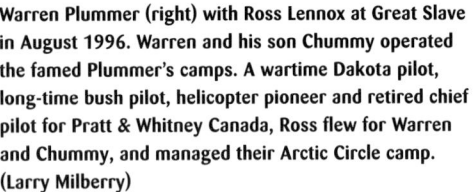

Warren Plummer (right) with Ross Lennox at Great Slave in August 1996. Warren and his son Chummy operated the famed Plummer's camps. A wartime Dakota pilot, long-time bush pilot, helicopter pioneer and retired chief pilot for Pratt & Whitney Canada, Ross flew for Warren and Chummy, and managed their Arctic Circle camp. (Larry Milberry)

(Right) Carl and Helen Klaenhammer of Wyoming fished from Plummer's Arctic Circle camp over the summer of 1996. They used their own Turbo Beaver (C-FOEE, ex-MGAS) for local transportation. (Larry Milberry)

his Goose. In the spring it rolled out, but things soon changed. For his first flight Warren was to go over to the lake to fuel and head north, but on takeoff an engine caught fire. Warren braked and the Goose went on its back. He got out OK, but the Goose burned. There was $40,000 worth of insurance, but somehow this was disallowed. All the effort to put the Goose on the line was wasted. Other twins used by the Plummers included Barkley-Grow BQM and a series of Beech 18s (HZA, MVS, PCL, etc.). Warren obtained the rights from a company in Florida for the 7850 floats used on the Beech. In an arrangement with Bristol of Winnipeg he sold many sets around the country.

A number of lodges were established on Great Bear Lake, the first at Conjuror Bay on the southeast shore. In 1959 this camp was moved by tractor train over the ice 150 miles northeast to Dease Arm. The new location was a few miles north of the Arctic Circle (some 400 nm north of Great Slave Lake Lodge). Here a gravel strip was put in for the Bristol Freighter and DC-3. In the early 1960s Plummer started bringing in fishermen on Transair DC-4s, then DC-6s. The strip was improved year by year, eventually to 6,000 feet. Meanwhile, Arctic Circle Lodge was opened about 60 nm south in the Hornby Bay region. Other camps on Great Bear Lake included Nealan Bay and Trophy.

The concept of fishing for grayling, and for lake trout of 40-70 pounds inevitably took another step. A camp was established near the mouth of the Tree River on the Arctic coast 230 nm northeast of Great Bear Lake Lodge.

Here 16-20 avid fishermen put up in tents to cast for Arctic char. About the time Tree River opened, Warren brought his son in as a business partner. Soon chartered DC-4s, DC-6s and Electras were bringing guests to Great Slave, using the old RCAF strip at Sawmill Bay. By the 1990s Plummer's Lodges were handling about 1,000 guests yearly. The season would begin in May, when staff arrived to open the camps (the ice was still in). Thereafter a new group arrived each Saturday on a chartered Canadian North Boeing 737, while the previous group departed. The last guests left early in September. A typical trip departed Winnipeg on August 3, 1996 with Capt Bill Brady of Canadian North in charge. En route a champagne breakfast with steak and eggs greeted the jovial crowd of 85 passengers. First stop was Great Slave Lake Lodge, two hours away. There a number of guests deplaned, supplies and equipment were offloaded, and the 737 took off for Yellowknife, a 20 minute hop (90 nm). Besides passengers going on to Great Bear, it carried those finished at Great Slave to

Yellowknife to await the 737's return from Great Bear (a couple of hours later), then board for Winnipeg. At Yellowknife some from Edmonton boarded for Great Bear. Finally, the charter headed for Great Bear Lake Lodge with 56 passengers. Once there, some guests stayed, others took the DC-3 to Trophy camp for their week of fishing. A week later Capt George Chivers of Canadian North arrived to bring in new guests.

Over the 1996 season Plummer's Lodges had a versatile air operation. First were the 737 guest rotation charters. Within the NWT Plummer's long-serving DC-3 C-FQHY connected Great Slave, Great Bear, Trophy and Tree River camps, and tied in to Yellowknife, to and from which it carried passengers, but also staples like propane, fuel in 45-gallon drums and groceries. Next were two PZL-powered Otters that took guests eight at a time on special fly-outs—to choice fishing spots. A Beaver and Ce.206 catered to small groups. At season's end the fleet went to Selkirk for winter maintenance and storage.

Aviation and Mining

The stories of aviation and mining from the 1920s are legendary. In those days shaky HS-2L and Junkers bushplanes, crewed by fearless pioneers, blazed the trails. Technology advanced—planes like the Norseman, DC-3, Twin Otter and Caravan led the way to new developments. Mineral prospects still entice—there are some 1.2 million square miles of Canadian Shield to explore for base metals, oil and gas. Only the surface has been scratched. There are several large mines in the NWT relying on aviation. These include Echo Bay's Lupin gold property at Contwoyto Lake, the Polaris lead-zinc mine on Little Cornwallis Island, and the Nanasivik lead-zinc-silver mine on Baffin Island. In 1991 the latter two shipped (by sea) 285,300 and 716,400 tons of concentrate. Properties in the Kitikmeot region south of Coronation Gulf were promising, but depended on establishing a port. The NWT and bordering provinces were thinking of new roads e.g. up the Hudson Bay coast as far as Chesterfield Inlet.

Another place where aviation and mining interact is in the Hope Lake district about 40 miles south of Coppermine. In the mid-1960s there was much exploration there. In 1968 about 40 companies were active and a 3,400-foot strip was in operation. In 1967 aircraft delivered 300-400 tons of supplies and fuel, using the ice. In addition 50 Bristol and 20 DC-3 loads were delivered around the area. In 1968 60 Hercules and 30 DC-4 flights delivered 1,650 tons, while another 150 came by Bristol and 100 by DC-3. With such activity the ATB granted PWA and NWT Airways rights into Hope Lake. In September 1968 the ATB considered NWT Airways' application to provide Class 3 irregular specific point service to a number of exploration sites. The points applied for would be served as two routes, one north-south serving Port Radium or Sawmill Bay (depending on the season), Hope Lake, Coppermine, Holman Island, Winter Harbour and Resolute Bay. Winter Harbour was on Melville Island, which the applicant proposed having as a staging point for a Panarctic project on Melville Island. The second route was east-west from Yellowknife, serving Hope Lake, Coppermine, Lady Franklin, Cambridge Bay, Spence Bay, Pelly Bay, Igloolik and Hall Beach.

The applicant suggested that the proposed Yellowknife service to Hope Lake, Sawmill Bay and Port Radium could be a frequent service in summer, perhaps two or three trips weekly. Coppermine could be included, when the airport was completed. Lateral service would operate twice a month. That to Holman Island would depend on the resource development taking place on Melville Island. The ATB decision noted that Port Radium was the site of a high grade silver mine being worked by Echo Bay Mines, served by an ice strip in winter and Sawmill Bay airstrip in summer. Echo Bay DC-4s carried silver concentrate, and about 1,000 passengers a year. Other exploration was under way in the area. In summer sport fishing was important; PWA and Transair served Sawmill Bay, mainly in the fishing trade. As for

Since the early 1950s Canadian geophysical companies have searched for minerals using aircraft with electronic equipment. This Piper PA-31 was on duty from Coppermine over the summer of 1996, seeking geologic formations that might hold diamonds. (Larry Milberry)

Holman Island the committee heard that NWT Airways carried 221 passengers and 11,000 pounds of cargo there in the most recent 18 months.

In defending its proposed east-west route, NWT Airways described it as a "community development" route, especially with the strong connection between Yellowknife and Coppermine. Its desire to serve Lady Franklin (a DEW Line site) was mentioned in the decision as: "Lady Franklin is located 70 miles from Coppermine... It is equipped with an airport, hangar and is capable of all-weather operations. Canadian National Telecommunications has a module for vertical communications and DEW Line communications. There is a small Eskimo settlement consisting of four or five families. The chief traffic to and from the point consists of communications technicians, who are rotated weekly, and equipment and supplies for the personnel and facilities located there." The ATB approved Holman Island on NWT Airways' unit toll Yellowknife-Melville Island run, but denied Resolute Bay. It denied the east-west plan, due to lack of business: "The application to serve the points Spence Bay, Pelly Bay, Igloolik and Hall Beach is therefore denied." Lady Franklin-Cambridge Bay was OK'd, but "The Licensee will be restricted ... from operating direct flights between Yellowknife and Cambridge Bay." Port Radium/

Sawmill Bay was approved, also Coppermine, Holman Island, Melville I., Lady Franklin and Cambridge Bay. No direct Yellowknife-Coppermine flights were approved in the application.

Echo Bay

Echo Bay Mines Ltd. formed in 1964 under Norman Edgar. Its first big project was to revitalize Eldorado Mines' old silver operation on Great Bear Lake, mothballed since 1960. It soon was processing 70-80 tons of ore daily. In 1977 it began developing a gold property known as Lupin, 250 miles north of Yellowknife. The viability of Lupin was proved in 1979-80 and an airstrip and mine were put in. By this time Bill Granley was managing Echo Bay's aviation department. A pilot and engineer, he had been in aviation since the 1950s. At first the company used its DC-3 and the good support from NWT Airways (DC-3, DC-6). Echo Bay soon upgraded to a Convair 640 and added a Bell 47 and Bell 206 for exploration. As Lupin grew, it needed heavier airlift. In 1980 it leased, then purchased, Herc CF-DSX from SEBJ. In its first year on the job DSX carried 47 million pounds of cargo, mainly from Yellowknife. It logged 1,990 hours and 2,083 landings. It also worked at Port Radium, where the old Echo Bay mine again was being decommissioned.

Echo Bay's DC-3 (ex-CanForces 12915) at Edmonton on July 31, 1975. It joined Echo Bay in June that year, thence moved to NWT Airways. Convair 640 CF-PWO (1976), the 727 C-FPXD (1984) followed the DC-3. (Les Corness)

By this time in the markings of Corp Air, Echo Bay's 727 C-FPXD loads at Yellowknife for Lupin on August 16, 1995. PXD started in 1968 with Trans International (N1727T), then went to PWA, and Echo Bay. It seated as many as 115, or worked as a 17.5-ton freighter. By July 1993 it had amassed 40,000 hours and 28,000 cycles. (Larry Milberry)

In 1984, by when it had flown about 7,000 hours for Echo Bay, DSX and the Convair were replaced by an ex-PWA 727. The 54-passenger combi flew a thrice weekly sked from Edmonton to Yellowknife and Lupin, where there were 250-300 at work. The same year a King Air 200 was added. Based in Reno, it served Echo Bay's Nevada and Colorado gold properties. In 1986 a Lear 35 was added for Edmonton, and a Bell 206L replaced the 206B. A Citation replaced the King Air in 1987 and a Twin Otter was bought the following year for the NWT. In this era Echo Bay flight ops came under the name Corp Air. In September 1991, under an agreement with Gulf Canada, PXD began bimonthly charters to Usinsk, Russia supporting joint venture petroleum developments. It routed via Iqaluit, Keflavik and Murmansk, turned around in two hours and backtracked to Keflavik, where a slip crew took over. Edmonton-Usinsk-Edmonton lapse time was 24.5 hours. In 1993 annual utilization on the 727 was 1,300 hours.

In 1990 Bill Granley was awarded the McKee Trophy. Under his direction, since 1975 Echo Bay's aviation department logged 38,000 hours, carried 160,000 passengers and hauled 82,000 tons of cargo. The company grew from a worth of a few million to $3 billion. Bill Granley fell ill in this period, passing away in February 1992. This was a blow to the Granleys, whose aviation members included brother Bud, an ex-RCAF fighter pilot, United Airlines captain, and airshow performer; Bud's flying children, two of whom (Chris and Ross) became CF-18 and Snowbird pilots; and Bill's sons, Brian (Echo Bay 727 pilot) and Darcy (Canadian Forces test pilot).

Diamonds on the Tundra

In the mid-1970s a few prospectors were turning over stones on the tundra north of Yellowknife looking for something few would have suspected—diamonds. There must have been something to this activity, for giant De Beers was on the scene. Eventually, independent prospector Chuck Fipke came up with some promising samples, after toughing it out around Lac le Gras since 1975, grub-staked by the small Kelowna company Dia Met Minerals. After years of piecing together the big picture, Fipke began staking. By April 1990 he was sure he was sitting on diamonds. He presented his findings to Broken Hill Proprietary Co. of Australia, one of the world's largest mining concerns. In August 1990 BHP agreed to finance efforts for a 51% interest. The 1991 drill season netted 81 diamonds; soon a staking rush was on by outfits big and small. Kennecott Canada moved in, but its samples didn't prove out. De Beers, too, seemed unlucky. This dampened enthusiasm and stocks plummeted. By 1995 the NWT rush faded. Of some 200 companies (mostly huckster outfits flogging penny stocks on the Vancouver Stock Exchange), only a few players remained. Meanwhile, BHP claimed to have five excellent kimberlite pipes—natural diamond depositories—all within 50 miles of each other. The pipes were beneath lakes, which would have to be drained for mining. Now came the next stage—clearing environmental and native rights hurdles. BHP spent $100 million getting ready and had prospects of spending another $500 million to bring a mine on stream. Its hope was to profit $60 in gem- and industrial-quality diamonds per ton, and to be processing 18,000 tons daily by year four of start-up. Early in 1997 native, government and mining interests came to an understanding, and mine development at Lac le Gras got underway. This was great news for those in air transport.

A Young Pilot's Career—It Can Be Tough

Young pilots usually face an uphill battle to get established. The economy controls the availability of jobs. If fishing, forestry or mining are down, jobs are hard to get. A general recession reduces the need for pilots in activities like sport fishing and hunting. If business overall is on the skids, the airlines have fewer passengers, tighten their belts and "de-hire" pilots. This drives young pilots away from commercial aviation to seek other work. Determined ones, however, hang in. In the 1950s-1980s many young commercial pilots, laid off or simply unable to find a first job, at least had the option of a military career. With the recessions of the 1970s-80s, many called on their local CanForces recruiting office and found a flying job in uniform. By the late 1990s, however, this option was gone—the air force had shrunk to a handful of squadrons.

For those who survive aviation's ups and downs, it's rarely smooth sailing. The pay traditionally is low; there are few "benefits". If there is little flying, operators usually offer an hourly rate. If there is lots of flying, they pay by the mile. Pilots and employers often don't see eye to eye, and some operators can be hard on new pilots. Himself a graduate of the school of hard knocks, the boss may feel his pilots should pay their own dues. A few operators still push pilots to operate unsafe planes, or fly in rotten weather. One pilot from a small airline told of a saying his boss had about airworthiness: "If you can start it, you can taxi it; if you can taxi it, you can fly it." When a new pilot departed on a passenger sked and had an engine failure, he turned back and landed. He reported by phone to home base, and was fired on the spot. As far as his boss was concerned, the pilot should have proceeded to base on one engine, so the plane could be repaired less expensively. A safe operator (as most are) would have congratulated his pilot for quick action in landing at the nearest airport. Pushing a pilot to fly in duff weather is another curse to commercial aviation. Such cases, with conclusions like "controlled flight into terrain", are common in the tons of accident reports in the National Archives.

There are always operators (or their tough base managers and chief pilots), who turn a blind eye when it comes to a plane being despatched grossly overloaded. One pilot, when asked if his company's crashed plane was overloaded, summed up things: "Probably definitely." Training and experience are other problem areas. Some operators will hire pilot after pilot, firing them before they get enough seniority to qualify for severance pay. One young pilot reported: "My boss wouldn't provide any training. He knew he was going to fire me before long anyway." He was hired by a small carrier, checked out quickly on the Navajo, and put on the sked as captain, even though he never had flown a Navajo. Eventually this company killed a load of passengers and was shut down. One wonders why district government inspectors don't shut down such companies before deaths occur.

Gus Kraus was an American who came to the NWT from Chicago in the 1950s to get away from it all. He married an Indian girl named Mary and settled at "the gates of the Nahanni". He trapped and prospected, but it's said that the real work was done by Mary. When the Nahanni became a national park, Gus and Mary moved to a cabin just outside the park. One of Gus's hobbies was keeping daily tabs on air traffic in the area. Shortly before he died in 1993, he passed his journal on to Joe McBryan of Buffalo Airways. Some of Gus's entries are shown in these sample pages.

Yellowknife airport looking from the terminal down Bravo taxiway on May 7, 1996. Some of the buildings beyond are NWT Air, Buffalo Airways, DND/RCMP, and Echo Bay/Shell FBO. Each airport has lore. In one winter incident from about 1960 a Luscombe owned by the duo of Avery and Markle took off from Yellowknife for the three-minute hop to the Old Town ski base. After 35 years no trace of the plane had turned up. The second photo shows the ticket and check-in counters at Yellowknife in August 1995. (Larry Milberry)

(Left) An aerial perspective from August 17, 1995 showing Yellowknife's Old Town and the main seaplane docks. Compare this photo with the ones from 35-40 years earlier that appear on page 794. (Larry Milberry)

The famous Wild Cat café (above) began as a beanery serving prospectors, miners, aviators, etc. in Yellowknife's original gold fever days. In modern times it catered to the tourists, with local fare like caribou burgers. More contemporary are places downtown like the Gold Range, a rough and tumble strip joint where, as one of the local bush pilots put it, "We're all ruddy heroes!" (Below) A summer hangout in the 1990s was Yellowknife's Bush Pilots brew pub, formerly used by operators like PWA and La Ronge Aviation. (Larry Milberry)

The shadow of a Buffalo Airways DC-4 brushes by the old Thompson-Lundren mine 27 nm north of Yellowknife. Somewhere in the foliage were the remains of an old Anson. Then, a tundra scene from June 27, 1993. An esker snakes across the landscape at 65° 58" 73' N lat., 109° 28" 72' W long. Glacial deposits like eskers were priceless sources of gravel with postwar defence and mining projects. (Larry Milberry)

(Right) A buzz around Yellowknife on August 17, 1995 with Bryan Currie in Great Slave Helicopters' Jet Ranger C-FGSD netted aerial views of Max Ward's famous Bristol CF-TFX permanently "airborne" on the road to the airport. (Larry Milberry)

(Below) Max Ward's Twin Otter C-FWAH at Yellowknife on July 19, 1992. WAH was bought by Wardair in 1969. In 1980 it went to Chevron Overseas Petroleum for West African operations. Max bought it in 1985 to serve "Rock Haven", his lodge 180 nm north of Yellowknife on Redrock Lake. In the second view Max manoeuvres a load for WAH. (Henry Tenby)

(Above) An RCMP Twin Otter departs Yellowknife on July 18, 1992. By 1997 this aircraft had been on police duty for 26 years. (Larry Milberry)

(Above) Merle J. Carter got into aviation in 1953 working as a helper on Ansons for old-time pilot Stan McMillan. The Ansons hauled for Carter's father, George, who ran a fishery. In 1962 Carter formed his own air service with Ce.180 KOW and Norsemen FUU. He added Lockeed 10 HTV, Beech 18 PJD and Otter IOF, but HTV burned in an accident at Birch Lake, NWT. Here Carter's Twin Otter KAZ waits at Hay River on July 13, 1992. (Larry Milberry)

(Right) The Short Skyvan was well-suited to the tundra. C-GKOA began in 1974 with the Mexican Air Force. In 1990-91 it was N53NS with North Star Air Cargo, then in April 1993 joined Imperial Airways of Yellowknife for diamond exploration work. Imperial soon folded and KOA moved to Air Tindi. Pilots Vicki Veldhoen and Barb Hilliard crewed it through 1995. (Larry Milberry)

(Left) In February 1992 Aklak Air became the first Canadian operator of the Do.228. The 19-seater first flew in 1981. Powered by Garrett TPE-331s, it was faster than others in its class, but found few sales in Canada. The 1996 acquisition of Dornier by San Antonio-based Fairchild brightened the North American future of the Do.228 and larger Do.328. This Aklak Do.228 was at Inuvik on July 14, 1992. (Larry Milberry)

Sturdy types like the Apache, Aztec and Islander served the Arctic well for many years. These Ursus Aviation Geronimos (hopped-up Apaches) were dormant at Fort Norman over the summer of 1995. (Right) A Lambair Islander visits Rankin Inlet on July 19, 1973. (Larry Milberry)

(Left) Helicopters are essential throughout the tundra and in many cases have replaced smaller fixed-wing planes. Here a Great Slave Bell 206B Jet Ranger waits at Coppermine as Pierrette Paroz hovers her Canadian Helicopters Bell 206L at Coppermine before setting off for the Lupin Mine far to the south. (Larry Milberry)

(Right) Bizjets frequent places like Iqaluit and Yellowknife for fuel or on business and sporting trips. Here a Falcon 50 owned by Rich Products of Buffalo, NY arrives in Yellowknife on July 3, 1993. (Larry Milberry)

(Below) An August 1992 view of the historic main hangar at Iqaluit. It was built by the USAF in DEW Line days. (Larry Milberry)

Ice reconnaissance Challenger 600S at Iqaluit on July 1, 1993. In early years types used for this important work ranged from the Fokker Universal to the Beech 18, DC-4 and Electra. (Larry Milberry)

Helicopters at Iqaluit. Aero Arctic SA341G Gazelle XTW (above), seen on July 1, 1993, entered Canada in 1977 with Apex Helicopters of North Battleford. The CanForces Labrador (right) of 103 RU, seen on August 14, 1992, was on detachment there, while working at Resolution Island and Lake Harbour. (Larry Milberry)

The annual sea lift under way at low tide at Iqaluit in August 1992. (Larry Milberry)

Froblisher Bay relics. The wreck of an old Bradley truck, then the DEW Line radar installation, abandoned in the late 1980s. A huge project later focused on cleaning up the Arctic and this old site was demolished. (Larry Milberry)

Inuit stone figures, or inukshuk, have been known in the Arctic for centuries. This one was seen at Kuujjuaq in 1992. (Larry Milberry)

More of the North

Red Lake, Ontario in a scene from July 16, 1995. The town was celebrating Norseman Days, so a dozen of the famous old bushplanes were on hand. Red Lake had first attracted commercial aviation with the gold rush of 1925-26 and, like many other Canadian communities, the mining-aviation pact remained strong ever since. For this year's celebrations, attending Norsemen were BTC, BTH, DRD, DTL, FQI, FUU, IGX, JEC, JIN, KAO, LZO, OBE, OBR, SAN and N45TG (ex-BSH). (Larry Milberry)

Some Northern Operators

Many operators did business in northern Ontario and Quebec in the 1930s—Eclipse Airways from Chapleau, Austin Airways—Sudbury, Starratt Airways—Hudson, General Airways—Noranda-Rouyn, etc. As the HS-2Ls aged, new Fairchilds and Norsemen replaced them. Then came the first modern twins—Barkley-Grow, Beech 18, Lockheed 10. In the 1950s the Beaver, Husky, Otter and Cessna 180; then the Anson, DC-3 and Canso brought further growth. Carriers from Red Lake to Chibougamau provided the traditional services—supplying Indian reserves, carrying trappers and sportsmen, doing mining and forestry work. A free enterprise was usually in charge. Growth and profits, though vital, were not necessarily his only goals—he might place service ahead of profits. Instead of rushing to buy new equipment, his pride might be keeping an old Norseman or Beech in pristine condition year after year. If he was Arthur Fecteau, however, he might insist on a Beaver or Otter fresh from the production line. The terms "local" and "third level" were applied to such carriers in the 1950s-60s. These had been around since TCA and CPA days, but seemed to have been overlooked by officialdom. While Ottawa grappled with airline policy, just what a local carrier was, and under which parameters it operated, was unclear, even though most functioned smoothly (compared to the tumultuous regionals). In the 1970s a new label appeared for the bigger local carriers—"commuter airline". In the 1990s Ontario and Quebec were served by carriers that were, it could be argued, similar to the pioneer bush operators, even though elements of big business were in the picture. The object of providing good service to small, widely-separated communities remained. Some companies, like Air Creebec, Bearskin Airlines and norOntair were well-established; others like Thunder Air were newer.

norOntair

In the 1960s the Ontario government decided to provide scheduled air service to remote northern settlements. Years of studies resulted in the formation of norOntair, launched in 1971 on a three-year trial. Three Twin Otters were acquired, the inaugural route being Sault Ste. Marie-Timmins-Earlton. The initial operating contract went to White River Air Services. It brought norOntair through the early years, then Bradley Air Services took over in 1974. Since it owned Voyageur Airways of North Bay, Bradley could offer integrated feeder service between the Soo and Ottawa. Air-Dale also had a turn operating norOntair, which grew steadily, although it was not profitable, considering the thin markets served. In 1984 norOntair became one of the original Dash 8 operators. Its first Dash 8 (C-GJCB) was named for James C. Bell of Austin Airways. Its other (C-GPYD) was named for Paul Yettvart Davoud, who had been a bush pilot, night fighter pilot, industry executive, chairman of the Air Transport Board, and norOntair

planner. The Dash 8s linked places like Timmins, Thunder Bay and Winnipeg with settlements where the Twin Otters flew.

As service by local operators improved, in 1993 norOntair dropped centres like Red Lake, Kenora, Dryden, Atikokan, Pickle Lake and Terrace Bay. CRA (then known as Canadian Partner) later reduced activities in the region, leaving the way clear for Bearskin and Calm Air to expand. Calm Air, for example, took over CRA's skeds from Winnipeg to Dryden and Thunder Bay in September 1995. By now traffic had grown—a far cry from the scanty revenue available in the 1950s-60s. Thunder Bay became a plum for any operator in the 1990s. Toronto-Thunder Bay was one of Air Canada's more profitable routes. Little wonder

The sight of norOntair's brightly-coloured fleet was common across Northern Ontario for 25 years. Here a Dash 8 and Twin Otter are seen at Thunder Bay on March 26, 1992. Then, Twin Otter NHB at Sudbury on April 12, 1991. By 1997 the norOntair fleet of two Dash 8s and four Twin Otters was working in Papua New Guinea with Milne Bay Air. Even at this time PNG's longest road measured only 30 km and no two towns were connected by land. PNG was a region that needed DHC aircraft if ever there was one. (Larry Milberry)

that CRA, having given up much of Northern Ontario, put the F.28 on that run late in 1995. Meanwhile, Ontario was reassessing norOntair. The fact that air service across the North was vastly improved, and that Ontario was facing a budget crunch, led to the shut down of norOntair. Its final flight landed at the Soo at midnight on March 29, 1996.

James Campbell Bell

Each region has a few aviators who leave a special legacy. One of the best known in Northern Ontario and Quebec was James Campbell Bell, many years the chief pilot at Austin Airways. A native of Copper Cliff, near Sudbury, he started flying in the early 1930s and studied at Parks Air College in St. Louis. In 1937 Austin Airways hired him to instruct in Sudbury. In WWII he flew in the BCATP, worked in Northern Quebec for CPA, then returned to Austin for the rest of his career.

James C. Bell shows off a respectable trout after a day's fishing with his friend Frederick Leary of Rochester, NY. They were at Rupert House in 1963. (Leary Col.)

In the 1950s-60s Jim Bell and his friend Rosemary Leary, Rochester, NY's first female police officer, exchanged some letters. Besides telling about aviation and the North, the letters are revealing about the enigmatic Bell. They show a sensitive side to a man usually characterized as pragmatic, if not cold. Some, who didn't know him well, used to characterize him as "old stone face", but Rosemary commented in 1994: "Without sentiment and emotion these letters would not have been written at all, nor so carefully saved, would they? For myself, even after all these years, the revived emotional freight is so heavy it would almost require a DC-3 to transport it! When I die Jim's letters to me, and the photographs, will no doubt be discarded. Then there may be nothing left of his words. Gone the comments about the lovely colour of the autumn leaves, about 'fool' aircraft, about eagerly awaiting the fall hunt,

about spring break-up on James Bay. In that sense, the letters do not belong to me alone. Their place is as a small part of Canadian aviation history, because they were written by the famous Jim Bell." Here are some excerpts from the letters to Rosemary.

October 24, 1956: Think you would enjoy this country. This past few weeks the leaves have disappeared, but were really lovely when they turned colour. We were doing quite a bit of work outside and it was most enjoyable. Got in a few days goose hunting down the mouth of the river, and a day up at Albany. It makes my heart bleed to think of you poor city people who can only go into the country on holidays and we have it all our lives.

November 28, 1956: You are probably the hardest lass in the whole of the USA to get to see. Here I go by in all directions, north, south, east and west and I can't get you to budge out of Rochester. No, sir! Could be that I'm such a suspicious looking character that you might have to bring your six shooter... would have come over to your home town for a short visit, but was so darned busy trying to find another C-47 that I just couldn't arrange the time. Actually, I only got a week's holiday that I spent on the deer hunt. I was travelling in vain all the time after that. Finally we bought one [C-47] through an agent and it is now being flown over from Iran... We are constructing a winter air strip here [Moosonee] and at the moment have a bunch of tractors working outside my window. Will be here for some time. I expect at least for the next two weeks or so till I get this thing in operation.

January 13, 1957: Spent New Year's at Cochrane and have been busy pretty well since then with one thing and another. The airplane we were getting through Hartford is still not ready for delivery and we are sort of in a spot not having it. From the looks of things it will be at least the first of February before it moves. The day after Christmas one of our Cansos [CF-FAR] went missing between Kapuskasing and Timmins. We had a full-swing search on our hands with 10 airplanes for two days until it was found. We were lucky in that the crew got away with superficial injuries, but the aircraft was a write-off and will probably be left where it is in the bush. The strip here is in fine shape now and will hold a DC-4 at least. We are flying our DC-3 and Canso night and day from here north, so things that way are going well.

Austin Airways did many good years of business with Bill Anderson, who ran fishing and hunting camps around Albany on James Bay. (Leary Col.)

February 8, 1957: We are working from the winter airstrip at Moosonee with a fair fleet at the moment. We have three DC-3s, two C-46s, a few Norseman, etc. working 24-hours a day, and will have until late March at least. Am living half the time with the Mitchells and the other half in the cabin by myself (hermit). Sort of a gypsy existence I guess, but have been at it for so long can't seem to do it any other way now. Have had the most beautiful winter weather here through January. Cold, but nice and clear. Has been down to 56 below, but no wind.

May 24, 1957: Since seeing you last in Rochester, have been roaming the country. The night I left I went to Oswego, or somewhere, and stayed in a motel. Left early, but still was late arriving in Montreal and got a bit of hell from the boss. Since then have been up to Moose for a while, then on a wild goose chase looking for a PBY. Covered the country from coast to coast, including Long Beach, California. Got so tired of travel.

July 3, 1957: It seems like it's rained every day for months, and this place is a real quagmire. The darned river is still at its spring flood level and the weather is something to make one wonder why we ever stay in a place like this. Have been working out here with the one Canso. Our two DC-3s are elsewhere—one up at Frobisher and the other at Winisk. Haven't been doing too much flying myself except for a week last month at Fort Chimo on

Canso CF-FAR down for the count between Kapuskasing and Timmins. The crew of Jim Hobbs, Jim Pengelly and Stan Cramer survived the crash of December 27, 1956, but FAR was abandoned in the bush. (CANAV Col.)

Ungava with one of the Threes. Was in the bush north of here for a couple of days staking claims. While out we got into some real trout fishing. Got several six-pound speckles and a couple of real big lakers. The flies in the bush just about send you, though!

August 27, 1957: Should have taken time at least for a short note ages ago, but have had my worries and have been on the road almost continuously since seeing you last. Sorry to hear that you have been under the weather. Have had the damndest summer myself. My tummy has been acting up. On top of incessant travel this has left me at a low ebb most of the summer. Growling at people most of the time and being real nasty. Mostly I think, though, our personal troubles are tiny ones compared to what some others go through. One of our engineers took suddenly ill this summer. It turned out he had a perforated ulcer. Then, last week one of the boys keeled over with a heart attack and was gone. So perhaps we are fortunate. Well, to get to more pleasant matters... the goose and duck season is only three weeks away! Then I'll be able to get the long boots on and away to the marshes! I can hardly wait. After that comes deer hunting. I just tolerate the rest of the year for the hunt.

January 30, 1960: Guess I am a real delinquent and should be censured for not writing. It's been the same old rat race. Was in the hospital in Sudbury for three weeks last April and lost 20 pounds. So was sort of taking it easy all summer. Did some flying, but the Canso was a bit on the heavy side for me. Confined my activities to pinch hitting for some of the boys. Am the original hard-luck kid for 1959 and hope this year is different. In the fall we had a good deer hunt—filled our count and got two moose as well. Then, at the tail end of the season, got a cold or some darned thing, then pneumonia. Got sick in the camp one night, and guess it was rough. I came to again a couple of days later in hospital looking at an oxygen tank. Fun, eh! Anyway, seem to be functioning normally again and working.

We had a fair operation going here again this winter, and it looks as though the area will boom again next summer. Have a DC-3, two

Norsemen, an Otter and a Beaver doing coast work 600 miles up both sides of the Bay. The weather has been for the birds this winter, though the temperatures have been ranging from the high 20s one day to 40 below the next. Haven't had too much snow this winter, though, and the ice on the river is over three feet thick. Should make for a real break-up this spring! Next summer I think we will have several trips away up into north Baffin, and late this spring are going with a DC-3 on a trip to all the Arctic posts of the Hudson's Bay Company. We will be gone at least three weeks on it.

November 21, 1960: Spent a week in the hospital [in Sudbury] for some investigating and the medics want to start doing some radium treatments for a fuzzy throat. Have me concerned that it might be rough. I'm somewhat upset and at a loss at the moment. Where do we go from here?

April 21, 1961: Have been to see more than enough of the medical people. The whole thing to me is sort of frustrating. Took the DC-3 back to Kapuskasing on Sunday. The crew have been off for a week or so, as they have to take it to Baffin on the 23rd of this month. I was flying it a bit while they were away. Still have a Norseman and a Beaver here on skis and will have a week or so, depending on the temperatures. It has been very warm for this time of year and think we will have to get off the river ice any time, if it stays the way it is.

August 29, 1961: I got away on a trip to relieve a crew we had at Frobisher Bay on ice patrol and get out on a tour with Bishop Marsh of the Arctic. Took him on what started out to be a five-day junket and ended up at ten on account of gale winds in the Pangnirtung area holding us up. Am still bugged by the throat and it leaves me like a bear most of the time. I am not really like that, and my poor co-pilot has begun to think that there are better ways of making a living than travelling with me. I am seriously thinking of packing the thing up after this fall's goose hunt. The country to the south of here has been burning up this year and our fire fighting Canso has really paid off. It's been a bind keeping crews on it, as it is hard work. Will be glad when it's over. Right now the New-

foundland government want us to go down and help out, as they are in a state of emergency. But we can't spare the airplane. Well, enough of this, and I hope to see you this fall, if you promise not to use your police special on me!

October 20, 1962: We are delayed here [Moosonee] somewhat longer this fall. It has stayed open surprisingly long, and even yet there is only the odd day of cool, frosty weather. Then it gets warm again. We are naturally getting anxious to get away, as it has been a long season. Five months, and at the end of it, I guess we have gotten a bit bushed! The Cansos have been sent south for the winter now, but we have some extra small aircraft to clean up inland work on the two coasts. These

Austin's water-bombing Canso CF-JTL in action in 1960. It used two external tanks each of 350 Imperial gallons. These were quickly refilled by landing on the water. Jim Bell, Joe Lucas, Jack Austin and Frank Russell collaborated to develop this system, which was used for years by the province of Quebec on its Cansos. (CANAV Col.)

will be here for another couple of weeks anyway, I imagine, unless we get a cold snap and get chased south.

Finished up the goose hunt in fair shape. We had a total of 80 hunters. In another year I expect we will run this figure to 100 and that would be about all the camp could stand without crowding. We closed the camp on the 10th and I went up for a night and cleaned up on what accounting there was with Bob and Bill Anderson. We are still horsing around with our staff house. It isn't finished yet, as we are stuck for power and setting up a diesel unit of our own to take care of the requirements. All the power locally is from the railway. After promising it to us, they backed down. We had to set up a power plant of our own.

We are starting to get some of the furniture in and it will not be too long before we are able to use it for the winter, at least. Haven't hired a cook, but expect to get a Finnish couple from the Timmins area to look after things for us. It will save a lot of headaches having a decent place to live and eat for a change and (I hope) be infinitely cheaper. Hope you had a good trip home and that the one goose arrived in good shape. I went up to the camp shortly after you folks left, but only hunted one afternoon for a couple of hours. Managed to shoot six waveys and just missed getting a shot at a bunch of Canadas that came sailing over. It's

Austin Airways Canso CF-DFB puts sportsmen ashore for some Arctic char fishing in James Bay country over the 1973 season. (Oscar Cecutti)

Austin through the Decades

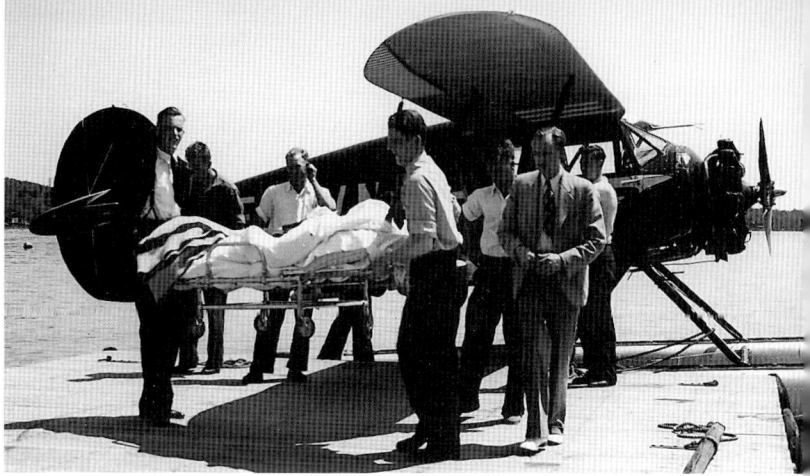

Austin Airways started with Waco CF-AVL in April 1934 and later had AVN, AWI and BDN. Here AVN is shown in summer and winter. Air ambulance trips by the hundreds were flown by Austin. When it came to such emergencies, the company set aside all other concerns. If a patient could not afford a flight, Chuck and Jack never quibbled about payment, but got their passenger to hospital as soon as possible. (Gord Mitchell, Rusty Blakey)

Austin Airways made good use of Bellancas and Fairchilds, including this Fairchild 51/71. It began in 1934 with bush pilot Johnny Fauquier, then served Austin from 1938. It ended in a crash near Nakina on June 17, 1949. That day Felix Cryderman, a wartime Typhoon pilot, mysteriously dove straight in moments after taking off. (Gord Mitchell)

In 1952 Austin purchased Huskies EIQ and SAQ. EIQ was lost when Nickel Belt Airways' hangar burned at Sudbury in December 1954. SAQ served till 1965, then was sold to Parsons Airways. (Neil A. Macdougall)

(Right) Austin 748 C-FMAA lands at Pickle Lake on August 22, 1979. The 748 carried an excellent load of freight and was liked by passengers—it was luxury compared to a Cessna or a Beech, and a bargain for operators compared to a new turboprop. 748s cost as little as $200,000 used, while $10 million would barely get one into the Dash 8 or Saab 2000 game. As one fan of the 748 put it, "For $10 million we can buy a lot of gas for our 748s." (Larry Milberry)

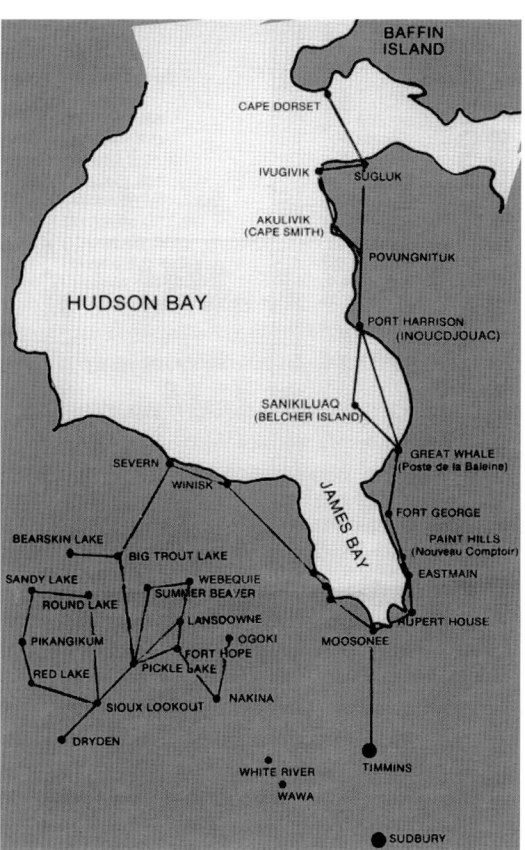

Austin Airways country in July 1979.

Feasting aboard C-GOUT—Capt Bob Isaacson prepares lunch as the 748 plods up Hudson Bay toward Cape Dorset. (Larry Milberry)

(Right) Austin Airways' famous letterhead from the 1950s remained in use into the turboprop years. (CANAV Col.)

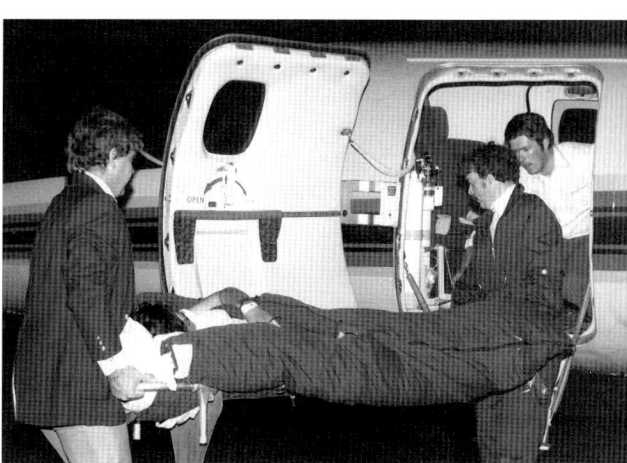

Pilots Bob Perkins, André Thibodeau and Ken Smith get a patient aboard an Austin Airways Cessna Citation air ambulance at Timmins in a scene from the mid-1980s. (Larry Milberry)

funny, but with the late season here, most of the geese are still around Hanna Bay, over on the east side. They will leave, though, with the first big north wind. All the camps are closed again, although there are still a few hunters coming in and hunting the river mouth.

January 20, 1963: Guess what I need is a blast every now and then to get going. Never seem to improve and have so many other things on the old noodle at the moment. It gets hard to get the time to sit down and write. Got your letter at Great Whale the other day. We had been there on an awkward operation since the 4th of January and things were anything but good. Had a DC-3 up there hauling freight and I just went up to make sure the thing got going OK. Stayed for two weeks and we had some pretty cold weather and plagued with all sorts of unserviceabilities with the aircraft. One thing after another and finally took it to Montreal on Thursday night and are just on our way back north now. Guess I owe you more than an apology for the neglect over the fall. I got out and was out for some time, but things sort of gang up. I didn't really go anywhere except to do a bit of deer hunting. Even that was cut short—old age must have caught up with me for sure. What with other things I just haven't had the enthusiasm to move. Our boss is due up this week and I am tempted to chuck the whole works, goose hunt and all, and find something else to do. As I mentioned before, I lost my medical a year ago December. Although it doesn't really affect my position in the company, this business of sitting and waiting for someone else to do a job is more than my patience and nerves can take.

February 5, 1963: Many thanks for the last letter, but must admit you are a mighty curious gal. All the questions! I would have to write a book to answer them. Will make an attempt. Have had, as most folks have, a few setbacks in my day. One's make-up is usually designed to take care of these things. Maybe mine isn't and that would be about the thing in a nutshell. Things, as I mentioned before, have bothered me for many years, probably all my life. Not one thing, but many. How I tackled it was to get drunk and disorderly at frequent intervals through the years. This is fine for a while, until the years start catching up. Then all sorts of things take place. One has to quit and I never did and probably never will. This is the reason I have avoided getting too close to you and hurting you. I am leaving one of our senior captains in charge of operations as a temporary measure for now. Hate to do this as it's a busy winter, and sort of an imposition to leave someone with a bunch of problems at a time like this. But am so darned tired I just can't do anything else.

Things actually have been going fairly well of late. We have had innumerable cold weather problems with equipment and what not, but these are normal for winter time. This winter has been, if anything, colder than normal. In another week or so it should start letting up and be a bit more livable. Maybe by that time the sun will shine again and make things look better. The boss is most cooperative in all this,

which is a blessing. At least I get something for 12 years of slaving in the sticks. Have two DC-3s working here at the moment and four other small aircraft. With at least another two months to go they should end up with a fairly successful operation.

July 6, 1964: The company have extended their northern run both ways for the next two months. They start from Timmins early on Monday morning and go right through to Cape Dorset on Baffin Island, this once-a-week and twice from Moosonee to POV and back. Business was never so good. Guess when I got out of there things really started to improve.

Jim Bell presided over a flock of young pilots. He hired them, nursed them along from dock boys to Cessna pilots then, progressively, to Beaver or Norseman. For those who persevered, the right seat in the Canso or DC-3 was their reward. Year by year the youngsters plugged along, except for those who didn't fit in, or who found jobs with bigger companies. Eventually, pilots like Jim Hall, Ray McLean, Bill Pullen and Jim Sheldon earned their Canso and DC-3 captaincies under Bell. While they became captains with major carriers, as apprentices they sometimes caused Jim Bell grief. Although he could count on them to do a good job, after hours it sometimes was another story. At Moosonee, for example, Austin Airways had privileges in the RCAF Officers Mess at the nearby Mid Canada Line radar station. More than once Bell's charges disgraced him in the mess, from where they eventually were banished. They moved to the Warrant Officers and Sergeants Mess, but, again, got the heave-ho. Thereafter, they did their drinking in the enlisted men's mess.

All the bush flying professionals who knew Jim Bell admired him. Fred Hotson met Bell in Sudbury soon after the war, when he (Fred) was flying a Husky for Nickel Belt. He found Bell to be "quiet, even self-enclosed". He didn't get to know Bell till later, when he was in Sudbury with a Lorne Airways Husky. Bell needed some help fixing his own Husky and approached Hotson. Through this contact Hotson learned what a fascinating and knowledgeable aviator Bell was. Wess McIntosh got to know Bell at HUT. Besides the DC-3, Bell flew the Canso. McIntosh described him as "an excellent boat man... he was quiet, but got along with everyone."

Austin Engineer

Besides famous pilots like Jim Bell, Austin Airways had many top air engineers and mechanics, beginning with chief engineer Frank Russell. John der Weduwen was another. Born in Zeeland, The Netherlands on November 20, 1923, as a young man in WWII he did farm work, then forced-labour in a milk factory. These were tough times—the Germans would confiscate anybody's food, and severe famine eventually hit the Dutch. Later, while the Germans still occupied parts of The Netherlands, the Allies brought relief. Reminiscing in 1995, der Weduwen recalled gathering food dropped by the RAF and USAAF.

Two famous Austin Airways engineers: John der Weduwen and Bob Kerr. (Der Weduwen Col.)

Der Weduwen got into aviation in 1946, joining KLM as an apprentice mechanic. His early types were the DC-3, Convair 240 and DC-4. After five years he earned his engineer's licence. He became a specialist on the Constellation and Super Constellation, staying with them for 10 years and ending as a foreman. In time der Weduwen was enticed by the idea of farming in Canada—farming was in his blood, but there was no land for him at home. Canada had nothing but land! In 1956 der Weduwen ended in the countryside near Trenton, Ontario. Soon he found the Ontario practice of mixed farming not to his liking. He took a job in a machine shop in nearby Batawa. This was dirty piece work at low pay, and he was obliged to speak in German on the job. Reading the Toronto *Star* in November 1956, he saw an Austin Airways advertisement for air engineers. He took the train to Kapuskasing for an interview. Once he heard der Weduwen's story, Austin engineer Tom Egett had one question: "When can you start?"

Der Weduwen started in Kapuskasing in January 1957. In March he went to Moosonee to spell off famed engineer Bob Kerr. His first trip to Moosonee was with Capt Ray Lejeune in an Austin DC-3. The one-week posting turned into a year. He worked on the Beaver, Husky, Norseman, Canso (in summer) and DC-3 (in winter). In 1996 he recalled: "During 1958 Ray Morin and I alternated between Kapuskasing and Moosonee. In the spring of 1959 Ray moved to Moosonee and I went with Ray Lejeune on Anson CF-DTW working inland from Fort George." Things ended with some damage to DTW due to slush and ice.

Over the summer of 1959 Cansos IZU and HVV (leased from World Wide), and another Canso (leased from Northern Wings) worked for Austin, mainly on Mid Canada Line resupply. It was a busy time, especially with HVV, which was not in the best of shape. Chief pilot at Moosonee was Jim Bell, who worked ceaselessly and taught much to the younger pilots. As they would fly along in the Canso or DC-3, Bell would quiz his co-pilots about systems, procedures, etc. With co-pilot Terry Monk he flew the sked with Canso JCV to Fort George, Great Whale, Port Harrison and Povungnituk ("POV"). The crew would over-

metal, but a DOT inspector disapproved and ordered a return to fabric.

In 1974 there were rumours around Timmins that Austin Airways would be sold. Some employees made an offer to buy, but Jack Austin said that there was no truth to the rumour. A few weeks later he sold to Stanley Deluce. This was disenchanting for some employees, who felt that too much had changed around the company. When Austin and Great Lakes Airlines of Sarnia amalgamated, head office moved from Timmins to London, where some employees did not wish to relocate. New people were hired, including a group of Chileans, who came along with some LAN Chile HS748s purchased by Stan Deluce. Thus did Austin Airways take on a new character. John der Weduwen left to crew on Bradley Brothers Drilling's DC-3. In 1978, however, he returned to Austin. In 1989 he retired as deputy chief inspector of maintenance. Through his years Austin Airways had grown rapidly, especially by adding the 748s. When it merged with Great Lakes, it become Air Ontario. Its mixed bag of types eventually gave way to a fleet of Dash 8s. Clearly, Stan Deluce had taken Austin Airways out of the bush, creating an airline for the times.

The Life of a Canso

In 1952 Austin Airways needed a big freighter to support exploration being done by Inco in the NWT, so bought its first PBY Canso, CF-FAR. This was a sound choice—as an amphibian it could reach isolated communities with three tons of cargo. Besides carrying a big load, it could tanker hundreds of gallons of fuel in its wing tanks (or in drums) and operate in all seasons. The airframe was rugged and easily supportable, and its P&W R1830s were the most reliable in their class. In 1960 Austin bought CF-JCV, a PBY-5A built by Canadian Vickers in 1944. It was RCAF 11054 till struck of strength in 1947. Thereafter it worked in Central America, returning to Canada in 1956 as CF-JCV with Dorval-based Eastern Canada Stevedoring.

In 1976 Austin sold JCV to Aero Trades of Winnipeg, a company serving northern native centres. In 1982 it moved to Air Caledonia in Vancouver. Owned by Peter Sperling, it specialized in coastal sport fishing charters. JCV carried 24 passengers non-stop to distant locations, one being the converted Norwegian whaler *Thorfin* in the Queen Charlottes. When a Convair took over this work, JCV was stored in Reno. In 1986 it was bought by the Catalina

night at POV, then retrace its route the following day. JCV, bought from Aircraft Industries to replace IHB, which had crashed at POV, could be configured for as many as 26 passengers and carried fuel, freight and mail.

Der Weduwen moved to Moosonee, staying from 1960-65; then relocated to Timmins. He spent considerable time as engineer on wheel-ski DC-3s flying on government contracts in the Arctic e.g. putting in fuel caches on Baffin Island for survey parties. This suited der Weduwen, who always enjoyed seeing the country. Another of his jobs (1969-71) was a geophysical survey over northern Quebec. The task was to help determine gravity characteristics, which the US needed to know about in connection with intercontinental ballistic missile launches. There were some troubles along the way. On January 9, 1964 DC-3 CF-ILQ, which was hauling fuel, crashed 20 minutes east of Rupert House. Pilots Bob Hamilton and George Charity were injured, but next day Lindy Louttit rescued them. A worse accident occurred at Val d'Or when DC-3 CF-AAC crashed in July 1970. Pilots Bruce McManus and Ray McLean, and der Weduwen were badly hurt, but their load of children heading to Fort George escaped. Der Weduwen had been going to Fort George that day to repair another

aircraft. At Val d'Or AAC took on full fuel, passengers and baggage. At the last minute a Cree chief insisted on boarding, even though there already was a full house. The captain relented, bringing AAC to maximum weight for departure. On takeoff there was an engine fire. The tower radioed McManus that AAC was smoking. Der Weduwen heard McManus say, "We're not going to make it!" then saw a wingtip slash into the trees. He blanked out momentarily, but as soon as the plane stopped, could hear the school kids. He clambered out to find McManus unconscious in his seat atop a wing. McLean was in the cockpit unconscious. An RCAF rescue crew arrived quickly and got everyone away. In spite of the potential for disaster, all survived. The crash was caused when the seal failed on the cowl gills jack on one engine. This allowed hydraulic fluid to spray onto the engine and ignite. McManus feathered the engine, but the plane stalled and crashed.

Like any professional in aviation John der Weduwen had his opinion about the types on which he served. He got to know the Fairchild Husky well and found it "a very nice machine". One complaint was the fabric panels atop the wing. Sometimes these would tear. Bob Kerr eventually replaced them with sheet

CP1960's route paralleled Hwy 400 and Hwy 69 en route to Sudbury, starting descent near the French River. Three hundred years earlier the French was *the* highway in the region, used by warriors, explorers, fur traders and missionaries. Adventurers from New France travelled by canoe from Montreal, taking weeks to reach Georgian Bay via the Ottawa, Mattawa and French. Now the river lay frozen and windswept, as HCP whistled passed

Austin Canso JCV visiting remote Pond Inlet on the top of Baffin Island in the 1960s. (CANAV Col.)

The new face of Austin Airways is seen in these two photos from August 1979. BAe748 C-GOUT was loading grocries at Kapuskasing. Citation air ambulance C-GRQA was at Timmins beside an Austin Ce.185. (Larry Milberry)

Austin Airways Fleet, December 1986

Type	Registration
Beech 99	C-GDFX, C-GEOI, C-GFKB, C-GFQC, C-GGLE, C-GGPP, C-GHVI, C-GJEZ, C-GQAH
Beech 200	C-GQXF
Cessna 402C	C-GBPE, C-GCGQ, C-GGPX, C-GGXH, C-GHGM, C-GHMI, C-GHMW, C-GHOE, C-GHOG, C-GHOI, C-GHOR, C-GIBL, C-GYHZ, C-GINR, C-GIQA
Cessna 500	C-GRQA
Cessna 501	C-GFEE
de Havilland DHC-6	C-GBOX, C-GDAA, C-GGAA, C-GNPS, C-GTYX, C-FZKP
Douglas DC-3	C-FAAM, C-FBJE, C-GWYX, C-FQBC
HS748	C-GFFU, C-GGNZ, C-GGOO, C-GMAA, C-GOUT, C-GQSV, C-GQTG, C-GQWO, C-GSXS

Safari Co., a French venture headed by Pierre Jaunet, whose idea was to operate tours between Cairo and Victoria Falls. JCV was fitted to carry 16 passengers in luxury. The first charter operated late in 1988 with Canadian pilots Ray Bernard and his son René. Various packages were offered, e.g. one lasting nearly a month that cost each tourist $15,000.

Sudbury and Timmins

In the 1970s modern turboprops began serving Northern Ontario and Quebec. A typical service began on January 22, 1992 at commuter Gate 7B at Lester B. Pearson International Airport. In the pre-dawn blackness Ontario Express Flight CP1960 waited for passengers for Sudbury, Timmins and points beyond. Just be-

Mining is the name of the game in Sudbury and by far the greatest player is Inco, one of the world's top nickel producers. This 1994 view shows the main Inco property at Copper Cliff close to Sudbury. As mining prospers in Sudbury, so does aviation. (Larry Milberry)

fore 0700 they filed down the corridor, across a wet tarmac, and onto their 48-seat ATR42 (C-GHCP). Capt Peter Boruta and FO Kevin Deslauriers started up for a 0705 departure, as flight attendant Wella Sparkes managed her 31 charges. HCP taxied to join the seven o'clock log jam, queuing as No. 13 for takeoff on R33. After a 15-minute wait takeoff power was set on its 1,800-shp P&WC PW120 turbines and HCP quickly nosed into low cloud, climbing for 16,000 feet.

unnoticed to land after a 56-minute trip. As HCP came to the terminal to join a pair of norOntair Twin Otters, an Air Ontario Dash 8 was taxiing for takeoff. Buried in snow near the terminal were the hulks of three Voyageur Airways ST-27s, rugged 19-seaters that for years had flown on many local routes.

The Sudbury-bound passengers deplaned into minus -21°C degree weather (it had been -8° at Pearson). Ten minutes later the ATR fired up, this time with 22 passengers. Takeoff was at 0834. HCP turned north on a 41-minute leg to Timmins, where it was -28°. The ramp was quiet, with an Air Creebec 748 and a Frontier/ Canadian Partner Beech 1900C. The ATR was refuelled and turned around as CP1961 for Sudbury and Toronto, due to leave at 0935. As it turned out, this was the last day of Toronto-Timmins ATR service. For the frequency of Ontario Express flights the ATR had proved to be too much airplane. The 30-seat Brasilia took over the following day. Matching the right airplane to a route was paramount for successful commuter operations, but the public often disagreed with the outcome. For years Sault Ste. Marie, Sudbury and Timmins had Air Canada DC-9 service from Toronto. When the DC-9 was dropped in favour of Dash 8s (Air Ontario) and ATRs (Ontario Express), there was an outcry. Travellers wanted the prestige of a jet. With commuter planes, however, they were ahead of the game with increased frequency, and there was little difference in block time to and from Toronto, all things considered.

Many passengers on Toronto-Sudbury-Timmins flights are business people. Those going to Sudbury (1997 population: 162,000) often are involved with the giant Inco and Falconbridge mining companies. Timmins passengers work with the Dome, Kidd Creek or Royal Oak mines. Others are in the endless variety of sales and service activities. On weekends many northerners fly to Toronto to shop and take in

Canadian Partner CP1960's crew on January 22, 1992: Capt Peter Boruta, FA Wella Sparkes and FO Kevin Deslauriers. Then, a view of C-GHCP at LBPIA on September 3, 1990. In the crowded commuter industry, Aerospatiale-Aeritalia, which built the ATR42, led in 1994. It sold 30 aircraft for a total of 435 since the ATR appeared in 1984. Like most twin-turbine commuters the ATR series progressed year by year, this process invariably depending on engine development. The original ATR42-200 appeared in 1984 with 1800-shp P&WC PW120 engines and a 34,700-pound MTOW. In 1996 the ATR42-500 was introduced with 2,400 shp PW127Es and 41,000 MTOW. (Larry Milberry)

(Left) Winter scenes at Timmins on January 24, 1992. BAe748s OUT and QTG were in storage. By 1987 most Austin Airways planes wore Air Ontario/Air Canada Connector colours. Then, Canadian Partner/Frontier Beech 1900C C-GFAC bound for James Bay. Frontier (parent company: Rog-Air) became a CAIL affiliate in early 1989. With the Beech 1900C, Frontier retired older types like the Aztec and DC-3. (Larry Milberry)

the big city's attractions. In winter others use Toronto to jump off to southern vacation spots. From Timmins there is much air travel northward. Routes pioneered by Austin pilots like Jim Bell and Gord Mitchell in the Fairchild 71 still are flown, but with none of the uncertainty of early days. In Austin's time flights were irregular. Sometimes the Indians and Eskimos greeting a Fairchild or Norseman never had seen an airplane. The occasional Hudson's Bay man, RCMP officer, government surveyor or missionary were the only other whites in James and Hudson Bay. Today things are different— skeds and charters are dispatched daily to the region.

In 1992 Timmins, a city of 47,000, was served by five sked carriers: Ontario Express, Air Ontario, norOntair, Air Creebec and Frontier. Voyageur of North Bay was also a regular visitor with King Airs on charter and ambulance work. Timmins also served many transient and corporate flights. General Motors visited with its Convair 580s, bringing personnel and equipment for its cold weather test base at Kapuskasing. Timmins airport was established after WW II on 2,280 acres; a new terminal was opened in 1955 to welcome TCA DC-3s, then Viscounts. For 1991 its modernized terminal served 185,000 passengers and 36,000 flights. Timmins has two runways: 03-21 (6,000 feet) and 10-28 (4,900). A main hangar, built by Austin Airways, was owned by Air Creebec; a second served the Ministry of Natural Resources, Ontario Provincial Police and Ontario Hydro. Frontier had a hangar for a B.1900, the Ministry of Health had one for Citation air ambulances FEE and RQA. Transport Canada managed the airport with about 50

The Akela Aircraft Repair base at Grassy Lake. Many such small operations, dotted across the North, serve hundreds of small commercial bushplanes. (Larry Milberry)

staff. The MNR Timmins operation was sizable, including 12 pilots in summer (eight in winter) and seven maintenance people; two Twin Otters, two Turbo Beavers, a CL-215 and a Bell 206B. The MNR also maintained a Twin

Otter for the OPP. In the early 1990s the base was flying about 4,000 hours yearly on fire fighting, fish and game law enforcement, fish planting, the annual moose count, surveys of goose nesting, beluga whale migration studies,

Rusty Blakey Heritage

Thurston A. "Rusty" Blakey about to depart on a local trip in a Ramsey Air Beaver (his favourite type) in September 1985. Born on December 10, 1911, he became Sudbury's best-known aviator. He worked for Sudbury Boat and Canoe in the 1930s, then Phil Sauvé of Austin Airways hired him in 1937. Before long Rusty was taking lessons from Jim Bell and flying the Fleet 7. He flew steadily until Austin sold its Sudbury float and ski base to Ramsey Air. Rusty was here for nearly 50 years. Along the way he trained many young pilots. In 1995 Brooks Hewstan, by then a BAe146 pilot with Air BC, recalled how Rusty took him over Shoo-fly Lake near Sudbury, asking what he thought of it for a Cessna 185. Hewstan considered it too small, but Rusty quickly landed and took off on the lake, giving the sprog some good pointers. On October 11, 1986 Rusty, returning to work after lunch, was involved in a cycling accident. He died the same day. A member of the Order of Canada and of Canada's Aviation Hall of Fame, he was further honoured by the Rusty Blakey Heritage Group, which instituted a local aviators' hall of fame. (Larry Milberry)

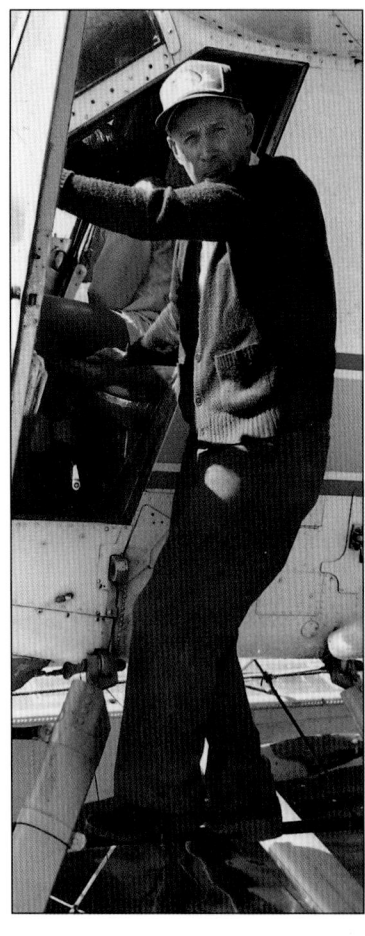

(Left) Rusty Blakey Heritage inductees Chuck and Jack Austin as young businessmen fresh from the University of Toronto in 1934. They had just accepted their first aircraft, a new cabin biplane from Waco in Troy, Ohio. Then, their famous chief pilot, Jimmy Bell, with a load of Christmas presents for the New Golden Rose Mine (CANAV Col.)

Members of the Rusty Blakey Heritage	
1988	Rusty Blakey
1989	Jim Bell
1990	Chuck and Jack Austin
1991	Max Ward
1992	Frank Russell
1993	Stan Deluce
1994	Tom Cooke
1995	Vi (Warren) Milstead
1996	George Theriault
1997	Russell Bannock

(Above) The great air engineer Frank Russell was the 1992 RBH inductee. Frank, also a member of Canada's Aviation Hall of Fame, was the first to do a full day's work for Austin Airways—in 1934 he opened the company's shack at the foot of Yonge St. in Toronto. Frank passed away on December 15, 1994. Here he's at Sudbury in August 1988 with his son David, a long-time MNR engineer. (Larry Milberry)

Rusty Blakey's wife Pearl attended the annual RBH evenings in Sudbury from 1988. Here she accepts a model of the Wardair A310 named in Rusty's honour. Beyond are RBH sponsor Risto Laamanen and inductee Max Ward. (Larry Milberry)

Max Ward was inducted into the RBH in 1991. Here he and his wife, Marjorie, are at Sudbury's bush pilots' monument at Science North. At Toronto in July 1990 Max Ward named Airbus A310 C-GDWD "T. Rusty Blakey" in honour of the famed bushpilot. (Larry Milberry)

(Left) Stan Deluce served on Hurricane fighters in WWII. After the war he formed White River Air Services in Lake Superior country. His first plane was a Fleet Canuck bought from Carl Millard in Toronto. He expanded yearly and his efforts led to contracts with norOntair, the takeover of Austin Airways, and the formation of Air Ontario and of Canada's leading charter airline, Canada 3000. (Larry Milberry)

(Right) Tom Cooke served on anti-submarine duties in WWII, sinking the German U-boat U-342 on April 17, 1944, then had a career with the OPAS. He was the 1994 RBH inductee. Here Tom was at the RBH celebrations in 1996. (Mike Valenti)

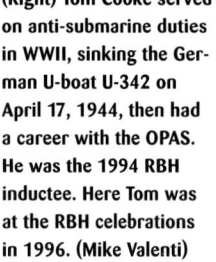

After his RCAF career George Theriault (right) founded a small tourist operation at Chapleau, near Sudbury. His sons continued operating Theriault Air Service after their father retired. George was the 1996 RBH inductee and is seen with Russ Bannock, who presented him with a model of the Beaver. (Larry Milberry)

(Left) Vi Milstead learned to fly in the 1930s. She held an exclusive job in WWII as a civilian pilot. She ferried all types of combat aircraft in the UK, from Spitfire to Mosquito to Lancaster. Later, she and her husband, Arnold Warren, flew for Nickel Belt Airways. One of Vi's student pilots there was Everett Makela. He later ran a bush air service and maintenance base (Akela Aircraft Repair of Grassy Lake). Here Vi, back from a 1995 flight in a Ce.185, poses with Everett (right) and RBH official James T. Miller. (Larry Milberry)

VIP transport and aerial photography. Typical was a survey in the winter of 1988-89 to collar 48 caribou west of Fort Severn with electronic devices. Later, the caribou were tracked by air to determine migration habits.

The Citation ambulances at Timmins in 1992 were under contract from Voyageur to the provincial Ministry of Health. Staff included nine paramedics and nine pilots. The Citations averaged about 770 flights yearly (1,800-2,000 hours). The base worked 24-hours a day with one aircraft ready, and the other on "mechanical stand-by". A typical call could send a Citation to Fort Albany for a patient destined for treatment in London, Toronto or Kingston. Attendants were community college graduates with two or three years of experience on road ambulances. At this time other Ontario air ambulances were: Bell 212—Sudbury, S-76—Thunder Bay, S-76—Buttonville, King Air 200—Sioux Lookout, Bell 212—Moosonee, Bell 222—Kenora, S-76—Ottawa.

Northern communities always are ready for the next cycle in the boom-or-bust stakes by which they live. In late October 1994 Inco announced a $75 million investment in a new nickel mine near Sudbury. On October 30, however, Canadian Regional announced the end of its Timmins service. Ten Timmins employees took early retirement, were laid off or moved to other bases. Two years later Royal Oak Mines gave local air carriers a shot in the arm by announcing an extension for its Timmins operations. One never knew what might happen next to influence air service.

The James Bay Coast
Upon arrival at Timmins on January 22, passengers on CP1960 destined for points north already were booked through on Frontier or Air Creebec. With Kapuskasing-based Frontier there were six passengers for CP1752—Beech 1900C C-GFAC, one of five in the fleet, crewed by Capt Kim Lessing and FO Stephen Hindley. Lessing had trained a few years earlier at the Brampton and Edmonton flying clubs and Confederation College in Thunder Bay, then flew in the bush. Hindley trained at Toronto Airways and flew with Pem-Air. Both were enjoying the Beech 1900C, while awaiting promotion to larger equipment and what that usually meant— a move to a bigger centre and a pay increase.

The crew praised the Beech 1900C, with its capacity for 19 passengers and 500 pounds of baggage (or 4,000 pounds in cargo configuration) and its 3 1/2-hour endurance. It was quick— 25 knots faster than the Jetstream, and a rocket compared to a Twin Otter. Lessing and Hindley soon had their engines humming, the -25° temperature being no problem for the start-anytime PT6s. FAC taxied, waited briefly for norOntair Dash 8 "P. Y. Davoud" as it arrived, then departed on R21. It climbed to 17,000 feet, comfortably holding 160 KIAS and levelling at 210. After 47 minutes and 166 nm FAC was on the ground at Moosonee, a busy Cree settlement (with its adjacent community of Moose Factory) and centre of regional government. En route FAC burned 800 pounds of fuel.

The ramp at Moosonee was quiet—Air

A typical scene with Air Creebec BAe748 on January 23, 1992. The 748 revolutionized aviation in the Canadian north. The type first appeared in Canada in the 1960s, but it was Stan Deluce of Austin Airways who brought it to prominence in the late 1970s. He added this ex-South African Airways 748 in 1983. Capt Marc Boisvert and FO Bruce Godby were operating it this day on the Timmins-Moosonee-East Coast milk run. It's shown being thronged by the locals at "YKU"–Chisasibi. Many of the passengers who boarded here were excitable first-time fliers. (Larry Milberry)

Creebec Ce.402 C-GIBL had just taxied out, and a Samaritan Air Service MU-2 had landed a few minutes earlier. Its crew—Tom Fairbank and Dave Moore, two lads who had started in the Air Cadets—were on a bi-weekly sked from Kingston for the Ontario Ministry of Health. Usually they carried nine of the local Cree, who had been out for medical treatment. Frontier's Beech left Moosonee for points north after a few minutes. It headed 68 nm down James Bay's west coast for Fort Albany, then made the seven-mile hop to Kaschechewan (one of the world's shortest skeds) and carried on 45 nm to Attawapiskat. Had the combination of snow and -40° at Attawapiskat not caused a landing gear snag, the flight would have been in Timmins as CP1753 at 1505. As it turned out it had to ferry to Timmins for maintenance. Air Creebec stepped to the rescue as its 748 C-GQWO was on a daily sked to the same centres.

The west coast of James Bay first saw an airplane in August 1924. Today's fast, high-flying turboprops had little in common with that old machine—Vickers Viking G-CAEB, which Laurentide Air Service had purchased in 1922. The sturdy Viking would earn its keep for a

decade. Its James Bay trip was a government contract. Besides pilot Roy Grandy and engineer B. McClatchey, an Indian Affairs man was along, dispensing annual treaty money to the Cree. This usually was done in summer by a canoe trip of several weeks. The Viking had started from Remi Lake near Kapuskasing. Grandy turned north at about 85 mph and hit the Albany River, tracing it to Fort Albany at the mouth. After finishing there, he flew to Attawapiskat. The return part of the 12-day excursion was to Moose Factory, thence cross-country to base at Grand'Mère, north of Trois-Rivières. What the Cree would have seen in that long ago time was a majestic biplane cruising slowly just above the trees. The sound would have been the loud brrrrr of a 425-hp Napier Lion. The Cree would have wondered what new gimmick the white man had come up with this time! These days, however, with turbo-props coming and going at Fort Albany and Attawapiskat, it was no big deal.

In the roomy Air Creebec 748 cockpit for the flight to Timmins were seasoned pilots Capt Ken Peddie and FO Robert Cook. Peddie had flown with Green Airways in Red Lake,

Tom Fairbank and Dave Moore of Samaritan Air Service at Moosonee on January 22, 1992. Their aircraft was MU-2 call sign "Halo". Their work took them around Ontario, but also to Florida to bring ailing "Snowbirds" home to Canada. (Larry Milberry)

Air Creebec's Ken Peddie and Robert Cook aboard 748 QWO going Moosonee-Cochrane-Timmins on January 22, 1992. (Larry Milberry)

and had nine years with Air Manitoba on DC-3s and 748s. Cook was from Awood Air and Bearskin. Their first stop was Moosonee. On leaving there, QWO was fairly light— 36,000 pounds (AUW 46,500). It was configured with 32 seats, with cargo forward. Thirteen passengers boarded and QWO was off at 1526 for Cochrane, 133 nm south. At 10,500 feet it levelled, cruising at 200 KIAS. A peek at the logbook showed that this old "bush plane" was a 21,180-hour veteran. The Air Creebec fleet included three other 748s, one a freighter, the high-time example having about 38,000 hours. QWO reached Cochrane at 1608 for a short stop, then flew 35 nm south to Timmins. On arrival there was a bit of activity — government Twin Otter C-FOPI was departing; a norOntair Dash 8 was in from Kapuskasing for Sudbury, Sault St. Marie and Thunder Bay; and an Air Ontario Dash 8 was turning around for Toronto.

Besides traffic on the west coast of James Bay, there is plenty on the Quebec side, where airplanes had visited for decades. On a 1929 trip Roy Maxwell of the OPAS flew a Moth to Rupert House to pick up ailing Anglican missionary Rev. Morrow, whom he took to Cochrane. From there Morrow reached hospital by train. Mrs. Morrow followed, getting to Cochrane by dog team in about two weeks. In a masterpiece of understatement she later wrote: "Travelling by dogs is not nearly so speedy... I have come to greatly appreciate the value of planes and radios." In January 1929 pilots T.M. "Pat" Reid, H.A. "Doc" Oakes and two engineers of NAME took a Loening from base at Sioux Lookout to Moose Factory via Longlac and Remi Lake. They pressed on down the east coast to Richmond Gulf to rescue several prospectors stranded earlier by freeze-up.

The sight of airplanes in the region remained rare for years. Gord Mitchell of Austin Airways flew a Fairchild 71 to Moose Factory in the spring of 1944. From here he did some local charters. Then Austin won a contract to fly a doctor to settlements on the Bay. Jack Austin and Mitchell did the flying with Fairchild 71 CF-BVI. This led to an ad hoc mail contract, when the Austin crew met a Post Office man on the east coast. He was stuck getting the mail moving—the CPA Fairchild he had contracted was broken. Meanwhile, Austin did some freighting for the HBC. The first trip was far inland from Fort George to Caniapiscau. A 1,400-pound load of flour was taken from Fort George down to Great Whale, and other trips were made to Factory River, Eastmain and Rupert House. When available, furs were back-hauled. All this showed Jack Austin that there was a future for aviation on the Quebec shore. Looking back in 1988 over his four decades with Austin Airways, pilot Jeff Wyborn discussed some memories of winter flying:

Slush was a condition which we often encountered on certain lakes. An aircraft might become mired in a combination of deep snow, slush and water layered in that order to a depth of 2-2$\frac{1}{2}$ feet. This invariably meant

Jeff and Muriel Wyborn the evening Jeff was honoured by his peers for his aviation achievements. This was at the old Airport Hotel in South Porcupine. Jeff had started flying with O.J. Wieben at Fort William in 1937. He joined Austin Airways in 1950, staying into the 1980s. Aviation wives like Muriel Wyborn or Pearl Blakey led amazing lives in their own right, often left alone for weeks or months, and worrying endlessly when their men were overdue. (Larry Milberry)

hours of hard slugging, jacking up the aircraft and cutting logs in the bush, these being laboriously dragged back to the machine and worked under the skis. In this process you could become soaking wet in temperatures that could drop to -15° or even -40°F. Then it might be necessary to spend the night preparing a path immediately in front of the machine, and waiting for snow conditions to improve before attempting your takeoff run. It was situations like this that made the glamour of bush flying more myth than fact.

On winter operations we regularly were faced with temperatures of -40°F, and conditions which dictated extra chores and precautions. A cover would be placed over the engine and sleeve-type coverings pulled over the wings. The former kept the weather out of the engine and the latter prevented a build-up of frost on the wing surfaces. Seemingly a small matter, even a thin film of frost on the wing could drastically reduce its lifting ability, and result in a difficult takeoff or even a serious mishap.

Planks or logs had to be inserted between the skis and the snow. If this were not done, the skis would freeze in place, and would have to be broken loose and cleaned thoroughly. Before takeoff, plumbers' blow pots filled with naphtha gas were used to heat the engine. We would light them, and when the blow pot's generator became heated, would open the valve, producing a hot blue flame. The pots had to be pumped up regularly to keep the pressure up. We would sit beneath the engine cover with two or three of these roaring away, with an extinguisher handy, watching for any sign of fire. Periods of such pre-heating varied from 25 to 45 minutes, depending on the outside air temperature and the wind-chill factor. During extremely cold spells we would remove the oil and battery, and carry them indoors where it was warm.

One January day I departed Moose Factory in the Norseman with mail and freight for Winisk and Fort Severn, stopping at Attawapiskat for fuel for the final 250 miles to Severn. Since gasoline was very expensive north of Attawapiskat, we used no more than was necessary. We had left Moose Factory on a clear, cold morning, with the temperature at -25°F. By the time we had reached the Winisk area it had dropped to -50°F. Then I noticed that the cylinder head and oil temperatures were at the lower end of the operating range, while the engine oil pressure was much above normal. These irregularities obviously were caused by the extreme cold thickening the oil and hampering circulation. Since I had no place to land, I pressed on towards Winisk, which was 15 or 20 miles ahead. Suddenly all hell broke loose. Engine oil leaking through the seals in the propeller spattered the windshield with a tar-like layer. With no vision ahead I was able to scrape enough congealed oil from the side window to reach Winisk, where I managed a rather tricky landing on the 1,200-foot strip of packed snow... When the oil seals gave, I was reminded of the old axiom: a chain is only as strong as its weakest link.

Since Austin Airways' beginnings on James Bay, aviation progressed from Fairchild 71 to Norseman, Beaver, Husky, DC-3, Canso, etc. In the mid-1970s Stan Deluce began a restructuring Austin Airways. He was convinced that the piston era was through. It was time to build larger markets with more efficient turbine aircraft. He purchased a fleet of 748s and brought many of LAN Chile's technical people to Canada as support staff. Austin's first 748 came from Air Gabon in June 1976. Thereafter, the company operated and brokered many 748s. It also bought several competitors, helped the 33,000-strong James Bay Cree establish their own airline in 1982 (Air Creebec, with the Deluces holding 49%), and took shares in Air Manitoba. Through a merger London-based Great Lakes Airlines and Austin Airways became Air Ontario, taking much of the commuter business between Windsor and Montreal/Ottawa, as well as Toronto-Timmins, and Thunder Bay-Winnipeg. Chuck and Jack Austin would have marvelled at what became of the little airline they had founded in 1934 with two little Wacos.

In 1988 the James Bay Cree under Chief Billy Diamond assumed ownership of Air Creebec, paying $20 million to Air Ontario. The Cree had the money—$137 million from the 1975 James Bay Agreement with Ottawa (Chief Diamond led those negotiations). Air Ontario was happy to leave Air Creebec, as it wished to focus on southern and mid-northern Dash 8 routes. The deal settled a long-standing issue—when Air Ontario moved from Timmins to London, many longtime Timmins employees reluctantly moved there. Those who kept their jobs under the Air Creebec buy-out now could return north. Next, the James Bay Cree formed an alliance with the 22,000 Nishnawbe Aski people based in Thunder Bay. An agreement also was made with Frank Kelner of Pickle

Chief Billy Diamond, who founded Air Creebec. (Larry Milberry)

Hubbub in the Great Whale terminal as three flights converge on January 23, 1992—Canadian North, Air Creebec and Air Inuit. (Larry Milberry)

Lake (Kelner Airways) and the whole organization was placed under a new name, Obi-Cree Airways.

No sooner had the deal been signed than Air Creebec started Val d'Or-Montreal with a Dash 8. It also took Air Ontario's northern routes from Timmins. The company which had begun with two Twin Otters now had eight 748s, two Beech 99s, a Twin Otter and a Dash 8. By January 1992 it employed 200. Air Creebec adopted the Rolls-Royce "18-60" mod for some of its Dart turbines. Although expensive, this offered savings. It included changes to the low pressure impeller, high pressure guide vanes and low pressure turbine blades. There were mods to the fuel control unit and other accessories. This improved fuel efficiency by 10%-14%, and reduced noise, but climb and cruise performance suffered. Air Creebec had three 18-60 engines by 1992. With fuel again relatively cheap, however, the incentive to convert more was low. The Dart continued giving excellent service, boasting a 6,000-hour TBO (for a DC-3's R1830 the TBO was 1,200 hours). Of Air Creebec's seven 748s in January 1992, the high time one was C-FSXS with 37,000 hours; low time was C-GQWO at 23,000. Two aircraft were in storage for the winter (OUT and QTG), SXS was at Springer's Airport near the Soo for a 750-hour check, FFU was awaiting the same check, freighter MAA was at La Grande for repairs after being hit by a fork lift truck, and GNZ and QWO were flying the line. Periodically since the mid-70s the fleet also earned revenue on international leases as far afield as the Indian Ocean or Caribbean.

An example of the 748's work at Air Creebec was F107 of January 23, 1992. The crew was Capt Marc Boisvert, FO Bruce Godby and FA Carol D'Amours. Godby was retired from the OPP after 36 years, 20 as a pilot. The aircraft was C-GGNZ, configured for 40 passengers. It left Timmins at 1355 with three passengers, climbed to FL170 at 150

Keen passengers wait to board Air Creebec F107/108 at Waskaganish. Then, flight attendant Carol D'Amours serving her passengers. Her long duty day included flying six hours—hard work for slim wages. Workers at smaller air carriers generally were not well paid. One FA mentioned in 1992 that of 14 who had started on her course 1$^{1}/_{2}$ years earlier only two remained flying. In 1995 Ottawa approved raising the ratio of passengers per FA from 40 to 50, making the FA's work all the more challenging. (Larry Milberry)

KIAS and cruised at 175 KIAS to Moosonee, landing after 50 minutes. Two passengers left and two (plus an infant) boarded after a 30-minute stop. For the rest of the day GNZ was in and out of east coast communities, each strip being like a bus stop in the big city—the 748 comes in, the passengers get off, cargo is thrown off, the next load boards and away they go!

First stop from Moosonee was Waskaganish, a reserve of about 1,900. As GNZ arrived, Darts whining, Air Creebec's Dash 8 C-FBVN was quietly taxiing out, half way through a day's work that would get it to Dorval. Waskaganish otherwise was quiet, with an Aztec and a Ce.172 freezing on the ramp. The town's most famous citizen, Chief Billy Diamond, came out to chat with the crew, and give a hand loading a heavy box. When GNZ left, it had 14 passengers and plenty of freight. Next stop was Eastmain, and the pilots had their work cut out—no sooner were they airborne than they were in thick cloud. So it would be for the rest of the day. Departing Eastmain there were 13 passengers, and getting into Wemindji took a go-around in cruddy viz and blustery winds. Most of Air Wimindji fleet was tied down on the ramp— Otter C-GCQK, wheel-ski Ce.206 C-GKJX and Ce.182R C-GIXN. Eighteen passengers left Wemindji for Chisasibi. All the coastal settlements had runways, but only Chisasibi had no building, not even a

waiting room. As GNZ landed, a fleet of pick-up trucks and skidoos waited. On shut down the 748's cargo door swung up, and baggage, freight and mail were tossed out. Passengers used the rear door. Those for Chisasibi hurried in -40° to reach one of the pick-ups, those for Great Whale (nine passengers plus a papoose) left their warm trucks and hustled aboard.

Great Whale is an ancient Hudson's Bay Company site that became a Mid Canada Line radar base. A paved runway was put in for C-119s, C-124s, etc., making this the region's best airport. It even had a large hangar. This stop could leave a traveller a little bewildered. Most in the 1990s still called it Great Whale, but to the Inuit it was Kuujjuarapik—so it

While the James Bay and Hudson Bay coasts are desolate and forbidding through the winter, in summer they offer spectacular beauty. Here is the east coast with a chain of islands by Richmond Gulf north of Great Whale in August 1979. (Larry Milberry)

appeared in the Canadian North schedule. The Cree, on the other hand, called it Whapmagoostui, listing it so in the Air Creebec schedule. To Quebeckers Great Whale is Grande Rivière de la Baleine. The neutral airport designator "YGW" seemed to simplify the matter. Was it confusion as to where one might be when visiting here that caused *The Canadian Encyclopedia* to omit this important place altogether?

When Flight 107 got into Great Whale, the terminal was abuzz. At least 200 were milling about, many in colourful native garb. There was a din of chatter, and tobacco smoke choked the air. A film crew had its kit stacked in the middle of the floor, and all sorts of non-native characters, mostly government flunkies—a lawyer, a teacher, various "consultants"—milled about. It was standing room only. Several flights were on the go, including Canadian North 737 combi C-GNDU "Spirit of Resolute". Soon it boarded passengers and left for Montreal. Meanwhile, GNZ refuelled. The pilots scurried around, sorting out their next

move. Normally, F107 would turn around as F108 after an hour and a half, and retrace its route. Now came new plans—a detour to La Grande to take up some slack due to a Dash 8 scheduling problem; but the detour was scrubbed—there were too many passengers waiting down the line.

GNZ left at 15:20 with 10 passengers and some return mail and freight. It winged south into fading light over the bleak-looking bay, featureless Canadian Shield. It landed within 30 minutes at Chisasibi. Here all but nine passengers left and the plane was mobbed by 37 newcomers, mostly school kids heading for a weekend hockey tournament. At 1645 the flight got into Wemindji, then was off again, FA D'Amours rattling off her flight safety spiel for the tenth time. Nobody paid attention. The hockey players had their faces glued to the windows checking the takeoff, and chatting excitedly in English. They were supposed to be talking French, but Quebec's hard-core language cops had failed to make the Cree subservient. The students' young teacher tried setting an example, but even he kept lapsing into *maudit anglais*. Meanwhile, the old timers conversed in Cree. The east coast run would be a delight for some crackpot sociologist cooking up new theories.

At Eastmain the hockey players made a quick run to the terminal for a pee and to buy goodies. Capt Boisvert waited a few extra minutes for the stragglers. Three new passengers boarded, including a graduate of McGill in Montreal, who had gone to India years earlier to work in "development". Now she was back in Canada consulting for the feds in health care matters. She was heading for Churchill, Manitoba for another session, then returning to India. At Waskaganish only three passengers stayed. The kids had reached their destination, so merrily poured onto the ramp and disappeared in the dark. Several forgot to take along their air sick bags — it had been a bumpy evening. The lady from India hurried into the waiting room to make sure her bags made the exchange from 748 to Dash 8—she was making the connection for Montreal. There was a panic as the station agent rushed to and from the 748 trying to reassure the excitable lady about her bags.

The final leg was across the bottom of the bay to Moosonee with 22 passengers, including another hockey team. Big guys this time. Some girls were also heading to Moosonee for broomball. Takeoff was gusty and the approach a roller coaster ride. Now all that remained was Timmins, where F108 bumped its way in at 2015 just as a blizzard howled across the airport and the duty manager closed the place down. The crew had been aloft for six hours, and were happy to be back, even if they found their cars buried in snowdrifts and coated in ice. This was, after all, Timmins in the dead of winter.

A Sked from Sudbury

By the early 1990s Canada's two big airlines had daily skeds to only one Northern Ontario community —Air Canada (DC-9s) and Canadian Airlines (737) served Thunder Bay. All other flights were with commuters, and service was excellent. In the 1960s a Viscount or two per day was considered adequate to Sudbury, North Bay, Timmins, Sault Ste. Marie and Thunder Bay. Later, things improved with DC-9s. In the 1970s Nordair 737s landed daily in the Soo. The main carriers eventually abandoned such service, e.g. Air Canada dropped the Soo in October 1986; but by 1992 there were many daily flights there—as many as 10 by Canadian Partner alone.

Since the 1980s Air Ontario was a key regional operator in Northern Ontario. Along with CRA and Bearskin it provided service from Winnipeg to Montreal. The 1997 fleet counted 17 Dash 8-100s and six -300s. Typical of Air Ontario's hundreds of daily flights was F1306 (Dash 8-300 C-GUON "City of London") operating Sudbury-Toronto on June 20, 1994. It originated in Toronto as F1305 that morning. Being a Sunday, the load was light— 14 passengers from Sudbury, one being Capt

In 1994 Air Creebec became the first Canadian operator of the Beech 1900D. It sought to improve passenger appeal by adding headroom—note the deeper fuselage compared to the B.1900C. In January 1997 the 500th Beech 1900 was delivered at Wichita, half were "Cs", half "Ds". Robert S. Grant shot this "D" at Val d'Or in February 1995. New technology like the B.1900D quickly supplanted earlier types like the Austin Airways B.99 (bottom), seen at Timmins in July 1985. (Larry Milberry)

Capt Isaacson and FO Thompson of Air Ontario F1305. (Larry Milberry)

Tony Jarvis of NWT Air, who had been holidaying in the Sudbury area.

The crew was Capt Bob Isaacson, FO Marc Thompson, Purser Dalia Pereira and FA Sandra Jackson. Isaacson, a native of Wawa, was one of Air Ontario's senior men. He had started flying with Stan Deluce's White River Air Service in 1959. Stan hired him in 1961 and he stayed on through Cessnas, Beavers, Twin Otters and 748s. By this time he had 24,000 hours, a third of them on the 748. By comparison FO Thompson was a sprog. He started in 1981 with Condor Aviation of Hamilton, finishing his commercial at Seneca College in Toronto. His first job was on the bottom rung—flying a light plane on aerial photography. Eventually, he got on with Voyageur, then Air Ontario.

F1306 taxied from Sudbury at 1058 hours (local) and was airborne in clear skies three minutes later at 33,000 pounds. It climbed over Sudbury and Ramsey Lake towards Georgian Bay, levelling at FL200 on the 1:04-hour, 188-sm route. Just past Collingwood F1306 began the descent to Toronto for a landing on R06R. From there, the crew flew to Cleveland before returning to home base in London. Altogether it was a soft day with light loads in fine weather.

Sault Ste. Marie

Another northern trip began on February 4, 1992 with the departure from Toronto of Ontario Express Brasilia C-GUOE, a 31-seater bound for Sault Ste. Marie and points west as CP1940. About 0910 UOE taxied to the de-icing pad at Terminal 3. It waited as Pem-Air's King Air C-GDAM was sprayed and inspected. Soon UOE was heading for R24R. Ahead were an Air Canada Airbus, a United 737, an Air Canada 747 and 767, a Metro, a Beech 1900C and a FedEx Caravan. Behind was a longer line—looming out of the fog were a 747 and a DC-10. This was LBPIA at rush hour again—40 minutes after start CP1940 finally lifted off and headed for FL210 on a course over Georgian Bay towns like Wiarton. In the days before the Trans-Canada Highway around Lake Superior these were famous for their shipyards, grain handling and Great Lakes passenger terminals. The flight carried along the Bruce Peninsula, over ice-coated Georgian Bay and Manitoulin Island. Being a fast turboprop

(cruising at 220 KIAS, or 300 KTAS), within an hour the Brasilia was descending. It touched down at 11:05.

Mid-northern Sault Ste. Marie (1997 population. 81,500) has an economy based on the sprawling Algoma Steel mills, pulp and paper, tourism, and St. Lawrence Seaway locks that connect Lake Superior with the lower Great Lakes. In 1924 it gained an early foothold in aviation with the OPAS. It built a hangar downtown on the St. Mary's River, from where its fleet fanned out each spring. The main tasks were fire patrol, aerial forest sketching and photography. The OPAS had many renowned aviators, mostly old WW I fliers. The Ministry of Natural Resources (formerly the OPAS) still was in the Soo. Its old hangar was home to the Canadian Bushplane Heritage Centre with such excellent exhibits as a Fairchild KR-34, Seabee, Beaver, Husky, Otter, ST-27 and Bell 47. Over the decades the Beaver was the OPAS' most important type. Ontario Lands and Forests director George Ponsford gave Phil Garratt the first OPAS Beaver order (four aircraft) in 1947. The

Husky competed, but lost. When senior OPAS pilot Frank MacDougall flew the Beaver in Toronto with DHC's George Neal, it is said that he was so impressed that the order was assured. Another view was that politics played a hand, with Ponsford refusing to spend Ontario tax dollars on the Quebec-made Husky, when Ontario made an equally fine plane.

The first OPAS Beaver (CF-OBS), was delivered April 26, 1948. Over the years the department owned 45 Beavers and 28 Turbo Beavers. Just before freeze-up most went to the Soo for maintenance. A few, however, were kept in the field for winter duties like the annual moose count. In the spring the fleet was test flown, then dispersed. In April 1951 pilots Tommy Cooke and Tony Shrive flight tested much of the fleet at the Soo—Shrive's log shows eight Norsemen and 20 Beavers for that month. About this time the OPAS began selling Norsemen, which soon were at work commercially with Austin Airways, Rusty Myers Flying Service, OCA, PWA and Transair.

The Otter and Twin Otter also made their mark with the OPAS. Over the decades 17 Otters served, some of which had crashes. CF-ODT, in the OPAS' original batch of 10, piled in around Kenora in 1961. Stan Deluce bought the wreck and, as one old timer later said, "rebuilt it by hanging a bunch of parts around it." ODT soldiered on for decades more. A second batch of seven, all ex-RCAF, came in 1978-79. The last in service was CF-ODX,

(Above) Sault Ste. Marie airport on April 11, 1991. The view is down R22 to the southwest. The MNR facility is closest, then the terminal in the centre. (Larry Milberry)

(Below) Canadian Partner's three types at the Soo on April 11, 1991: BAe Jetstream 31, ATR42 and Brasilia. These mainly connected the Soo with Toronto. The Brasilia operated to Thunder Bay and Winnipeg. (Larry Milberry)

Two of the important types in the CBHC collection. CF-OBS was the first OPAS Beaver and remained airworthy into 1997. Note the small water bombing tanks atop the floats. C-FDKG represents the Seabee as flown in the early 1950s by Air-Dale of Sault Ste. Marie. (Larry Milberry)

(Right) The MNR operated one BK117. It was good for fire fighting, its rear clamshell doors making for easy loading of equipment. This scene was at Science North in Sudbury on June 18, 1994. The BK117 eventually was sold—it was expensive operating it as a fleet of one. (Larry Milberry)

(Below) The Soo was the historic base for the OPAS, then the MNR. In the 1990s the main types supported were the Turbo Beaver, Twin Otter, Jet Ranger and CL-215. Here a Turbo Beaver is seen during a mod program over the winter of 1991-92. (Larry Milberry)

The well-known crest of the Ontario MNR has flown over Northern Ontario for decades. (Larry Milberry)

OPAS Fleet in 1991

Type	Hours Flown	Passengers	Cargo (pounds)
CL-215	1,181.2	164	574,701
Twin Otter	3,698.3	9,430	1,113,767
Turbo Beaver	5,053.0	9,999	553,609
Jet Ranger	2,849.0	4,413	294,300
BK-117	248.7	761	55,766
King Air	743.9	1,966	30,255
Navajo	142.6	2	590
Beaver	19.4	30	1,935
Totals	13,936.1	26,765	2,624,923

which retired in May 1990. ODU joined the museum in the Soo. In 1971-72 the MNR converted six ex-RCN Trackers to 800-gallon water bombers. Engineer Knox Hawkshaw of Field Aviation designed the bombing system and DHC did the fuselage mods. The Tracker was a good performer and at home on rough 3,500-foot strips. It could reach a fire quickly and knock it down before it grew. Before long, however, the MNR reviewed its fire fighting philosophy. This centred on directly attacking a fire; while the main use of the Tracker was to flank a fire—laying retardant on its fringes to control spreading. The ministry continued to

favour direct attack. Meanwhile, some were perturbed by how the Tracker forced a dependency on fixed bases. For 50 years Ontario had had a freewheeling operation, working off handy lakes and rivers. When fires were too far from strips, the Tracker was handicapped compared to a float plane. Most of Ontario's worst fires occurred in the northwest, with its abundance of lakes perfectly suited to water-borne ops. Another factor weighing against the Tracker was the cost of chemical retardants. In 1980 the MNR sold its Trackers to Conair. Saskatchewan then bought them. The OPAS took delivery of its first Twin Otters (CF-OEG, CF-OEQ) in 1967. These Series 100 aircraft were traded for Series 300s in 1969. OEG went to NASA as N508NA; OEQ was with Kenn Borek/Harbour Air in 1997.

On February 4, 1992 many MNR aircraft were at the ministry's new hangar at Sault St. Marie airport. Three CL-215s were inside, others were outside buried in snow. Two Turbo Beavers and a Bell Long Ranger also were present. One Turbo Beaver was getting airframe mods and a PT6A-27 in place of its standard -20. The rest of the fleet was in the field. At this time the MNR employed about 140 people, including 65 pilots (22 on the

CL-215). The latter had winters off, as their services were needed only during the fire season. The 1992 MNR fleet included nine CL-215s, seven Twin Otters, 13 Turbo Beavers, five Jet Rangers, a BK-117, two King Air 200s, a Navajo and a Beaver. The latter was CF-OBS. It was kept for historical purposes. Figures for the Turbo Beaver fleet in 1990-91 showed typical use. 6,710.2 hours were logged including: fish and wildlife 40.4%, inter-ministry (e.g. support of law enforcement) 15.9%, aviation fire management 11.1%, extra fire fighting 9.7%, forest management 9.6%, recreation areas 5.2%, and land management 2.9%. The CL-215s were used nearly 100% on fires. Helicopters flew more in fire management than Turbo Beavers, and less on inter-ministry work. Twin Otters and Turbo Beavers were less likely to be used water bombing, at least for big fires.

Other operations at Sault Ste. Marie in the early 1990s were Sky Services, Clarm-Aire, and Algoma Steel. Sky Services maintained norOntair's two Dash 8s and two Twin Otters, and provided maintenance to northern operators like Nakina Outpost Camps, which had a Twin Otter, turbo Otter, Caravan and Beaver. Clarm-Aire provided basic and advanced flying instruction for students at Sault College, one of several colleges supported by the Ontario government. In 1992 it had six Ce.152s, three Ce.172s and three PA-30 light twins flying about 7,200 hours annually. Sault College graduates earned an aviation technology diploma and a multi-engine IFR commercial rating. Algoma Steel's corporate aviation division also was at Sault St. Marie airport, although it soon was to be disbanded. From the 1950s it flew a Mallard and DC-3, then a Gulfstream and King Air.

An MNR Career

Since the original WWI veterans founded it in 1924, hundreds flew for the OPAS. Here was a band of aviators with an affinity for the northland. Newspaperman Bruce West wrote of them in his 1974 book *The Firebirds*. Thomas Cooke was one of the post-WWII firebirds. Born in Goderich, Ontario in 1920, he grew up with no thought of flying. In 1935 he joined the Royal Bank, where he plodded away till 1939. That year, keen to help with the war effort, he visited an RCAF recruiting centre to enquire about a clerk's position. The desk sergeant asked if he'd like to be a pilot. "I guess so", replied the naive lad, and he promptly was sworn in. A few months later Cooke was training on Finches at London, then progressed to Camp Borden. After checking out on Yales, he flew Harvards and Nomads under famed OPAS pilot George Phillips.

Cooke's first posting was instructing on Harvards and Cranes at RCAF Station Dauphin, then he flew at a bombing and gunnery school on the Anson, Battle and Bolingbroke. Next came a navigation course at Summerside, PEI. This led to anti-submarine duties with 162 Squadron. There were patrols from East Coast detachments, then the squadron transferred to Iceland in January 1944. Things heated up—the North Atlantic was thick with enemy U-boats and 162 Squadron was soon in combat. On April 17 Cooke was piloting Canso 9767 on patrol when U-342 was sighted. Cooke attacked, but the Germans sent up heavy flak. His nose gunner raked the submarine with .303 fire, then Cooke dropped three depth charges. U-342 faltered and slipped below the surface—a few minutes later it exploded violently to become 162's first kill.

Cooke was posted home to 124 (Ferry) Squadron. Soon he was seconded to an experimental budworm spray program using four RCAF Cansos. The job was to cover 106 square miles near the Lakehead. The requirements seemed tough—lines 10 miles long by 200 feet had to be accurately laid down. At first these were marked by a Norseman dropping strips of toilet paper, but Cooke improved on the process. He had detailed photogrammetric strip maps made up. Using these and a drift sight, a pilot could navigate accurately. Cooke recalled: "Each crew had a one-mile strip and that's all they sprayed during the whole period. They did as many as three runs before spraying, to make sure it came out right at the other end. A pilot picked out all kinds of landmarks—little swamps, dead chicos, crows'

nests, and used these to perfect the navigation. We normally flew our lines at 103 knots." Tom Cooke left the RCAF in February 1946 and returned to banking, but soon the OPAS called him. It had bought a Canso for $15,000 and needed a spray pilot. The pay of $500 a month looked good, compared to the bank's $175. For spraying, one of the Canso's two main 730-gallon tanks carried the chemical, the other carried fuel. Cooke finished the job, but George Ponsford wooed him away from the bank once and for all in August 1946 with an offer of $3,000 a year. One of his routine jobs became flying Ponsford around. "I got a few good bush flying tips from the old man", recalled Cooke.

When Tom Cooke came to the OPAS there had been some water bombing experiments. In 1944 Cam Crossley tried bombing with the OPAS Fairchild KR-34. The Beaver was later tried with five-gallon, water-filled paper bags. Neither approach was much good, so one winter, when the pilots spent most of their time sitting around the hangar at the Soo, Tom Cooke and one of the engineers sketched out a new system—a 90-gallon tank in the Beaver cabin with a dump hatch that the pilot controlled with a simple handle. There was no water pickup system, so the tank would be filled by landing on a lake and using a portable pump. In 1952 the tank system was evaluated against water bags, but Cooke realized that it couldn't get the water out in a concentrated mass. When George Ponsford heard about the tests, he ordered his boys to knock off their games. Nothing further happened until 1957, when OPAS engineer George Miles was authorized to build a water drop system for the Otter. At this early date Miles suggested putting the water inside the floats, but everyone poo-pooed that.

The next idea came from George Gill, who suggested tanks mounted atop the floats. These would be emptied by rotating them 180°. For refilling, fixed probes would scoop up water in a few seconds as the pilot planed over any lake. "We talked about this idea in the morning," noted Tom Cooke, "and by the afternoon were building the first tank." By now George Ponsford was supportive, perhaps because he had heard of similar experiments in Minnesota. The tank system was evaluated at the Soo. At first there was trouble with the probes—they were too long and impeded take-off. Soon the bugs were out and the tanks became standard in the OPAS for years until replaced by a built-in float water drop system.

In the spring of 1947 Cooke had moved to the summer base at South Porcupine to fly the Norseman, Beaver and Otter in succession. Years of experience convinced him that the Otter was the best of the bunch, followed by the Norseman, then the Beaver. In 1960 Cooke moved to Pembroke, flying a Beaver year-round. In 1961 he went with George Beauchene and Otter CF-ODL to help with fires in Newfoundland. After a few days on the job the boys were perturbed to find metal filings in their oil. Cooke reminisced: "We kept plugging away until one morning, soon after take-off, my oil temperature went off the clock and the pressure fell to zero. I popped down

An OPAS Twin Otter drops water from its floats—the idea of fire-fighting had come a long way since the days of dropping paper bags filled with water, and Tom Cooke saw it at every stage. (Halford Col.)

onto Island Pond near Gander to check the oil. I didn't have any! A Ce.180 flew over five gallons, then I took off. The problem recurred and I quickly returned to the same lake. There was a big fire raging nearby, so I anchored on the lake and we planned an engine change. That had to be done ashore and George Willoughby got the job going. He hired a crew of local Newfies. Eventually the fire came through and they had to push the plane back onto the water and spend the night in it. George's crew eventually finished and ODL returned to Ontario."

Tom Cooke got to know everyone in the OPAS. His years were full of memories, such as how Frank McDougal, one of the senior pilots, would entertain the folks at night with expert fiddling. George Phillips was another friend. "I knew George 39 years," recalled Cooke. "Both he and Frank were pioneers, great bush pilots and wonderful men all round." Cooke's engineer George Holmberg also was special. His son Joe later took over his job. In 1964 Cooke became forest protection supervisor at Chapleau, and the following year was OPAS supervisor in the Soo, a position he held until retiring in 1977. He was one of the early members of the Canadian Bushplane Heritage Centre at the Soo, and in 1995 was inducted into Sudbury's Rusty Blakey Heritage aviation hall of fame.

Thunder Bay

At the end of a day at Sault Ste. Marie on February 4, 1992 it was time to board norOntair Dash 8 "P.Y. Davoud" for Thunder Bay. This was F105 crewed by Capt Dave Weber, FO Wayne Gorecki and FA Vicki Zeppa. It took off at 1905, climbing in perfect weather to FL160. The lights of Thunder Bay appeared on the horizon at about 65 nm. The flight started down at 1959 and touched down 1 hour, 1 minute and 1 second after start-up at the Soo. Thunder Bay (1997 population: 114,000) has been a busy transportation centre since the days when 18th century "coureur de bois" fur traders used the area as a meeting place. In the next century, by when the twin cities of Fort William and Port Arthur had been founded here, the transcontinental railroad came through. Once the prairies opened to immigration and agriculture,

(Left) The heart of Thunder Bay's prosperity is shipping prairie grain, with forest products next. When this overhead view was taken on April 11, 1991, it was slack season at the Lakehead—the Great Lakes-St. Lawrence Seaway was closed for the winter. (Larry Milberry)

(Below) A different angle on Lakehead aviation. This gaggle, stored for the winter on the old Port Arthur waterfront, was seen from a passing 447 Squadron Chinook helicopter. Included are an MNR Otter, a Beaver, some Cessnas, Pipers and the same ST-27 that appears on page 631. (Larry Milberry)

A Thunder Airlines Navajo at the Lakehead on May 1, 1996. This operator came onto the scene in 1994 after some entrepreneurs under Ken Bittle took over defunct Awood Air. A graduate of Confederation College, Bittle had been with Patricia Air Transport, Austin Airways and Air Ontario before joining Air Manitoba in 1990 as president. Thunder Air tapped the traditional Kelner Airways market, opening a base at Pickle Lake. It also did courier runs, e.g. Thunder Bay-Hamilton. For 1997 it had three King Air 100s, three Grand Caravans, and a Short 330, the first Canadian example with a rear cargo door. (Larry Milberry)

Charter & Schedule Information:
Pour reseignements sur les vois reguliers et nolises:

(map of northwestern Ontario showing communities: Fort Severn, Bearskin Lake, Big Trout Lake, Angling Lake, Sachigo Lake, Kingfisher Lake, Kasabonika, Muskrat Dam, Webequie, Sandy Lake, Round Lake, Wunnumin Lake, Summer Beaver, Lansdowne House, Deer Lake, North Spirit Lake, Cat Lake, Pikangikum, Pickle Lake, Fort Hope, Red Lake, Geraldton, Dryden, Sioux Lookout, Kenora, Manitouwadge, Timmins, Atikokan, Marathon, Wawa, Fort Frances, Thunder Bay, Elliot Lake, Sudbury, North Bay, Sault Ste. Marie, Ottawa)

(Above) Communities served by Bearskin Air in 1996. Note the use of Cree syllabics.

(Left) A Bearskin Air Metro boards passengers for a March 26, 1992 flight from Thunder Bay. A norOntair Twin Otter has just arrived. For 1997 Bearskin, which had 320 employees, served 36 communities between Winnipeg and Ottawa. Its 32 aircraft made 160 daily flights, creating about $35 million in sales. (Larry Milberry)

these cities prospered as trans-shipping ports. The railroads carried in grain from the prairies, then it was loaded onto freighters to go down the Great Lakes to market. The first airport for the twin cities was Bishop's Field, opened in March 1929. It fared badly in the Depression—its licence lapsed in October 1935. A new airport was begun in August 1938. It prospered through the war with EFTS, and Canadian Car and Foundry activity. TCA began international flights from here in July 1946 with service to Duluth, from where passengers could reach Toronto via Chicago or Cleveland in eight hours. Scheduled service to Winnipeg began in 1947. In 1948 the airport was given a new name—Lakehead. In 1970 Fort William and Port Arthur amalgamated as Thunder Bay. In 1991 there was a sod turning for a new terminal,

then an official opening on February 18, 1994. By 1997 the airport was handling about 500,000 passengers yearly.

Bearskin Air

Thunder Bay's most prominent airline in the 1990s was Bearskin Air. It was founded in 1962 by John Hegland, a trader at Bearskin Lake, 400 nm northwest of Thunder Bay. His company was Bearskin Lake Air Service and started with Ce.180 CF-HJN. In the spring of 1962 he moved to Big Trout Lake, a larger centre, and added Ce.180 CF-JEN, then (1965), a Beaver. Hegland lost his life when the Beaver crashed. Local pilot Henri Boulanger and trader Albert Cone took over, but Cone sold his half to Harvey Friesen of Saskatoon. The company flew its first sked between Big Trout Lake

and Sioux Lookout. In 1977 Boulanger sold to Friesen, whose brother Clifford, Karl Friesen (no relation) and Rick Baratta also became partners. The fleet expanded to the Aztec, Beaver, Beech 18, Navajo, Norseman and Otter, and started serving more Indian settlements around Big Trout Lake. In 1978 a maintenance hangar was opened in Sioux Lookout. On July 17, 1988 Bearskin celebrated 25 years in business without serious injury or fatality to a passenger.

By 1992 Bearskin had 225 employees, about a third being pilots. It was carrying about 100,000 passengers yearly and about a half-million pounds of freight. It operated 60 scheduled flights daily to 27 communities. Charters made up 10-15% of business. Skeds ranged from Minneapolis to Winnipeg to Big Trout

(Above) A Bearskin King Air overnighting at Thunder Bay on February 4, 1992. Originally for the corporate market, King Airs worked their way into the commuter and courier worlds as they aged. (Larry Milberry)

Lake. The fleet was five Aztecs for summer MNR fire patrol contracts, two speedy Piper T-1040s, five 12-passenger King Air 100s and nine 14-seat Beech 99s. Two 19-seat Metro IIIs were added in March 1992. With its rugged airframe and PT6s the Beech 99 was ideal for Bearskin. It was a money-maker. Although it dated only to 1968, many had more hours than a lot of 50-year-old DC-3s. For Bearskin's fleet on February 5, 1992 the high time B.99 was C-GHVI (153rd off the line at Wichita) with 31,453 hours. The pressurized Metro was faster and had better overall performance than the unpressurized B.99. It added passenger appeal to those travelling north of Thunder Bay. The first arrived soon after Bearskin successfully opposed norOntair's proposal to extend Dash 8s to communities in Bearskin's market. Ironically, Bearskin was working for norOntair at this time, operating a Navajo and two Twin Otters for it at Thunder Bay.

On April 3, 1993 Bearskin, using new Metroliners, took over CRA's Ottawa-North Bay-Sudbury-Sault Ste. Marie-Thunder Bay service. In March 1996 norOntair ceased operations, the victim of government belt-tightening. It had been losing $5 million a year, so was a ready target for government cost-cutters. Its routes were taken by local carriers. Air Creebec, for example, took Earlton, Hearst, Kapuskasing, Kirkland Lake and Sudbury. Bearskin added Wawa and Elliot Lake. By 1997 Bearskin's fleet included seven Metros, eight B.99s, four King Air 100s, plus five Aztecs and six Ce.337s for fire patrol. There were about 200 flights daily and business was increasing Thunder Bay-Winnipeg and Winnipeg-Red Lake. At 420 nm Sault Ste. Marie-Thunder Bay was Bearskin's longest route.

F.28 to Thunder Bay

For 1997 Canadian Regional Airlines had a streamlined fleet of Dash 8-100s, Dash 8-300s, ATR42s and F.28s. The 65-seat F.28 was especially important with 18 working coast to coast. One of its premier routes was Toronto-Thunder Bay with five dailies on the schedule. CP1991 operated on May 1, 1996 with Capts Mark Whitehead and Earl Dyck. Their F.28 was C-FCRU, a 39,823-hour veteran. Pushback at LBPIA's Terminal 3 was at 0830 with 38 passengers. CRU taxied at 62,632 pounds (max brake release: 66,500 pounds) and held behind CP923 (737 C-GCPX) at R24R. For a flaps 11° takeoff, V1 (the top speed for a rejected take-off) was 131 knots, V2 (bringing

the nose up for liftoff) was 138. There was no delay departing and CP1991 was airborne at 0840. It climbed at 260 KIAS, levelling at FL310 at 0857. Time along the 509-nm route would be 1:31 hours—the F.28 is no speed demon, cruising at a modest M.70.

Capt Dyck had begun flying in 1983 with Central Airways at Toronto Island. He got his first job flying for Vic Pipoli of Batchawana Bay Air Service north of the Soo. From there he migrated to Rogair in 1986 flying from Port Loring and Nakina; then moved to Frontier Airlines. His types through these years were the Ce.185, Beaver, Otter, Aztec, Navajo, DC-3, Twin Otter and Beech 1900C. Eventually, Lloyd Rogerson sold Frontier to CRA. Most of his pilots went to CRA—Dyck got onto the F.28 in 1995. Capt Whitehead was a Seneca College grad. He began instructing at Toronto Airways, then flew a Ce.310 for Soundair at Winnipeg. Next stop was Air Toronto on the Jetstream and Metro.

The trip to Thunder Bay was, as usual, up the Bruce Peninsula, by Sault Ste. Marie and across Lake Superior. At 0950, while 63 miles from Thunder Bay airport, CP1991 started down. As it came across the Lakehead there was a good view of the ocean-going freighters lying against the ice, waiting for the last of it to melt or blow offshore, so they could get about the business of loading prairie grain. CP1991 now set up its approach at 140 KIAS. At 1009 it touched down on R25 at 127 knots weighing 56,000 pounds. Air Canada DC-9 C-FTLM had just cleared the runway intersection for the terminal and a Calm Air Saab 340 was touching down on R30, so things were busy. Soon CRU discharged its passengers and turned around for Toronto. The crew would complete 2½ such trips today, then overnight at Thunder Bay.

Pickle Lake

A typical day at Bearskin Air began before sunrise, as maintenance readied the fleet. From Thunder Bay there would be flights to places like Pickle Lake, Sioux Lookout, Round Lake, Webequie, Kasabonika, Kenora and Minncapolis. On February 4, 1992 Flight 353 prepared for the daily run to Pickle Lake, a town of 1,000 people 195 nm north. F353 (King Air C-GKAJ) had four passengers plus freight. The ramp was cluttered with aircraft as it taxied—two B.99s and two King Airs of Bearskin, a Frontier B.1900C, two norOntair Twin Otters, an Air Ontario Dash 8, Canadian 737 C-GRPW and Air Canada DC-9 C-FTMA. KAJ taxied to R07 behind Frontier and took off at 1119. It climbed to 11,000 feet, cruised on at 200 KIAS, and landed at Pickle Lake at 1235. Bearskin's nerves immediately were tested—all points westward were weathered out. Much of the day's schedule was kaput, but winter ops in the region are like that.

Pickle Lake is a historic aviation centre. It was a commercial fishing hub for years, especially with O.J. Wieben. Hooker Air Service, run by Dave and Horace Hooker, flourished there. It began in 1947 and took over Transair's local routes in 1967. By 1970 Hooker had a Beaver, an Otter, Bellanca 66-75 and two Beech 18s. This day there were all the Bearskin skeds, plus activity with local operator Kelner Airways. On the tarmac were a Transport Canada Twin Otter and Gold Belt Air Transport Otter CZO. Buried in the snow was another Otter—it had hit the trees the summer before—and Air Manitoba C-46 C-GTPO. TPO's pilot had tried aborting takeoff after losing control. He ground-looped avoiding a drop-off at the end of R27. It turned out that the crew had attempted takeoff outside the accelerate-stop limits for 4,700-foot R27. Also at Pickle Lake this day were two Mid West Helicopters B.206s and a B.205 on a drilling contract.

Within an hour on February 4 most of Kelner's fleet had come and gone at Pickle on the daily grind, mainly with cargo. Caravans C-FKAL, KDL and KSL were busy with grocery flights, as was 748 KTL. KAL had just been out to Summer Beaver (115 sm) and Kingfisher (58 sm). BAe 748 LTC was tankering fuel oil between Pickle and North Spirit, 7,000 litres a trip. A third 748 was in Thunder Bay being converted to a freighter. It

A Kelner 748 C-GTLC at Pickle Lake on February 5, 1992, while busy tankering fuel. TLC served the German government 1969-84, then came to Canada in 1986 via the US. Kelner was taken over in 1992 by First Nation interests and re-named Wasaya Airways.(Larry Milberry)

Kelner chief pilot Bruce Whitly with James Allinson. (Larry Milberry)

A Kelner Caravan at Big Trout Lake on February 5, 1992. DHC originally toyed with the idea of a single-engine turboprop, but rejected it–the company was too engrossed in its glamorous Dash 7. The small turboprop migrated to Cessna, where design was completed. The prototype flew as the Caravan in December 1982. With a 600-675-shp P&WC PT6-114A it took naturally to back-country markets. In 1995 Soloy and P&WC combined to develop a Caravan with a pair of PT6-114As, the so-called SDP (Soloy "Dual Pac"). It had a six-foot stretch and gross weight upped to 10,500 pounds. (Larry Milberry)

was Frank Kelner who introduced the Caravan to Canada. By 1992 he already had owned nine. His aircraft were logging about five hours a day in winter, more in summer. Each averaged 2,400 hours yearly. With all his equipment, be it aircraft or fork lifts, it was his policy to keep them as new as possible. Thus, he was pleased to find two low-time 1980-model 748s in Niger. These had been in mothballs for two years, one with a mere 2,300 hours, the other 2,800—virtually new machines at an excellent price. LTC came from defunct Inter City Airlines of Oshawa and had 10,000 hours by early 1992. Each 748 was logging about 160 hours monthly.

The Caravan's longest usual leg from Pickle was to Big Trout Lake, 170 sm north. Pilot James Allinson started KDL for a grocery flight there at 1413 on February 4. As he taxied, "Super Cargomaster" KSL came back from a trip to Summer Beaver, and Transport Canada Twin Otter C-GDCZ took off ahead. Allinson got airborne from R09 at about 85 knots and climbed to 5,000 feet, where KDL settled down at 145 KIAS. Allinson was typical of young 1990s bush pilots. At 23 he graduated from Sault College in 1989. Earlier he spent several years in the Soo with 155 Air Cadet Squadron. He remembered his Sault College instructor, Bernie McComisky, an ex-RCAF fighter pilot from Sabre and Voodoo days: "He wouldn't give me an inch or let me get away with anything. There were no first names—it was 'Sir' or 'Mister'. But I couldn't have had a better instructor." Allinson noted that of the 10 who graduated with him, two were at Awood Air, two at Pem Air, and one each with norOntair (Twin Otter), First Air (Twin Otter), Collingwood Air, and basic training in the air force. One was unemployed and one had died in a crash the summer before, flying a Travel Air for Perimeter. After graduating, Allinson found co-pilot work on a Newfoundland Labrador Airways Ce.402 in Deer Lake, then was hired by Kelner. There he was piling up the hours—1,100 a year was not unusual for a full-time Caravan pilot. On this trip he had 2,680 pounds of groceries for the native co-op in Big Trout. He touched down there on R14 at 1520, pulled onto the ramp and was met by the co-op truck. As it goes in the North, the pilot set to work offloading—every box, then the mail. Meantime, Bearskin's second daily

B.99 came in to a flurry of activity as skidoos and pick-ups roared out to greet passengers. Minutes later it taxied in a cloud of snow kicked up by the props. The only other aircraft present was a Ross Air Beaver on a Bell Canada contract. At 1542 Allinson fired up, made a 300-foot takeoff roll and set course for Pickle, landing at 1652. The smoothness of this operation left no doubt that the small turbine was the way to go in the North. The Caravan already had done much to edge out the piston Otter. On the other hand the turbo Otter conversion won its own followers, who touted its ruggedness, compared to the Caravan's supposed "tinniness" (if one thing is certain in aviation, a pilot's view of his airplane automatically will be refuted by the nearest pilot flying a rival type).

Sioux Lookout

At day's end at Pickle Lake on February 4, 1992 the weather cleared enough for a B.99 to get seven passengers the 45-minute hop to Sioux Lookout. This is the region's busiest airport. During the day its waiting room bustles, mostly with First Nation frequent fliers—the Indians seem forever on the wing. This day much of the crowd was in bright winter apparel. It was quiet, with people chatting, munching goodies and downing cans of pop (the pop- and snack-dispensing machines were working overtime gobbling up "loonies",

(Right) Welcome to aviation in Sioux Lookout. (Larry Milberry)

(Below) Besides Slate Falls, Knobby's was another of Sioux Lookout's busy summer operations. Here its Otter ODL returns from a charter on July 14, 1991. (Larry Milberry)

Canada's one-dollar coins). A few travellers sat around waiting to get back to Thunder Bay or Winnipeg. One was a serviceman in from Winnipeg to fix some equipment in the local hospital. Some local cops waiting for a flight were trading stories about partying and how much they could drink. There also were three rough-looking bushwhackers off a hardrock drill and anxious to get south.

At this time Sioux Lookout had a Bearskin maintenance base, a Ministry of Health hangar with King Airs, an MNR operation and several smaller businesses. Being winter, numerous float planes sat in the snow drifts, awaiting break-up. Not so likely to get airborne in the spring was a collection of derelict Beech 18s and other worn out machines. Downtown on the lake little was doing. Planes were hauled up, some with motors pulled for overhaul. February 6, 1992 saw only a Beaver, a 185 and a 206 on the ice at Slate Falls Airways and Knobby Clarke's place, awaiting any business that might crop up. Around the corner sat Glenn Tudhope's FBA-2C, wings covered against the elements. It had been quiet for the Found. Each winter Tudhope noticed a drop in

business. This he linked to the modernization of facilities in the area—new runways and nav aids everywhere, the advent of the versatile Caravan, and road improvements.

Kenora Rescue

March 28, 1979 was a day to remember in Northwestern Ontario. Pilot Jerry Krushenski was flying Aztec ASK to Kenora on a medevac, when his engines overheated in icing. He alerted Kenora FSS. Before long he lost one engine, then the other. He couldn't make the airport, so Kenora called the nearby MNR office on Lake-of-the-Woods, asking about the local ice strip. MNR pilot Robert S. Grant briefed the FSS and the details were relayed. In marginal weather Krushenski spotted what he thought was the ice strip and made an approach. But he was lining up on a winter road, not a runway. It didn't matter, for he crash-landed short, and on the only part of the lake with any open water. The Aztec slid across the ice into the water. Krushenski, Dr. Yvonne Kason and nurse Sally Irwin were lucky to get out—the plane went straight to the bottom. Patient Jean Peters, strapped to a stretcher, was lost. Krushenski and Kason managed to swim ashore, but not Irwin.

About the same time that Krushenski had departed Sioux Lookout, Heli Voyageur pilot Brian Clegg was ferrying east in a Bell 206L. Unable to get into Kenora airport because of weather, he put into the MNR base and was chatting there with Bob Grant. Suddenly the FSS called. The duo immediately scrambled. Grant, knowing the area, guided Clegg along in the Bell. Eventually they spotted skid marks in the snow and saw Irwin floundering. As Clegg hovered, Grant tried grabbing Irwin, but she slipped away several times. He finally got her hands on the skid, then stood on them, while maintaining his precarious perch on the skid.

Clegg hovered towards some ice and Grant got Irwin partially aboard. Minutes later she was delivered to Kenora hospital. Clegg and Grant then returned for Krushenski and Kason to complete the impromptu rescue. Rarely in Canadian aviation history had such an operation been so successfully completed, so quickly. A number of awards were dispensed in the months that followed. Bob Grant, for example, received the Royal Canadian Humane Association award, The Star of Courage from the Canadian government, a silver medallion from the Ontario government, and the prestigious Carnegie Hero Fund silver medal from the US.

Red Lake

The 1925 Red Lake gold strike entrenched aviation in Northwest Ontario. The area boomed even through the Depression. Commercial operators hauled equipment, fuel, dynamite, and groceries; and carried passengers from big time Toronto mining executives to prospectors, geologists, miners, construction workers, machine operators, traders, cooks, missionaries and hookers. Winnipeg was the focus of great activity. From there travellers and supplies moved northwest, first by train to

Lac du Bonnet, thence by air to Red Lake. Hudson, on the CNR, was a similar jump-off point. A key innovation came in 1936 when Canadian Airways put a speedy 10-passenger Lockheed 10A on Winnipeg-Red Lake. It was anxious to introduce a fleet of these modern airliners coast-to-coast. CF-BAF, the second Canadian L.10, was adapted to skis and proved a grand success on the busy route. Captains Herbert Hollick-Kenyon and Z.L. Leigh flew BAF. Both were renowned— Hollick-Kenyon had flown in the Antarctic with the American explorer Lincoln Ellsworth. Leigh had barnstormed on the prairies, operated an air service in Atlantic Canada, flown with Canadian Airways on the Mackenzie, and was Canada's first instrument-rated airline pilot. Red Lake would remain dependent on aviation, water transport and portages in summer, and ice roads in winter till 1947, when Highway 105 was opened from Vermillion Bay, about 110 miles south.

In the 1990s Red Lake remained busy in aviation. Activity focused downtown on Howey Bay, and at Cochenour, about 12 miles out, where a modern airport and float operators were found. The area's historic carrier was Green Airways, a solid old firm run by its

(Above) When gold was discovered near Red Lake in 1925, prospectors and developers were quick to use airplanes to reach their claims north of the transcontinental railroad. First came Jack Elliot with his JN-4s and Doc Oakes with the famous Curtiss Lark. Soon more practical types appeared, like this speedy Canadian Airways Stinson Reliant, seen in 1936. Today the Cessna 185, Beaver and Otter dominate the scene. (Red Lake Museum)

(Right) Red Lake's famous symbol is the Norseman. This example was part of a 1945 USAAF order. It came to Canada in 1954 for Ontario Central Airlines. Eventually it ended as a hulk at Kuby's Aircraft in Kenora, from where it came to Red Lake for restoration and display in 1992. Over its career CF-DRD logged 7,109 flying hours (296 days) in the air. (Larry Milberry)

(Right) Norm Wright brings Sabourin's Twin Otter DMR to the dock at Couchenour on July 15, 1991. Then, Sabourin Norseman JEC getting away on a trip the same day. (Larry Milberry)

(Above) Pilot Robert S. "Bob" Grant of Thunder Airlines returns to base after a day of medevacs to Fort Frances and Kenora with fellow pilot John Mercer. One of Canada's leading aviation journalists, Grant also authored popular books like *Bush Flying: The Romance of the North*. (Larry Milberry)

(Left to right) Pilots old and new—Al McNeill, Eddie Johnson and Steve Preston flew at Sabourin in 1991-92. Steve, the son of an airline captain, is shown in Ce.185 C-GDSJ during a Norseman Days on July 25, 1992. Typically, the veterans in a company brought the sprogs along step-by-step. Once they had put in their time as dock hands and general "gophers", the beginners got onto the Cessna. A season or so later they progressed to the Norseman, Otter or Beech 18. This system prepared them to move up to local carriers like Bearskin. (Larry Milberry)

founder's sons. The largest local carrier was Sabourin Lake Airways—large, but on its last legs by the mid-1990s. It dated to the 1970s, when formed by fishing lodge operator Ralph Webb. A variety of planes was used, including Beech 18s PCX, PFC, TBD. These did the usual freighting and fish hauling. Sabourin was taken over by Art and Norm Hegland, brothers of John. Art sold to Norm in 1977, who sold to the local Indians in 1987. Sabourin had two main activities: serving Indian reserves and other centres in the region; and catering to summer tourists. Over 1991 the fleet included Twin Otters C-FDMR and C-FISO, B.99s C-GJCC and C-GJKO, Navajo C-GYYJ, Otters C-FVQD and C-FPHD, Norseman C-FJEC, Ce.185 C-GDSJ and Ce.206 C-GIKR. By the winter of 1991-92 a leased Twin Otter and the Navajo were gone. One Otter and the Norseman were laid up for maintenance, and the other singles were on a reduced schedule. Much of Sabourin's summer flying was to Sabourin Lake Lodge, 30 minutes from Cochenour by Twin Otter. It was established in the 1950s by Ralph Webb, but was owned in the 1990s by Ron Williams of Winnipeg. It offered 48 guests the finest in facilities and fishing. A typical group flew out in the Twin Otter the afternoon of July 23, 1992— eight businessmen from Arkansas, who had arrived that day on a Cessna Citation.

Sabourin's pilots were widely experienced. Eddie Johnson had started flying in Winnipeg in 1954, working for such operators as Northland Airlines, Ilford-Riverton, Selkirk Air and Transair. From 1972-89 he was with the MGAS, finishing on the CL-215. By the end of the 1992 summer season he had about 21,000 hours, and still was spending several weeks each winter working his trap lines around Lac du Bonnet. Al McNeill was another 20,000-hour veteran, with most time on the Otter (11,000). He earned his wings in 1962 at Bradley Air Services and got his first job with Superior Airways on the Ce.180 and Norseman. He later came to Red Lake for Green Airways. January to December 1991 McNeill logged

In the summer of 1995 Sabourin DC-3 C-FBXY had 450,000 pounds of cargo at Red Lake for local Indian reserves. At about 7,000 pounds a trip, the task was daunting. Here BXY loads on July 13, surrounded by stacks of cargo, which had sat outside day after day in all kinds of weather. About this time Air Manitoba turned up with a shiny 748 freighter and put up a storage shed—not a happy sight for Sabourin, which went out of business a few months later. BXY was in the USAF till 1958, then flew with several civil operators till coming to Fort St. John, BC in 1971 for Knight's Rathole Drilling. It was later with North Caribou, in whose colours it remained during its Red Lake lease. (Larry Milberry)

1,104 Otter hours—a good year's flying for a bush pilot. Twin Otter pilot Norm Wright came to Red Lake in 1972 from the US, where he had been in the navy as an aero engine tech. Through the 1990s he flew DMR on floats in summer. In winter it was on wheels for routine trips to places with a runway or ice strip. For service into rougher areas the wheels were replaced by tricycle skis. Wright sometimes had a young pilot flying with him, teaching him the basics of the Twin Otter. In 1991-92 Pat Kalist was sometimes lucky to fly with him. He was slowly learning the ropes and, in the stagnant job market of the early 1990s, was content with any job. He spent most of his days as a dock hand and gopher. Clark Griffith also was at Sabourin. He had had various bush jobs, then flew for Skywalker. Later he flew Electras with NWT Air, but these were sold. Griffith crewed on the last one—a service to Calgary on February 1, 1991. Then it was back on the job-hunting trail, the tale of woe of so many in aviation. This brought him to Sabourin (where he had worked earlier). For 1991 he flew the Norseman, then took over the Ce.185 after

freeze-up. In the spring of 1992 Sabourin acquired a DC-3 from North Caribou Flying Service in Fort St. John, BC. Clark Griffith and Lyle Griffith flew it through the year. Three years later they were still at it. Soon after Sabourin went out of business in 1996, Wasaya stepped up activities in Red Lake, using the 748 and Caravan. Thunder Air also took an added share of the business.

Green Airways was formed in 1956 by George H. Green. To him the future in local aviation would be in tourism, not minerals, forestry or commercial fishing. For years his trademark was Stinson SR-9 Reliant CF-BGN, which had come to Canada in 1937. It flew with the OPAS until Green bought it in 1950, but eventually was lost in a fire, which spread from a burning church. BGN's remains were donated to the WCAM in Winnipeg, but later moved to the Canadian Bushplane Heritage Centre. For 1991 Green Airways had three Otters, two Beavers, two Cessnas, a Beech 18 and a metalized Norseman. The Beavers had extended cabins (two feet more usable cabin space) retrofitted in Alberta by Big Bird Sky

Bob Green (co-owner of Green Airways with his brother Jack) with company pilots Joe Sinkowski, Scott McAllister (later with Air Tindi), Kris Manchip (later Bradley), Ken Olson (later Bearskin) and Steve Wall (later Harbour Air) on their dock at day's end on July 15, 1991. Then, typical load going aboard Green's Otter CF-ODJ (landing beyond is Norseman OBJ, piloted this season by Steve Preston). The PZL Otter uses 7850 floats compared to a standard Otter's smaller 7170s. The numbers that designate a float refer to the gross weight accommodated. (Larry Milberry)

Farmers. Green had about eight pilots in summer, three or four in winter.

Red Lake is home to other aviation activities. In 1993 there was service from Canadian Partner, Bearskin and norOntair. Kelner often appeared with a Caravan or 748, and Perimeter had courier runs. Red Lake Air Service and Wild Country Air worked their own niches. Red Lake Seaplane Service, run by Whitey Hostetler, provided a host of technical needs for aircraft from Cubs to DC-3s at its Howey Bay base and at the airport. Winter was the busy time—dozens of aircraft were serviced at Whitey's to be ready as soon as the ice went out in April. Wild Country was at Cochenour. Till late 1991 its mainstay was leased Norseman CF-OBR, an OPAS veteran with more than 50 years of service. Company owner Bill Cousineau, whose father had flown from Red Lake since WW II, got Beech D18 C-FBCC from Big Bird Sky Farmers. It originally served in BC and had logged about 10,000 hours. At this late time (the prototype flew in 1937) the Beech 18 was in resurgence. One in good condition cost about $100,000. On wheels it could carry 2,800-pounds at 140 KIAS. A Caravan carried the same load 25 knots faster, but cost $1.2 million.

Beech 18 pilots, including many who graduated to the major airlines, always sing its praises. To some this was the best bush plane, but it also had detractors. A spate of crashes in the 1960s-70s gave it a bad name, but analysis showed that these often involved low-time or careless pilots. Typical was the October 10, 1975 fatal crash of CF-NCL in the northern BC bush. The DOT commented: "After a delayed departure because of low ceilings and visibility, the aircraft took off ... [it] was grossly overloaded and was not equipped with anti-icing or de-icing systems." To some the Beech 18 suddenly was your standard death machine, reminding us of goofy stories that the Martin B-26 and Lockheed F-104 were flying coffins. Any professional who flew these always stood by his machine, praising it to the heavens. A frequent story was that a Beech 18 was not safe on one engine; but with R985 Wasp Juniors it couldn't have better engines. Of course a heavily laden Beech (like any twin) would have trouble maintaining height on one engine. A typical case occurred on July 28,1978. The pilot of Superior Airways Beech 18 CF-ZQH

The Green Airways base at Red Lake seen from Twin Otter DMR flown by Norm Wright. (Larry Milberry)

After graduating in geology from Queen's University and spending several years in the field with Canadian Oxydental Petroleum, Dave Robertson started flying at Bonavair in Ottawa in 1980. He joined Green Airways in 1981 and had accumulated 7,000 hours by the summer of 1995. Here he hefts an outboard motor aboard Otter ODJ on July 13, 1995. His trip took him from Red Lake to tourist camps at Shabu Lake and Birch Lake. ODJ's PZL engine made takeoffs easy. It also had STOL-modified leading edges to improve handling in cruise and on approach, although pilots could be cagey about giving this mod too much credit. By 1997 AirTech of Peterborough, Ontario had completed 13 1000-hp PLZ Otter conversions. Then, Gary Miller of Red Lake Seaplane Service prepares an R1340 Wasp for Norseman JIN during routine maintenance on July 15, 1991. (Larry Milberry)

had to shut down his right engine. ZQH (a wheel plane) would not hold altitude and ditched in Pikitigushi Lake, Ontario. All six aboard made it ashore using lifejackets. Norm Wright logged many Beech 18 hours before moving to the Twin Otter. In his view the Beech was a safe machine with ample power. It was responsive, and a nice IFR aircraft. A bonus was its ability to carry a hefty load of ice. This said, C-FBCC crashed in 1992. Wild Country then acquired Super 18 CF-KEL. It no longer had CF-OBR, which was with Thunderbird Lodge in Manitoba. Instead, it had an enlarged fleet headed by KEL plus Otter CF-ITF, a Ce.206 and a Navajo. In 1995 Sabourin, which had been on shaky ground for some time, closed its doors.

Over the years an airplane can suffer different kinds of abuse. Bushplanes historically fly overloaded. In the long run this leads to shorter engine and airframe life. Then, there are things like hard landings, even over-stressing in rough air. Drinking and flying is another old problem that can go hard on both planes and people. On October 25, 1977 a pilot took off from Snow Lake, Manitoba in Beech 18 CF-PCK with a load of fish. PCK landed hard at destination, bounced, overturned and sank. The pilot, whose blood alcohol was above the legal limit, drowned.

A number of retired aviators lived in Red Lake, one being Jake Siegel. He was born in Toronto in 1914 and got hooked on aviation when Harry E. Tegart set up a shop near the Siegel home to renovate JN-4s. One day Siegel met Tegart at the old airport on Dufferin St. Tegart encouraged him to hang around and do odd jobs. Thus did Siegel occasionally scrounge a flight. His first was in an OX5 Swallow. He took some flying lessons from Ken Smith, soloed in Fleet CF-ANC in 1932, and soon had a licence. Since a new pilot needed 50 flying hours to get a commercial licence, Siegel took an engineer's job with mining man Howard Watt. This got him down the St. Lawrence, where one job was ferrying wood cutters by Fox Moth between their homes in and around Matane, and their jobs on the North Shore. Next, Siegel got on with the Hennessy brothers, who had Bellanca AMO working between Pickle Lake and Dog Hole Bay, a 26-mile freight run. Ken Smith was the pilot and Siegel the engineer. In 1939 Smith joined Starratt, taking Siegel along. Here Siegel finally got a flying job. He was given Moth AGX to fly Ike Richie of Ontario Hydro around to camps. In the fall Siegel was freighting on the 40-mile stretch between Gold Pines and Uchi Lake with Fokker AJB. The war started and Starratt pilot Frank Brown left for the RCAF. This meant that Siegel could move to Fairchild 82 AXL.

There was lots of flying through the war. One project was building hydro dams in the Lac St. Jean watershed. Jake Siegel and Charlie Robinson went there in Fairchilds AXG and AXL to work on the Manouan dam. Siegel was next posted to Sioux Lookout; then became a staff pilot on Ansons at No. 5 AOS in Winnipeg. This was valuable experience with a

Jake Siegel in Red Lake in July 1995, wearing his WCAM cap. For all his flying, Jake's favourite bushplanes were the Fairchild 82, Norseman and Otter. (Larry Milberry)

lot of night and instrument flying. Later Siegel returned to Sioux Lookout. One day in 1944 Punch Dickins and Stan Wagner of CPA were in town on an inspection. Base pilot Art Shady took the opportunity to complain about Siegel, who was summarily fired. Getting fired is not a great occasion in a bush pilot's life, but Siegel landed on his feet. Old-time OPAS pilot Giff Swartman heard of Siegel's plight and remembered that he "owed him one". Swartman contacted George Ponsford and got Siegel into the OPAS. He was posted to Oba Lake with Buhl CF-OAR, a machine he disliked. Only rarely could he get it to 1,000 feet agl and the machine dropped like a stone as soon as power was chopped for a landing. Siegel considered the Buhl's heavy Canadian Vickers floats the cause of this. The next two years Siegel was at Pickle Lake on Fairchild 71 OAP. In subsequent years he moved to the Norseman, Beaver and Otter (his favourite types) and retired in 1978. He settled in Red Lake and bought a PA-18, which he flew till 1992. Jake Siegel treasured many memories. In 1995 he told of a flight from Ilford, Manitoba with HBC factor Mr. Bland. He flew Bland to his post at York Factory and was astounded to find the place still packed with ancient trade goods. There were old ship's lanterns and hundreds of muskets in perfect condition. On a roof was an ancient telescope, used to study the horizon for approaching vessels. When Siegel departed, Bland made him a gift of an airplane carved from ivory by an Eskimo artist.

Flight Safety in the North

Northerners have lived with air accidents since the 1920s. Bad weather, usually coupled with a pilot's poor decision-making, has been the traditional cause of these mishaps. On other occasions accidents were unavoidable, as with structural failures. A number of accidents are case studies, one centring on Fairchild 24 CF-FXT. On September 2, 1951 it disappeared in James Bay country. Aboard were Dr. Albert H. Hudson and Bill Barilko, a well-known hockey player for the Toronto Maple Leafs.

Returning to Timmins after fishing north of Fort George, they stopped for fuel at Rupert House. Although advised to wait overnight for better weather, Hudson pushed on into stiff headwinds. Nothing further was heard of FXT. A September 2 telex from the search coordination centre at RCAF Station Trenton outlines what was known:

Report of special flight to Rupert House 02 Sept by OPAS Beaver a/c with RCAF representative. Flight of Fairchild CF-FXT. Left Rupert House 1615-1630 EST. CF-FXT refuelled Rupert House. Wind was east surface 15-18 a/c had difficulty on takeoff... Weather was muggy. Humid day. Carrying fish on pontoons. Vis poor approx 6 miles. Mr. Wheeler advised pilot that weather was poor and suggested that he stay all night. Pilot answered It's OK I'm used to flying in this weather. Indians and missionary stated that a/c looked to be doing stunts. Appeared to have 300-400 ft when a/c disappeared over tree line. A/c appeared so close to point of stall that it was dropping a wing. It was stated that Barilko showed hesitation and appeared reluctant to fly. A/c took off in easterly direction and set approx course 260 degrees magnetic. Was established at Moose Factory that wind was 220/50 mph. Summer resort close to track did not hear or see a/c. Pilot and pass were sober. After viewing takeoff surmised a/c would soon return. Expert advise is that if a/c spun in little or any a/c would remain visible. Impossible for Indians to walk on muskeg terrain even with snowshoes.

With this rough summary began a huge search. The Maple Leafs put up a $10,000 reward for locating the lost men, and the media revelled in rumours that Hudson's flight had something to do with stolen gold. After several weeks the search ended without a trace. In the following years people flying in the area often would scour the muskeg for FXT. Finally, on May 31, 1962 a Bell 47 pilot north of Cochrane caught a glint below. He inspected and found a wreck that soon was identified as the Fairchild. The DOT's report noted: "The aircraft struck the trees in a 50° nose down attitude at an angle of bank of 25° to the left. It was apparent from the damage to the propeller that the engine had been either under power at the time of impact, or windmilling at high rpm. The elevator trim tab was found in the almost full nose up position. The trim control in the cockpit corresponded to this position. There is no evidence available to explain this trim setting, which is considered abnormal to cruise flight. There was insufficient evidence to determine the cause of the accident."

Another sad case is that of Austin Airways Otter CF-MIT. On September 4, 1976 it departed Moosonee for Timmins. En route it flew into wires at Little Abitibi Canyon, killing all aboard. A newspaper report noted: "for the ten most recent deaths in this desolate place there is no marker, only a patch of blackened hillside." The report mentioned that since 1970 there had been 520 air accidents and 107 crash deaths north of the line between Ottawa and

Georgian Bay. But Austin Airways had had but four fatalities between 1934 and this time. Evidence indicated that MIT should not have departed. One passenger, fearful of conditions, had refused to board at the last minute, prudence that saved his life.

Ce.402B CF-EIA crashed at Deer Lake, Ontario on March 24, 1980. The pilot attempted takeoff from the 3,000-foot slush-covered runway, but never got airborne. Six died, three were injured. Contributing to the disaster, EIA was 500 pounds overloaded. Twin Otter C-GTLA of Austin Airways crashed at Lansdowne House on November 23, 1983. Arriving in bad weather from Webequie, it landed short and burned. Four died and the pilots and another passenger were injured. Investigators found that the crew made the flight without the weather details. There were whiteout conditions on approach, and the cargo had been ill secured. The passengers had not been briefed about safety matters.

Many DC-3s have been lost in the North. C-FBJE crashed in bad weather at Pikangikum on November 1, 1988 killing the pilots. BJE had been on a cargo flight from Red Lake, when it crashed a half-mile west of the airport. On May 12, 1977 there was double trouble at Pickle Lake. A Patricia Air Service DC-3 crashed into the lake on takeoff. A PAS Beaver went to the rescue. Pilot Gary Linger landed with a helper. They hauled the co-pilot aboard and tried diving for the pilot. Unable to find him, they took off for base. His eyes being soaked with gas and oil after the rescue attempt, Linger could barely see. He crashed on landing, but all three survived. In another DC-3 tragedy Air Manitoba's C-FADD crashed near Pickle Lake May 11, 1987 after hauling fuel to Big Trout Lake. Cause? The port wing failed in normal flight due to a fatigue crack in the centre section lower wing skin. Radiographs taken about 300 flight hours earlier showed the crack, plus missing fasteners and a rivet. This was not properly reported and no corrective action was taken.

Ice helped kill the pilots of Farfard Aviation's DC-3 C-FBZN in Quebec. BZN was at La Grande on February 28, 1989 hauling fuel to Bienville. It failed to arrive and was found only a mile from the end of the runway at La Grande. The port engine crankshaft had failed. This may not have been recognized instantly, since the rear cylinder bank kept operating (delivering no power to the prop). BZN stalled and went in steeply; 40% of the wing area carried 3mm-5mm of ice. There had been no icing off the end of the runway, so the ice must have been left from the morning, when the crew swept the wings. Such ice would have increased stall speed and aggravated drag with stall onset. Add to that a windmilling prop (as was the case—the crew had not secured the engine upon failure). This was a formula for disaster. A sad detail was that de-icing fluid was available at La Grande. The loss of BZN ended a long career. It had started as USAAF 43-48029. It rolled out in July 1944 and became RAF KG745. It reached Dorval July 20, where it operated with Ferry Command. In

May 1946 it started with CPA, staying for a decade, then joined Trans Labrador Airlines. In 1963 it was sold to Northern Wings and spent the rest of its days in Quebec.

Another weather-related DC-3 crash involved CF-NNA. It struck the 150-foot NDB tower at Sachigo Lake on January 19, 1989, while carrying fuel from Red Lake to a drill site near Sachigo Lake. The pilots were landing in poor visibility, when NNA hit the tower. The starboard propeller tore off and pieces of the tower were imbedded in the fuselage. A wheels-up landing was made, injuring the captain and a passenger. Of 17 full fuel drums 15 broke their tie-downs and flew ahead, trapping two passengers. After a few minutes everyone got out through the cockpit hatch. Transport Canada later classified the accident as avoidable—all the captain had to do was return to Red Lake, instead of attempting to land in rotten weather. TC stated in one publication, "The 'push on' attitude exhibited, but widespread in bush and Canadian Arctic operations, is an endemic and persistent problem in Canadian aviation."

One day the pilot of a Beaver with six passengers set down on James Bay in weather. A strut broke on landing, so the plane was unserviceable. The pilot contacted an overflight, which relayed his distress. Seven stranded people would spend the night on the ice, but they had no survival kit. Thanks to Canada's excellent SAR system, a break in the weather, and whatever other fortunate circumstances came into play, at 0145 a SAR C-130 from 424 Squadron (CFB Trenton) was over James Bay. In difficult conditions it located the Beaver using night vision goggles, dropped survival gear, then put four SARTechs on the ice by parachute. They got the seven under cover and provided hot drinks. A 424 Labrador arrived at 0615 and took the survivors to Moosonee. Had the weather deteriorated, all seven could have perished, all for the operator's lack of concern about winter survival. Another Beaver pilot was moving passengers to a new fishing camp. He removed his rear seats, squeezed in all the kit and passengers, took off, struck trees on climb-out and crashed into a lake. The pilot, crushed against the instrument panel by the cargo he failed to tie down, drowned. The aircraft was determined to have been in good condition and the weight and balance in order. Why did it not climb above the trees? Why, in the first place, had the pilot illegally loaded his plane, not briefed his passengers about safety, or even provided seat belts? No one can be sure, but it is known that he had little sleep the night before, and had been using cocaine a short time before he died.

A tragedy which nobody could have escaped took place on May 1, 1995. At 1328 hours CDT that day Bearskin Metro C-GYYB (F362) and Air Sandy Navajo C-GYPX collided in clear conditions 12 miles northwest of Sioux Lookout. All aboard died. The Metro had departed Sioux Lookout for Red Lake on an IFR flight plan at 1300 with two pilots and a passenger. The Navajo had left Red Lake on a VFR flight plan at 1323 for Sioux Lookout

with the pilot and four passengers. At 1315 Winnipeg Centre advised Sioux Lookout FSS (flight service station, no radar capability) that F362 was estimating arrival at Red Lake at 1332. At 1327 F362 called the FSS to advise that it was 14 nm out. A minute later the FSS operator heard the sound of an ELT. Simultaneously, a Bearskin B.99 in the area reported a flash and falling debris. Efforts to contact Bearskin 362 failed and a search began. An MNR helicopter crew quickly identified a fire in the bush as Air Sandy, then reported debris in nearby Lac Seul (soon confirmed as from F362). Weather at the time included scattered cloud at 6,500 feet and visibility 15 miles. At the moment of collision the Metro was descending at about 230K; the Navajo was in level flight accelerating to 180K. In a follow-up MIT study it was determined that there was a low probability (13%) of the Air Sandy pilot detecting the Metro to allow for the 12 seconds the average pilot would need to see it at 1.4 nm (considering colour scheme, sun angle, cloud, snow cover in the background, closing speed of about 410 knots, etc.), assess the situation and react. Not until about four seconds before impact would both pilots have had a 50-50 chance of seeing each other. It appears that the Navajo pilot saw the oncoming Metro and banked sharply to port. At such a late moment this increased the Navajo's cross-section, making it a bigger target.

In its assessment the Transportation Safety Board lamented that formal training in recognizing in-flight collision geometry and reacting properly to the threat of collision were not required in pilot training. It also noted that neither plane had, nor was required to have, TCAS (traffic-alert collision and avoidance system). The Metro and Navajo each carried GPS, a precise, reliable nav system, but one that can increase the chance of mid-air collision on busy airways with a steady flow of traffic on reciprocal courses. In the aftermath Transport Canada stepped up educational programming regarding pilot diligence in keeping an eye for conflicting traffic, and in urging pilots using GPS to offset their tracks at least a mile to starboard. Bearskin instituted a policy whereby its planes would not exceed 150 knots within five miles of busy Sioux Lookout airport.

Any accident is hard to take, but another in the Red Lake district was poignant. Soon after takeoff from Red Lake on July 1, 1985 a passenger aboard Ce.414 N4639G issued a Mayday, saying that his father, the pilot, had been incapacitated and that nobody else on board was a pilot. Kenora radar control tried to pass along assistance. Fourteen minutes later the passenger radioed that he could not control the plane. It crashed 20 miles from Kenora, killing all four aboard.

Tragedy at Dryden

One of Canada's infamous weather-related disasters involved Air Ontario Flight 1363—Fokker F.28 C-FONF. On March 10, 1989 ONF was city-hopping from Toronto to Winnipeg. For passengers on this well-travelled run it could be a pleasant day, so long as they enjoyed flying, the weather was fine and nobody was in a rush. It also could be a dreary day, as the plane stopped and started—Toronto, Sault Ste. Marie, Thunder Bay, Dryden, Winnipeg. Weather-wise, in summer there could be thunderstorms; in winter... who knows? On this day there were snow squalls and icing by the time ONF was at Dryden. As passengers deplaned and others boarded, ONF's surfaces gathered a coat of snow. Capt George Morwood and FO Keith Mills surely would have discussed this, especially since Morwood always had been a "safety first" captain. To de-ice or not to de-ice, would have been the question. Nobody knows what was said, but it is known that the F.28 was behind schedule, and its APU was on the fritz. That obliged Morwood to keep an engine running on the ground. ONF couldn't be de-iced without risk, since a blast of fluid might dowse the engine. Without an APU to restart, F1363 would be stuck if that happened.

Things were going wrong, those little things that grate on a crew. ONF taxied. Now came another snag. A light plane was inbound, but having trouble with the weather. ONF sat till it got down. This gave time for more snow to accumulate. What did the pilots think? That the snow would blow off once they got rolling? The flight proceeded. Fully fuelled, carrying 69 people and freight, ONF hardly needed a lift penalty from snow and ice on the wings. Even the snow on the runway would do its bit to slow acceleration. ONF groaned into the air, but mushed and wouldn't climb. Within seconds it slashed through spruce and poplar off the end of the runway, crashed and burned. It was a tough crash to pull through—ONF had cracked up in the bush on a cold and blustery day. There was a long delay before rescuers arrived on skidoos. Twenty-four died, others performed heroic acts of rescue. Among those lost were the pilots and FA Katherine Say.

The crash of F1363 spawned a public inquiry headed by Justice Virgil Moshansky. It focused on air regulations, company operating procedures, personalities, etc. As to procedures Moshansky noted that Air Ontario, then in a period of rapid growth, only recently had acquired the F.28. Its crews were learning the plane—Morwood and Mills each had less than 100 hours on type. Moshansky's four-volume report was presented on March 26, 1992. In the end nobody could say why Capt Morwood took off. Moshansky would not accept the old cliché—"pilot error". To do so, he said, "would have been to ignore the larger issue of what factors may have caused a highly experienced captain with some 24,000 hours of flight time to make such an error in judgment." Rather than laying all the fault on the captain, Moshansky questioned government regulations and the way Ottawa had begun deregulating the airlines in the 1980s, but didn't keep pace with the process. Ironically, just a few days before Moshansky's report appeared, a US Air F.28 failed to get airborne in icing at New York's La Guardia Airport. Such accidents continue to plague aviation, and the airlines will keep up their campaign to reduce them.

Ill-fated F.28 C-FONF at LPBIA. In 1988 it came to Canada from France, where it had been F-GEXT. It crashed at Dryden March 10, 1989, when take-off was attempted in the worst of icing conditions. (Larry Milberry)

The best approach will be training, especially regarding pilot attitudes. Technology will also help. An example is CWDS, or "clean wing detection system", first used on an Air Atlantic BAe146. Designed by Air Atlantic and Instrumar of Newfoundland, CWDS differentiates between snow, ice, de-icing fluid, water, etc. on the upper wing and tail. This information is displayed on the flight deck. In October 1990 Air Ontario flight attendant Sonia Hartwick, the only crew member to survive the Dryden disaster, was honoured for her good work in helping passengers after the crash. She received the 1990 Gordon R. McGregor Trophy, presented annually by the Air Force Association of Canada. It is awarded for "outstanding and meritorious achievement by a Canadian in the field of air transportation."

Norseman Days

Even though more than 900 were built, by 1960 the Norseman was fast fading. With wooden wings and fabric covering it was difficult for operators to maintain. All-metal types like the Beaver gradually dominated. The sight of a Norseman rotting in a grown-over lot was common once the cheaper ex-US Army Beavers and Otters were available. Yet, the Norseman would not go quietly. There always seemed to be one doing a day's work for a lodge or other small operator, mainly in Northwestern Ontario, their final stronghold. Inevitably, the nostalgia bug started infecting Norseman fans. People like Gordon Hughes of Huntsville (later of Ignace), Ontario turned out superb restorations. One by one examples started rising from the ashes. Red Lake became "Norseman Capital of the World" and some locals felt the time was right to make something of this. In 1988 the Red Lake business community sponsored a waterfront clean-up. This grew into a plan for some new parks. Considering Red Lake's roots, it seemed natural that one be dedicated to aviation. With government funding Moore-George Associates of Toronto developed a theme park around the Norseman. For 1992 Red Lake planned to dedicate its Norseman Heritage Park (two years earlier Red Lake hosted its first Norseman fly-in). A hulk was located in Kuby's aircraft salvage operation in Kenora and was shipped to Red Lake Seaplane Service. The plane was CF-DRD, Norseman No. 831, originally delivered to the US Army Air Corps on May 4, 1945 as UC-64A No. 45-41747. It went surplus to a US

operator in August 1946, then was purchased in 1953 by Barney Lamm of OCA.

Over the winter of 1991-92 DRD was restored, then launched on June 28 for a last taxi on Howey Bay. Meanwhile, Singleton and Associates of Toronto designed a graceful arch upon which DRD was mounted in time for a July 24-26 Norseman "bash". Eleven Norsemen attended, some local, others from as far away as Lake Winnipegosis and Minnesota. All were docked at the head of Howey Bay. The weekend guest of honour was retired engineer Bob Noorduyn, son of the Norseman designer. He dispensed Norseman lore and made the point that the Norseman was really a third generation design. His father's first design had been the Fokker Super Universal. Next came the Bellanca Skyrocket, finally the Norseman. The first were rugged, practical types, so Noorduyn put their best features into his ultimate design. Another point he made was that the Norseman was the first Canadian bushplane designed for floats, unlike predecessors, which were land planes modified for floats. Noorduyn recalled how every Norseman his father sold came "with a free paddle, and shotgun". He also told of his father's postwar plans for a recreational two-seater, and for a "twin Norseman" using such Harvard parts as outer wing panels. These ideas did not gel—the industry slumped at war's end and Noorduyn returned to the US.

Other visitors to Red Lake were Barney Lamm from Winnipeg; Air Canada 767 captain Ken Reppen, who had started in aviation as a dock hand with Austin Airways; Al Williams from Edmonton, who had flown DRD for Taylor Airways of Lynn Lake; and Joe McBryan and Peter Austin, who spent nine hours flying down from Yellowknife in a Ce.185. McBryan brought two original DRD logbooks, most of the signatures in them being Al Williams'. Jim Olson flew 220 miles from Steep Rock, Manitoba in Norseman CF-IGX. The 1943 model had 4,300 flying hours. Rollie Hammerstedt of Redditt, Ontario also attended. Over the years he had worked on many Norsemen, and helped develop the metalizing process. This started with the bellies. Next, the fuselage between the doors was metalized. Finally, Norsemen OBE and JEC were fully metalized. Like everyone else who visits Red Lake, Jim Olson wandered into the Lakeview Restaurant. There he perused the menu. "Norseman Burger" caught his eye. His selec-

Bob Noorduyn, Jr. (left) was born in 1922 and spent much of his youth in Montreal, where his father was developing the Norseman. He received his training at Parks Air College, then followed his father into aeronautics. In the 1950s he was vice-president of Noorduyn Norseman Aircraft Ltd. in Montreal, but spent most of his later career at North American Aviation in Columbus, Ohio. Bob is pictured at the 1995 Norseman Days festivities with fellow Norseman history buff Bruce Gowans of Calgary. In the second photo, Al Williams, who flew Norseman DRD for several years, attended the original Norseman Days in July 1992. (Larry Milberry)

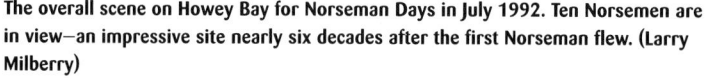

The overall scene on Howey Bay for Norseman Days in July 1992. Ten Norsemen are in view—an impressive site nearly six decades after the first Norseman flew. (Larry Milberry)

This is how DRD reached Red Lake in 1991 for restoration. (Larry Milberry)

(Right) Norsemen JIN, DRD and GLI on July 23, 1992. JIN was ex-RCAF 2482. It also served companies like Austin Airways and Labrador Mining and Exploration before joining Red Lake Airways in 1988. In former years GLI , seen here with Red Lake Airways, worked for companies like QCA, Air-Dale and Kryo's. In the second photo, Joe Sinkowski lands KAO on Howey Bay on July 16, 1995. (Larry Milberry)

tion was made with the comment, "If I came to Red Lake and didn't try a Norseman Burger, I'd hate myself."

The Lakeview is to Red Lake what the Wildcat is to Yellowknife. There all the gossip in town can be heard—talk of mining and unemployment, of partying, screwing and salvation. One laid-off miner was overheard chatting to his buddies about entering a government retraining program. "I can't decide," he began. "Should I go in for a brain surgeon, or maybe an astronaut?" Another diner, when criticized about his beer-belly, crowed, "Hey, it took me a lotta years to get her into this good a shape!" Bad jokes are the curse of places like the Lakeview. Diner No. 1: "Howze a woman like a quart of milk?" Diner No. 2: "I dunno." Diner No. 1: "Keep her around too long and she's sure to go sour." At another table the talk

Norseman CF-SAN at Red Lake for Norseman Days 1995. Originally delivered to Saskatchewan in 1946, it later did the rounds of bush operators till a take-off prang at Fort Simpson in December 1981. Joe McBryan of Buffalo Airways bought the wreck in 1993 and had it restored at Wetaskiwin. SAN flew again on July 2, 1995, just in time to make Norseman days. Beyond is CF-BTC. From 1968-92 it was owned by Zane Palmer, who kept it in nearly-original condition. Following Palmer's death in 1992 , BTC made a forced landing on Cree Lake north of La Ronge, when fabric began tearing from the wings. J.W. "Whitey" Hostetler bought it from owner Randy Daoust and sent in John Blaszczyk and Joe Sinkowski, who made temporary repairs, then ferried BTC to Red Lake. (Larry Milberry)

The cockpit of a working Norseman–CF-JEC on July 15, 1991. Then, Joe Sinkowski flying Norseman CF-KAO (Chimo Air Service) on a July 15, 1995 sunrise trip. Born in 1950, Joe took geology at Cambrian College in Sudbury. He started flying in 1972 and got his first job flying at Pickle Lake for O.J. Wieben. Those were the days before airstrips throughout Northwestern Ontario, so there was steady work year round. Joe could get 800 hours a year without any trouble. He moved to Green Airways in 1977, but left in 1991 to work for Whitey Hostetler. He continued flying for local operators like Chimo Lodge and Viking Outpost Camps. (Larry Milberry)

was of the MNR. One fellow summed it all up, at least in his mind: "You go into that goddam MNR office down there and ask a question to four different people. You get four different answers." Elsewhere, a fellow was dreaming about moving on, maybe getting out west. So it went—all the topics, all the viewpoints, not many answers, but good therapy for everyone.

Each year one or two "new" Norsemen take flight, rebuilt from available hulks. Buffs scour forest and tundra for wrecks, some of which have been landmarks for decades. One by one these re-appear as factory-fresh aircraft, the only guaranteed original part being the data plate. Besides the move by buffs to restore wrecks, others have a more practical bent. By 1990 the price of a Beaver exceeded $200,000. An operator in the market might consider a Norseman rebuild. In 1996 he could have one for perhaps $160,000 and have more performance and payload than a Beaver. Besides airworthy examples, by 1997 there were various Norseman museum pieces and restoration projects. With the occasional new Norseman still turning up, it wasn't likely that we would see the demise of this old favourite.

Canada's Surviving Norsemen in 1997

Reg.	s/n	Notes
1. CF-BFT	17	547962 Alberta Ltd, Fort Smith, Alberta
2. C-FBHZ	N29-13	Art Latto Air Svc, Savant Lake, Ontario
3. CF-BSB	N29-15	Bolton Lake Lodge, Winnipeg, Manitoba
4. CF-BTC	29	J.W. Hostetler, Red Lake, Ontario
5. CF-DTL	57	Gordon H. Hughes, Ignace, Ontario
6. CF-ECG	N29-43	Gogal Air Service, Snow Lake, Manitoba
7. CF-ENB	324	Grass River Lodge, The Pas, Manitoba
8. CF-FOX	340	Randy R. Doust, St. Albert, Alberta
9. CF-FQI	364	Viking Outpost Air, Red Lake
10. C-FFQX	625	Osnaburgh Airways, Pickle Lake, Ontario
11. CF-FFUU	74	Birch Lake Lodge, Red Lake
12. CF-GLI	365	Gogal Air Service, Snow Lake, Manitoba
13. C-FGSR	N29-47	Stewart Lake Airways, Vermillion Bay, Ontario
14. C-FGUE	542	Sioux Lookout Fly-In Camps, Sioux Lookout, Ontario
15. CF-IGX	141	David Lindskog and R.W. Polinuk, Winnipeg
16. C-FJEC	469	Sabourin Lake Airways, Cochenour, Ontario
17. C-FJIN	CCF-55	Chimo Air Service, Red Lake
18. C-FKAO	636	Chimo Air Service, Red Lake
19. C-FKAS	367	G.R. Arnold, St. Albert
20. C-FLZO	535	Wrong Lake Airways, Winnipeg
21. C-FOBE	480	Green Airways, Red Lake
22. C-FOBR	N29-35	Wrong Lake Airways, Winnipeg
23. CF-SAN	N29-29	Buffalo Airways, Hay River, NWT
24. C-FSAP	231	Nueltin Lake Air Service, Thompson, Manitoba
25. C-FUUD	224	G.L. Crandall, Ponoka, Alberta
26. C-GRZI	175	Grass River Lodge, The Pas, Manitoba

Museum Display Norsemen

Reg.	s/n	Notes
1. CF-AYO	1	Canadian Bushplane Heritage Centre (wreck only), Sault Ste. Marie, Ontario
2. CF-BSC	N29-17	Canadian Museum of Flight and Transportation, Langley, BC Sold in May 1997 to Jennings River Airways.
3. CF-DRD	831	Town of Red Lake, Ontario
4. CF-EIH	94	Alberta Aviation Museum, Edmonton
5. CF-JDG	538	BC Air Museum, Sidney, BC
6. CF-MAM	N29-26	Petro Canada Building, Calgary
7. CF-SAM	N29-27	Western Development Museum, Moose Jaw, Saskatchewan
8. RCAF 787	136	National Aviation Museum, Ottawa

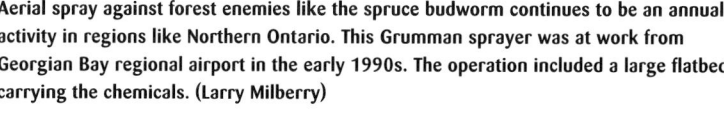

Aerial spray against forest enemies like the spruce budworm continues to be an annual activity in regions like Northern Ontario. This Grumman sprayer was at work from Georgian Bay regional airport in the early 1990s. The operation included a large flatbed carrying the chemicals. (Larry Milberry)

The North has many independent air operators. Some are involved in growing and harvesting wild rice; others, like Bob Buckler of Loon Air, harvest minnows for live bait. Here he works with his minnows at his Ear Falls location south of Red Lake in July 1991. Then, Bob's PA-18 Super Cub equipped with wing-tip "lift boosters". He used it to retrieve 1,000 dozen minnows daily along his 120-mile trap line. (Larry Milberry)

(Right) For decades there were no runways in the North, but with WWII many airports were built, boosting aviation in the postwar era. In the 1970s Ontario began building airstrips with navigation beacons at remote Indian reserves. This high-level view shows the new gravel strip at Sandy Lake, Ontario in the mid-1970s. Such strips brought large planes like the DC-3 and 748, causing the decline of float and ski operations. Operators like Green Airways, who once supplied the reserves, now depended largely on summer tourist business. Although some mineral exploration still went on, this often was supported by the more versatile helicopter. (CANAV Col.)

Quebec and Ontario:
The Regionals Compete

Regional airline competition in Quebec is as intense as anywhere. CRA partner Inter-Canadien and Air Canada partner Air Alliance were the big operators in 1997. Air Nova, Air Atlantic and Air Ontario served Dorval from outside the province, while local carriers like Air Alma and Air Montreal covered niche markets. Each worked hard to attract business and there were constant changes in fares and schedules aimed at outfoxing the competition. Inevitably, carriers stepped on each other's toes. Early in 1993, for example, Air Montreal extended service east to Baie Comeau and Îles de la Madeleine. This annoyed Inter-Canadien, which noted: "Normally they operate as a courier, but also carry passengers at very low rates." On February 1 Air Satellite lowered some fares and increased some frequencies,

(Above) Although not often thought to be, communities on the lower St. Lawrence River are as northerly as Timmins or Kenora, so share their characteristics when it comes to wintry weather. As with places like Timmins, river communities were served by the main commuter airlines and various smaller outfits. This view was at Baie Comeau (49˚7' N. lat.) on November 18, 1992. Present was Air Montreal SA-226TC Metro C-GIQG, offloading cargo. It had started in Jugoslavia in 1979, moved to Belgium (1986), France (1987), then joined Quebec's Intair in 1989. The Metro was popular with Intair on passenger skeds in pre-ATR days. An Inter-Canadien ATR42 is farther down the ramp. (Larry Milberry)

(Right) The "big guns" in Quebec regional operations in the 1990s were Air Alliance (Air Canada connector) and Inter-Canadien (CAIL connector). This Air Alliance Dash 8 was at Sept-Îles on November 18, 1992. The Inter-Canadien ATR42 was at Mont Joli the following morning. In the 1990s the 46-seat ATR42 was Inter-Canadien's Quebec workhorse. The company originally ordered five in 1988. The ATR's ability to operate with passengers, as a combi, or as a freighter made it perfect for the job. Its turboprops were ideal for severe winter ops. Passengers were spoiled by its spaciousness, quiet interior and speed. Once they had sampled the ATR they were reluctant to book on the Beech 99, Metro, Aztec, etc. With Air Alliance operating Dash 8s, Inter-Canadien could offer nothing less, and in 1994 went one better by placing the F.28 on prime Quebec routes. (Larry Milberry)

e.g. Baie Comeau to Sept-Îles with a $129 one-way fare compared to Inter-Canadien's $163. Air Satellite, however, had a battle to win customers on its smaller, noisier planes that had no in-flight catering. In 1993 there also was news that giant Air Transat was offering a $30 bus fare from Lac St. Jean/Saguenay to Quebec, where passengers connected to southbound flights. This annoyed Inter-Canadien, which lost some Bagotville-Quebec fares to the bus. Inter-Canadien was pleased with its $99 Bagotville-Dorval weekend return fare—about 800 tickets were sold in a month. A Christmas 1992 seat sale throughout Quebec was a success. Meanwhile, frequencies were reduced Quebec-Dorval and Dorval-Ottawa. Bad news came with the announced closing of a mine in Chibougamau. Inter-Canadien also feared additional layoffs by the Iron Ore Company of Canada in Sept-Îles and Wabush.

In Ontario the same kind of competition prevailed. Canadian Partner noted Air Ontario's reduced frequency early in 1993 to places like Sudbury and North Bay. In January it offered an incentive fare Red Lake/Dryden-Toronto of $339 to offset reduced service in Northwestern Ontario. A public uproar from local politicians and travel agents led to a

quick reinstatement of frequency. At the same time Canadian Partner increased fares by 5%. This was matched by Air Ontario, norOntair and Bearskin, but not by Air Creebec. Meanwhile, Air Ontario in its employee newsletter *Wingspan* discussed coming changes at Canadian Partner: "They will be after us more than ever; their survival depends on it. We must make it our first priority to ensure that they do not succeed." At this time Air Ontario had about 66% of the regional commuter market and wanted to keep it that way: "We must remain focused on customer service and we must continue to examine our product to ensure that it meets and exceeds the expectations of our customers." Clearly, customer satisfaction was the ultimate goal of the commuters in this fiercely competitive market.

As of January 11, 1993 CRA frequencies were reduced Toronto-Sault Ste. Marie, Sudbury-Timmins, London-Toronto, North Bay-Toronto and Dryden-Thunder Bay. A new 0605 Timmins-Toronto (via Sudbury) departure was introduced to replace the 2100 non-stop. The company explained, "This allows better connections and same-day return for business travellers." Other CRA information concerned a new 500-job joint venture at Kingston by

Distek Energy that might bring some business. The Falconbridge and Inco announcement of six-week Christmas furloughs in Sudbury, and that Falconbridge would lay off a further 200 was bad news. Thousands of miners would be out of pocket and less likely to fly from Sudbury than drive. More bad news hinted at 1,500 new layoffs from Algoma Steel in the Soo.

Air carriers always are keen to have the best equipment. Cost efficient, fast and comfortable aircraft are the goal. Air Alliance, Air Nova and Air Ontario all used Dash 8s. This left Canadian bringing up the rear with Jetstreams and Brasilias until the CRA merger. Soon CRA/Inter-Canadien rationalized their fleet, and gave ATR42s the run of Ontario and Quebec. In February 1994 Air Ontario had 21 Dash 8s to CRA's seven Ontario-based ATRs. Air Alliance had 11 Dash 8s to Inter-Canadien's eight ATRs. CRA was still the underdog, but happy with market share and with slowly winning new customers. Keeping travel agents current is vital to the regionals. When CRA introduced the ATR42, it gave every Ontario travel consultant the chance for a familiarization flight. They wanted to show the ATR's appeal compared to the Jetstream. To CRA this was good business, for 65%-75% of their sales came through agencies. Two big CRA/Inter-Canadien events took place late in 1994. On November 14 it began non-stop Toronto-Quebec service with the F.28. The F.28 also went on Montreal-Quebec-Sept Îles and Montreal-Sept Îles non-stop on December 6. At the same time Wabush and Abitibi gained F.28 service. An Inter-Canadien newspaper ad was headlined "Montréal-Sept-Îles en Jet", something that must have riled Quebec's fanatical language police, who strive to "purify" Quebec by driving out the use of English words.

Important to all the carriers is public relations and advertising. Each airline supports local and regional festivals, charitable, educational, sporting and arts events, etc. Radio, TV and newspaper advertising is vital, as are promotions with car rental agencies and hotels. Everything that can help a company hold or improve its slice of the market is explored to the fullest. The airlines also struggle for perfection in on-time arrivals and departures. A late airplane draws calls and letters from customers, who usually finish by saying how they are going to give their business to the competition next time. This pressures airlines to keep aircraft in top shape; and ticketing, baggage and ramp teams giving 100%.

Always a struggle for the regionals is keeping equipment in shape while reducing maintenance costs. How can one have the best of anything, while pinching pennies? The use of advanced concepts in areas like finances has helped achieve high maintenance levels. In its 1994 maintenance planning statement CRA listed its areas of interest: reducing engine costs through enhanced engine management, i.e. involving suppliers to a larger degree in management programs, decreasing inventory, increasing revenue from the sale of parts, lowering the time to do aircraft block checks, running human relations courses, and improving on-time performance. The block check was a means of bringing an airplane down after a period of days for all the checks that usually were done more frequently, e.g. overnight or on weekends, bit by bit. CRA stated: "As the fleet ages, this will allow us to do more in the way of major refurbishment and repairs."

Sept-Îles

The main centre on the Quebec North Shore is Sept-Îles, also known in years gone by as Seven Islands (1997 population: 25,000). Sept-Îles was incorporated as a city in 1951, although its site near the mouth of the Moisie River was used for centuries by Indians and early European traders, fishermen and explorers. By 1900 only about 800 Acadians and Indians lived around here, but the opening of a pulp operation at nearby Clarke City in 1908 brought a population surge. A major boost came 1948-54 with construction of the railroad to the iron ore deposits in Labrador. Sept-Îles' population pushed towards 30,000 in the 1960s. Today it serves chiefly as a deep water port exporting iron ore, and as a regional service and government centre. In the 1980s Sept-Îles began producing aluminum.

Since the early airmail flights of the 1920s,

aviation was important to Sept-Îles. Carriers like Canadian Airways served it in the 1930s and Quebec Airways' de Havilland Rapides were well known there later. From the 1950s air service grew. Trans-river connections improved to places like Mont Joli and Rimouski; up-river to Baie Comeau, Quebec and Montreal, and downriver as far as Blanc Sablon. Northern Wings became the key local carrier after WWII, then TCA and Quebecair began skeds in the 1960s. By the mid-1990s Sept-Îles was a busy regional hub, with daily skeds in all directions—north to Goose Bay, south to places like Mont Joli, east to Blanc Sablon and west to Quebec and beyond. In 1993 the main operator was Alexandair, run by former Quebecair pilot Jean-Marc Roy. Air Alliance and Inter-Canadien each had daily Dash 8 and ATR42 skeds. The ATR was especially busy below Sept-Îles, usually as a combi carrying passengers, mail and freight. When the river was frozen, the ATR sometimes took full loads of groceries (9,000 pounds) down the shore. Until a road is built beyond Sept-Îles, communities downstream will depend on river boats in summer and freight flights in winter.

A typical flight from Sept-Îles was on November 18, 1992. Six passengers boarded Alexandair's 30-minute sked to Port Menier, the only permanent settlement on 140-mile long Anticosti Island. Three of the passengers

(Right) Pat Roy and Pierre Corbeil crewed the Alexandair Metro for the trip downstream to Port Menier on November 18, 1992. Roy had been flying F.100s earlier, but the demise of Intair sent pilots like him scrounging for jobs. Those who got to fly even a Metro were doing well in a tight job market. Three years later the airlines were booming and couldn't find enough pilots! As to Alexandair, it formed in 1987 from Air Brousse (1983) Inc., taking over Propair's North Shore operations and seven of its aircraft. (Larry Milberry)

were Americans up to hunt deer. The aircraft was Metro II C-GQAP flown by Pat Roy and Pierre Corbeil. Jacques Cartier was the first-known European to visit Anticosti, arriving in 1534. In 1884 a land company went bankrupt trying to settle the island, and Anticosti was auctioned for $101,000 to an English entrepreneur. He went broke and French chocolate king Henri Menier bought Anticosti as a private hunting preserve. In 1896 he brought in 220 Virginia white-tail deer. These multiplied into the thousands.

In 1926 Consolidated Bathurst purchased Anticosti for its pulpwood. The population grew to 3,000. But forestry declined, then ceased in 1972. By 1994 Port Menier had about 175 full-time residents. The permanent population was being phased out by denying habitation rights to young residents, once they were old enough to attend school on the mainland.

As QAP taxied to the small terminal at Port Menier, a Conifair DC-6 with its load of hunters and their kill was firing up for St. Hubert. A strip at the east end of the island was being served at this time by an F.27 of Les Ailes de l'Île. A Navajo of Trans-Côte was also on the

Much of Sept-Îles' traffic is attracted by heavy industry—iron ore shipping and aluminum manufacturing. This view shows the main iron ore docks backed by the city in November 1992. By this time mining had dwindled. Even Schefferville, Canada's iron ore mining capital and a model planned community in the 1950s, was a ghost town. The locals in such communities prefer sked flights to the long drive to places like Quebec and Montreal. The loosening of constraints on the regional carriers led to the opening of many new Quebec routes; e.g. in 1988-89 Air Alliance began Montreal-Boston, and Air Nova introduced Quebec City-Fredericton. (Larry Milberry)

(Below) Aviation Québec-Labrador Bandeirante C-FGCL arrives at Sept-Îles from Lac St.-Jean on November 20, 1992. This company still operated the venerable Aztec, which was favoured for charters to small, remote places. (Larry Milberry)

(Above) The ramp at Sept-Îles the morning of November 18, 1992 included a Trans-Côte Navajo, Alexandair Metroliner and Air Alliance Dash 8. Air Canada connectors were on strike in early 1997. This was a bonanza for CAIL associates like Inter-Canadien, and smaller carriers like Air Creebec—to the dismay of striking pilots, they took all the Air Canada business they could handle. The strike concerned the connector pilots' demand that their seniority list be merged with the main Air Canada list, e.g. placing long-time Dash 8 pilots ahead of Air Canada pilots flying much heavier aircraft. Needless to say, Air Canada pilots opposed the proposition and a strike ensued. (Larry Milberry)

A Laurentian Air Service Beech 99 departs Sept-Îles for Schefferville on November 18, 1992. (Larry Milberry)

(Left) Air Satellite Twin Otter captains Christine Vaillancourt and Nancy Lemelin ready for a trip from Sept-Îles on November 19, 1992. Their destination was the isolated hydro base at Poste Montagnais, where about 100 people were stationed. (Larry Milberry)

A Trans-Côte Navajo ready for a trip from Sept-Îles, then Navajo pilot Jean Duchesne (left) with aviation writer Ken Swartz en route Sept-Îles to Mont Joli in Alexandair's Navajo C-GQAI on November 19, 1992. Two passengers were carried out and six back. (Larry Milberry)

ramp, as was a La Ronge Twin Otter on contract from August to December. Pilot Robert Burns was supplying five hunting camps. Another local flight from Sept-Îles was to Mont Joli aboard Trans-Côte Navajo C-GQAI. The 45-minute route followed by pilot Jean Duchesne was up the North Shore to Godbout, then across the St. Lawrence. He delivered his two passengers, then waited about 15 minutes before boarding six to return.

Mont Joli was a Conifair base at this time. Conifair had a fleet of DC-4s, DC-6s, Constellations and Convair 580s specializing in spring budworm attack, and hunting charters to Labrador and Anticosti. A number of DC-4s and DC-6s were present in November 1994, including Conifair's last Lockheed Constellation. C-GXKR had been derelict for some years (following an engine failure), but still looked imposing. In 1993 the Constellation Group USA of Scottsdale, Arizona arrived to change two engines on XKR and get it ready to fly. A

crew headed by Capt Frank Lang flew XKR (now N749VR) away on September 9, 1994. First destination was Presque Isle, Maine where weather and a snag with No. 3 engine forced a two-day layover. N749VR now made a 4:46-hour transit to Dayton, Ohio, then carried on to Fort Smith, Arkansas. Next day the ferry terminated at Avra Valley, near Tucson. There it parked beside N494TW, another ex-Conifair sprayer, resplendent in its MATS paint job. Work now began preparing for a flight to Holland, where the Dutch Constellation Association awaited XKR's arrival.

Nordair: Profile of a Quebec Regional

One of Quebec's great airlines was Dorval-based Nordair. It dated to February 25, 1946—the day Roberval-based Mont Laurier Aviation was incorporated. With Roberval-Fort Chimo rights, it began Canso operations in July 1947. In 1953 Mont Laurier was taken over by Boreal Airways of St. Felicien. Boreal, formed

in 1947, was serving the bush with a Norseman and Husky. Boreal then was bought by Fred T. Briggs and Carl Burke of Maritime Central Airways. They were keen to follow up by absorbing Mont Laurier, mainly to gain access to Fort Chimo. Once the merger was made, it began building routes in northern Quebec. Contracts were won on the DEW Line and Mid Canada Line. On May 24, 1957 Boreal and Mont Laurier merged totally as Nordair. The first service was on May 24, when a DC-4 operated Montreal-Roberval-Fort Chimo-Frobisher. Nordair focused on the DEW Line—contractors were earning 80¢ per mile flown. Capt J.S. "Pat" Patterson of Nordair recalls the joys of navigation in the Arctic in DEW Line days:

Most aircraft had only magnetic compasses and directional gyros, with precession rates far in excess of what would be considered minimal today. North of 60°N, magnetic compasses are

Wabush Mines' corporate Gulfstream (C-GWAM) at Sept-Îles in November 1992. It was on a regular sked from Dorval. Besides this regular visitor the Iron Ore Company of Canada and the Quebec North Shore and Labrador Railway each kept an F.27 at Sept-Îles for corporate travel needs. (Larry Milberry)

Lac St-Jean/Saguenay is a major industrial region developed in WWII with American money to manufacture aluminum. At the same time a fighter base was built at nearby Bagotville to defend the area. CPA linked the region with Montreal and Quebec using the Barkley-Grow, Boeing 247 and DC-3. After WWII the aluminum complex and base expanded. Air Canada served the region with the Viscount, then the DC-9. The great Alcan complex is seen in a June 1994 photo; then, a view of the airport at the same time—the small civil terminal is beyond the military flightline. (Larry Milberry)

(Right) Helicopters are vital along the North Shore and in Quebec's hinterland. Here a Viking Jet Ranger and AStar and a Coast Guard Jet Ranger sit at Sept-Îles on November 20, 1992. About 20 helicopters were in the area on such tasks as hydro-electricity and railway line patrol, mining and hydrological work, and carrying mail to communities isolated during freeze-up. (Larry Milberry)

(Right) Viking AStar C-FIOC cruises near Sept-Îles; then, pilot Michel Durant. A veteran pilot from the French Army, he came to Canada in 1963 to fly for companies like Autair, Northern Wings and Okanagan. With Okanagan he flew offshore in places like South America, India and Ireland. (Larry Milberry)

(Left) Sept-Îles, Dorval and Deer Lake, Newfoundland are some of the places with flights to Goose Bay. On January 22, 1993 this Provincial Airways Ce.208B Cessna Caravan "Super Cargomaster" was at Goose Bay. All-cargo Caravan service was introduced by Provincial in 1988— 14 communities were served, and an overnight courier sked was introduced: Goose Bay-Deer Lake-St. John's, Stephenville-Sydney-Halifax. Once development began of the vast mineral deposits at Voisey Bay, near Nain, Labrador, aviation in the area hit new levels. Labrador Airways and other regional operators expanded overnight and new companies appeared (Frank Kelner set up a base at Goose Bay as the first Canadian operator of the Pilatus PC-12). (Larry Milberry)

practically useless. The easiest way to check headings then was to ascertain the local time zone, set one's wrist watch to that time, and point the hour hand at the sun . Between 6:00 AM and 6:00 PM, half way between the hour hand and 12:00 was true south. Between 6:00 PM and 6:00 AM, half way between the hour hand and 12:00 was true north. Strangely enough this primitive method worked quite well. A heading within 5-10° accuracy was possible

Arctic flying requires pilots and mechanics who are independent and adventuresome. They adapt to rough 'n ready conditions; and to the daunting weather and long winters. Many are young and go north for the experience. The job makes or breaks them, but most survive to move to other branches of aviation. It is no surprise that most have their sights on a career "down south", once they have been schooled in the Arctic. A few, however, enjoy the Arctic and stay for a lifetime. Capt Patterson logged

more than half of his 28,000 flying hours above the Arctic Circle. Nobody would know more about the region, so we can depend on the accuracy of his description of Arctic weather:

Winter in the Arctic is a phenomenon in itself. A large high pressure area usually covers the whole area and is called just that—the Arctic High. This brings clear conditions with light wind. Except for the bitter cold, often as low as

-45°C, and the eternal darkness, flying conditions are generally quite good. When a blow does come up, it is hard to predict. Often the differential in pressure will be as much as 20 millibars in a distance of 100 miles, and the winds will be calm. In other cases, under similar or even less differential, the wind will be gusting 30-40 knots.

Visibility is reduced in blowing snow to ⅛ of a mile or less. In these cases an Arctic station will go on "white alert", wherein only essential people are allowed outside, and only in pairs. When winds are higher, in the 60-70 knot bracket, with visibility down to a few feet, a red alert is called and nobody goes outside. Once the blow is over, the task of digging out begins, as entire buildings can be drifted under. Strangely enough, landing in a white out of blowing snow is quite easy. The depth of the blowing snow is only about 50 feet, so from the air, runways and runway lights are clearly discernible. Taxying to the ramp is the difficult part. Takeoff, of course, is out of the question.

Patterson also had another way of navigating on the DEW Line—flying from one plane wreck to the next, but that wasn't so accurate—some of the pilots had been off course when they crashed! There were no reliable maps of the region for some time. Thus did the tried and true astro compass become a vital tool in navigating. Improved maps eventually appeared, due to the intensive program to photo-map the Arctic using RCAF Lancasters, and high-altitude Spartan Mosquitos. Radar was available in newer aircraft and was the best in navigation equipment until such systems at Doppler, Loran, INS and GPS.

In April 1960 Nordair took over the heavy aircraft division of Wheeler Airlines and sold its VFR operation to Wheeler Airlines (1960) Ltd. This gave Nordair access to the important centres of Val d'Or and Great Whale. On July 1, 1960 Nordair took over the DEW Line lateral resupply between Cape Dyer in the east and Barter Island in the west. It also conducted the annual fuel lift from Hall Beach to remote DEW Line sites. This busied as many as six C-46s. Flying around the clock, each could log as many as 18 hours a day. Later in 1960 the Class 3 licence of Sarnia Airlines was purchased and in January 1961 Nordair commenced daily service Windsor-Sarnia-London-Toronto using a DC-3. This was the first incursion into the Ontario sked market of a Quebec-based airline. The Sarnia Airlines licence was upgraded in May 1961 and service was expanded to include Oshawa, Kingston and Montreal. This was known as the Seaway Route. In April 1961 the 36-seat Dart Herald took over from the DC-3, but poor sales forced cancellation of the Seaway Route that August.

On August 6, 1961 Nordair operated the first scheduled flight to the High Arctic. Capt J.S. Patterson flew a DC-4 off the runway at Dorval at 2300 hours, headed to Resolute Bay via Frobisher Bay. The flight arrived at "Rez" 16 hours later. Patterson recalled that they authenticated several dozen letters with the Resolute Bay postal stamp for collectors around the

Nordair used the C-54 from 1958, CF-IQM being its first. The C-54 established Nordair as a long-range DEW Line carrier and remained in service into the 1970s. IQM was one of the last C-54s built—it was delivered to the USAAF in October 1945. It came to Canada for Wheeler in 1956 and was flying from Yellowknife for Buffalo Airways 31 years later. Here it was at Dorval on February 19, 1966 in a classy-looking scheme. (Larry Milberry)

world. The flight was back in Dorval at 0900 on August 8. In July 1962 Nordair introduced a Montreal-Frobisher Bay-Cape Dyer-Hall Beach-Resolute Bay sked. Meanwhile, Nordair was interested in MCA. In 1963 that was settled when EPA absorbed MCA. Late that year Nordair entered the ITC market with DC-6A CF-NAB. DC-7C NAI replaced it the following June. It arrived in May but soon was dealt to PWA, when four Super Constellations were purchased in November 1964. These were for Atlantic and Caribbean charters, but also worked the Arctic. Pat Patterson noted: "The Constellation, owing to high maintenance costs, was never known as a big money maker with the airlines, yet did yeoman service in the north. There were the usual maintenance problems, but not even Lockheed envisaged Super Connies flying into -50°C. An on-time departure or arrival was an event, but late or not, no flight was ever cancelled." Brian Harris, writing of Nordair in *Propliner* (Summer 1980) described an especially hard time with Connie NAM, which was operating Nordair's last overseas service for 1967:

No problems were encountered until the last flight, which was routed Montreal-Amsterdam-Dusseldorf-Gander-Montreal. No fewer than five engine failures occurred! The first was on the approach to Amsterdam. An engine change was required, and it was decided to make a three-engine ferry to Hamburg, where Lufthansa was to complete the work. A second engine was shut down on the way to Hamburg, where CF-NAM landed on two. The following day, during a test flight after both engines had been changed, one of the new engines failed. This was rectified and CF-NAM was then ferried to Dusseldorf to collect a German group travelling to Montreal. By this time the aircraft was two days late, and the German press was quite interested, to say the least. Just after takeoff from Dusseldorf, another engine failed. The aircraft dumped fuel and returned to the field, where it was found that another engine change was necessary... Further engine problems developed an hour or so before landing at Gander. Rather than perform the usual precautionary engine shut-down, the captain elected to keep the engine turning at minimum revs so as not to worry the passengers any more. It did

not do the engine any good, but probably prevented a full-scale riot. A replacement Super Connie was immediately despatched from Montreal, and the passengers had only a short wait at Gander. They eventually arrived in Montreal 3½ days late. Their return flight to Dusseldorf was subcontracted to Wardair.

Nordair entered the jet age in November 1968, leasing some Convair 990s from Modern Air Transport in the US. These were operated on Atlantic charters. That November Nordair accepted its first 737 and soon had five (three leased from United Airlines). The 737 brought the first modern jet service to the Arctic and gave Nordair attractive equipment as it expanded ITCs to the Caribbean and Florida. It also proved an ideal Arctic freighter. Flying from Dorval it could haul 33,000 pounds to Fort Chimo, or 28,000 to Frobisher. For 1968 Nordair carried over 47,000 passengers. With its designation by the ATB in 1969 as the regional airline for Western Quebec, and Southern Ontario, sked service was inaugurated between Montreal and Hamilton that April, Montreal and Ottawa in May 1971 and on to Pittsburgh from Hamilton in May 1972. Connecting the two steel cities seemed natural, but equipment eventually had to be downgraded to the FH-227; the route finally was terminated on account of poor loads. Now the FH-227 found a home in the Arctic. Beginning in October 1971, two were dedicated to DEW Line lateral re-supply, logging 160-170 hours monthly to places like Cape Dyer, Hall Beach and Cambridge Bay from Frobisher. Reliable Dart engines, a large cargo door and a 15,000-pound payload made the FH-227 suitable for this work. A third FH-227 stayed at Dorval for skeds to Chibougamau, Val d'Or and Matagami.

For Arctic 737 operations Nordair had Boeing design a gravel-deflecting modification for the nose gear to keep stones away from the engine intakes. Vortex dissipaters were installed under the engine nacelles to keep gravel from the area just in front of and below the intakes. Protective teflon applied to the belly completed the gravel kit, which proved its worth in succeeding decades. The first revenue flight of a Nordair 737 to a gravel strip was on April 30, 1969—Montreal-Resolute Bay.

Nordair Convair 990 N5615 at Prestwick during a one-month lease from Modern Air Transport in June 1968. The 990 first flew in January 1961 but only 37 were built. This example started with American Airlines in May 1962, was sold to MAT in 1968, and went for scrap in 1986. (Wilf White)

Lapse time on this now-routine return trip was 14 hours; air time 10:40. Nordair later pioneered with 737s on ice strips. Its first landing was at the Strathcona Sound lead-zinc mine at Nanisivik, where a 7,000-foot strip was prepared on 10 feet of clear blue ice. In April 1976 Nordair began operating onto Nanisivik's new 5,300-foot gravel strip. In 1977 Capt J.S. Patterson, writing in *Shell Aviation News*, described a routine 737 trip north:

A typical flight north would have Montreal centre clear you to the Frobisher airport centre-stored routing, to maintain FL290. FL330 would be more desirable, but some fuel must be burned off before that altitude can be reached. Nordair in living memory has never sent a flight north that wasn't at maximum gross weight! Out of range of Montreal VOR [VHF Omni Range], a powerful beacon at Chibougamau is picked up, and DME [distance measuring equipment] from the TACAN [tactical air navigation] there gives a constant ground speed check. Next, Nitchequon is tuned in. The interesting thing about this place is that it is the geographic centre of Quebec, Canada's largest province. Twenty minutes north of Nitchequon, Fort Chimo VOR is picked up, and thereafter it's strictly airways with complete VOR coverage all the way to Frobisher.

Now something funny occurs. To begin a descent into Frobisher, clearance must be obtained from Goose Bay. Communications are good, and generally the clearance is relayed through Frobisher radio in a couple of minutes. There are times with freaky weather situations where communications break down. The descent clearance is delayed. On occasion an aircraft will be hung up at FL330 over Frobisher Bay waiting for descent.

Fuel is always precious, but it's a hell of a lot more precious when your nearest alternate is three or four hundred miles away. It has always struck me as ludicrous that a clearance must be obtained from more than 600 miles distance. With the amount of traffic plying Arctic skies, a new system will have to be set up for control.

Out of Frobisher northward, the VOR radial is good for about 200 nm at FL310. Now it's time to convert from magnetic to true heading. Out comes the faithful astro compass, and a

sun, moon or planet shot is taken. Both magnetic gyro compasses are switched to DG [directional gyro], and Hall Beach VOR is tuned in. At 310 there is a blank area for about 50 miles until Hall Beach is picked up. Generally you are within a degree or two of the correct radial to Hall Beach. A check-in with Hall Beach radio gets you all the latest actuals [i.e. weather] for your destination.

North of Hall Beach, still flying on a true radial, one can check bearing and distance with the DEW Line. Although this is not part of their job, they are very accurate and quite obliging, especially if one of the lady flight attendants happens to use the mike. Before Hall Beach is out of range, Resolute VOR can be picked up. If the destination is Resolute Bay or Nanisivik, descent clearance to FL290 is obtained from Edmonton, more than a thousand miles away. It sounds silly, but it's true. Below 290, pilots provide their own separation. At Resolute, a normal ILS is carried out. If the destination happens to be Nanisivik, a beacon let-down is all that is available at present. The return flight is a reverse of the northbound flight. True headings until 200 nm south of Hall Beach, then magnetic the rest of the way south.

Nordair sell refreshments on all northern flights, and this has become a bit of a problem. Construction workers aren't exactly normal drinkers, and orders of a double whiskey with a beer chaser are common. Fortunately, few passengers get out of hand—the word spread very quickly in the north that trouble makers will be dropped off at the next stop, into the waiting arms of the RCMP. Besides, a frosty look from a flight attendant usually cools the most ebullient passenger.

In 1969 the Twin Otter and Short Skyvan began work from Frobisher under the name Nordair Arctic Ltd. (which also used the Beaver, Canso and Turbo Mallard). In early 1972 Nordair won a five-year ice patrol contract from the Department of the Environment and that March purchased Lockheed Electras CF-NAX and NAZ. These were equipped by Canadair with advanced instrumentation for such duties as mapping and photographing ice conditions in the Eastern Arctic and down the East Coast. For summer ops patrols flew from

Frobisher. In winter they moved to Gander or Summerside. A third Electra was acquired in August 1972 for general duties. (On March 31, 1977 NAZ was at Summerside when an Argus patrol plane crashed on landing. It ploughed into NAZ, wrecking it. NAZ's fuselage later was used to rebuild Nordair Electra C-GNDZ.)

By 1972 Nordair was operating various licences; e.g., Class 1 and 8 were sked licences, while Class 2, 3 , 9-2 and 9-3 were non-sked. In 1969, 1973, 1974 and 1976 respectively the "Connie", DC-4, C-46 and DC-3 left Nordair. In September 1972 Nordair began serving the James Bay region with initial service to Val d'Or and Fort George. In June 1973 Matagami and La Grande were added as the James Bay hydro project grew. Quebecair, using BAC 1-11s, made inroads on some of these routes. Nordair battled in Ottawa through the early 1970s to serve points like Sudbury, Sault Ste. Marie and Thunder Bay, but Air Canada and Transair always opposed this and Ottawa would reject Nordair's claims. It was always Nordair's contention that mid-northern Ontario was rightfully its territory (and not Air Canada's or Transair's) as far as regional theory went.

Nordair's ITC activities were boosted in October 1974 with the first of several DC-8s. By 1976 it had six 737s, a DC-8-61F, three Electras, three FH-227s, two DC-3s, a Turbo Mallard, a Skyvan and three Twin Otters. It also ordered eight Dash 7s to use on a Toronto-Ottawa STOL service, but the order was later cancelled. Nordair' departments at this time included flight operations, skeds, charters and tours, engineering and maintenance, quality assurance, finance, purchasing, public relations and advertising, and labour relations. 1974 was an important year for all the regionals, for Ottawa relaxed regulations on ITCs. No longer did the price of an ITC package have to equal at least 115% of the normal sked rate, and the stipulation that ITC tickets had to be purchased at least seven days in advance was reduced to three days.

Nordair prospered steadily since it began. The advent of the 737 saw its Arctic statistics leap from 31,329 passengers/7,000 tons of cargo in 1972, to 123,354 passengers/13,000 tons in 1976. Late that year the Air Transport Committee approved Nordair's application to serve Quebec City on its Montreal-Fort Chimo run. However, to appease Air Canada and Quebecair, no passengers destined to Quebec City could be carried from Montreal. At the same time the ATC turned down Quebecair's request to add Fort Chimo on its Sept-Îles-Wabush-Schefferville service. Thus, it could be seen that the regionals continued dogfighting for business in their respective back yards.

While some airlines suffered strikes, Nordair had few labour troubles and was in the black. Even so, in 1977 owner James Tooley offered Nordair for sale—Tooley was tired of dealing with French language regulations imposed by Quebec and Ottawa. If a buyer could not be found, Tooley simply could fold and sell off piecemeal. Great Lakes Airlines of London was interested, but could not put a good enough offer on the table. Transport Minister Otto Lang preferred a buy-out by Quebecair with maybe some EPA involvement, but neither was interested. Nordair then approached Air Canada, which made a cash offer. This was accepted in December. Following hearings through early 1978, where many parties objected to the offer, usually on the grounds that it would restrict competition, Air Canada acquired Nordair. Opponents forced a second look at the deal and in October 1978 Otto Lang ordered a compromise. Air Canada was to divest itself of Nordair within a year. The federal election of May 1979 saw the Liberals tossed from office; the Progressive Conservatives under Joe Clark were installed (new transport minister: Don Mazankowski).

The Nordair matter was as contentious as ever. James Plaxton of Great Lakes Airlines (who had taken over the financially shaky airline in 1975), Alfred Hamel of Quebecair, and other Quebec-Ontario business interests, aligned to try a shared Nordair takeover with the view of amalgamating all three carriers. An opposing syndicate included Harry Steele of EPA, the Algoma Central Railway of Sault Ste. Marie and Quebec interests. André Lizotte, formerly a Quebecair director, now of Nordair, formed a third group that included the Inuit Makivik Corp. The Clark government fell in the election of February 1980 and was replaced by the Liberals—before the next Nordair sale

was finalized. The new government let the issue simmer, waiting for results of a federal referendum in May 1980 on the question of Quebec independence. Transport minister Jean Luc Pépin then announced that Ottawa favoured the Hamel consortium. This was bad news for Nordair employees, most of whom were anglophones. They knew that under Hamel French would be the working language. Negotiations with Hamel collapsed in August 1980. A few weeks later Quebecair took over a large block of Nordair shares. In a move so typical of the machinations of Canada's regional airline industry, Air Canada now proposed Nordair buying Quebecair! Nothing came of this, but Pépin decided not to enforce the order for Air Canada to sell Nordair. In 1980 Nordair withdrew from the overseas charter market, its DC-8s going to Nationair.

In 1981 the Deluce family took a 50% share in Great Lake Airlines, changing the corporate name to Delplax's Air Ontario (reflecting the names of the two partners—Stan Deluce and Jim Plaxton). Talks between the Ontario and Quebec transport ministers now led to new plans for amalgamating Quebecair and Nordair. Finally, in late 1982, after Pépin had rejected Ontario-Quebec bilateral talks, Quebecair was liquidated and replaced by a new francophone company, owned by Air Canada and Quebec, to take over Quebecair's routes. More months of discussion followed; in

mid-1983 Quebec finally purchased the common stock of Quebecair.

Nordair was strikebound for the second half of 1982 and lost money for the year. Now Innocan, a group partly owned by Air Canada employees (through their pension fund) offered to buy Nordair. Delplax counter-offered. Transport minister Lloyd Axworthy favoured Innocan, as this would keep control of Nordair in Quebec. Innocan took over in 1984. In September 1984 the political climate changed again with the ouster of the federal Liberals by Brian Mulroney's PCs. This led to Quebec (owner of all Quebecair shares) offering to buy Nordair. Meanwhile CPAir made its own offer. Each bidder upped its offer as negotiations progressed. By early 1986 CPAir held the majority of Nordair shares. Nordair and CP Air schedules now were integrated; it was a bonus that there was fleet commonality with the 737. In December 1985 Nordair, Conifair and Avitair formed Nordair Metro, a Quebec feeder operation with CV 580s. A few months later soaring Nordair Metro bought Quebecair (which then re-formed as a Convair operator on behalf of the SEBJ). Soon Nordair's 737s were painted in CP Air's colours and its reservations were switched from Air Canada's to the CP Air's system. The last official Nordair flight operated on January 24, 1987.

The mainstay of Nordair was the Boeing 737, which was used with equal success on DEW Line, inter-urban, and holiday charter runs. C-GNDD was at Toronto's Terminal 1 on December 29, 1979. It went new to EPA as C-GEPB in 1975, but soon was leased to Wein Alaska, Aloha, then sold to Guinness Peat Aviation. It was leased to Nordair in 1978 as NDD, returning to GPA in 1982, and last was heard of in Chile in the mid-1990s. (Dave Thompson)

Commuters to Couriers

Southern Ontario's Local Carriers

Since WWII various local carriers have linked centres along the Great Lakes-St. Lawrence Seaway. Besides being a general convenience for people travelling from town to town, they also delivered passengers to Malton or Dorval to catch long-distance flights. One carrier was Sarnia Airlines. In the 1950s it ran Apache, Beech 18 and Dove skeds, mainly for the petrochemical industry in Sarnia, linking it to Toronto, where companies such as Imperial Oil had offices. At the same time Southern Provincial Airlines of Toronto Island Airport operated planes as big as the DC-3. Such companies competed with the railroads and interprovincial bus lines, which offered frequent, affordable service, so long-term survival was a gamble. Sarnia Airlines and SPA soon faded. Undaunted, new carriers occasionally would appear. One was Royalair of Montreal, which introduced Montreal-Pembroke-Peterborough-Toronto-St. Catharines service in August 1968. A DC-3, Lodestar, Riley Heron and Twin Otter were used. First-year traffic of 25,000 passengers was predicted, but business did not materialize. Royalair ceased operations in December 1969, its interests being assumed by Quebecair.

As long as local carriers stayed on their turf, they were hardly noticed by the larger ones. As the regionals grew, especially with jets in the late 1960s, they were sometimes happy to let the locals take lesser routes. When Air Canada retired its Viscounts in 1974, it no longer could serve certain communities. Its smallest airliner now was the DC-9. Some markets and airports (notably on the prairies) that were OK for Viscounts were not for jets—their runways were too short. If a local carrier wanted to take over cities like Windsor and London, Air Canada might concur. In 1960 Sarnia auto parts manufacturer Holmes-Blunt set up a corporate aviation division. It evolved in 1967 into Great Lakes Airlines, serving Sarnia and Toronto. Using two DC-3s, it carried about 10,000 passengers in year one. Two Convair 440s were bought in 1970 and London was added as an

After WWII small operators often used war surplus planes like this Cessna Crane seen at the old Hamilton airport. Such equipment soon was replaced by more modern types like the Sarnia Airlines' Dove (below) at Malton on October 10, 1959. Carrying 8-10 passengers at about 180 mph, the Dove was an ideal mini-airliner. Standard Doves used 380 hp-400 hp D.H. Gipsy Queen engines. (Jack Whorwood, Merlin Reddy)

The chief personnel at Great Lakes Airlines were John Blunt—president, George Capern—operations, and Chris Frost—chief pilot. An RCAF veteran, Frost earned a George Cross in WWII and flew Sabres postwar. (via David Frost)

(Above) Royalair's 16-passenger Riley Heron over suburban Montreal. Harold N. Miller was president of Royalair. The Riley conversion used 290-hp Lycoming IO-540s compared to standard 250-hp Gipsy Majors. Flying at 200 mph (faster than a DC-3), it was speedy enough for Royalair's Montreal-Toronto run. (via Gord McNulty)

(Left) The first Great Lakes Airlines Convair at Sarnia in June 1970. (via David Frost)

(Below) City Express inaugurated Toronto-Montreal STOL service with Dash 7s in September 1984, more than a decade after the concept first was touted by Ottawa. Besides skeds, City Express ran charters; one of its Dash 7s flew 111 relief trips in Sudan. But things started deteriorating for the company— airplane paint jobs got shabby, interiors got ratty, employees were asked to work without pay. Air Ontario moved to Toronto Island with a new terminal and a fleet of spiffy Dash 8s. City Express could not compete. On August 1, 1990 its Dash 8s were seized by Mutual Life Assurance. This scene dates to March 1987. (Kenneth I. Swartz)

intermediate stop. The Convairs had a big impact and for 1973 GLA carried about 80,000 passengers. New skeds were added from Toronto to Ottawa via Peterborough, and to Waterloo between London and Toronto, but soon were abandoned. GLA was on the verge of folding in 1975, when new management took over. The 440s were replaced by turbo-prop 580s. Eventually, GLA formed an alliance with Austin Airways and Air Ontario was born.

In the early 1970s Otonabee Airways of Peterborough (later called Air Atonabee) and Pem-Air of Pembroke made inroads along the Ottawa-Toronto corridor, initially by agreeing to serve places like Peterborough and Kingston with the Cessna 402, Beech 18, Navajo, DC-3 and ST-27. In time they convinced Ottawa to let them drop the unprofitable intermediate stops and concentrate on Toronto.

Joseph D. Csumrik, who headed Otonabee, expanded service to Toronto Island Airport (TIA), where he negotiated a monopoly situation. In 1984 he sold to Victor Pappalardo for $6 million. Pappalardo renamed the company City Express and began expanding. This was after a flock of would-be interurban companies had fought for years for rights to a federally-promoted Toronto-Montreal STOL service. In 1982 Ottawa awarded the route to City Centre Airways, a paper airline and an Air Ontario adjunct headed by James Plaxton. It dilly-dallied about starting up, so Ottawa withdrew its offer in April 1984, opening the route to all comers.

Building on Csumrik's hard work, City Express got off to a good start, operating from TIA to Ottawa, Montreal and Newark, and pushing east to Quebec City and Sept-Îles. Its tried-and-true ST-27s were supplemented by

Canadian Partner/Ontario Express commenced operations July 15, 1987 with service to Kingston, London, Sarnia, Toronto and Windsor. Ottawa, Sault St. Marie and Sudbury were added August 3; 75,000 passengers were carried the first year. Hamilton-Ottawa-Montreal Jetstream service began early in 1989, but ended October 1, 1991 due to poor loads. The Jetstream was a stalwart with Canadian Partner, Air Toronto and Air BC in the 1980s. As it came onto the used plane market in the mid 1990s, it found favour with smaller operators, e.g. Samaritan Air Service, which converted one for air ambulance work. The company eventually reached west to Winnipeg. PWA had a 49% share in OE. Here a Jetstream 31 departs Mount Hope for Ottawa on June 11, 1991. (Larry Milberry)

two Dash 7s, then four Dash 8s. The number of passengers rose from 24,000 in 1984 to 300,000 in 1989 by when there were 50 flights daily. Everything looked rosy, then City Express lost its monopoly. Air Ontario moved onto TIA with Dash 8s, a new commuter terminal and more attractive schedules. Intair of Montreal and Canadian Partner also sought entry to the TIA market. Business at City Express nose-dived and the company could not even pay its employees. Its fleet deteriorated; the mortgagor foreclosed in February 1991, City Express owing $45 million. In June 1994, however, Pappalardo re-entered the fray. He had retained one Dash 7, which he launched on charters under the Trans Capital Air banner.

Canadian Partner

Canadian Partner was formed in 1987 to tie smaller Ontario communities to CAIL's overall system. This was CAIL's answer to Air Canada's 75% purchase of Air Ontario. Toronto was the focus for both these ambitious commuter firms.

Canadian Partner became a member in CAIL's reservation system, its worldwide Empress Lounges, the Canadian Plus frequent flyer program and flight scheduling and joint marketing schemes. Canadian Partner was the operating branch of Ontario Express, a holding company headed by Ronald L. Patmore. It built a fleet of BAe Jetstream 31s, then ordered six ATR42s and five Brasilias. In 1989 Beech

Toronto-Buttonville Airport in July 1971. Various attempts over the years to develop it into a commuter airport for Toronto had failed by 1997. Since this photo was taken most of the open fields have been developed for residential and commercial use, as developing Toronto pushed the rural-urban fringe farther and farther out. (Larry Milberry)

A Jetstream 32 in maintenance at Toronto in July 1989. With the demise of Air Toronto it migrated to the US as N840JX, serving Atlantic Coast Airlines. (Larry Milberry)

The Embraer Brasilia operated with Canadian Partner, but only from 1991-93. The speedy turboprop filled a niche between the 19-seat Jetstream and B.1900C, and the 40-50-seat Dash 8 and ATR, but somehow didn't fit in economically and was dealt away. This scene was at Thunder Bay on March 26, 1992. (Larry Milberry)

1900C operator Frontier Airlines of Kapuskasing was absorbed. This gave Canadian Partner access to James Bay. Airplane types were matched to markets; each morning the fleet spread out among Windsor-Montreal-Ottawa, Toronto-James Bay, and Sault Ste. Marie-Winnipeg. In 1989 Montreal and Ottawa Dash 8 service was launched from Toronto-Buttonville Airport. Traffic did not materialize, so the Beech 1900C took over. Business remained marginal—in early 1993 Toronto-Buttonville service died. Laurentian Air Service of Ottawa stepped in with King Air 200s, but this failed as well.

In May 1991 Ontario Express Ltd. absorbed Air Toronto, an upstart Jetstream operator formed by Brian Child and run by Stephen Smith to serve niche markets in bordering US cities. In 1974 Child had formed Soundair, a flying school in Wiarton. In 1983 he acquired DC-3s IAZ, KAZ and QBI to operate night courier runs for Emery between Montreal and Hartford; and for Airborne Express between Toronto and Buffalo. Child's interests soared in the mid-1980s. They included Executive Jet Canada with Lears, and Soundair Express/ Commuter Express (later renamed Air Toronto) with the Metro II, Merlin IV, Convair 580, F.27 and DC-4. DC-4s C-FGNI and C-GCXG came from Aero Trades of Winnipeg. GNI worked the Emery contract from Dorval. Soon a DC-4 was needed in Toronto, so C-GPSH was bought from Basler. In 1986 the DC-4s were flying nightly to Emery's hub in Dayton, Ohio from Montreal and Toronto. The plum of Child's holdings became a classy charter company, Odyssey International. It operated the 737 and 757 under direction from former Wardair executives like Tom Lewis and Brian Walker. In 1987 Air Toronto became an Air

Canada Connector, but in April 1990 Child's empire crumbled under the weight of $65 million in debt. As Air Toronto was profitable, the receiver let it continue. PWA Corp/CRA and Air Canada fought for ownership. The courts finally approved its sale to the former early in 1991. About this time CRA took 100% of Time Air, 70% of Inter-Canadien and 45% of Air Atlantic.

Canadian Partner began slashing routes— James Bay, across mid-northern Ontario, to Red Lake, Hamilton, all the Air Toronto routes, etc. (For James Bay this was welcomed by Air Creebec, and VisionAir started up to serve this market). There were wide layoffs at Canadian Partner. Everyone in Air Toronto was chopped, then most in the parent operation at LPBIA. In 1993 all 26 Jetstreams and Brasilias were disposed of, and replaced by a small fleet of ATR42s. A new round was beginning in the cycle of a type entering service with much fanfare, then being replaced and moving out of the limelight. Jetstream 31 C-FCPE, for example, had come in 1988 to Canadian Partner. In 1995 it joined Northwestern Air of Fort Smith. While it no longer was suitable for major southern work, it was an improvement in the NWT over types like the Navajo or Beech 99.

With fleet modernization and layoffs CRA hoped to stop the hemorrhaging of capital and put its feeders on a profitable basis. By 1995 it was even more streamlined and in a better position to survive. A new regulation allowed a single flight attendant for a 50-seat aircraft (the previous limit was 40). Thus, on the Dash 8-300 the crew was reduced from four to three (two pilots, one FA). CRA considered the saving worth $2 million annually. The ATR42 left Ontario in favour of the Dash 8; only Inter-Canadien now operated ATRs. Otherwise, a strict business plan ensured savings across the company e.g. meal service was reduced; ramp services were out-sourced at some stations; crew layovers, travel and training costs were tightened; maintenance looked for business servicing other operators' planes, and parts sales outside CRA; purchasing for CRA, Inter-Canadien, Air Atlantic and Horizon Air of Seattle was done jointly where possible; some properties were sublet or sold; and aircraft painting was brought in-house (to Calgary), saving some $20,000, for example, on each F.28 repaint. Other Canadian airlines were as serious about savings, and developed their own new business plans, e.g. Air Ontario realized the value of Air Toronto's old service to places like Harrisburg and Columbus, and reinstated them.

Toronto's Airports

In November 1936 Toronto set up a committee to report on suitable sites for a new municipal airport. F.D. Tolchard of the Board of Trade, E.L. Cousins of the Toronto Harbour Commis-

Charles Willard made Toronto's first flight in 1909. By 1915 a flying school was in business on Toronto Bay–it ran this advertisement in *Canadian Motor Boat* in June that year. The aircraft is the Curtiss E "Sunfish", flown by Theodore C. Macaulay. A modern airport opened on the bay in 1939 and remained a valuable part of Toronto's waterfront into the late 1990s. (Ontario Archives S15751)

FLYING IS THE MOST ATTRACTIVE SPORT OF THIS AGE

TORONTO AVIATION SCHOOL

WE TEACH you to be a Pilot or an Aviation Mechanic, positions which command excellent salaries. FLYING BOATS, SEAPLANES and AEROPLANES—everything pertaining to the skilful operation of these wonderful craft, fast coming into general use—will be taught by our school by men of wide experience in aviation. All those desiring to enter the school should make application at once. Call, or write for particulars to

W. A. DEAN, Room 21, Bank of Toronto Building, 205 Yonge Street, Toronto

(Above) A close-up of Toronto Island Airport from the mid-1950s. The slips for the old cable ferry are seen either side of the western gap linking the harbour and Lake Ontario. All the buildings were built in the late 1930s and during WWII, when the TIA was a Norwegian training base. Then, an overall eastward view taken May 17, 1971 from radio station CFRB's Twin Commanche CF-AKC. The airport's proximity to the central business district is apparent. (Toronto Harbour Commission Archives PC14/3130, Larry Milberry)

(Right) A view from HMCS *Ottawa* showing the TIA ferry slip and terminal during some local improvements about 1988. The ferry *Maple City* is about to shove off for the mainland. (Larry Milberry)

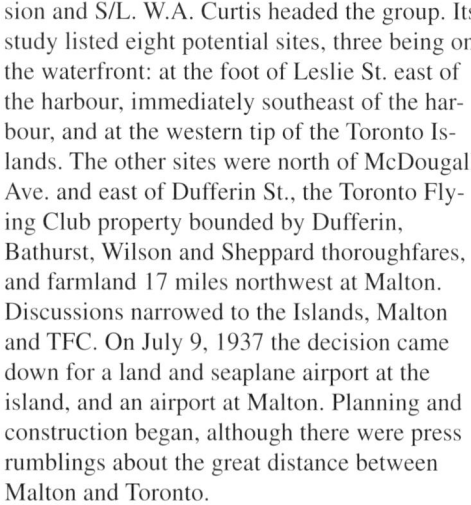

sion and S/L. W.A. Curtis headed the group. Its study listed eight potential sites, three being on the waterfront: at the foot of Leslie St. east of the harbour, immediately southeast of the harbour, and at the western tip of the Toronto Islands. The other sites were north of McDougall Ave. and east of Dufferin St., the Toronto Flying Club property bounded by Dufferin, Bathurst, Wilson and Sheppard thoroughfares, and farmland 17 miles northwest at Malton. Discussions narrowed to the Islands, Malton and TFC. On July 9, 1937 the decision came down for a land and seaplane airport at the island, and an airport at Malton. Planning and construction began, although there were press rumblings about the great distance between Malton and Toronto.

The island required a huge dredging effort. Nearly 2,000,000 cubic yards of fill were needed to extend the natural sandbar deposits. Besides runways, retaining and break walls, and a seaplane ramp were needed. Malton was straightforward and opened on August 29,

1938, the arrival of an American Airlines DC-3 being the highlight that day. The first plane landed (officially) at the island on February 4, 1939. The site was christened Port George VI Airport in honour of the King, who was visiting Canada on a royal tour that year. In 1939 Toronto leased Malton to the federal government for $1.00 a year. The Toronto Harbour Commission operated the island for the city.

While the island was to have been Toronto's primary airport, WWII changed this. Malton was ideal for the war effort—training and building airplanes. No. 1 EFTS and No. 1 AOS opened there in mid-1940, and all the infrastructure for such large operations was built. Meanwhile, National Steel Car, which had built a plant at Malton in 1938 for Lysander production, expanded, receiving contracts to build Hampdens, Ansons, etc. It boomed after December 1941, when an order came for Lancasters. None of this activity would have been suitable for the island, which was too close to the city core. Instead, it became a fly-

ing training base for the Royal Norwegian Air Force. Meanwhile, TCA made Malton its main base. No commercial activity was drawn to the island till after the war, when a few small operators like Central Airways moved in.

In the late 1940s and early 1950s domestic and trans-border travel, scheduled trans-Atlantic service and the extension of TCA service to Bermuda and the Caribbean guaranteed steady growth at Malton. In 1957 it was handed over to the DOT in exchange for federal funds to improve TIA, which, in 1962, became an entity of the Toronto Harbour Commission. TIA's age-old problem was its isolation from the mainland—access was only by boat. For a generation this was an ancient cable ferry, then the modern *Maple City* was bought to replace it in the 1960s. It ran in daylight hours about every 15 minutes, but this discouraged widespread use of the airport. On rough-weather days, ferry service sometimes was curtailed. Meanwhile, the building of a "fixed link" to the island was steadily opposed by a small group

TIA scenes from the late 1950s-early 1960s. The Austin Airways DC-3 was on a charter with prime minister John G. Diefenbaker. Austin Airways had its head office on the second floor of the terminal beyond. The Alouette, the first turbine helicopter in Canada, was on a demonstration tour. Southern Provincial's Lodestar was being barged away for scrap after being damaged by Hurricane Hazel. The RCAF C-130 could have been training or delivering cargo. (Toronto Harbour Commission PC14/10227, '7908, '4225, '8585)

(Below) In the 1990s Air Ontario Dash 8s were the key commuter airliners using Toronto City Centre Airport (as it was known in the 1990s). Points served were between Windsor and Montreal, and south to the US. (Larry Milberry)

(LBPIA). Four years later construction began on Terminal 3 using private funds, a first for a major Canadian airport. Terminal 3, opened in early 1991, was built mainly for CAIL and American Airlines, although other carriers gradually were added. In March 1997 Ottawa agreed to contribute $185 million toward renovations at LBPIA, including a 130-gate terminal to replace Terminals 1 and 2 (79 gates). It was noted that LBPIA supported 96,500 jobs and generated $9.7 billion annually. By this time a new north-south runway parallelling R15-33 was nearing completion at an airport that was taxed to the limits. Meanwhile, the lands reserved for Pickering airport still were held by Ottawa in anticipation of requirements in the mid-21st century.

Toronto Tragedies

Toronto's worst aviation accident involved Flight 621 near Malton on July 5, 1970. F621 (CF-TIW, a new DC-8-63 with only 453 hours) was inbound from Montreal with 109 aboard. Capt Pete Hamilton (a WWII bomber pilot and senior CALPA man), FO Donald Rowland and SO Harry Hill were on the flight deck. At 0809 Hamilton was bringing TIW across the threshold of R32 on a normal approach for its 208,000 pounds. At about 60 feet he asked the FO to arm the ground spoilers—on touchdown they would deploy automatically. This did not happen. Just as Hamilton asked for arming, he

of anti-airport activists. It wasn't till the early 1990s that Toronto City Council finally agreed to a bridge across the narrow gap to the TIA. By 1997 there still had been no move to build the bridge. Thus, for half a century Toronto, intimidated by vociferous, special interest opponents, squandered its chance to have a vibrant downtown airport.

Meanwhile, Malton Airport grew steadily. In the late 1950s construction began on a modern aeroquay, a marvel at the time. Prime minister Lester B. Pearson officiated at the opening on February 28, 1964. Other aeroquays were planned, but not built on account of the

expense. Instead, construction began in 1969 of Terminal 2 and plans went forward for a new airport northeast of Toronto—the so-called Pickering airport. A public outcry put the kibosh on Pickering in 1975. Terminal 2 opened in June 1972. In less than a year it was the home of all Air Canada flights. Other operators used Terminal 1. Both terminals underwent major improvements, but traffic kept growing. A new ATC centre was opened in 1981.

As the years passed, the name Malton Airport disappeared, replaced by Toronto International Airport. In January 1984 it became Lester B. Pearson International Airport

A view of Malton's passenger terminal about 1955 with an American Airlines CV240 and a TCA North Star waiting. Well-wishers crowd the observation deck. For 10¢ one could pass through the turnstile and spend hours on the deck enjoying the activity as Convairs, Britannias, DC-6s, Super Connies and Viscounts came and went and many other interesting types like CF-100s passed in review. (Bob Bolivar)

shouted, "No, no, no!" Rowland inadvertently had pulled the lever back to deploy the spoilers in flight. This killed lift, so TIW dropped (the DC-8-63's maximum safe descent rate for landing was 10 feet per second; it was estimated that TIW reached 24, but Hamilton arrested this to 18). Rowland was heard on the cockpit voice recorder (CVR) to say, "Sorry, Pete," just as the gear crunched down. This subjected the airframe to 5g (five times actual weight) at its centre of gravity. Hamilton already was going-around, but, unknown to anyone, No. 4 engine had torn off, and a fuel tank had ruptured. As TIW climbed, the tower asked: "Do you wish to come in for an immediate on five right?" Hamilton replied, "Oh, we'll go around. I think we're all right." The crew cleaned up the plane—gear up, flaps 25°.

Moments after TIW passed the outbound end of R32-14, the tower advised that R32 was closed due to debris; it gave Hamilton vectors for R23L. At 1 minute 47 seconds into the overshoot the crew realized that No. 4 was not delivering power. Less than a minute later, three explosions hammered the starboard wing and No. 3 engine and a section of outer wing fell off. (At this time the driver of a DOT ground vehicle radioed: "That 8 just landed... he, ah, knocked off an engine and his right wing may be on fire... He just lost another engine. Looks like it's going to crash in Malton Village.") TIW started to pitch down. FO Rowland said, "Pete! Sorry!" Someone in the cockpit made the last comment of the day— "We've lost a wing." It was all over—52 minutes after leaving Dorval, TIW buried itself in a field 8 1/2 miles from the threshold of R32 and 2 1/2 miles from its centre line. Air Canada F254, a Viscount inbound from North Bay, witnessed the last moments, then reported to the tower, "Yeh, 621 has, ah, crashed... He sure did, he went down in flames."

The board of enquiry looking into the accident determined that captain and FO had differing views about arming the ground spoilers for landing. Hamilton liked to arm them once on the ground, suspecting that in-flight arming had led to the crash of a CPA DC-8 at Tokyo on March 4, 1966. Rowland preferred arming the spoilers on the flare—moments before touchdown. Air Canada's DC-8 manual stated that spoilers could not be deployed in flight. Its pilot operating manual taught arming the spoilers at the before-landing check, i.e. at 1,000 feet agl. This day, the "spoilers armed" check was omitted—the pilots had agreed a few minutes from landing to arm on the flare, even though it was Hamilton's landing—he seems to have agreed to Rowland's technique this time, and a few minutes before landing said, "All right, give them [spoilers] to me on the flare." There wasn't any frustration in his tone, and some chuckling was heard at this time on the cockpit voice recorder.

Like many accidents where "pilot error" is freely used by the news media, the crash of TIW was a complex affair. There was a difference of opinion on the flight deck, but Air Canada itself was ambiguous about ground spoilers. Before the accident there was no warning in its operating manual that ground spoilers could be deployed in flight. One statement read, "They [ground spoilers] do not extend in flight," implying a safe system, but in CPA's DC-8 manual there was a warning: "The spoilers can be extended in flight by manual selection of the lever to the extend position. Do not apply any rearward pressure on the spoiler lever when arming the spoiler." Eastern Airlines' DC-8 manual included the warning, "Movement of the spoiler lever toward the extend position should never be attempted in flight." KLM had a similar warning.

Other details were cited by the board of enquiry as contributing to the accident. The board criticized the dangerous simplicity of a) arming, b) deploying the spoilers. Instead of these being mechanically separate and sequenced actions, it was possible for a pilot to carry out one function, while intending the other. Also, on DC-8-40s, the spoiler actuating lever required a harder pull than on the DC-8-50s or -60s. Other details questioned by the board of enquiry included the absence of an indicator to show that an engine had fallen off, and of a g-meter to show how hard a landing was. Also studied was how the fail-safe bolts on the No. 4 engine pylon didn't work as advertised—they did not shear cleanly nor in sequence. Instead, they tore the lower wing plates away, bursting the fuel tank. Bare electrical wiring then likely ignited spewing fuel, causing the explosions.

The argument has been made that Capt Hamilton should have landed TIW after touchdown, instead of going around. The board said he did the right thing. Even before TIW hit, he had applied go-around power, rotated the nose, and called for gear up. It would have been wrong to alter this plan. The normal thing, given what Hamilton knew about the condition of TIW, was to go around. A comment in Justice Hugh F. Gibson's *Report of the Board of Enquiry into the Accident at Toronto International Airport ... on July 5, 1970* notes: "As the tape recorder clearly indicates from the words used and the tones of their voices, the captain and first officer were of the opinion that no substantial damage had been done to this aircraft and that they would be able to make another circuit and land safely." Ultimately, the board of enquiry was hard on Air Canada, stating: "The Air Canada ground training school staff, until July 5, 1970, did not know the ground spoilers on the DC-8s could be deployed in flight, when the undercarriage was down, by manually pulling aft the actuating lever in the cockpit, whether or not the lever was armed. Why the operations staff or the engineering staff or both... did not communicate this information to the ground training school staff long prior to July 5, 1970 is difficult to understand, especially when the whole matter of when and how these ground spoilers could be deployed in flight was the subject of wide debate and discussion among Air Canada pilots." Douglas also was chastised for having equivocal and inaccurate statements in its DC-8 flight manuals. The Department of Transport was criticized for not inspecting Air Canada and CPA DC-8 flight manuals and for failing to compare them with Douglas' edition for errors or confusion. The board of enquiry listed every shortcoming it could, as it was meant to do. It was there to shake up Air Canada, pilots and the entire industry, in the hope that there never would be a similar disaster. Meanwhile, the system for using DC-8 ground spoilers was not changed; but pilots were given better training in its use.

Another tragic day for Toronto was January 12, 1987, when a Trillium Air Islander, coming into Toronto Island from Niagara Regional Airport, ran out of fuel and ditched in Lake Ontario about a mile short of R33. It quickly sank, leaving pilot Rolf Pahl and passenger Hank Emson floating in their life jackets in frigid waters. A harbour police launch and a light plane immediately started searching, but the weather was grim and the lake rough. A Ranger Helicopters A350, used on the Air Canada shuttle between downtown and LBPIA, was at its waterfront base when word came of the ditching. Pilot Andy Stevens, Bob Gareh and Jason Pearce took off. They had no life jackets and no SAR equipment, other than a length of rope.

The Ranger crew spent half an hour scouring the waves till they spotted the men. Gareh dropped the rope to Emson and twice hauled him onto the skid, but he slipped back. An air ambulance BK117 arrived, so the A350 pulled away. It and another light chopper stood by to help. Meanwhile, a 424 Squadron Labrador was despatched from CFB Trenton, 90 miles to the east. Shortly after noon paramedics Brian Hershey and David Clarke got Emson onto the BK117 skid and held him there while pilots Alex Gonier and Dan Seyewich shuttled to hospital. By then the police launch arrived and pulled Pahl from the lake. One of the helicopters directed it to a spit where the Labrador waited. It took 15 minutes to transfer Pahl. SARTechs Cpl Pat Mercer and Cpl Dan Massana tried reviving him as they flew to Sunnybrook Hospital. There the trauma team worked on the two men. Only Emson pulled through.

Air Ontario—a Proposal

Since the 1980s "Air Ontario" was a household name between Winnipeg and Montreal. But the name was used earlier. In 1971-72 a group headed by former Air Canada vice-president Howard C. Cotterall proposed a regional carrier to be called Air Ontario. Also in the company management were Sturrock Sadler, formerly vice-president of administration with Air Canada, and James T. Bain, Air Canada's head of engineering. Air Ontario was to serve points that Air Canada wished to abandon—Timmins, SSM, Sudbury, North Bay, London and Windsor. Cotterall stated that the new company would not seek subsidies, nor object to the services being provided in Ontario by Transair and Nordair. Air Ontario applied for a licence describing itself as an "intraprovincial" carrier, a classification not recognized in Ottawa. Ottawa feared that Ontario was now trying to establish its own regional, since the 1969 regional airline policy had designated Nordair and Transair as Ontario's carriers.

In January 1972 Air Ontario announced its order for eight refurbished BAC-111 airliners from American Airlines. It stated that it would begin service that summer. In March the matter was raised in Parliament, with Minister of Transport Don Jamieson replying to a query from the member for Algoma (Maurice Foster): "As things stand... and as I indicated yesterday, Air Canada will be holding discussions with the government of Ontario with the view to providing more in the way of what might be described as regional or hub-and-spoke services. Given the existing regional air policy, of course, the company to which the honourable member refers [Air Ontario] would not be entitled to apply for services within Ontario or anywhere else unless that policy were changed." On May 12, 1972 Ottawa, in a statement from CTC President J.W. Pickersgill, scuttled Air Ontario's plans. It stated that Air Canada was doing an adequate job in Ontario. If it was to withdraw from them, the routes would be turned over to the two designated regionals.

With the end of the Air Ontario scheme Ontario pursued a plan for regional service by subsidizing several Ontario operators: Air-Dale, Bradley Air Services, On Air and White River Air Services. These operated Ontario-owned planes under the norOntair banner. With provincial subsidies, they automatically were in a good position when applying for a route, making it tough on competitors. This argument against norOntair was made in 1977 by the Canadian Association of Primary Air Carriers. norOntair operated into 1996, then folded. By that time Ontario was being well served by the new Air Ontario and numerous smaller carriers like Bearskin Air and Air Creebec.

Local Carriers in Atlantic Canada

Local service also was important in the Maritimes, mainly when EPA acquired 737s. Not all its markets would support these 109-seaters, so it didn't automatically object if a local carrier such as Atlantic Central Airlines appeared with a Beech 99, Navajo or ST-27 to serve small centres. Thus, as the regionals expanded and focused on big cities, southern holiday charters, even trans-Atlantic business, the locals blossomed. They upgraded to attract passengers, and became more aggressive in marketing. Their pleas for new licences before the Canadian Transport Commission often were successful. By the 1970s some seemed as "regional" in influence as the regionals themselves had been a few years earlier. Through most of the 1970s the average annual revenue

increase was greater for local carriers than for the nationals or regionals.

It was a matter of time before some of the bigger local carriers began challenging the regionals. And why not? Ottawa had never defined the borders. EPA reacted to the growth in local traffic by forming its own local carrier, Air Maritime. Using 748s, it tapped the market beyond hubs like Halifax, Saint John and St. John's. In May 1984 it opened trans-Newfoundland service to St. John's, Gander, Deer Lake and Stephenville. Air Maritime trimmed in-flight service in order to hold or cut fares. But the company was short-lived. As the sole such operator in the area, it could survive; but in 1986 two Dash 8 operators appeared—Air Nova (Air Canada connector) and Air Atlantic (CAIL connector). Each would reap the benefits of interconnected schedules, routes, ticketing and baggage handling. When inundated by frequent Dash 8 service, Air Maritime quickly folded. EPA described its loss as "an inevitable result of deregulation", adding: "Our strategy will be to form an alliance with Air Atlantic to protect this feed traffic and thereby maintain the viability of our full jet operation."

By 1990 Canada's Atlantic airlines no longer were the "local carriers" of the 1970s and 1980s. They had taken routes discarded by Air Canada and CP Air and extended their influence to destinations as distant as Montreal, Ottawa and Boston. In the mid-1990s, however, Air Atlantic was floundering. Expansion had become too much of a good thing. It had too many planes flying too frequently on thin routes, trying to compete with Air Nova (which had about 65% of the regional market). British Aerospace was forced to step in, when Air Atlantic defaulted on payments for its BAe146s. It took an ownership interest and provided five 27-seat Jetstream 41s, while reducing the more costly Dash 8 fleet. In 1996 IMP purchased Air

Atlantic and imposed a tighter business plan. The two-month Air Nova strike of early 1996 was a short-term boom to Air Atlantic. Each week it lasted, Air Atlantic reaped millions in "found money".

The Commuters Mature

Since the 1950s much changed for the good in the commuters. The industry that started shakily with war surplus Ansons and Cranes, was flying state-of-the-art Dash 8s by the mid-1980s, and regional jets were being discussed. At the same time the US FAA was proposing sweeping changes. This followed a decade with dozens of commuter accidents. Carriers were operating two main categories: 10-19 seaters like the Navajo and Beech 99, and 30-70 seaters like the ATR72, Dash 8-300 and Saab 340. As stated in April 1995, the new regulations would include such things as: all airline dispatchers to be FAA certified; commuter lines having the same flying limits as big carriers (smaller carriers had been allowed to work pilots 1,200 hours yearly, compared to 1,000); use of the "Age 60" rule for commuter pilots—at age 60 a pilot would retire; establishing daily duty time limits for maintenance personnel; certifying all 10-19-seat aircraft in the transport category; upping flammability standards in commuter planes for items like seat cushions; installing weather radar on all 10-19-seaters; requiring tighter regulations about carry-on luggage; installing smoke masks in the cockpits of small commuters; and adding TCAS on all such planes. The FAA recommended a 10-year period for implementing these standards. In the mid-1990s some operators felt that the money required to improve their fleets would spell the end to the 19-passenger airliner. Companies like Chautauga immediately began replacing them with bigger planes. The FAA was serious about all these changes; eventually Canada would follow.

The Beech 99 gave excellent service in the era when small carriers began replacing old piston types like the Aztec, Ce.402 and Beech 18 with their first turboprops. This Atlantic Central Airlines commuter was the 37th Beech 99, built in 1968 for Time Airlines in the US. It was at Fredericton on August 8, 1975, a few weeks after coming to Canada. It served ACA to early 1977, then returned to the US. It flew for Columbia Pacific Airlines till it crashed near Richland, Washington on February 10, 1978. (Larry Milberry)

(Left) An Air Nova BAe146 lands at Ottawa in April 1990. With such modern equipment Air Nova and Air Atlantic brought the best of commuter service to the Atlantic region. (Andy Graham)

Air Taxis and Commuters

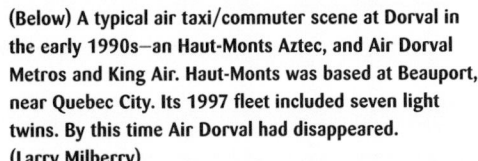

(Left) Many small types operated on air taxi and flight training duties in the 1990s. Even a few forty-year-old Piper Apaches survived. Mont Joli-based Avion Taxi operated three as late as 1987 and still had one (C-GUVT) 10 years later. (Below) The Moncton Flying Club Piper Seneca was seen on July 6, 1978. (Ken Swartz, Larry Milberry)

(Below) A typical air taxi/commuter scene at Dorval in the early 1990s—an Haut-Monts Aztec, and Air Dorval Metros and King Air. Haut-Monts was based at Beauport, near Quebec City. Its 1997 fleet included seven light twins. By this time Air Dorval had disappeared. (Larry Milberry)

(Below) The Beech 18 remained popular with local carriers into the 1980s. This example was at Kingston, Ontario on February 15, 1975. The DC-3 was just as irrepressible. This Pem-Air machine was at Toronto International on April 15, 1973. By 1997 these types had disappeared from Canada's airlines except in the North. (Larry Milberry)

Air Cargo

Air cargo, now such huge business, was a sideline for TCA after WWII; but airlifts like Seven Islands-Knob Lake, and the DEW Line brought several companies to prominence in the field—Dorval Air Transport, HUT, World Wide Airways, etc. Occasionally a new company would appear, one being Lome Airways of Toronto. It had a Class 9-4 licence: international, non-sked, cargo only. With the Knob Lake airlift, business partners in the US and UK sent a Tudor freighter (CF-FCY) to Canada to take part. For a fee it registered, then operated, the Tudor under Lome's licence. The plan was to carry cement from Seven Islands to Menihek, where a dam was being built. The British captain made a proving flight, then wisely decided that Menihek was too marginal for a Tudor.

The Tudor then looked for work at Malton, making its first flight from there on December 8, 1952. Crewed by Capt A.L. Firmin, FO D.T. Brooker, radio operator W. Matlashewski, and engineer F. C. Brobyn, it operated to Nassau with 18,000 pounds of Canada Packers meat. It returned to Malton with perishable fruit and vegetables. This was the first four-engine Canadian freight charter. In early 1953 TCA chartered FCY for a weekly trans-Atlantic freight contract. The first flight (in February) would carry Red Cross supplies for Holland, where disastrous floods had struck. On the eve of FCY's departure the DOT grounded it for lacking approved anti-icing equipment

(i.e. de-icing boots) to deal with the winter conditions prevalent on the Atlantic. Further, FCY's cockpit heating system was inoperative. Before long FCY returned to the UK for good.

Another far-out plan was devised by Donald E. Merriam, operating as Tradewind Air Transport of Canada Ltd. He applied to operate from Montreal on a "tramp freighter" basis, and was opposed by TCA, CPA, Curtiss-Reid Flying Service, World Wide Aviation, Global Aviation Ltd., and Wheeler Air Lines. The ATB noted: "It has been made clear to the Board that the purpose of this application was intended to provide a service of heavy freight air transport initially within Canada, but to be extended to serve any part of the world where business is offered... Those who opposed the application

(Right) Turboprop commuters like the ATR42 bridged a huge gap, bringing the latest in technology and banishing most older types. This Intair ATR42 was at Ottawa in April 1990. (Andy Graham)

(Left) By far the finest of the commuter turboprops is Canada's Dash 8. More than 500 were in service by 1997 and production was on-going. In Canada the Dash 8 served the main commuter airlines—Air BC, Air Ontario, Air Alliance, Air Atlantic and Air Nova. These Air Ontario Dash 8s were at LBPIA on August 18, 1991. (Larry Milberry)

Many US commuter carriers cross into Canada every day, from Horizon Air of Seattle to Comair of Cincinnati to Business Express of Westport, CT. In this case a Precision Airlines (Northwest Air Link) Do.228 was at Quebec City on August 18, 1991. (Larry Milberry)

By the mid-1990s many Cessna Caravans operated with courier companies in Canada. This one of Windsor, Ontario-based Simo Air was at LBPIA on March 4, 1995. Simo Air ceased doing business upon the deaths of the owners in a Cheyenne crash near Killarney, Ontario. In early 1997 Cessna delivered the 850th Caravan. (Larry Milberry)

Skycraft of Oshawa was founded in 1977 by Charlie Robson, whose many interests included a Lysander spray operation in the early 1950s. Skycraft offered flight training, passenger skeds and charters, and served the auto industry in air freight. Shown are its Short 3-30, Embraer Bandeirante and DC-3. Skycraft ceased operation in the mid-1990s. Jan Stroomenbergh recorded this scene at Oshawa on July 23, 1989.

did so on the grounds that they... were fully capable of carrying out such a service, but that with the exception of Trans-Canada Air Lines they adduced no evidence that they had carried out such a service in the past, or were in a position to carry it out at present unless and until they acquired suitable aircraft." While TCA had suitable aircraft for such work, it was clear that there was little business to justify a dedicated airplane.

The board felt that this application should get a chance, so approved it in February 1951, stating that "the granting of this licence might well act as a stimulus to this branch of carriage by air." The licence was "restricted to the employment of aircraft with a disposable load in excess of 10,000 pound." The stipulations of

the day included: passenger liability of $20,000 per seat; property damage $5,000 per accident; and public liability of $20,000 minimum for one person, and $40,000 minimum total per aircraft; and $5,000 bond as security against "any necessary rescue work". The decision was contingent upon Tradewind beginning within six months. Six month later Tradewind hadn't entered the market; it pleaded with the ATB for an extension, claiming that, due to the Korean War, it couldn't locate a suitable plane. This was agreed to, but in February 1951 the licence was revoked "on the grounds of non-prosecution." Thus ended an airline that only ever existed on paper.

Later in the 1950s J.W. Earl Crane of Kitchener applied to fly fish from Newfound-

land towns and outports to the US Northeast coast, returning with fresh farm produce; and from Newfoundland to a fish processing centre in Amherstburg, Ontario, returning with farm products from a number of Lake Erie ports. Crane planned to use Solent flying boats. His application was before the ATB in April 1961, and was accepted despite opposition from operators like EPA and Sarnia Airlines. The ATB commented: "While the board does not share the optimism of the applicant with respect to the profit potential of the proposed operation, it is satisfied such a service would be in the public interest and that the applicant should be given the opportunity to test the potential of the traffic." Crane's scheme did not get airborne, but as late as 1972 he still was active, and promoting CL-44s on a Newfoundland service. For this he was granted a licence in January 1972, but the scheme died.

Overnight courier flying, also so big today, had its Canadian beginnings in November 1969, when Canadian Helicopters proposed 17 routes from 12 locations carrying overnight cheques to data processing centres. This would keep business moving—millions in cheques would not be "dead money" overnight. The CHL idea was not implemented. Then, in June 1977 Alan Newnham, formerly of Pan Arctic Oil, proposed a Calgary-based, trans-Canada courier called Uranus Air Systems. Equipment would be three Howard 500s, a type that carried 6,000 pounds of cargo at 350 mph. This is thought to have been the first Canadian application for a FedEx-type courier service, although Worldways applied for a similar service about the same time using two Lear 35s. Uranus applied for Vancouver-Calgary-Edmonton-Regina-Winnipeg-Thunder Bay-Toronto-Ottawa-Montreal with one flight east and one west Monday through Friday. This would tap the small package market that needed better service than offered by the sked carriers. Gradually, overnight courier business caught on. CanAir Cargo became a major player. It was formed in December 1989 by Dan Goliger from the leftovers of bankrupt Tempus Air of Hamilton. By 1993 CanAir had six CV580s.

A typical CanAir night run was Flight 501 (call sign "Trader 501") of July 5, 1993. Convair C-FBHW started in Vancouver with Capt Jim Rogozynski and FO Joe Davidson. Each held a university degree and had graduated with a pilot's licence in 1983, Jim from the Brampton Flying Club, Joe from Toronto Airways. Jim had flown earlier with Tempus, Joe with Air 500. Trader 501 operated Vancouver-Calgary (410 nm) from 0123Z-0254Z carrying five standard containers. It was ready to roll from Calgary at 0320 and was airborne 10 minutes later off R28 at 116 KIAS. It climbed at a steady 1,000 fpm, levelling at FL210 at 0351. The night was quiet as the flight ploughed on for Winnipeg, a trip of 652 nm. It arrived at 0541, but there were signs of starter trouble. As the groundcrew off-loaded, the pilots called maintenance in Toronto for words of wisdom. In the end all was well and Trader 501 was airborne again at 0740 (an hour late). It was 836 nm en route to Toronto, and a gruelling trip it would be. Few flying jobs are more stressful than trans-Canada courier runs in old prop planes. Along the way the pilots chat a lot and take turns catnapping, sipping coffee and munching sandwiches. They listen in to the various centres en route to relieve the boredom. The late-night airwaves can be entertaining. Passing through Chicago's zone this night, for example, there was a quiz underway. Someone in ATC was asking esoteric questions to crews aloft, e.g. "Name the four US presidents who are buried outside the USA". Correct answers merited applause. Daytime frequencies are too busy for such fun.

As Trader 501 proceeded, word came to divert to Hamilton. Its airport finally had given up on becoming a major Toronto satellite. Now

Hundreds of courier flights operate daily in Canadian skies. Kelowna Flightcraft CV580 C-GKFY was at Halifax on September 16, 1988 after a flight from Toronto. Merlin 2B C-GSWJ was at Hamilton on June 18, 1993, heading for Ottawa. Each year since the 1970s this industry carried greater amounts of cargo. At the same time Canada Post, having driven its good customers away by not providing for their needs, carried much less. (Larry Milberry)

it was building its reputation as a cargo and courier hub. A few minutes back from Hamilton, pilots Rogozynski and Davidson shook themselves into full consciousness and descended into a hazy Southern Ontario for an IFR approach. Two miles back, with the sun about to pop over the horizon, the runway appeared. Trader 501 landed—nine hours after leaving Vancouver (6.7 hours aloft). The pilots were bushed and couldn't wait to get home. What lay ahead? More of the same, but in the tight job market of the 1980s and early 1990s, a job was a job. Where else could a young pilot fly a big turboprop and log a thousand hours a year?

Another busy courier fleet a LBPIA was Jetall. Formed in 1987, it operated with Metros and Merlins, then added 580s and 737s. The first 580 (C-GKRF) came in March 1988 and later was joined by three others. KFR was the prototype Canadair CL-66, having been converted to Elands in 1959. It was in the RCAF from 1961-67, then became a CV580 with

AVENSA. It returned to the US in 1975, serving various operators before becoming KFR in 1985.

On December 1, 1990 a Jetall Metro was involved in a $16-million heist at Dorval. Included in the loot were 16 gold bars worth $6 million. The Metro had been to New York on a Brinks run. When it landed at Dorval at 0348 local, a five-man team of armed robbers awaited. As a diversion they set off a bomb on another part of the airport, then drove a stolen garbage truck through a fence onto the tarmack. The Jetall pilots were ordered to stay in their seats as the loot was transferred into two small trucks, which soon sped away. By mid-1997 there still had been no word about the crooks or the loot.

While Jetall seemed to be screaming ahead, in 1996 it suddenly went bankrupt. Some linked this to hastiness in getting into jets. Perhaps the lesson of Trans Provincial Airlines had not been learned.

CanAir Convair 580 C-FIWN on March 4, 1995 at Toronto after a night's work, as a corporate Piper Cheyenne departs. (Larry Milberry)

Airline Developments

Air Canada 747-433 Combi C-GAGM at Toronto, first on July 31, 1993 in original colours; then, repainted, on September 1, 1995. The first -400 flew in April 1988. Some of its distinguishing features compared to the popular -200 series were: AUW increased (833,000 to 870,000 pounds); wing span increased (195' 8" to 211'); range increased (6,900 to 7,600 nm); and cockpit crew reduced (two pilots + flight engineer to two pilots). (Larry Milberry)

Air Canada in the Nineties

The 1980s and early 1990s were crisis years at Air Canada. Stung by debt, high interest rates, rising fuel costs, the demands of labour, fierce competition, etc., it lost billions. Along the way it fought a brutal dogfight with arch-competitor CAIL. In some periods each was losing at the daily rate of $1 million. Nobody knew how long this could last, although the feeling was that Air Canada could outlast CAIL. In February 1992 Hollis Harris took over Air Canada from Claude Taylor, who joined TCA in 1949 as a reservations clerk, rose to president in 1976, then became chairman in 1984. Harris, a former Delta (1987) and Continental (1990) head, arrived as Air Canada announced a 1991 loss of $218 million. One of his first moves was to stir the Orient pot. In May 1992 he told travel agents: "The travel market between Canada and Japan is growing 8% to 10% a year. Yet only two carriers fly the route [from Canada]—Japan Air Lines and Canadian Airlines. Right now up to 40% of the traffic moving between Canada and Japan is flowing over US gateways. That creates jobs in places like Chicago and Los Angeles. We want those jobs in Vancouver, Toronto and Montreal. Gaining access to Japan is a key to our ability to implement a broad Pacific strategy. Without Japan

other routes in the region aren't viable for Air Canada on a stand-alone basis."

In early 1992 Hollis Harris predicted that by year's end there would be only one major Canadian airline—his. A few months later Air Canada offered to buy CAIL from PWA Corp. Air Canada was accused of making the offer under pressure from Ottawa, but denied the allegation. CAIL's biggest fear, should there be a takeover, was that Air Canada would cannibalize CAIL's routes and fleet, leaving 6,000-10,000 out of work. Even though Air Canada promised that this would not happen, and that CAIL would continue operating under its own name, CAIL rejected the offer. It had another concern—a suspicion that Quebec politicking was afoot. If Air Canada could win CAIL, this would bring a host of new jobs to Montreal, headquarters of Air Canada, while stripping the West of the same jobs. If this happened, the ever-wily federal government would garner Quebec votes in the next federal election. CAIL's suspicions in this regard were well-founded historically—whenever a federal election loomed, Ottawa made lavish gifts and promises to Quebec.

Even into the mid-1990s Air Canada could not make a penny. In the first half of 1993 it laid off 250 pilots, bringing to 6,000 the

number let go since 1988, when Air Canada had 24,000 on the payroll. Yet Air Canada and CAIL continued buying new fleets, adding routes and getting into an overcapacity jam that became ludicrous. On busy routes like Toronto-Montreal there was so much capacity that A320s and 737s flew almost empty. Neither carrier would blink as the tidal wave of red ink grew.

In January 1993 the US DOT approved a deal whereby Air Canada and Dallas-based Air Partners bought 55% of Continental Airlines. This gave Air Canada access to Continental's markets in Mexico, South America, Africa, Asia and the Mid East. In July 1993 Air Canada, Continental and Air France formed an alliance including 600 aircraft serving 75 million passengers and 400 destinations. In this triangle arrangement, the partners began feeding each other's hubs in Toronto, Houston and Paris. Air Canada's 28% stake in troubled Continental originally was worth US$235 million. Its shares soared from $11 to $60 by the time Air Canada sold the last of them in 1996. Further funds were raised, when Air Canada sold two 747-400s to GE Capital Corp for $300 million in July 1993, then leased them back. It raised $240 million earlier by the sell-lease-back of five A320s. Such funds often were used to retire debt, pay day-to-day expenses, or raise credibility with suppliers. Sell-lease back was used commonly by airlines starting in the late 1980s. Following privatization, Air Canada started developing a new corporate image. On December 1, 1993 it officially adopted a giant red maple leaf for the evergreen tail of each of its 103 planes.

On March 22, 1994 Air Canada released the following statement: "Air Canada and PWA Corporation are pleased to have resolved all remaining matters regarding Gemini and the AMR transaction... Gemini or the successor company will continue to maintain hosting for and provide services to Canadian's Pegasus system until its transfer to American Airlines Sabre system in November 1994. All litigation between PWA and Air Canada regarding Gemini and the AMR transaction is now concluded... Air Canada has acquired ultimate ownership of Gemini." Gemini soon disappeared into a new Air Canada computer reservation system operation known as Galileo Canada. The Air Canada-CAIL arrangement became effective on March 15, 1994. So ended the monumental battle between Canada's two major airlines—CAIL had struggled for years to extricate itself from its Air Canada-Gemini agreement in order to merge with American Airlines and switch to its Saber reservation system. Most of the 700 jobs held previously under Gemini were preserved under the new Air Canada organization.

In May 1994 Air Canada announced an order for 25 111-seat A319s (compared to the 131-seat A320). The A319 would replace the DC-9, modernization plans for which had been announced in December 1993. These plans now were shelved, but the DC-9 would remain indefinitely, considering the boom in air travel that materialized after 1995. A great advantage in A319s would be the interchangeability of crews between the A319 and A320. The aircraft also used the same engines, spares and simulators. Also in the fleet planning department, in early 1994 Air Canada decided to reactivate several L.1011s from desert storage. These were for Montreal-Toronto-Vancouver, and Toronto-LA services. This was necessary due to a shortfall in capacity resulting from increased prospects in India, Korea and Japan.

With the high costs of fuel and of financing new equipment in the 1980s, the leading airlines found it increasingly difficult to function independently. Thus began the trend of two or more companies forming alliances, the benefits from which included coordination of schedules and pricing, joint marketing (e.g. merging frequent flyer incentives) and integrated travel agency commission programs. In the long run such alliances could save billions by coordinating fleet purchases, maintenance, training and computer systems. With schedules, the big benefit of such alliances was in code-sharing, whereby one airline could list the other's connecting flights in its schedule, as if these were its own flights. Another benefit was that each partner fed business into the other's hubs. One Air Canada alliance was with Korean Airlines. On May 16, 1994 Air Canada began B.747-400

(299 seats) service to Seoul from Toronto. Under a code-sharing alliance Air Canada and Korean offered six flights weekly. Next, Air Canada won Toronto-Vancouver-Osaka, inaugurating service in September 1994 with long-range (6,700 nm) A340s. In an agreement of August 1996 it agreed to let All Nippon Airways sell seats on Air Canada's daily flights to Osaka.

The A340 was ideal for the Pacific, having excellent capacity—a maximum of 33 LD3 containers or 11 pallets to 45,000 pounds, plus 284 passengers. Air Canada became the first North American user of the A340. Its Osaka coup hurt CAIL, which considered the Orient its special preserve. It was widely claimed that Air Canada won Osaka when it dropped its lawsuit to keep CAIL from withdrawing from

Air Canada A340s at Vancouver in February 1997. C-FYKX is nearest. A Nanaimo-bound Pacific Coastal Beech 99 (Air Canada connector) is doing its run-up in the foreground. (Jan Stroomenbergh)

Gemini. Air Canada continued lobbying for Hong Kong. This was vigorously opposed by CAIL, whose CEO Kevin Jenkins was peeved—CAIL recently had invested $500 million in new 747-400s for Hong Kong. He complained, "You just can't plan a business if the rug gets pulled out from under you." Ottawa gave Air Canada rights to Hong Kong in March 1995. Service was inaugurated from Vancouver on December 20. Hollis Harris commented that day: "Let the competition begin and may the best airline win." Meanwhile, CAIL was granted rights to Germany, previously the preserve of Air Canada. At the same time CAIL got Malaysia, Vietnam and the Philippines.

Open Skies

Open Skies was an airline philosophy periodically discussed in the US and Canada. In August 1939 there was a bilateral Canada-US air transport agreement: "Having in mind the desirability of mutually stimulating and promoting the sound economic development of air transportation between the United States and Canada, the Parties to this Arrangement agree that the establishment and development of air transport services between their respective territories by air carrier enterprises holding proper authorizations from their respective Governments shall be governed by the following provisions...." These included that overflights could be made by whichever nation over the US, Alaska and Canada and their territorial waters. There could be non-stop services through each others' air space, stops in each others territories, airworthiness

understandings, etc. Such agreements developed steadily as trans-border routes grew.

In 1992 Canada and the US began serious talks about bringing down barriers in trans-border flying. The object was to make it easier for Canadian carriers to serve US communities, and vice versa. With NAFTA soon to be in place, this was not viewed as a big problem, although Canada argued for a 10-year phase-in plan for US airlines gaining free access to Canada, while Canadian carriers should have immediate privileges in the US. Naturally, the big labour unions, ever fearful of any kind of progress, strenuously objected to Open Skies, whining that "Canadian carriers could never win in an open border policy." Fortunately, their negativism was ignored, and in February 1995 Canada and the US signed the Open Skies agreement. It opened all Canadian points to US carriers, except for Toronto, Vancouver and Montreal. Authority for them had to be sought through the US DOT, but this was assured after a short waiting period. Air Canada quickly introduced several new routes, including Montreal-Atlanta, Ottawa-Chicago, Toronto-Denver and Vancouver-Houston. Fourteen new slots were made available at LaGuardia for Air Canada and CAIL; and 10 in Chicago. Before long trans-border business was booming, fleets were growing and jobs again were plentiful in the airline industry.

On the home front Air Canada entered the world of the regional jet with a December 1993 order for 24 Canadair CRJ-200 regional jets. The US$17-million Canadair carried 50 passengers on stage lengths out to 2,000 nm. This showed its commitment to the domestic and trans-border markets, something that was engendered by the Canada-US Open Skies agreement. Air Canada and CAIL now were flying unfettered to dozens of new US destinations that best suited a smaller jet, one that was more economic per seat-mile than a DC-9. To entrench its domestic and trans-border interests further, in October 1994 Air Canada took 100% control of Air Ontario and Air Alliance. Early in 1997 it added several more Canadairs. After more than a decade of losses, Air Canada finally announced some earnings—$129 million in 1994, $52 million in 1995 (CAIL continued losing money in the same period). It had been a long, painful road—when Ottawa sold Air Canada in 1989, its shares each were worth $12.00. When Harris handed over to

Hollis Harris and Lamar Durrett, who headed Air Canada in the 1990s. Both had been top men at Delta and Continental. (Air Canada)

Lamar Durrett in 1996, they were listed at $5.10.

Air Canada's pilots made history in 1995 by leaving 4,300-member CALPA. This came about due to Air Canada pilots disputing a move in CALPA to merge pilot seniority lists among Air Canada, CAIL and smaller airlines. This could give a pilot flying a plane like a Beech 1900 seniority over a junior Air Canada pilot on something like a 767. The dissidents formed the Air Canada Pilots Association with 1,600 members. CALPA had been formed by TCA [Air Canada] pilots in 1937.

In May 1996 Hollis Harris, whose salary in 1995 was a reported $700,000, turned over the Air Canada presidency to fellow American, Lamar Durrett (another veteran from Delta and Continental). By early 1997 Air Canada seemed to be flying high after decades in the doldrums. It was fresh from posting a 1996 profit (on sales of $4.88 billion) of $149 million, most gleaned from the sale of its Continental shares. For the first time in its history, Air Canada earned more (52%) from non-domestic sales. Even so, its share of the domestic market climbed slightly to 57%, this in spite of serious competition from WestJet, Greyhound Air, Canada 3000 and others. In February 1997 CEO Durrett announced that 1,100 new employees would be hired by Air Canada, noting that "A large percentage of that will be flight attendants to support the increase in airplanes that are coming this year." He expected 22 new

planes for a year-end fleet of 160 airliners—the largest in company history. In June 1997 Air Canada inaugurated thrice-weekly Toronto-Osaka non-stop service.

The Air Charter Roller Coaster

The story of Canada's charter airlines dates to early post-WWII days. Companies like Curtiss-Reid, Maritime Central, Nordair, PWA, Wheeler and Worldwide tested the market, especially overseas possibilities. Starting with war surplus DC-4s, then hand-me-down DC-6s and Super Constellations, those that survived were flying jets by the early 1970s. Wardair set the standards. Operators came and went, depending on organization and management, and the ups-and-downs of the world economy. Failures sometimes occurred when poorly-run or shady tour companies, to which airlines sometimes were tied, collapsed (in 1972 twelve Toronto travel agencies dealing in airline charter tickets folded). Another cause of airline failures was a fuel crisis. Traditionally, European and American companies running Arab oil fields guaranteed supplies of cheap fuel. In 1973 however, the Arabs decided to reap more of the profits for themselves. They restricted supplies. In early 1973 oil had been selling at US$2.60 a barrel. By the end of the year the price had soared to $11.65. It hit $34.00 in 1981. Airline fares rose, driving millions of passengers to cheaper modes of travel.

In the 1970s new Canadian charter names appeared— Holidair, Nationair, Ontario World

Air, Vacationair, Worldways. One by one they disappeared. Others arose—Air Club, Air Transat, Canada 3000, Royal Air. The industry finally seemed stable and a global economic upsurge encouraged travel. Besides overseas business, the number of charter passengers travelling between Canadian cities soared from 20,000 in 1988 to 800,000 in 1994 and continued growing. Worldways was one of Canada's leading charter companies of the 1970s-80s. Formed in Toronto in 1975 under Roy Moore it began with corporate Lears, and a DC-4 hauling freight and fuel in the North. Next came two Convair 640s, used especially for sports charters. The Lears and Convairs also did courier work. In 1981 Worldways entered the international market with an ex-Ontario World Air 707. Business grew and two more 707s were added. The Convairs were sold and Worldways took over the former Wardair hangar at LBPIA. Four ex-CPAir DC-8-63s were added. Contracts were taken in 1983 to operate Hercules CF-DSX for Echo Bay Mines and a Convair 640 for Petrocan in Newfoundland.

In 1985 Worldways leased a 707 to the DND for a domestic sked between Shearwater, NS and Comox, BC. In 1986 DC-8s began operating Trenton-Gatwick-Lahr for the DND. Meanwhile, Worldways sold its 707s and added two long-range L1011-500s. The DC-8s were fitted with hush kits to permit operations to US destinations, where noise restrictions were increasing. Suddenly, Worldways folded, perhaps due to overly-enthusiastic growth. Montreal-based Nationair immediately filled the vacuum; but soon failed as well. With one charter carrier after another failing, it was

Worldways was one of Canada's leading charter airlines in the 1980s. Here its L.1011 C-GIFE lands at Toronto on September 3, 1989. It was originally with Pacific Southwest Airlines. After PSA folded, the aircraft sat in the Arizona desert, then went to Aero Peru in 1979. Worldways bought it in June 1985. Then (left), Worldways DC-4 CF-KAD at Malton in November 1975. It had worked as far afield as Afghanistan before coming to Canada in 1957 for PWA, Transair and Kenting. Hereafter it was in Egypt with Hercules Global Airlines, then moved to Texas in the 1980s. Compare KAD's tail logo with later versions on the CV640 and L.1011. Convair 640s C-FPWT and PWY served Worldways. PWT was at Toronto on June 6, 1976. It began in 1952 as a corporate CV340 (Arabian American Oil). In 1966 it was converted to a CV640, served 1967-76 with PWA, then joined Worldways. 1982-89 it was in the US as N862FW, flew 1989-90 with Canada West Air as C-FCWE, again was N862FW and ended in a disastrous crash in Senegal on February 9, 1992. (Larry Milberry)

Worldways L.1011 C-GIES on pushback at Toronto's Terminal 1. Also an old PSA and Aero Peru liner, it was taken over by the Royal Bank of Canada when Worldways folded in 1991. Later it was used for humanitarian relief by a US church organization. (Andy Graham)

some aging airliners, like Royal Air's 727s and Tristars, but these were well-maintained, and flown by top crews. Air Buses and 757s steadily were taking over— Canada 3000 set new standards by starting with the 757.

Canada 3000

Canada 3000 was formed in Toronto in 1988 in a joint venture among Air 2000 of the UK, John Lecky of Calgary, Adventure Tours and the Deluce family. Lecky was a successful businessman, former Olympic rower, and organizer of the 1988 Calgary Winter Olympics. The Deluces were an aviation-minded family based in Timmins. Family head Stan Deluce had bought Austin Airways in 1975, then took control of various smaller air carriers. This led to the Deluces having considerable sway in companies like Air Ontario, Air Manitoba and Air Creebec. By 1997 Canada 3000 had bases in Montreal, Toronto and Vancouver. It had a core of ex-air force pilots including veterans from ATG, but also many from CF-18s. Among the latter were pilots like Bob Wade and Ian Struthers, who had flown on top fighter meets like William Tell, done airshow and test flying and, in Struthers' case, commanded a squadron. The air force lost some of its best experience when these pilots left, but Canada 3000 cashed in. By 1994 about half its 125 pilots were ex-military.

A typical Canada 3000 charter operated on June 23, 1993. This was Flight 355 (Boeing 757 C-FXOO) going Toronto-Calgary return with Capt Ted Ryczko and FO Moe Storozuk. Aboard was 41,890 pounds of fuel and a 19,400-pound "traffic load". Push-back from Gate "K" at Toronto's Terminal 1 was at 1541 local time. First the right, then the left engines were fired up, and Ryczko taxied to R6R. He waited for a DC-9 on final, lined up and was airborne at 144K at 1552. Aboard F355 were the two pilots, eight cabin crew and 226 passengers. The cabin crew soon were busy, while the flight deck began with a hold at 7,000 feet—it was 1600 before F355 was cleared to FL230. It climbed over the Orangeville area, where the FO could pick out his farm, then arched northwestward toward Owen Sound and Wiarton to start the run up the Bruce Peninsula in CAVU conditions. Clearance came for FL350, then for FL390, where XOO levelled at 1628. Here it sat comfortably at a computer-controlled Mach .799. Captain Ryczko pointed out that present fuel flow was about half that of

amazing that there were volunteers to take their places, yet there always were.

With the recession of the 1980s and early 1990s airlines were able to pick flight and cabin crew from a large, educated and experienced labour pool. In a tight job market people would go outside their usual fields to work hard for deflated wages. As to pilots and flight engineers with the charter lines, they were different from those with the trunk carriers. Many were ex-CanForces personnel, some pensioned off after 20 years. Others were younger ex-military men who gambled on an early release for an airline career. When Worldways and Nationair were at their peaks, there was a joke that they really were subsidiaries of CanForces Air Transport Group at CFB Trenton. Many ex-ATG crews with Worldways and Nationair commuted to Toronto from their Trenton-area homes.

Air force people brought valuable experience—years of top training with thousands of hours on the 707, C-130, etc. Some argued that military discipline was another asset. To management, the ex-air force types were perfect, since they worked happily for less pay than Air

Canada or CAIL pilots. A major attraction for the pilots was that they could rise quickly to captaincies. Within two or three years a pilot in his early thirties could move to the left seat. At Air Canada or CAIL such a promotion would take 15-20 years. But many were rudely awakened in the 1980s, when the charter boom waned. Worldways, Nationair, then Wardair folded. Some ex-air force pilots found work with other carriers, including in the Far East. A few returned to uniform. Others left aviation. Now charter flying seemed to stabilize, even though two new companies folded almost as soon as they started. Newfoundland-based Triton tried entering the market in 1993 with a 737 equipped with hush-kitted engines. It became entangled in government red tape and disappeared. Atlantic Island Airlines of PEI also faded within a few weeks in this period. By 1997 Air Club, Air Transat and Royal Air of Montreal, and Canada 3000 and Sky Service of Toronto were Canada's leading charter airlines. Their markets were Europe, the Caribbean and Latin America, and US sun and gambling spots; there also were daily flights among Canadian cities. Airlines still were operating

Canada 3000's Boeing 757 C-FXOF at Toronto on July 31, 1993. The 757, which used the tried and true fuselage cross-section of the 707, 727 and 737, first flew in February 1982. The 757 became one of aviation's most solid types. There was no accident till December 20, 1995, when an American Airlines 757 crashed in Colombia—the cause being purely human. Canada 3000's first revenue flight operated Toronto-St. Petersburg, Florida on November 14, 1988. Canada 3000 shared its 757 fleet with Air 2000. It was world-famous as an efficiently-run, profit-making carrier that offered excellent service at affordable rates. (Larry Milberry)

a 707, although he hastened to comment that aviation was still aviation, in spite of a thousand technological improvements.

After 1:36 hours F355 was abeam Dryden in Northwestern Ontario. Capt Ryczko pointed out this pulp and paper mill town on the Trans Canada Highway west of Thunder Bay—it was his home town. When he made his announcement to the passengers, the FO mused that the captain deserved an award as Dryden's top PR man. Ryczko had started flying in the RCAF in 1966 and was one of the few trained from the outset on the Tutor jet trainer. This was an ill-conceived plan—putting recruits off the street straight into a jet, instead of the usual program—starting on the basic Chipmunk trainer. The RCAF reverted to its tried-and-true scheme after most of the Tutor students failed. Ryczko, however, survived and was posted to No. 2 Maritime OTU to train on the Neptune, thence to fly the Argus for 5,000 hours. In 1975 he moved to ATG at Trenton on 707s (437 Squadron). Many considered this the air force's plum posting. Ryczko spent most of his time on domestic and overseas skeds, but did numerous VIP trips, including September 9-20, 1984 with Pope John Paul II. He made round-the-world flights and many trips supporting the army—Op Snowgoose to Cyprus, trooping to and from Australia with the army, etc. He carried Vietnamese boat people to Canada in 1978-79 and recalled how he once brought a load to Dorval. The newcomers boarded buses for a reception centre in Longueuil. Ryczko flew his 707 back to Trenton, drove home and flipped on the TV news to see a happy gang of Vietnamese kids having a snowball fight in Longueuil. In the spring of 1986 Ryczko left the air force to fly briefly with Worldways, then returned to Trenton as a Class "C" reservist. He also did some flying for ACS, a Mirabel-based cargo carrier with DC-8s. His final move was in August 1988 to Canada 3000. By June 1993, when it had 109 pilots, Ryczko held pilot seniority No. 2.

F/O Storozuk followed a different route into the cockpit of C-GXOO. While he mined nickel underground in Thompson, Manitoba in 1970, he trained at Connie Jeff's flying school. Ilford-Riverton Airlines hired him to fly bush-planes, then Storozuk moved to Cross Canada Flights in Ottawa. Persistence got him further up the ladder—right seat, then captaincy on DC-3s for Pem-Air. After three years he progressed to Convair 580s at Great Lakes Airlines. It evolved into Air Ontario and Storozuk moved to Dash 8s, then jumped to a succession of charter companies—Vacationair (737), Odyssey (737) and Worldways (727). One after another these folded. In 1992 Storozuk joined Canada 3000.

At 1700 hours (local) F355 started its programmed decent to Calgary. At 1711 it was cleared to contact arrival and descend to 8,000.

Canada 3000's Capt Ted Ryczko and FO Moe Storozuk (above) en route Toronto-Calgary in 757 C-FXOO on June 23, 1993. Shown in Toronto-Vancouver Canada 3000 cockpit scene is Capt Chris Kay, in charge of 757 C-FXOF on July 31, 1993. He had been in the air force flying T-33s and Challengers before entering the airline world. Then, Gerry Graham on the flight deck of XOF. (Larry Milberry)

Six minutes later it was cleared to 6,500. Calgary ATC was busy and F355 was number nine to land. At 1721 and eight miles back the FO selected 15° of flap. A minute later the captain called for the landing check. At 1724 the flight broke out of cloud at 5,700 feet and crossed the outer marker a minute later. The wind was gusting to 18 as the 757 crossed the threshold of R28 at 145 knots and the FO called the final altitudes—"50 feet, 30, 10". At 1727 local F355 touched down. When the pilots shut down, XOO had 11,905 pounds of fuel left. Toronto-Calgary was scheduled as a 3:21 hrs flight, 3:31 going back. This made for a nice day's work for Capt Ryczko. He preferred such runs. Others liked the international routes, but Ryczko had had his fill of the global airways. At this stage he preferred getting home at day's end, to spending the night in some distant hotel.

Canada 3000 F643 (Boeing 757) of July 31, 1993 operated Toronto-Vancouver. On the flight deck were Capt Chris Kay and FO Gerry Graham. Running the back end was purser Karen English, who hadn't worked in the airline business until joining Canada 3000 in 1988, a leap she made from her job with the Toronto Humane Society. Among her staff were several who had battled through the charter lines, jumping from one sinking ship to the next until getting on with Canada 3000.

Capt Kay started flying in Midland, Ontario at age 17 and joined the air force in 1977. His first posting was to Cold Lake to spend three years flying a Dakota on odd jobs for the base—medevacs, passenger skeds to Edmonton, courier flights, airshows, etc. From here he went to 414 Squadron at North Bay. He joined Canada 3000 in 1989. Gerry Graham had started flying in 1981 at the St. Catharines Flying Club. As a young pilot he worked in the dreaded night-time courier game, hauling bags of mail, bank cheques, etc. for Soundair and

Nighthawk before getting a civilized job with Canada 3000.

At 1013 hours F643 departed LBPIA at 149 knots. With 227 passengers and 10 crew it weighed 227,100 pounds. At 1038 local it was level at FL350 and settling in for the 4:42-hours, 1,861-nm trip. A weather check at 1058 informed Kay and Graham that turbulence was disrupting flights over Lake Michigan. Some had to divert for fuel and safety to Duluth. F643 skirted the trouble, but still got a rough ride for a half hour. Then it was clear sailing across the US Great Plains and over the Cordillera until Kay started down for Vancouver at 1133 local. Nineteen minutes later F643 landed on R26. At shutdown the lapsed-time clock indicated 4:46 since push-back.

A return flight (F444, C-FXOO) from Vancouver followed on August 12. Capt Roger Clark and FO Tom Hamilton were ready for engine start at 1744 local. By the time XOO was pushed back it had been on the ground from Toronto a mere 1:17 hours. In that time the groomers had been through the cabin to inspect each seat for litter, spills, etc. Seat pouches were rearranged with their in-flight magazines, safety info cards and barf bags. The lavatories, recently used by the 211 Toronto-Vancouver passengers, were cleaned and their tanks pumped. Fresh water was replenished. The galley was cleaned and restocked by the caterers. Everything again was perfect for the comfort of 208 Toronto-bound passengers. Outside, service techs inspected XOO, checking fluid levels, tire condition and a dozen other details. 44,100 pounds of fuel were aboard for the trip east. The pilots, meanwhile, were reviewing their flight plan. Soon the passengers came down the jetway and boarded to find their seats and stash their carry-ons. In a rare happening, a mother and her newborn were put off minutes before taxiing. For safety, newborns were not allowed to fly commercially. Although cabins are pressurized to

6,000 feet (comfortable for children and adults) such thin air could cause a newborn distress.

At 1754 F444 was off into a clear evening sky—clear except for a canopy of smog over Vancouver. The auto pilot routed XOO over the Lower Fraser Valley passed suburbs like Richmond, Delta, and Surrey. Lush dairy lands fell behind, giving way to rows of north-south mountains. The Fraser, once a vital canoe route for explorers and traders, turned north at Hope and disappeared in the mountains. In 1808 Simon Fraser became the first white man to descend this great waterway to the Pacific. Another 50 years and a rush of fortune hunters swarmed its banks searching for gold. Next, the Fraser canyon was adopted as a natural route for Canada's transcontinental railroad pushing through from the Prairies. So went the evolution of transportation in the region. The pilots enjoyed the scenery—mountains, glaciers, rivers, forests, the ever-scenic finger lakes of BC's southern interior. They had the best seats in the house. Back in the cabin there was the usual hubbub of meals and drinks being served, videos running, passengers jostling up and down aisles on trips to the lavatory, children fussing.

Capt Clark was Canada 3000's senior pilot next to Capt D.V. "Dusty" Thompson, head of operations. Clark had learned to fly in 1959 at the Moncton Flying Club, then joined the RCAF. His first tour was on Sabres with 430 Squadron in France. Next he instructed at Moose Jaw on Harvards and Tutors, then flew C-130s and 707s at Trenton. He left the air force in 1980 for Ontario World Air, then flew with Wardair, Worldways and Nationair, before joining Canada 3000 at its start in August 1988. His initial months were in training, and waiting for a dispute to be settled (partner Air 2000 of the UK held more control in Canada 3000 than Canadian regulations permitted). Canada 3000 finally got the go-ahead and took

delivery of its first Boeing 757 (C-FOOB) in November.

Tom Hamilton had launched into aviation at the Calgary Flying Club in 1979. He instructed there for a time, then flew for a small resource company before getting a job with Soundair on night bag runs. This led to an FO position when Soundair subsidiary Odyssey International got 757s. When Odyssey folded, most of its pilots went to Canada 3000. As to today's airplane, C-FXOO had been delivered new to Canada 3000 15 months earlier. At this time Canada 3000 was flying its 757s at or near record daily utilization rates. Its policy was to acquire new aircraft, fly them to the max, and replace them on a five-year cycle. One pilot described this as a sound approach, since the 757 seemed to love being in the air. The more it flew, the fewer snags it seemed to have. F444 worked its way eastward. Regina-at-dusk slid by the port wing. Next came the lights of Brandon, two hours en route. Winnipeg was on the nose, probably getting rained on, as that's all that had been happening there through summer.

Meanwhile, XOO was being shadowed by Canada 3000 F456—Airbus C-GVXA under Capt Chris Ludwig. It stayed at the same Mach number some miles off XOO's port wing all the way to let-down at Toronto. XOO sailed on toward Thunder Bay and 3:18 hours after takeoff was over Sault Ste. Marie. Eighteen minutes later the crew was talking to Toronto Centre and starting down. Clearances were given to 7,000, then 3,000. VXA got the same instructions. F444 landed at 00:55, five minutes before LPBIA's 01:00 curfew. The two gleaming Canada 3000 airliners docked moments apart at adjacent gates to let off their 377 bleary-eyed charges. The passengers soon congregated at the baggage carousel to await their luggage. There was a flurry of activity outside as drivers picking up friends and relatives jostled for parking spots. A line of limos appeared from nowhere, collected those needing transport, and disappeared into the maze of highways leading toward Toronto. On the tarmac the night shift got to work cleaning and servicing XOO and VXA. Terminal 1 was now quiet as the morgue.

Canada's Major Airline Fleets, June 1997		
Company	*Aircraft*	*Quantity*
Air Canada	Airbus A320-211	34
	Airbus A319-113	8
	Airbus A340-313	8
	Boeing 747-100	3
	Boeing 747-200C	3
	Boeing 747-400C	3
	Boeing 767-200/300	29
	Canadair CL-600-2B19	
	Regional Jet	26
	MD DC-9-32	32
Canadian Airlines International	Airbus A320-211	12
	Boeing 737-200	44
	Boeing 747-400	4
	Boeing 767-300ER	11
	MD DC-10-30	9
Canada 3000	Airbus A320	6
	Boeing 757	6
Air Transat	Boeing 757	4
	Lockheed L1011	13
Royal Airlines	Airbus A310	3
	Boeing 727	7
	Lockheed L1011	2
Air Club	Airbus A310	4
Skyservice	Airbus A320	1
	Airbus A330	1
CanAir	A320	2
	Boeing 737	6
	CV580	8
First Air	Boeing 727	6
	BAe748	8
All Canada Express	Boeing 727	6
	CV580	2
Greyhound Air	Boeing 727	7
NWT Air	Boeing 737	3
	Lockheed L100-30	1
WestJet Airlines	Boeing 737	5
Vistajet	Boeing 737	1
Canadian Regional Airlines	Fokker F.28	18
	DHC-8-100	10
	DHC-8-300	14
	ATR42	10
Air BC	BAe146	5
	DHC-8	17
Air Ontario	DHC-8-100	19
	DHC-8-300	6
Air Nova	BAE146	5
	DHC-8-100	13
Air Alliance	DHC-8-100	8
	Beech 1900D	5
Air Atlantic	BAe146	3
	Dash 8-100	3
	Jetstream 41	5
Inter-Canadien	Fokker F.28	1
	ATR42	8

A March 22, 1997 newspaper ad showing the diverse routes and bargain fares offered by Canada 3000.

Air BC's main types in the early 1990s were the BAe146, Jetstream and Dash 8, all three of which appear in this July 8, 1989 Henry Tenby scene at Edmonton Muni.

(Left) Air Transat 727 C-GAAD originally was Air Canada C-GAAF, delivered in March 1975. It joined Air Transat in June 1991 to serve its hemispheric routes. AAD (181-seat configuration) was at Mirabel on July 29, 1994. In charter service such planes, readily available at bargain basement prices, could earn millions for the owners. By the year 2000, however, such old jets would need new or hush-kitted engines in order to meet Stage 3 noise level standards. Otherwise, they would be banned from the world's major airports. (Larry Milberry)

(Right) In 1990 the DND signed Nationair to fly passenger skeds previously flown by Worldways. This was done to free ATG's 707s for more operational duties. As a 437 Squadron A310 boarded at Lahr, Germany for Nairobi on March 15, 1993, the last DND Nationair flight was waiting to return to Trenton. The company was broke and few in the industry were surprised. Its DC-8 fleet had become infamous for breakdowns. Nationair cabin crew were so shabbily treated that they walked off the job in September 1991. The company locked them out and continued operations using scabs. A mediator was appointed to assist in negotiations, but Nationair rejected his report that recommended a 53% pay increase for the strikers. There was no sympathy for Nationair when it folded after its horrendous DC-8 crash in Saudi Arabia. (Larry Milberry)

(Left) Courier specialist CanAir entered the passenger charter field in the 1990s, first with 737s, then A320s. Here its 737 C-FNAP taxies at Toronto on July 13, 1996. (Larry Milberry)

(Right) Crownair's DC-8 CRN (ex-Air New Zealand) at Toronto in May 1989. Crownair formed in February 1989. It folded a year later, when its Vancouver-based financial backer withdrew support. CRN moved to Florida as N42920. In December 1994 it was seized by the DEA for drug involvement. (Andy Graham)

(Below) Another of Canada's many defunct charter airlines was Calgary-based Advance Air Charters. Its DC-8 C-FHAA is seen stored at Vancouver in October 1994 after a charter for the rock group UB40. HAA served Alitalia 1968-81, then a host of lesser carriers. (Larry Milberry)

Millardair

For nearly 40 years Toronto-based Millardair was prominent among Canada's charter airlines. It had planes ready to fly any time the phone rang, and its reputation for service brought many loyal customers. At a peak in the early 1970s Millardair had four Beech 18s, a Citation, six DC-3s, eight DC-4s, a Hansa Jet and several twin Pipers. Company owner Carl Millard was born on November 28, 1913 on a farm near Foldens Corners, a hamlet in the Ingersoll area of southwestern Ontario. His mother was related to England's Shuttleworth family, which later helped preserve British aviation history. His father was distantly related to Joseph Brant, Canada's famous Mohawk chief. As a boy Millard often cycled to Tommy Williams' farm. A WWI aviator, Williams often hosted Sunday barnstorming. One day Millard bought a ticket and went joy riding with Art Leavens in a Waco. This whetted his appetite for more flying.

When he was 16, Millard traded his father a cow for an old Model-T truck, which he used to haul cordwood. Meanwhile, he was aware of a disused water mill at nearby Centreville. He thought that he could get the mill working again, so negotiated with owner Charlie Downing. He agreed that Millard could run the mill rent free for three years. Millard cleared out the tail race by hand and restored the dam. He installed bins and a grinder, and convinced Baird's Machine Shop in Woodstock to repair the old turbine. When it got cold, Millard, who had moved into the mill, borrowed a chick hatchery stove to keep warm. Soon he had wagons waiting to off-load grain and cart off product. He also produced feed and meal for

local sale. Before long Millard was able to buy a 1929 Pontiac coupe. By the end of his lease, he was eyeing the old Thompson Hotel in Ingersoll. A company had been using it to store hearses, one of which fell through the floor one day. The place was condemned. Millard talked to one of the Thompsons. She agreed to his offer to repair the building in exchange for two years of free rent and an option to buy. Millard repaired the floor with salvaged lumber. For $200 he bought an old 50-hp Mack truck engine. A shop in Tilbury modified it to run on natural gas, and Millard then used it to power his new mill for four years.

One day Williams asked for Millard's views about aviation as a career, then gave some simple advice: "I could tell you about a lot of guys who got killed flying, and about all the others who went broke. Forget about flying and put your money back into your business." Millard listened, but was undeterred. His friend Wally Siple had Aeronca C-2 CF-BIG, which a fellow had pranged at St. Catharines. Millard bought the wreck for $150 and trucked it to his

Carl Millard outside his feed mill in the early 1930s, then with his personal Citation at Malton in 1989. (Millard Col., Larry Milberry)

father's farm. With Bill Minzatuk's help he soon had BIG looking good. With a new prop fitted, Millard taxied it around. This attracted a few locals. One dared him to take it up. Millard had taken a few lessons in Siple's E-2 Cub, so he poured on the coal and soon was looking down on the countryside. But he had to get down soon—it was dusk. After three or four dicey tries he succeeded, but a wheel hit a groundhog hole and BIG flipped. So ended Carl Millard's first solo.

Millard repaired BIG and used it flying around southwestern Ontario. One day Tommy Williams looked it over. He checked the rudder, yanked it upward and it came off—there was no pin securing it! Williams suggested, "You should get this fixed before it falls off and hits somebody." Millard later sold BIG to Bill Elgie of Ingersoll, who eventually cracked it up. In 1997 aviation collector Bob Stewart of Oshawa still had its remains. Millard now took some flying lessons from Fred Gillies in Toronto and got his licence. In September 1939 he paid J.R. Crapp of Hamilton $600 for Curtiss-Reid Ramble CF-CEA. He got to know Hamilton's aviation celebrities —Gord Rayner, Sr., Joe Reed, Don Rogers, Ernie Taylor. Millard was a successful businessman by now—he had a mill, a feed store, a Nash dealership, and was earning spare money joy riding, charging $1.00 a ride, $2.00 for a loop. One day in 1938 he had a look at Toronto's Island Airport. Construction was under way, but he landed beside the runway. Manager Ed Johnson shooed him away, but Millard had made the first of millions of landings at the new airport. He next took a commercial licence. For the examination each student had to do right and left spins, and a spot landing. Since he was used to his heavier Rambler, Millard was edgy about doing the test in a Toronto Flying Club Moth. Colonel Doug Joy of the DOT was present, and eased Millard's mind by asking, "Why don't you fly your own plane?" After passing the test, Millard took an instructor's rating at RCAF Station Trenton.

At the time he took his private, Millard had a medical with Dr. Tice in Hamilton, who gave him a limited (A-2) ticket on account of a family history of asthma. This displeased Millard, who wanted to apply at TCA. ("You may as well apply to be an astronaut today, as try getting in to TCA in those days," Millard reflected in 1997.) When the time came for the medical for his commercial, Millard went to Dr. Cargill in London, who was not known as overly strict. Cargill asked about medical problems. The question of asthma wasn't discussed and Cargill passed Millard, but needed to sign his licence (with the restriction stamped in it). Millard had "forgotten" to bring it, so Cargill decided to approve him for an A-1 commercial.

One day early in the war Millard's phone rang. Walter Leavens wondered if he would come to London to fly Ansons. Millard hesitated. He wanted to write TCA about a job. A few days later TCA's reply came. It included a pass to Winnipeg for an interview. Millard decided on TCA and was soon at Malton awaiting a stand by seat on a Lockheed 14. The plane, however, was full, and he spent the night on a couch in the manager's office in the Malton House. After breakfast he walked the mile back to the TCA terminal, hoping for a seat that night. Then, due to a no-show, he got on the noon flight. The captain was Jock Barclay, with FO Gordon Haslett and they made the trip via North Bay, Kapuskasing and Armstrong. A few hours after reaching Winnipeg, Millard heard that the night flight from Toronto had crashed at Armstrong, killing everyone.

Millard's interview went well and he joined other trainees in TCA's rigorous apprenticeship—working two weeks in each of the Winnipeg shops—engine, instrument, radio, sheet metal, etc. Ron George ran the operation, and Millard concluded that TCA could not have had a better man. He was tough, but that was what a fledgling airline needed. Jim Meekin, in charge of Link training, was similar. He told the lads, "This is where you're going to be a pilot for TCA, or go back where you came from." Soon Millard and three others began flight training at Lethbridge under Howard "Sandy" Sandgathe. Then he was posted on the Windsor-London-Toronto-Ottawa-Montreal run with captains Art Hollinsworth and Gerry Avison. Their Lockheed 14s carried only essential passengers—it was wartime, so each passenger had to be approved before a ticket was issued. As a first officer Millard got little opportunity to fly. As he recalled, "We first officers were mainly recognized when we made a mistake."

The Lockheed 14 was not the easiest plane to fly. It spun viciously, had troublesome engines (later replaced with R1830s), and the wing slots had been faired over to prevent them from icing. Without slots the Lockheed had reduced low-speed performance. In those days, when a plane had its annual C of A renewal, it had to be flown and spun. On one such flight Sandy Sandgathe had a Lockheed 14 go on its back. To recover, he had to over-stress the plane, bending a wingtip 90 degrees before regaining control.

After 14 months Carl Millard was promoted to the Toronto-Montreal-Blissville-Moncton-Halifax run. Next came his captain's course at Lethbridge. Back at Malton he finished upgrading with Frank Young, a well-liked captain who had started with the Toronto Flying Club. Young was a fine pilot, but usually flew with seat belt unfastened and legs crossed. As Young flew smoothly in for a bad weather landing, Millard used to wonder if he was ever going to uncross them! Young was a tough instructor. On instrument takeoffs (under the hood), trainees could expect him to cut an engine as soon as the gear was up, then cage (shut off) the gyros, so that all a pilot had to go on was his turn-and-bank indicator. The student would battle the asymmetric Lockheed, using all his strength on one rudder pedal. One day Millard asked Young what he would do if a trainee lost control at that point. He coolly replied, "He'd soon lose the other engine." At this time, a fresh TCA captain was earning $400 monthly, rising to $500 after six months on probation, then by $50 for each year served.

Once promoted, Millard flew Toronto-New York return twice daily, with a 36-hour New York layover. Pilots stayed at the Forest Hills Inn and had time to enjoy the city. Millard would haunt Seventh Avenue with its stores full of interesting junk. While at a boat show in New York, he met the owner of the Churchward company, which was marketing a $4,000, all-metal, 28-foot cruiser with a Chrysler engine. Millard took a dealership. He sold a number of boats around Ontario, but a 30% import tax killed this venture.

Millard spent eight years flying to New York, first on the Lodestar, then the DC-3 and North Star. He flew the inaugural Toronto-New York with the North Star, a plane he always enjoyed, although he had his reservations—mainly about the Merlin engine. His sense was that Rolls-Royce was trying to push out too much power. Early on the Merlins were cruised at 1,100 hp, then, to reduce wear, TCA reduced that to 1,000, finally 900. The Merlin needed a block check (tear-down) every 400 hours, a complete overhaul every 800 (TCA's R1830s went 2,000 hours between overhauls). To reduce brake wear, TCA had a policy of taxiing North Stars on Nos. 2 and 3 engines. If a news reporter saw one coming to the terminal on two engines he was likely to write a story about a North Star making yet another emergency landing. This delighted the opposition Conservative party, which used such stories to discredit the ruling Liberals, who had launched the North Star in the first place.

While with TCA Carl Millard had a business buying and selling small planes. One deal came with Fleet, which was winding up the Canuck project. Millard was offered the last 50 Canucks at $2,500 each. He didn't bite, and the price fell to $1,500, plus $200 in taxes. Then he bought the last 28 Canucks. He sold them one by one for $2,500. The Wongs of Central Airways, and Sunny Dale of Sault St. Marie bought, as did the Windsor Flying Club. One day Stan Deluce of White River met Millard to look over a Canuck on floats at the Island Airport. He had his three young boys with him, and explained that he was a railroad engineer hoping to start a small tourist business. The price was $2,500 but Deluce hedged. Millard told him that if his plans petered out, he'd take back the Fleet in a year for $2,000. Deluce agreed and in time became a major northern operator. Meanwhile, Millard was taking in other aircraft repossessed by the Industrial Acceptance Corp.

In the postwar era Canada allowed the import only of new general aviation planes. This policy was protectionist, but the US was swamped with cheap planes for which keen Canadian buyers waited. Carl Millard decided to challenge the restriction. He bought a Seabee in Pennsylvania, flew it to Ottawa, then arranged to meet the finance minister. Millard explained that the Seabee was still under warranty and had flown only 50 hours. The minister appreciated this point and waived the usual customs rule. Thereafter, Millard regularly brought in used planes—Bellancas, Cessnas, Ercoupes, Pipers, Seabees, Swifts, Taylorcrafts, Widgeons, etc. He also bought several new float-equipped, 150-hp Super Cubs

Della and Carl Millard picnicking on a Northern Ontario lake, which they had reached with one of their Canucks. (Millard Col.)

($10,000 each) from Ted Hiebert (Piper's largest distributor) at Teterboro Airport, New Jersey. These he would fly off the runway using an improvised dolly, while a strong lad ran along for the first few seconds holding the tail in flying position. Millard also enjoyed visiting the Bellancas in Newcastle, Delaware. They were genuine types, always refining their designs. At the time the affordable, wooden-wing Model 14-13 Cruisair was their popular type. With a 150-hp engine it could keep pace with a Bonanza, which was touted as the fastest four-seater, but had the price tag to go with it.

One day Pete Lazarenko called Millard from Winnipeg offering a Stinson Reliant for $7,000, but with a time expired engine. It was "buy now" over the phone or forget it. Millard accepted, took a Super Connie to Winnipeg and headed home in the Reliant, going non-stop to Blind River for fuel. There Carl L. Mattini, who had bought a Seabee from Millard, greeted him at the dock, looked over the Reliant and asked, "How much?" "I'll settle for $8,000", was the reply, and the deal was made. Next day Mattaini drove Millard to Sudbury to catch the TCA DC-3 to Malton.

Millard was interested in a Beech dealership, so visited Olive Ann Beech in Wichita. In those days one had to purchase a new aircraft to qualify for a dealership, so Millard bought a Model 18 off the line, registered it CF-EPJ and quickly talked to an old customer, the McFarland Construction Co. of Picton, Ontario. The latest corporate plane in Canada was the DH Dove, but it was easy to show that a Beech 18 was more airplane. McFarland was convinced and took EPJ, trading its Widgeon. Flown by McFarland pilot Bob Byers, EPJ arrived in Picton with the company's hand-picked paint job, radios and interior. Another customer was Inco. Millard met its pilot, George Cranmer, to talk over the pros and cons of a corporate plane. Cranmer discussed this with Inco vice-president Ralph Parker, who liked the idea, but wanted a demo. Millard arranged a flight from Toronto Island to Sudbury. While Cranmer flew, Millard and Parker talked. Below stretched the railroad north—Parker mused about all the hours he spent down there. By the time they reached Sudbury he was sold on Millard's idea. Soon Inco had a beautiful Beech 18—CF-HHI. It was one of only 31 Model D18Cs with 525-hp Continental R-9A engines compared to the usual 450-hp P&W R985s; but the Continentals were cantankerous. HHI's previous owner had traded it to Larry Rausch of Teterboro, who retrofitted R985s. He traded it to Millard in 1953 for a new Bonanza, worth about $25,000. Millard added further improvements before delivering HHI to Inco, where it served many years.

In 1963 Millardair was founded, beginning with a 150-mile Malton-Sarnia Beech 18 sked. Carl's son, 21-year-old Wayne, headed the airline. He had already been flying more than three years and had accumulated about 3,000 hours ferrying planes for his father and doing charters with the Apache, Goose, Widgeon, etc. Meanwhile, Carl's wife Dell was playing a key part in the administrative side of the business. Sarnia was Ontario's petrochemical capital, but most company head offices were in Toronto. It made sense that officials fly between the two centres. Bill Moon had this idea about the same time and formed Sarnia Airlines with a Dove. Both operations eventually disappeared, mainly because of Sarnia's poor airport. The

runway was prone to dangerous crosswinds. One night Carl Millard was coming in there in bad weather. He hadn't heard that a 20-foot drill rig had been erected at one end of the runway. Had it not been for Gary Kaiser of the local flying club, who radioed about the obstruction, Millard likely would have ended his flying career that night. Airport improvements eventually came. In 1961 John Blunt formed Great Lakes Airlines at Sarnia with DC-3s and Convair 440s.

Carl Millard looked at other ventures in the 1950s. One prospect was the Czechoslovakian Aero 45, a light twin similar to the Apache. Bill Stonehouse of Kitchener brought one to Canada and asked Millard to engineer some improvements, including improved brakes and heater. George Clark of Fleet was interested in the Aero 45, but Fleet was involved in US military contracts. In Cold War days the US looked askance at doing business with the communists, so pressured Fleet it forget about Aero. Millard also incorporated several improvements to the standard Apache. Another interesting deal involved an ex-TCA L.14. Most of TCA's fleet had gone to Latin America in the late 1940s, but Millard bought one (CF-TCN) on a take-it-or-leave-it basis for $5,000. He then offered it for $6,000 to Bill Sheppard of Windsor, whose brother, Gordon Fairley, had hotel interests in the Bahamas. Fairley was using a Goose to commute south and accepted the L.14. Millard checked out his pilots, Hugh Carlisle and Chuck Spurgeon, but when Fairley's Bahamas investments grew shaky, he asked Millard to find a buyer for the L.14. Texan Nat Kalt bought it for $5,000, then found that the FAA would not certify it because of more than 1,000 mods done over the years by TCA. Kalt asked Millard to peddle the Lockheed and it went to Central Northern Airlines of Winnipeg, where it served for years before going to G.H. Godsall's Commander Aviation of Gananoque, Ontario.

Millard also dealt in aircraft salvage. One day a Royal Gull crashed in Bellwoods Lake west of Malton. He bought it from the insurers for $5,000, spent some days locating it on the lake bottom, towed it ashore, then offered it to Leavens Brothers for $10,000. Clare Leavens

met him at Bellwoods Lake. He hedged, asking for Millard's rock bottom price. The answer was $7,500, at which point Leavens pulled a cheque from his pocket, already made out for the amount!

In 1962 Carl Millard bought Goose CF-OIA from Grumman for passenger charters. Ontario Premier John Robarts frequently used it. Eventually, OIA went to Warren Plummer of Sioux Narrows. He spent the winter readying the Goose for tourist operations. That spring Plummer lost OIA on take-off; he asked if Carl could find him another Goose, the sooner the better. Millard's interest in Grumman products led to friendship with Mac McKinnon of Sandy, Oregon. Over the years Millard converted several Widgeons using McKinnon kits. McKinnon then introduced a four-engine Goose. In 1959 it was at a Las Vegas trade show along with new types like the Convair 540, DC-8, Electra and F.27. Millard was pleased when McKinnon asked him to spend the week demonstrating the Goose. The routine was to make a four-engine pass, come around on three, then on two. Tony Le Vier and Fish Salmon were doing the same with the Electra. On the show's last day Millard overheard them on the radio asking for an extra three minutes to finish their show. The tower agreed and the Electra made passes on four, three and two engines. Then it came around on one! As the crowd watched, the Electra slowly became uncontrollable and disappeared over a small ridge. Everyone expected a cloud of smoke. Lucky for the crew, however, straight ahead was Nellis Air Force Base, where Le Vier made a judicious landing. Later he told Millard that the feathered engines would not re-light, something they hadn't faced before. Millard brought the Goose back to Malton to look for a sale, but none transpired. At $150,000 it was too expensive, and, at 12,000 pounds (8,750 for a standard Goose) lacked performance.

In 1956 Carl Millard heard from Earl Brownridge of American Motors Corp., which had a plant in Brampton, near Malton. He needed some auto parts delivered quickly to AMC in Kenosha, Wisconsin. This was done, and Millardair was suddenly in the auto parts freight business. This specialty became syn-

Millardair's famous Toronto base in a view from February 1972 showing Beech 18s and a DC-3. These hangars were built in WWII a half-mile south of this location. Then, a view from August 1962 showing one of the old 350- by 160-foot hangars being moved to its new home along the taxiway in front of Field Aviation's hangar at the north end of the airport. Few thought this feat possible, but Millard paid a contractor $25,000 for the job after agreeing that it might not work out. (Gordon Schwartz)

onymous with Millardair—its fleet henceforth spent most nights connecting auto industry hubs with outlying suppliers. About 1960 Millardair started Canada's first licenced instrument flight training school, using Apaches, then Twin Comanches. Instructors included a such experienced pilots as Wayne Millard, George Morwood and Archie Vanhee.

In February 1966 Millardair bought its first DC-3, CF-WCM. After WWII it had served the Champion Spark Plug Co. of Toledo, Ohio. Late in 1966 Jim McBride of Midwest Airlines approached Carl Millard looking for a DC-3 to serve the hydro site at Gillam, Manitoba. Millard flew a few demo trips. It was just before Christmas and hundreds of construction workers were keen to get home. Millard captained the plane. McBride sold tickets at Gillam as the workers stepped aboard. For the first trip Millard had to draw the line, when passengers started boarding with heavy tool kits—the kits had to stay behind. WCM remained with Midwest for nearly six years, then made its way south of the border.

The Beech 18

Always keen on the Beech 18, Carl Millard jumped at an offer from George Clark of Fleet to go to France in 1958 to see a Beech 18 with Turbomeca Astazou turbines. Clark thought the Astazou might be suitable for the RCAF. The two boarded a Pan Am 707 at Idlewild Airport, NY on what was Millard's first jet flight. He knew that the plane was in good hands—the captain departed on his own terms, but touched down in Paris precisely on schedule. Millard and Clark visited Dassault and Turbomeca, and flew the turbine Beech; but it was not such an attractive proposition—the engines were too far forward and too noisy. Millard later was in Texas picking up a Mooney, for which he had a dealership. Coming home he stopped by Dee Howard's in San Antonio and told him about Canada's new PT6 engine. They talked about a Beech 18 conversion and next day visited their pal, Ed Swearingen. He and Howard had been Braniff Airlines mechanics and were keen on conversions—Howard was doing Ventura corporate planes, and Swearingen was converting a Queen Air to his prototype Merlin. Millard suggested that he drop his proposed round windows for big "TV-screen" windows for the Merlin, and go for PT6s.

Millard put Howard in touch with Peter Thompson, who ran Aircraft Industries in St. Jean near Montreal. Millard, Howard and Thompson met with Canadian Pratt and Whitney president Thor Stevenson. As they talked, Stevenson was called to the phone and chatted away about some PT6 deal. When he hung up he told his friends that it was Ed Swearingen—he had just ordered two PT6s. The Merlin flew in 1965. Production ensued, although most models had Garretts, not PT6s—Garrett had offered better terms. Aircraft Industries built one turbo Beech 18—the

Jobmaster. By the time it appeared, other conversions had a grip on the market. Ed West, who had worked on many aircraft including Howard Hughes's "Spruce Goose", had placed the first PT6 order in 1963. West had a jump on Dee Howard and Aircraft Industries. His turbo Beech 18 (the Westwind), along with conversions by Gordon Hamilton of Tucson, led the market.

Although Carl Millard always was interested in the turbo Beech, he never operated one. He had faith in the stock plane, and couldn't rationalize the cost of conversion for some improved performance; but he was interested in improving the stock Beech. There had been wing failures, which he discussed with Dave Saunders, an ex-RCAF pilot, who had turned to design and modification. With the crash of some Beech 18s, the US Post Office decreed that the type no longer would carry mail. Dee Howard and Gordon Hamilton had

Millardair DC-3 at Malton on August 9, 1969. In 1988 McDonnell Douglas formed the Supplemental Inspection Document Working Group to provide the latest in DC-3 data and support to operators of some 2,500 DC-3s airworthy worldwide. This proved an excellent concept. Operators could call or fax McDonnell Douglas for hard-to-find information (including technical drawings) at reasonable rates. Millard noted, "We couldn't have had better service if we owned a new airplane." (Larry Milberry)

made some improvements to the Beech 18 wing, but Millard and Saunders discussed further mods to restore the plane's reputation. Saunders prepared drawings for a strap to strengthen the attachment of the outer wing to the main spar. In the late 1960s a strap kit was made in Millard's hangar and retrofitted to CF-WGP. This was done with the DOT co-operating. Millard and Saunders flew WGP with and without the strap, using strain gauges and a g-meter. To prove the viability of the kit, WGP had to be stressed to 2-g. Without the kit, data showed that the wing, at 2-g, was severely strained; with it, it was solid. Saunders began marketing the kit; hundreds were sold. In the early days Carl Millard had told Saunders that he could "make a million" with the strap. One day Saunders happily reporting that he indeed had sold $1 million worth of kits. He later designed wing strap kits for the King Air and Aero Commander, and STC'd other mods to

make various aircraft safer. Millardair's Beech 18s mainly flew auto parts, although one was an air ambulance. The Ontario government stopped using it for supposed safety considerations. Various "safer" Ontario air ambulances crashed in subsequent years, including turboprops. The Beech 18's good safety record continued.

The DC-3

By 1970 Millardair's included 10 DC-3s, five configured for passengers. There was lots of work with clubs like the Rotarians and Kiwanians, hockey teams, corporations and political parties. There were some booms when Air Canada or the air traffic controllers were on strike. In January 1972 an ATC strike grounded the sked airlines, costing them millions a day. This was good for smaller carriers, which got extra charters. In one strike Millardair flew a General Motors marketing team across Canada. It also offered aerial field trips to schools. A typical geography trip left Malton and took in such features as the Scarborough Bluffs, Toronto waterfront, Credit River from mouth to source, Hamilton harbour and steelworks and the Niagara Peninsula, all in about an hour. The trip was a bargain at $325 and a great experience for the students, few of whom had been in an airplane. Some 10,000 students experienced a Millardair field trip. During one Ontario election campaign all three political parties used Millardair. Each morning their DC-3s fired up to fan out across Ontario. Towards the end of the campaign Millardair put a special cake aboard the Bill Davis (Conservative) and Bob Nixon (Liberal) planes. The cakes were to be served on the homebound leg and were appropriately decorated "Good luck, Bill" and "Good luck, Bob". It was a real laugh when the cakes were served on the wrong planes!

The cargo DC-3 had a natural home with Millardair. The 1960s-70s were the heyday of the Canada-US auto pact, which brought prosperity to Southern Ontario. Each night DC-3s and Beech 18s crisscrossed the airways with Canadian-made components for US plants. There also were special charters for the DC-3. One took Carl Millard to Mexico for six dolphins destined for Marineland in Niagara Falls. Carl would fly non-stop to Toronto with a long-range DC-3 (1,600 gallons of fuel). To accommodate the dolphins, which were attended by experts, he would stay below 4,000 feet; but weather forced a stop in Texas. US Customs there took a jaundiced view of the operation. When Millard came out to the plane in the morning, he found it chained to the tarmac and the dolphins being off-loaded. Four were deposited in the chemical-ravaged waters of the local harbour. After negotiations the flight continued with the remaining animals. On another trip, Millardair flew a gorilla from Toronto to Pittsburgh for a zoo.

When it first started to grow, Carl Millard

hired experienced pilots on salary. In time he found that this was unwise—sometimes, when a pilot was needed for a rush trip, he was unavailable. Millard scrapped salary and introduced mileage. The more they flew, the more pilots earned. There was another problem—most experienced pilots were creatures of habit—they were not adaptable to Millardair's needs. This led to some costly abuse of equipment. Millard standardized on youth. He found that lads straight from their first job, such as instructing at a flying school, were best. They could be taught, and most loved to fly—anytime, anywhere. Over the years many of Millardair's young pilots and mechanics lived in right in the company hangar. They were despatched as calls came in for a load to go here or there, and got home by sunrise.

An added incentive was that new pilots had to pay for type check-outs. A DC-3 rating cost $3,000. Millard decided on this system, when he realized that many pilots were using him to get free ratings, then were leaving for greener pastures. When charged for check-outs, sprog pilots were more serious about their work. This system did not prevent pilots from leaving Millardair as soon as they got work elsewhere. Millard's system of charging young pilots was not new, but was controversial. The press liked to characterize such operators as squeezing poor lads of their last pennies. Such a portrayal was totally false. The system was widely used in the 1990s by much larger airlines, e.g. US commuter giant Comair charged prospective pilots even to be interviewed. Its advertisement in the March 1997 *Professional Pilot* invited applications from new pilots with 1,000 hours, but noted a fee of $325 per interview. In the end, most pilots appreciated what Millardair did for them. When the DC-3 turned 50, about 30 of them took him to dinner and presented him with a plaque which read, in part, "In appreciation for many fine starts."

As its DC-3 fleet grew, Millardair introduced a number of refinements. The DC-3 was famous for its unreliable heating system. Millardair engineered one that worked, got it STC'd and sold it to many operators. It also fitted the fleet with "Dr. Hoerner wing tips". These shortened each wing by 18 inches. While no lift was sacrificed, hangar storage was facilitated. Millardair looked at two potential DC-3 "replacements" in the 1970s. One was a conversion from John M. "Jack" Conroy of Air Technical Services Corp., Van Nuys, California (also Aero Spacelines of Santa Barbara). He, Carl Millard and Dave Saunders discussed a DC-3 with three PT6-45s. Conroy earlier had tested a DC-3 with Rolls-Royce Darts. He foresaw a market for turbo DC-3s, but not with the fuel-guzzling Dart. His PT6 demonstrator flew in November 1977 as the "Tri-Turbo Three", but the FAA denied an STC— since Conroy had added an engine, the plane had to be certified as a new aircraft. Unable to finance this, Conroy abandoned the project. He died of cancer shortly after.

Years earlier Douglas had built a better DC-3—the Super DC-3, first flown in 1950. It had a new wing and tail, Wright R1820s and a

A time exposure of Millardair C-117 C-GDOG at Toronto. (Tony Cassanova)

gross weight of 31,000 pounds (4,000 pounds heavier than a typical DC-3). Capital Airlines tried the new plane, but the project would have died had the US Navy not converted 100 of its R4D-5s and -6s (C-47s) to Supers. These were redesignated R4D-8 (later known as C-117s). They served till the early 1980s. While visiting Tucson in the 1980s, Carl Millard noticed many storage C-117s. They were for sale for about $25,000 each. He and engineer Bob Raynard test flew one and liked it. Performance was better than a DC-3—45 knots faster and able to carry 10,000 pounds of cargo over the same distance that a DC-3 carried 6,000. He started replacing DC-3s with C-117s, which served into the 1990s. As C-117s joined the fleet, DC-3s were sold for about $150,000 each. CF-DTV went to a Mexican operator, another went to Alaska. Two or three later turned up in the drug business. One was impounded at Allentown, Pennsylvania, while heading for Canada carrying dope.

The C-117, of which Millardair had eight, flew all sorts of charters. One day a Mohawk chief called, enquiring about freighting cigarettes from Montreal to Indian reserves in southern Ontario. His cargo normally moved along Hwy 401 between Montreal and Windsor; but the Ontario Provincial Police had been pulling over his trucks for spot checks, looking for cigarettes smuggled from the US. Loads would be impounded on suspicion, the chief would fight charges in court, but the cigarettes would spoil in the meantime. He wanted to try an airlift and wave goodbye to the police. In the late 1980s a number of C-117 loads were carried from Dorval to Ontario reserves. One day the OPP visited Millard to suggest that he was implicated with smugglers, but he heard nothing further of this. The airlift stopped when OPP highway inspections declined.

The Hansa
In 1970 Carl Millard decided to spend $550,000 on an HFB320 Hansa jet, a German design with a forward-swept wing. First flown in April 1961, it was certified in 1967. The Hansa, of which only 46 were built, carried two pilots and 12 passengers. Gross weight was 20,300 pounds compared to 16,800 for the seven-passenger Jet Commander (same engines: GE CJ610-9s). Although a bit underpowered and short on range, the Hansa was popular for charters and the Ontario government was a regular customer.

Carl Millard's favourite Hansa story con-

cerned its first charter. He and co-pilot Walter Kellor were flying Newfoundland Premier Frank Moores and several of his officials from St. John's to Montreal. While they passed Moncton in clear air at FL370, the right engine fuel filter warning light illuminated, indicating an iced-up filter. The circuit breaker was pushed to send de-icing glycol through the filter, but the light stayed on. Next, the fuel pressure warning light came on and the engine quit. Millard asked Moncton for weather—the ceiling was 300 feet, visibility half a mile in freezing rain. He requested a clearance and that the runway be sanded. Passing FL250, the second engine quit and instruments like the flight directors failed. Only some stand-by instruments remained. On the way down Millard relayed some weather to a worried Apache pilot trying to make Saint John (he crashed short of the runway). Passing the outer marker at 190 knots, Millard asked for the sander to clear the runway. He broke out at 200-300 feet and landed uneventfully. His biggest worry had been that the windshield could ice over, but it was thick enough to hold enough heat to stay clear. The cause of the incident was traced to water in the fuel taken on in St. John's. In 1974 Carl Millard purchased a Citation and sold the Hansa (in 1997 it was N171GA of Grand Aire Express in Monroe, Michigan). Millard took an American Airlines ground school and flight training course on the Citation before taking delivery. On the day of his check ride the DOT pilot arrived and they boarded the plane. Millard, however, was flabbergasted that the DOT's man couldn't read some of the flight instruments. Millard called off the check ride and the DOT sent its man on a Citation course.

Millardair and the DC-4
As the auto parts business boomed, Millardair purchased its first of eight DC-4s in 1976. Most were bought for about $30,000 from US military surplus at Davis-Monthan Air Force Base in Arizona. They were ferried a short distance to Ryan Field, where Aero American (known locally as "the Mayo Clinic for old planes") brought them to FAA standards. Next they went to Sebring, Florida for paint. Millardair DC-4s RYY and RYZ were ex-Royal Danish Air Force. Max Ahmad of New York owned them, but re-sold them "as is, where is" to Millard. They were sitting in a gravel pit at Vaerløse Air Force Base. Millard sent engineer T.A. Joe to work on them, assisted by two young Central Technical School graduates. En-

A Millardair DC-4 arrives home in December 1982 carrying thoroughbred horses. Then, a look at off-loading the passengers. (Larry Milberry)

gines and parts were shipped from Toronto on Lufthansa and within a few weeks the first aircraft was ferried to Toronto by Wayne Millard, arriving with no snags.

The DC-4 carried 18,000 pounds, was faster and had more range than a DC-3. To optimize the fleet, each got a new instrument panel. The forward bulkhead was moved ahead four feet; the electrics were converted from a hodge-podge of systems to all-DC (direct current). The Millardair charter rate at this time for a DC-4 was $2,000 hourly ($1,000 for a DC-3, $450 for a Beech 18). Besides hauling auto parts, the DC-4s flew race horse charters. Wayne Millard once took 12 horses from Toronto to Los Angeles. To ease the loading of horses Millardair introduced an STC'd mod—the DC-4 cargo door sill was raised 18 inches. Millardair DC-4s also airlifted the instruments and accoutrements of the Toronto Symphony Orchestra on tours of the Arctic and West Coast; and two DC-4s completed a magneto-metric survey over the North Atlantic.

Bureaucratic Hassles

Carl Millard had ongoing struggles with the DOT in the late 1970s. After training some pilots on the DC-4, he called the DOT to arrange check rides. He was told that so-and-so would do the rides, but Millard knew that this pilot was not qualified. The closest to the DC-4 that he had flown had been the North Star, and that was 20 years earlier. He turned down the DOT pilot. A new one was assigned, but Millard insisted that he take a ground school offered by a seasoned DC-4 captain. This man observed the DOT check pilot's lack of knowledge, even though his log showed captain time on types like the DC-4 and 707. Further research showed him to be an impostor. Millard documented all this, then met with a DOT official, who agreed to provide a qualified DC-4 check pilot.

One week Millardair underwent a DOT audit; some records were seized by the RCMP. A number of criminal charges were laid regarding a Millardair pilot with (supposedly) a lapsed medical, engines (supposedly) operating beyond overhaul schedules, etc. Millardair engaged a lawyer well-versed in aviation matters. The case climaxed in a 3 1/2-day court session. Besides defending itself successfully on all 10 charges, Millardair, using evidence surreptitiously left in its front office, illustrated that the DOT itself was negligent.

In more than 40 years Millardair had few safety problems. One involved a hockey charter that got into trouble landing at the Soo—the gear would not come down. Carl Millard had his pilot fly to Toronto, where a safe wheels-up landing was made. The company's only fatal crash involved pilot Tom Mulroney, flying a night Aztec charter from Gore Bay on Manitoulin Island to Toronto with five passengers. Minutes after takeoff he returned, but crashed near the runway. Four aboard died. During a lawsuit against Millardair, it was revealed that the pilot had suffered a heart attack and was trying to return to Gore Bay. This was revealed through testimony from a pathologist and a survivor. Testimony showed that the runway lights had been turned off when the Aztec departed. This, it was explained, was standard practice at small airports. But the DOT publicly claimed that runway lights at many such airports were on through the night. If a disabled pilot could not contact the radio attendant at a place like Gore Bay to request that the lights be turned back on, he would be on his own in finding the runway. In this case, the attendant had left the airport office when the Aztec departed, so nobody was available to turn on the lights. Millardair won the case.

Phasing Out

In the late 1980s Millardair was winding down and seeking buyers for its fleet. The DC-3s had been sold, and a buyer was found for the ex-Danish C-54s. In 1989 they went to Interocean Airways of Mozambique. In the early 1990s the C-117s were sold in BC. Two things helped Carl Millard decide to leave the airline business. Since the 1980s ATC at Toronto had imposed limits on the number of flights it would handle—no more than 70 an hour, only four being for general aviation. Even though there were only two busy times in the day at Toronto, this regulation applied throughout the day. All flights had to operate according to slot times, prearranged through the airport flow control centre. This was not practical for a charter carrier, whose phone rang at all hours with customers needing short-notice transportation.

Operators at Toronto were used to the power of the ATC union. Through the 1970s it had tied up traffic by imposing one hold after another. Even

when there was light (if any) traffic, the tower might force planes to hold before landing. Word of this spread throughout North America and drove away much business, mainly from corporate operators, who found other places to do business or to hold their conventions. Facing the slot time rule, Millardair tried to arrange an appointment with the Minister of Transport, but he was never available, and didn't return calls. For a time Millardair continued operating without slot times. With each departure the company was fined. Priority customers accepted this being added to their invoices. Before long, however, ATC began denying takeoff clearance to flights lacking slot times. Millardair was effectively grounded as of May 31, 1990. Meanwhile, the transport of auto parts was influenced by the North American Free Trade Agreement, which came into effect January 1, 1993. US carriers could operate more freely in Canadian skies. Detroit's Big Three auto producers began using US companies—Beech 18s, Convairs, and DC-6s at Detroit Willow Run airport began monopolizing the business. By 1995 the only flying being done at Millardair was with Navajos. Meanwhile, it began an interesting new phase—it built a modern hangar, which it leased to Canada 3000 for the servicing of Boeing 757s and A320s.

Wayne and Carl Millard, with Wayne's son Dellen. Wayne's first flight was as an infant. He learned to fly as soon as he could reach the controls. Once licenced, he worked for his father, often ferrying light planes around Canada and the US. He earned his multi-engine and instrument ratings, and at age 21 got on with Air Canada. In 12 years there he flew types like the Viscount, Vanguard, DC-8 and L.1011. Meanwhile, he squeezed in plenty of flying with Millardair. Later, Wayne flew the 757 for Canada 3000, and added ratings on the 747 and 767. (Millard Col.)

Corporate Aviation

The Grumman G.159 Gulfstream became one of Canada's top corporate aircraft. CF-MUR began in Canada with Massey Ferguson in 1960. It later was sold to the Iron Ore Company of Canada, with which it is seen at Toronto Island on October 30, 1971. (Larry Milberry)

Since earliest times aviation has been important in Canadian corporate affairs. Canada's first bush planes belonged to resource firms seeking to modernize, expand and increase profits. The St. Maurice Forest Protective Association and Imperial Oil Ltd. formed aviation departments, the one for mapping forest reserves in Quebec, the other for developing oil properties in the Mackenzie River valley. Since those times corporate aviation kept pace. In the 1930s entrepreneurs like J.P. Bickell, John David Eaton, K.C. Irving and Harry McLean did much to further corporate aviation, using planes like the speedy Beech 17 and amphibious Goose. From 1939-45 corporate flying was curtailed by wartime conditions, but a postwar boom followed. Markets surged, driving business forward. The airlines

couldn't meet every demand—the frequency of interurban flights was limited and thousands of US and Canadian communities were without sked service. For many business travellers taking a train, bus or U-drive car didn't always satisfy the need. The solution lay in corporate flying. A glut of war surplus aircraft did much to fill the gap for businesses needing a plane. In Canada, large corporations could pick from a selection of refurbished aircraft. The DC-3 was favoured by companies like Abitibi Paper (CF-ITQ), Algoma Steel (GJZ), Avro Canada (DJT), Canadair (DXU), Canadian Breweries (CBL), Duplate Canada (JUV), T. Eaton Co. (ETE), Ford (ORD), Goodyear (TDJ), Impe-

rial Oil (ESO, IOC), Interprovincial Pipelines (BZI), K.C. Irving (KCI), Miron Frères (HGD), and Ontario Paper (IKD). The Lodestar was a favourite among oil and gas companies, which wanted something faster than a DC-3— BA Oil (BAL, BAO), Home Oil (EAE), Imperial Oil (TDB), Pacific Petroleum (IYS)—but jazzed-up Lodestars also served companies like Canada Packers (CPK), Canadian Pacific Railwas (CPA), Mannix Construction (TDI), Massey Harris (TDG) and Noranda Mines (TCV).

Meanwhile, Beech still offered the Model 18 — $75,000 new. It was popular with companies carrying smaller loads on shorter routes. In the 1950s Beeches were operated by companies like Alnor Construction (GJS), Anthes-Imperial (ANT), Canadian Breweries (HOP), Curran-Briggs (MCH), Gardner Steel (FUP), International Harvester (LPC), McNamara Construction (IJI), and Sea Breeze Manufacturing (KLI). Carl Millard bought a new Beech 18 off the line (CF-EPJ) and quickly sold it to an old customer, McFarland Construction of Picton, Ontario. Over the decades Carl Millard would buy and sell many corporate Beech 18s.

A number of new designs entered the market in the late 1940s. From Britain came the Dove. Powered by two 400-hp Gipsy Queens, it first flew in September 1945, being Britain's first postwar civil design. It accommodated 6-8 passengers and was as fast as the Beech 18. Several Canadian companies bought Doves, including Federal Equipment, Imperial Oil, Massey Harris and Shell Oil. Grumman amphibians also were popular after the war—they were excellent for companies that liked treating executives and VIPs to fishing trips. The Goose flew with companies like BA Oil (BAE), J.P. Bickell (BKE), Canada Veneers (BZY) and MacMillan Bloedel (IOC, HUD). For companies on a budget, there was the

(Right) The speedy, comfortable Beech 17 was one of North America's popular executive planes of the 1930s. John David Eaton's is pictured in Toronto Bay. He kept CF-BKQ from 1938 to 1946, when it began a career with several bush operators. (Best Col.)

(Below) Two of Canada's well-known corporate DC-3s were Eaton's CF-ETE and Ford Motor Co's CF-ORD. Here ETE was at Malton; ORD was at Mount Hope. The DC-3 was the most desirable Canadian executive plane till it was toppled from its place by the first of the executive jets in the mid-1960s. (Larry Milberry, Jack McNulty)

The Lodestar was another classy executive plane of the 1950s. Here Eaton's CF-ETE is at Cartierville, fresh from refurbishment by Canadair in the late 1940s. ETE had a brief career. It caught fire in flight one day. The crew set down quickly, but their lovely airplane burned to destruction. Then, Calgary-based Hudson Bay Oil and Gas Co.'s CF-INY firing up at Malton on October 6, 1962. (Canadair Ltd. D4624, Larry Milberry)

(Below) The Dove served many Canadian companies in the 1950s-60s, but quickly disappeared when the bizjets arrived. Perhaps the last Canadian Dove was C-GEDT of Canadian Voyageur Airlines of Fort Frances, Ontario. EDT was seen at Dryden on August 24, 1976. (Larry Milberry)

(Below) The Widgeon, Goose and Mallard all were important corporate planes in the post-war years. Widgeon EHD, seen at Malton about 1950, was owned by Sheridan Equipment of Leaside, Ontario, a major dealer in heavy construction equipment. (NA PA191779/McNulty Col.)

(Below) Two of the early postwar light twins in Canada were Beech B95 Travel Air CF-LEN (1958) and Cessna 310 CF-LEU (1959). Over the years they had various owners. LEN, for example, once was owned by Del Bodkin Motors of Toronto, LEU by National Aviation Consultants of Toronto. Both were shot at Malton, LEN on May 14, 1966, LEU on March 24, 1972, when it was owned by Torrid Oven and Equipment. (Larry Milberry)

smaller Widgeon; for those with money to spare there was the Mallard. It was perfect for tycoons like J.P. Bickell, K.C. Irving and E.P. Taylor.

Not all businessmen could afford, nor did they always need, a DC-3 or a Beech 18. For them there were the new light twins. This market was the domain of Aero Design and Engineering, Beech, Cessna and Piper. The typical light twin was an IFR-equipped 4-5 seater. The elegant Aero Commander 520 first flew in 1948, and was an instant success. Beech offered America's first postwar light twin in

November 1949—the Model D50 Twin Bonanza. From Piper came the PA-23 Apache (first flight: March 1952), while Cessna flew the Ce.310 (January 1953). Beech added to the picture in 1956 with the Model 95 Travel Air. Many Canadian firms operated small twins. The owner of a one-man engineering company could make an intercity flight, do his business and be home for dinner. Even huge corporations found the light twin perfect for a hundred and one jobs. Trans Canada Pipelines, for example, used Apaches and Aztecs for pipeline patrols, basing planes at airports across Canada.

A typical light twin was Canada's first Aero Commander, CF-GBG. The 32nd "520" off the line, it reached Canada in 1952 for G.H. Godsall Equipment of Toronto. In 1957 it was registered to Commander Aviation. By 1970 it was being flown by Cross Canada Flights of Ottawa on charters and multi-engine training. As late as 1988—3$\frac{1}{2}$ decades after it came off the line — it was with a small company in St. Albert, Alberta. Besides the small twins, other businesses operated single-engine planes. The Beech Bonanza and Cessna 195 were the high-end types favoured in the early postwar years.

Pioneer Operator—The Hudson's Bay Company

One of Canada's historic corporate flight departments was run by the Hudson's Bay Company. Founded in 1670, it always used the best means available to get around. As soon as aircraft went north, it started chartering them to move men and supplies among its far flung posts. In the 1930s it purchased Beech 18D CF-BMI. It carried management on visits to posts to get a first-hand picture of business in furs, etc. In earlier times such trips took months, even years of heavy slogging. With a plane they took a few days or weeks. HBC air engineer Duncan D. McLaren wrote of operations with BMI (excerpt below) in the March 1941 issue of *The Beaver*. Although not informative regarding business affairs, it gives a good picture of the challenges in northern aviation (although flying was luxury compared to travelling by dogsled and snowshoe, the HBC's usual means of winter travel at this time):

As an example of the conditions met in winter flying, here is a short description made up from the log of our recent flight in BMI through northern Manitoba and Saskatchewan. We left Winnipeg at 1025 on the morning of January 6, having been delayed an hour and a half by fog. The fog continued over Lake Winnipeg, and prevented us from getting to either Cedar Lake post or The Pas; so we landed on northeast Cedar Lake to await better visibility. After two hours, and a cold lunch, we took off for The Pas with the ceiling at 300 feet, and landed there at half past three. We were held there all next day of account of fog and couldn't get away until 1130 the following morning. Soon we reached Cedar Lake, and landed at the post at noon. After a three-hour stay we got back to The Pas in mid-afternoon. Again we spent the night there, and next day left for Cumberland House, where we arrived after a 25 minute flight. It was overcast but the visibility was 10 to 12 miles and the ceiling was 3,000 feet.

At Cumberland House we stayed over night and left there at two o'clock the next day, taking with us post manager F. Reid. Fifteen minutes later we came down at Pine Bluff. We stayed there three quarters of an hour and then flew to Channing, half an hour away. When we landed there it was snowing. We spent the night and left at ten o'clock the next morning, January 11, taking along mail for South Reindeer and Lac du Brochet. During the take-off we ripped one of the ski bottoms on the ice. We arrived at Pelican Narrows in 25 minutes. The dull day made the snow very difficult to see.

At Pelican Narrows we were held up on account of the weather, but got away on January 13 at nine o'clock. The flight to South Reindeer post took 35 minutes, and we left there the same day for Lac du Brochet, an hour away. On this day we hit some pretty cold weather. It was 48° below zero at South Reindeer, but only 40° below at du Brochet. Flying north on Reindeer Lake we saw a few hundred caribou on the ice. We saw more of them, or perhaps the same ones the next day when we attempted to fly to Stanley. The weather closed down about 115 miles south of du Brochet and we had to return. It was 52° below at du Brochet that morning.

Next day we had more luck and hit Stanley about noon, bucking strong headwinds on the way and seeing more caribou. It took us only 20 minutes to fly the next day from Stanley to Lac la Ronge. On the following day, January 17, we flew against strong headwinds to Green Lake. On the 18th we made a half-hour trip to Beauval, where we spent the night, and in the morning flew back up to Ile à la Crosse. On this trip the dull light again made it difficult to see the snow. The next day we took in three more posts—Buffalo Narrows, Buffalo River and Clear Lake. Landing at the first was difficult owing to tractor ruts and holes in the ice

made for fishing. At Clear Lake, where we arrived at 2:30, the weather was closing in with snow squalls. We stayed there all night, leaving for Pine River in the morning, and at three o'clock in the afternoon came down at Ile à la Crosse again. From there, the next day, we took a hop over to Montreal Lake, where we were again grounded by bad weather.

Leaving Montreal Lake on the 24th, we arrived at Prince Albert at 9:30 in the morning. This was our coldest day—the temperature at Meridian Cabin in the bush was 55° below, although the post thermometer showed 43°. On arrival at Prince Albert it was estimated that the Montreal Lake temperature was at least 60° below. These low temperatures gave us trouble with the engine controls. Fifteen minutes out of Montreal Lake the temperature inside the plane was still 20° below, and coming down to Prince Albert the throttle on one of the engines froze in the idling position during decent. We therefore had to make a single engine landing. The Prince Albert temperature was 46° below.

Bad weather then forced us to spend the next day on the ground, but we got away before nine o'clock the following morning and headed for Winnipeg. The weather reports indicated clear weather all the way, but about 180 miles east of Prince Albert we ran into a cold front and encountered low weather. We turned northeast and flew along the edge of the front. An attempt to cross Lake Winnipeg and get into Berens River was frustrated. We were forced back and landed at Norway House at noon. Next morning we got away to an early start, arrived at Fort Alexander (at the other end of the lake) in an hour and a half, and after spending three hours there we got back to Winnipeg on January 27 at two o'clock.

Imperial Oil

Trailblazing down the Mackenzie in 1920 showed Imperial Oil that airplanes were the key to northern development. Even at this early date it foresaw a market in aviation products. From the early 1920s it began caching fuel and lubricants in the North for users who soon would need them. By the 1990s Imperial Oil still had the lion's share in this market and yearly was selling some 100 million gallons in Canada. Thomas Mayne "Pat" Reid, was Imperial's first aviation products manager. Born in Ireland in 1895, he served on RNAS flying boats in WWI. In 1920 he joined Handley Page Air Transport, flying airliners between London and the Continent. He emigrated in 1924,

(Above) The Hudson's Bay Company operated many types, from pre-war Beech 18, to Goose, DC-3 and Twin Otter. Here Goose HBC was at Winnipeg in May 1966. (Andy Graham)

(Right) Imperial Oil's G.1159 Gulfstream II CF-IOT at Malton on March 28, 1972. Along with the Falcon 20, it replaced the last of this company's propeller types, the Gulfstream I. Passengers now travelled long distances at more than 500 mph in quiet, spacious cabins. In the 1990s IOT was operating in Panama. (Larry Milberry)

joined the OPAS, then moved to NAME, where he operated in the Arctic on mineral exploration and explored the east coast of Hudson Bay. In January 1929 he and Doc Oakes took two aircraft to the Richmond Gulf on Hudson Bay to rescue 13 stranded prospectors. That August Duke Schiller and his prospecting party were stranded in the tundra, when Loening C-GATM ran out of gas. They set out to walk to Baker Lake, but Reid spotted them from Fairchild C-GATL, landed, and flew them to base. There were other Arctic adventures before Reid joined Imperial Oil in 1931. That summer he was tour leader for the Trans-Canada Air Pageant, flying Puss Moth CF-IOL. The next year he was assigned to co-ordinate sales of Imperial Oil products nation wide. This was very much Reid's domain—by now he was well-known among Canada's aircraft and engine builders, air carriers, pilots and engineers, insurers, people in government, and the RCAF. He worked tirelessly to promote Imperial Oil, often piloting its Beech 17 and Stinson Reliant on sales trips. He was honoured with the 1942-43 McKee Trophy. On April 8, 1954 he, his wife and 33 others on a TCA North Star died when it was struck near Moose Jaw by an RCAF Harvard. In 1973 Reid became a member of Canada's Aviation Hall of Fame.

Aviation boomed in the Canadian North after WW II. There were huge projects like the railroad to Knob Lake, the hydro project at Kemano, and the DEW Line. To keep in touch with customers all over the country, Imperial bought Lodestar CF-TDB from TCA in 1947 and had Canadair refurbish it. DC-3s ESO and IOC, based in Edmonton and Toronto, became company stalwarts. So did Dove BNU, Goose IOL and Mallard IOA. The Mallard served from 1949. One day in 1955 pilots Bruce Middleton and Larry MacKinnon got into severe icing, something especially dangerous in the Mallard. Imperial Oil immediately got rid of IOA and it went to PWA. On August 3, 1955 it was operating near Kitimat when it crashed into a mountain side.

For exploration in Alberta, Otters IOD and IOF were acquired, and aircraft routinely were chartered. In 1954 Convair 240 IOK became Imperial's premier aircraft. It came from China, where it had been with Civil Air Transport. Imperial Oil also ran an in-house travel bureau booking thousands of employee trips yearly on its planes, but also on the airlines, railroads and bus lines. In 1961 the Lodestar and Convair retired. Gulfstream IOM was acquired, then Falcon 20 ESO. F-27F IOG was added in 1965 and Electra IJR was bought in 1968 to support refining at Norman Wells and exploration in the Beaufort Sea. IOG was sold to Mobile Oil in 1972. Twin Otters IOH, IOJ and IOK were acquired in 1969-70, then a King Air 200 was added (all based in Alberta). Gulfstream II IOT joined the Toronto fleet in 1972. Smaller types operated in the 1950s-80s were the Apache, Aztec and Beaver. By 1997 Imperial Oil still was in corporate aviation, 77 years after its first involvement.

Massey Ferguson's modified Lodestar CF-TDG idling at Malton on June 29, 1960. (Larry Milberry)

The Massey Fleet

In 1947 the Massey Harris farm machinery company formed a flight department at Malton with ex-TCA Lodestar CF-TDG. Soon TDG was speeding company officials from one place to the next, visiting suppliers and dealers, many in small agricultural centres not served by airlines. Massey Harris appreciated the idea of corporate aviation and in 1954 Dove GYQ was added for shorter trips. In 1957 the company became Massey Ferguson and in 1959 Howard Super Ventura CF-MFL joined the operation. It was the speediest corporate plane in Canada, but Massey Ferguson added Gulfstream MUR in November 1960. As an industrial giant, it had the cash for one of these million-dollar beauties. MFL went to Dominion Tar and Chemical of Montreal. The Dove was sold in 1961 and in 1965 D.H.125 SDA replaced the Lodestar, which went to Execaire.

Massey Ferguson was now an all-turbine operation. With each advance of the bizjet world it would move up. Late in 1967 Falcon 20 WRA was bought and the Gulfstream went to HUT of Montreal. By this time Massey Ferguson belonged to Conrad Black, who brought it into his conglomerate, Argus Corporation. In 1968 he renamed his flight department Sugra. In 1972 SDA was replaced by D.H.125 AOS. Sugra managed it for the Canadian Imperial Bank of Commerce. In 1981 Massey Ferguson's huge Toronto complex was boarded up. So it remained for years until bulldozed and turned into a housing development. Argus Corp now had more lucrative interests, if Sugra was any indicator—

CF-AOS gave way to Challenger 600 C-GCIB. Argus absorbed Labrador Mining and Exploration and Norcen Energy Resources; their flight departments joined Sugra. By early 1982 Sugra had the Falcon, Challenger, LM&E's D.H.125 CF-HLL, Norcen's Saberliner CF-NCG and Black's Challenger C-GWRT, although HLL and NCG were sold. When Norcen moved to Calgary in 1985, it acquired Gulfstream II C-FNCG. Meanwhile, the aging Falcon was sold. So was Norcen (1987), resulting in NCG coming to Toronto with Sugra. By this time Sugra was disenchanted with its Challengers—there was too much downtime, so they were sold. C-GCIB was replaced with a BAe 125-800 with the same registration. Other companies like Dominion Stores and Standard Broadcasting used the Sugra aircraft as shareholders. In the mid-1990s the Sugra fleet was GII C-GNCG and a Bahamian-registered GIV VP-BHG.

Algoma Steel

The Algoma Steel Company of Sault Ste. Marie, owned by Sir James Dunn, needed its own flight department. Located in Ontario's mid-northern region, a 350-mile drive from Toronto, Sault Ste. Marie had no scheduled service until TCA opened its Great Lakes route in 1947. Even then a DC-3 flight a day was all that TCA offered. A major steel company had to have better connections with its suppliers and customers, so in 1946 Algoma opened a flight department with Norseman CF-BSH. In May 1947 it added DC-3 CF-GJZ, and in July 1948 swapped it for McIntyre-Porcupine Mines' DC-3 CF-GHL. Canadair rebuilds, each

Algoma Steel's Gulfstream ASC lands at Malton on July 1, 1969. This aircraft remained with the company till it closed its corporate aviation department in the early 1990s. ASC then was sold in the US for parting out. (Larry Milberry)

had a three-seat sofa, two reclining seats forward and four reclining seats in the rear. Allan Coggon, who flew for Algoma from September 1952 to January 1978, noted an odd bit of DC-3 trivia: "The 'biffy' in CF-GHL remained in its virginal state while in Sir James Dunn's service. He died in 1956." Algoma used its DC-3 mainly on runs to Montreal, New York, Charleston, Chicago and Fort William. There were occasional trips to Vancouver, Edmonton, New Brunswick and Florida. In March 1964 Algoma purchased Gulfstream I CF-ASC. Over the years it also operated a Beech 18C, an Aero Commander 680E, Beavers CF-AOP, CF-GBN, Turbo Beaver CF-ASA, and King Air E90 CF-ASD. By the mid-1990s Algoma Steel was in tight financial straits, so was forced to close its corporate aviation department. By this time Sault Ste. Marie was well-served by commuter carriers making dozens of flights daily with top equipment like the Dash 8.

Aviation and Mining

Mineral exploration and aviation became intertwined in the early 1920s. Developers like Consolidated Mining and Smelting recognized the airplane as the key to opening remote districts in the Canadian Shield and Cordillera. There were many smaller operators like Bobby Cockeram, Norb Millar and Howard Watt. As president of Prospector Airways, Cockeram qualified for a commercial licence in 1930, then earned his air engineer's licence, when he was 49. In 1933 Colonel D.G. Joy, Inspector of Civil Aviation, noted of him: "Learned to fly about four years ago, when considerably older than the usual flying pupil. Has used aircraft in his companies and explored about half of Canada from Winnipeg to the Arctic, and Montreal to Hudson Strait. Exceptionally good record of quiet, careful and efficient operations without accidents." An unknown commentator added: "Realizing the extent of the yet unexplored mineral areas distributed within the not readily accessible northern stretches of Canada, he early appreciated the value and uses of aircraft for primary geological and topographical reconnaissances, general scouting, photography for subsequent study of general rock structures, besides a means of transportation to aid the prospector in opening up these areas. In 1928, as an observer with two others of his organization, he spent the season flying around the Barrens and country surrounding Reindeer, Wollaston and South Indian Lake areas scouting their mineral possibilities. In the subsequent prospecting seasons of 1929, '30 and '31, while flying his own Moth seaplane, his operations were confined to the known mining areas in Quebec and Ontario south of the transcontinental [railroad]. The use of aircraft enabled him easily to investigate many out-of-the-way properties."

Another mining man dedicated to corporate aviation was M.J. Boylen of Toronto. In 1959 he bought Goose CF-IFN, an old US Navy JRF-5. After the war it had been the VIP plane for Cuban dictator Batista. When dictator Castro took over Cuba, Batista fled to Miami in his Goose. It became N2720A, then was

sold to Boylen. In July 1966 Reg Phillips became Boylen's pilot. Phillips noted, "One day we could be transporting a load of rugged prospectors and the next have New York financiers in their pinstripe suits." The captains of industry always are linked to top politicians. On occasion Boylen loaned IFN to Newfoundland's famous premier, Joey Smallwood, during election campaigns—the Goose was perfect for getting around Newfoundland and Labrador. Phillips also flew Boylen to mine properties in places like northern New Brunswick. The many "local" entries in his logbook turn out mainly to have been short trips from Malton airport to Fort Erie, where Boylen enjoyed afternoons following thoroughbred horse racing.

Boylen also used IFN in aerial mineral surveying. He had a custom magnetometer installed—a compact unit that fit in the nose. Phillips noted: "It was designed by Hans Lundberg but was owned by Lloyd Leach, as was the four-channel spectrometer. Leach had worked for Lundberg, and when the company shut down, bought the equipment. Lloyd installed it and acted as navigator and technician on most of our flights." There was no external "mag" or electronic cables as usually seen on aero survey planes. The mag measured the magnetic fields and the spectrometer measured the different types of radiation." One intensive survey was flown from Sept-Îles and Wabush over July 1967. Phillips flew on 26 days (94 hours logged) that month, mostly off the water. One day he flew eight hours. In January 1968 IFN was in Las Vegas for several weeks of surveying. One morning Phillips came out to fly and found a note on IFN's windscreen. It was from an old crony from Austin Airways days, Gil Hudson. Parked beside the Goose was a Mallard that Hudson was flying for a US company.

Before he died in 1969, M.J. Boylen had Phillips sell the Goose. Phillips was contacted by Charles Blair of Antilles Air Boats in the Caribbean and a deal was made. Blair, whose flying boat collection included the only Sikorsky VS-44A, kept IFN mainly for private use. When he died in a crash, his widow, actress Maureen O'Hara, stored IFN for several years with Fred Frakes in Texas, then sold it in the late 1980s, when it had more than 12,000 flying hours. Phillips later did some mining consulting in aircraft operations for remote areas. He also bought and sold airplanes and had a placement service for pilots looking for work. In 1972 he joined the DOT in accident investigation and air regulations enforcement, then retired in 1994.

Another company sold on corporate aviation was Steep Rock Iron Mines Ltd. of Toronto, which had mines at Atikokan in Ontario's Rainy River District. Over the winter of 1937 high grade iron ore was discovered on the bed of Steep Rock Lake, 135 miles west of the

Steep Rock's Tony Shrive in 1996. (Larry Milberry)

Lakehead. The lake was drained and the ore bodies developed as open pit mines. The railway division point of Atikokan quickly grew to accommodate 6,000 people. The ore went by rail to Fort William, where it was transferred to ships. Annual production reached 3,000,000 tons by 1956.

Steep Rock had business all around North America. It enjoyed excellent rail service, and Highway 11 from Fort William reached Atikokan in the midfifties. But a corporate airplane would facilitate business, especially since head office was in Toronto. The nearest air connection was with TCA from Fort William. The schedules did not always suit company plans, so in the early 1950s Steep Rock purchased Beech D18S CF-EPJ from Carl Millard. The chief pilot was Borden Fawcett, an ex-miner who had flown in the RCAF during the war. In 1960 Steep Rock purchased Super 18 CF-TAB from Timmins Aviation. It offered greater speed and comfort than earlier versions. With TAB came a new pilot, A.N. "Tony" Shrive. He was born in Hamilton in 1925. His father, an RFC observer in WWI, worked in radio sales and service, then joined the BCATP in WWII. Sons Norm and Tony followed him into aviation. While earning his private pilot's licence at the Hamilton Aero Club, Tony worked as an apprentice mechanic at No. 10 EFTS at Mount Hope. In 1942 he joined the RCAF and was posted to No. 10 for elementary training on Tiger Moths. He earned his wings at No. 16 SFTS at Hagersville, near Hamilton, and spent the war flying Ansons at places like Rivers, London and Charlottetown. When he left the service, he worked at his father-in-law's garage, but kept his eyes peeled for a flying job. Stan Welsh of Matachewan near Kirkland Lake, Ontario had bus and trucking operations, but wanted to diversify into bush flying. He offered Shrive $40 a week and a dollar-a-flying-hour. Shrive started on March 6, 1947, initially flying J3C-65 CF-DGR, then Stinson 108-2 CF-FJX. Business focused on prospectors and tourists. Besides flying, Shrive kept DGR and FJX fit.

Shrive next took a job with Borden Fawcett, who was operating Northland Aviation at South Porcupine. It had Aeronca 11BC CF-DUP, Seabee CF-DKO and Waco ZQC-6 CF-BDW. Business involved lumbering, mining and tourism. A year later Shrive was hired by the Department of Lands and Forests. He was posted to Remi Lake near Kapuskasing with Beaver OCL and started flying May 4, 1949. This was a seven-days-a-week job with lots of variety. The pay was average—about $3,500 a year, but the government provided summer accommodation for Shrive, his wife Ruth and their children. After five years, Shrive rejoined the RCAF. He took the instructor's course at Trenton, then instructed at Penhold and Centralia on Harvards and Chipmunks. He finished his five-year, short-term service contract flying Expeditors with 424 Squadron at Mount Hope. While there he heard that Anthes-Imperial of St. Catharines, which owned Beech D18S CF-ANT, was looking for a pilot. The company was in the furnace and cast iron pipe business and had widespread offices and plants. Shrive took the job, flying ANT initially on October 3, 1958. ANT, based at St. Catharines, operated throughout the continent. Although it was convenient having ANT in St. Catharines, there was a drawback—the airport was VFR only. If the weather was down, ANT was grounded or, if inbound, had to divert to Buffalo or Toronto. In April 1959 Anthes-Imperial purchased an E18S (also ANT). The earlier Beech went to the Hunting Survey Co. as CF-LLF.

While Shrive was in the air force, Borden Fawcett started Steep Rock Iron Mines' aviation department. Needing a pilot, he solicited Shrive, who accepted. He made his first flight in Beech TAB on March 28, 1960. Head office was 1,000 miles from Atikokan. Company president M.S. "Pop" Fotheringham, however, preferred living in Atikokan. Almost every week he commuted to Toronto, and made frequent trips to iron and coal industry cities like Cleveland and Chicago. TAB was good for this work—it was fast (160K), comfortable and reliable. A drawback was that it was unpressurized—it seldom flew above 10,000 feet, so could get stuck in weather.

Eventually TAB came due for a costly overhaul. At this time the owners talked to Commander Aviation in Toronto. Tony Shrive had gone to school in Hamilton with Dan Seagrove of Commander. They talked about the merits of replacing TAB with an Aero Commander, a faster plane that handled icing better than a Beech. Aero Commander 680F CF-SRG was bought; about a year later it was replaced with Turbo Commander CF-FEO, which cruised at about 280 mph. Unlike the Super 18 it was pressurized, so could get above the weather. By late 1967 Tony Shrive had made the Atikokan-Toronto run about 800 times. Then came word that, under a new president, Steep Rock Mines was closing its flight department. Tony Shrive heard through Irv Watson of Commander Aviation that DHC was looking for a pilot, arranged an interview with Dave Fairbanks there and was soon was doing ferry and sales demo work with Twin Otters and Buffalos.

Canada's First FBO

By 1950 corporate aviation in the US was well-served by FBOs, or fixed base operators, like Butler Aviation. All major airports had at least one FBO. When a plane arrived, whether a Bonanza or a Convair, it could taxi to the FBO, where attendants awaited. Passengers would be shuttled from the ramp to a comfortable lounge. Limousines or taxis to the hotel, if not standing by, would be arranged. A conference room was available for meetings, in case passengers were in a rush. Meanwhile, the plane was fuelled and serviced. If necessary, minor snags were rectified. If passengers were returning the same day, the crew could wait in comfortable surroundings. Flight planning, weather services and catering all were avail-

Super 18 CF-TAB in service with Steep Rock Iron Mines. (Gordon Schwartz)

Turbo Commander FEO, seen at Malton on May 13, 1967, replaced Beech TAB. Being fast, pressurized and turbo-powered, it was a great step ahead of the Beech. (Larry Milberry)

able. In Canada at this time there were no FBOs. When a corporate plane arrived, it was up to the crew to arrange everything. John Timmins decided to do something about this. In 1958 he set up Canada's first FBO—Timmins Aviation at Dorval.

Born in Montreal in 1928, John Timmins took flying lessons at Cartierville with Curtiss-Reid Flying Service. At age 17 he had a private pilot's licence. He also had a job as a Joe-boy at RAF Ferry Command. This provided useful hands-on experience and the opportunity to meet people in the business. Timmins bought an interest in Curtiss-Reid Rambler III CF-BIB. He barnstormed with it, but the venture was short-lived. On September 26, 1946 BIB had engine failure taking off at St. Jean, and crashed. Timmins' injuries kept him on the ground for about a year. Next he took a job in Timmins, the town named for his grandfather Noah of mining fame. This time he was working for a fellow called Wolf, a construction man with an air service on the side—Northland Aviation of South Porcupine.

On August 8, 1949 Timmins had just taken off in Waco CF-AWK. It was a hot day and he had three passengers with lots of gear. Within moments the engine quit and AWK came down like a stone. Timmins crash-landed on a street in town. His worst injuries were sustained when his burly passengers trampled him in the rush to get clear. For flying overloaded, Timmins was grounded by the DOT for a few months. Then HUT hired him to fly the DC-3. After two years he went to CPA at Dorval.

Soon after starting there he began having sinus troubles, so left for a low-level contract on a Spartan Canso. This allowed him to fly without sinus trouble. Within six months his condition improved and he returned to HUT.

In 1954 John Timmins and Don McClintock took over flying HUT's Lodestar, CF-CPA. It had been the CPR's corporate plane, used mainly by president N.R. "Buck" Crump. Now the big wheels in the Iron Ore Company of Canada took it on business trips to New York, to coal and steel centres like Pittsburgh and Cleveland, etc. John's relative, Jules Timmins, was a frequent passenger, as he was a key figure in attracting capital to keep iron ore devel-

John Timmins in his Avline office in Sarasota, Florida in January 1996. The posters on the wall honour the Cowboy Junkies, a group of famous musical performers founded by John's children. (Larry Milberry)

D.H.125s Registered in Canada in 1972

Registration	Serial	Imported	Owner
CF-ALC	25087	1966	Aluminum Company of Canada, Montreal
CF-ANL	25042	1965	Anthes Imperial, St. Catharines, Ont.
CF-BNK	NA746	1970	Execaire Aviation, Montreal
CF-CFL	NA741	1970	Churchill Falls Labrador Corp, St. John's
CF-DOM	25018	1965	Execaire Aviation
CF-DSC	25086	1965	Distillers Corp, Montreal
CF-HLL	25034	1965	Labrador Mining and Exploration, Toronto
CF-IPJ	25053	1965	Interprovincial Pipeline, Edmonton
CF-KCI	NA702	1968	Irving Oil, Saint John, NB
CF-MDB	25075	1968	International Jet Air, Calgary
CF-NER	NA714	1969	Execaire Aviation
CF-OPC	25016	1965	Canada Packers, Toronto
CF-PQG	25036	1965	Government of Quebec, Quebec City
CF-QNS	25152	1967	Ontario Paper Co., Thorold, Ont.
CF-SDH	NA724	1969	Shell Canada, Toronto
CF-SEN	25027	1965	Execaire Aviation
CF-SIM	25039	1965	Robert Simpson Co., Toronto
CF-TEC	NA754	1970	Execaire Aviation
CF-WOS	25159	1968	Gulf Oil Canada, Toronto

opment moving in Labrador and Quebec. The HUT airlift ceased when the railroad opened in 1954. As John Timmins described it, "HUT was the first airline ever replaced by a railroad." Most of the fleet was sold, but DC-3 CF-FBS was outfitted with a VIP interior to replace the Lodestar. FBS operated years from Dorval under the Hollinger Consolidated Gold Mines banner till replaced by an F-27.

Flying the Lodestar and DC-3 was valuable experience for John Timmins. It familiarized him with corporate aviation in the US and with FBOs. He met those who converted planes to corporate configuration e.g. at places like Remmert-Werner in St. Louis. He was aware of the dearth of services in Canada—of coming into Dorval or Malton on a blustery, wet day and being assigned a parking spot on some bleak tarmac; having to battle with officialdom to get a taxi to the plane; having to tie down, scrounge for fuel, etc. This was a shame, for corporate aviation was expanding in Canada. John Timmins entered the FBO field by leasing land at Dorval and erecting a $300,000 hangar, designed to accommodate a DC-6. His first employees were old HUT men—pilot John Weston and engineer John Luty. Vic Bennett was company secretary; Jack Graham was general manager. The company was modelled after Butler Aviation with a lounge, Shell Oil concession and services like making hotel reservations and providing ground transport for passengers and crew. The first airplane to visit was Massey Harris Lodestar CF-TDG (Capt B.G. Smith). The second was a four-seat Morane-Saulnier Paris, perhaps the first corporate jet to visit Canada.

Soon Timmins Aviation branched into a line where it would win acclaim—remodelling older aircraft as corporate executive planes. The renowned New York designer, Charles Butler, was hired, and a design office was set up at Dorval. The first conversion was Goodyear Rubber's DC-3 CF-TDJ. In August 1958 Grumman introduced the world' first turboprop business plane, the Gulfstream. Timmins won the Canadian distributorship. Several Gulfstreams were sold in 1959-60 to companies like Home Oil of Calgary, and Imperial Oil, J.F. Carruthers and Massey Ferguson of Toronto. Gulfstreams were delivered to Dorval "green" (no interiors) for custom finishing. Activities increased soon after John Timmins attended the 1962 Farnborough Airshow with DHC's Phil Garratt. There they viewed the new D.H.125 bizjet. Timmins was taken aback when Garratt offered him the Canadian distributorship, which he accepted. One day Timmins and Bennett were musing about how many of the $700,000 D.H.125s might be sold in Canada. Timmins thought, "15 in five years". This seemed reasonable for what was understood about the market for bizjets in those times. Timmins sold 15 in two years, including two on the same day. This could happen since neither Lear nor North American was earnestly marketing bizjets in Canada. The first D.H.125 in Canada was Ontario Paper's CF-OPC. It was finished by Timmins and delivered in 1964. Meanwhile, other interior contracts were won, including for RCAF VIP Cosmopolitans.

While business boomed, John Timmins wondered about expanding. He looked at the design of aircraft galleys—something to give a unique product line. Its first galley (for the

BAC 1-11) was displayed at the Paris Airshow. Aer Lingus ordered several for its BAC 1-11s. These were the days when B.737s, DC-9s and other jet liners were pouring from production lines. Soon Timmins was swamped. Typical was an order from United Airlines for 110 B.737 galleys. Ironically, this success caused losses—the company was not geared to mass production. Timmins Aviation was pushed to the brink. When the opportunity arose in 1968, it was sold to Atlantic Aviation of New Jersey, headed by Henry Dupont.

John Timmins had a small Imperial Oil FBO at Malton run by Stu Irving. He now envisioned a trans-Canada chain of such basic FBOs, a concept he had not exploited originally. He discussed possibilities with Charlie Dunn of Texaco and the concept of Texaco Sky Service was born in Canada. Timmins now approached Lear about marketing its bizjet and Learjet Canada was set up under him. He also took on the Saberliner and Model 690 Canadian markets for Rockwell International. Such activity took him to 1976, the year Russ Bannock, then president of DHC, invited him to be vice-president of marketing. Timmins joined DHC at the peak of Dash 7 marketing, then carried into the Dash 8 era. This involved him to 1984, when he left DHC to form Avline, a company specializing in leasing and re-marketing pressurized regional airliners from Metros to Dash 8s. What drew Timmins to this field was the scarcity of companies with a knowledge of the commuter industry and leasing. He had a firm understanding of the airlines, their needs, aircraft available, and how to write leases. Much of this came from his years at DHC, but the beginnings dated to when Timmins Aviation leased its first aircraft—DC-3s to DEW Line contractors. This was a school of hard knocks, for Timmins knew nothing about leasing. Those were the days when deals still were made on a handshake. When he leased a DC-3 to World Wide Airways, the main stipulation was that it be returned in such condition as dictated "by normal wear and tear". The DC-3 came back a near wreck, with the undercarriage locked down, held firmly with 2 x 4s! When Timmins complained, his customer (Don McVicar) said, "Hey, you said normal wear and tear. This is what we consider to be normal wear and tear." Timmins decided to show more interest in lease writing.

In January 1996 Avline had 36 airliners on lease including 10 Dash 8s. Over the years John Timmins found the market for used regional airliners to be steady. While bigger carriers were susceptible to economic instability, the regionals weathered hard times well. In serving more specialized markets, they rarely discounted seats. Fleets were tailored to needs, so load factors were higher than for mainline carriers. The typical passenger was a business traveller, who was not as concerned about ticket price as the casual or holiday traveller. In the mid-1990s used planes like the B.1900C, Jetstream and Metro represented good value and held their prices, since they were not being replaced by new designs. The cost of introducing new types discouraged airframers from

Steven B. Roman of Denison Mines travelled in this beautiful GIII or his King Air 200. CF-SBR was at Malton on March 28, 1972. In 1996 SBR was noted as N48EC with Eastman Chemical Co. in Tennessee. (Larry Milberry)

new ventures—they had the capability, but knew that the 19-seat operator would balk at $5 or $6 million, when used planes were under $2 million.

Corporate Ups and Downs

Corporate aviation always is susceptible to economic times. When times are good, flight departments are busy and may grow. In tough times departments will shrink, perhaps fold altogether. Corporate flying hours in Canada dropped about 25% from 1981 to 1982. This resulted from a global recession, which also forced industry to tighten its belt. Companies had less business, so less of a requirement for executive travel. Thus, Falconbridge and Inco, Toronto-based mining multi-nationals, sold their Gulfstream IIs. Inco still had other aircraft, but Falconbridge opted for chartering and using the airlines. At the same time real estate giant Campeau Corp sold one of two Sabreliners. MacMillan Bloedel, hurting from a slump in wood product sales, got rid of a Westwind and mothballed a Goose. At the same time hard times hit the major airlines, so they were obliged to reduce service. For the corporate traveller this aggravated matters, making the company plane look good again. An interesting solution for some was the sharing of corporate planes among several firms.

Ambiguity about running a corporate fleet convinced many that it was more sensible to charter from specialists like Montreal-based Execaire. It started in 1964 with Beech Baron CF-EXA, and built up a fleet of bizjets and turboprops. Besides chartering, Execaire managed some corporate flight departments e.g. D.H.125s CF-BNK for the Bank of Montreal, and CF-DOM for Domtar. Some bizjet owners applied for licences to charter their aircraft to others, thus subsidizing the cost of their flight departments. Short-term results of the 1980s recession included a glut of used aircraft (at excellent prices for the few who were buying), and many pilots and AMEs looking for work. In spite of tougher times, however, some companies maintained their bizjets. Sometimes the operators were so wealthy that their planes were as much the CEO's toys as they were business assets. This was the case in the early 1980s with Denison Mines of Toronto. Its

Gulfstream III and King Air 200 were available on a moment's notice from CEO Steven B. Roman.

Flight Safety

In modern times, with flight safety at such a high level, there is little concern about how many senior company people board the same flight. Accidents are almost down to zero, but the age-old question is still asked: "Do the benefits of corporate aviation outweigh the risks?" Most of the time the question is academic, and the answer is a quick, "Yes". But accidents bring more practical considerations and companies re-assess policies about corporate travel. In March 1997, for example, four top men with Basler Turbo Conversions (including famed company president Warren Basler) died when two company planes (a DC-3 and a Bonanza) collided in flight in Wisconsin. Another such accident occurred in BC on April 27, 1963. That day R.E. Lewko was flying circuits in Cessna 140 CF-GIH at Penticton, an uncontrolled airport in those days. Meanwhile, Frank Sibley took off in Argosy Oil's Aero Commander 680E CF-OJL. Aboard were Argosy employees with family members. Sibley took off towards the north, then turned south. As Lewko was turning at 1,000 feet to land, the planes collided, killing all eight aboard. Neither pilot had seen the other in time to turn away.

One of the worst accidents in Canadian corporate aviation occurred near Frobisher Bay on February 27, 1974. Involved was Brethour Realty Services' Saberliner CF-BRL, piloted by Lloyd Nichol and Bernard Lawrence. Brethour made frequent trips overseas, which got it into corporate aviation. Company owner Max Buschler had asked Carl Millard about buying a Sabreliner, to which Millard replied, "You're playing on the edge with a jet." He had suggested that no jet at that time was really suitable, but something like a Convair 580 would do. Brethour did not want a propeller plane and bought BRL. It was inbound from Europe on February 27 and cleared by Moncton ATC to land at Frobisher, when the beacon there failed. BRL was not informed of this and the beacon stayed off for 49 minutes. Back-up equipment failed to keep it running automatically and the beacon was not switched to an alternate power source. Without a beacon to guide it, BRL had to descend to search for the airport. By so doing it burned excess fuel. The airport was not located and BRL's tanks ran dry. It crashed, killing all nine aboard. Lawsuits following the disaster led to a number of settlements favouring the plainiffs, including one for $1.2-million, the largest such settlement in Canada. In 1994 Carl Millard recalled that on the night of the crash he was called by the Toronto *Star*. It wanted a jet to fly a news team to Frobisher. They mentioned that a business jet had crashed. Without being told the details, the thought crossed Millard's mind, "It's the Brethour jet." Another corporate jet accident occurred December 9, 1977. Churchill Falls Labrador Corporation's HS-125 went down at night while approaching Churchill Falls. All eight aboard died. The crash of Rockwell 690A N400N on December 1, 1990 claimed the lives of several employees from Madill Ltd. of Nanaimo. So badly effected was the company that it closed its flight department.

Corporate Aircraft Sales

Gordon Schwartz of Toronto spent part of his aviation career buying and selling corporate aircraft. Born in 1919 in Toronto, he graduated in aviation from Central Technical School in 1937. He apprenticed at the Toronto Flying Club, then located on the northeast corner of Dufferin St. and Wilson Ave. in North Toronto. When the war began, the TFC became involved training RCAF pilots and air engineers. Schwartz became chief engineer, when Gord Rayner moved to Malton to open No. 1 EFTS. When the TFC closed, as DHC expanded and took over the airfield, Schwartz became head of aircraft overhaul at Nos. 20 and 21 EFTS at Oshawa. There was lots of work—the school had 72 Tiger Moths.

In 1942 Schwartz joined the RCAF and was posted directly to the Test and Development unit at Rockcliffe in maintenance. He rose to flight engineer. T&D was involved in such work as testing de-icing systems, special antennae, ski installations on a Hurricane, armament, night lighting, and all sorts of components. It also tested aircraft with odd engine installations—a Stranraer, a Fairey Battle and a Bolingbroke with Wright R1820s; a Bolingbroke with twin Wasp Juniors, and a Cornell with an opposed six-cylinder Lycoming. T&D had a Ventura with a malfunctioning right engine. It was ferried to St. Hubert for Canadian Pratt & Whitney to check. After 10 days of work CP&W corrected the problem. During a subsequent test flight at St. Hubert the left engine quit on approach. The Ventura crashed, injuring Schwartz and his pilot. Once he partially recovered, Schwartz got a medical discharge. He took a job at Chisholm Manufacturing, a Toronto company making aircraft components. After the war he returned to the TFC, where he again took over from Gordon Rayner, who went to Maritime Central Airways as chief engineer.

After the war British American Oil of Toronto bought Goose CF-EXA from War Assets and hired Smitty Pruner and Gordon Schwartz as crew. The plane was refurbished by Grumman and put to work on corporate duties. Through 1947-48 there were trips to BA headquarters in Tulsa, and to such far-flung areas as Alaska and the Lower St. Lawrence. One day BA received an offer from Nickel Belt Airways to operate EXA on BA's behalf. In the process Pruner and Schwartz left the company. This was a mixed blessing for Schwartz, who had been away from his family much of the year on the Goose. He settled down and bought a service station in Downsview. Business was good, but the job was just as bad as corporate aviation—he worked 16 hours a day, seven days a week.

Schwartz kept in touch with aviation through the 1950s. Although he had spent all of his time till then in maintenance, he also had several hundred hours as a flight engineer. He decided to get into flying. He went to Les Baxter, an instructor with the TCA Flying Club, soloed in two hours, and obtained his private pilot's licence. He started doing some ferrying for Carl Millard, mainly delivering light planes from the US. He upgraded to twins and by 1960 had his multi-engine IFR commercial rating. From 1954-60 he was active in COPA and owned a Piper Tri Pacer, which he flew all over Canada and the US to aviation events. In 1960 Schwartz sold his service station to BA Oil. There was lots of work through Carl Millard, and assignments upgrading local aircraft owners to instrument standards. Schwartz would accompany pilots, who were not fully at ease by themselves, but wanted to make longer cross-country trips. He renewed his acquaintance with Collin Campbell about this time. A former TCA captain, Campbell was operating a North Star for International Air Freighters, flying between Toronto and Havana. He took on Schwartz to assist flight crew Willy Milne and Stan Hegstrom in radio work and other duties on their long trips. While farm animals and essentials like wire and tires were carried from Toronto, the North Star would come home with such loads as frozen lobster tails for Canada Packers. There were also a few passengers. The Cuban trips were non-stop—under a trade embargo against Cuba such flights faced seizure if they landed in the US. ATC was cool towards any Canada-Cuba air traffic. An example was when the IAF crew might ask Charleston or Pittsburgh for Toronto weather and get a reply back like, "Why don't you ask your buddy Castro?"

Gordon Schwartz's career with Fleet and the Super V is told earlier in *Air Transport in Canada*. When he left Fleet, he was asked to join Commander Aviation Ltd. It had been started by Gordon Godsall, a heavy equipment dealer with an interest in aviation. Godsall bought an early Model 520 (CF-GBG) in Buffalo in 1954. In 1955 Aero Commander of Oklahoma City struck a deal for Godsall to be distributor. His company, Commander Aviation Ltd., began in Gananoque, near Kingston, Ontario. As it expanded, CAL won a contract with Westinghouse in Hamilton for corporate transportation. They bought a new Model 680E (CF-LAC) with radar and IFR equipment. CAL's first employee was Irv Watson, an AME

from DHC who also did some flying and became operations manager. When CAL moved to Hamilton with a new airplane and a contract to operate it, they needed a full-time professional captain, so hired John Weston. Mo Servos, a TCA captain, knew Watson and Godsall and was asked to assist with the Westinghouse contract. CAL grew slowly and in 1962 moved to Malton. One of the new pilots was Dan Seagrove.

By 1963 Godsall was disenchanted with aviation and wanted out. Seagrove and Watson had no money, but saw an opportunity. Through personal mortgages and signing promissory notes, they persuaded Godsall and Aero Commander to turn the company and its distributorship over to them. Schwartz joined CAL in mid-1964 as vice-president of sales. As Canadian distributor CAL sold all Aero Commander models, which included the 500B, 560F, 680F and 680FL Grand Commander. The line was well-received because of its all-weather capability and excellent stability for camera and EM work. CAL sold new and used aircraft to meet price requirements and accommodate trade-ins. A charter licence helped introduce potential customers to the product line. By this time CAL was based in the Field Aviation hangar at Malton.

The introduction of the 707 and DC-8 in 1959 threatened the importance of the corporate aircraft as a long distance resource. Executives now had their pilots fly them to the nearest main airport, where they boarded big jets for long trips. The manufacturers of corporate aircraft saw the writing on the wall and started designing the first of the corporate jets—DH the 125, Lear the Learjet, Lockheed the Jetstar, Rockwell International (at its Aero Commander Division in Bethany, Oklahoma) the Jet Commander, and North American the Sabreliner. CAL was unique in Canada—it always bought its own demonstrators. To finance an $800,000 jet took courage, but CAL went ahead. About this time Al Lowe, a WWII pilot and successful Toronto businessman with a variety of holdings, bought 50% of CAL. He owned a 680F, which he used on business trips. Moe Fraser, CAL's chief pilot, Lowe and Schwartz went to the factory to take the course on the Jet Commander. They had flown piston-powered aircraft all their lives at lower altitudes, cruising at 180K, approaching at 80K, climbing at 1,000 fpm. It was a major transition to come to terms with jets climbing at 3,500 fpm, cruising at up to 41,000 feet at 420K, and approaching at 110K. As Schwartz recalled, this was intimidating and it took about 50 hours to catch up with the airplane. Once one did, however, the Jet Commander was nicer to fly than its piston-engine relatives.

CAL took delivery of its first demonstrator (CF-SUA) in mid-1965. It was still the sole Canadian distributor with its own demonstrator. A vigorous sales campaign followed and CAL learned quickly the cost of financing, operating and marketing business jets. The first sale was to the Acres Company, a Toronto engineering firm involved with the Churchill Falls hydro power development. Acres needed

Jet Commander CF-SUA at Dorval on February 21, 1966. SUA later became N91669, although it was noted in 1996 as withdrawn from use at Gilroy Aircraft in California. (Larry Milberry)

reliable, high-speed, long-distance transportation. Further sales followed to such companies as Avco Finance, the Bank of Montreal, International Nickel (two), Kenting Aviation of Toronto; Central Trust (Moncton), Mother Tucker's Restaurants (Winnipeg), Lundrigan Construction (Newfoundland) and IMP (Halifax). Competition was fierce and closing a sale could take as long as a year.

In 1969 CAL was flying high. It built a hangar and moved out of Field Aviation. Also in 1969 Rockwell International bought North American Aviation and found itself the manufacturer of two business jets—the Saberliner and the Jet Commander. The US Justice Department ordered it to sell one. The Jet Commander was sold to Israeli Aircraft Industries. Schwartz immediately went to New York City to meet with IAI officials and found them anxious to keep their distributorships intact. Since they had taken the entire project, they found themselves with inventory that had to be moved. Before any other distributor made a sale, Schwatrz sold the first Jet Commander for IAI to Lundrigans.

In the early 1960s, when other manufactures were designing and building jets, Beech decided against that and entered the turbine market, re-engining its pressurized, piston-powered Model 88 Queen Air with CP&W PT6 turbines. Thus was the King Air born, and Beech never looked back. Although the aircraft was slow (215K cruise), had low cabin pressure and pressurized off only one engine-driven compressor, it was attractive to Beech followers. Aero Commander countered by turbinizing its pressurized, piston-powered 680FLP Grand Commander. TPE331 engines, of which Garrett had built thousands as APUs, were used. Garrett felt it would be simple to add a gear box to the TPE331 and put a prop on it. The engine started out troublesome, but as experience was gained, became reliable. The 680T series (575-eshp Garretts delivering 4 psi pressure differential cruising at 240K) developed into the 690 series (710 eshp and 5psi pressure differential cruising at 280K). Eventually the series had 1,000 eshp and cruised over 300K.

True to its policy CAL ordered a demonstrator. In mid-1966 Schwartz went to the factory, took the course and brought the first Turbo Commander into Canada (CF-SVJ). He

started immediately on an eastern tour through to Newfoundland and was successful in selling the demonstrator to Diamond Construction in Fredericton, a previous Aero Commander operator. In the ensuing years Turbo Commanders were sold from coast-to-coast to companies like Irving Oil (Saint John), the Province of New Brunswick (Fredericton), Avco Finance (Toronto), Steep Rock Iron Mines (Atikokan) and Perini Construction.

When Seagrove and Watson formed their partnership in 1963, Seagrove became president. Over the years he proved an astute businessman as well as a capable leader. He formed, financed and controlled various divisions of CAL, ran them autonomously and installed their own presidents. He formed Navair, an avionics sales and repair company run by Jack Grose. In the early 1970s, with a downturn in business (and Schwartz having some medical problems), CAL relinquished its exclusive Aero Commander distributorship. In 1975 a partnership was struck with John Timmins and a separate company (Timmins Aviation Ltd.) was formed. Timmins, the president, brought the Canadian Saberliner distributorship with him. He operated from Montreal, Schwartz from Toronto. They did well the first year, selling several Saberliners, Aero Commanders and other aircraft.

In 1976 Russ Bannock returned to DHC as president, after Ottawa bought DHC. Timmins joined as vice-president of marketing and sales. Schwartz remained with Seagrove, but their interest in aircraft sales was waning. In 1977 Seagrove spun off the charter portion of CAL into a separate company. Frank Smith, formerly general manager of Kenting and a past-president of Nordair Ontario, became president of CAL Air Charter. When Schwartz was looking for assistance in 1978, Smith took over Timmins Aviation as president. They worked well as a team, but in 1979 Bruce Sully, who owned Business Air Service, a charter company in Goderich, bought CAL Air Charter and Smith went with the move. At this point Seagrove decided to fold up Timmins Aviation. In 1979 Seagrove bought out an aircraft parts company—High Line Aviation. Schwartz now tried his hand at component sales. Late in 1981 he resigned after 18 years with his old friend and associate, Dan Seagrove. Seagrove bought out Watson and

when Al Lowe died suddenly in 1982, he bought back his portion, becoming the sole owner of CAL. He previously had sold the hangar to Field Aviation and later sold them Navair. He sold High Line to an American firm, keeping CAL as a shell, and retired from active business.

In 1982 John Timmins called Schwartz with an offer. De Havilland was taking in used airliners as it sold new Dash 7s. Timmins needed an expert to re-market the older planes. Schwartz took a six-month contract and initially had six ex-Rio Airlines Metroliners to move. These, and later sales, led to some complex arrangements. For example, two Twin Otters were taken as trades when Air BC bought a Dash 7. Schwartz had them refurbished in Victoria, then sold them to a Delaware holding company and financed them through a Chicago house. They then were operated by a separate company in Hawaii. Schwartz was starting to get busier than he wanted, so had his friend L.L. "Slim" Jones join him. Jones, a WWII Sunderland pilot, and a pioneer Ferry Command pilot, had been with DHC in aircraft sales and was highly regarded. Jones and Schwartz worked well together. In 1984 Ottawa decided to sell Canadair and DHC. For the interim it installed Joel Bell as its watchdog over the body running the two companies—CIDC. Bell was tough and, as someone put it, "scared the pants off everybody at de Havilland." Soon Schwartz found that the fun was going out of his job. He resigned after $2^1/_2$ years of interesting work and left Jones in charge. Meanwhile, John Timmins started a company in Toronto to broker commuter planes. Schwartz worked there for two years, during which he developed a manual giving the minutest details for every aircraft type on the commuter market. He was also active as a corporate aviation consultant for companies requiring air transport. Unfortunately he had further medical problems, so resigned for the third time from aviation after a career that spanned 50 years. He became active in the CAHS. One day some of his friends presented him with two framed photos. One showed him as a lad refuelling the DOT's new Beech 17 CF-CCA in 1938. The other showed him 52 years later gassing up the same plane, which by then had been restored by his old friend Mo Servos. Few in aviation could say that they had had a happier career.

Corporate Types through the Years

(Right) Thousands of small planes worked in the Canadian business world over the decades. This Seabee was with Bloxom Brothers and was shot at Toronto Island about 1955. (Al Martin)

(Left) A Piaggio P.136 Royal Gull at Malton in 1957. This was an exotic type that appealed to the corporate operator interested in sport fishing and resorts at water's edge. But the Royal Gull was expensive and demanding to fly. This example was lost in a crash soon after Al Martin photographed it.

(Below) Beech seemed to have a new light twin every year. The six-to-eight-seat, 7,700-pound (AUW) B.65 Queen Air was introduced in 1958. About 400 were sold, mainly as corporate planes. Powered by 340-hp Lycomings, it cruised at about 180 mph. This one was at Uplands on June 11, 1972, where it served on business charters with Cross Canada Flights. (Larry Milberry)

(Above) The Beech B.60 Duke was a pressurized four-to-six-seater introduced in 1966. With 380-hp Lycomings it could hit 250 mph. This example was at Toronto Island in August 1969. It was owned by the Ontario government and used to speed top officials around the province. (Larry Milberry)

(Above) Beech's greatest success story since 1960 is the King Air. The program started by modifying B.80 Queen Airs to take PT6 turbines. First flown in 1964, the King Air remained in production more than 30 years later. These King Airs were with Transport Canada at Uplands on April 12, 1991. (Larry Milberry)

(Above) A host of light, twin Cessnas did superb work in corporate aviation since the 1950s. This Ce.411 was at Malton on April 19, 1969. Many such planes remained in use to the turn of the century. (Larry Milberry)

(Right) While large companies preferred the DC-3, Lodestar or Beech 18, others used more exotic types like converted wartime bombers. This Douglas A-26 served Canadian Comstock. Especially modified by On Mark in the US, it was faster than all standard corporate planes (300 mph), was comfortable and quiet. Al Martin photographed it at Malton in 1963.

The light business jet quickly replaced planes like the DC-3 in corporate service. The greatest successes were the Lear Jet and the Cessna Citation. Descendants remained in production at century's end. Here an early Lear was at Malton on February 20, 1968. CF-CPW, the second Citation built, was there on February 27, 1972. CPW was owned by Canadian Pratt & Whitney, which supplied thousands of JT15D engines for Citations over the decades. (Larry Milberry)

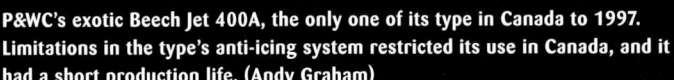

P&WC's exotic Beech Jet 400A, the only one of its type in Canada to 1997. Limitations in the type's anti-icing system restricted its use in Canada, and it had a short production life. (Andy Graham)

For the 1990s Beech marketed the aesthetically pleasing Starship. Powered by PT6s, it drew glances wherever it went, but production was limited. It was almost as costly as a small jet, but noisier and slower. This US visitor was at LBPIA on August 13, 1991. (Dave Thompson)

(Right) One of the early corporate jets was the Lockheed 1329 Jetstar, which first flew in September 1957. The two prototypes used twin 4,850-pound thrust Bristol Orpheus engines, but subsequent models had four 2,400-pound thrust P&W JT12s. With a maximum weight above 43,000 pounds (22,000 for a D.H.125), the 10-passenger Jetstar was a heavyweight. This one was at Malton on September 1, 1958. The prototype was N329J, No. 2 was N711Z. All other Jetstars had four engines, so this is either No.1 or No. 2 with a change of registration. No. 1 ended with the Pacific Vocational School in Vancouver, BC. (Paul J. Regan)

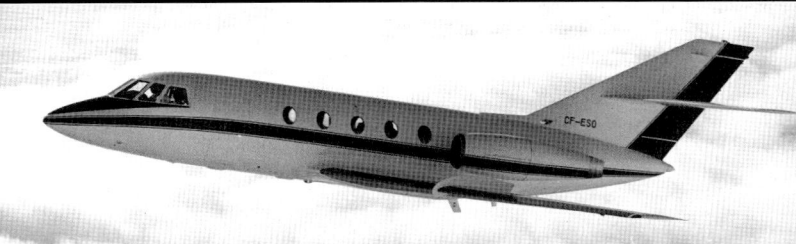

(Above) Imperial Oil's Falcon 20 in a nice in-flight pose. It joined the company in 1966, being 46th off the line. In the late 1990s it was with Audeli Air Express in Spain as EC-EHC. (Halford Col.)

(Above) Canada's mark in the bizjet world has been made with the superb family of Canadair Challengers. After rocky times in the early days of the program, the Challenger became the choice of corporations wanting the best. Here Challenger 601 C-FBEL of Execaire/Bell Canada Enterprises delivers passengers at Uplands. (Larry Milberry)

(Above) Canada's biggest corporate jet in the late 1990s was Boeing 727 C-GPXD of CorpAir, the Edmonton-based company supporting Echo Bay operations at places like the Lupin Mine in the NWT. Henry Tenby took this view of PXD on September 27, 1994.

(Right) From the Challenger 600 of 1978 came the Bombardier Global Express, the corporate jet for the turn-of-the-century. Here it's seen during roll-out ceremonies at Downsview in August 1996. (Larry Milberry)

Overseas Adventures

ures like chaff, flares and radar warning receivers), troop seats and machine guns were added, while ASW (anti-submarine warfare) gear was removed. The flotilla left Halifax on August 24, arriving in the Persian Gulf on September 27 after intensive work-ups en route. Canada expanded its involvement with a composite squadron of CF-18s. The first of 26 fighters left CFB Baden-Soellingen for Doha, Qatar on October 6.

To sustain its Sea Kings and Hornets, ATG went into high gear. Its main thrust (October 1-16) was Op Scimitar—carrying 1,100 Canadians and 1,200 tons of cargo to the Gulf. This involved everyone at Trenton and Edmonton, including 200 aircrew; plus 150 support people at Lahr (Canada's main Gulf War air head), in the Med and in the Gulf. During Op Scimitar a 707 or C-130 was being despatched every three hours. When the CF-18s flew to Doha, they were accompanied by a 707 tanker. Both 437 Squadron tankers were committed to Scimitar, but one was unserviceable most of the time. For liaison a Challenger, crewed by 414 Squadron, then 412, was based at Canadian HQ in Bahrain. Frequent Herc and 707 flights from Lahr sustained Canada's Gulf War contingent.

Delivering aid to Somalia typified Air Transport Group's efforts in the 1990s. Here an ATG Hercules offloads at Mogadishu on March 20, 1993. CanForces personnel, equipment and supplies moved through here daily, whether upbound to Belet Uen or downbound to Nairobi or Mombasa. In this view, the berm beyond separated the airport from the Indian Ocean. (Larry Milberry)

Air transport in the RCAF made great strides after WWII, moving from the Dakota, Goose and Lockheed 10 to the North Star, C-119, Expeditor, Hercules, Cosmopolitan, Yukon, 707, Buffalo, Challenger and Airbus. The heyday of it all was the Cold War, when Air Transport Command, headquartered in Trenton, supported Canadian fighter squadrons in England, France and Germany. At the same time ATC regularly flew United Nations and humanitarian missions. Following unification of Canada's forces in 1968, ATC became Air Transport Group. ATG disappeared in July 1997, a victim of world politics and fiscal belt tightening. With the Berlin Wall down, the need for military force adequate to fight the USSR no longer was so urgent. Canada joined nations everywhere in paring its forces. Then, Canada, like all Western nations, had spent itself into penury by the 1990s, so could not afford the impressive forces of earlier years. With the demise of ATG air transport ops devolved to the wing level, with no overall commander.

ATG in the 1990s

ATG in the 1990s focused on UN and humanitarian tasks. In case of flood or earthquake it usually was first on the scene with Red Cross assistance. If there was strife in the Persian Gulf, Yugoslavia or the Horn of Africa, ATG's hard-pressed planes and crews were front and centre. At home a number of squadrons specialized in search and rescue. Hercules supported the army on exercises large and small, and re-supplied Arctic bases like Alert. 707 tankers logged thousands of hours over the

years on aerial refuelling (AR). Meanwhile, all the training for which ATG was famous continued at bases from Comox to Edmonton, Winnipeg, Trenton, Greenwood and Gander.

The first big challenge to ATG in the 1990s was the Gulf War. When Iraq invaded Kuwait on August 2, 1990, Canada quickly joined a UN coalition determined to restore Kuwaiti independence. On August 10 Ottawa committed two destroyers and a supply ship to the Persian Gulf to help enforce a trade embargo against Iraq. Eight Maritime Air Group Sea Kings were modified for the flotilla—FLIR (forward-looking infrared), GPS (global positioning system), ECCM (electronic counter-counter meas-

Operation Preserve

More typical of ATG's activities than the Gulf War was Operation Preserve, which began in August 1991. It was mounted to help starving Ethiopia, a country of about 50 million with 80-90 tribal and clan groups and as many dialects. The area is geographically diverse. In the southeast is desert; elsewhere are rugged highlands with plateaus to 10,000 feet. Vegetation ranges from harsh desert to lush greenery. The economy is mainly agrarian, 90% of the people living on the land. Nomadic herding is the main activity, but grains and cash crops like coffee are grown.

After the September 1974 overthrow of

One of the eight Sea Kings modified for Persian Gulf ops visited Hamilton, Ontario on June 14, 1991. Its many mods included nose-mounted FLIR and an IR missile warning receiver (under the tail boom beneath the orange ELT). Over the door is an electric cargo winch. (Larry Milberry)

Tanker 704, seen at Trenton on May 20, 1994, was the mainstay of Canada's CF-18 contingent in the Persian Gulf. Note the wingtip refuelling pods. Then, a CF-18 ready for action at Doha as 704 returns from a long AR mission. (Larry Milberry)

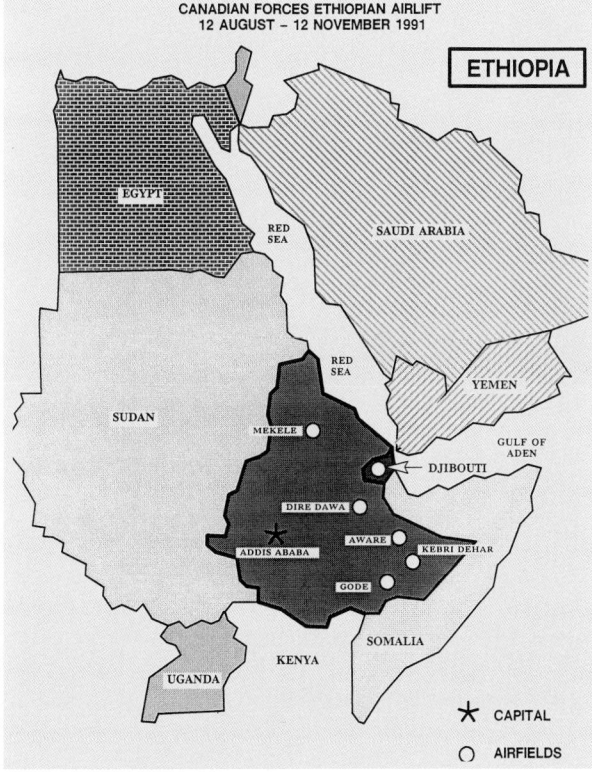

Challenger 144610 of 412 Squadron's detachment at Bahrain on January 11, 1991. It served in communications and liaison from October 1990-April 1991, logging 248 hours. (Larry Milberry)

(Below) Canada's involvement in Ethiopia in 1991 included flying relief to the centres shown, plus to Jijiga (a short distance east of Dire Dawa).

King Haile Selassie, Mengistu Haile Mariam took over (1977). The despotic king's downfall resulted from the airing of a British television documentary. Jonathan Dimbleby secretly had visited Ethiopia during the 1974 famine; his show alerted the world to the horror. The king, meanwhile, tried to hush up the trouble, and

denied aid to his own subjects. He was arrested the day after the documentary aired, and eventually died in prison. With $4 billion in military aid from the USSR and Cuba, Mengistu brought some stability, but Ethiopians still suffered. They went from being under a despotic king, to a ruthless Marxist dictator with a superbly-equipped military. The new regime squelched traditional customs. Canadian journalist Eric Margolis wrote that it "confiscated farmers' seed stocks, collectivized farms and outlawed private farming. In other words, it destroyed its own agricultural system in order to break resistance to communism among conservative farmers." Ethiopia's youth was sent to be cannon fodder in a futile war with Somalia, another dictatorship that, like Ethiopia, alternately was a vassal of the USSR or the USA.

Famine hit parts of Ethiopia in 1984. It was aggravated by the deliberate destruction of the rural infrastructure. This made tools, seed, fuel, transportation, etc. impossible to obtain. There was unrest in Eritrea and Tigre in northern Ethiopia; in 1952 Eritrea had been accorded autonomy by the UN, but Haile Selassie, with US support, annexed it in 1962. Disgruntled Eritreans launched a brutal guerrilla war which lasted until 1991. One government tactic in dealing with the north was forced resettlement. Russian planes carried 600,000 unwilling Tigreans to Wollo in the south. Western aid groups helped finance the operation in which thousands died in concentration camps. At this

Canadian Hercs load food relief on the taxiway at Djibouti on November 30, 1991. Then, an aerial view of colourless, parched Ethiopia between Djibouti and Jijiga an hour-and-a-half by Hercules to the south. Once agriculturally productive, Ethiopia was ruined by a lethal combination of military madmen, perverse politicians and prolonged drought. (Larry Milberry)

899

ATG's headquarters at Djibouti—rudimentary, but it did the job. Then, 0400 and ops already is alive. Crews are flight-planning for pre-dawn departures. (Larry Milberry)

time Canada was the largest Western aid giver, but much of this aid was siphoned off by Mengistu.

In 1988 Mengistu negotiated peace with Somalia, but the revolutions in Tigre and Eritrea continued. Some last-minute concessions from Addis Ababa, including a plan to disband collective farms, came too late. The Tigrean-dominated Eritrean People's Democratic Front ousted Mengistu in May 1991. He accepted asylum from fellow communist Robert Mugabe of Zimbabwe. By this time an estimated 6.5 million Ethiopians were starving and more than 1 million tons of food aid urgently was needed. Canada responded with Op Preserve. Canada knew the region—through the 1980s it had given Ethiopia $480 million. In July 1991 Canada sent a recce team to the region to determine the best location for an ALCE (airlift control element). Djibouti, formerly French Somalia, was chosen. This tiny trading state had been independent since 1977, but remained a de facto French protectorate.

Deployment from Canada for Op Preserve began on August 7, 1991, with the first relief flights flown on the 12th. ATG last had worked in Ethiopia in 1988, when three Hercs spent three months delivering 7,000 tons of food. For 1991 the ALCE at Djibouti airport comprised a small cluster of tents and sea containers. Looks can be deceiving—the ALCE was the epitome of efficiency. Its 64 personnel worked in administration, air movements, communications, flying, medicine, ops, servicing and supply. Staff came from all across ATG and worked alternate days, a shift being 15-18 hours. People worked till the job

was done, whether it meant smooth sailing or major aircraft snags. Life in Djibouti generally was enjoyable. The Canadians lived in the Sheraton with all the amenities. Downtown it was rough and ready, but there were many restaurants with everything from camel stew at Assamo's to pizza and superb seafood. One pizzeria displayed snowshoes on the wall. Years earlier the owner, a retired Foreign Legionnaire, had been on exercise in Northern Quebec, where he collected the snowshoes as a souvenir.

ATG support crew rotated home every four weeks, so four different groups got to serve during Op Preserve. Aircrew did a two-week tour. A twice-monthly Herc from Lahr ferried people back and forth, and carried mail, spares, beer and other essentials. Communications were first class, with AM, FM and HF radios, satellite link, and local telephone. An adjacent French air base was a bonus—the French assisted with equipment, tools, parts, and intelligence. The Op Preserve Hercs were a mixed bag—326 was a 1966 model with 30,433 hours logged as of December 1, 1991. Next came 333 with 19,326 hours, then the relatively new 337 with 9,446. Typical of operating difficulties was with 337, which was down for several days at Djibouti. It needed an engine change (unserviceable fuel control unit). The techs managed to complete this formidable job in 12 hours, using bare-bones facilities. There were four crews of five to fly two Hercs (the third was a spare). For a day's work crews rose at 0300, were at the airport by 0500 for briefing (Hercs were loaded the night before), and airborne into a black sky for ar-

rival at destination at sunrise. Thus could every minute be squeezed from the day. The last trips would land home at sunset. Early in Op Preserve there were other Hercs on the job at Djibouti—Ethiopian Airlines, the Spanish Air Force, Southern Air Transport. These disappeared as contracts expired. An Ethiopian Herc went for good—into the side of a mountain en route to Djibouti. As all three crew were in the rear, people speculated that the pilots had gone back to help with a problem. They left the Herc on autopilot, unaware that it was in a slight descent.

All day long truckloads of food arrived at the airport from Djibouti port, where supplies were stockpiled. A UN official said that a typical day's operations with the Canadian Hercs cost $12,000, or about $118 per ton. He praised the operation: "Without the Canadians we'd be buggered." The UN brought in food by ship, managed offloading, storage, bagging, trucking, and tasked the Hercs based on UN WFP (World Food Program) and UNHCR (High Commission for Refugees) assessments. Ops were initially into northern sites, then focused in the east and south at Addis Ababa, Dire Dawa, Gode and Jijiga. Dire Dawa and Jijiga were only about 150 nm from Djibouti, but air was the only way in—guerrillas had sabotaged roads, bridges and the 486-mile Djibouti-Addis Ababa railroad. Eight ATG Hercules loads went daily to Ethiopia. The UN co-ordinated the airlift, food being distributed in the field by non-government organizations (NGOs) like Save the Children and CARE. Op Preserve's personnel were widely experienced. Most were young, but there were a few veterans. Capt Rolly Tassé, loadmaster on Capt John Pedneault's crew, was one of only three commissioned "loadies" in ATG. He had joined the RCAF in 1955. He flew 1,500 hours on C-119s before going to C-130Bs in 1962.

(Left) The crew of Herc 130333 at Djibouti at day's end on December 1, 1991. Capt John Pedneault (aircraft commander), Capt Rich Pittet (FO), Capt Rolly Tassé (LM), Maj Tom Whitburn (Nav) and Sgt Don Levins (FE). (Larry Milberry)

(Below) A Herc loads at Djibouti for Ethiopia. (Larry Milberry)

Although he worked on numerous types, he had most of his time on Hercs—6,200 hours. Over the years he had been on his share of detachments—to the Congo on UN peacekeeping, 46 days in starving Biafra, in Peru on earthquake relief. Recently he had done a Gulf War tour. Op Preserve was nothing new to Tassé, as it was for the sprogs, some of whom had never been abroad. S/L John Barrass of Capt Ross' 436 Squadron crew brought another kind of experience. An RAF exchange nav, he had been on Hercs during Ethiopian relief in 1985. At that time food was being air-dropped to Tigre, each Herc delivering 16 x 2,000-pound pallets. These were pushed out at 15 feet and 125 knots—dicey flying. Barrass explained how such ops go hard on the equipment. Clean-up and repairs (including inspecting under the floor, where hundreds of pounds of grain would accumulate and germinate) later cost about £250,000 per aircraft.

Local labour was employed at Djibouti. Men worked reluctantly, especially as the day heated, and they got chewing khat—the local drug of choice. It often took 1 1/2 hours to load a Herc (44,000 pounds of food), but when it arrived at destination it was met by keen helpers, who could empty a Herc in 12-15 minutes. They charged aboard to heft the sacks, while rapping out some cheerful tunes. One day the Canadians were amused to hear what the boys were rapping. It seems that Ibriham had been wooing a local Jijiga damsel. He had spent his last penny, but failed to impress her. Next morning his buddies made his disappointment the topic of their chant.

Op Preserve had its frustrations. There were the khat chewers, then there were delivery snags between port and airport caused by troublesome foremen and clapped out trucks. Downpours could flood the roads—Djibouti had no storm sewers. Daily temperatures of 50°C didn't help. Several Canadians came down with gastro-intestinal disorders, keeping the medical officer busy. Aircraft also took a beating, operating at max weights into rough spots like Jijiga. Security could not be ignored. There were belligerents in Djibouti. One day the Canadians were warned to avoid downtown—guerrillas had infiltrated to stir things up during election campaigning. A fire fight downtown left dozens of casualties. Meanwhile, Ethiopia was full of armed clansmen. Thugs from Somalia roamed the desert looking for trouble. Security at the airstrips was not guaranteed, although UN officials usually had a good daily assessment of the situation. The Canadians had no trouble, but one day 61 locals died in a shoot-out in Dire Dawa. A serious problem was the hijacking of food. At Jijiga three of nine truckloads brought in one day by the Hercs "went missing"

Op Preserve Flying Schedule, November 30, 1991					
Herc	To	ETD*	ETA	Fuel	Load
326	Jijiga	0515	0715	18K	19** corn
333	Dire Dawa	0530	0730	18K	20 wheat
326	Jijiga	0815	1015	18K	19 corn
333	Dire Dawa	0830	1030	18K	20 wheat
326	Jijiga	1115	1315	18K	19 corn
333	Dire Dawa	1130	1330	18K	20 wheat
326	Jijiga	1415	1615	18K	19 corn
333	Dire Dawa	1430	1630	18K	20 wheat
* local times ** metric tons					

between the airstrip and a nearby compound. This could have been by pre-arrangement, or by armed hijacking. The worker lamented this, but rationalized that at least someone was getting to eat.

A week-long visit to Ethiopia provided a first-hand look at Op Preserve. It began November 27 by catching UN Flight 6165 (Herc 323—AC Capt Dave Ross, FO 1Lt Pete Stolz, Nav S/L John Barrass, FE Sgt Mark Theaker, LM MCpl Serge Pelland) at Stuttgart. UN6165 carried passengers, supplies and 8,000 pounds of clothing, mainly seconds donated to Ethiopia by Tilley Endurables of Toronto. First stop was Iraklion, Crete for fuel, then it was on to Djibouti, where the flight touched down in withering early morning heat on the 28th. The passengers were soon at the Sheraton, where a briefing by ALCE staff included info about various routines, e.g. a 2030 curfew, highly recommended considering the chances of after-hours shoot-outs. Other tips related to pickpockets, dehydration, the incidence of AIDS among local prostitutes, and the dos and don'ts of eating out. Next morning the flying began. We made three trips that day—two to Jijiga (144 nm south), one to Dire Dawa (135 nm). Loads were usually 800 x 55-lb, or 640 x 55-lb sacks of corn meal, depending on fuel required (18,000-22,000 pounds); or 400 x 110 lb or

320 x 110 lb sacks of wheat. The Hercs were lightened by removing such things as external fuel tanks. This reduced basic weight to 75,000 pounds (usually 82,000).

At 140,000 pounds Herc 326 (call sign "Canuck 36", Ross crew) was airborne for Jijiga at 0545 at 110 knots. It climbed on a 45-minute leg, levelling at 14,000 feet. Soon it was bumping down Jijiga's rough strip. As it shut down, hundreds of people came out of the morning mist. The FE lowered the ramp and unloading began. Bags of corn were stacked aboard NGO trucks, as shivering onlookers watched. Canuck 36 soon fired up. A makeshift Transafrik chart showed Jijiga's runway (03-21) as 7,450 x 150 feet. The Canadians knew that only 5,400 was usable. At 0700 the Herc was airborne for Djibouti at 88 knots, this time at 92,000 pounds. It went to 15,000, the level strictly designated by Djibouti ATC.

Trip two was in Herc 333 to Dire Dawa with a 429 crew: AC Capt Vince Schurman, FO LCol Skinner, nav Capt Frank Costello, FE WO Stephane Guy and LM MCpl Mike Robinson. Skinner was the new CO of 413 Squadron, sent out for some hands-on flying before 413 started Herc ops. Along the route were miles of dry, gray fields, where crops

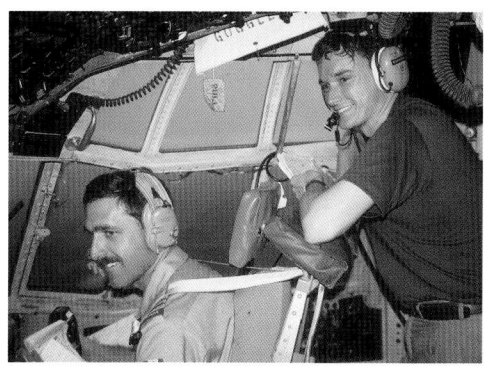

LCol Marc Dumais and S/L John Barrass en route Dagahabur on December 2. (Larry Milberry)

(Above) "Canuck 36" at Jijiga on November 30. The off-loaders hop to it as the locals (right), trying to keep warm in the chilly morning air, look on. (Larry Milberry)

(Left) At Jijiga on November 30 John Barrass, Mark Theaker and Dave Ross check an artillery shell over which their Herc nearly taxied. (Larry Milberry)

Herc 333 at Dire Dawa on November 30; and a photo of Capt Vince Schurman at work on the same trip. Vince lost his life in a Hercules crash at CFB Wainwright on July 22, 1993. (Larry Milberry)

once had been abundant. On December 16, 1991 Mark Abley wrote of Dire Dawa in the Montreal *Gazette*: "The new Ethiopian government has been working without great success to establish full control here... Dire Dawa once nestled below green hills; now those hills are bare. Desert threatens to invade the town..." Dire Dawa airfield was military. Rumour was that its encircling anti-aircraft sites remained active. On approach to land on 10,000-feet of concrete, dozens of MiG-15s, 17s, 21s and 23s could be seen. On the ground there was time for a quick recce. An unserviceable Aeroflot Mil 17 sat near our Herc, its crew apparently stranded, waiting for spares. None of the fighters was serviceable. Apparently, the Soviet pilots and techs had gone home, taking their manuals, without which MiGs were useless in Ethiopia. A trip to Dire Dawa next day was with Maj Wayne Davidson's 429 Squadron crew: FO 1Lt Jim Bertrand, nav Capt Gary Stone, FE Sgt Steve Huffman, LM MCpl Paul Brochu. The return to Djibouti began with a fly-by for the boys in the control tower. This allowed for a few aerial photos of the MiGs. Then we headed out low level, blasting up dry river beds, scattering goats, thousands of which dot the landscape like confetti. The country was mostly bare— having no fuel for their stoves, the people had stripped it of anything burnable, a practice that hastened desertification.

On December 2 several journalists flew to Jijiga aboard 326 with a crew headed by LCol Marc Dumais, ALCE commander and CO of 436 Squadron. The FO was Capt Jacques Dufort, the rest of the crew being Dave Ross' boys. It had been raining at Jijiga, so a fly-by was in order to inspect the runway. Someone observed, "We'd be hard pressed to find anything to drag us out of the mud if we got bogged down." We landed. As the Herc offloaded, the journalists and a few others piled into Toyotas to go cross-country to two refugee camps. With EPRDF men (Eritrean People's Revolutionary Democratic Front) riding shotgun, we bumped along for 35 miles, passing devastated villages, one apparently destroyed by clansmen from the next village settling some score. Armed men loped along near the road, always waving and smiling. At one point our EPRDF "guns for hire" leapt from the Toyotas to fan out ahead, AK-47s ready. They sensed an ambush. Instead they found a convoy of water trucks wallowing in a mud hole. We pushed on and met a tank piled high with

Gedebursis gunmen guarding their turf—even locals had their private armour. More Gedebursis came past in a new Land Rover. How had the fellows come by a brand new vehicle? They had stolen it from Save the Children, removed the roof and bolted on a machine gun.

Our destination was Dherwenaje, three miles from the Somali border. Just inside this camp was a mountain of food; overseeing it, some toughs. Here was the end-of-the-road for tons of food delivered by ATG. It seemed odd that Dherwenaje's supplies were not controlled by the NGOs. Something happened once food left Jijiga—it was no longer the UN's nor the NGO's. The camps were run by heavily-armed local mafia. The NGOs went to them as beggars. A squabble at the stockpile led to a bloodbath one day. On all sides in Dherwenaje were the

Jim Bertrand and Gary Stone of Wayne Davidson's 429 crew. (Larry Milberry)

ERPDF guns-for-hire on the road from Jijiga to Dherwenaje on December 2.

rough little wattle huts, shelter for about 115,000. Most were northern Somalis from Hargeisa, a city of 300,000 until Siad Barre's southern army levelled it beginning in 1988 (the same year hostilities with Ethiopia ended, letting Barre turn his tanks and MiGs on his countrymen). Life at Dherwenaje was squalid, but the death rate was down to a few per month, thanks to Op Preserve. The camp presently was using 131 sacks of food daily. It had no doctor, only four NGO nurses, who this day were glum. They had heard about their bogged-down water convoy. We watched the feeding stations at work and saw some of the hopeless cases, the pits that served as latrines, and makeshift facilities like the TB clinic and open kitchens. There was even a "corner store" where pop, chocolate bars and cigarettes were sold to anyone with cash.

The return to Jijiga was through rain squalls. A consolation was that, as we started back, the water trucks were coming towards camp. We stopped briefly while the EPRDF chitchatted with a few guerillas—Somalis on a raiding or recce party, but the EPRDF didn't care. By now Jijiga was a sea of mud. As the day darkened, a Herc approached. It had a careful look, then turned north, the crew apparently thinking better about a landing. But just as it almost had disappeared, it turned. Soon it was on final. Herc 326 landed as short as possible, sending up a rooster tail of mud. It cocked to port, but stayed on the strip. It cut through flocks of large birds that swooshed in all directions, seemingly on the attack. A number caromed off the Herc. With judicious reverse and braking 326 soon was turning onto the apron. Next came 333 and it too made a good "splashdown". When inspecting one of the downed birds, a crewman assessed the situation: "This fellow is a mess— he's left his fuselage and gear all over the runway."

We boarded the Hercs, some for a direct flight home in 333, others routing via Dagahabur, 80 nm

to the south, where an air-strip recce was requested by the UNHCR. 326 taxied and the crew briefed for a soft field takeoff, the AC commenting, "We land back here only if we really, really have to." On the way south there were storms on all sides. Dagahabur needed food, but was its runway safe? Dumais and Dufort made two low passes. As we circled, we could see the townsfolk hurrying out, no doubt hoping for food. Now 333 set up for a touch-and-go to get the feel of the runway. The crew could tell that it was too wet and soft, but would be fine once it dried out a bit. We made a final pass as S/L Barrass videotaped the strip.

Our visit to the Horn of Africa ended on December 4. We departed for Larnica in Herc 334 (UN6166, Davidson crew). From Larnica the Pedneault crew took charge to Lahr. Aboard 334 were tons of freight. Since the ALCE was about to close, everything possible was crammed aboard—333's duff engine (4,500 pounds), a refrigeration unit (3,500 pounds), a heavy set of floodlights, some screw jacks, spare Herc tires (each 300 pounds) and 1,500 pounds of fluids. At Lahr we caught Nationair 757 C-GNXU to Ottawa.

Op Preserve continued until December 12, when redeployment to Canada began. It had delivered 808 loads—1,600 tons of food. All went smoothly and much was learned. What

Stockpiled food brought to Jijiga by C-130. From there it was trucked to camp, where local toughs (two shown, at right) doled it out as they decided. Then, the object of the world's concern through the 1970s, 80s and 90s— some of the gravely malnourished of Ethiopia patiently await their next handout. (Larry Milberry)

Capt Schurman speeding away from Jijiga after an early morning delivery on December 2, 1991. Then, his dicey landing in the mud of Jijiga that afternoon. First he's crabbing, then he has the landing under control. Besides the mud, the Herc had to battle through some very big, very dumb birds. Inset, Theaker and Dumais inspect a "kill". (Larry Milberry)

(Above) Pigeons clear the way at Djibouti as a Herc returns from Ethiopia on November 30. Then, Cpl Kim Tucker marshals it to the ramp. She was one of the 64 ALCE people who kept Op Preserve rolling. (Larry Milberry)

(Right) Herc 333 loads at day's end to be ready for its next pre-dawn mission. (Larry Milberry)

(Below) Changing an engine at Djibouti. Then, getting a Herc engine ready to ship home. Sgt Ron Mifflin (driving) talks shop with Capt Glenn Watters. (Larry Milberry)

happened after Canada left Djibouti? Some letters from UN aid worker Tracey Buckenmeyer working in Jijiga provided an update. On January 27, 1992 she wrote, "Another airlift has started today, using only one Ethiopian Airlines Hercules. We haven't had rain since the last airlift, so I am sure it will begin to rain soon! The security situation is still poor and no one is quite sure what to do about it. We still manage to get to the camps in convoys with EPRDF escorts, but there is always some sort of incident on the road, be it bandits or some 'political' group doing its thing. It's a bit nuts, but no one is ready to pack it in yet."

On May 31 Tracey noted: "Security-wise things are much better now after a terrible time two months ago when two of my colleagues were killed by bandits. The food is moving pretty well to the camps, though we still go with the EPRDF." Her news of November 29 was brighter: "I'm now in charge of repatriation—getting the Somalis to go home... By the turn of the century all Somalis will be in Somalia, all Ethiopians in Ethiopia and off the in-

ternational dole. I can dream, can't I? Ethiopia is much quieter. We are moving freely to the camps without military escort and movement along major roads is safe." Things were so good that the UN was starting to divert aid from Jijiga towards Mogadishu, where a new crisis was brewing.

Somalia Goes Down the Drain

In 1991 Siad Barre, military dictator of Somalia since 1969, was overthrown by the United Somali Congress. On January 27 he fled to Kenya. He had ruled like a despot, but this kept Somalia, a nation of traditionally warring clans, together. When a region did not toe the line, Barre crushed it. He did this 1988-90 to the northern Isak people. His military went in and ravaged cities like Hargeisa. It was surrounded, then plastered with heavy artillery and aerial bombardment (using Chinese-made F-6 fighter bombers flown by South African mercenaries). Along the way Barre's men sewed perhaps a million mines, making Hargeisa an urban time bomb for years to come. Most of its survivors fled to camps in Ethiopia.

In late December 1991 there was an opportunity to visit Hargeisa. This came while chat-

ting in a Djibouti bar with UN official Gary Perkins. He mentioned that a UN Caravan was going to Hargeisa the next day. There were some empty seats for CANAV, and writer Mark Abley and photographer Michael Mahoney of the Montreal *Gazette*. Perkins offered a vehicle, driver and guide for the visit. Caravan 5Y-ZBZ, on lease from a Nairobi charter company, had arrived in Africa in October after a 46-hour ferry from Wichita via St. John's, Birmingham and Luxor. On December 3 we made the 55-minute flight southeast to Hargeisa. The Caravan landed long on Hargeisa's runway—the first 1,000 feet were mined. At the terminal was a Red Cross Caravan. Its engineer was Montrealer Anil Patel. He had graduated from Confederation College in Thunder Bay and apprenticed with Bearskin Airlines before going overseas.

We set off to see the city and interview whomever we could. Hargeisa was slowly coming back together. An estimated 70,000 had returned from Ethiopia. Unlike places like Dire Dawa, it felt safe. An ordinance kept all weapons out of town. The only ones we saw were recovered mines and a heavily armed "technical" (a truck decked out with weapons) that roared through town. The city centre was abustle. As gasoline was hard to get, donkeys were hauling everything, including beat-up drums of drinking water, which came from miles away. Hargeisa was friendly. A young father, who had lost an arm fighting, invited us into his ramshackle house, where several fami-

With love from Somaliland, whose people declared independence on May 18, 1991. These were some of the scenes in Hargeisa on December 3, 1991: restoring the main mosque, boys at play in the rubble that once was a proud city, local public transportation, water hauled from miles away, the baker making his rounds. Then, some of the ordnance recovered by the UN's mine-clearing operation; and two mine hunters disabled on the job. (Larry Milberry)

The Caravan from Djibouti on its daily 133-nm flight. The pilots were Tony Pettinger, Tad Watts and Ramesh Pashavaria, who were crew training this day. Then, Chinese F-6 fighter-bombers (licence-built MiG-19s) left behind by Siad Barre's men after they devastated Hargeisa. General Mohamed Hersi Morgan orchestrated this mayhem, later fought to save the Barre regime, and survived to take part in post-UNOSOM Somalia. The F-6s were flown by white mercenaries from countries like South Africa. (Larry Milberry)

lies lived. At a busy intersection about a dozen men sat at battered desks poking at equally battered typewriters—the local "word processing" contingent. Business was good—line-ups for each typist. There were a thousand incredible sites, but only a day to take them in.

Soon we had to hurry back to the airport. Over its battle-scarred terminal building flew a fresh green-and-white flag, that of newly-declared Somaliland. No one had recognized this state, but proud locals were anxious about displaying their flag. Around the airport were several abandoned MiGs. A fellow in tattered clothes reported that he had been a MiG pilot and was keen to resume flying. Good luck, we thought. After an inadvertent stroll through a minefield to take some photos, we left for Djibouti. There we were chastised and fined by immigration for having made an unauthorized cross-border trip.

Op Deliverance

The United Nation's largest peace keeping and humanitarian operation took place 1993-95. This was UNOSOM—UN Operations in So-

malia. Billions were spent trying to bring peace among warring factions and end widespread starvation. Since time immemorial war and famine plagued the Horn of Africa. This manifestation began in 1988 in the north—former British Somaliland. When that area sought independence, Siad Barre's men devastated and looted the region. In mid-1992, his southern stronghold collapsed midst inter-clan fighting. He and his henchmen fled to Kenya, Canada, and wherever else they could find easy refuge. Countless civilians were slaughtered or starved, especially in the countryside. Hoodlums took the country apart, going so far as to tear up overhead wires, tin roofs and cast iron sewer pipes for shipment to scrap dealers in South Africa. Thus did Somalia's so-called warlords sell their own country for scrap. Finally, the US intervened. In late

1992 it landed forces in Somalia to begin restoring order .

The warlords faded into back alleys of Mogadishu and into the countryside. The US planned to stay in Somalia till the following May, then hand over to the UN. Canada joined the US multinational coalition Unified Task Force (the US later handed UNITAF over to the UN, where it became UNOSOM). ATG already had an ALCE at Nairobi—as early as September 1992 its C-130s were flying on Op Relief from there to Mogadishu for the WFP and ICRC (International Commission for the

Red Cross). Op Relief ended February 28, 1993. Meanwhile, the USAF was carrying aid on Op Provide Relief and Op Provide Hope. Another Canadian element during Op Relief comprised Sea Kings, operating from HMCS *Preserver* off Somalia, first to the port of Bosaso, then to Baledogle, 50 nm northwest of Mogadishu. *Preserver* was also a refueller for other UNITAF vessels; and a haven, where soldiers, after weeks in the desert, could have a shower, a decent meal and relax overnight.

An overall formation ran all Canadian activities—Canadian Joint Forces Somalia, tasked to secure a large humanitarian relief sector (HRS) centred on Belet Uen, 162 nm north of Mogadishu. CJFS comprised the Canadian Airborne Battle Group (850 men in Belet Uen), the 93rd Rotary Wing Aviation Flight (six Twin Hueys and 100 men in Belet Uen), CJFS HQ (80 people at the US Embassy compound in Mogadishu), support elements of 25 at Mogadishu and 10 at Nairobi (Jomo Kenyata) airports, and a small R&R contingent at Mombasa. The ATG element was designated

the Canadian Air Transport Detachment Somalia (CATDS). Its task was designated Op Deliverance.

CATDS operated three Hercs daily to Belet Uen: one direct from Nairobi, and two from Mogadishu, although this routine was flexible. Most flights were freighters carrying IMPs ("infantry meals prepared", also known as MREs—"meals ready to eat"), some fresh food, bottled water, fruit juice, soft drinks, beer, combat supplies, etc. Otherwise, the way to Belet Uen was overland from Mogadishu on mined roads that claimed two Canadian armoured vehicles. Mines even threatened airplanes—a civil-

ian Hercules was blown apart in Sudan when it triggered a mine. A program was begun January 20, 1993 to provide four days of R&R to each Canadian with 30 days in-country. The first R&R flight was on January 20, 1993 to Nairobi.

Staff at the CATDS worked alternate days and were quartered in downtown hotels. For daily ops aircrew were picked up at the Safari Club Hotel between 0300-0630 hours for the 20-minute drive to Jomo Kenyata International Airport (JKIA). There they got busy with flight planning, weather and intel briefings, and pre-flighting the aircraft (which had been serviced and loaded the night before). All ops were tasked through a central US-managed office in Mombasa (Unified Task Force HQ). A typical mission (CF320, passenger sked with Herc 320) began March 19, 1993 with an 0630 pick-up at the hotel. The op order specified takeoff at 0830 for Mombasa, thence to Mogadishu and Belet Uen, and home on the same route. The 435 Squadron crew was: Capts Derek Stobbs and FO Randy King (pilots), Capt Greg Illchuk (nav), Sgts Don Levins and Steve Hull (FEs), and Cpl Jim Beaugrand (LM). All were green to Somalia except for King, who had served on an earlier tour. Besides their Herc, Nos. 332 and 337 were on the ramp. Only two normally were despatched, the other being a back-up.

CF320 started at 0820, backed from its parking spot using reverse pitch, and taxied at 123,000 pounds. It was airborne on time, climbing for FL190. The course was east for Mombasa 235 nm away on the Kenyan coast. Mount Kilimanjaro soon appeared 55 nm to starboard, its 20,000-foot peak poking above cloud. Descent began at 0915. As CF320 swooshed across the threshold at Mombasa, it joined a scene crammed with planes of all types, mainly involved in relief operations. CF320 took on fuel and 28 passengers returning from R&R, then headed up the shore of the Indian Ocean toward Mogadishu, 503 nm dead north. Within a half hour the green of Kenya

Somalia bound. Airbus 15002 boards at Lahr for Nairobi on March 15, 1993. It departed at 0815Z with troops and supplies for the UN's Somalia peacekeeping operation. Takeoff weight was 320,000 pounds, including 120,000 of fuel. (Larry Milberry)

The UN flightline at Nairobi in a typical scene from March 31, 1993 showing Canadian, Swedish and US Hercs. Then, a view of the Canadian ALCE, including a spare Herc propeller and engine. ALCE business was done from the trailers, sea containers and tents beyond. Finally, a load is prepared at the ALCE for shipment to Belet Uen. (Larry Milberry)

(Right) Herc crews gather at the ALCE for the morning intelligence briefing. Information about the condition of isolated strips, possible threats from guerillas, etc. wasn't always available to ATG. Any versatile intelligence officer made it his business to chat up Southern Air Transport crews, whether in the hotel bar or on the ramp, to glean tidbits. SAT was the most experienced freight operator in Africa, so its people knew more than anyone about daily conditions. Most SAT crew were veterans of the USAF, CIA or both. Then, Capt Jim Kinnear, Lt Biff Jones and Capt Jean Houle of 436 Squadron check paper work before a March 20 trip to Somalia on Herc 320. (Larry Milberry)

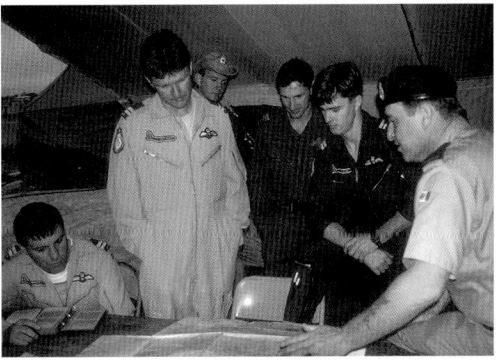

(Left) Maj Don Thain was ALCE commander in Nairobi in March 1993. Capt Rod Lanning (right) was his deputy. As a rule they flew desks at the ALCE, but on this occasion they were at Mogadishu on March 24, 1993 as an augmentation crew. (Larry Milberry)

(Right) Capt Jim Kinnear in the co-pilot's seat during a liesurely flight back to Nairobi from Mogadishu on March 22. The 2.2-hour flight was a full house—77 were on board. (Larry Milberry)

(Left) A Royal Canadian Dragoons recce squadron provides security with one of its Bison armoured personnel carriers, while Herc 320 turns around at Mog on March 20, 1993. Personnel going to or from the airport and the Canadian compound used APCs—one never knew when some kook might toss a grenade. (Larry Milberry)

(Right) The busy TAMS shack at Mogadishu with WO J.R.C. Tremblay (left) at the door with one of his troops. (Larry Milberry)

Mogadishu had a great variety of aircraft movements. The Tu.154, carrying an out-of-date registration, was in from some ex-USSR republic on March 20. It was game to have shown up. After all, this was not a functioning country. The airport only worked because the Americans made it work. The other side of the fence were guerrillas with SAM-7s and itchy trigger fingers. RNZAF 727 NZ7271 (ex-United Air Lines) was delivering supplies on March 31. In the foreground is a Kuwaiti C-130. Turbo DC-3 N5156T was at Mog on March 24. (Larry Milberry)

(Right) Belet Uen on March 20, 1993 by when the population was estimated around 40,000. It had been much higher, when thousands of Ethiopians had sought refuge there earlier. The bridge was built by Canadian troops. (Larry Milberry)

(Below) A view from a Twin Huey of the Canadian base at Belet Uen with a door-mounted 5.56mm C9 machine gun as foreground. From this camp fewer than 1,000 Canadian troops effectively kept the peace over their 30,000 square-kilometer HRS and brought more than a semblance of normality to the lives of thousands. Then, a close-up in camp. The last of this force was out of Somalia by June 17, 1993. Although the Canadians did an outstanding job in Somalia, a media circus developed over the indiscriminate actions of a few members of the Canadian Airborne Regiment. A huge scandal developed. A public enquiry indicated dubious leadership in the military top brass and weaknesses in the DND from the minister on down. Morale throughout the CanForces plummeted. Sadly, ordinary Canadians soon forgot what great work the majority of their troops had done on the Somali mission. This resulted from the public's willingness to be led by the nose by a media that sensationalizes news, highlighting the negative aspects of a story at every opportunity. (Larry Milberry)

(Above) Under Capt B. Ladouceur of 429 Squadron, Herc 320 arrives in a dustball on March 28, 1993 to delivers supplies and personnel to Belet Uen. (Larry Milberry)

(Left) Don Levins, Greg Illchuk and Derek Stobbs inspect prop blade damage at Belet Uen. Over the decades, relief operations in Africa caused millions in damage to ATG's Hercules fleet. Then (right), Levins, Stobbs, King and Illchuk (outside) get their Herc cranked up. The sooner the better. This day the thermometer at Belet Uen hit 56°C. (Larry Milberry)

faded to the sand of southern Somalia. It was all desert as the Herc passed Kissmayo at 1°N. lat. Descent into Mogadishu, crowded with air traffic, was under USAF radar control.

CF320 did a "straight-in" onto the north-south runway, taxied to the north end and onto a general purpose ramp. UNITAF aircraft were everywhere—Italian Chinooks, Mongooses, H-3 s and G.222s; RNZAF Andovers; a Saudi Herc, USMC Cobras and Hueys, as well as An-26s, King Airs and Caravans in UN service. In a corner was part of the once-powerful Somalia Air Force—derelict MiGs, Islanders, Hueys, P.166s, etc. CF320 was marshalled to a spot where the Canadian TAMS (temporary air movements section) people got busy offloading and fuelling. The flight quickly was sent on its way and soon was descending for Belet Uen. Arrival included a low pass to check for camels and donkeys, which often cluttered the runway. Stobbs and King then brought their Herc around for a "max effort" landing (full reverse and brakes) on the 5,670-foot strip. Reversing created a huge dust ball. The pilot reduced reverse artfully, keeping the dust ball just behind the engine intakes, then pulled onto an even dustier offloading area at the end of the runway. About 30 passengers deplaned, including some US Special Forces troops, who were met by an escort of heavily armed vehicles. "Welcome to Belet Uen" meant two things—heat and dust. As soon as the doors opened, a blast of hot air gushed into the Herc. Daily temperatures of 50°-55°C were the norm. Then came the fine, ankle deep dust that rose like smoke with each step. The visitor could take all this, but he felt for the peacekeepers stuck in Belet Uen on six-month tours. One of them made the comment, "This place is so hostile that even the butterflies have fangs."

One of the passengers on CF320 was a European lady working for an NGO. She described how, six months earlier, warring clans had wrecked Belet Uen and its thriving economy. There was no food and hundreds starved to death every day. Since the Canadians set up their HRS, the region was on the way to recovery. The locals were grateful for the Canadian Airborne—the warlords were keeping their heads down. The market had reopened and there was lots of food. The Canadians had built a bridge across the local river, were volunteering in clinics, had helped restore schools and were training some local police. In the case of the bridge (steel spans bolted together), as soon as it was erected, some entrepreneurs started taking it down to sell for scrap. When they heard this, the Canadians restored the damage, but this time welded all the bolts.

After an hour on the ground CF320 loaded about 40 passengers and got airborne at 102,000 pounds. More boarded at Mogadishu, most going to Mombasa on R&R. It was 1750 hours by the time the flight finished the day's work at Nairobi. It had logged 8.3 hours, but the duty day had been twice that. On shutdown the crew reported on snags, checked into Ops, finished the paper work and took the bus downtown. After a sweaty and dusty day they looked forward to cleaning up and having a meal in one of Nairobi's many excellent restaurants. These were things that the Canadians in Belet Uen could only dream of, as they ate their dreary rations before bedding down under canvas with their nocturnal companions—scorpions and camel spiders.

Twin Hueys

The 93rd RWAF, of which the vanguard arrived in Somalia on March 1, was a utility outfit doing recces, medevacs, etc. Five USAF C-5s carried it in from Trenton. It took over the duties performed by the HMCS *Preserver* Sea Kings (*Preserver* had set sail for home on March 10). The core of the 93rd (89 people) was about 70 personnel from 427 Squadron from CFB Petawawa. The others included several air reservists like Capt Ken Jones, a pilot from 2 Tactical Air Wing in Toronto. LCol Ken Sorfleet commanded the 93rd. He had joined the air force in 1969 and had flown with 408 and 444 Squadrons, worked in NDHQ, attended Staff College in Australia, was chief of staff and flew Twin Otters at Northern Region HQ in Yellowknife, etc. before taking over 427. The 93rd's Twin Hueys on March 28 were Nos. 111, 129, 141, 142 and 148. Aircraft 150 was being returned to Canada following a hard landing at Belet Uen on March 14.

A key task for the 93rd was ensuring the movement of food and supplies throughout the Canadian HRS. The unit flew its first patrol on March 5—a recce from Mogadishu-Belet Uen return. Two days later three Hueys flew to Belet Uen, and on the 8th the first operational

(Above and left) Twin Hueys of the 93rd RWAF operated from the Canadian encampment, 6 km from Belet Uen airfield. For Somali ops they were fitted with extra armour plating, the APR39 radar warning receiver, ALQ144 infrared jammer and a missile approach warning system. This raised basic aircraft weight from 6,500 to 7,000 pounds. During their stay in Somalia there was no need for any of this equipment. The only damage suffered was to aircraft 135150. (Larry Milberry, John McQuarrie)

(Below) The crew on Twin Huey 111 on March 28 were Capts Eric Manchester and Yves Grenier (pilots), MCpls Pierre Blair and Mario Giasson (FEs) and Sgt Gilles Paradis (observer). Cpls Mike Muranetz and Jason MacKinnon of the Airborne Regiment (tan outfits) were aboard to do some practice firing. (Larry Milberry)

trip was logged—a recce in support of the Royal Canadian Dragoons. The 93rd began its first deployment on March 13, when three Hueys flew to Cadaado on a five-day recce looking for technicals. They also met with various local clan leaders. A particular concern to the 93rd was the supporters of warlord Aideed, who held sway in the country northeast of Belet Uen. Because of such threats the Hueys carried a door-mounted light machine gun on either side. Although two Canadian Bison armoured vehicles were wrecked by land mines in this HRS, there is no record of damage from hostile fire to the Hueys.

A typical training flight from Belet Uen was flown on March 28 in 135111 to Matabaan, where the flight engineers would practice firing the C-9 machine gun. Two gunners from the Airborne also were along to train. The aircraft commander, Capt Yves Grenier, had been in the forces since 1974. He was a hard core fighter jock with 1,000 hours on the CF-5, but also had logged 1,200 hours instructing on the Tutor. He had 1,000 hours with 444 Squadron in Kiowas, had served on 427 in 1989, gone to Honduras with the 89th RWAU, then returned to 427. His co-pilot was Capt Eric Manchester, who had come to 427 in January 1990 from the Basic Helicopter School at CFB Portage. The half-hour flight to Matabaan was NOE—"nap of the earth" i.e. as low as possible, and at 100 knots. The country was hilly desert, with prominent termite mounds as high as 20 feet. The range for this day's gunnery was an old battlefield where Somali and Ethiopian tanks had fought it out in the 1970s. When Huey 111 arrived, the crew had a host of targets on which to fire. The Huey started its passes and the gunners fired away, using up one ammo box after another before it was time to return to base.

Operation Lifeline Sudan

Other routine Herc flights included one on March 21 (CF401, Herc 320) carrying a 26,000-pound forklift to Mogadishu for the TAMS; March 21 (CF401, Herc 337) with 32,000 pounds of canned soft drinks for Mogadishu; thence to Belet Uen with 20,500 pounds, mainly bottled water; and March 22 (CF400, Herc 332), an airstrip recce to Dousa Mareb on the Ethiopian border. The latter trip was made at low level. On nearing destination, the Canadians received a briefing from "Red 433"—a Swedish Air Force Herc. It reported that the strip was in good condition and that two helicopters and two armoured vehicles (all Canadian) were present. Dousa Marreb had a fine 7,000-foot gravel runway, but it took four approaches to get in, due to camels on the runway. Once down, a number of passengers and a local clan chief came aboard.

On March 30 UN320 (Herc 336) made a long trip from Nairobi to Mayen Abun, Sudan, via Lokichoggio, a refugee camp in Kenya near the Ethiopian border. Sudan, which had been closed to most western aid on account of civil war, was a mess. The war had displaced millions; populations of entire communities, such as Kongor, had been slaughtered. In late March

Khartoum approved 20 UN flights to needy communities in south central Sudan, starting March 29. On the next day Herc 336 took off with 26,000 pounds of grain. Capt Gerry Stark landed on the rough 3,300-foot strip at Mayen Abun (9° 08' 1" N lat., 28° 21' 9" E long.) and the Herc immediately was surrounded by hundreds of locals keen to share in the bounty. The stop was an ERO—engine running offload. This was a precaution against not being able to get re-started due to mechanical trouble, and a good idea in general, lest a hasty get-away be required.

The food quickly was offloaded. So anxious were the locals, that fights erupted around the rear ramp, obliging the neighbourhood enforcers to wield their long switches. No sooner was the last sack tossed off than dozens of Sudanese rushed forward with sick and wounded, who were stretched out on the floor of the Herc. This was not in the plans, and the FE, consulting over his headset with the captain on the flight deck, negotiated with a Sudanese doctor. As fast as the wounded had come aboard, they were taken off and the Herc departed. It had been a tense few minutes. Had the locals somehow taken offence, the Canadians, who had no arms, could become instant hostages—or worse. Back at "Loki", Herc 336 took on a load for Leer, southeast of Mayen Abun.

Besides the work done in East Africa by ATG, other Canadians and Canadian-made aircraft were busy there in the 1990s. From Loki the UN flew aid to famine-stricken regions on Operation Lifeline Sudan. Hard at work from Loki's dusty strip were Hercules, King Airs and Caravans, but also two Buffalos and a Twin Otter. The Buffalos were operated by Avior Technologies of Miami, and Ethiopian Airlines. Avior's was C-GDOB, a 1982 model,

the 108th example off the line. It was the sole DHC-5E Transporter, an upgraded civilianized version of the standard military DHC-5D. DHC was unable to find a market for the expensive "E", so converted it to a "D". It went south in April 1984 for the Ecuadorian army. It sat in storage at Las Vegas through the late 1980s, but in September 1990 George Neal of de Havilland ferried it to Field Aviation in Calgary. There it again became an "E" and was sold to Avior. Crewing DOB in April 1993 were pilots Emmanuel Anassis and Andrew Burton from Canada, and engineer Philip Cooper of New Zealand. Maintenance was being managed by Ron Edwards, who had joined DOB after 23 years in the Canadian Forces working on Buffalos with 424 and 442 squadrons and 116 ATU. With an all-up weight of 49,200 lbs (standard Buffalo, 41,000) DOB could carry up to 15,500 pounds on its 180-360 nm runs from Loki. A well-kept machine, as of April it had a mere 2,300 flying hours since new. The pilots expressed their satisfaction with its performance on the region's limited airstrips—it got a hefty load into pretty well any strip in Sudan. The second Buffalo (ET-AHJ, c/n 102), was one of a pair of "D"s sold to Ethiopia in 1981. Wess McIntosh and Lou Sytsma had delivered it as C-GBXL from Toronto to Addis Ababa May 29-June 4, 1981. In the typical way Canadians have of covering the planet, after the delivery McIntosh went directly to Abu Dhabi to spend June 9-14 doing check rides with local pilots on the Twin Otter and Dash 7, then returned to Toronto. By this time AHJ, looking a little rough, was flown by an Ethiopian crew as a workaday "trash hauler".

The Twin Otter at Loki was N910HD of California-based Air-Serv International, a 12-plane company (two Twin Otters, 10 Cessnas)

Herc 336 at Loki after its run to Leer on March 30, 1993. It carried huge Canadian and UN markings under the wings and tail to let rebels know that it was not a Sudanese Air Force plane The crew was Capts Gerry Stark and Paul Anderson (pilots), André Gagnon (nav), Sgt Al Allaire and Cpl Dave Hutchinson (FEs), and Sgt John Harries. Herc 336 was an H73 model that had served earlier in the UAE. (Larry Milberry)

(Right) Getting ready to go around a third time at Dousa Marreb, 85 nm northeast of Belet Uen. (Larry Milberry)

(Below) Sudanese tribespeople mob the Canadian Herc at Mayen Abun, hopeful of something to eat or a flight to hospital in Loki. (Larry Milberry)

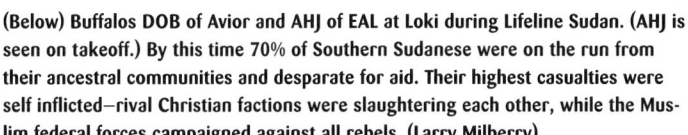

(Below) Buffalos DOB of Avior and AHJ of EAL at Loki during Lifeline Sudan. (AHJ is seen on takeoff.) By this time 70% of Southern Sudanese were on the run from their ancestral communities and desparate for aid. Their highest casualties were self inflicted—rival Christian factions were slaughtering each other, while the Muslim federal forces campaigned against all rebels. (Larry Milberry)

(Above) Canadians Don Cressman and Steve Urquhart at Loki in March 1993. According to Air-Serv, it logged 14,906 hours and carried 5,642 tons of cargo/33,724 passenger with two Twin Otters in Sudan and Kenya from January 1, 1985 to April 30, 1992. Like most Twin Otters N910HD (lower right) had been around. It rolled off de Havilland's Toronto line in July 1970 as the 289th Twin Otter and was delivered to Miami Aviation as N623MA. The same month it was listed with Oasis Oil Co. In 1974 it was flying with National Oil in Libya as 5A-DBD. In 1978 it became F-GBDE with Air Alpes, was N48MK 1978-80, then came to Canada as C-GOPR for Kimba Air, Shirley Air Service and La Ronge Aviation. In January 1983 it became N910H with Amoco Madagascar Petroleum, and finally joined Air-Serv in December 1987. It had worked from Arctic tundra to jungle to desert, and flown something like 3.5 million miles! (Larry Milberry)

specializing in relief operations in African. It employed several Canadians. Albertan Don Cressman and New Brunswicker Steve Urquhart were on N910HD, serving about a dozen refugee camps. The morning flight of March 30 was typical— flying 15 UN workers to Juba, north of the Uganda border in Sudan. N910HD was logging about 200 hours monthly on a 7-days-a-week operation. Since it began flying in the area in 1989, it had averaged

downtime of only two days a month. The second Air-Serv Twin Otter flew from Khartoum. Canadian-built airplanes were elsewhere in the region. Air Kenya Twin Otters worked skeds and carried tourists on safaris. Kenya Air Force Buffalos flew on military and police duties. In 1993 a US civil Caribou was hauling freight from Nairobi's Wilson Airport. Beavers and Turbo Beavers of Kenya's Desert Locust Control were spraying insecticide in the back country.

Kenya Field Report

In January 1994 Emmanuel Anassis of AirCo Ltd. described some of the Kenya-based relief flying activities:

During the first two weeks of November local relief operators lost several aircraft. The first accident involved a C-46. It went on its nose landing at an airstrip called Akot. It was the end of the wet season and the ground was still

(Left) Southern Air Transport delivering a Herc load to *Médecins sans Frontières* at Kismaayo. A few days later MSF had to leave the area–the Somalis even resented doctors. Since the 1970s famines, earthquakes, floods and civil wars made millions for sharp businessmen in air transportation–disaster means big bucks to such specialist operators. (Larry Milberry)

As a UNITAF convoy sped through Kismaayo, the camera caught a sudden drama. A bully was after a young girl. Willing urchins were moving in with their knives and stones to get in on the fun. What was the outcome? The convoy didn't stop. The neighbourhood was too dangerous and this was an incident all-too-common in a brutal society. (Larry Milberry)

soft. My hat goes off to the guys flying the old airplanes—we have a hard enough time in the Buffalo, which was designed for this sort of thing. A Simo Airways Caravan lost its nose gear and prop, and dragged its left wing tip in the dirt after it tried to land at Akot to get the C-46 guys out. Another Caravan ran into a train after it aborted take-off. The aircraft was a write-off. A turbine DC-3 from South Africa lost control during take-off and crashed into a South African Caravan, which, in turn, slid into an Air Serv Caravan. The DC-3 and Caravan were completely destroyed, but no one was seriously hurt. [The DC-3T prang involved some recklessness—the captain had invited a local pilot, who did not fly DC-3s, to try his hand at a takeoff. The results were predictable.]

The Ethiopian Buffalo was almost lost when it had a double engine failure on climb-out. Earlier in the year a Canadian C-130 was damaged at Loki. The pilot seemed to have stalled at about 50 feet and dropped it onto the runway very hard. The rear fuselage showed a fair amount of skin wrinkles. The flight engineer gave the aircraft a serious walk-around with a lot of head shaking. We seem to be the only guys not to have had an incident, however, we have many close calls. On one occasion we had to evacuate a gravely ill person from Nasir. It was the height of the wet season and it had been raining for several days. A bordering river had flooded across the airstrip. After several passes, my co-pilot and I decided to try it. The approach was normal, bringing the Buff to the stick shaker on touchdown. Not wanting to use reverse for fear that it would cause the air-

craft to skid sideways on the slippery black cotton soil (just like black ice), I held the nose up to aid in aerodynamic braking. Everything looked fine until the nose wheel touched. At that point the world disappeared! Throughout the roll we could hear the engines surging violently as they ingested water and mud. As the mains began to dig in, the aircraft fishtailed down the strip. When we finally stopped, the Buff was 90 degrees to the left of our landing direction. We had landed in 12 inches of slop. We managed to medevac our patient to hospital. He survived only because of our tough airplane. It is to the Buff's credit (and to those who designed it) that we could operate year round safely and efficiently.

C-46 Woes

Jeff Schroeder of Winnipeg flew the C-46 for Relief Air Transport. Generally the aircraft were serviceable, and carried a great deal of cargo for World Vision. One day, however, Schroeder had trouble. An engine failed on C-FIBX while departing Loki on March 23, 1994. With 14,000 pounds of freight for Ui, Sudan the C-46 would not hold altitude. Schroeder did a straight-

ahead, wheels-up landing. The crew survived, but IBX was a loss. Schroeder also was involved in the incident at Akot. While landing there, TXW's wheels dug in to soft, deep sand, went on its nose, then came down heavily on the tail. Schroeder called Loki on HF, but was told that a rescue plane could not be sent. They settled in for the night in an area occupied by hostile tribesmen. About midnight some tribesmen pounded on the C-46. The crew emerged to an invitation to join a campfire party. Next day a Caravan landed with a repair party, but it sank in and was damaged. Loki was advised and suggested that the crew hoof it to the next strip, about 50 miles distant across the desert, and await transportation. This was declined and later that day a second Caravan arrived. Repairs began and the stranded airmen got back to Loki. A few weeks later the C-46 flew out, but the repaired Caravan collided on takeoff with a local train that passed by only every few months!

The End of UNOSOM

"Somali warlord Mohamed Farah Aideed's militiamen swept into Mogadishu airport yesterday, chasing away looters and filling the void left by retreating United Nations peace keepers." So began a news report of March 2, 1995. After more than two years of US and UN efforts to save Somalia from national harakiri, there was little left to say. The final scenes as

---

Somalia in 1991: A Few Facts and Figures

Population – 6.7 million
Per cent of population, pastoral nomads – 70
Per cent of population, agricultural – 30
Life expectancy – 56
Literacy rate – 24% at age 15
GDP – $1.7 billion
Per capita income – $210 per year
Arable land – 2%
Chief export – livestock (40% of GDP, 65% of exports)
Crops – sugar, bananas, sorghum, corn
Per cent GDP from industry – 5
Usable airports – 46 (8 with paved runways)
Independence gained – July 1, 1960
Aircraft used by SAF – An-2, An-24, An-26, Bell 204, Bell 212, BN-2B Islander, C212 Aviocar, Do.28, Fiat G-222, J-6 (MiG-19), Hunter, Mi-4, Mi-8, MiG-15, MiG-17, MiG-21, SF-260W, Yak-11 Moose.

C-46 5Y-TXW "Ancient Lady" at Nairobi with Relief Air Transport in 1993. (Donald A. Rogers)

An Italian Army CH-47C Chinook chugs across the airfield at Mogadishu on March 24, 1993. Next to the Americans, the Italians sent the largest contingent to UNITAF. (Larry Milberry)

An Australian Navy Sea King leaves the 5,700-ton landing craft HMANS *Tobruk* off Mogadishu on March 25, 1993. *Tobruk* had just refuelled from *Pecos*. (Larry Milberry)

(Above) Far to the south of Belet Uen, Kismaayo was not in the Canadian HRS. The US Marine Corps made a landing there on March 26, 1993, seeking to stabilize the area in view of warring factions. Here an HMM-263 Squadron CH-46 arrives on March 25 aboard the oiler USNS *Pecos*; as USS *Wasp* sails behind. *Wasp*, a new 844' x 106' assault carrier, sailed from Norfolk on February 23, arriving off Somalia March 23 as part of the 24th Marine Expeditionary Unit. Its complement was 12 CH-46s, four CH-53s, four AH-1Ws, three UH-1Ns and six Harriers. It was about to come alongside *Pecos* for 450,000 gallons of fuel. The Americans invariably are in the forefront of relief operations like Somalia. When it comes to generosity in places like Somalia or Rwanda, they have no equals. Predictably, they suffer the heaviest dose of rotten criticism in the media. (Larry Milberry)

(Left) A Sikorsky Black Hawk of the 10th Mountain Division (Fort Drum, NY) departs Mogadishu on March 25, 1993. The Americans did a super job of quelling clan rivalry in Somalia, and facilitated food distribution to millions of starving people. All this cost them 43 killed in action and accidents. Black Hawks and Cobras got involved in the worst UN action in Somalia, when the US decided to ferret out warlord Aideed. Pushing the Somalis proved rash and signalled the beginning of the end of US involvement. It all came to a head on October 4, 1993, when 13 Americans died in a shoot-out in the dusty streets of Mogadishu. (Larry Milberry)

USMC LCAC hovercraft (payload 75 tons) on the beach at Kismaayo on March 26. The Belgians recently had taken over this HRS from the Americans. At this date UNITAF comprised 27,000 personnel. Participants included Australia, Belgium, Botswana, Canada, Egypt, France, Greece, India, Italy, Kuwait, Morocco, Nigeria, Pakistan, Saudi Arabia, Sweden, Tunisia, Turkey, the UAE, the USA and Zimbabwe. There were more than 13,000 Americans in Somalia. Italy, Canada and Australia were the next biggest partners. (Larry Milberry)

The only fighter planes involved in the Somali operation were the Harriers aboard *Wasp*. This one was preparing for ops off Kismaayo on March 25. (Larry Milberry)

the US and Italian rearguard boarded boats and helicopters for their offshore fleet were of Somalis doing what they had done best for years—loot. The end of the Canadian Airborne Regiment came the same day at CFB Petawawa. The tough, dedicated and effective Airborne had gone to Somalia to run the Belet Uen HRS. It did a good job under horrendous conditions; but got into trouble with the media after a few bad apples tortured and murdered a local Somali, who had infiltrated their camp. Adding to this, an opportunistic publisher released a video tape showing some Airborne hijinks—traditional hazing ceremonies, harmless, goofy sorts of things, but disgusting to a naive public. The media fell upon the Airborne like vultures. Minister of National Defence David Collenette, no fighting man, took the easy way out. With prime minister Jean Chrétien's blessing, he ordered the Airborne disbanded, admitting that his decision was pure politics. His military chief, General de Chastelaine, refused to stand up for his men and slunk away from the spotlights. One Canadian editorialized: "Perhaps the Prime Minister would like to impose the same type of rules in dealing with our elected Members of Parliament. For instance, should an elected member of his party commit an unfavourable act, which would upset the public or bring discredit upon the country, then he should dissolve Parliament and call an election. Anything less would be cowardly. I have much more respect for an individual who is willing to die for his country than one who is only willing to feed at the trough and waste away the taxpayer's money."

Other African Adventures

Over the years several Canadians crewed on foreign Hercules, mainly in Africa. James L. Gillespie of Edmonton was with Compania de Diamentes de Angola. He had joined the RCAF in 1958, flying Sabres with 427 and 421 squadrons, instructing on Harvards and Tutors, doing a tour on Voodoos with 416 Squadron, flying Hercs at 435 Squadron 1981-84, when he left the air force for Angola. The main work was supplying the diamond mine at Dundo (in the interior desert) from an air head at Luanda. Gillespie's L-100-30 Hercules D2-EHD flew about 520 hours a month using three crews. On one occasion it set a C-130 world record by logging 574 hours. One flight turned into the most dramatic of Gillespie's career. On February 10, 1986 he was captain on D2-EHD with a Filipino crew: FO Noel Navarette, FE Tito Acelar and LM Paddy Padilla. They had off-loaded at Dundo and were returning to Luanda. The airfield was engulfed in fog out to 2-3 miles, with a ceiling of 300 feet. Nonetheless, the Herc departed at 0542Z. Eight minutes later there was an explosion in the starboard wing. Shrapnel tore into the fuselage. In 1994 Gillespie reminisced about this:

On the first indication of trouble I initially thought the No. 4 engine turbine wheel had disintegrated for whatever reason. Immediately, the master fire warning light illuminated, as did the light in the No. 4 fire pull T-handle, followed by the No. 4 nacelle overheat light. The engine was shut down in accordance with checklist procedures. After the first fire bottle was discharged, a quick check confirmed a serious fire. The second bottle was fused, with no apparent effect. So much for my thoughts about proceeding to Luanda where we had a maintenance facility.

We now declared an emergency, stated we were setting up for an immediate landing on R23, and requested that the fire trucks be called —we had to get on the ground ASAP. I banked sharply to the left and rolled out 180° to my departure track. The entire area was obscured by fog. For navaids there was only a weak NDB [non-directional beacon] a few hundred yards from the runway, with no published approach procedure. I mentally calculated our position in relation to the field and began a timed descent to set myself up at the approach end of the runway, hopefully to be in a position to land. I set the radar altimeter at 50', hoping the cloud base was still 300'. The fog had by now thickened— when we entered it at 1500' AGL the reflection from the 200' of flame we were trailing from the wing was startling.

There would be no going around from this approach—the loadmaster was giving updates on the fire, including comments about pieces falling off the wing. My plan was to land, even if we couldn't find the runway. We broke out at 150' and the FE spotted the runway at 9 o'clock for 1 1/2 miles. I banked sharply left 180°, then right 90° using 60°-90° of bank and was surprised that we didn't drag a wingtip. We passed over a village with thatched huts and wondered if we might set them alight.

We lined up with the runway after avoiding two antennae to the right of our approach. We were a mile back on final and below 100' with wings level when I secured No. 3 engine, which was now burning fiercely. Control was getting difficult with the two engines out and much of the starboard aileron gone. People below were scattering as we were far too low for a normal approach. Our touch-

Jim Gillespie in air force days. (DND)

down, six minutes after the explosion, was welcomed and I had to remind myself not to use reverse or we would have resembled a pinwheel.

Our FE was outstanding through the entire ordeal and had completed all checklist items as quickly as I had called for them. He had shut down Nos. 1 and 2 engines as the wheels touched, and the aircraft was fully secured by the time it came to a smoking halt. The crew exited, each with a hand-held fire extinguisher. As we rounded the nose we saw the starboard wing sag to the ground. By this time the flames, fed from what remained of 8,000 pounds of fuel that had been in the starboard wing tanks, were higher than the Herc's 36-foot tail. We emptied all the extinguishers, then the fire department arrived. The fire was doused just as the chemical truck squirted its last bit of foam.

*After the fire we counted 27 shrapnel holes in the fuselage. At the time of the explosion, the FE had been standing near the centre of the cargo bay and could hear things zinging passed his ears. He escaped injury, and no damage was done to our hydraulics or electrics. The aircraft was repaired with minimum equipment in a mere 120 days. The engines were removed and both wings replaced. Nos. 1 and 2 engines were rein-*stalled and two new ones mounted on Nos. 3 and 4 positions. The work was done with skilful use of a forklift and a small crane. The aircraft was then flown to Luanda where things were properly rigged.*

What happened to EHD was nothing to do with Lockheed or Allison. In 1975 the Portuguese had quit their long-running civil war with Angolan rebels. They went home, but soon Angolan's marxist government was embroiled in a civil war with a US-supported group known as UNITA, which had good stocks of CIA-supplied Stinger surface-to-air missiles, as many as 6,000. Another former Canadian Herc, CF-PWN of PWA, transported Stingers to Angola for use against government aircraft. As a rule UNITA did not harass Hercules, since they brought food for the locals. On this occasion, however, it was different—a SAM brought down EHD; but it was not certain who fired it, since the government also had SAMs. Someone suggested that the

Herc EHD shortly after its hairy landing at Dundo. (Gillespie Col.)

Angolans had hit their own plane.

Lloyds of London covered repairs to EHD—CDA's policy included loss or damage from hostile action. Lockheed flew in spares and EHD was flying four months later, logging 530 hours in its first month. Jim Gillespie, who had logged 1,400 hours during his year on EHD, returned to Edmonton to spend seven years flying Twin Otters with 418 Squadron. In November 1991 he retired as 418's CO and took a job flying a corporate Jetstar II. As to EHD, on January 22, 1993 it had just departed Luena, when a rocket-propelled grenade hit, tearing off No. 4 propeller. The prop went under the wing, opening No. 4 fuel cell. Capt Bonzo Von Haven, ex-USAF and the world's high-time C-130 driver, immediately turned back. Unaware of the extent of the damage, he asked his Angolan co-pilot to confirm that No. 4 prop was feathered. The FO looked out, but instead of reporting the propeller gone, told the captain, "It's feathered!" Von Haven got EHD down. A few days later it made a three-engine ferry to Benguela, then flew to Portugal for full repairs.

In 1988 former Canadian L-100 CF-DSX (now S9-NAI) went to Transafrik, a São Tome-based operator with as many as seven Hercules that serviced diamond mining. As the Angolan civil war expanded, Transafrik took on more government work, including military contracts. Thus, the civil Hercs became fair game for UNITA. In a short period three were hit by ground fire: D2-EHD, S9-NAI and CP1564 (Bolivian Air Force on lease to Transafrik). NAI was lost at Luena on April 8, 1989, while tankering 20,000 litres of fuel in an aluminum tank. The safe approach to evade SAMs was a spiral from altitude, but this day the Portuguese captain was dragging NAI in low and slow, making himself a fat target. Once hit, NAI pancaked into the scrub, the crew escaping through the captain's cockpit window. CP1564 was flown by Edmontonian Graham Page McPhee, an ex-PWA pilot. On March 16, 1991 it was operating Luanda-Cafunfo on a mining job. While at 17,000 feet (lower than it should have been), it was hit by two SAMs. The forward fuselage broke off, carrying the four crew to their deaths.

(Below) Bolivian Herc CP-1564 at Benguela. Page McPhee of Edmonton was shot down in this aircraft. (Donald A. Rogers)

(Right) CF-DSX later served in Angola as S9-NAI. This is how it ended after the SAM attack of April 1989. (Donald A. Rogers)

(Left) St. Lucia Airways Herc J6-SLO at Khartoum. It started in 1966 with Zambia Air Cargoes as 9J-RBW. After three years it joined PWA as CF-PWN. While in for repairs after a mishap at Eureka, it was stretched to an L.100-20. It stayed with PWA till 1984, then was sold to St. Lucia Airways, where it was dubbed "Juicey Lucy". Most of its work now was hauling CIA weapons from Zaire to Angola for the UNITA rebels under Jonas Savimbi. In 1987 J6-SLO became N9205T ("Grey Ghost") with CIA contractor Tepper Aviation. On November 27, 1989 Bud Peddy, who owned Tepper, was landing at night at Jamba, Angola from Kamina, Zaire. Something went awry and N9205T crashed, killing seven. (Donald A. Rogers)

Herc HPW spent considerable time in Angola with both its Canadian operators. Here it is in PWA colours at Prestwick on March 3, 1979; then with NWT Airways at Benguela in 1983. Over the winter of 1996-97 HPW went through major overhaul ("D-check") at CAE in Edmonton, so was ready for years more work around the world. (Maxwell Col., Donald A. Rogers)

Disaster over Western Sahara

In 1988 Don Lumsden of Dawson Creek, BC was hired by T&G Aviation of Phoenix to fly a DC-7C on a US aid program in Africa. A veteran ag pilot, he had started at the Winnipeg Flying Club in 1957, flew in the bush, then got into ag and forestry work with Skyways of Langley, BC, learning the ropes on the Stearman and TBM. For the African job two aircraft were to spray against locust in Senegal and Morocco. Lumsden left on November 13, 1988 for Senegal, where he flew N284, an old Northwest Airlines plane.

The contract, under US government humanitarian funding, soon ended in Senegal. On December 8 Lumsden left for Morocco, this time in N90804, previously G-AOIF of BOAC (photo on p.289). The DC-7Cs flew north together. The day was clear and everything routine—the flights had been planned 24 hours

earlier in accordance with regulations. Overflights were approved by Mauritania and Morocco; ATC in Madrid, the controlling agency, had all the details. The sprayers entered the area between Mauritania and the Western Sahara at 11,000 feet. Suddenly both were hit with SAMs and started burning. N284 lost a wing and crashed, killing all five aboard. N90804 took its SAM in No. 1 engine, which caught fire and fell from the wing. This opened the wing and the fire quickly died. The crew pushed on and landed 1:55 hours later on an abandoned airstrip near Ifni, south of Agadir. The locust sprayers, in Africa to help the ordinary people, had been used for target practice by Polisario desert guerrillas. Once again those involved on humanitarian duties had learned that their help often is not appreciated in Africa. On December 14 Lumsden was home. His African venture had been short, but not very

sweet. He was happy to get back to work with his Avenger and Ag-Wagon in the safe environs of northeast British Columbia.

A Rwandan Odyssey

Not long after it left Somalia, ATG was recalled to Africa. In April 10, 1994 it started flying aid to Rwanda, a nation in a part of East Africa cursed by centuries of tribal war. Matters culminated in 1994 when one tribe attempted to exterminate another. This followed such horrors as the October 1993 assassination by Tutsis of Burundi's Hutu president; and the April 1994 missile attack on Rwanda's presidential jet, wherein the country's president, Juvenal Habyarimana, was killed along with the president of Burundi. This led to total collapse in Rawandan society.

Rwanda is a small country (1993 population 8.5 million) in the lakes region where the River

Nile begins. For centuries it was fought over by the Hutu and Tutsi. The militaristic Tutsi, although a minority, usually dominated, and treated the Hutu as social underlings. The first whites in this area arrived in 1894. Germans, then Belgians, colonized it. The Belgians favoured the Tutsi when it came to privileges like education. Following national independence from Belgium in 1962, the Hutu took power, and it was the Tutsis' turn to be persecuted. Thousands fled to bordering countries, where they organized a guerrilla army, the Rwandan Patriotic Front. In 1990 the RPF began making incursions into Rwanda. Meanwhile, the government launched a racist campaign over the airways, inflaming the Hutu against the 15% of Rwandans who were Tutsi.

When the president was assassinated, Rwandan radio blamed the Tutsi and called for revenge. Nothing short of genocide would suffice. As many as a million were slaughtered within a few weeks, most by ordinary citizens intimidated into turning on friends and neighbours. Kigali, a city of 300,000 early in 1994, dwindled to almost zero. The UN had a small force present, the UN Assistance Mission in Rwanda, formed in 1993 following RPF incursions, commanded by Canadian MGen Romeo Dallaire. UNAMIR had no mandate to intervene. Its patrols often had to stand by as innocents were killed. In one case a group of Belgian soldiers was butchered by the Hutu. Meanwhile, USAF C-130s and C-141s led in the evacuation of thousands of expatriates from Kigali to Nairobi. Soon there were few UN troops in Kigali, and the Hutu slaughter rolled on.

Canada moved quickly to help get people out of Kigali. On April 5 a MAMS (mobile air movement section) team arrived. Canada's ALCE at Ancona, Italy was split, half going to Nairobi. A C-130 left Ancona on April 8 for Bujumbura, Burundi. On the 10th Maj Ken Pfander took the Herc to Kigali to begin Canada's part of the UN evacuation—Operation Scotch. The Hutu army was still in charge, and the airport was under Tutsi fire from the surrounding hills. Even though a Canadian Herc was nearly hit (June 5), flights continued, ATG being the only airlifter involved in this dicey operation. The RPF routed the Hutu army, which fled to a southwest enclave defended by the French and across the northwest border to Zaire, where the city of Goma became a refugee centre crowded with Hutu in terror of Tutsi reprisals. Now the need was to aid the very people who recently had been exterminating Tutsi. In July the UN seemed paralyzed—it was able to deliver only 20 of 78 planned flights in the first week of its airlift plan. The United States initiated its own humanitarian airlift, air-dropping its first aid on July 24. Meanwhile, the Tutsis established an interim multi-tribal government. In Canada the DND contracted with Toronto's SkyLink Aviation, which organized a fleet of Russian, Ukrainian and American freighters to move aid from CFB Trenton—18 trips with An-124s, 12 with IL-76MDs and eight with 747s. The first left for Entebbe on July 26. ATG kept up daily

C-130 flights from its ALCE at Nairobi, and used Airbus and 707 flights to support this air head. The ALCE comprised 62 personnel from all essential trades.

The first An-124 (UR-82008) was loaded at Trenton on July 29 with an 85-ton cargo of water purification units, diesel generators and runway sweepers. The sweepers were for Kigali; the purification units for Goma, where refugees were contracting cholera by drinking from polluted Lake Kivu. The An-124s used were improved "Dash 100s", which met regulations for operations in the US. The "One-twenty-four" first had flown in December 1982 and entered production in 1984. With an all-up weight of 405,000 kg it eclipsed the C-5B Galaxy (349,000 kg) as the world's largest operating aircraft (the larger An-225 was grounded by 1994). There were about 50 An-124s; demand for them kept production going into the mid-1990s. Aircraft 008's crew of 21 was headed by Valeri Shlyakhov, chief pilot of the Antonov Design Bureau in Kiev. Co-captain was Anatoli Khroustitski, a senior Antonov test pilot.

Tipping the scales at 308 tons, UR-82008 taxied at 1400 hours (local). It backtracked on R06 for takeoff on R24. After lining up, a four-minute engine run, mandatory for an An-124, was completed: two minutes at 40-50% power, two at 70%. The pilots then advanced the throttles and 008 rumbled away, eating up 7,000 feet of runway before lifting off. Capt Shlyakhov then made a sharp one-eighty and climbed across the Bay of Quinte. Course was set for Mirabel, where the required fuel (Jet A) was available. After an hour, the crew set up for a long final on R24 behind a Lufthansa Airbus. They approached at 280 kph, touched like a feather at 1555 hours, and were marshalled behind An-124 RA-82047 of Volga-Dnepr, which was freighting between Amsterdam and Prince George.

UR-82008 was airborne again at 1900 on a 7.5-hour flight-plan for Tenerife in the Canaries. Soon after takeoff the back-end crew set to work preparing salads, and cooking on a small electric stove. The larder was well-stocked, as was the bar. A feast was soon on the table, and some serious toasting commenced. Afterwards,

everyone stretched out on wooden bunks to sleep their way to Tenerife. At 0730 (local) 008 landed and the crew soon was downtown for a day's rest. The third leg began with takeoff at 2124 for the 5.2-hour flight to Lagos. After re-fuelling, it was off to Entebbe—5.5-hours. Touchdown was a 1105. Entebbe was abuzz with relief activities, with aircraft like the AN-12, C-130, CH-53E, Il-76, KC-10 and VC-137. Assisted by ATG MAMS personnel, 008 soon was empty. It departed at 1725 for Mombasa for fuel. A stormy landing was made 1.5 hours later. The crew laid over for two days at the Nyali Beach Hotel, then departed at sunrise on August 2, laden with 138 tons of Jet A — enough to get it to Kiev for some urgent servicing.

The Rwandan relief operation, based in Nairobi, included dozens of flights each day. The largest presence was the USAF with its C-5s, C-130s, C-141s and KC-10s. These were co-ordinated through a busy ALCE at Mombasa. A trip of August 3 was typical. It involved Atlant/Jove Aviation's Il-76 (UN Flight 392), which operated most days to Mogadishu. This day it carried 13 tons of grain to Mog, then, ironically, took on 28 tons of surplus infantry MREs. A year earlier Mogadishu was the focus of an airlift. The airport was jammed and the circuit busy enough that pilots were twitchy about conflicting traffic. This day it had nothing going on. The only other plane on the main ramp was a New Zealand DC-3. UN392 soon was turned around with eight pallets of MREs (now referred to locally as "meals Rwandans won't eat", a comment about the quality of infantry field rations). UN392 climbed to 7,600 metres for the 1:35-hour flight, cruising at 750 kph indicated. To a first-time Il-76 passenger, it was clear that this was an excellent airplane—rugged and utilitarian, although one detail took some getting used to: it was not easy to tell who was in control—both pilots appeared to be steering!

The Il-76 landed at Kigali at 1015. It was a surprise to see how quiet the airport was, for all the news had been of the great Kigali airlift. By now, however, the action had moved to Goma. The only aircraft present were Canadian and RAF C-130s, a USAF C-5, a Luftwaffe

CanForces Herc 337, carrying relief supplies for beleaguered Rwanda, arrives at Kigali on August 6, 1994. A Russian Il-76 is in the distance. The hangar across the field once housed Rwanda's presidential flight. Now it was the base for UN helicopters—Twin Hueys on contract from Canadian Helicopters Ltd. (Larry Milberry)

An-124-100 UR-82008 on the ramp at CFB Trenton on July 29, 1994. It was loading heavy equipment, including diesel motors and a water purification plant. Built in 1985, this was one of the earliest 124s. Before this Rwandan contract it had been hauling space equipment to Brazil for NASA and airport jetways from Utah to Bogotá. Other recent loads had been a GM locomotive from London, Ontario to Dublin, and 216 ostriches from Namibia to Gatwick. (Larry Milberry)

Capt Shlyakhov en route Trenton-Mirabel on July 29. Then, offloading at Entebbe on July 30, following the Trenton-Mirabel-Teneriffe-Lagos odyssey. (Larry Milberry)

Transall and a civil Boeing 727. UN392 departed empty from Kigali at 1440 and covered the 412 nm to Nairobi in about an hour. By this time ATG was operating to Kigali, Goma and Entebbe. Its ALCE was in the same spot as the year before, when all eyes had been on Somalia. The routines were the same—about 0400 hours the servicing crew went to the airport. An hour later the flight crews arrived. On August 4 C-130E No. 315 was tasked as UN6763 to operate Nairobi-Kigali return (twice); C-130H No. 337 as UN931 Nairobi-Goma return, Nairobi-Kigali return. Each crew did a morning trip; had hot dogs, sandwiches, pizza and soft drinks at the ALCE; then flew again.

UN931 was flown by a 436 Squadron crew under Capt John Stevens. He was keen about Op Scotch, but also talked of a recent Maple Flag exercise, where he had flown a simulated special forces paradrop mission. His Herc covered a hostile route with a SAM (surface to air missile) site or unfriendly fighters at every turn. It got through unscathed. At the debrief, when Stevens wondered how, someone commented, "That is the first time anybody here has ever asked why he *wasn't* shot down." With such sophisticated training, aircrew like Stevens were ready for whatever Rwanda had to offer. UN931 was airborne at 0810 with 31,000 pounds of high energy biscuits, bottled water and boxes of rubber boots. It laboured up to 22,000 feet, then cruised the 426 nm route at about 260 KIAS, passing over Tanzania's rolling farmland, Lake Victoria, then rugged eastern Zaire.

Reports from Goma were not what pilots like. The first Canadian Herc there (August 3) had spent two hours in a hold, then returned to base on low fuel. The crew bitched that night in the bar about the hold and rumours of an unsafe runway. There was so much traffic at Goma that planes were stuck on the ground and stacked overhead! This day Stevens got in—he pushed the nose down at 1004 hours (an hour's difference between Nairobi and Goma), 27 miles back. It was a standard letdown, not the steep, tense combat entry at Sarajevo or at Kigali when people were lobbing mortars. At 15,000 feet the crew briefed for landing (airport elevation is 5,220 feet), noting an approach speed of 146 knots, threshold at 120. At 1006 UN931 was cleared to 10,000 feet to hold, then OK'd to land.

This part of Africa is in perpetual haze from tropical air, smoke from slash-and-burn farming, and volcanoes. Two miles back the runway appeared. That was reassuring, except that there were people and animals wandering all over it! This, it turned out, was the norm. Refugees had encamped on both sides of the single runway and, with no attempt by the Zaire army to control things, wandered freely around. The Herc pressed on, the laid-back folks on the runway parted just in time, and UN931 arrived at 1013. Goma offered a rare collection of air-

(Left) The Il-76 has been a reliable workhorse in Africa through the decades, whether supporting violent civil wars as in Angola, or on humanitarian airlifts. Here Atlant's Il-76 disgorges cargo at Mogadishu on August 3 before loading MREs for Kigali. (Larry Milberry)

Capt John Stevens on the way to Goma on August 4 in Herc 337. His crew included Capt Doug Hinton (FO), Capt Steve Camps (Nav), Sgt John Day (FE), MCpl Ron Guest (LM, in helmet), Sgt Serge Pelland (LM) and WO Don Drennan (ASO, dozing behind Guest). Stevens and Hinton were enthusiastic fliers, who had learned to fly on civvie street while in their teens. Stevens, with eight years in the air force, learned at Toronto's Seneca College. Hinton graduated from Mount Royal College in Calgary, did some bush flying in Yellowknife, then got into the air force. (Larry Milberry)

planes: 707, 727 and 737, C-130, Caravan, DC-3, DC-8, Gazelle, Il-76, Jet Ranger, King Air, Partenavia, Puma, Transall, Twin Otter. The ATG Herc was down just long enough to allow for some hectic photography—an hour after touchdown, it was firing up. It taxied and took off passed groups of local scamps playing chicken by darting as close as they dared to the on-rushing Herc. On descent to Nairobi 1.7 hours later one of the crew commented, "I don't see what the big hassle was about Goma." Somebody agreed: "Been there, done that. You know!"

On the afternoon trip Herc 337 (same crew) operated to Kigali as Canuck 37. Aboard were 19 passengers, including an MRP to rescue Herc 315, unserviceable at Kigali. Just before taxiing Canuck 37 heard that 315 was running OK, so the MRP deplaned. The run to Kigali took 1:50 hours. The only thing unusual about the afternoon was two of the return passengers—large Alsatians belonging to the local Sabena agent. When he had fled in April, he could not take the dogs, so released them to fend for themselves. More than three months later he found them. Like many others dogs, they had survived by feeding on human corpses. Somehow they eluded being picked off by RPF and UN soldiers, who used wandering dogs for target practice. The two dogs didn't make a sound. Seemingly petrified, they pressed themselves to the tarmac. Their master had to carry each onto the Herc.

Besides freight Op Scotch carried hundreds of passengers e.g. peace keeping soldiers from Bangladesh, Ethiopia, Ghana and Uruguay. There were also representatives from NGOs like Care, the Red Cross, UNICEF and World Vision; and government officials. The UN handled bookings and nobody boarded without its say-so. In a place like Rwanda a person is in the UN's good books, or else he is in trouble. One occasionally met foreigners unable to get a flight because of strict procedures. A three-man Czech TV crew spent days stranded at Kigali. Lacking the UN's blessing, they had to camp at the airport. Finally they approached the aircraft commander of a Canadian Herc, who used his discretion to fly the Czechs to Nairobi. With the memory of what had happened in April still fresh, one Canadian crew visibly was relieved one day when the new president of Rwanda failed to turn up for a flight to Nairobi. There could still be the odd SAM "out there" and someone with a score to settle.

Canadian Hercs occasionally departed Kigali using special ops tactics—low level and fast (described by one crewman as "our sky and ground departure" for the energetic yanking and banking involved). This gave a chance to see Rwanda close up. As advertised, it was scenic—verdant hills, rivers and lakes, prosperous-looking farms (coffee comprised 90% of Rwanda's exports). Only one ingredient was missing, and that made for a spooky feeling— there were next to no people. They were either dead (if Tutsi) or gone to Goma (if Hutu).

Don't worry, be happy! With the fringes of Goma airport crowded with refugees and security being easy, the locals took to strolling on the runway any time they pleased. Here a King Air approaches, the pilot knowing that the pedestrians in front will saunter off before he runs them over. In the second view an Italian Transall roars away from Goma to the amusement of the amblers. The local scamps often enjoyed a game of "chicken", standing on the runway as long as they dared as planes roared straight at them. (Larry Milberry)

(Left) During the Rwanda airlift many types of aircraft were seen daily at Goma airport. Here an ancient Russian Il-18 and a Shabair 727 sit on the tarmac on August 8. (Larry Milberry)

Corpses floated downriver towards Lake Victoria —a low-level tour by Herc was also a lesson in horror. There was no telling what might happen next on Op Scotch. On August 8 UN931 was at Goma awaiting takeoff. It fired up at 1741, but had to wait for arriving Partenavia, Transal, Twin Otter and Ce.206 traffic, then for a USAF Herc to depart. Next, UN959, a Shark-nosed Herc from Pope AFB, was on final. It appeared out of the haze in a rakish nose-down angle and too close to make a good landing. It overshot. By this time the ATG crew was getting a bit cheesed. UN959 began a wide go-around, as if everyone else at Goma had all day. Someone aboard UN931 wondered out loud, "Are these guys on a trainer? Are they circling Zaire?" UN959 finally landed, touching short on the two- mile runway, yet missed its turn-off and rolled to the far end, before backtracking. UN931 was happy to get away and forget about its aggravations. It proceeded to Nairobi low level, too low to raise Kigali tower for a routine position check. A high-flying USAF Herc relayed UN931's call. The Canucks were in a kibitzing mood by now and had told the USAF radio Kigali that they were the lead plane in a C-5 Galaxy three-ship. "Will be overhead Kigali for the break in 10 minutes," was the message. Amazingly, the tower took this seriously and went into a tizzy. UN931 split its collective gut, then shut up and disappeared into the twilight as someone mumbled, "That's us... making friends for Canada wherever we go."

In mid-September several flights were made from Dar es Salaam to Bujumbura. On September 16 a 437 Squadron Airbus delivered supplies and equipment directly to Kigali. The ATG airlift ended on September 27, although other Canadians remained in Rwanda on UN duties. ATG logged some 1,200 flying hours on 312 flights on Op Scotch, carrying nearly 3,000 tons of cargo and 6,340 passengers.

One conundrum of the West providing aid to the Third World is that it often goes unappreciated. Western aid workers regularly are abused, even murdered. A small taste of African gratitude came at the end of Op Scotch, when an ATG Herc delivered supplies to Kigali. As a favour to Belgium, the ALCE commander in Nairobi agreed to bring out the remains of three Belgians killed in the original uprising. A crew member recalled:

Being obliging Canadians, we accepted the request. Upon arrival at Kigali we were greeted by a delegation, who oversaw the loading of the three coffins. I learned that two of the dead had been young newly-weds, and the third was a teacher. Apparently they were killed about April 7 but, because everyone had evacuated, there was no one around to take care of delivering them home to their families.... The loadmasters strapped the coffins to a pallet at the rear of the aircraft and we prepared to depart for Nairobi, where the Belgians had a C-130 waiting to transport the coffins homeward. Requesting clearance from the tower, we were asked if we had all the necessary paperwork. Thinking that they were referring to our flight plan, we answered in the affirmative and prepared to taxi. All of a sudden a Rwandan (one of the self-appointed generals)... ordered the local fire truck to park across our nose.

We waited while our aircraft commander tried to sort things out, before shutting down. It turns out that in Rwanda things must be done "properly". For every dead body there must be a certificate signed by the Minister of Health! Well now, we pondered, had the minister issued certificates for the 500,000 poor souls whose bodies went floating down the river during the heat of the conflict? Being diplomats we did not ask that question aloud.

During the two hours while we awaited "authority" to leave, one of the Rwandans asked to whom he could write in the Canadian embassy in Nairobi to secure landing fees for our flights into Rwanda! One only had to look out in any direction across the crowded Kigali ramp to see how ludicrous this question was. Piled all over the place were sacks of food, water purification systems and chemicals, piping, etc... basically all one would need to start up a small civilization. Every item had been donated by the West, yet this clown wanted us to pay for the privilege of being the charity that we were to Rwanda. We remained calm, obtained our "authorization" in the form of a hastily scribbled note from some Belgian official, promised to be good and made a speedy departure.

Relief workers, many with Canadian T-shirts, load supplies at Goma for the outlying refugee camps. (Larry Milberry)

Boeing 707 9Q-CMD. An old Varig machine, its Zairean owner bore one of the great names—Business Cash Flow Aviation. One day its cash hurtled off the runway at Goma to become a permanent landmark. In the distance a Syrian Il-76 offloads Red Crescent supplies. (Larry Milberry)

Crowds throng the streets in suburban Goma on August 4, 1994. Then, a view of the activity around the French field hospital at Goma airport. (Larry Milberry)

The passenger terminal (left) at Kigali's Gregoire Kayibanda airport was badly shot up by Hutu and Tutsi gunslingers. The tower was trashed and emergency equipment had to be flown in to make it serviceable. Over the summer of 1994 it was providing rudimentary daylight air traffic control to a steady flow of aircraft. Here Capt Roland Lapointe (CFB Comox), USAF Capt Bill Mahoney and MCpl Yvan Poisson (CFB Bagotville) comprise the ATC shift on August 6, 1994. A Canadian airfield support detachment had put the airport back into service on August 1. (Larry Milberry)

(Right) Herc 337 delivers a new Toyota to the UN in Kigali on August 5, 1994. The control tower is across the tarmac. Many Canadians applaud their nation's efforts at international mediation, which date to the Arab-Israeli war of 1956. At least as many wonder about the cost and ultimate usefulness of this work. On the CBC program "On the Line" of April 2, 1995 retired LGen Lewis Mackenzie commented cynically about Canada's efforts in troubled places like Rwanda: "It's all we have, hockey and peacekeeping. Don't take them away from us." (Larry Milberry)

The "We've got that goin' for us" Jim Bertrand crew gets into high finance after delivering some soft drinks and pizza to the troops at Kigali. (Larry Milberry)

Lots of well-worn 707 and DC-8 freighters still were making a living in Africa in the 1990s. Liberian 707 EL-JNS was at Kigali on August 5, 1994. Having begun in 1963 with American Airlines, it still was active 31 years later. (Larry Milberry)

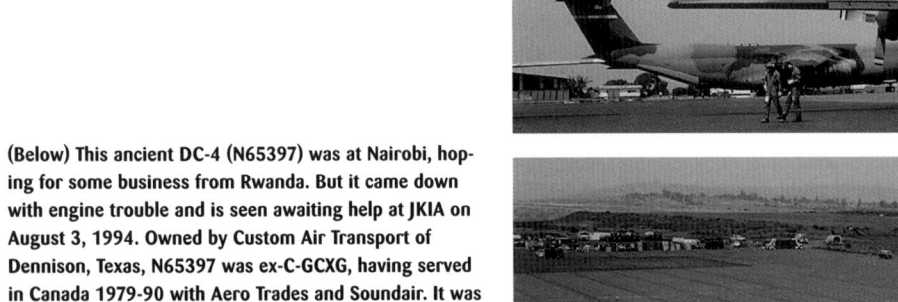

(Right) Heavy metal on the ramp at Kigali—an RAF Herc, a USAF C-5 Galaxy and a Luftwaffe Transall. (Larry Milberry)

(Below) This ancient DC-4 (N65397) was at Nairobi, hoping for some business from Rwanda. But it came down with engine trouble and is seen awaiting help at JKIA on August 3, 1994. Owned by Custom Air Transport of Dennison, Texas, N65397 was ex-C-GCXG, having served in Canada 1979-90 with Aero Trades and Soundair. It was sold by Soundair's receivers for $5,000. In March 1997 it resurfaced in a mention in John Wegg's intelligence monthly *Airline and Commercial Aircraft Report*—the old clunker now was in Zaire with Air Transport Office as 9Q-CLM. (Larry Milberry)

(Above) An Aeroflot Il-76 delivers a hefty load to Kigali on August 6. This ancient freighter still was being improved late in life—in 1995 the stretched Il-76MF was unveiled. (Larry Milberry)

Canadian Helicopter's Bell 212 C-FRWL on the presidential ramp at Kigali as C-FRWI flies out with Canadian techs to check an airport beacon. RWI had just flown in from Kabale, Uganda with the crew of Norm Robichaud and Stan Binns (pilots) and Chris MacKay (radio dispatcher). CHL won many offshore contracts in the 1990s. Others included six machines for UN use in Kuwait/ Iran; and 11 for peacekeeping support in Cambodia. (Larry Milberry)

Home sweet home... CHL's crew in Kigali in August 1994: engineer Jim Hannam and pilots Murray Cheslock and Len Crocker. Crocker had begun flying in 1965 and spent two years with Ontario Central Airlines on Norseman CF-DRD. He moved to Halifax and worked with the Department of Lands and Forests, flying a Beaver and King Air. In 1972 he got a helicopter ticket and spent several years on a Hughes 500C for the Nova Scotia Power Corp before moving on to companies like Viking, Great Slave and Pegasus. (Larry Milberry)

Capt Mike Biehl of 429 Squadron checks his charts, while operating Kigali-Entebbe on August 5 as CF6763. His crew comprised Capt Gary Moore (FO), Capt Howard Tetzlaff (Nav), WO Stephane Guy (FE), MCpl Ben Fraser (LM), MCpl Steve Frotter (LM) and MCpl Lorne Herrick (ASO). (Larry Milberry)

Sgt Pelland relaxes for a few minutes while en route Goma. (Larry Milberry)

(Left) Dave Hutchison and James Pierotti man the starboard observation blister during a "sky and ground" (combat egress) departure from Kigali on August 5. Capt Bertrand was at the helm. Crossing Lake Victoria the Herc buzzed by a small island fishing community. (Larry Milberry)

Lunch time at the ALCE in Nairobi. Capt Doug Hinton, Maj Brian Jossul (ALCE deputy commander), Capt John Stevens and Capt Steve Camps take a break after the morning trips to Goma and Kigali. (Larry Milberry)

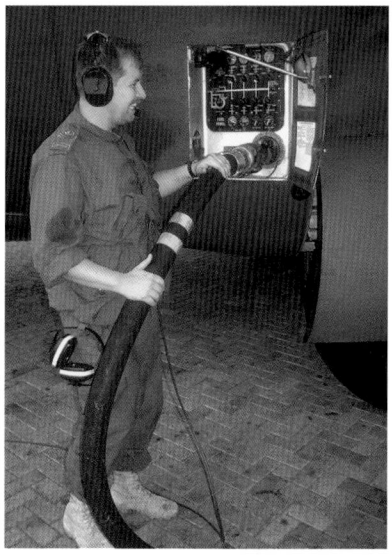

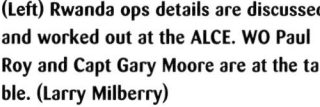

(Above) MCpl Ross Windeler refuels Herc 337 at the ALCE on August 4, 1994. (Larry Milberry)

(Left) Rwanda ops details are discussed and worked out at the ALCE. WO Paul Roy and Capt Gary Moore are at the table. (Larry Milberry)

The Bertrand crew at Nairobi after the day's work of August 8. They had done Nairobi-Kigali in the morning and Nairobi-Goma after lunch for about eight hours of flying: Capt James Pierotti (Nav), MCpl Dave Hutchinson (LM), Capt Rob Butler (FO), Capt Jim Bertrand (AC), MCpl Denis Culver (LM), MCpl France Dufort (ASO) and (in front) Sgt Marc Kovacic (FE). As a lad Rob Butler earned his private pilot's licence at the Campbell River Flying Club. Then he joined the air force and worked nine years as an instrument-electric tech before remustering to pilot in the spring of 1991. (Larry Milberry)

Besides ATG other "Canadian content" was noted in East Africa in 1994. NWT Air's famous red-and-white Hercules arrived in Nairobi on August 28, having routed St. John's-Shannon-Athens-Jeddah-Nairobi-Kigali to deliver vehicles for Care Canada. It then sat at Nairobi hoping for some ad hoc charters. With the tarmac covered with military transports and Russian "heavy metal"—Antonovs and Ilyushins—not much developed and NWT Air departed for Yellowknife on September 11. Also at Nairobi in August were C-46s 5Y-IBX and 5Y-TXW (formerly C-GIBX and C-GTXW). Operated by Relief Air Transport, they were down from Lokichoggio, their base since September 1993, hoping for work. Todd Lamb, formerly of Air Manitoba, managed the operation, using Canadian crews. Other word was that one or two civil DHC-5 Buffalos (with several Canadians) still were working from Loki. Some Twin Otters were operating on airline skeds, tourist charters and relief work in Burundi, Kenya, Rwanda, Somalia and Zaire; as were a Caravan and two King Airs of SkyLink. An interesting relic at Nairobi was C-54D-DC N65397, a 1945-model originally delivered to the USAAF. After a water bombing career in the US with Aero Union, it worked in Canada. After damaging its undercarriage, it rotted for more than a year at Dorval, then was purchased by US interests. It now was registered with Custom Air Transport of Texas. It sat all summer at JKIA awaiting an engine change and by late August had yet to fly a trip. Meanwhile, Custom Air's 727, bearing Air East Africa colours, was busy flying Nairobi-Bukavu with aid.

The An-124

The An-124 Ruslan was developed to carry nuclear missiles within the USSR, including on rough strips. It entered service in 1985 and Antonov in Kiev built 17. The line then closed, but re-opened in 1993. About 35 more were turned out by 1996 at the Aviastar plant in Ulyanovsk, south of Moscow. The An-124 was introduced to the West in 1989 through an arrangement between AirFoyle of the UK and the Antonov Design Bureau in Kiev; then through HeavyLift (also a UK freight broker) and Volga-Dnepr Airlines. Antonov and Aviastar were obliged to meet certain Western standards with the An-124. In time organizations like the US FAA were satisfied and the An-124 began appearing at Western airports. Almost instantly there was demand—there was no other such airlifter.

The An-124 has a payload of 165 tons, although 132 tons is the normal limit. Full fuel totals 236 tons. The hold has 750 m³ of useable space, and has self-contained fore and aft cargo handling. An integral crane can move items as heavy as 22 tons within the hold. For gross-weight takeoffs minimum runway length is 9,850 feet; maximum landing weight is 364 tons. There is no fuel dump system. In the event of a technical problem at maximum take-off weight, an An-124 would have to hold for four hours to bring weight down for a safe landing.

By 1995 the An-124 allegedly had captured half of the world's $150,000,000+ worth of outsized and charter cargo business. The biggest operator was Volga-Dnepr, formed in 1991 and owned by Aviastar. In 1995 it claimed sales beyond $50 million. Its six aircraft monthly were logging 200-250 hours each, and four additional machines were on order. Antonov's fleet of six 124s, like Volga-Dnepr's, was booked at most times. The secret to this success seems to have been alliances with Western brokers—AirFoyle (in 1989) and HeavyLift (in 1991). The brokers themselves were associated since 1989. Most work originated outside the CIS. A typical trip was made in July 1992 from Winnipeg—an An-124 carried farm combines, grain storage bins, a grain dryer and other agricultural equipment to Chelyabinsk in Russia. This was reportedly the biggest air cargo load ever handled at Winnipeg.

In 1994 Aeroflot established a fleet of three An-124s, and Rossya, a one-time Aeroflot branch, introduced two. Transcharter started up with a lone An-124. The Russian Air Force was offering its fleet of 26 aircraft for charter. The success of operations encouraged the CIS to expand in air freighting and to start integrating services using smaller types to feed cargo hubs. In the mid-1990s the An-124 could be described as a moneymaking machine for the CIS. Jobs were being paid in US dollars, expenses in rubles, leaving lots of room for profits.

While the An-124 impressed everyone, it had its Achilles heel. Its D-18T engines were unreliable. Time-between-overhaul was low, and fuel consumption was unacceptable. Lost revenue caused by this led to retrofitting Western engines. In July 1996 the first An-124-130 was rolled out at Ulyanovsk with GE CF6-80 engines, and the need for a smaller crew—only four on the flight deck plus four engineers/loadmasters. On earlier models a crew of about 20 was normal. To 1997 the An-124 had a fair safety record. One crashed on a test flight—the hinged nose door ripped off in a dive. In October 1996 another crashed approaching Turin, where a cargo of luxury cars for the UAE and Brunei awaited. The doomed machine was operated by Aeroflot on behalf of Ayaks Cargo Airlines of Moscow. In early 1997 a Volga Dnepr An-124 set a world record for a single load (volume-wise), carrying a BAe Nimrod fuselage from Kinloss, Scotland to Bournemouth in the south of England.

Typical Kigali Movements, August 7, 1994					
Mission No.	Carrier	Aircraft	ETA	From	ETD
UN800	NGO*	An-12	08:00	Dar es Salaam	09:00
UN475	Canada	C-130	08:30	Nairobi	09:00
UN931	UN	C-130	09:15	Nairobi	09:45
UN831	UK	C-5	09:45	Brize Norton, UK	13:00
UN947	NGO	King Air	09:45	Entebbe	10:15
UN825	Germany	Transall	10:00	Nairobi	10:30
UN830	US	C-130	11:00	Entebbe	12:00
UN950	US	C-130	12:30	Goma	15:35
UN950	Canada	Il-76	13:00	Cairo	16:00
UN951	Canada	Il-76	13:15	Cairo	16:15
UN801	NGO	707	13:30	Cairo	16:00
UN475	Canada	C-130	13:30	Nairobi	14:00
UN800	NGO	An-12	15:00	Dar es Salaam	16:00
UN831	US	C-130	16:00	Entebbe	17:00
UN802	UK	C-5	18:15	Diego Garcia	21:30
UN803	UK	C-5	23:59	Brize Norton	03:15

An-124 Specs	
Wing span – 240.5 feet	
Length – 226.8	
Height – 68.2	
MGTOW – 864,200 pounds	
MLW – 727,525	
Payload – 264,560	
Max Fuel – 214,000	
Price – $25 million US	

(Above) A Volga-Dnepr/AirFoyle An-124-100 offloads at Kigali on August 6, 1994 as a USMC KC-130F of VMGR-252 (NAS Cherry Point) taxis. For Rwanda the An-124, C-5 and 747 provided the heavy airlift from overseas, while the C-130, Il-76 and An-12 were the regional workhorses. (Larry Milberry)

SAR and Other Developments

No. 111 Search and Rescue Unit Lancaster FM215 at Winnipeg in August 1959. The Lancaster was Canada's primary long-range SAR aircraft in the postwar years. The last served at Torbay with 107 RU till replaced by North Stars in 1963. A few years earlier FM215 had been on photo duties with 408 Squadron (see page 443). (B. Pidverny)

Search and rescue (SAR) is one of the great specialties in Canadian aviation. It dates to the earliest days, when pilots in rickety HS-2Ls or Fokkers were down in the bush or on the tundra. The hunt for the McAlpine Expedition still rates as one of Canada's greatest aerial searches. Advances in SAR came with WWII—the first para-rescue specialists were trained and the first specialized SAR equipment and dedicated flying units appeared. The pressures of war, however, meant that, if it wasn't found within a few days, a missing crew was given up. Episodes of wartime (air-sea rescue, as it then was termed) and postwar SAR appear in earlier chapters of *Air Transport in Canada*.

In the 1990s Canada's main SAR resources were four squadrons: 442 at Comox (Buffalo and Labrador), 424 at Trenton (Hercules and Labrador), 413 at Greenwood (Hercules and

In the interwar years SAR was an ad hoc matter. If a plane or ship was missing, a search might be arranged with local resources. There were no special SAR equipment or personnel. Matters suddenly improved with WWII, when there was a pressing need for SAR. The Sikorsky S-51, the RCAF's first helicopter, brought a much-needed advance in SAR. Here No. 9606 from Vancouver works at the site of B-25 No. 5246 of 406 Squadron, which crashed in the Rockies on February 16, 1953, killing all aboard. Note the rescue hoist atop the S-51's cabin. (DND)

(Left) SAR often proves futile, as with the 406 Squadron B-25, but often there are happy results. In June 1949 veteran Arctic pilot Ernie Boffa and mechanic Fred Riley were down with their Anson between Holman Island and Coppermine, 160 miles north of the Arctic Circle. Six RCAF and various civil planes searched four days before the duo was found. (DND)

(Right) Although many ships' crews have been lost off Canada's coasts, hundreds have been saved by CanForces SAR operations. This lifeboat-full of sailors adrift in the Atlantic must have had their spirits soar when an Argus from Greenwood dropped rafts and supplies. In most cases surface vessels complete the rescue, but sometimes helicopters hoist the survivors aboard. In the second case, a Sea King is involved in a 1973 medevac off the east coast. Maj Jean-Marie Rouleau, a medical officer from CFB Shearwater, is being lowered aboard the trawler *Cape Nova* to assist an ailing seaman. (DND)

(Left) Larger helicopters replaced the S-51 in RCAF service, first the S-55, then the S-58 and H-21. These did outstanding work till replaced in the mid-1960s by the CH-113 Labrador. This H-21 was at the site of CF-104 12712's crash north of Montreal on November 25, 1961. Canadair test pilot Bruce Fleming had to eject when he got into a deep stall. (Canadair Ltd. 29318)

A 442 Squadron Labrador training near CFB Comox in 1992. The "Lab" was Canada's frontline search and rescue helicopter since 1963 and 13 remained in service in 1997. Empty weight was 13,500 pounds; but with fuel (3,000 pounds), SAR kit, tools and spares (1,300), and crew (1,200), a Lab hit 19,000 pounds. In 1996 Ottawa deferred selecting a Lab replacement for at least another year. It was likely that the Lab, refurbished time and again, would see the year 2000 in CAF service. (John McQuarrie)

Labrador) and 103 at Gander (Labrador). There was secondary support from other Hercules squadrons (435 at Winnipeg, 429 and 436 at Trenton), 440 Squadron at Yellowknife (Twin Otters) and squadrons with Twin Huey, T-33s, Challengers, Sea Kings, etc. Air force SAR was well-supported by the Canadian Coast Guard, police agencies like the RCMP, provincial air services, CASARA (Civil Air Search and Rescue Association), and the international SAR satellite network, SARSAT.

West Coast SAR

Canada's West Coast is the busiest search and rescue region. It includes vast stretches of the Pacific Ocean, thousands of miles of coast with endless bays and inlets, mountain ranges (the Cordillera geographic region), and challenging flying weather. The SAR squadron in charge of this huge area in the 1990s was No. 442. It began in WWII with Spitfires and Mustangs. Post war it flew Vampires, Sabres, Expeditors and Otters as a Vancouver-based auxiliary squadron. For most of the postwar era SAR on the coast was carried out by 121 Composite and 123 SAR flights at RCAF Station Sea Island (Vancouver). They had formed in 1946-47, using the Hudson, Canso, Lancaster, Dakota, S-51, H-21, T-33, Albatross and Otter over the years. In 1953 these units merged as 121 Communications and Rescue Flight. In 1958 No. 121 C&R Flight became 121 Composite Unit ("121 KU"), and in 1964 it moved to Comox, where the latest SAR helicopter, the CH-113 Labrador, entered service that year. In 1968 121KU became 442 Transport and Rescue Squadron. In 1970 it added three Buffalos.

Taskings at 442 Squadron include coverage of the Victoria search and rescue region, general transport, Buffalo and Labrador operational training flight (OTF), support for the Skyhawks parachute demonstration team, operation of the CF sea survival training unit, and support for CASARA and PEP—Provincial Emergency Program. For 1995 442's fleet included Buffalos 115451, 452, 456, 457, 462 and 465; and Labradors 11310, 12, 16 and 18. The complement of 216 personnel included 24 pilots and four navigators.

442 Squadron Flying in 1995
Total hours – 4,953
RCC taskings – Buffalo 639 hours, Labrador 611
Transport and training – Buffalo 1,801, Labrador 1,008
OTF – Buffalo 182, Labrador 531

For 1995 No. 442 was involved in 292 SAR launches. These involved searches for aircraft and vessels, civil aid, humanitarian work, and taskings where the source of distress was not identified. The squadron finished 1994 on the last day of the year, when Labrador 312 and Buffalo rescued an injured crewman from a tug 100 nm offshore. The next SAR tasking came on January 3—a call for emergency food and water for a sail boat in distress 120 nm offshore. The drop by Buffalo 456 was good. The same day Buffalo 462 located a Hughes 500 destroyed in a crash on Cradock Glacier near Chilko Lake. Four survivors were stranded at the 6,500-foot level. Survival gear was dropped and two SARTechs jumped in to set up camp. Next day Labrador 312 made a confined area landing to pick up the party, then flew to Powell River.

On January 9 Lab 310 medevaced four Rivers Inlet people who had ingested some type of toxic root. They were taken to hospital in Port McNeill. At 0233 two days later Buffalo 451 was despatched to Masset to search for a medevac Learjet (SAR Jorgensen) missing with five persons on board (POB). Two Labs and several other resources went to work at first light. After three days 442 was stood down. Two bodies and some wreckage had been found. Later the main wreckage was located in 250 feet of water near Masset. A subsequent report concluded that the pilots had not changed their altimeters from cruise to approach setting, causing the plane to fly into the water during a night approach.

A search for a missing light plane with two POB (SAR Holl) began January 12. No. 442 dedicated 46.7 hours over four days. A USN P-3 and many vessels took part for a total of 431.7 search hours. One body was found and a surface vessel discovered the wreckage on the sea bottom in Barfleur Passage. On January 22 Lab 312 made a confined area landing on Mount Washington to rescue an injured skier

and ferry him to Campbell River. There was lots of work on February 11, when Buffalo 457 took off in darkness searching for the fishing vessel *Pacific Bandit* 35 nm south of Tofino. Using NVGs (night vision goggles) the crew located a raft with two POB. Lab 318 picked them up, the fishing boat *Ocean Star* saved another man (later hoisted by the Lab), but a fourth was not found. The squadron recorded: "All three persons were treated for mild-moderate hypothermia with electric blankets, casualty bags and warm beverages, and transported to Tofino hospital." Most of 442's 15 launches in March proved to be false alarms or were noted as "Nothing found".

On April 8 pumps were dropped by Buffalo 465 to the foundering 34-foot *Ichiban*. One pump could not be recovered by the fishermen, and they were unable to get the second to work. CCG *James Sinclair* arrived to take *Ichiban* in tow. Two days later Buffalo 465 was despatched to search for three crew from the abandoned *Hilikum*. One person was saved, another found dead. April 13 a lady, badly injured in a fall, was medevaced to Victoria. A typical false alarm occurred April 21, when Lab 316 picked up an ELT (emergency locator transmitter) signal near Port Renfrew. This was traced to a helicopter shut down on a logging road. As soon as 312 arrived on scene the ELT ceased.

On April 24 Ce.185 CF-CJS sank off Princess Royal Island after a flight from Chilliwack. Two passengers swam ashore. An R-22 helicopter spotted the pilot, then he disappeared, leaving his floatation device on the surface. Lab 312 searched the area to no avail. May was busy with 37 taskings, a third being futile ELT searches, e.g. on May 18 the squadron reported: "ELT was homed to a garage that had an aircraft parked inside." In another case it noted, "a Coast Guard helo in the hangar at Shoal Point was the culprit." SAR Fillipone of May 19 involved a rock climber who fell into a crevasse on Chief Mountain. Local ground rescuers extricated the climber, whom Lab 310 flew to a waiting ambulance in Squamish.

On June 7 Lab 310 investigated a small aluminum boat with an unconscious man aboard. A helpful citizen towed the boat ashore and the Lab landed nearby. The occupant was a drunk who had passed out. One ELT proved legitimate. On June 13 (SAR Sundeman) Buffalo 462 homed on the signal between Whitehorse and Inuvik, located a crashed plane and, with the assistance of a Bell 206, confirmed that the sole occupant was dead. An odd case on the 19th involved an aircraft hulk trucked from Denver to Vancouver as a film prop. Someone failed to notice that it contained an active ELT. Lab 310 finally pinpointed it in the warehouse, where the film was being shot.

On June 20 Lab 318 assisted FV *Courageous*, capsized at Edith Point. Five survivors were rescued by the Coast Guard. Another was thought trapped in the boat, so RCC authorized a dive. One of the SARTechs was pinned by heavy machinery and lost his air tank. A second diver cleared the mess and got his mate to the surface. The missing crewman wasn't found. On July 20 PA-23 C-GGNA reported that it was lost and low on fuel. Buffalo 456 was despatched, intercepted GNA and escorted it to Campbell River. SAR Boot ended with bad news, when RCC learned that the missing PA-12 was found north of Smithers with three POB dead. On September 1 Lab 310 flew about 100 nm offshore from Port Hardy to take on a gravely injured ship's crewman. The Lab then returned to Port Hardy from where Buffalo 451 rushed the patient to Victoria.

SAR "C-FEBX" commenced September 28 —a Turbo Otter on approach to Campbell River airport with 10 POB had failed to land. Buffalo 462, en route Comox-Williams Lake, quickly pinpointed an ELT signal eight miles west of the airport near the local NDB site. The weather was duff and night had fallen, so 462 returned to Comox, a few minutes away, and the SARTechs transferred to Lab 318. Weather grounded 318 at Campbell River hospital for about an hour, then it departed. Using NVGs, a SARTech spotted the wreckage. Three SAR-Techs were lowered to the site. There were two survivors; one was hoisted and flown to hospital. Good teamwork saved an injured helo pilot down near Salmon Arm on October 16. Involved were Buffalo 465, which dropped SARTechs; a Bell 412 from CFB Cold Lake that was transitting the area; a first aid expert who hiked to the site; and an AStar helicopter, which long-lined two SARTechs to the crash.

A large search commenced November 28 after a twin-engine Rockwell 700 with five POB disappeared between Calgary and Hillsboro, Oregon. Buffalo 462 began with a night electronic search. Poor weather restricted searching next day. A search HQ was set up at Cranbrook and aircraft from 442, 417, CASARA, the USAF and US Army arrived. The search continued in poor weather for seven days, but nothing was found. On this search 442 provided two Labs and four Buffalos, flying 119.2 hours. Its last tasking of 1995 occurred December 10-13. This was SAR Dalewood Provider—the search for survivors from a capsized fishing boat around Cape Beal. One survivor and two dead crew were taken from the water. In September 1995 442 took part in ATG's annual SAREX at CFB Greenwood. The main SAREX categories were parachute accuracy, medical, search, rescue, maintenance and team spirit. 442 won in the parachute category. March 6-8 the squadron launched a long-range Buffalo trainer to Alaska and the Aleutians. A southern trainer in October included delivering humanitarian supplies to Mexico. Op Hurricane took a 442 Buffalo to Eureka, Alert and Tanquari Fjord in June to haul fuel. Skyhawks activities included 68 shows at 28 sites

That Others May Live

Over the decades since WWII many RCAF/CF SAR personnel sacrificed their lives. There have been tragedies in training, and others on ops. Both pilots of a No. 121 Communications and Rescue Flight Canso died in a crash near Vancouver in July 1955. On April 23, 1966 Albatross No. 9302 crashed into a mountain near Hope, BC. Only one of six aboard sur-

Buffalo No. 457 of 442 Squadron lands at Namao on May 22, 1994; then, some nose art on a 442 Buffalo in February 1987, indicating one in-flight birth, and two bird strikes. (Larry Milberry)

(Below) Buffalo colour schemes from earlier days as recorded by Les Corness. No. 459 was at Gatwick on May 31, 1968; No. 460 was at Namao on June 12, 1982.

vived. On November 2, 1971 a 429 Squadron Dakota was at Cape Perry in the NWT searching for a light plane. When it crashed during an equipment drop, seven died. On October 15, 1980 Hercules 130312 was in northern Quebec searching for a lost helicopter. It stalled in a turn and crashed, killing eight of the 10 aboard. In April 1988 Capt Mike Erickson died while on a night search off Vancouver Island for a missing vessel. Rotten weather had kept 442 Squadron grounded, but Erickson volunteered to fly an electronic search. While flying low in IFR conditions, his CF-18 slammed into a cliff.

The Boeing Vertol CH-113 Labrador has been Canada's chief SAR helicopter since 1963. Although upgraded over the decades,

one could not expect trouble-free operations. This was brought home on April 30, 1992 while Labrador 11311 was searching 15 nm southeast of Bella Coola, BC, where hikers Jennifer and Bob Kovacs were lost. The Lab had hoisted three searchers and was hovering over a glacier at the 5,400-foot level, when an engine failed. The pilot tried to save the heavily-loaded machine. He knew there was a SARTech below. If the Lab crashed, he likely would be crushed. The Lab manoeuvred to gain speed and lift by flying down the side of the mountain, but it was too late. It crashed, rolled and burned. Cpl Philip L.C. Young was hurled out and killed. Everyone else aboard was injured.

In a preliminary report the DND Directorate of Flight Safety noted: "The poor single engine performance of the CH-113A Labrador is well understood. The nature of the SAR role demands that the aircraft be flown in flight regimes that preclude the option of a single-engine flame-out. The aircraft is operated in this role on the statistical improbability of a power loss occurring in such importune circumstances. In this instance the crew had everything going against them—a high density altitude, a large number of personnel on board [nine], and no safe landing area." In spite of this severe criticism of the Labrador, it had not suffered a fatality in perhaps 100,000 SAR flying hours over 28 years. Relative to other helicopters the old Lab still had a good safety record. Shortly after the crash, the formal search for the Kovacs ended. Relatives continued searching. Before the summer was out, their team located the couple, who had died in an avalanche.

The Atlantic Region

In the late-1990s two SAR units were covering the Atlantic region—413 Transport and SAR Squadron at CFB Greenwood, and 103 Squadron at CFS Gander. 413 arrived at Greenwood from Summerside, following closure of that base on June 10, 1991. It had two C-130s, two Buffalos and three Labradors. On June 24 the C-130 took over as the primary fixed wing type and the Buffalo was retired. For 1991, 413's flying hours included: Buffalo 1,042, Hercules 1,655, Labrador 1,639. There were 1,630 hours on SAR taskings. The first SAR call-out in 1991 was on January 4, when a Buffalo and a Lab co-operated in a medevac—a sailor with appendicitis was hoisted from his boat 60 nm east of Sydney. A major action took place January 11-15. The 830-foot bulk carrier MV Protektor had signalled distress 220 nm southeast of St. John's. The 33 crew abandoned ship. 413 despatched Buffalo 459 and Lab 304 to St. John's via Sydney and St. Pierre. Meanwhile, 103's Labs were grounded in Gander by weather. It soon was determined that Protektor had sunk with no survivors.

Next, 413 was occupied with the Spanish trawler San Eduardo Chao caught in the same storm as Protektor. It was found January 12, laden in ice, its bridge windows smashed, radios out and five crew injured. A sister ship managed to get San Eduardo Chao in tow for St. John's. On the 14th Halifax RCC was asked to airlift the injured captain from a point 180 nm east of St. John's, just outside the Labrador's range. The squadron diarist noted, "The mission was quickly calculated and, if everything went exactly right, the helo could just complete the mission based on the reported winds and the fuel on board." Lab 304 launched from St. John's, preceded by Buffalo 459. Aboard Rescue 304 were Maj Bob Cuthill (aircraft commander), Capt Kelly Freitag (co-pilot), Sgt Walt Levesque (flight engineer), and MCpl Rob Walker and Cpl Mario Michaud (SARTechs). On Rescue 459 were Capt Craig Kennedy (aircraft commander), Lt Fab Micoli (co-pilot), Capt Denis Ouellet (nav), MCpl Frank Soos (FE), and MCpl Bob Mondeville and Cpl Derek Curtis (SARTechs).

Near San Eduardo Chao sea conditions were intimidating—35-foot swells in winds to 45 knots. The situation was described by the diarist: "It was still about sea state eight! Without power, and under tow, turning the Eduardo Chao, so that the wind would favour the helo doing a bow hoist, would risk having the boat roll over... As it was, she would roll up to 30 degrees and pitch some 35 feet at the bow all in a random pattern due to the wind on the bow and the swell from the port side."

Rescue 304 realized that the bow was the only safe area for hoisting, but hovering would have to be with the wind on the Lab's port side. This was too dangerous, so the Lab had to await calmer conditions—it flew back to St. John's. It refuelled and the crew briefed for a return flight, getting airborne at 1520 local. Flying top cover (back-up) was Buffalo 457 with Capt Rick Green (aircraft commander), Capt John Kordich (co-pilot), Capt Martin Mattes (nav), MCpl Conrad Wilson (FE), and MCpl Bob Best and Cpl Brian Keefe (SARTechs). Arriving on scene, Rescue 304 found the winds at 30-35 and the seas as bad as ever. The Lab set up over the ship's bow. By observing swell patterns, the crew determined the best moment for lowering a SARTech: "The crew was in agreement that they could at least give it a try with a degree of safety and probability of success. The deck was still covered in clear ice and was treacherous. Even the crew of the Eduardo Chao were reluctant to venture out, and the SARTechs would probably be left to their own devices." Soon two SARTechs were aboard ship. The Lab orbited. It was a struggle to get the injured ready for hoisting. As the wind howled and the vessel pitched, the Spaniards and SARTechs were brought aboard the Lab. Now came a 1.5-hour flight to shore into a 30-knot wind. A dicey landing was made in the dark at the St. John's hospital.

Naturally, 413 was plagued by false ELTs. The first of these in 1991 came on January 19—the signal was traced to an aircraft at Yarmouth airport. On January 25, 413 flew top cover, while a 103 Lab hoisted crew from a foundering vessel off Deer Lake, Newfoundland. The first air incident of 1991 was on February 3-4. A Buffalo and Lab searched for an overdue aircraft; it was found near Matane with the pilot dead. On February 8-9 this scenario was repeated, when a Caravan was found in Labrador. Throughout February and March there were many medevacs with patients having heart trouble, failing kidneys, difficult pregnancies, broken bones, newborn problems, etc. April 23-24 a 413 Buffalo and Lab helped in the search for a 60-foot tug missing between North Cape and Gaspé. Life rafts, a Zodiac and four bodies were found from the Patricia B. McAllister. Only one survivor was found. Two other marine distresses occurred on June 5. Rescue 302 was tasked to aid the fishing vessel Little Darren 115 nm off Yarmouth. En route it was diverted to assist the Christopher R., taking on water. The Lab stayed by this vessel until CCG Clark's Harbour arrived, flew ashore to refuel, then made for Little Darren, where a dicey hoist of an injured seaman was made in

15-foot seas. On June 14 the first 413 SAR Hercules operation was flown—a search for a wind surfer, but nothing was found. On the 24th Herc 314 was despatched to a sailboat 325 miles south of Halifax. There it replaced a US Herc, and vectored another vessel to help.

On July 2 a 413 Herc dropped fuel to a light plane out of gas on Anticosti Island. This good Samaritan operation was offset the next day when the search for Piper Cherokee C-GQDA ended with the wreck being found with the four POB dead. Numerous taskings were carried out through the summer to aid recreational boaters. In one case a canoe with four on board was swamped on Bras d'Or Lake, Cape Breton. A Lab had to airlift two of the canoeists to hospital—one with hypothermia, another with heart failure (the SARTechs revived him on site). When the fishing vessel Mirabel I caught fire on August 9 off Stephenville, a 413 Herc dropped a SKAD (survival kit air droppable). The crew was picked up at night by a 103 Lab as the Herc illuminated the area with flares. On the 13th two 413 Labs hoisted seven tourists trapped along a cliff by a rising Fundy tide. Somewhat greater excitement occurred on September 19 as Lab 304 was lowering pumps and SARTechs to the J. Venture 100 nm off Bridgewater, Nova Scotia. The vessel suddenly caught fire. Cpl House and one fisherman were lifted aboard, but Sgt Maltais and the other fisherman had to leap overboard and swim from the inferno before being taken on board the Lab.

October 18-22 was a busy time for Halifax RCC—a major search was under way for a T-33 from VU-32 (CFB Shearwater). Hercs, Labs, Auroras, Sea Kings and T-33s took part, but no trace was found of the jet and its two occupants. The busiest 413 tasking of the year was October 30-November 3, when Herc 322 from 435 Squadron crashed near CFS Alert in the high Arctic. The job was to airlift 100 flares to Alert. Rescue 305, fully fuelled and weighing 159,000 pounds, was airborne from Greenwood at 2110 local. It arrived over the crash site 7.5 hours later and dropped two flares without getting any visual results. It stayed overhead until Rescue 342 refuelled at Alert, then recovered at Thule after 9.8 hours aloft.

After a five-hour crew rest at Thule, 305 again took off for the crash site which, by then, was known (info passed on by Herc 342). A flare was dropped without results. After much effort 305 was able to contact the ground search party and assist it through an Arctic storm by lighting the way with flares. Occasionally it would return to the crash site, drop a flare and orbit to obtain visual references. After being on site about two hours, 305 made the first sighting of Herc 322 (this was at 0330Z). It was able to assess surface winds and terrain conditions, then returned to aid the ground party, which reached the site at 0410Z. By this time USAF Herc R102 was illuminating the crash site and Rescue 305 was tasked for a SARTech drop. R102 was to illuminate the DZ, but this seemed unsafe to R305. It decided to illuminate the DZ itself. The first flare was dropped, but onto unfamiliar terrain. R305 then realized that its compasses were out and climbed to reset them with a star shot. Meanwhile, R102 attempted to drop its SARTechs, but could not get visual contact in the area.

At 0535Z R305 dropped flares. As the squadron diarist recorded: "The site was re-acquired and a target marker light was dropped at 1,500 feet AGL (2,500 MSL), with 1,000-foot free-fall on the next flare pattern, the drift light was dropped directly over the crash site. Due to poor visibility near the ground, strong surface winds and weak-intensity lights, neither of the lights was seen again. After having dropped over 90 flares that night, it was possible to determine a very accurate wind direction. Timing for the drop was attained from the drift of the flares and the on-board GPS. The crash site appeared to be situated on a 2-km by 1-km flat snowfield valley. The crew agreed that an into-wind drop on a known outbound heading from on top the crash site was the best choice for the SARTech insertion."

The drop required three runs for the SARTechs and one for equipment. The flares were dropped between cloud layers at 5,500 MSL with a 500-foot drop setting. With the flares gone a descending turn and pre-drop pattern were flown as an IFR procedure with the pilot flying vectors given by the co-pilot, as the nav ensured terrain clearance with radar mapping. The SARTechs exited through the right door. On each jump two flares were dropped in a left climbing turn 15 and 30 seconds after the last jumper departed. The first stick of jumpers went in at 0545Z and included six men—a medical team and on-scene commander. Next, five SARTechs descended, followed by a supply drop with three toboggan camp kits and two bundles of B25 survival kits. All drops were on the DZ. There were a few injuries among the jumpers, and only one toboggan kit was saved—the others were carried away in 35-knot winds. On the ground the jumpers from Rescue 305 had a formidable job, as indicated in the squadron history:

The wreckage was searched by three two-man teams because of the vast area it covered. Survivors were heard calling from the tail section and one team proceeded into the tail. The survivors indicated that there were two other survivors buried in some wreckage 100 meters behind the tail section, so two teams went to search for them and found them 10-15 minutes later. Initial findings were that there were 13 survivors, four dead and one missing... Initial diagnosis of the patients was very rudimentary because of the severe cold and lack of shelter. The first priority was getting the survivors shelter, so all of the parachutes were used to close off the tail section and to put on and under the casualties, who were all hypothermic, wet and lying on metal. After the first toboggan was recovered, the tent was set up and the two survivors buried in the snow were dug out, splinted and moved inside. Both had major injuries.

Rescue 305, airborne 10.7 hours, landed at Alert as soon as it had dropped its equipment— it had minimum fuel for safe diversion to Thule. It was followed at the crash site by

R342, which dropped six SARTechs but did not provide illumination for the jumpers once they leaped out. The SARTechs landed long, and out of sight of the wreck. They had some injuries and were about two hours working their way back. R342 now dropped two four-man survival kits, one toboggan kit, medical supplies and SKAD kit survival bundles. By late morning all casualties were treated and ready for transport to Alert. It was late afternoon, however, before a Twin Huey arrived to begin ferrying casualties and the dead. It made three trips, then an S-61 made two, carrying SARTechs and other rescue personnel.

The wreck of Herc 322 near Alert. Then, the monument erected in memory of those lost in the tragedy, including Capt John Crouch, aircraft commander. (DND)

SAR Childers

A most astounding SAR operation took place in January 1992. Jerry Childers and Steve Raynor had set off from St. John's for Reykjavik in a Beech Bonanza. About 2000 hours (local) they had engine trouble. The mood on the Bonanza must have been of sheer terror. The pilots put out a May Day. Overhead, an Alitalia flight picked it up, relaying the position to Gander Area Control Centre. Halifax RCC was informed; it advised Childers to activate his ELT and update his position before ditching. A few minutes later Childers and Raynor faced their ultimate nightmare—their engine quit 250 miles east of Gander (50°1'N 48°32'W). At 2024 Childers radioed his position.

Already RCC was in motion. Two CF-18s (433 Squadron) from Goose Bay, a C-130 (413 Squadron) and an Aurora (415 Squadron) from Greenwood, a Fisheries and Oceans King Air from St. John's, and CCG vessels in the area were alerted. From New York came information on other vessels in the area. According to regulations, the Bonanza carried sea-survival equipment, immersion suits included. The pilots may have thought to themselves, or even joked as they loaded this kit, that it would not be of much use should they ditch—the North Atlantic was nothing but a graveyard for sailors or airmen in distress at this time of year. As they descended, the pilots placed their raft between them and hoped for the best. They ditched successfully, but the Bonanza immediately turned nose down in the heavy seas. Only the air trapped in the tail kept it afloat. After a struggle Childers and Raynor got out and scrambled into their raft, closing its cover against a wind. At least they were alive, but would anyone find them before the cold drained them of life?

At 2215 the CF-18s were overhead, but were not picking up Childers' marine ELT. Neither was SARSAT, which rarely missed a signal. It was later determined that the ELT was cracked and disabled by water. This normally would have spelled the end for the survivors—how were they to be pinpointed in the darkness? The CF-18s had to leave; the King Air took over. Soon the Aurora arrived, and the two planes divided the search area. Herc 306 from Greenwood was also in the search. Aboard were Maj Marvin McCauley and Maj Price (pilots), Capt M. Mattes (nav), Sgt Miller (FE), MCpl Norris (LM) and MCpls Mondeville and O'Reilly (SARTechs). At this point the nearest ship was 60 miles away; nobody was hopeful of a happy ending. By now the survivors had been about four hours adrift. There were signs of hypothermia, but Childers was still sharp enough to pick up what he thought was the sound of an airplane. He looked out and there, dead ahead, were lights in the sky! Certain that it was homing on his ELT, he might have been happy to sit by; but wisely, took the initiative of firing a flare— the crossing of Herc and raft had been pure coincidence. No flare, and the Herc would have swooshed overhead, knowing nothing of the raft below. The flare caught Maj Price's eye. He suspected it was the Aurora, but a quick call confirmed otherwise. "We've got them," announced Price. The sighting was recalled by the Herc navigator: "It was a red flare approximately 10 nm on the nose. The wave of emotion that swept throughout the cockpit was beyond description. That moment definitely proved to be the highlight of my SAR tour, and probably one of the most moving moments of my life."

Herc 306 dropped two two-bundle SKAD

SAR Scenes

Since the 1960s 424 Squadron at CFB Trenton met the SAR requirements for much of Ontario and Quebec. For many years it flew the Labrador and Buffalo. Labrador 308 (left) was on exercise in the Bay of Quinte with the Coast Guard in June 1991. No. 424 was the first SAR squadron to upgrade from Buffalo to Hercules. One of its Hercs is seen at Trenton in October 1994. (Larry Milberry, Tony Cassanova)

The air-droppable, major air disaster (MAJAID) kit used by SAR Hercs. This example was in permanent readiness at Trenton in the 1990s and included everything from skidoos to tents, medical supplies and rations. (Tony Cassanova)

Although not dedicated SAR resources, 440 Squadron's four Twin Otters were available for SAR duties in the 1990s. John McQuarrie photographed No. 809 (aircraft commander Capt Brian Degere) north of Edmonton. Mike Valenti caught No. 806 landing on tundra tires at Namao in May 1992.

SAR training at CFB Trenton in 1994. (Tony Cassanova)

(Bottom) Over the years the CAF operated 10 CH-118s, 50 CH-135s and 100 CH-146s. They gave valuable service since the first CH-118s appeared in 1968. This type equipped Base Flights at Moose Jaw, Cold Lake and Bagotville till the CH-146 took over in the mid-1990s (Base Flight Moose Jaw disbanded). CH-135s served with 424 Squadron in the 1970s and at Base Flight Goose Bay. These utility units later were re-designated 421 Squadron (Goose Bay), 417 (Cold Lake) and 439 (Bagotville). Here CH-118 No. 104 departs Cold Lake on May 26, 1993; while the CH-146 is seen in a snow quall at Yellowkife on May 4, 1996. Note that each is equipped with the Canadian-designed wire-strike protection system—the knife-like projections atop and below the nose. (Larry Milberry)

kits to Childers and Raynor, but they made no effort to retrieve them. They were relatively safe in their raft, but also were sapped of energy. The best thing to do was to direct the nearest boat to the site. This was carried out by the Aurora; about two hours after initial contact, the fishing vessel *Zandberg* arrived to pluck the exhausted Bonanza crew from the swells. The Atlantic should have been their final resting place, but great SAR planning and coordination; the professionalism, dedication and bravery of many aviators; and good fortune turned tragedy into salvation.

SAR Childers was not the only incredible SAR in modern times—Christmas 1979 was special for the pilot of Mooney M20J N3757H. Three days earlier he had ditched in the Atlantic 800 miles east of Newfoundland. He managed to deploy his dinghy, was spotted by SAR aircraft three hours later and picked up by the Soviet vessel *Ukshakov* 10 hours after ditching. This incident proved the wisdom of flying well-equipped over open water. The pilot carried a dinghy, sea survival suit and personal ELT (in addition to the one on the aircraft). He used all these to best advantage and lived to celebrate Christmas at home. If even one of these items had been missing, he would not have survived. The two people aboard Cessna 402 C-GDTW were not so fortunate. They departed Goose Bay on January 19, 1981 for an 8.5-hour flight to Reykjavik. They were heard from 4:35-hours later, but failed to arrive in Iceland. No trace was found of aircraft or crew.

No. 103 Squadron in 1996

In its annual résumé for 1996, 103 SAR Squadron reported a unit strength of 97 personnel including 13 pilots, eight flight engineers, 11 SARTechs and 59 in maintenance. Three Labradors were on strength. They flew 1,592 hours, including 557 on SAR operations. There were 81 SAR missions: 24 marine distress, 23 marine medevac, 12 land medevac, six air distress, five ELT, six missing persons, two aid to civil power and three SAR patrols. The first notable tasking for 1996 occurred on January 29—a medevac from Belle Island to St. John's. Hereafter there were many call-outs, a percentage of which were inconclusive due to factors such as weather and aircraft unserviceabilities. Typical was the case on April 10 of which the

squadron diarist noted: "Rescue 303 was tasked to investigate an overdue sealing dory, approximately 20 nm offshore from St. Anthony. Nighttime en route weather deteriorated and combined with aircraft equipment unserviceabilities, forcing the crew to return to Gander and the mission to be called off." Floods around Flat Bay brought 103 to the rescue on February 18; 25 people were airlifted to safety. An April 11 search for three overdue seal hunters was cut short, when the men were located by CCG *Humphrey Gilbert*. A week later several missing sealers were located and transported by Rescue 304 to Main Brook. At the same time 103 was on a high degree of readiness as the lobster season opened. On April 20 R303 stood by a foundering fishing vessel till CCG *Anne Harvey* arrived to assist.

On May 4 a call came from the high seas. A seaman was severely injured 300 miles off shore. R303 was despatched with a Herc from 413 Squadron, hoisted the patient aboard and returned to St. John's. On May 6 pumps were lowered to a vessel in distress 80 nm off shore. R303 had to depart with minimum fuel, but a 413 Herc kept station till CCG *Cygnus* arrived. On May 30 R305 flew to Kuujjuaq to search for a missing light plane. After two days the plane was located with no survivors. In a night SAR 200 nm southeast of St. John's R305 hoisted a seaman, while a 413 Herc illuminated the scene with flares. Often bad weather complicated SAR efforts, sometimes making it impossible to locate a vessel. On July 17 R305 flew 140 nm off shore where a seaman required assistance. Several approaches were made looking for the ship, but R305 had to return to St. John's empty-handed. Three hours later another effort was made, this time successfully.

In July 1996 the Saguenay area was devastated by floods caused by a combination of heavy rains and poorly-built dams. Among the many SAR resources tasked to assist in the disaster was 103, which logged 30 hours, evacuating 142 people in distress. An ELT hit on July 22 sent R305 to the west coast of Newfoundland, where a light plane was found crashed on a hillside with no survivors. On August 10 a missing hiker in Terra Nova Park led to R303 being scrambled. The hiker was found dead the same day. On September 9 an ELT caused R306 to be scrambled, but the signal was false. It was

traced to a vessel at a fish plant on the Burin Peninsula. Three days later R303 investigated a small boat going in circles in Long Harbour. No trace was found of the two occupants. On September 17 a hunter suffering heart failure was medevaced from Buchan's Junction to Grand Falls by R303. The next day R303 medevaced two men injured in a mine explosion, carrying them to St. John's. Using searchlights and night vision goggles, R306 assisted in finding two missing dorymen on Tweed Island on October 1. R305 and R306 departed for Goose Bay the same day on SAR Walker—a missing Beaver. A 103 SARTech spotted an oil slick on the eighth day, then divers confirmed a sad ending to the story—the Beaver lay 100 feet below the surface with all aboard dead.

On November 6 a single-engine plane reported fuel transfer problems 200 nm east of St. John's. R305 was despatched and escorted the plane to St. John's. November 13 saw R305 tasked to northern Labrador, where a 444 Squadron Griffon was missing. It arrived as a 413 Labrador was rescuing the Griffon crew, who had ditched in bad weather. R305 remained in the area to support the accident investigation team. The last tasking of the year for 103 came the day before Christmas—a bulk carrier needed assistance 80 nm off the west coast. Bad weather en route forced R301 to abort the mission.

Besides its usual operations, 103 participated in many activities. In one case its personnel undertook to clean up the old Sabena DC-4 crash site—St. Martin-in-the-Woods. On September 18, 1996 a new memorial was erected there through the efforts of CFB Gander. Various members of 103 took part in SAR exercises from the Canary Islands to Russia and Hungary; and 103 hosted the 1996 SAREX. It won the Diamond Trophy (overall SAR proficiency trophy) for the second year running. It also was awarded the trophies for team spirit and for the medical exercise event.

Demise of the CFE

The fall of the Berlin Wall brought great changes overnight. Nations that had been totally oppressed since as early as 1917 felt a bit of sunshine for the first time. But the collapse of Soviet communism had side effects. Without the old restraints that day-to-day communism

A typical scene at Lahr in the 1980s when this was Canada's major NATO air transportation centre. Over the years 707s, Hercs, Cosmos, Dash 7s and a host of other types were handled daily. Here 13702 of 437 Squadron refuels before departure for Ottawa and Trenton. (Larry Milberry)

put on individual freedoms, crime swept unfettered across the former USSR. New mafias arose. In one case, an entire nation—Albania—was ruined within a few weeks by old-fashioned flimflam artists. The West also suffered difficult changes in this period. With no more "evil empire" to defend against, NATO and other alliances started downsizing militarily. Canada was part of this movement. Within a few years its once powerful European forces were dismantled. Baden-Soellingen and Lahr, Canada's two German bases, closed. Infrastructure and equipment was sold to local bidders and the last Canadian came home.

Canadian Forces Europe (CFE) paraded for the last time at Lahr on July 13, 1993. It was a nostalgic affair with German, French and American troops swelling the ranks. CFE HQ disbanded on July 31. The last CF-18s already had left Baden, which then became a Lahr detachment. Only 400 service and civilian employees, and dependants remained at Baden, while Lahr had about 800. This compared to a CFE peak during the Gulf War of some 18,000. Lahr's airfield closed October 1, when ATC services ended. By this time greater Lahr had a population of about 36,000. The Canadian pull-out hurt. From taxi drivers to shop and hotel keepers, everyone seemed gloomy—the economic rug had been pulled from under the local economy. The future of the *flugplatz* was undecided; there were rumours of it becoming a regional airport, an industrial park or a housing complex for Germany's economic refugees. Last-minute activity at Lahr included CADC's cash-and-carry sell off of anything that wasn't nailed down. People came from miles around to buy furniture, office equipment, kitchen fixtures, etc. A Dutchman showed up with $1 million to buy 50 10-ton trucks and dozens of armoured personnel carriers (wanted mainly for their diesel engines).

On July 26, 1993 the last CFE Cosmo trip was despatched from Lahr by No. 1 ATU (formerly 5 AMU); the last CFE Hercules on August 2. In September the twice weekly Air Canada 767 service flights were curtailed. The occasional CF or UN Herc came through in September, mainly supporting former Yugoslavia. On September 1, 1993 two Hercs were present. CF410 (Herc 315) under Capt Doug Hinton (436 Squadron) was preparing to leave for Shearwater via Shannon with Sea King spares. CF412 (Herc 327) under Capt Tony Milani (429 Squadron) was in from Zagreb heading for Prestwick with 27,000 pounds—soldiers' personal kit. Milani's crew included Capt Dave Wrathall (FO), Capt Martin Dufort (Nav), WO Frank Payeur (FE) and MCpl Jean-François Lapointe (LM).

CF412 departed for Prestwick at 1600Z. Seventeen minutes later it levelled at 18,000 feet; it chugged along

With CF412 on September 1/2, 1993. First, an overall view on the flight deck en route Lahr-Prestwick. Capt Tony Milani (AC), Capt Dave Wrathall (FO) and WO Frank Payeur (FE) are in their seats, while the loadie, MCpl Jean-François Lapointe, stands. Then, Milani checks his charts as the flight passes 12,000 feet. Finally, Milani, Wrathall, Lapointe and Dufour arrange for ground services at Prestwick. (Larry Milberry)

till cleared 55 nm back to descend. It slid onto R31 at 1908Z, two minutes ahead of ETA. Once-booming Prestwick was by now a quiet spot, so the local fuelling company was happy to see a Canadian Herc. The crew laid over in the nearby village of Ayre, where many Canadians had been stationed during the war with famed RCAF squadrons like 401 (Hurricanes) and 410 (Mosquitos). A pub crawl ensued and at one hotel the crew was beset by a coach-load of senior ladies, who regaled the ATG "heroes" with wartime tales.

CF412 started the next morning at 1118Z. It was airborne 11 minutes later off R13 and climbing for FL160. It started west roughly along the 56th parallel, then gradually arced south once beyond the 35th meridian. It was 1,858 nm to Gander, so there was lots of time at 280 knots for relaxed chitchat. Every topic under the sun eventually comes up on such a trip. Today things started out all about clouds—there had been some beauties as the flight left the Scottish coast. As the Herc swished along on the cloud tops, one of the boys mused, "Wow, you don't get to do this VFR!" As always there was plenty of discussion about procedures and manuals, various characters past and present in ATG, women, sports, good spots to eat and drink in whichever country, good or not-so-good books to read, etc. Meanwhile the paperwork proceeded apace, mainly at the nav's table and on the FE's clipboard. Everyone was keen on the weather (especially the winds), course, fuel burn and speeds. Today's flight was off to a good start, with a 10-35 knot push (tail wind) for the first three hours.

On the chitchat frequency CF412 eavesdropped as crews gossiped back and forth. Two American Airlines pilots were yakking about their hero, news commentator Rush Limbaugh. Meanwhile, fellows on the Herc razzed each other good-naturedly, enjoyed their hot meals, and filled up on coffee. They talked to Boston, then Montreal ATC centres and heard of rotten weather at Gander, with gusts to 43 knots. On arrival, however, everything was fine. CF412 landed after 6.3 hours aloft. While the Herc refuelled, the crew wandered around an empty Gander terminal. Then CF412 set off to do battle with some heavy weather on the final leg (1,005 nm). The closer it got to Trenton, the worse the weather. One cell after another forced the Herc to dipsy-doodle, and fly ever lower. 75 miles back the crew talked to Toronto Centre, then contacted Trenton tower. It started down 41 miles back and landed smoothly on R24 after a four-hour slug. The Herc pulled onto the tarmac in front of the AMU and soon the crew split and headed home. The folks at the AMU started offloading the hundreds of soldiers' boxes, and servicing began its checks. So ended one of ATG's last flights from Lahr.

Aircraft Disposal

The military periodically disposes of surplus equipment. The greatest period for this was in the late 1940s, when thousands of aircraft were available for scrap or resale through the War Assets Disposal Corporation. Many small flying services got started with an Anson or Crane bought for a few hundred dollars. Bigger companies like Canadair and DHC profited by refurbishing DC-3s and Cansos to sell around the world. The 1970s-80s was another period of disposal. Through the Crown Assets Disposal Corporation the air force trimmed its fleets drastically, especially of obsolete types like the Expeditor and Dakota. Dozens joined commercial operators. By the late 1980s few Canadian military aircraft were left for disposal. AIRCOM, however, was under pressure to rationalize its fleet—it was too costly having many small fleets, e.g. two Dash 7s. Soon types like the Otter, Dakota, Dash 7 and Tracker retired. All found eager buyers. Several Dakotas went to Basler in Wisconsin for conversion to PT6s. The Otters were snapped up by bush operators. The Trackers went to Conair for its water bomber program. The collapse of the USSR also hastened Canada's military downsizing. This was the so-called peace dividend—if Canada no longer had the USSR as an enemy, it hardly needed such a big air force. The defence budget could be cut and the money spent elsewhere. In this period aircraft were offered for sale (e.g. CF-5s, some of which were sold to Botswana); others, like many CF-18s, were placed in storage.

In 1994 some of the last surplus CanForces aircraft were on the chopping block—seven Cosmopolitans, four Buffalos and three Twin Otters. Rumours of the demise of the Buffalo had been heard since the 1980s. It was increasingly expensive to maintain, and downtime was excessive by the early 1990s. On the other hand the VIP Cosmopolitans recently had been overhauled and fitted with EFIS (electronic flight instrument system) cockpits by Kelowna Flightcraft. But Ottawa decided that it no longer could justify this high-maintenance fleet. As for the Twin Otters, they too fell to budget cuts, even though they were a valuable, low-maintenance type.

The Buffalos for sale were Nos. 115453 (airframe hours: 13,654), 115485 (15,784), 115459 (15,991) and 115463 (14,944). In 1993 they were stored at Mountain View, near Trenton. Attempts to sell them that year failed. Eventually they found buyers, later being heard of in Zimbabwe, up for resale. The Cosmos were the survivors of 10 RCAF aircraft, two of which eventually were scrapped, another destroyed by fire at Dorval. They were built by Canadair in 1960-61: 109151 (airframe hours: 16,877), 109152 (18,861), 109154 (18,746), 109156 (21,866), 109157 (21,950), 109159 (20,884) and 109160 (21,325). The fleet was stored at CFB Trenton in mid-1994 and was described in a CADC document as "maintained in flyable preservation state... except that the gaseous oxygen is not being recharged." A DND summary of June 29, 1994 described the Cosmo fleet as "considered to be flyable", although a special note was made about 10915: "has been susceptible to unusual airframe vibrations in the past."

Twin Otters 13803, 13808 and 13809 were offered in September 1994 with 18,326, 20,478 and 17,802 hours respectively. They had joined the air force 1971-73. Mods in 1986 increased their AUW to 14,000 from 12,500. Many upgrades were made over the years, e.g. in 1988 the aircraft (according to CADC) "were modified to provide enhanced crash-worthiness by increasing the energy absorption capability at the attachment of the wing rear spar to the fuselage to retard wing movement when the aircraft is subject to abnormal deceleration. Aluminum wing attachments were replaced in 1987 with improved steel adapters. In 1991 the top and bottom horizontal stabilizer centre hinge location was reinforced with a doubler to prevent skin cracking." The Twin Otters found ready bidders among bush operators.

Aerial Refuelling

With Operation Rhine Prosit in January 1993 Canada pulled its CF-18s from Germany. This ended its NATO air support role, begun with the formation of the Air Division in 1952. Nos. 410 and 411 squadrons took their Sabres overseas then aboard ship, but the third squadron, No. 439, ferried to England from Ottawa via Goose Bay, Greenland and Iceland. This was a bold move with single-engine fighters, but not exactly pioneering work—during WW II thousands of aircraft made the trip. Some, however, never arrived, so the RCAF planned carefully for the 439 move. Over the next 40 years Air Division Sabres, CF-100s and T-33s flew back and forth across the Atlantic thousands of times without the luxury of aerial refuelling.

When Canada decided to join in the defence of NATO's northern flank, it committed its CF-5s. Equipped for aerial refuelling, they made their first Atlantic crossing in June 1972 (Exercise Long Leap). The short-range CF-5 required five or six refuellings from accompanying 437 "Husky" Squadron CC-137s (as Canada oddly designated its 707s). Five CanForces 707s logged about 180,000 flying hours over 25 years. The idea for a tanker version was sparked by the NATO commitment, the necessary mods being done by the CanForces, Boeing and Beech. A satisfactory system was ready in 1971—the Beech Model 1080. Aircraft 13703 and 704 were converted. With delivery of the first CF-18s in 1982 the

Canada's Military Twin Otters 1971-97			
Tail No.	c/n	Delivered	
13801	303	July 1971	440 Squadron, Yellowknife 1997
13802	304	July 1971	440 Squadron, Yellowknife 1997
13803	305	Aug. 1971	To N774A, to C-FUGP 3-95, to C-FMOL 3-95
13804	306	Sept. 1971	440 Squadron, Yellowknife 1997
13805	307	Sept. 1971	440 Squadron, Yellowknife 1997
13806	308	Sept. 1971	To C-FSLR 5-94, to N776A 12-94, to C-FTXQ (Ptarmigan Airways) 1-95
13807	309	Sept. 1971	Crashed in Rocky Mountains 14-6-86
13808	310	July 1971	Destroyed in Indian Air Force attack on Kashmir 5-12-71
13809	382	Nov. 1973	To N677A 12-94, to C-FUGT 3-95, to Bradley Air Service

commander with a hundred-and-one details to cover—everything from arranging quarters and ground transport, to getting the latest weather and oceanic clearances, and liaising with 437 and 439 Squadrons, Air Transport Operations Centre (ATOC), Fighter Group, etc. The 3,451-nm route from Trenton to Baden-Soellingen was covered in 6.8 hours. When the crew deplaned, there was a sense of foreboding on the base that once had been the workplace for 2,500 Canadians. Fighter ops had ceased the previous December; only a few training missions had been flown since. The last of Baden's T-33s had left for Canada on April 6, 1992 (Capt Ab Lamoureux, Maj Roger Arseneault). With a plan to reduce to a skeleton staff of 150

tankers were busier than ever. In May 1984 the first CF-18s flew to CFB Baden-Soellingen using AR (Exercise Rhine Hornet). In June 1985, 409 Squadron took up residence there with its Hornets. Since then Hornets crossed routinely. During the Gulf War 26 were shepherded to and from Qatar by 437 Squadron.

Operation Rhine Prosit, months in the planning, got under way from CFB Trenton with the departure of tanker 703 on January 16, 1993. Aboard were 24 crew and 18 passengers. The mission commander was Maj Chris Beaty, 437's USAF exchange officer. Also aboard was Capt Rusty Wright, the detachment

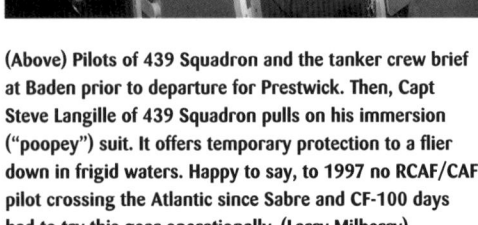

(Above) Pilots of 439 Squadron and the tanker crew brief at Baden prior to departure for Prestwick. Then, Capt Steve Langille of 439 Squadron pulls on his immersion ("poopey") suit. It offers temporary protection to a flier down in frigid waters. Happy to say, to 1997 no RCAF/CAF pilot crossing the Atlantic since Sabre and CF-100 days had to try this gear operationally. (Larry Milberry)

(Right) Sgt Terry Odell at the FE's station during Op Rhine Prosit on January 16, 1993. Then (below), Sgt D.M. Knockleby (FE) at the air-to-air observer's station on January 18, controlling fighters coming on and off the starboard refuelling boom. (Larry Milberry)

(Above) Tanker crew flight plan at NAS Keflavik on January 20, 1993. From the left are Capt Todd Guenther (navigator), Maj Kevin Keogh (navigator), Capt Lee Williams (pilot), Maj Chris Beaty (mission commander) and Capt Rusty Wright (detachment commander). Williams was typical of 437's experienced crew. Before joining 437 he had flown Buffalos and Hercs, instructed at Moose Jaw and worked in the Rescue Coordination Centre at Trenton. For 1992 he logged 400 hours on the 707, flying to places like Brussels, Djibouti, Islamabad, Minsk, Nairobi and Zagreb. (Larry Milberry)

(Left) The pod techs inspect their equipment during downtime at Keflavik on January 22, 1993. (Larry Milberry)

at Baden by August 1993, there was no jubilation at 439 as it prepared to fly its Hornets home.

On January 18 pilots involved in Rhine Prosit flew an AR training mission. This was done in a tight 30-mile racetrack pattern with the tanker at 250 KIAS at 29,000 feet. Hour after hour Hornets rendezvoused to fly the carefully briefed profile, then returned to base to "hot refuel" (i.e. engines running) and change pilots. This got everyone ready for the business of flying 24 Hornets 3,500 miles to Goose Bay. That afternoon there was a mass briefing to discuss the departure for Canada. Next day the last Air Division Hornets blasted off. There were some nostalgic fly-bys, including one over downtown Baden-Baden, then the fighters set course for Prestwick, eastern staging base for Rhine Prosit. Capt Tom Jackson had to return to base with a pressurization problem, but the rest was routine. Tanker 703 followed. As it taxied, it passed one HAS (hardened aircraft shelter) after another. The bombproof HASs, each of which once housed a Hornet, now sat empty.

The tanker made the 1.7-hour flight to Prestwick to deliver groundcrew and equip-

A 437 Squadron tanker with its refuelling equipment deployed for an AR exercise on January 12, 1994. (Tony Cassanova)

ment, then refuelled and pushed on to US Naval Air Station Keflavik in Iceland, two hours away. The fighters overnighted in Scotland. The first four launched in early morning darkness on the 20th to rendezvous west of Iceland with 703 at 32,000 feet at 400 KIAS. Each took 10,000 pounds of fuel (total capacity was 17,200 with three external tanks), then flew with the tanker till off the tip of Greenland.

There the tanker turned for Keflavik. Meanwhile, Auroras from CFB Greenwood tracked back and forth below. These were the "duckbutts", or stand-by rescue aircraft, equipped with SKADs ("survival kits air droppable"). In the event of a fighter pilot ending in the ocean, the Auroras soon could be on site with SKADs and vector ships to the site.

Several AR trips were made as the Hornets

A Hornet four-ship departs the tanker after refuelling over Germany on January 18, 1993. Then, a close-up of Hornet 764 in splashy 439 colours as pilot Tom Chester prepares to take on fuel. (Larry Milberry)

(Above)) Tanker 704 landing at Trenton on September 10, 1983. The last CanForces 707s were retired from service on April 1, 1997. All five went to the USAF. (Larry Milberry)

(Left) AIRCOM acquired four KC-130H tankers in 1992. Operated by 435 Squadron at Edmonton (then Winnipeg) they supported Canada's CF-18s on domestic ops, but lacked the strategic capability of the 707. Here No. 339 sits at LBPIA on September 6, 1992. Note its AR pods outboard of the engines. (Dave Thompson)

(Right) Hercules tanker "Oiler 38" (435 Squadron, LCol Peter Nodwell aircraft commander) refuels 425 Squadron CF-18s near Bagotville on July 15, 1993. The AR Herc carries 90,000 lbs of fuel, 23,315 in a quick-change 19' x 6' aluminum tank in the hold. Each 170" x 34", 1,343-lb underwing AR pod contains 78 feet of 2-in. hose. Maximum refuelling rate is 300 gal/min. The Herc tanker has an operating envelope of 188-250 knots. (Larry Milberry)

flew the ocean in fours. The tanker had two crews that alternated morning and afternoon. On January 21 tanker 704 arrived from Trenton to speed up operations. There were some weather delays at Prestwick and Keflavik, and a ground handling hold-up that caused four Hornets to overnight at "Kef", but things generally went as planned. Early on January 22 tanker 704 landed in Goose Bay behind four Hornets. All refuelled, then 704 took off with four more Hornets for Cold Lake. Other Hornets already had terminated at Bagotville. Tanker 703 followed into Goose Bay. Operation Rhine Prosit was in its last hours. Forty years of Canadian fighter operations in Germany were history. Not a shot was fired in anger by any of Canada's fighters. In everyone's heart was a feeling of a job well done. After all, the Iron Curtain had collapsed. This was a good way to win a war.

Zagreb

Another successful ATG operation was supporting UN peacekeeping in former Yugoslavia. On Canada's part this included the Hercules, 707 and Airbus. In time this op exceeded the Berlin Airlift in scope. In the spring 1993 ATG *Newsletter* Maj D.M. Duggan wrote of ops into Sarajevo, Serbia. At the time ATG's airlift control element was at Zagreb, Croatia. Because of tensions there it later moved across the Adriatic Sea to Ancona, Italy:

Flying into Sarajevo remains challenging. Since the beginning of the UNHCR airlift op-

0500 local. Aircrew arrive at 0530 to begin preflight checks, complete flight planning and, most importantly, receive their update intelligence briefing. The Canadians are assigned three slot times daily into Sarajevo from the UNHCR operations centre in Geneva. Loads typically weigh close to 35,000 pounds and comprise dry food, canned food, water, condensed milk, blankets and clothing. Most of the cargo is transported to the ALCE by truck; however, other quantities are airlifted by huge military or civilian transports. It is not uncommon to see a civilian B.747, USAF C-5 or Russian Antonov offloading more aid, all of which is destined for Sarajevo Airport and is relayed throughout Bosnia-Herzegovina...

Crews routinely depart Zagreb at 0615 to meet their first Sarajevo arrival time of 0730. Ground operations are normal at Zagreb with the exception of ensuring the CMS equipment is serviceable and personnel flak jackets are on board. After departing the airhead, crews contact Zagreb radar for normal air traffic control flight following. Then they contact "Magic", a NATO AWACS aircraft for information on all airborne traffic within Bosnian and Yugoslavia airspace.

The UN-directed routing is via the Adriatic coastline to a VOR at Split. Crews perform a combat entry check at this point, ensuring all mobile personnel don helmets and flack jackets. After passing Split VOR, the crew sets an easterly course for Sarajevo. Approximately 35 miles from Sarajevo the crew contacts the French-controlled airport tower to receive

nearby shelling, all of which is a constant reminder of why we are there. These few moments spent waiting for our loadmaster to state "unloading complete", are moments spent watching Serbian and Muslim stronghold positions just 800 meters away with small arms at the ready.

Once offloading is complete, little time is spent preparing for take-off... With an empty CC-130 and full power applied, acceleration is less than what a fighter jock would experience, but nonetheless responsive to our needs. Approximately 2,000 feet later the venerable Charlie 130 is streaking skyward... Occasionally flares are discharged during this critical phase of flight and all the while the crew maintains a vigilant climb schedule and thorough lookout for any signs of aggression. After a safe altitude is achieved, the crew resumes a normal climb profile, re-acquires AWACS and Zagreb control and continues the mission back to Zagreb for another load...

Based at Ancona, across the Adriatic from Split, ATG's one-Herc detachment shouldered the load with the RAF (a Herc) and the Luftwaffe (a Transall). Their theatre of operations was not a friendly place; during the UN effort an Italian Transall was shot down, and a UN IL-76 badly shot up by some of the local kooks. This explained why the Ancona aircraft carried CMS, flares, and armour plate in the cockpit and at rear crew stations. When over Yugoslavia, crew wore helmets and flak vests. On November 17, 1994 the Canadians were reminded of the importance of following war zone procedures—their Herc was hit by small arms fire as it departed Sarajevo. The crew had just finished a mission that brought to 25,000 tons the humanitarian relief carried to Sarajevo by ATG.

Help for former Yugoslavia—an ATG Herc and 707, an RAF Herc, a German Transall and a Russian IL-76, all involved in humanitarian missions. In the 1990s the world did what it could to ease the misery of people in this region, but, with heavily armed lunatics, drug dealers, gun runners and mass murderers running it, there was little hope of long-term peace. Many UN peacekeepers, including Canadians, gave their lives in this frustrating operation. Canada remained involved into 1997. (DND)

eration, ATG has used one CC-130 fitted with countermeasure system (CMS) equipment for missions into Sarajevo. This equipment allows aircrew to detect if their aircraft is being targeted by radar-emitting sources or if it is being fired upon by infrared-seeking weapons. Quantities of chaff and flares are loaded for each mission, which further enable the crew to take appropriate action if engaged by enemy sources.

Due to the hostile nature of this operation, ATG employs crews who have received advanced tactical airlift (ATAL) training. As the operation continues, the same crew members are required to return for second, third and even fourth tours. Although taxing for the individuals involved, the frequent rotations back in-theatre create increasing experience levels and margins of safety and effectiveness...

A typical mission begins each day with ALCE personnel arriving at the airfield about

weather and airfield condition updates. Once it is established that the airfield is open, the crew commences either an IMC (instrument) or VMC (visual) approach. During VMC penetrations the crew strives to maintain a high altitude to minimize the time exposed to small arms fire. Periodically, humanitarian airlift aircraft are hit by small arms fire during their approach into Sarajevo. Although it is more difficult to navigate into the airfield during degraded weather conditions, an added measure of security for the crews is provided by the frequent cloud cover and persistent fog since those on the ground cannot see the incoming aircraft.

Once safely on the ground... one engine is left running for precautionary measures since there is absolutely no ground support equipment to assist with engine starting... awaiting offloading, crews can hear and sometimes see

The End of LAPES

Hercules 130321 ("Trucker 5") departed CFB Edmonton at 0955 hours July 22, 1993 for a LAPES training exercise at nearby CFB Wainwright. Capts Mike Allen and Vince Schurman, the pilots, were on their first LAPES trainer, having just arrived on 435 Squadron. Trucker 5 was observed entering the LAPES mode and flaring, but it continued to

Canada's Military Transport Squadrons, May 1997		
Sqn.	Location	Aircraft Types
103	Gander	CH-113 Labrador
402	Winnipeg	CC-142/CT-142 Dash 8
412	Ottawa	CC-144 Challenger
413	Greenwood	CC-130, CH-113
424	Trenton	CC-130 Hercules, CH-113
429	Trenton	CC-130
435	Winnipeg	CC-130
436	Trenton	CC-130
437	Trenton	CC-150 Aurora
440	Yellowknife	CC-138 Twin Otter
442	Comox	CC-115 Buffalo, CH-113

descend, striking the ground nose first. It slid about 650 feet, hit an embankment and broke into three main parts. The wing landed atop the cockpit. The pilots, two loadmasters and an observer from the Canadian Airborne Regiment died. The navigator, a non-flying pilot on instructing duties, and two flight engineers survived. Three of the survivors had been standing on the flight deck.

The DND Directorate of Flight Safety concluded that Trucker 5 was not ideally set up for the LAPES and hit the ground at a high rate of descent—1,000 fpm. So much damage was caused that a go-around was impossible. The DFS commented: "LAPES is one of the most demanding tactical manoeuvres flown with the C-130 aircraft. A successful load delivery requires coordination between the pilot flying the aircraft and the right seat pilot who releases the load. With a planned release height of five to eight feet, there is very little tolerance for error. To be successful, an overshoot must be commenced immediately any time it is recognized that the profile is not progressing normally." In another LAPES mishap all aboard died at Edmonton on November 16, 1982 when the load hung up on Trucker 2 as it attempted a drop.

Canada's Original Hercs: The C-130Bs

Tail No.	TOS	SOC	Notes
10301	5-10-60	29-3-67	435 Squadron. Traded to Lockheed as N4652. Sold to Colombia January 1969 as "1003". Ditched off New Jersey 16-10-82 when crew became lost and ran out of fuel.
10302	5-10-60	29-3-67	435 Squadron. Traded to Lockheed as N4653. Sold to Colombia January 1969 as "1001",
10303	5-10-60	29-3-67	435 Squadron. Traded to Lockheed. Sold to Colombia January 1969 as "1002". Crashed August 26, 1969.
10304	5-10-60	29-3-67	435 Squadron. Crash-landed near Borden, Sask. 15-4-66 when cargo door blew off in flight. Written off and salvaged. Components traded to Lockheed in deal for C-130Es.

Canadian Forces C-130 Crashes

Tail No.	Sqn.	Date	Location	Misc.
1. 10304	435	29-3-66	Borden, Sask.	Lost cargo door in flight, crash landed, no fatalities.
2. 130309	436	27-4-65	Trenton, Ont.	Crashed on night training flight (runaway trim control), all fatal.
3. 130312	436	15-10-80	Chapais, Que.	Stalled while turning to check a target while flying low-level on SAR Ryan. Eight of 10 aboard died.
4. 130318	435	29-1-89	Wainwright, Alaska	Crashed short of runway while landing in ice fog.
5. 130322	435	30-10-91	Alert, NWT	Crashed at night when crew failed to monitor their approach.
6. 130329	435	16-11-82	Namao	Crashed when LAPES load hung up part way out the door, all fatal (seven).
7. 130330	435	29-3-85	Namao	Mid air collision with 331. All fatal.
8. 130331	435	29-3-85	Namao	Mid air collision with 330. All fatal (10 aboard both aircraft).
9. 130321	435	22-7-93	CFB Wainright	Crashed during LAPES exercise. Five fatal.

JUNE 10, 1990

R.C.A.F.

DEDICATED TO THE MEMORY OF MEMBERS OF 435 AND 436 SQUADRONS, WHO, IN WAR OR IN PEACE, GAVE THEIR LIVES IN THE SERVICE OF THEIR COUNTRY

C. F.

(Left) These monuments at CFB Edmonton honour the men lost with 435 and 436 Squadrons, especially the crew of Trucker Two, killed during LAPES training. (Larry Milberry)

(Below) Canada's C-130 Hercules had innumerable adventures around the world. The E-models began with the RCAF in 1965, so were into their fourth decade of service in 1997. On April 1 that year a ceremony at CFB Trenton commemorated Hercules 130315 attaining 40,000 flying hours. Its missions over the years sometimes were dangerous, e.g. it was damaged by hostile fire at Kigali early in the Rwandan genocide. Here it was at Nairobi on August 4, 1994. (Larry Milberry)

Herc 330 demonstrating LAPES at Abbotsford in August 1977. The drogue is deployed to pull out the main chutes. In the second photo Herc 326 has just made a LAPES delivery during training at Trenton on June 27, 1992. (Kenneth I. Swartz, Larry Milberry)

(Left) Herc tails in a line at Namao on May 22, 1994. (Larry Milberry)

Members of Canada's Skyhawks parachute demo team leap from a Herc during training at CFB Edmonton in the early 1990s. Then (below), ready for action, members of Canada's elite Canadian Airborne Regiment sprint from a Herc at Petawawa during war games. (John McQuarrie)

(Below) Hercs of 435 Squadron and a Twin Otter of 440 at the AMU at Namao on May 16, 1992. The AMU recently had opened, but within three years the air force left Namao (except for 408 Squadron) and the AMU closed. (Larry Milberry)

The scene at Namao on May 7, 1994 as 435 Squadron celebrated its 50th anniversary. Many original members from Burma days attended. (Larry Milberry)

(Right) A dramatic view of Hercs during a Talex (tactical airlift exercise) at CFB Trenton. (John McQuarrie)

(Below) Trenton, on the Lake Ontario shore about 100 miles east of Toronto, was the traditional home of air transport in the RCAF and CF. This view of the base was taken on September 24, 1993. The town straddles the Trent River where it empties into the lake. The river mouth was first settled in the 1790s, but a town was not begun until 1853. It prospered in the lumber industry, then in munitions during WWI. After the war investors tried turning Trenton into Canada's film industry capital. That failed and little happened until the early 1930s, when land east of town was deleveped as the RCAF's most modern station. WWII brought a boom to the local economy. Thereafter, Trenton was the RCAF's focus for air transport. Activity peaked when 426 Squadron moved there from Dorval in 1959. (Larry Milberry)

(Above) Herc 323 refuels at Iraklion, Crete on November 23, 1991. It was en route Lahr-Djibouti during humanitarian relief operations in Ethiopia. Then, Herc 325 loading a CanForces twin Huey at Tegucigalpa, Honduras on November 27, 1990 during the UN peacekeeping mandate there. The Twin Huey was being returned to CFB Gagetown. (Larry Milberry)

(Left) A C-141 Starlifter delivers freight to CFB North Bay on April 25, 1980 as the DOT's Viscount CF-GXK taxis. In 1965 there was a rumour that the RCAF would acquire 10 C-141s. Les Edwards wrote that year, "Having recently flown from Trenton to Cyprus and return in a CC-106 Yukon, I can readily understand the requirement for jets if the RCAF is to remain current in the air transport business." The cost of a C-130 in 1966 was noted at $2 million compared to $10 million for a C-141. The RCAF had a contract in July 1967 for four C-141s, but Ottawa dillydallied and Lockheed closed the C-141 line to make room for C-5 production. (Dave Thompson)

937

On April 10, 1970 the first 437 Squadron Boeing 707 combi (CF designation CC-137) was delivered to Trenton; 27 years later the last of the five-plane fleet retired. Over the decades the 707 did outstanding work with ATG, flying passengers and cargo, and doing aerial refuelling. In this scene from January 27, 1987 a CC-137 lets off its passengers at Trenton. For the retirement of the 707, tanker 703 was specially festooned with a giant Husky. (Larry Milberry, Tony Cassanova)

Airbus A310 15004 of 437 Squadron at Trenton in April 1997 with its new cargo door and paint job. Formerly C-FNWD "Jack Moar" of Wardair/CAIL, 15004 was purchased by the Canadian government in February 1993. (Tony Cassanova)

(Above) Capt Daryl Smith and Maj Tom Adkins were the pilots aboard 437 Squadron Airbus 15002 on March 15, 1993. Closest is CAIL engineer Don Gil. When ATG acquired its A310s, a staff of more than 60 CAIL people moved to Trenton to maintain them. Every flight was accompanied by two seasoned CAIL men who were de facto FEs. (Larry Milberry)

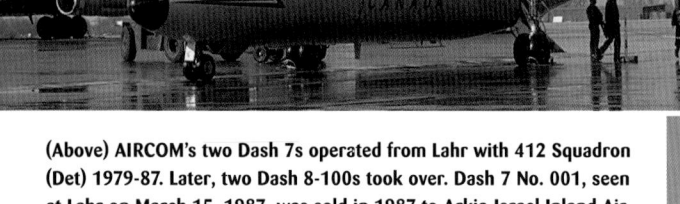

(Above) AIRCOM's two Dash 7s operated from Lahr with 412 Squadron (Det) 1979-87. Later, two Dash 8-100s took over. Dash 7 No. 001, seen at Lahr on March 15, 1987, was sold in 1987 to Arkia-Israel Inland Airlines. (Larry Milberry)

(Right) In the late-1990s AIRCOM had six Dash 8s—two transports for general duties and four especially equipped for training navigators. They all were based at Winnipeg, where the CF Air Navigation School and 402 Air Reserve Squadron operated them. Here nav trainer 804 lands at Winnipeg on January 9, 1993. (Larry Milberry)

(Left) Seven Dassault Falcon 20s served 412 and 414 Squadrons and the CF Airborne Sensing Unit 1967-90 as VIP, EW training, and remote sensing planes. In 1990 they were retired and Challengers took over their duties. Dave Thomspon photographed the ASU's spiffy-looking Falcon at Trenton in January 1972.

Challenger 614 of 412 Squadron in its attractive VIP colour scheme. It was at Quebec City on August 18, 1991. Then, No. 610, an electronic warfare trainer with 414 Squadron, in its workaday military scheme. (Larry Milberry, John McQuarrie)

(Below) The Bell Kiowa (i.e. Bell 206 Jet Ranger) was Canada's light military helicopter 1971-96. It replaced the last of Canada's Hiller UH-12 Nomads and Cessna L-19 Bird Dogs. There originally were 74 of the Bells, then 14 were added in 1981 for the Basic Helicopter School at Portage la Prairie. The Kiowa was retired in 1996, the last giving way to new CH-146 Griffons. These 408 Squadron Kiowas were at Namao on May 22, 1994. (Larry Milberry)

(Above) A German Army Bo.105 chases a 444 Squadron Kiowa during NATO training in southern Germany. (John McQuarrie)

(Right) Base Flight Moose Jaw CH-118 Twin Hueys at home on July 11, 1992; then (below), Capt Dwayne Lovegrove of 417 Squadron (formerly Base Flight Cold Lake) ready for a CH-118 mission on April 26, 1993. A rough and tumble bunch, "rotorheads" like Lovegrove will always tell you that they have the best job in the world. (Larry Milberry)

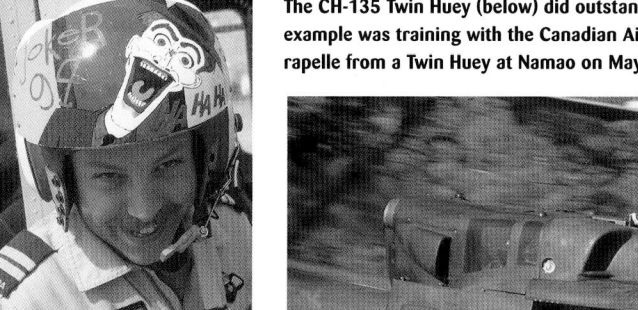

The CH-135 Twin Huey (below) did outstanding work since the first of 50 was accepted in 1971. This example was training with the Canadian Airborne Regiment at Petawawa. In the second scene, troops rapelle from a Twin Huey at Namao on May 22, 1994. (John McQuarrie, Larry Milberry)

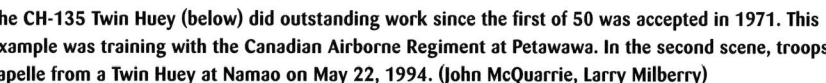

(Above) This S-55 had been RCAF 9624 and did some of the early Mid Canada Line work with 108 (Comm) Flight. It became CF-JTE in March 1965 and had a long civil career thereafter. Andy Graham photographed it at the Dominion Helicopters base on Hwy 400 at the King Side Road (north of Toronto) in November 1966. The RCAF also operated six S-58s 1955-71. Turbo Tarling shot 9632 at Cold Lake in June 1958.

(Right) In 1963-64 Canada acquired 18 Boeing-Vertol CH-113s, 12 for the Army (CH-113A Voyageurs), six for the RCAF (CH-113 Labradors). These were the Army's first medium-lift helicopters. For the RCAF they meant the end for its well-worn S-58s and H-21s. Here Voyageur 412 attracts a crowd during a stopover in a park near Beamsville, Ontario, where it was delivering prime minister Pierre Trudeau. In time the CH-113s/113As were converted to SAR specs and known as Labradors. Of the original machines, 13 remained on CF strength in 1997. (Hugh Halliday, Larry Milberry)

With the Voyageur, the CF's old H-21s were sold. A few then saw commercial use, mainly in logging, but their second wind didn't last. This example was on a dump in Fort Simpson in August 1979. (Brent Wallace)

Chinook details from 447 Squadron in 1991: one of three 500-Imp.-gallon rubber fuel bladders for cross-country trips; then, one of the Chinook's Lycoming T55-L-11C turbine engines. The "11C" was a hybrid that eventually was difficult to support with spares. (Larry Milberry)

(Above) The Boeing Vertol CH-147 Chinook had a productive career in Canada since entering service in 1974, but fiscal cuts forced its premature retirement in 1991. Canada had nine Chinooks over the years. With their special mods they were dubbed "Super Cs". Mods inluded a 28,000-pound hook for slinging, an external hoist, auto flight control system (for maintaining heading), speed trip system (to keep the aircraft in trim at high speed, when Chinooks tend to fly nose down), stability augmentation system (three-axis auto stability), CPI (crash position indicator) and T55-L-11C engines (unique to CF Chinooks). The Chinook served with distinction with 447 Squadron at Namao and 450 Squadron at Uplands. In this John McQuarrie photo, a Chinook is about to sling a light truck during exercises at CFB Wainwright.

The three 447 Squadron CH-147 Chinooks about to depart Edmonton on April 8, 1991 for storage in Mountain View. All seven surviving CF Chinooks soon were sold to the Netherlands. They went first to Boeing in Philadelphia to be upgraded to CH-47D standards (an option turned down by Canada). In this view, aircraft 007 is nearest with the crew of Maj Krayer and Capt Jule (pilots), MCpls Orcutt and Medland (FEs) and MCpl Langlois (loadie). At 28,800 pounds (empty), a CH-147 weighed about as much as a loaded DC-3. (Larry Milberry)

The Turbine Helicopter in Canada

Helicopter history falls naturally into piston and turbine eras. Piston types appeared in the 1930s, the first practical commercial designs appearing post-WWII from companies like Bell, Bristol, Hiller, Kaman, Kamov, Mil, Piasecki and Sikorsky. By the mid-1950s helicopters were accepted, in spite of their obvious limitations. Then the lightweight turbine engine appeared, giving the helicopter new importance.

Okanagan Helicopters of Vancouver kept on the leading edge of rotary-wing flight. It ordered Canada's first turbine-powered verticraft—the 70-seat Fairey Rotodyne, first flown in late 1957. Its giant four-blade rotor was driven by blade-tip jets supplied with compressed bleed air from two 3,000-shp Eland turboprops. The Elands (with standard propellers for forward thrust) were on stub wings. In January 1959 the plane set a rotorcraft world speed record—185 mph. Okanagan applied to operate Rotodynes to Vancouver, Victoria and Seattle. When delivery slipped, it filed with the ATB to fly a twin-engine helicopter instead. The ATB noted: "This application will be proceeded with if the route studies now being made and the economics of the helicopter under consideration satisfy your Directors that the operation can be conducted without direct subsidy." In 1962 Fairey dropped the complex, expensive Rotodyne. It would be another generation before Agar's interurban concept reached fruition.

The first practical turbine helicopter was France's Sud Alouette II, which flew in March 1955 with a 400-shp Turbomeca Artouste (successive versions had more power). The Alouette II entered service with the French Army in 1956; about 1,000 were on order by 1962. The first Canadian example was CF-KMW, acquired by Autair of Montreal in August 1958 (lost in a ground mishap in May 1959). Next came CF-JMC, bought by the Ontario government in June 1959; it crashed that September. Other Alouette IIs came to Canada, e.g. in 1964-65 for Bullock of Calgary and Lac St. Jean Aviation of Quebec City. Alouette IIIs (790-shp Artouste) CF-CAW, 'X and 'Z went to the DOT 1965-66. Allegedly bought in connection with France's bid to sell Sud Caravelles to Air Canada, they served the Canadian Coast Guard for more than 20 years, mainly supporting marine navigation on the BC coast. (Upon retirement they went to trade schools as classroom training aids.) The

The Bell 204B, certified in April 1963, brought Canadian commercial helicopter operations into the turbine era in a big way. With its 1,100-shp T53 engine it could lift 3,000 pounds, better than an Otter. This example (C-GTNP) was working on a 2,000-acre fire near Fort Good Hope, NWT in July 1986. (R.S. Petite)

A Kamov KA-26 at the CIAS on August 30, 1969. The first Russian helicopter in Canada, it was one of the rare light piston twins. Ross Lennox flew it in Montreal; he liked its handy rear-loading arrangement, although it was noisy. Idiosyncrasies included long landing gear oleos (making it useless in sand, snow, swamp, etc.), and drooping rotor blades when at rest. No KA-26s were sold in Canada, although hundreds served Aeroflot. Three decades later the Russians still were trying to make inroads in Canada, the mighty KA-32 turbine logger being their first hit. (Larry Milberry)

(Right) One of the great SA318C Alouette IIs. CF-EDG originally came to Canada in 1972 for Northern Helicopters of Abbotsford, BC. Here it was in May 1977 with Darvill Copters of St. Albert, Alberta. Then, Quasar SA316 Alouette III C-GSHF at Edmonton Muni on August 5, 1977. When Northern Helicopters folded in 1975, Calgarian Ralph Hancock bought its assets and formed Quasar. By late 1977 it had Alouettes, Bell 47s, Bell 206s, Hughes 500s and Sikorsky S-55Ts. It subsequently was purchased by Dan Dunn, a former Okanagan vice president. It introduced the Bell 214ST in the Beaufort, had Bell 214B-1s on logging and construction, and supported the Polar Continental Shelf. In the 1980s Quasar fell on hard times and folded, its assets becoming the basis of Campbell Helicopters. The bottom photo shows a CCG Alouette III at Vancouver. (Kenneth I. Swartz)

The DOT's S-62 at Prince Rupert. This type made its mark in SAR with the US Coast Guard. In Canada it soon was eclipsed by the S-61, which offered the twin-engine reliabilty needed in Arctic and offshore flying. (NA/McNulty Col. PA189940)

(Above) The original Bell XH-40 from which came the legendary "Huey" and Bell 204B. (Bell Helicopter Corp 217234)

(Left) Bow Helicopters' Bell 204B CF-BWR working a fire near Lac laBiche, Alberta. It came to Canada for Bullock in 1965 as one of Canada's first 204s. (Paul Kristapovich)

Alouette revolutionized Canadian helicopter operations, especially in the Arctic, where turbine engine starts in the coldest of weather were simple. Even so, it was expensive to buy, and customers needed time to adjust to higher charter rates. Being a European product, the Alouette met resistance from loyal Bell, Hiller and Sikorsky operators, who were wary about product support for a European design. They did not appreciate how early Artoustes had to be shipped to France for service.

The next practical turbine was Sikorsky's S-62, flown in May 1958. With the rotor and transmission of the S-55 and a 1,050-shp GE T58 it carried two crew and 11 passengers at 100 mph. A watertight hull and semi-retractable undercarriage (in outrigger floats), provided versatility, but such a helicopter could be justified only for specialized jobs. The first in Canada was CF-OKA, bought by Okanagan for $310,000. In January 1961 it was delivered from Sikorsky to Canadian Pratt & Whitney in Longueuil; then flew to Goose Bay for training and to relieve Okanagan's three S-58s for overhaul. After one of these crashed, OKA remained in Goose Bay till March 1962, then returned to Sikorsky in exchange for S-58 OKS.

In July 1961 S-62 CF-CAH was delivered to the DOT/CCG for servicing marine nav aids and lighthouses, mainly from Prince Rupert. Its reliable turbine and amphibious nature were reassuring on the long over-water flight to the Queen Charlottes. In time the CCG needed greater over-water and rough weather capability. The ideal replacement was the S-61 (civil HSS-2 Sea King, two T58s) first flown in March 1959. In March 1966 the CCG accepted S-61N CF-CGF. Meanwhile, Okanagan acquired an S-61 for its first North Sea oil support job.

Of the new commercial turbines the Bell 204B was the greatest success. The original 204 began in 1954 as the US Army XH-40. Its design was based on the lessons of the Korean War. It could handle troops, stretcher cases and sling loads; had the rugged Lycoming T53 turbine, and was transportable by C-119, C-130, etc. Bell pilot Floyd Carlson flew the prototype in October 1956. Pre-production models (HU-1s, hence the nickname "Huey") were delivered to the Army in September 1958. In April 1960 Bell flew the HU-1B with an im-

proved rotor system and uprated T53, allowing an 8,500-pound gross weight. It was offered commercially as the 204. In 1964-65 Associated, Autair, Bullock and Okanagan bought the first Canadian examples to serve oilfield drill camps. It was clear to drillers that the 204 could ease their work, lifting heavy machinery and getting it across rivers, muskeg and rough terrain. Big Indian Drilling of Calgary had a two piece drill matched perfectly to the 204. Its power base was 3,600 pounds, the drill frame, 3,000. In April 1970 *Rotor and Wing* described a typical move:

Once a drilling hole has been completed, the machinery is dismantled and the drill frame section is hooked to the Bell 204B, which lifts it off and carries it to the next location. The power base section is then lifted and flown to the next location and set in position, as identified by a flag. The helicopter returns for the drilling crew which, upon arrival, levels the power base. Then the mast section is picked up by the helicopter, moved over the power base and guided into position by the drilling crew. An air hose is connected, engines are started, and the drilling gets under way. The move from one shot point to the next has been made in as little as 6 1/2 minutes.

The RCN and RCAF also entered the turbine helicopter age. In 1963 the RCN accepted its first Sea Kings, eventually taking 41. All but the original four were assembled by CP&W. These replaced the HO4S (S-55) on ASW, SAR and utility duties. Meanwhile, the Canadian Army got its first Vertol CH-113A Voyageurs and the RCAF its first CH-113 Labradors. The Voyageur was the Army's first tactical helicopter, while the Labrador was for SAR and utility work. These outstanding types remained in CanForces use to the turn of the century.

Okanagan Spreads Its Influence
Through 1964 Okanagan's 60 helicopters flew more than 20,000 hours. It won a two-year, 12-

Sea King action aboard HMCS *Bonaventure* in Halifax on May 21, 1968. Aircraft No. 01 was one of the original four that Sikorsky built for the RCN. No. 05 was the first built at Longueuil by Canadian Pratt & Whitney. Of 41 RCN/CF Sea Kings, about 30 survived into the late 1990s, including these three. Although primarily an ASW resource, the Sea King often served as a general purpose transport. (Turbo Tarling)

(Right) Okanagan's Bell 206B YKZ at Vancouver in March 1978 with S-58T OKG beyond. (Brent Wallace)

machine contract with the BC Forest Service. New bases were opened in Nanaimo and Fort Nelson. It had long term contracts at Kemano (Alcan) and Goose Bay (Pinetree Line). The latter went for six years, mainly using S-58s. That year Okanagan absorbed Pacific Helicopters. On the downside, gradual closure of the Mid Canada Line reduced helicopter demand. In February 1965 Okanagan helped at the Granduc Mine, north of Stewart, BC, when an avalanche hit. An S-58 was the first rescue aircraft on site. It evacuated 100 survivors and 28 bodies despite bad weather. Two $400,000 Bell 204Bs were purchased in 1965. Dave Alder, Dave Ramscore, Jim Reid, Jim Ritson and Roy Webster were some of Okanagan's original 204 pilots, while Hank Ellwin and Cary Carstison were engineers. A major push was made into aerial construction. The 204B's 4,000-pound external capacity was ideal in the mountains, when constructing microwave towers, power lines and ski tows. A typical machine in 1965 hauled concrete for the Whistler Mountain ski lift, slung equipment and diesel fuel in seven-drum loads to a Vancouver Island mine site, and set 3,000-pound poles on a hydro project near Kamloops. Such markets had been developed by the S-58. Okanagan used this type to prove its own power line stringer on a job at Seaton Lake, BC. With the closing of the Mid Canada Line, government-owned H-21s and S-55s were sold. Okanagan took three S-55s. In 1965 they planted telephone poles across the face of the giant Hope landslide; and helped at Smithers during floods on the Bulkley River. Okanagan revenue hours for 1965 exceeded 28,000.

By early 1966 Okanagan pilots and engineers evaluated three new light turbines—the Bell 206A Jet Ranger, Fairchild Hiller FH1100 and Hughes 369. All three had been developed for a US Army light observation helicopter requirement. Winner of the competition was the Hughes, but each manufacturer then adapted its design for commercial markets. Three $115,000 Jet Rangers were ordered for 1967, and an FH1100 was leased for geophysical survey work. In December 1967 Okanagan ordered 10 FH1100s, with an option for 20. In February 1968 its fleet of new FH1100s arrived at Vancouver en masse.

Power line construction, mostly with Bell 204Bs, totalled 414 miles in 1967. This was the S-61's first commercial job in Canada. Pro-

ductivity was high—68 aluminum towers as heavy as 1,400 pounds were set in one day on a 659-tower job around Alice Arm, BC. An S-61 set 14 towers in the Pine Pass on the Peace River line, while many miles of wire were strung there using 204Bs and S-58s. Heavy lift prospects prompted Glen MacPherson, head of Okanagan, and chief pilot Don Jacques, to visit Russia to evaluate types like the giant Mi-10 aircrane. The Sikorsky S-64 aircrane also was studied. Okanagan lost money in 1968—revenue hours dropped from 31,751 (1967) to 26,681, mainly because of less forest fire activity.

In 1969 Okanagan ordered 11 Jet Rangers to replace its FH1100s. The 206As featured removable door posts to accommodate a litter kit, permitting air ambulance duties from logging, mining or construction sites. This brought an end to Okanagan's flirtation with the FH1100, crashes of which had claimed three Okanagan pilots. Soon Bell 206B Jet Ranger IIs with the 400-shp Allison 250 C20 engine (317-shp in the 206A) were the norm. While 69 Jet Rangers were with Canadian operators in 1971, a year later there were 137. The Hughes 369 entered Canada in 1969.

Mackenzie Delta and Arctic islands oil-gas exploration fired the market for such types. New operators in this period included Bow Valley, Haida, Kenting, Liftair and Viking.

By now Okanagan had a network of licenced bases. Prince George (1961) was the earliest. The ATB realized that these cost an operator a lot of money, so protected them from competition. Bases opened in the 1960s-70s included Campbell River, Calgary, Cranbrook, Fort St. John, Fort Simpson, Goose Bay, Inuvik, Kamloops, Montreal, Smithers, Tofino and Whitehorse. The bread and butter for the BC bases was the BC Forest Service contracts. By the early 1970s still more helicopter firms appeared, including Air Alma, La Vérendrye, Nahanni and Trans Quebec.

Ross Lennox: Helicopter Veteran

Ross Lennox was one of Canada's helicopter pioneers. Born in Winnipeg on December 25, 1922, he joined the RCAF as a lad, making his first flight on August 8, 1942 in a Tiger Moth at No. 19 EFTS in Virden, Manitoba. He won his wings at No. 10 SFTS at Dauphin, then was posted to instruct at Claresholm, Gimli and Comox. He got overseas at war's end, but his only ops were a few trips on 436 Squadron Dakotas. Lennox's first peacetime flying was in a Waco owned by O.J. Wieben. He then joined the

Ross Lennox (right) with fellow helicopter pilots Ralph Heard (left) and Bruce Best. They were in Carl Millard's hangar in Toronto for the 1989 launch of *Power: The Pratt & Whitney Canada Story.* (Larry Milberry)

Air Transport Syndicate in Flin Flon on Cub DSH and Stinson 108 EXO. The syndicate had a nickel property at Mystery Lake. Inco took an interest in it, which led to the great mine at Thompson. Lennox's next opportunity came with Hudson Bay Air Transport of Flin Flon. It had Norsemen BFT and BFU and recently had added a Mallard. In March 1949 it bought Crane FQP at Moose Jaw. FQP operated on MacDonald Brothers floats between Channing Lake at Flin Flon, and Island Falls. Lennox then took it to the Yukon to serve exploration camps. In 1952 Beaver GYU, then Otter GBX, were purchased.

Convinced that too much time was being wasted in the field caring for sled dogs and pack horses, HBAT president Rosco H. Channing studied the potential of helicopters. He sent Ross Lennox to Okanagan to train, his first flight being with Fred Snell in Bell 47 FZX on February 13, 1953. He graduated April 25 with about 75 hours on Bells FDN, FJA, FSR and FZX. Now Channing ordered an S-55 and Lennox went to Sikorsky for conversion. Jim Chuters taught him the ropes May 4-8, 1953 and on May 9 Lennox and mechanic Art Wilson set off in S-55 CF-HAB from Bridgeport. They routed Warren (Ohio), Toledo, Chicago, Minneapolis, Grand Forks, Winnipeg, Saskatoon, Fort Nelson and Teslin, arriving May 25. Lennox now operated HAB in BC and the Yukon supplying prospectors. By December 8, when HAB was laid up for winter maintenance at Flin Flon, he had 367 hours on type. On May 7, 1954 he ferried HAB back to the Yukon.

In 1958 Ross Lennox joined Okanagan on the Mid Canada and DEW line. A job in October 1962 took him to Cape Dyer on S-55 CF-JLP. In February 1963, after he was recommended by tech rep Jock Graham, he joined United Aircraft of Canada, working mainly on the RCN Sea King. His first flight was on May 10 at Bridgeport on Sea King 4001. He soon test flew Nos. 02, 03, and 04. He flew the first Longueuil-built Sea King (05) on April 9, 1964. This was a booming time as UAC established its Sea King line, and delivered aircraft. RCN pilots like Larry Ashley, Seth Grossmith, Joe Sosnkowski and Larry Zbitnew were Lennox's main pilot contacts on the program. They were top young men—all attained high rank, or moved to important positions in industry. Lennox did much production test flying, mainly on aircraft being overhauled or modified. There was also test flying to prove equipment installed on the HO4S and Sea King. By the end of 1969 Lennox's log showed: S-55—4,350 hours, Sea King/S-61—1,335, S-64—

Ross Lennox flew the S-55 for Okanagan on the Mid Canada Line. At top is the hangar at the MCL site near Bird, Manitoba. Lennox was involved in the most unusual MCL event in this period. On March 15, 1960 a Lockheed U-2 spy plane of the 4080th Strategic Reconaissance Wing landed on Wapawekka Lake, near microwave Site 715 in the area of La Ronge, Saskatchewan. The U-2, designed in 1954 under Clarence L. Johnson, first flew at Groom Lake, Nevada in July 1955 with test pilot Tony Le Vier. It was soon operational on CIA missions over the USSR—the first (July 4, 1956) overflew Moscow. This U-2, flown by Capt Roger Cooper, was officially on a high altitude sampling program (HASP) mission, studying the stratospheric distribution of nuclear fallout. This involved taking samples through the fuselage pod seen near the port engine intake. Conflicting data indicate that this U-2 was heading for Minot, North Dakota after photographing a Soviet ice island. As Cooper flew south his engine quit at about 72,000 feet. As he glided down he spotted the buildings at S715 and landed nearby. Cooper had difficulty slogging through heavy snow, but a Bombardier from S715 arrived. S-55 CF-JTE (pilot Pender Smith) arrived from Site 700 at Cranberry Portage (S/L Clive Dean-Freeman commanding), near Flin Flon, about 100 miles east.

Smith flew Cooper to Cranberry Portage for a meal and a bit of a party. By this time Ross Lennox arrived from Flin Flon; next day he took a security detachment to the U-2 in CF-JTB. A USAF T-33 landed at The Pas to take Cooper to Minot. On day three repairs were made to a damaged outrigger on the U-2 landing gear. Then a USAF lieutenant colonel arrived. He asked that only 1,000 feet of snow be levelled for takeoff, strapped in, blasted off downwind and disappeared into cloud for Minot. Pender Smith later received a medal from the US government for helping out. Here Cooper's U-2 sits on Wapawekka Lake. (Ross Lennox)

371, Bell 47—353, Hiller 12E—146, S-58—18 and KA-26—0:30. He also had 5,904 fixed-wing hours.

In April and May 1970 Lennox was in Prince Rupert training CCG crews on S-61N CF-CCG. Back in Longueuil he flew UAC's speedy Beech 18 PT6 test bed CF-ZWY-X, and its King Air and Citation. In May 1973 he flew the S-58T for the first time, then added the test bed Viscount. Lennox retired in 1982, his last flight being on P&WC's Citation II on December 7 going Vero Beach-St. Hubert. By this time his log showed: Citation—2,971 hours, Sea King/S-61 2,620, S-64—594, Viscount—404, Beech 18—202 and Lear 36A—23. In 1983 he returned to the bush, flying a PZL Otter for Plummer's Camps in the NWT. He also flew the S-64 for Ericson Air-Crane, a company he represented in Canada into the 1990s.

Trans-Atlantic Sikorskys

In 1964 Okanagan Helicopters and British European Airways combined under International Helicopters to serve Shell Oil drill rigs in the North Sea. Each partner purchased a 26-passenger S-61N for carrying passengers and cargo to rigs as far as 150 miles offshore.

While BEA shipped its S-61 by sea from New York, Okanagan sent CF-OKY "Morning Star" on the first trans-Atlantic ferry of a commercial helicopter. The 4,500-nm trip would save $26,000 in shipping. No helicopter had made an unescorted Atlantic flight. (In July 1952 two USAF H-19s routed across the North Atlantic escorted by a B-17 Flying Fortress. In 1963 CH-3B "Otis Falcon" crossed with C-130 escorts.)

Okanagan's ferry would be with no requirements that were "not a part of normal operations or within the capabilities of the twin-engine helicopter." UAC's Ross Lennox, Thomas Scheer (co-pilot) and Keith Rutledge (engineer) of Okanagan, and UAC engineer Tom Harrison were the crew. OKY flew from Stratford, Connecticut to Longueuil on May 7 with Sikorsky pilot Dick Stephansky. The next few days involved preparations such as evaluating the 250-gallon ferry tank carried in the cabin.

At 0836 on May 14 OKY was airborne from Longueuil for Baie Comeau. From there it flew to Knob Lake, a turbulent leg, so speed was reduced from 110 to 95 KIAS for a more comfortable ride. Freezing rain beyond Knob Lake grounded OKY till May 17, when it reached Fort Chimo and pushed on. Lennox noted in his diary: "Approaching Hudson Strait, we climbed to 6,000 feet to clear the clouds, and this was maintained across the low range of mountains to Frobisher. A heavy snow shower was encountered at Frobisher after let down, but we held to the west until a clearing allowed us to land without difficulty."

Soon after takeoff from Frobisher on May 19 fog encroached, and Cape Dyer reported being in weather. OKY pushed on with the option of returning to Frobisher. Next day came a flight to Sondrestrom. May 21 found OKY cruising at 11,000 feet across 300 miles of Greenland ice cap. Sea Bass DEW Line station passed below. Lennox noted: "Both TWA and MATS were heard carrying on amusing radio conversations with personnel manning this outpost. Apparently this was usually the case—it helped to pass the time and monotony." OKY landed at Kulusuk near Big Gun DEW Line site. It departed on May 22 for the open sea. Of this leg Lennox reported: "For navigation the two ADF beacons of Big Gun DEW Line site in Greenland, and Reykjavik were used, along with two fixes obtained from radar picket aircraft. The Reykjavik beacon was excellent, as it could be picked up over the entire leg. Our dead reckoning navigation was within one mile of the last fix given by the radar ship when 95 miles out of Reykjavik. Flight planning out of

The Flight of CF-OKY, May 1965			
Leg	Distance (nm)	Time	Fuel Used (Imp. gal.)
Longueuil–Baie Comeau	321	3:12	419
Baie Comeau–Knob Lake	336	3:30	411
Knob Lake–Ft. Chimo	205	2:10	315
Ft. Chimo–Frobisher	355	3:45	452
Frobisher–Cape Dyer	249	2:25	300
Cape Dyer–Sondrestrom	260	2:50	220
Sondrestrom–Kulusuk	338	3:50	520
Kulusuk–Reykjavik	404	3:45	461
Reykjavik–Hofn	240	2:50	360
Hofn–Vagar (Faeroes)	255	2:25	309
Vagar–Prestwick	420	3:30	425
Prestwick–Gatwick	325	3:00	375

CF-OKY was the first S-61 to fly the Atlantic. Here it is at Longueuil before leaving on its epic trip. OKY's pioneer trip proved the feasibility of trans-ocean helicopter flying. Such flights later became routine. OKY's flight was specially commemorated by carrying 30 No. 10-size covers. Each was signed by the crew and bore stamps from Canada, Greenland, Iceland, Denmark and Great Britain, which were cancelled in Kensington, UK on May 31. (Lennox Col.)

OKY at Reykjavik on May 22, 1965 for fuel and an overnight stay. Engineer Keith Rutledge stands in front. In 1997 this S-61, long since in British registry, still was doing offshore work in the UK. It was owned by British International Helicopters, by then a CHC subsidiary. Then, Norwegian S-61N LN-ORH at Fort Chimo on July 28, 1966. (Ross Lennox)

Lennox later made recommendations for those attempting such trips e.g. that pilots have instrument ratings and sufficient hours on type, that up-to-date information about radio facilities be available, that maps of four, or eight miles-to-the-inch be used (not the 16 miles used on OKY), that flights be limited to summer, and that a less cumbersome internal ferry tank be used. Engineer Harrison's final report listed no mechanical snags. For him the most useful items carried had been a light, gas-powered pump for refuelling from drums, and a 12-foot aluminum ladder to reach the tail rotor.

Through the spring of 1966 Ross Lennox planned two S-61N ferries to Norway for Interessentkapet Helibuss. Helikopter Service A/S of Oslo would use them to support the Esso North Sea drill rig *Ocean Traveler*. The first S-61 (LN-ORE) left Sikorsky for Montreal on July 1, the second (LN-ORH) on July 15. Aboard ORE were Lennox plus Norwegian co-pilot Kjell Bakkeli, and engineers Per Listerud (Norway) and Richard Smith (Sikorsky). Their route from Montreal included Baie Comeau, Knob Lake, Fort Chimo, Frobisher Bay, Cape Dyer, Sondrestrom, Kulusuk, Reykjavik, Hofn, Vagar, ending in Stavanger, Norway on July 13. With a maximum of 4,460 pounds of fuel in the mains and 2,000 in the auxiliaries, ORE had an average fuel burn of 125 Imp. gallons per hour. This gave a 6:46-hour endurance. Gross weight per leg was 21,000 lbs. After delivering ORE, Lennox returned the Montreal to join the crew of ORH.

S-61 to Thailand
In early 1974 Okanagan took an S-61 job to service gas fields in the Gulf of Thailand for Union Oil. Since the machine (CF-OKM) was urgently needed, it was ferried from Canada. After serving Mobil, OKM was idle in Halifax. Okanagan changed its engines and loaded it with long-range equipment and survival gear. Pilots Don Jacques and Bill Janicke, and engineers Dan Lemire and Chuck Taylor comprised the crew. In the first week of February 1974 they left Halifax for Sept-Îles, then flew to Frobisher Bay, making most of the journey on instruments at night. There was icing between Baffin Island and Sondrestrom. Over the Greenland ice cap weather forced OKM to 12,000 feet. While at Kulusuk, it made a side trip to Angmaggssalik to deliver Christmas mail delayed by local aircraft trouble.

OKM next made Reykjavik, Vagar and Ab-

Kulusuk was 3:50 en route, and actual time was 3:45, so navigation was no real problem on this leg." Twenty-three years later he added, "OKY was equipped with only one low frequency ADF for navigation. We had a second ADF receiver, but it was an extra, kept in our box of helicopter spares. All navigation (dead reckoning) was by relative ADF bearings being plotted on a mercator chart. We were never able to raise the picket ship *Mike* off the southeast tip of Greenland either by VHF or HF. However, the single contact we made with a Constellation radar picket aircraft was very reassuring, as we were on top of cloud at 6,000 feet at this time. Our position was one mile out, as confirmed by theirs as they flew by us visually."

OKY reached Hofn, Iceland from Reykjavik, where weather grounded it for two days.

For May 27 Vagar in the Faeroes was on the flight plan. The first 200 miles was flown at 300-700 feet on account of weather. Next day it was on to Prestwick: "Landfall was made on the north coast of the Outer Hebrides, right on track, and the ETA at Stornaway was within a minute. A strong tailwind helped on the last part of the Prestwick leg down the west side of the Scottish coast. This was an enjoyable leg with altitudes of 1,500-3,000 feet being held over very interesting terrain. On arrival at Prestwick the aircraft was met by a strong contingent of radio, press and TV personnel." On May 29 OKY went south through industrial smog that kept it 1,000 feet AGL. The flight terminated at Gatwick, where OKY was handed over to International Helicopters. The offshore operation began on July 9. Ross

S-64 Skycrane N6959R at work over Cape Dorset in August 1969. Then, pilots Lee Ramage and Larry Pravicek with N6959R in the Arctic in 1969. Years later they visited Toronto for several weeks of construction on the CN Tower. (Ross Lennox)

erdeen. In Southend, near the mouth of the Thames, the Arctic survival gear was left behind and OKM flew to Nice, Rome, Brindisi, Athens, Iraklion, and Beirut. Military activity in northern Iraq forced it south via Damascus, then Qaysumah, Bahrain and Dubai. The final legs were to Karachi, Bombay, Nagpur, Calcutta, Rangoon, Bangkok and Songkhla, Thailand. The trip included 98 flying hours and covered 12,959 miles, the longest ferry by a commercial helicopter. There were stops in 16 countries over 30 days. Four months later pilot Bill Janicke, co-pilots Roy Webster and Pierre Bock, and engineer Dan Lemire returned OKM to Newfoundland in a record 13 days (11 flying days). The only weather encountered was a sand storm over Saudi Arabia.

The Skycrane in Canada

Canada's annual Arctic sea lift involved various federal departments, with commercial agencies supplying ships and manpower. Traditionally, lighters moved cargo ashore, a slow process that invariably damaged goods. Delays from rough seas, winds and ice were common. Other than doing some ice patrols and slinging, helicopters had been used little, but for 1969 the DOT experimented with an S-64 Skycrane. It listed its goals in a memo of July 7, 1969:

1. To test the feasibility of the ship-shore heavy lift helicopter system as the primary marine cargo delivery system for supply of settlements in the Hudson Bay and the Eastern Arctic.

2. To test the feasibility of unitizing general cargo destined for the settlements in containers, gondolas and pallets to a weight of between 16,000 lbs and 20,000 lbs.

3. To investigate the problems involved in loading, stowing and handling unitized cargo both at Montreal, on board ship and ashore at the settlements.

4. To assess the capabilities of the S-64A for operations in high latitudes, with a minimum of support facilities, and to investigate the problems involved in ferrying the aircraft from settlement to settlement in support of the sealift.

A Sikorsky document of July 14, 1969 added further details:

The cargo vessel Sir John Crosbie will depart Montreal July 25 and will sail seven days to

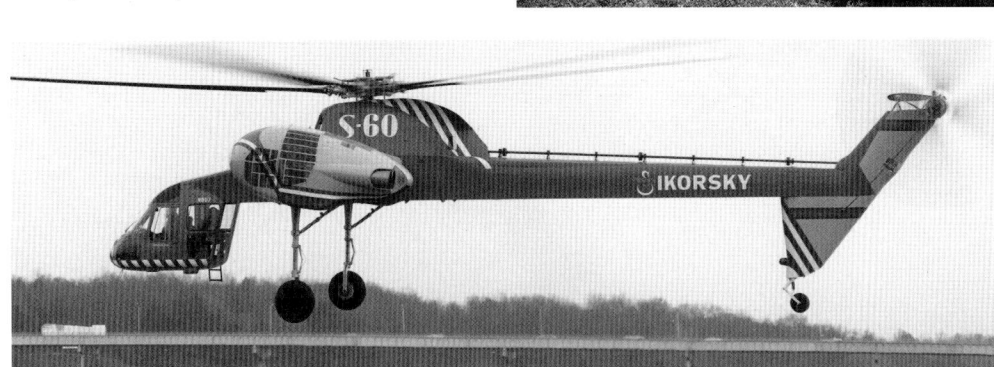

The prototype Sikorsky S-60 Skycrane quickly evolved into the turbine S-64. (Best Col.)

Cape Dorset, arriving August 1. Crane 002 [N6959R]will intercept the Crosbie at Cape Dorset August 1. At Cape Dorset 750 tons of varied cargo will be offloaded. The Crosbie and the crane will then rendezvous some 180 miles west at Coral Harbour where 250 tons will be offloaded.

Accompanying the motor vessel Crosbie will be the Canadian Coast Guard ship Raven which primarily acts as a back-up vessel to the Crosbie. Aboard the Raven will be the fuel requirements for the crane consisting of 34,000 gallons of JP5 which we require for offloading at both sites. The schedule calls for unloading at Cape Dorset in two days and approximately ¹/₂ day at Coral Harbour. In just 24:38 hours of flying, N6959R is to offload 1,296 tons directly to villages and construction sites at the two settlements. (Crosbie was a package freighter not designed for containerized or palletized cargo; its 72-foot mast was an obstruction, so it was poor for helicopters.)

The prototype Sikorsky S-60 crane flew in 1959 with two R2800 engines. Trials led to the CH-54, flown in May 1962 with the T37 turbine (military JT12D). Although the instigator of the CH-54 had been the German military, only the US Army ordered. The first of 107 was delivered late in 1964. The CH-54 did outstanding work in Vietnam. Ten civil S-64s were built on speculation.

N6959R's Work in 1970

Site	Dates	Tons Delivered	No. Flights	Hours Flown
Resolute Bay	Aug. 20-25	903.9	224	33.7
Arctic Bay	Aug. 26-27	298.8	74	6.9
Pond Inlet	Aug. 28-30	551.0	133	14.6
Clyde River	Aug. 31-Sept. 2	695.3	153	15.6
Frobisher Bay	Sept. 5-7	413.3	116	20.9

For experience and publicity Sikorsky's Industrial Crane Marketing branch was keen to send the S-64 north. It assigned N6959R with pilots Lee Ramage and Larry Pravicek; and maintenance men R. Sousa, R. Szyja and W. Schmedlin. Ross Lennox was seconded from UAC. Sikorsky's fee for the job would be $80,000. Carrying a detachable support pod and operating on a 30-day DOT permit, N6959R ferried from Stratford via Montreal. There it tested a 20-foot sea container and the cargo cage to be used. July 25-27 it flew Quebec-Baie Comeau- Sept-Îles- Wabush-Knob Lake-Fort Chimo-Cape Hopes Advance Bay-Frobisher Bay (1,750 nm). On July 29 it flew to Cape Dorset and Coral Harbour to work, returning to Frobisher on August 8 and setting off for Montreal next day.

Recommendations after the job included basing the helicopter aboard ship (to co-locate fuel, flight crew and maintenance personnel). Emphasis was on cargo being prepackaged in loads of 8-9 tons; and on ships having more than one hold (one could be readied, while the other was serviced by helicopter). For safety, radios had to be reliable, and cargo handlers

N6979R working at George River in July 1971. It is about to hoist a load from the forward hold of *Fort Chambly*. (Ross Lennox)

(Right) Views of N6979R in the Arctic in 1971, first aboard ship with its utility pod, then ashore, clean. (Ross Lennox)

trained in slinging. A new lift using the S-64 was planned for 1970. Called Operation Skylift, it involved the 450-foot, 5,947-ton Canada Steamship Lines MV *Fort St. Louis*. It had four wide hatches, hydraulic elevators, and deck space for the CH-54.

Deck cargo comprised 19 vehicles, some 50-foot house trailers, bottled gas, and the Skycrane (with pod). Below decks were fuel in 45-gallon drums, telephone poles, lumber, plywood and general cargo. Cargo would be slung off the deck by cables or chokers; or from below decks on pallets and in 4,400-pound, 8 x 20-foot steel cages. In 1993 Lennox recalled that cage weight totalled an inordinate 25% of disposable load. The job began on August 7, when the S-64 left Stratford for Longueuil. Next day it flew to the Montreal docks, where its pod was lifted onto *Fort St. Louis*... It then flew aboard and the ship sailed on the 3,000-mile trip to Resolute Bay. On August 19 it was joined by the CCG icebreaker *d'Iberville*. Next day they anchored off Resolute Bay. By August 25 the S-64 had delivered 1,315 tons to eight sites around Resolute Bay. The average lift was 5.38 tons; average hook-up time over the ship, 3:09 minutes. On August 25 the last 82.5 tons were offloaded, then 28 tons of cargo and forklifts were slung aboard. That evening *Fort St. Louis* sailed for Arctic Bay, Pond Inlet, Clyde River and Frobisher Bay. It left Frobisher on September 7 with the Skycrane, and tied up in Montreal five days later. Operation Skycrane was history's largest such replenishment. N6959R recorded an availability rate of 96.3%, although there

were snags—at Pond Inlet two sites were unsafe, one being too muddy for the forklifts, the other too close to buildings. At other times cargo was inadequately marked for speedy delivery.

In 1971 the S-64 operated in Ungava and Hudson bays with *Fort St. Louis* and *Fort Chambly*. The crews were: Ross Lennox (front pilot) and John Holt (aft pilot); and Bob Lenzicki (front) and Charlie Evans (aft). Flying began July 20, when Lennox ferried N6979R from Stratford to Montreal. *Fort St. Louis* had sailed the day before. On July 21 N6979R lifted off (pod attached), flew an hour to Quebec for fuel, to Baie Comeau and Sept-Îles and by late afternoon made Knob Lake after flying 7:45 hours. The forecast for July 22 was not promising, but Lennox flew to Fort Chimo, then George River. For the next three days the crew relaxed. Lennox noted: "Beautiful weather, temperature 75-80. Went fishing (15 fish, mostly salmon, the rest char)... Moved aircraft to refuel area and took pod off... Took walk—poor village and terrifically low tide, villagers only able to go out on high tide. Good area for caribou, some polar bear, seal and excellent fishing. No sign of ship or word. Went fishing and caught six char... Slept a lot and read S-64E manual. Few martinis."

On July 26 the supply ship anchored four miles off George River for the first lift. Work was done in four-hour shifts with crews alternating. Offloading began July 28. *Fort Chambly's* diary noted some of the activity: July 29—"0837 Skycrane approaches, operations resumed. Clear weather, temp. 60F. Slow

work preparing optimum loads in confined deck space, and in No. 4 hatch. Offloading progressed rapidly all day with stops for meals, fuel and maintenance. Final lifts were four complete school sections for assembly ashore, each 20,000 lbs. 2305 MV *Fort St. Louis* heaving anchor to depart for Fort Chimo". July 31—"0315 Anchored in river 19 miles from Fort Chimo. The outboard motors on the workboat are inoperative. Radio-telephone to Fort Chimo to send canoes to take shore gang to village. 0900 Skycrane commenced lifting forklift trucks ashore. 1600, hove anchor and moved one mile closer. 18 miles was the closest deemed prudent for this size of ship in this river. This had a profound effect on the operational plan, the itinerary of which called for a distance of one mile. There would be an appreciable delay and a considerable increase in helicopter fuel consumed. A decision was made to purchase fuel at Fort Chimo to conserve the fuel on board for the remainder of the voyage. 2016, Skycrane lands on deck.. Approx. 70 tons offloaded in 11 flights." Work so far had been fairly smooth except that some trailers were heavier than the Skycrane's 20,000-pound capacity, so had to be lightened.

On August 2 Lennox noted from aboard *Fort St. Louis:* "Continuous grind all week with tonnages running from 120-150 a day. Load indicator approx. 2,000 high, so true tonnage is less by two tons/hr flown. Averaging 9-11 hrs flying per day." For August 7 he wrote: "Good day, 150+ tons... Back on deck by 9:30 PM and sailed at 10:30. Heavy ice contacted during night, but arrived on mid-tide at Payne

Bay at 1030 AM on August 8. Most of cargo offloaded by 10:00 PM and finished on morning of August 9." The ship sailed to Koartak, arriving at 1900. Flying began with eight loads slung before fog curtailed operations. Next day the first load was flown at 0710, then fog intervened. This had been ongoing and concerned Lennox, who reminded his diary that the contract was to end August 13, but eight outposts remained. Next came Wakeham Bay, seven hours sailing (90 miles) along Hudson Strait. The ship anchored a mile off shore at 0030 on August 11. Flying commenced at 0800: "Low overcast and fog, but good flying for us. Landing zone excellent. Good loads to start with (50 tons/hr). Then slacked off."

Fort St. Louis arrived at Sugluk on August 12. There was trouble getting the ground crew ashore on account of a rocky approach: "By the time a landing made and site picked (had to get electrician to lower wires), it was 9:00 AM. Wind really blowing... Very, very hard flying, however the 259 tons moved and a/c back on by 9:30 PM... Headed out to sea and when in the open, the rolls and pitches were so bad, plus the very poor job of lashing down the helio, that it became apparent we could lose machine overboard. Entirely our fault for not overseeing and putting on more cables." Foreseeing trouble, Captain Lacey took shelter until about 0400 on August 13. Next stop was Ivugivik, where Lennox predicted a quick job moving 158 tons. They sailed at midnight for Povungnituk. Heaving seas damaged the S-64 pod. Fort St. Louis finally anchored off POV at 1600 on the 13th. Lennox mused: "Amazing that the last two nights we haven't lost either the pod or helicopter—a lesson learned."

For August 14-15 the supply ship was stuck outside POV's anchorage due to swells. Lacey pulled about 30 miles off in the lee of some islands. Lennox noted: "Everybody feeling fed up and with news of an extra 1,030 tons on *Fort Chambly*." On August 16 the ship anchored 6.5 miles off POV. By evening the S-64 delivered 180 tons. Day two also was productive and by 2000 the helicopter was lashed on deck. Lennox wrote: "Did some trading and buying last night for a few carvings. Should have bought a beautiful otter with fish for $20.00."

Next came Port Harrison, where both ships anchored on August 18. The S-64 got 235 tons ashore the first day: "Worked both ships all day and averaged 62.5 tons/hr. Ships ³/₄ and 1 mile off shore. Excellent sites... Port Harrison is by far the nicest of the Eskimo villages and quite large. So many carvings one couldn't believe, but all are in Co-op and prices too high... Be in Great Whale tonight around 9-11, so should be ready in morning. 850 tons should take 2¹/₂ days or less. Only 5 miles off shore and nice weather." At Port Harrison *Fort St. Louis* discharged the last of its load. Now the airlift transferred to *Fort Chambly*. *Fort St. Louis* was supposed to sail for Montreal. However, it still had 9,000 gallons of helicopter fuel, which would be needed for offloading *Fort Chambly* cargo at Great Whale. Great

Whale had a fine drop zone at the end of the runway. For the 20th Lennox had 35-45 knot winds: "Very rough, especially hovering over ship. Started refuelling on ship, which I think is quite hazardous, especially with rolling... Ended up in evening with only 395 tons off and ship only 1.5 miles from drop zone." For day two at Great Whale the weather was poor. The morning of the 22nd was spent on maintenance; good tonnage was hauled that afternoon. Mechanical trouble forced the Skycrane down. August 24: "Water so rough *Chambly* couldn't launch lifeboat to take mechanics ashore. Both ships steamed for shelter and overnighted behind whale-shaped islands. Tied up side-by-side late at night and transferred fuel and food between ships. Moved pilots' gear over and now quartered on *Chambly*—5 in wireless room, but not too bad... Mechanics ashore early in morning and fixed aircraft. Back flying by 1000... finished last loads night of August 25."

August 26 the Skycrane was in Fort George, where dust at the landing zone, and black flies were a bother. Loads included pallets of cement up to 16,800 pounds. These did not sling well and flew at only 70 knots. On the 26th the S-64 slung 285 tons, 281 on day two, 503 on day three. Fort George was wrapped up on August 30; the show moved up the coast to Paint Hills to an anchorage 10 miles from town. Soundings were made on September 1 and the captain of *Fort Chambly* moved two miles closer. About 280 tons were flown the first day. Weather delayed the finish till 1800 on September 3. Next day Lennox took a Beaver to Fort George, overnighted, then went by DC-3 to Amos, and by Air Canada to Montreal. The Sikorsky ferried south, landing in Stratford on September 4. It logged 241.6 hours on the 47-day charter. Sikorsky billed $296,724, while the ships cost $997,000. Overall, it had cost $230.74 to deliver each ton of cargo. This had been the fastest and most efficient such Arctic resupply.

Clearly the Skycrane was affordable and speedy in Arctic work. However, in 1972 the lowest bid ($140/ton) was accepted from a barge company. One barge later grounded in James Bay and its cargo had to be rescued by a

79-trip S-64 operation (N6960R). The S-64 burned nearly 11,000 gallons over 32:40 hours, adding considerable cost. This led Sikorsky to muse: "A very interesting concept presents itself if we consider using the heavy helicopter in conjunction with an ocean barge. The extensive open deck of the ocean barge, and the readily accessible cargo conjures up a speedy discharge of cargo which is the very essence of the helicopter's efficiency."

In October 1971 Hydro Quebec faced a crisis. Three James Bay survey camps needed fuel, but weather kept floatplanes from moving it before freeze-up. If the fuel couldn't be supplied, Hydro would have to pull its men from the bush. It needed to get 1,771 drums from Matagami to Nemiscau (110 miles) and from Fort George to Carbillet Lake (94 miles). Jock Graham of UAC calculated that pallets of 32 drums could be slung by S-64 on the 110-mile leg, flying at 90 KIAS, returning at 115. On the 94-mile leg, 35 drums could be slung. The rate on the S-64 was $3,000 per hour, and the job could be completed in 138 flying hours. Work proceeded and took two months. Ross Lennox flew the first month with Charlie Evans, the second with John Holt. Following this job the S-64 went to North Bay, then to Barrie, where a slinging demo was put on for Ontario Hydro.

Offshore

By 1964 Okanagan realized that international sales offered better prospects than the fickle Canadian market. Its first such work was a nine-month land use survey in East Pakistan, using a Bell 47J flown by Jack Milburn under the auspices of Canada's foreign aid program. Next, Okanagan entered the offshore oil market in the North Sea with British European Helicopters, but within a year sold its interest. In 1966, however, Shell ordered the 16,800-ton drill rig *Sedco 135-F*. In June 1967 service to it began with a Tofino-based Bell 204B on floats. An FH1100 assisted as needed. As drilling shifted north, the 204 moved to Sandspit in the Queen Charlottes. For all its effort (14 wells) Shell found nothing promising, so exploration ceased in May 1969. That year, however, drilling began in earnest off Canada's east and northern coasts. From July to October

The Skycrane delivers drums to Nemiscau during the emergency fuel haul from Matagami in the fall of 1971. (Lennox Col.)

Puma C-GUMP aboard the drill ship *Ben Ocean Lancer* on July 3, 1978 with pilots Paul Dorfman and Bernard Dasté; then a view of the ship the same day. In 1997 GUMP was operating with a Brazilian company. (Larry Milberry)

Okanagan supported the Aquataine Co. in Hudson Bay. Since its first S-61N hadn't arrived, Okanagan leased one in the US. It operated the longest nonstop, over-water helicopter sked—220 miles east from Churchill to the rig. In September 1969 S-61N CF-OKP began a multi-year Shell contract, supporting a rig off Halifax. With this, Okanagan became the first Canadian helicopter company operating under Instrument Flight Rules. On April 30, 1970 Don MacKenzie and Jim Reid were lifting off in S-61 CF-OKP from a drill rig off Nova Scotia. When they lost tail rotor control, MacKenzie made an immediate landing on the helipad. Although OKP was badly damaged in a rollover, there were no fatalities among the 16 aboard. For this good show MacKenzie was named "Pilot of the Year" by the Helicopter Association of America. OKP was recovered and rebuilt by Sikorsky.

In 1971 Okanagan acquired Universal Helicopters of Carp, Ontario, which had operations in eastern Canada. This brought an offshore contract starting that April—S-61N CF-DWC began serving an Amoco Canada rig. On one occasion Amoco sent DWC to fly an ailing man from a ship 175 miles off Newfoundland. Also in 1971 Okanagan operated from Cartwright, Labrador serving a Tenneco rig using an S-58T with long-range tanks and floatation gear. Okanagan signed a five-year contract with Mobil Oil Canada. For this it acquired S-61N CF-OKM from Sikorsky for $1.4 million. Work began in September 1972 from St. John's, then moved to Halifax. Besides regular supply and passenger flights, helicopters were used to spot for icebergs that might threaten the rigs. Special minima were approved for approaches and departures, as activity grew through the 1970s. Offshore approaches to a drill platform were approved for 150 feet radar altitude and a half mile visibility. ADF, VOR and ILS were used for primary navigation, but certain operations allowed airborne radar (e.g. to identify a rig), transponders and microwave landing system. By mid-1974 Okanagan was serving four east coast rigs with three S-61Ns, and additional ships with two smaller helicopters; but by year's end only one rig was left in Canadian waters, as exploration shifted to traditional properties, mainly in Alberta.

Bullock and Bow

Calgary-based Bullock Wing & Rotors Ltd. was formed in the late 1950s by brothers Bruce, Evan and Kurt Bullock. Bow Valley took over Bullock in March 1968, when Kurt Bullock died in a flying accident. At this time Bullock had 21 helicopters, 12 being turbines. It had been specializing in oil and gas exploration, with nine helicopters working the Arctic. In addition to the usual services, it moved drill rigs and seismic equipment. It had bases in Golden, BC, Yellowknife and Dawson Creek, where forestry, mining and heli-skiing were of interest.

In 1969 Bullock won a Pan Arctic contract away from Okanagan. That March it accepted its first Bell 205A-1. This machine started in seismic operations for Pan Arctic on Melville Island. Now Bullock became Bow Helicopters, named for its parent firm, Bow Valley Industries, which was active in gas and oil, coal and uranium. The outlook for 1970 improved with increased oil and mining exploration in the Arctic. Bow's aviation income increased to $2.7 million from $1.8 million a year earlier, although it still recorded a loss. Winter operations focused on heli-skiing in the Bugaboo Mountains of BC, using 204s and Alouettes. This was on behalf of heli-skiing pioneer Hans Gmoser. Bow made a profit in 1971. Work included a 204B constructing ski lifts in the Rockies; another worked in Labrador on the Churchill Falls hydro electric project. There

was a contract for a 205 in the Arctic islands and a lease on a 212 (the first in Canada) for northern work. In 1972 Bow had 10 turbine and five piston helicopters. Sales were up over 1971, but profits were poor due to high operating costs.

Bow added two Bell 212s and two Jet Rangers in 1972, while disposing of older types. Twin engines now were required for offshore operations over Arctic waters and for slinging, such as rig assembly and transmission tower erection requiring prolonged hovers. Results were disappointing in 1973. A surplus in medium helicopters curtailed revenue flight hours, and a wet summer reduced forest fires. In 1974 Bow had 14 helicopters in BC, Alberta and the NWT. By this time work in the Arctic and in BC mining was sliding. In 1975, however, there were jobs erecting transmission towers, supporting coal exploration, oil drilling and heli-skiing. By 1976 margins had shrunk midst growing competition and costs could not be passed on to customers. In 1977 Bow was supporting Dome Petroleum's drill ships in the Beaufort Sea, using 212s with floats. The fleet increased to 21 and a profit was shown. In 1978 a $3.9 million Sikorsky S-61N was added to support Dome. But gross margins were affected by increased costs, including foreign exchange losses on replacement parts from the United States. Key activities continued to be in petroleum, power line construction, forestry and heli-skiing; but, in spite of future potential,

Bow Copters Bell 204B C-FBWR, seen at Castlegar, BC on August 29, 1978, was one of the first of its type in Canada. It originally belonged to the Bullock brothers. In 1997 BWR was with Campbell Helicopters. (Brent Wallace)

in late 1979 Bow sold its helicopter operation, the assets going to Okanagan. The profit to be made disposing of its helicopters looked like a better deal.

More of the 1970s

At first it was difficult marketing the new, expensive turbine helicopters. Besides their price tags, the traditional market centred on the one- and two- passenger Bell 47 and Hiller UH-12E; an atmosphere had to be created for the three- or four-passenger turbine. The transition finally happened. At the end of 1969 there were 94 civil turbine helicopters in Canada: six private, 15 government, the rest commercial. There were 17 Alouette IIs and IIIs, 10 Bell 204Bs, five Bell 205A-1s, 38 Bell 206A Jet Rangers, 19 Fairchild-Hiller FH1100s, three Hughes 369HS Model 500s, and two Sikorsky S-61Ns. Okanagan had the largest turbine fleet—17. By 1972 the number of light turbines almost doubled over 1971, increasing 90% to constitute 47% of the total. Jet Rangers were most numerous, moving from 65 to 137. Alouette IIs grew from 20 to 35, the Hughes 500 fleet reached 29, but only six FH1100s remained. Vought Helicopter Corp delivered its first five-place Aérospatiale SA341G Gazelles and SA315B Lamas, the first Gazelle (CF-CWN) and Lama (CF-DGD) going to Canwest of Calgary in 1972. At this time Canwest was owned by John Pridie and Don Wederfort. They began with a Bell 47G, then bought

Alouette II CF-TQZ with a $120,000 loan from California businessman Tom West. TQZ went to work down the Mackenzie Valley and earned $800,000 in its first year.

In 1972 the S-55T (Garrett AirResearch turboshaft) entered service. The first in Canada (CF-JTF) went to Don Crowe's Nahanni Helicopters. Conversion was done by Deltaire Industries of Richmond, using a kit from Aviation Specialties of Mesa, Arizona. The S-62A reappeared in 1972 after 10 years—Okanagan bought CF-OKS from KLM Helikopters. Meanwhile, the piston S-58 and Brantly faded, and Bell 47s fell from 240 to 220, the first decrease since 1947. Average annual utilization per Bell 47 dropped 23% to under 400 hours. Hiller UH-12E numbers fell to 21.

Most growth in 1972 was in Quebec and the NWT, where hours increased 28% and 14%. Quebec represented 32% of domestic demand, the James Bay project being the key. Oil and gas exploration in the Mackenzie Delta and Arctic islands kept the NWT busy—49,367 hours (19% of Canada's total) were flown there in 1972 (18,570, or 12%, in 1968). Meanwhile, BC dropped from its historic number one slot, to third. Little mineral exploration was done in BC—a provincial NDP government was in power that espoused stiff mineral royalty taxes. Overall, Canada's fleet increased by 20% in 1972. Revenue hours rose by 6%, although average utilization per helicopter fell 11.5%, suggesting excess supply. Even so, in this era of a

regulated industry, operators earned a 12% return on investment. One newsworthy happening in 1972 involved an Okanagan Bell 204. On March 13 the freighter Vanlene ran aground approaching Vancouver. Salvage conditions were difficult, but, once masts and other obstacles were removed, Okanagan began salvaging the cargo of 150 Japanese cars by long line. Pilot Grant Soutar recalled that retrieving the cars in the third level was tricky. That hatch was only half the size of the upper one. Soutar and fellow pilot Roy Webster had to line up carefully, then make a rapid and continuous ascent, lifting each car straight through three decks, something Soutar likened to "squeezing a wet bar of soap."

Besides offshore work, the S-61 was good for general duty. In the fall of 1974 CF-OKB helped construct towers for BC Hydro's 500 kv line through the Fraser Canyon. Sections of 38 towers were lifted, then held in position as crews bolted them together. This job also involved the Bell 204, 205, 206 and S-58T. In November 1977 Okanagan began a special IFR S-61 job. Luscar Sterco Ltd. contracted it to shuttle miners between an open pit mine at Coal Valley and Edson, Alberta, each point having MLS equipment. The S-61 flew seven days a week, moving 32 passengers per trip on the 39-mile, 50-minute run (288 passengers a day, 2,300 hours a year). Such initiatives suited president John Pitt, as he moved Okanagan from its historic dependency on forestry into new fields. Through the 1970s Okanagan was

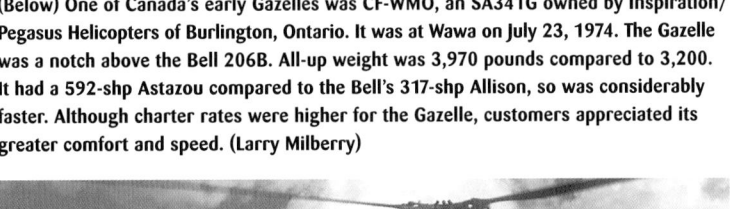

(Below) One of Canada's early Gazelles was CF-WMO, an SA341G owned by Inspiration/Pegasus Helicopters of Burlington, Ontario. It was at Wawa on July 23, 1974. The Gazelle was a notch above the Bell 206B. All-up weight was 3,970 pounds compared to 3,200. It had a 592-shp Astazou compared to the Bell's 317-shp Allison, so was considerably faster. Although charter rates were higher for the Gazelle, customers appreciated its greater comfort and speed. (Larry Milberry)

(Above) The Jet Ranger dominated the light turbine market in Canada from the late 1960s. C-GOXX belonged to the Ontario Provincial Police. It was shot at Toronto's Spadina Avenue heliport in December 1975 as a vintage Bell 47G departed. (Larry Milberry)

Two medium-lift Sikorsky turbines in Canada in the 1970s were the S-55T and the amphibious S-62. Although never in great numbers, they did excellent work. Athabaska Airways' fleet of S-55Ts worked for years on fire suppression in Wood Buffalo National Park. Here a Quasar S-55T sits at Delta, BC in September 1977, while Okanagan's S-62 hovers at Vancouver International in June 1976. (Andy Graham, Kenneth I. Swartz)

Okanagan S-58T CF-OKG heli-logging near Toba Inlet on the BC coast in November 1978. Powered by the P&WC PT6T TwinPac, the S-58T was greatly improved over the piston S-58. Empty weight, for example, fell from 8,275 pounds to 7,575, allowing a greater payload (5,400 pounds). With the natural safety factor of the TwinPac the S-58T was ideal for construction and logging. In passenger configuration the roomy cabin accommodated 16. (Brent Wallace)

involved in several takeovers. One of the most important involved Ontario-based Dominion-Pegasus with about 50 helicopters. It joined Okanagan in early 1975.

The S-58T

The idea of an S-58 with PT6s was discussed in the mid-1960s, when the US military was having performance trouble with its H-34s in Vietnam. J.F. "Jock" Graham of UAC proposed a conversion, but received no encouragement from Sikorsky. Later, however, Sikorsky realized that there was a market offshore for a revitalized S-58. It acquired a stock of H-34s from France, Germany and Israel. A prototype, converted to the PT6 TwinPac, flew in August 1970. Canadian customers were keen. Alpine, Okanagan and Ontario Hydro were first to buy. The S-58T, which carried a 5,000-pound payload, proved superb. Several went straight to work moving diamond drills in BC, hauling fuel in the NWT, planting telephone poles in Ontario, etc. Okanagan's first was its veteran S-58D CF-LWB, converted in Vancouver in 1971 along with H-34s OKE, OKG. Their first job was fire fighting near Prince George. They also did some experimental heli-logging, and served a Tenneco rig off Cartwright. So successful was year one that S-58Ts OKR and OKW were added in the spring of 1972; and LWE, OKH and OKO a year later. In August 1972 Okanagan won its first long-term international contract, a job in Greenland supporting Cominco's Black Angel lead-zinc mine at Marmorilik. One, then two, IFR S-58Ts operated over a 300-mile route from base in Sondrestromfjord. Besides miners, everything from food to furniture and dynamite was carried. The contract lasted 18 months and 3,000 flying hours. For 1974 Canada's S-58T fleet logged 3,266 hours of which 860 were international. Okanagan had a number of S-58T mishaps over the years. Veteran pilot Don Jacques died when LWE crashed on Grouse Mountain, BC on September 16, 1976. Although little wreckage was left after the fire, it appeared that mechanical failure was the cause. The S-58T stayed with Okanagan till 1980, when the last were sold in the US.

Overseas Growth

Okanagan's big push internationally began early in 1974. One S-58T went to Gabon in the summer of 1974 to fly 100 miles offshore to a drill ship. For a few months it also operated from Mauritania and Morocco. During the summer it flew to the border of Gabon, Equatorial Guinea, and Cameroon, carrying the head of state to resolve a border dispute. Later it ditched offshore, claiming the lives of several passengers. A push began in early 1975 with contracts in Burma and Thailand. By now Okanagan was active in 10 locations overseas. S-61N OKM worked in Thailand after its his-

toric flight from Halifax. Meanwhile, two S-58Ts were with Total Oil at Burma; and two 212s were in Burma for Esso. Another 212 completed a job for Tenneco at Thailand and moved to the Philippines for Amoco. An S-61N was in Guyana and Suriname on offshore for Shell and Elf Oil. A Bell 205 was supporting pipeline work in Peru's Amazon. Other flying was in Zaire, Cambodia, and Bangladesh. For 1975 20% of Okanagan revenues originated overseas. That year it purchased five 212s and an S-58T for $5.5 million. Orders for 10 S-76s were placed with deliveries starting in late 1978. The S-76 would accommodate 12-13 passengers, cruise at 125 kts, and fly 500 miles. It promised to be a good hot-temperature performer and an ideal S-58T replacement. Fleet additions through 1976 numbered 16 Jet Rangers, four 205s and two S-61Ns. Two 212s flew from Thailand to Egypt to support Esso drilling in the Red Sea. A 212 and a 206A were in the Bay of Bengal for Texas Pacific Oil. From Bombay an S-61N operated for the Oil and Natural Gas Commission of India. This was the start of 10 years of business in India.

Okanagan was perhaps the first Canadian helicopter operator in the southern hemisphere—it sent two UH-12E sprayers to New Zealand in the mid-1960s. United Helicopters of Calgary went to Fiji in the mid-1970s with a Hughes 500C and a Lama for aerial surveying. The first large venture came in 1977, when Okanagan teamed with New Zealand Helicopters to support drilling 160 miles off Invercargill. S-61N C-FDWC was cocooned in Van-

The S-76 first flew in March 1977 and was offered with the PT6T, Allison 250-C30 or Arriel 1S. Along with the Bell 212, it replaced the S-58T with Okanagan. Since April 1979 Okanagan used S-76s offshore, especially in Thailand and China. Here one appears in a typical setting—departing the drill ship *Scan Queen*. Then, C-GIMT of CHC over Toronto, on an air ambulance contract to the Ontario Minisitry of Health. (Grant Soutar, John McQuarrie)

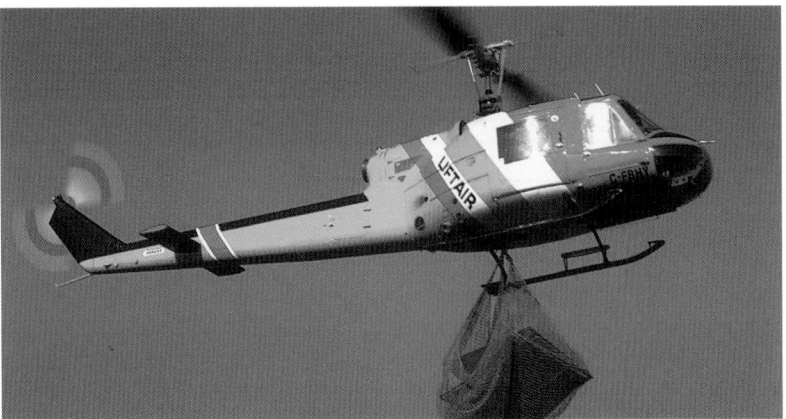

Rarely have government and industry combined to produce such a superb product as the Bell 212 Twin Huey. In this case the recipe for success was: one airframe (based on the tried-and-true UH-1 Huey), a power plant (PT6T TwinPac), funding (Canadian taxpayer) and a launch customer (the Canadian military). The prototype flew in 1968. The 212 launched the PT6T and evolved into the more advanced 214. Hundreds of Twin Hueys were built for military and civil customers. Helicopter and engine remained in production into the 21st century, having brought billions of dollars to Canada's economy. Here Okanagan Bell 212 SQM practices water bombing at Fort Simpson in July 1985. (Brent Wallace)

Liftair's Bell 204B C-FBHY slinging cargo in May 1980. (R.S. Pettite)

John Pridie (right) with ERA (Alaska) pilot Tom Mockler during an overseas AN-124 flight carrying Super Pumas to Greece. (Larry Milberry)

couver, loaded aboard MV *Hobart Star* and shipped to Auckland. Also in 1977 a Jet Ranger was on contract in Haiti. That year the company made its largest expenditure ($11 million), purchasing Associated Helicopters and some new equipment. In June 1977 Ottawa announced that the Mackenzie Valley Pipeline would be shelved, pending settlement of native land claims. This was a major blow to helicopter operators.

Okanagan contracts were completed in 1978 off Ireland, Bangladesh, and New Zealand. In early 1979 the company was active in the North Sea from Aberdeen, in the North Atlantic from Agidar, in the Persian Gulf from Abu Dhabi, in the Arabian Sea from Bombay, in the Gulf of Thailand, in the South China Sea from the Philippines, in the Indian Ocean from Australia, in the South Atlantic from Rio de Janeiro, and in the Orinico River delta. A new international operations division formed in 1979 under J.C. Jones to manage overseas work previously overseen by the Vancouver and Montreal offices. At this time Australia kept two S-61Ns and four S-76s active. Five S-76s were added during 1979 and an NWT Air Herc transported a newly-overhauled 212 to Bombay and an S-76 to Thailand. Orders and options for 23 S-76s were placed in 1979.

Although Okanagan was busy, by 1980 it was not in the best of health. It was suffering from over-expansion and outmoded operating strategies. New helicopters now cost as much as $5 million—gone were the days of the $30,000 Bell 47. To improve its fortunes,

Okanagan had to lower costs, abandon unprofitable activities, and win more jobs. In its favour were its leading domestic and international positions, and its strength in operating, management and field personnel. Solutions were found. Engineering support was concentrated in Vancouver. The Montreal hangar was leased out. Increased use was made of in-house engineering capability and 20 helicopters were sold.

Liftair, eventually a subsidiary of Simmons Drilling of Calgary, was another Canadian international pioneer. Early contracts involved oil exploration in Tanzania and the Caribbean. In 1981 five Liftair Hughes 500s sailed for the Antarctic aboard a German research vessel. After launching the two 500s ashore with an advance party, the ship was holed by ice. The 500s helped in the rush to save as much as possible before the ship sank. In 1982-83 Liftair won another Antarctic contract with the Germans. In the same period three of its 500s went to Egypt for such work as sightseeing. One was wrecked while being delivered by an Egyptian truck. After arrangements generally soured, the Canadians decamped for Israel.

Canwest of Calgary also worked overseas. One of its co-owners/pilots was John Pridie. In 1965 he came to Canada from England, having such helicopter experience as spraying olive groves in Greece with a UH-12. At first he worked in the west for Foothills-Klondike Helicopters, then flew an FH1100 in Toronto for radio station CFRB (flying the famed traffic reporter Henry Shannon). While on a fire in the NWT in 1972 he had to land a Bell 205

hurriedly in a muskeg, when his windshield suddenly melted during a water drop! In 1973 Pridie and Wederfort went to Ethiopia on a two-year Tenneco oil exploration job. Canwest provided two Bell 47Gs and a Lama. While in transit by sea, the Lama disappeared for several weeks, turning up later in a warehouse in Mombasa. The Lama was ill-fated—it ended being blown up during an attack by Eritrean guerillas on a Tenneco camp. Wederfort was kidnapped and held for several months. A Bell 47 sent to try rescuing him also was destroyed. Pilot Grant Wyatt was captured, a nurse hostage killed. In the midst of this, Canwest's bank foreclosed on its loans and the company was sold to Air Log. Wederfort later formed Liftair; Pridie and Gordon Oliver formed Peace Helicopters. When the recession came, Pridie got out of Peace. In the early 1990s he was flying on exploration and relief work in Yemen and Somalia.

Oil and Gas in the Eighties

For years companies like Imperial Oil, Shell, Dome and BP explored sedimentary formations in the NWT for hydrocarbons. From the late 1950s-early 1960s Bell 47s supported ex-

The Bell 47 did much of the pioneer oil exploration in the high Arctic, starting with geodetic surveying in the early 1950s, then supporting field camps for the next 35 years. At left North Mountain Bell 47G3B2 CF-XFW is seen at Watson Lake on June 2, 1979. Then, Shirley's Bell 47J C-FKYS. The "J" was a four-seat version of the two/three-seat "G". (Kenneth I. Swartz, R.S. Petite)

The Hiller UH-12 also did solid work in early Arctic oil and gas exploration. The UH-12 was returned to production in California in 1994 by a new company, Hiller Aircraft Corp, under Stanley Hiller's son Jeffery. By this time about 1,000 original UH-12s remained in use. Like many Bell 47s, some had been converted to light turbine power, using kits developed by the US company Soloy. Here is a turbine UH-12—Joe Soloy was at the controls at Abbotsford in August 1979. (Kenneth I. Swartz)

The Bell 206 changed the helicopter scene overnight and immediately was adopted by the oil industry. Customers now demanded nothing less than a turbine. Here Trans North Turbo Air's Bell 206B, a 1972 model, was at Ross River in the Yukon on June 26, 1980. (Kenneth I. Swartz)

ploration down the Mackenzie to the Beaufort Sea, and east-west across the Yukon and NWT. In 1961 Peter Bowden Drilling and Dome Petroleum drilled a winter well on Melville Island, proving that exploration could continue year round in the high Arctic. Meanwhile, since the 1950s US oil companies had been acquiring rights in the Prudhoe Bay area east of Point Barrow, Alaska. Seismic surveys began. Atlantic Richfield picked a site in shallow waters at Prudhoe Bay and began assembling more than 3,000 tons of supplies to meet a drilling deadline of April 1966. It considered moving this down the Mackenzie from Hay River. Instead, an airlift was begun from Fairbanks using the C-46, C-82, Super Constellation and newly-certified L.100 Hercules. Setting up the site called Susie No. 1 took 142 flights over 21 days. The well was drilled for 10 months, but no petroleum was discovered. Drilling continued through the region and on March 13, 1968 Atlantic Richfield brought in the largest oil discovery in the USA—reserves of 9.6 billion barrels. This set the stage for construction of an 800-mile pipeline from the North Slope to the ice-free Alaska port of Valdez.

Prudhoe Bay encouraged Ottawa to get involved in Mackenzie Delta/Beaufort Sea exploration. It strengthened the resolve of the Panarctic Oil partnership, which had formed earlier at the encouragement of Calgary entrepreneur J.C. "Cam" Sproule. In 1969 a base camp was set up at Rae Point on Melville

(Right) A typical Beaufort Sea offshore drill platform scene. An Okanagan S-61 is delivering supplies by sling to the *Molikpaq*. (Marc Gilbert Col.)

Island. Setting up a drill took an average of 140 Hercules flights, so drills were moved overland by Cat train, when feasible. Panarctic sustained its camps by air, including with its own Electra and Twin Otter, and dozens of helicopters. Drilling took place on artificial islands that had wide beaches to absorb Beaufort storms. Dome Petroleum used a fleet of drill ships. Ice islands also were used. Sea ice was flooded and built up as a pad to support a rig and an adjacent ice strip was built for the Herc and Electra. By 1985-86, when oil prices tumbled, Panarctic had spent $750 million and drilled 174 wells. Although no great find was made, more than 173 trillion cubic feet of gas reserves were proved. Meanwhile, the Western Energy Accord was signed by Ottawa and the western premiers. While this helped the waning petroleum industry in Western Canada, it ended assistance to drilling in northern and offshore frontiers.

Through this era, the helicopter industry prospered across Canada. Sealand of St. John's, a company formed with a VFR fleet after Craig Dobbin acquired the licences of GEM Air in Winnipeg, bought AS330G Puma C-GUMP. Associated, Okanagan, and Universal were supporting rigs off Newfoundland, where six oil companies were spending $200 million. Dome had four drill ships in the Beaufort—*Explorers 1-4* . These were supported from Tuktoyaktuk by Bow's S-61N, several Bell 212s and a Bell 214B. When Bow closed its doors, Dome hired some of its pilots and engineers to create its own helicopter division, and acquired an S-61N, two S-76s and a BO105. Edmonton-based Shirley Helicopter (formed in 1962 by Ralph Shirley, who had a Ford car and truck dealership), filled some of the vacuum left by Bow in 1980, operating 15 fixed-wing aircraft and 60 helicopters, including 212s from Tuk for Dome. Twin-engine medium helicopters became more common as Alpine, Highland, Pacific and Quasar acquired 212s. Panarctic used 212s, S-61Ns and sometimes an S-64 constructing its ice island airports and floating drill rigs. Its annual airlift was legendary, with stories of S-58s, then S-61s and 212s doing night slinging in perpetual darkness.

The Venture and Hibernia finds of 1979 expanded offshore prospects for Nova Scotia and Newfoundland, giving work to S-61Ns, S-76s and Pumas. To carry the fuel to reach rigs nearly 200 miles offshore, an S-61N could take only five passengers, so companies turned to the long-range Aérospatiale AS332C (short-fuselage), then AS332Ls (stretched) Super Pumas. The first in Canada were AS332Cs delivered to Associated, Okanagan and Sealand in 1981-82. Sealand leased theirs (C-GSLH) from PHI for a two year contract with Petro Canada, initially off Nova Scotia. Okanagan's (C-GQRL) started with Dome. Associated's (C-GQYX) flew from Frobisher for Elf Aquitaine between Greenland and Baffin Island, later moving to Saglek, Labrador, thence to Halifax for Home Oil. The first of Sealand's $30 million order for six AS332Ls began arriving in late 1982 for Atlantic Canada and Brazil.

Early in 1982 Kenting introduced a pair of

The AS332 Super Puma was the answer to Canada's offshore helicopter needs. While the S-61 was out of production and scarce, the new Super Puma was available. It had the speed, range and payload to get the job done. Here Sealand AS332L C-GSLK sits aboard *Sedco 710* off the east coast. Then, C-GSLJ of CHC at St. John's in April 1991. It later was HC-BNB in Ecuador. (Halford Col., Brent Wallace)

(Above) Associated's AS332C C-GQYX at Edmonton Muni in February 1983. (R.S. Petite)

(Left) C-GHJZ-X was the first BO105 registered in Canada. Bill Loftus demonstrated it in 1973. Here it was in Vancouver that March 20. (Brent Wallace)

412s and several TwinStars in the Beaufort supporting Esso Resources Canada. This lasted two years, then Okanagan got the business and the 412s went to Thailand for a resources survey. In 1983 Okanagan introduced two 214STs on the east coast. Quasar acquired one of its

own. Along with a 212 it worked for Gulf Canada in the Beaufort. Meanwhile, Bristow Helicopters of the UK sought a 49% stake in Okanagan Helicopters. Announced in October 1982, it would take till December 1984 to win Ottawa's blessing, but only after a change of government. By then five AS332Ls leased by Okanagan from Bristow in the UK were arriving in St. John's. The Sealand-Okanagan rivalry intensified. A string of discoveries in the Beaufort and east coast spurred exploration. Thirteen rigs were active in July 1984, down from 16 a year earlier, but most of the extra Super Pumas got other work. Signs of a downturn in the offshore came when Ottawa confirmed that it would not renew exploration grants once they expired in late 1985. Soon after the first Arctic oil was shipped by tanker, world oil prices collapsed. What exploration budgets remained, moved to western Canada. Okanagan returned its AS332s to Bristow; Sealand sought work for its in South America.

Aérospatiale

Aérospatiale of France has had a major influence on the helicopter industry in Canada. This began in the 1950s with its ancestor, Sud Aviation. Modern-day Aérospatiale got its North American foothold in 1969, when Vought Helicopter Inc., a subsidiary of LTV Aerospace, formed at Grande Prairie, Texas to market helicopters. Within two months Alouettes had been delivered to 30 customers. Canadian Gary Robb joined VHI in sales after holding senior positions at Shirley and Haida Helicopters. Years later he related some of his experiences to Kenneth I. Swartz.

I sold the very first Gazelle, CF-CWN, Serial No. 1 to Canwest Aviation of Calgary. I sold dozens of Alouettes my first year on the job. Lots of people wanted to buy helicopters. I'd sell them as many as they could afford, then look around for an investor like a doctor or lawyer, or someone in real estate. I'd sell them a helicopter which they could lease to an operator. That was back in the golden days of the 40% capital cost allowance for depreciation. You'd claim the depreciation and find that the value of the helicopter would appreciate 20%. One of the things I did was size up investors who could buy more helicopters than the operator could. I knew a lot of them in the oil industry, like Evan Bullock and Ed Darvill.

Vought soon added the SA315B Lama and SA341 Gazelle. In 1974 Aérospatiale took over VHI, becoming Aérospatiale Helicopter Corp. That year it sold its first SA330 Puma to Petroleum Helicopters in Louisiana. The single-engine AS350 AStar was introduced in 1978, then the twin-engine AS365, and AS355 TwinStar in 1979. With the AS350 AHC established a major facility at Grande Prairie, supporting AStar sales and the June 1979 order for 96 HH-65A Dolphins for the USCG. These were to be assembled and Americanized with Lycoming engines. By 1981 there were 715 AHC machines in service—20% of the US market. A sharp downturn in the market followed that hit all major manufacturers.

Following success with the SA330 Puma medium helicopter, Aérospatiale launched the AS332 Super Puma at the 1975 Paris Airshow. The Super Puma had more power (1,800-shp Turbomeca Makilas), better range, and improved crashworthiness, e.g. reinforced cockpit and cabin floor, main landing gear attachment frames and tail structure. The undercarriage was designed to absorb some crash impact, and could be kneeled for stowage or transport. (this proved useful decades later when Super Pumas started travelling the world in AN-124s). The prototype Super Puma, a modified AS330, flew in September 1978. The first production AS332C appeared in early 1980. The lengthened AS332L flew in October 1980. It offered increased gross weight to 18,080 pounds. Petroleum Helicopters placed the first Super Puma order—three AS332Cs and three AS332Ls, but when its first AS332C was delivered in November 1981, it went immediately to Sealand as C-GSLH for a two year contract with Petro Canada supporting the *Bow Drill 1* off Nova Scotia. Sealand also took PHI's first AS332L in February. Sealand subsequently ordered six "Ls", all of which were in service by the end of 1983. Okanagan took two "Cs" in April 1982.

More of the Puma

In 1975 a SA330G Puma was on tour across Canada. When Rocky Mountain Helicopters of Utah backed out of a deal to buy it, Heli Voyageur of Val d'Or took it on consignment. Re-registered C-GUMP, it sat over the winter, then came an offshore contract with Total Eastcan. At the end of the Vietnam War, many American and Vietnamese helicopter pilots came to Canada. Some were hired by Heli Voyageur for James Bay, including ex-US Marine Paul Dorfman, the first Heli Voyageur pilot checked out on "GUMP". He and Rob Freeman were trained by Aérospatiale while in Labrador. Bernard Dasté, an ex-French military Puma pilot, was the third pilot on GUMP. There was lots of offshore work for GUMP for the next two years from Halifax, Inuvik and Goose Bay.

While he was a policeman in Mississauga, Ontario, Rob Freeman would hang around Toronto International Airport, enjoying the action during time off. His dream was to get into aviation. He finally left policing, enrolled at Canadore College in North Bay and, in 1974, graduated as a helicopter pilot. He joined Heli Voyageur to fly a Bell 47G in James Bay. In July 1978 Freeman left Heli Voyageur and joined Sealand, which was about to purchase GUMP. The sale was pending when Heli Voyageur got an unexpected request for the Puma for a rush job overseas. When Freeman called Val d'Or to make arrangements to pick up GUMP, he was told that it was on the high seas heading for West Africa! Not surprisingly there was bad blood between the two companies after this.

The Puma was not overseas for long—the contract didn't prove a money maker. In a few months it was flown to Marseilles after having a lot of technical problems in Africa. In early 1979 GUMP returned to Canada aboard a container vessel. It was not well cocooned; the sea was so rough it even washed a container overboard. "I went down to Halifax to fly the aircraft from the docks to Halifax airport," recalled Freeman. GUMP was in sad shape, covered in sea salt and oil. Once it got to the airport, it was grounded for three months for maintenance.

Craig Dobbin saw the offshore starting to grow, so went looking for equipment. First he picked up GUMP, then bought GMNP new from Aérospatiale in Texas in July 1979. Meanwhile, Okanagan, which was short of helicopters so couldn't take advantage of offshore work, was playing hard ball in Ottawa, stalling Sealand's IFR operating certificate. In the end Sealand and Okanagan got together. Each would contribute a pilot to each helicopter, and the mixed crew would fly together, using Sealand's Pumas and pilot expertise, and Okanagan's contracts. Paul Dorfman and Rob Freeman now were back together on GUMP, flying for Petro Canada from Goose Bay and elsewhere in Labrador. In 1980 off season, PHI leased Sealand's Pumas for the Gulf of Mexico and Canada's east coast.

In the early 1980s Petro Canada decided to try the S-76. Sealand acquired C-GSLA and GSLE through PHI. The flying involved carrying a lot of tools and other heavy cargo, but the S-76 lacked cabin volume; oversized equipment had to be slung. After two years the S-76s disappeared. In 1981 Dobbin decided on the Super Puma. It proved ideal for long-distance flying—it had a huge cabin, and with external tanks endurance exceeded three hours. The Super Puma was certified in 1981 with more range and payload than an S-61N. Freeman recalled, "In October 1981 we picked up the first aircraft at Grand Prairie. This was interesting, since Bob Suggs of PHI wanted to be first with a new Super Puma. Ironically, we checked out on type at PHI's main base in Lafayette, Louisiana. On November 17 a huge Petro Canada party was scheduled in conjunction with the new rig *Bow Drill 1*. Everything was set, including cocktails for thousands. The company wanted the Super Puma there. The weather was poor and we ran scud from Lafayette half way to Halifax with the IFR installation in the aircraft not quite complete.

"We landed in a field about 10 miles south of Halifax and cleared customs right there. The fly-by was scheduled for 2:00 PM and Transport Canada finally signed off the paperwork at 1:15. Ex-U.S presidential pilot and Aérospatiale flight operations director Jim Creighton and I were flying. We made a bow to the crowd from three angles. Thus did Petro Canada unveil a new rig and helicopter on the same day." Freeman stayed with Sealand until July 1987, moved to Ontario Hydro as chief pilot, then joined Transport Canada doing flight safety programs, then check rides and inspections.

In January 1984 Okanagan began operating its first AS332L offshore from St. John's. In March it added two more. About this time Sealand accepted its 8th Super Puma. In early 1984 six were working off the east coast and two were on lease in Brazil. Three more Super Pumas were due for delivery in 1984, with two more on option. In January 1988 Canadian Helicopters Corporation, came into being through the merger of Sealand Helicopters, Toronto Helicopters and Okanagan Helicopters. By this time CHC had 11 Super

Pumas—five in Canada, four in South America, one in the USA and one in New Guinea.

Towards a Fall

Almost every region was busy in the late 1970s; more than 100 new helicopters were being imported yearly. Okanagan remained dominant. For logging, offshore and overseas it had 16 Sikorsky S-61Ns through 1979, including some of the last off the Sikorsky line before production ended in 1980. Okanagan recently had bought Bow (an S-61N, three 212s, a 214B-1, an AStar and some Jet Rangers). Bristow Helicopters was courting Sealand. Highland Helicopters, then Okanagan, placed the first Canadian orders for a Boeing 234, although these machines were not delivered. Northern Wings merged its helicopters with LaVérendrye, Maple Leaf Helicopters formed through the merger of Geddes Contracting and Deltaire, and Midwest Helicopters was sold to its employees by PWA. In 1979 VIH introduced the Bell Long Ranger II and turbine Soloy-Bell 47T in Canada. Viking, a 50% subsidiary of the Algoma Central Railway, ordered

Bow was one of Canada's first Aérospatiale AS350 operators. C-GBVS was at Calgary on August 25, 1978, soon after delivery from Texas. (G.L. Marshall)

16 Hughes 500Ds to add to its fleet of 50. Okanagan accepted its first two Sikorsky S-76s in April, shipping them to Thailand and Australia. Bow, Helicraft and Maple Leaf introduced the AS350 AStar with the Lycoming LTS-101. Another 80 AStars were on order from Canadian firms, including Canada's first Arriel-powered AS350Bs for Frontier. Olympic Helicopters of Montreal became the only Canadian company with the single-engine AS360C Dauphin (lost soon after delivery). The Royal Thailand Police fleet (six Bell 204Bs and nine FH1100s) was imported for re-sale by a Calgary group. Meanwhile, James Bay continued and, as if to show the industry's daring, Trans Quebec and Bow were running an experimental "Helibus" shuttle between Dorval and Mirabel using an S-61N. Overseas, Associated had a 212 offshore in Morocco. Okanagan had contracts in Australia, while other orange and white choppers worked from Abu Dhabi to the Philippines, Thailand, and Brazil.

Just as the stage seemed set for greater things, along came a worldwide recession. As fast as the industry had blossomed, it collapsed. Western operators suffered the first setback, when oil and gas suddenly cut exploration in reaction to Ottawa enacting the 1980

National Energy Program. The NEP placed the price of Canadian oil below world levels, favoured grants to Canadian-owned oil companies (discouraging offshore exploration), etc. Aggravating matters was industry deregulation. After years of filing long-term tariffs with the ATC for contracts of 30 days or more, operators now were free to set their prices, but now no one knew what competitors were bidding. This sent rates tumbling—instead of the 20% annual growth rates of the 1970s, demand went flat. With many helicopters, but too little business, declining revenue hours were the story. Operators cut rates to generate cash flow and pay debt. Although 1980 revenue hours increased over 1979's by 16% to 541,049, they slowed to 2% in 1981, then plummeted by 26% in 1982 and a further 17% in 1983 to 340,110 revenue hours. This loss of 200,000 hours would have kept 335 helicopters flying about 600 hours each yearly. Exports of helicopters to the US, Asia, and Europe helped reduce the surplus.

Canada's 1981 commercial fleet totalled 959, about the same as the previous year, but only six new helicopters were registered for the year. For the first time since at least the mid-1960's the number of machines declined in 1982. Hours dropped by 25.6%. On a per helicopter basis, average annual utilization fell 23.7%, to 439 hours, something not seen since the early 1960s. BC resumed its place as the largest Canadian helicopter market in 1980, replacing Quebec, where James Bay was slowing. Alberta was the second domestic market in 1982. Although oil and gas work was down, numerous forest fires kept Alberta busy. In face of recession, some companies merged, others folded. In December 1981 John Lecky's Montreal-based Resource Services Group, which increased its interest in Okanagan in 1973, gained control of the company. In February 1982 it took an additional 63%, raising its share to 98%. Soon Okanagan was wholly-owned.

Shirley of Edmonton entered receivership in March 1982. Operations ceased in May. Heli Voyageur of Val d'Or halved its James Bay fleet and diversified by going offshore. Silver Grizzly left Prince Rupert in 1982, when the Japanese market for unprocessed logs tightened. By early 1984 Okanagan had about half the 900 employees of 1982. There were fewer bases and helicopters. Other than China, India and Thailand, offshore was quiet. It didn't help when world oil prices fell in 1985-86, further discouraging exploration. The fleet bottomed at 945 aircraft in 1986, then gradually came back. According to Ottawa's Aviation Statistics Centre, Canadian helicopter operators logged 297,346 hours in 1988, about equal to the mid-1970s. BC was the largest market, accounting

for a third of domestic activity. Then came Quebec, Alberta, Ontario, the NWT, Newfoundland, and Manitoba. The greatest increases were in Manitoba (22%), BC (17%) and Alberta (8%). Cutbacks in the offshore meant fewer hours for Newfoundland, Nova Scotia and the NWT. As of June 30, 1989 there were 1,019 commercial helicopters registered in Canada.

Overseas in the Early 1980s

During the 1980s there was an overseas shift from offshore support to drill rig moves and seismic surveys ashore. Okanagan expanded to Australia (S-61, S-76A) and China (S-76). It also won contracts to train Chinese to fly the S-76A and, later, the Bell 214ST. After more than 10 years in India, Okanagan was the last foreign helicopter operator there before helicopter support was nationalized. Okanagan was the first foreign company invited back after foreign oil companies grew unhappy with the safety record and performance of government-owned helicopters. Meanwhile, Thailand remained important.

Sealand ventured into South America in 1981 with SA330J C-GSLA flying offshore from Belem, Brazil. Later work saw two Super Pumas at Belem. An offshore contract in Africa saw a TwinStar and Super Puma cross the Atlantic on a drilling rig. A storm blew the Super Puma on its side. A subsequent disagreement over the insurance payment contributed to the end of Sealand's partnership with Petroleum Helicopters Inc. New Sealand contracts were won in Trinidad and Colombia in 1985. AS332C C-GSEM worked with a joint venture company on a pipeline in the Andes, replacing a PHI Bell 214ST. It aided in rescue efforts during the eruption of a local volcano. In 1986 Sealand formed Helican, a joint venture company supporting exploration by Conoco in the Ecuadorian rain forest. Helican's fleet grew to five Super Pumas and a Bell 212 before exploration declined in late 1989.

Viking Helicopters

Viking Helicopters was founded by Larry Camphaug of Ottawa, and named for his nordic heritage. From a lone Bell 47G, Viking expanded across Canada. In 1973 it won an offshore oil contract in Tanzania. A UN World Health Organization contract (1974-79) supported smallpox inoculation in Ethiopia. Other WHO work was in West Africa, spraying for "river blindness" disease. This involved 10 Hughes 500Cs and three Turbo Porters. Viking kept this work till 1986, when it went to US operator Evergreen. Viking's first international petroleum contract was in Sudan between 1977-85. It was in Madagascar 1984-86, supporting Amoco Oil seismic surveying; then it went to Turkey on seismic work for Esso.

Northern Wings absorbed Autair in the late 1970s , then merged with LaVérendrye in 1978. Provost Corporation, active in trucking and buses, acquired LaVérendrye in 1980, Viking in 1981. In the fall of 1983 it merged it with LaVérendrye. Operations centred at St-Clet, west of Montreal. When Provost took

Viking Bell 205 C-GVIK practices slinging in the back field at company headquarters at Carleton Place, Ontario on December 3, 1976. A Viking Bell 47G and some Hughes 500s complete the scene. (Larry Milberry)

over Trans Quebec and Trans Canada Helicopters in early 1987, HQ shifted to neighbouring Les Cèdres. In 1989 Viking was active in Gabon on a seismic program for Conoco. This involved four TwinStars and two Bell 212s based at Libreville. With tall trees surrounding jungle camps, pilots delivered loads from 200 foot long lines. On September 11, 1989 Viking was purchased outright by CHC. With 63 helicopters it was Quebec's largest operator and the second largest in Canada. It remained an independent CHC unit till merged with Canadian's Eastern Division in 1996.

Heli Logging

Logging brightened in the late 1970s. As road construction costs soared, inaccessible timber could be harvested economically by helicopter. To equip its new OK Heli Logging division, Okanagan purchased an S-61L and short-body S-61A from Boise Cascade, and two S-61Ls from New York Airways. Erickson S-64Es were logging on the north BC coast with subsidiary Silver Grizzly, as were a Columbia Helicopters Vertol 107 and Evergreen S-64. By 1981 Okanagan had three S-61Ls in BC and 55% of the BC heli-logging market. Each helicopter was slinging between 17 and 35 cunits of timber hourly (one cunit = 100 cubic feet). US operators Evergreen, Erickson and Siller Bros. also were active, using S-64Es. Erickson had two S-64s flying from an old freighter north of Prince Rupert. They were supported on and off by an Alouette II, Hiller UH-12E, Jet Ranger, Bell 204B and S-58T. Whonnock Industries (which later became Helifor) was further south with a Columbia Vertol. Elsewhere, the Bell 214B was a popular timber hauler with Alpine, Can-Arc, Okanagan, Quasar and Transwest. Late in the decade Can-Arc, Coulson Aircrane and VIH acquired logging S-61Ns.

In the early 1990s VIH was looking for an S-61 heli-logging replacement. Having heard about the Kamov KA-32, it sent Ken Norrie and Bill Ross to Papua New Guinea to evaluate it. This type first appeared in 1981 as a commercial offshoot of the military KA-27. Like earlier Kamovs it had counter-rotating blades, so needed no tail rotor. This afforded great stability, an asset when hovering to select logs

from a tall stand. Two 2,235-shp NPO Klimov TV3-117 turbines were more than ample for lifting 11,000 pounds. The 6,000-fpm rate of climb was timesaving when logging steep slopes. At all-up weight the Kamov still had a 10% power reserve, so pilot stress was lessened. VIH quickly imported a KA-32. It came with several pilots and engineers. It logged successfully on Vancouver Island into the mid-1990s.

In 1987 Super Puma C-GQRL appeared on the BC logging scene. It had gone new to Okanagan in 1981 for offshore work. Eventually it went back to Aérospatiale in Grande Prairie from where it was acquired by Hydra Management Ltd. for heli-logging. BC operations began in July 1987. On October 3 it was working a few miles north of Campbell River. The crew had just refuelled and was starting to log with a 150 foot long line. QRL was seen to turn downhill with a load, when people heard a

"twang", followed by an explosion and fire in the engine and main gearbox area. The load was jettisoned, but main rotor RPM decreased rapidly and the helicopter crashed and burned. The event took just seven seconds and claimed the lives of both pilots.

Investigators found that the main gearbox flexible mounting plate had failed from fatigue just after the helicopter had picked up the logs. This let the gearbox rotate, misaligning the drive shafts from both engines to the gearbox. The drive shafts contacted and failed. The right engine oversped, then shut down automatically when the overspeed protection circuit cut in. The left engine then oversped because the circuit does not allow both engines to be shut down. It failed because of centrifugal overload of the first stage power turbine disk. Now rotor rpm decayed and the pilots did not have sufficient time to autorotate. It was learned that the true number of high-load cycles applied to the mounting plate during heli-logging was not considered in the manufacturer's determination of the service life of the plate. The operator did not detect cracks in the plate during inspec-

CHC S-61L C-FOKB logging in Howe Sound, 30 miles north of Vancouver on October 26, 1988. Then (above), a Canadian Air-Crane S-64E near Port Alberni on May 11, 1992. The helicopter allows loggers to reach previously inaccessible stands and work with a bit more discretion than in clear cutting, causing less environment damage. (Brent Wallace, Kenneth I. Swartz)

VIH's KA-32 RA31000 over Abbotsford in August 1989. It later returned to Russia. In April 1997 two others started work on Vancouver Island. (Brent Wallace)

tions before the accident. Shortly after the crash, Aérospatiale issued an airworthiness directive reducing the service life of the plate. Logging operations now required a newly-designed part and the replacement of other cycle-sensitive drive-train components.

While heli-logging grew, the more traditional role in forestry continued. After nearly 200 helicopters were called out on fires on two consecutive summers in the early 1980s, Alberta reconsidered its initial attack (IA) strategy and expanded its use of medium helicopters on season-long contracts. BC developed rap-attack (rappel attack) crews. Frontier Heli-

copters was acquired by Conair Aviation in the late 1970s and started to specialize in retardant delivery systems. Besides operating 205s and a 212 in Canada, it swapped helicopters with the National Safety Council of Australia, and operated seasonally in southern France and Spain in summer. Later Frontier added markets in Italy, Portugal, Chile and Mexico. Another fire fighting tool, SEI Industry's Bambi Bucket, became an international success since first used on Canadian fires in the early 1980s.

Scheduled Services
Since the days of Bernard Sznycer and the Grey Gull, operators in Canada dreamed of the practical helicopter sked service. Natural limitations kept the helicopter out of this field till the 1980s, when Helijet Airways of Vancouver opened service between Vancouver and Victoria with an IFR Bell 412, then built a fleet of

S-76s. While full profitability was elusive, in 1990 Helijet entered its fourth year, flying thousands of passengers, showing what was possible with the right routes, fares, equipment, financing and heliports. In 1997 Helijet added Victoria-Seattle skeds. Flying time would be 34 minutes ($92 fare) compared to the two-hour ($66) hovercraft sked.

In December 1989 Canadian Helicopters opened skeds to Whistler Mountain, Vancouver Airport and Vancouver Harbour Heliport. This helped raise the profile of the helicopter sked. Meanwhile, Ranger/Air Canada in Toronto, Cougar in Halifax, and Bow/Trans Quebec in Montreal all flirted with skeds in the 1980s, but were thwarted.

Emergency Medical Services
A growing helicopter market since the 1960s, especially in the US, has been emergency medical service (EMS). No argument was needed to sell the idea of helicopters rushing highway traffic victims to hospital, other than the dollars-and-cents of it. Helicopter EMS, both privately and publicly funded, crept into Canada in the 1980s. The former was represented by the Southern Alberta STARs program, operating an MBB BK117 on contract from ALC Airlift in Calgary. In contrast, Ontario's government program was part of an in-

Two of the world's most successful downtown heliports. Vancouver's (right), a concrete floating structure with full lighting and fuelling services, appears in a November 1991 view. Operated by Helijet, it was financed by Ports Canada. Then (below) Victoria's heliport in November 1995. Such West Coast facilities resulted from years of good work by dedicated committees like Vancouver's Harbour Heliport Society. (Brent Wallace)

(Above) After starting operations with an IFR Bell 412, Helijet Airways built a fleet of 12-passenger S-76s. The example at the top, seen at Vancouver, was specially painted for the 1994 Commonwealth Games held at Victoria. Then, HJV over Whistler on November 29, 1991, while Helijet was operating a short-lived Vancouver-Whistler sked. (Brent Wallace)

(Right) Ranger Helicopters had a contract operating an Air Canada shuttle between LBPIA and the Cherry Street heliport on Toronto harbour. AS350B C-GATX is shown at Cherry Street in May 1987. This service was short-lived—the heliport was in a remote and seedy neighbourhood, and the time saving benefits were marginal. (Kenneth I. Swartz)

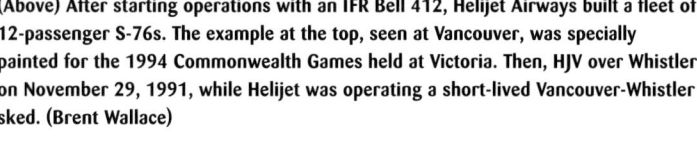

(Above) Typical EMS machines. A patient transfer from Huisson Aviation Bell 212 C-GAHZ at Moosonee on January 22, 1992. Then, ALC Airlift/STARS BK117 C-GDGP at Calgary in October 1992. STARS (Shock Trauma Air Rescue Society) was a privately-funded organization. (Larry Milberry, Brent Wallace)

(Left) Canada Coast Guard BO105 C-GCHW at Tofino on April 9, 1992, (Larry Milberry)

(Below) CHC AS350 AStar C-GRGJ over Vancouver in the early 1990s. This five/six-seat type gradually replaced earlier Jet Rangers, being preferred for its greater roominess, comfort and speed. First flown in 1974, within 20 years there were some 2,000 worldwide. Then, 212 C-FMPZ at Mike Wiegele's Heli-Ski Village on February 14, 1993. As many as six Bell 212s and three AS350s worked here at a time in the 1990s. Some skiers are being briefed before the day's fun on the choicest of slopes. (John McQuarrie, Brent Wallace)

tegrated system including contract helicopters (Bell 212, S-76), fixed-wing air ambulances, and designated trauma centres. BC's EMS program, e.g. in Prince Rupert, proved a success with a VIH 212, then a 222. By early 1990 EMS expanded to Quebec, when Heli-Max based a BO105 at St. Hubert .

Government Air Services
In the 1980s helicopters were appearing in greater numbers in government fleets. The Canadian Coast Guard introduced 16 BO105CBSs through the decade. This eliminated most CCG single-engine, over-water ops. CCG hopes for a large EH101-class helicopter faded when plans for Canada's huge *Polar 8* icebreaker were killed. The RCMP added some Jet Rangers and Long Rangers in the 1980s. Plans for medium helicopters to carry emergency response teams were discussed until arrangements were made to use CanForces Twin Hueys.

In 1981 Nova Scotia acquired a Bell 212 for fighting forest fires, then added an MD500E. Alberta bought a Bell 222UT for utility work in the north. Ontario's Ministry of Natural Resources bought four Long Ranger LIs and a BK117 for fire suppression. Based in Sudbury, the BK117 was used with IA crews. In 1988 Quebec added an executive Bell 222. Since the

number of provincially-owned helicopters used in law enforcement and EMS still was small, this market was expected to be active through the 1990s. Calgary became the first Canadian city with a dedicated police helicopter.

Mergers
In late 1986 Sealand Helicopters acquired Toronto Helicopters. Early the next year it made an offer to the Resource Services Group and Bristow Helicopters for their Okanagan shares. This led to the creation of Canadian Helicopters Corp. CHC was the product of 10 years of mergers and consolidation. Besides Okanagan, by the early 1990s it included Associated, Canadian, Bow, GEM, Heli Littoral, Heli Voyageur, Horizon, Lakeland, LaVérendrye, Maple Leaf, North Star, Northern Wings, Ranger, Sealand, Toronto, Trans Canada, Trans Quebec, Vernon and Viking (as well as their predecessors).

The Mid-Nineties
On June 30, 1993 there were 1,541 civil helicopters registered in Canada, 75% being commercial. With 250 machines Canadian Helicopters was the largest operator (and the second largest civil helicopter fleet outside the USSR). Northern Mountain had about 50, Highland

and VIH about 40 each. Many upstart companies appeared, usually with a single Jet Ranger or R-22, but the occasional Bell 47G and UH-12E still seemed suitable for the beginners. Only 62 commercial helicopters had summer fire contracts over 1993. However, the price of wood products was high, keeping heli-loggers busy. Environmental concerns also pointed the way to selective logging by helicopter, although environmental groups soon were targeting the heli-loggers with their patent vigour. Several S-58Ts and S-61s, two S-64Es, two Vertol 107s, a Boeing 234 and a Kamov were BC's primary logging machines. Meanwhile, gas and oil prices were high, so heli-seismic crews were busy in Western Canada. Northern Mountain, for example, had three Bell 212s on this work. Prospectors were busy in the NWT, where the diamond rush was going full tilt. To meet demand, Great Slave Helicopters of Yellowknife expanded its fleet in 1993 by nine Jet Rangers, three MD500Ds, two AS350Bs and a 204. EMS services were well-established by now, e.g. Ontario had four S-76s and two Bell 212s under contract to Canadian Helicopters at Sudbury, Thunder Bay and Toronto. Heli-skiing was growing in BC and sightseeing at Niagara Falls was lucrative. Overseas, Canadian companies were involved in UN work, fire

An overview of Bell Helicopter Textron Canada of Mirabel with a Taylor Energy Long Ranger III overhead. Then (left), a pair of Long Ranger IIIs in May 1991 before ferrying to the US for the Los Angeles Water Power Commission. In 1997 BHTC delivered the 4,400th Jet Ranger. Counting OH-58s and Long Rangers, 9,000 of this helicopter had been built since the first example, the experimental OH-4, was demonstrated to the US Army in 1964. (BHTC/Jean Huneault)

fighting and mineral exploration. Historic Liftair was one of the companies that went out of business in 1993.

Among the major Canadian helicopter stories of the mid-1990s was the EH Industries EH101, a medium military type being developed in the UK and Italy. Canada ordered 40 to replace its aging Chinooks, Labradors and Sea Kings, but the $3-billion deal became a political hot potato. The Liberals promised to axe the deal, if they were elected in the 1993 federal election. Once elected, they quickly kept their promise. While the EH101 project went down the drain, the CanForces were encouraged by a promise of 100 new Bell CH-146 Griffons. As to BHTC, it expected to deliver 208 helicopters in 1993, about the same number as 1992

A Helicopter Industry
By 1981 Canada, with the second largest civil helicopter fleet in the non-communist world (after the United States), had the highest ratio of turbine-to-piston machines—80%. With at least 100 helicopters being imported annually there was renewed interest in a domestic manufacturing industry. The Department of Industry, Trade and Commerce formed the Helicopter Inter-Departmental Committee to see if military and CCG requirements would support an industry. Its report noted the Canadian commercial fleet growing at 8% yearly. The fleet would reach 1,995 by 1992, and 2,860 by 2000. Some US$1.5 billion would be spent on new helicopters.

Other projections, based on conditions of the early 1980s (before the impact of the coming recession and the fall in oil prices), suggested that Canada's civil and military market had too many models and weight classes to provide the volume to support manufacturing.

It suggested that Canada seek a new helicopter program with worldwide appeal. This was not a new concept. In the 1970s Bell, P&WC and Ottawa had co-operated to develop the P&WC PT6T TwinPac, and Bell Model 212/UH-1H. This proved most successful, with thousands of engines and helicopters sold. Now the most promising concept recommended was a twin-engine helicopter with a maximum take-off weight between 4,410 and 7,500 pounds. A second was for heavy twins suitable for the military and for offshore support. A third idea suggested the design and manufacturing of high-value components, such as helicopter transmissions, drive shafts and rotor systems. Any undertaking would capitalize on P&WC's turboshaft engines.

In December 1982 Ottawa asked leading manufacturers for proposals, specifying that they would be expected to create an export-oriented industry. There would be no link between any new industry and government orders, but industry was interested, especially with the prospect of government financial support. Provincial governments commissioned studies, anxious to win new jobs for their regions. Aérospatiale, Bell, MBB and Sikorsky submitted proposals. Aérospatiale was interested in a Canadian outlet for its single and light twins and viewed its medium AS332L Super Puma as a replacement for Canada's aging military helicopters. Light twins were MBB's focus. First flown in 1967, its BO105 was considered the first practical light turbine twin. Sales were made to the German army, foreign militaries, the oil industry, and EMS operators. Developed with Kawasaki, MBB's larger BK117 served a slightly upscale market. Having a Canadian component manufacturer would give MBB a base from which to seek

government orders. Sikorsky's main interest lay with its SH-60 Seahawk as a Sea King and Labrador replacement. But the manufacturing base it had in Canada recently had been dismantled, when P&WC closed its Helicopter and Systems division.

The Bell Deal
In October 1983 Ottawa, Quebec, Bell Helicopter Textron Canada (BHTC) and P&WC announced a $514-million agreement to build a light twin-engine helicopter for world-markets. P&WC would provide the power with its PW200 series of light turboshafts. Ottawa offered much of the cash to get the project moving, most being used to establish facilities near Mirabel airport. The agreement covered BHTC manufacturing the Bell 400 Twin Ranger, and co-development of the Bell 400A and composite-construction 440. All would have the same rotor and dynamics system. P&WC would launch a new technology engine for which Ottawa offered $100 million in repayable loans (development cost: $252 million). Ottawa would provide more than $162 million towards the Bell program. Quebec would chip in $110 million. Bell envisioned 40% of the Model 400 being manufactured in Canada, climbing to 75% of the 400A and 85% of the 440. Bell's facility in Texas would manufacture components.

The Twin Ranger was development of the Long Ranger using twinned Allison 250-C20Bs, and the rotor and dynamics system of the US Army OH-58D. A proof-of-concept version (modified Jet Ranger) was due to fly in spring 1982, followed by first flight of a prototype by May 1984, and certification in a year. Asking price was $700,000-$800,000. Bell claimed more than 80 $10,000 deposits from customers. Through late 1982 it studied the 440 with two P&WC engines, development of which had started earlier when Bell challenged P&WC to develop a high technology 400-shp engine (a decade earlier P&WC had proposed a similar turboshaft for the Bell 222, but lost to Lycoming). P&WC's new engine would use half the parts of the PT6. Subsequent discussions defined the requirement for a 500-shp engine with a combining gearbox.

In the winter of 1983 Bell began recruiting employees for Mirabel. About 200 engineers, mechanics, manufacturing specialists and a few pilots were sent to Texas to train. The first executives for Mirabel, people like president Jim Schwabe, came from Bell in Texas. Some CanForces flight test specialists were recruited from AETE. The prototype 400 flew at Bell in Texas on July 4, 1984. The second prototype was scheduled to fly in December 1984. Certification was on track for late 1985/early 1986. By now the factory in Mirabel was going up on a 151-acre site. The first assignment was to manufacture 256 Bell UH-1 tail booms for the US Army. This provided experience and was a stopgap till the 400 started. Of course, BHTC had its sceptics. Some felt vindicated in this as recession hit and the helicopter industry stumbled. Orders for existing types and the Twin Ranger fizzled. In July 1986 BHTC, Ottawa and Quebec agreed to skip the 400, and jump to the PW209T-powered 400A and the all-composite 440. BHTC also would expand into research, design, development, production and product support, and become home for 206B and 206L-3 Long Ranger production. Bell in Fort Worth concentrated on US military projects and retained overall direction of commercial programs at Mirabel.

The first truckloads of Jet Ranger tooling and fixtures reached Mirabel in August 1986. Work began on near-complete helicopters from Texas—installing interiors and painting to customer specs. Of the first 100 deliveries Japan led with 29, followed by the US (26) and Mexico (16). Production of the Long Ranger III commenced at Mirabel in early 1987, the 212 in mid-1988; the 412 in 1989. The 500th Mirabel delivery was a Long Ranger III for Air Logistics. To accommodate increasing demand, BHTC added a preflight hangar and expanded the paint shop. It also subcontracted considerable sheet metal production to AWSM Ltd., a local business; and a second shift was added for painting, and some sub-assembly and bonding. Employment rose from 675 in September 1988 (when the 100th helicopter was delivered), to 850 in January 1989 (when the 250th was delivered). BHTC reported sales of $130 million by the end of 1989. By November 1991 600 helicopters had been built, 99% for export. Production passed 800 late in 1992. BHTC was by now the largest civil helicopter plant in North America. But this rosy picture would not last. The onset of recession in 1991 and the Iraqi invasion of Kuwait caused new orders to dry up, leaving BHTC with surplus capacity. A four-day week was instituted (instead of a 20% cut in jobs), although engineering kept busy on future products. Demand for helicopters remained weak, especially for 212s and 412s. Long Ranger sales suffered when Evergreen and PHI cancelled substantial orders. Bell, however, still expected the sale to Ottawa of 100 Model 412s. Its engineering talents were used on other projects. The DND contracted studies associated with the CF Light Helicopter Program, and work was done for EH Industries (Canada) as part of its proposal to build EH101s in Canada. EHI envisioned Bell taking a role in assembly and flight test, but demand for space at Mirabel made the idea unworkable.

Model 230

In early 1989 Bell ended production of its twin-engine 222 at 182 machines. Yet it was convinced that a market existed for this helicopter with its 3,000 to 3,500-pound payload. It considered re-engining the 222 with the P&WC PW200B, but had no launch customer. The CanForces and Coast Guard, however, were looking for a multi-mission twin. While neither program came to fruition and P&WC needed more time to develop the PW200, Allison had the 700-shp Model 250-C30G/2, an enhanced version of the Long Ranger III engine. It put out greater power on hot days or in the event of OEI (one-engine inoperative). It also had better fuel consumption. Here was the solution to Bell breaking from the Lycoming LTS101 (the 222's weak point). At the 1989 NBAA convention in Atlanta, Bell announced its new design, the Allison-powered Model 230. It would be the first turbine helicopter developed in Canada. A Bell-Ottawa-Quebec agreement was signed in October 1990. Plans included deliveries 29 months after go-ahead.

Parallelling the 230, BHTC adopted new management practices to improve efficiency. Inspired by Japanese "total quality" management, president, V-Ps, directors, engineers and workers regularly gathered to deal with the nuts and bolts of the 230 program. Bell also introduced "concurrent engineering". Previously, engineering would design a helicopter, then release the drawings to manufacturing and tooling. Invariably, they would conclude that the tooling was too expensive, or that such and such a part couldn't be built. All steps were now being done concurrently with departments working together from the outset. Tooling received advanced drawings and feedback came before design was frozen. In the end there were few radical changes.

Engineering for the 230 began in March 1990, peaking in late summer. In the fall two early 222s arrived for conversion to 230 prototypes. The lead ship, C-GBLL, was the sixth 222 prototype, used for 222B (corporate) and 222UT certification. The second, C-GEXP, was the third 222UT. The first step in building the prototypes was removing their LTS101s. Most changes were structural, e.g. a new engine deck, fire walls, intake cowl and engine cowls. The prototypes started coming together in April, with engines installed on the lead aircraft. Twelve men from Mirabel's customization department did the modification work, supervised by engineers from Bell's New Product Development Centre in Texas. Fabrication of the single-piece cabin and tail was contracted to Fleet Aerospace in Fort Erie. Fleet used modified 222 tooling to build the sheet metal cabin, with titanium engine deck parts supplied by Bell. Cabins reached BHTC by truck. Installation of engines, transmissions, landing gear and avionics was done in final assembly.

The prototypes were unveiled in August 1991, with first flight that month. There were many tests: torsional stability, engine vibration, drive shaft vibration, oil systems, fire extinguishing, power availability, cooling, noise, snow, etc. Work progressed on schedule and on September 11, 1991 Bell invited 160 guests for

The Bell 230 prototypes at Mirabel on September 11, 1991. There were 40 built of this type. Then, a prototype 430 flying there in December 1995. (Kenneth I. Swartz)

Bell 430 Specs

Length: 44' 1"
Width: 11' 3"
Max gross weight: 9,300 pounds
Useful load: 4,015
Engine power (takeoff): 715 shp (times 2)
Engine power (max continuous): 613 shp (times 2)
Max cruise speed: 135 kt
Range: 350 nm

the official 230 rollout and first public flight (C-GEXP). On hand was a large Japanese delegation from Mitsui & Co., the 230 launch customer, and Taiyo Kogyo, the largest 222 operator. Test pilots Don Wolfe and Leo Melin did the flying (first flight had been on August 12). A month of hot-and-high trials followed at Leadville, Colorado. Coming home the 230 visited some EMS operators as well as Fleet. Bell already was considering a 230 follow-on—the 430 with a stretched cabin, four-bladed main rotor and optional engines.

The first production 230 (C-GAHJ) flew on May 23, 1992. In August it ferried to the US for completion. Demo flights began in earnest. Prototype C-GEXP spent the first half of June flying over the Gulf of Mexico from a base in Louisiana. Offshore platform landings and takeoffs were made as part of tests to get certified for Category A elevated heliport operations. This provided data for fine tuning the gross weight flight envelope and determining approach techniques to elevated heliports. Meanwhile, C-GBLL flew to Bakersfield, California for noise certification tests.

BHTC deliveries for the first half of 1992 included 48 Long Ranger IIIs, 33 Jet Ranger IIIs, 12 Model 412s, four Model 212s, all for export. Next to the US, Germany was the largest market. Other customers included Algeria, Angola, Australia, Brazil, Czechoslovakia, Japan, New Zealand, Singapore, Sweden and Turkey. This brought Canadian production to 762 since the first 206 delivery in October 1986. The various 206s led with 279 Jet Rangers and 373 Long Rangers. In the end only 38 Bell 230s were delivered, going to countries like Germany, India, Korea, Mexico, South Africa and Turkey. Corporate buyers included Joseph E. Seagram & Sons Inc. in New York, and Textron Inc. of Providence. Several went for aero medical use. Now the 230 prototypes became 430s. As such they were stretched 45 cm, re-rotored (from a two- to four-blade system), and re-engined (from the twinned Allison 250-C30G2 to the -C40B with 10% more power). Rotor blade life increased from 6,000 hours in the 230, to 10,000; engine TBO grew from 1,750 hours to 2,250. Bell in Texas designed and built all 430 rotor and dynamic components, and did transmission bench tests and component fatigue tests. BHTC handled airframe, avionics and program management, and flight test. Its 430 staff was boosted by 30 engineers loaned by Texas.

The first 430 (C-GBLL) flew at Mirabel on October 25, 1994. The second (C-GEXP) flew December 16. On August 4, 1995 the first production machine flew (C-GRND). C-GBLL

made its last flight in September, then became a ground test airframe. As such it logged hundreds of hours at full power, while firmly anchored. Pilots did not relish this work, but it was indispensable in general development and for certification. Bell 430 No. 4 (C-FWQV) left Mirabel in November to have an EMS interior installed in Texas, before attending the 1996 Heli Expo show. In February 1996 aircraft No. 5 went to the Asian Aerospace show in Singapore. Meanwhile, there were hot-and-high trials in Colorado with EXP, while RND went to La Grande in northern Quebec for cold weather trials.

Canada's UTTH

The CanForces utility tactical transport helicopter (UTTH) arose from the need to replace an aging fleet of CH-118s, CH-135s and CH-136s, the first of which dated to 1968. When the UTTH selection was made in 1992,

118 old helicopters remained in service. The replacement was BHTC's civil 412CF, designated CH-146 Griffon by the military. By choosing the basic 412CF, millions were saved designing to military specs. The UTTH was modified as required, aircraft being fitted with or having provision for specific avionics/radio, NVG lighting, armoured floor and crew seats, spotter's window, litter kit, C-130 air-transportable kit, external hoist, cargo hook, WSPS, searchlight, heavy-duty cabin heater and rotor brake, etc. Self-defence would include radar warning receiver, missile approach warning system, chaff and flares. All such equipment could be added or removed to suit the mission.

The $750-million contract for 100 CH-146s was signed in September. The deal included customizing aircraft for missions, a flight simulator, air and ground crew training (630 personnel), and technical support. BHTC pilot Bob Trimble flew the first CH-146 at Mirabel

Model 412s and a 230 (far corner) near completion in a scene from November 17, 1992. Then, a typical flight test scene showing "green" helicopters—a 412 landing and a 212. (Larry Milberry, Kenneth I. Swartz)

One of the new multi-role CH-146 Griffons for the Canadian Forces. This superb type replaced more than 100 aged Kiowas, Hueys and Twin Hueys. (BHTC)

on April 30, 1994. Handover of the first example was on March 15, 1995. The first CH-146s went to AETE in Cold Lake for trials; the first unit to equip was the OTU—403 Squadron at CFB Gagetown. Next, three Griffons each went to 417 Squadron (Cold Lake), 439 (Bagotville) and 444 (Goose Bay). The first operational tac-hel squadron to convert was 430 at Val Cartier, Quebec (16 aircraft). Nos. 400, 401, 408 and 427 squadrons followed, with final deliveries in 1998.

The first Canadian-built Bell 407. Early Canadian operators of the 407 were Abitibi Helicopters of Quebec and Aerial Recon of Alberta. (BHTC)

The Model 407

BHTC had some 1,800 employees by 1995. That year it turned out 230 helicopters, accounting for 52% of worldwide deliveries. The 1,500th BHTC helicopter rolled out in early 1996. Besides the 430, BHTC also now had the 407 for which $50 million was budgeted, including a $9 million repayable loan from Ottawa. The 407 (5,500-pound AUW) featured the Long Ranger cabin plus the four-blade rotor system of the OH-58D and 790-shp Allison 250 C47, a derivative of the 430's C40. The 407 began in January 1994. A proof-of-concept prototype was built in Texas—N407LR, a modified Long Ranger III. It flew in April 1994 and logged 220 hours before being grounded. The project benefited from work done on the 400, which was the first BHTC project using the OH-58D rotor system. Now Mirabel began building the first 407—C-GFOS flew on June 29, 1996. Soon two test machines were at work on everything from cold weather trials in northern Quebec to hot-and-high trials in Colorado. Improvements over the Long Ranger IV included a wider cockpit and cabin, more power for takeoff, higher payload, greater range, and a top speed of 147 mph—11 mph faster. The first production 407 flew in October. Meanwhile, Bell was starting the Model 427, a twin (6,000-pound AUW) based on the 407.

Eurocopter

The roots of today's Eurocopter Canada, the second of Canada's modern helicopter manufacturers, date to the early 1970s, when MBB of Germany and Boeing of Philadelphia co-operated in designing the BO105 light twin turbine. The first example was demonstrated in Canada in 1973. In 1974 Bill Loftus, formerly of companies like Autair and Helisolair, became the Canadian sales rep. He and fellow pilot Steve Rickets toured aircraft D-HDBF (later C-GHJZ) across Canada that summer. Unfortunately, this was about the time that activity in places like the Beaufort was slowing. Besides, the BO105 was more than three times the cost of a single-engine Jet Ranger, FH1100 or Hughes 369. Also, while Boeing was proud of the BO105, potential customers complained about it being cramped for passengers. Before long, a stretch was engineered, and more powerful Allison 250-C20s offered. Meanwhile, the BO105 lost a US military competition and Boeing's interest in single-rotor technology faded. It dropped its MBB affiliation.

In 1979 an MBB support company was set up in Philadelphia. Bill Loftus, still the only Canadian BO105 rep, became the first MBB Helicopter Canada pilot. In 1980 he was joined in sales by R.L. "Buck" Rogers, a veteran RCN pilot. Jim Grant, an ex-CanForces officer from the CF helicopter procurement office in Ottawa, also joined.

On December 13, 1983 MBB, Fleet Industries, Ottawa and the Province of Ontario agreed to establish a facility in Fort Erie. This was MBB Helicopter Canada; its big project became the BO105LS, a variant for "hot and high" regions. The LS used the uprated transmission from the military BO105, and a 550-shp Allison. The first flew in Germany in 1986, Canadian production beginning the same year. Aircraft were built up from German components, although some parts were made in Fort Erie. Sales were made in California, Indiana, Mexico and Peru, but only 36 were delivered by mid-1992, when Eurocopter reduced staff from 110 to 64. Some staff were seconded to MBB in Germany, where design was proceeding on the twin-engine BO108, a version which would use the P&WC PW206B.

Boeing Canada

The names Piasecki and Vertol were famous in the helicopter world in the 1940s-60s. In time they disappeared under the Boeing umbrella. Boeing originally came to Canada in May 1929, the principals being William E. Boeing of Seattle and Henry S. Hoffar and Charles

(Right) Bill Loftus, one of the longest-serving members of Canada's helicopter community, brought the BO105 to Canada. (Kenneth I. Swartz)

(Left) Eurocopter's BO105LS assembly line at Fort Erie in March 1987. Then, the first LS with P&WC PW200 series engines. This testbed led Eurocopter to choose the PW200 for its EC135 production model. (Kenneth I. Swartz)

A 442 Squadron CH-113 Labrador at Abbotsford in August 1977; then, a 450 Squadron Chinook slinging a Voodoo from Uplands to Rockcliffe for the NAM on July 19, 1984. These types were overhauled and modified regularly over the decades by Boeing in Arnprior. (Kenneth I. Swartz)

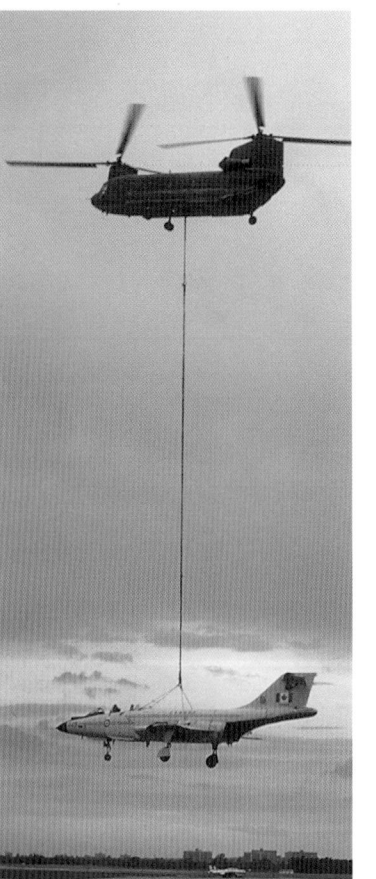

Beeching of Vancouver. Work began in the former Hoffar-Beeching Shipyards in Vancouver. Boeing Canada's first activity was building C-204 flying boats and 40H-4 mail planes. The indigenous Totem flying boat was turned out in 1932; thereafter Boeing was stuck in the Depression. Its recovery coincided with European war mongering. In 1937 Boeing built some Blackburn Shark torpedo planes for the RCAF. Once the war started, it expanded quickly with contracts for Anson components. A plant was erected at Vancouver airport to build PBY-5s. Satellite factories sprang up making parts for the B-29, Mosquito, Norseman, etc. There was some RCAF aircraft overhaul, and a contract to upgrade TCA L.14s. Employment peaked at 10,000.

Late in the war Boeing vied to build a new TCA/RCAF transport, but that went to Canadair. It sold its Sea Island plant to CPA. After WWII it developed subcontracting plants at Arnprior, Ontario, and Winnipeg. Arnprior had been a BCATP station, then the NRC's flight test centre. By 1954 Piasecki HUP-3s and H-21s were entering RCN and RCAF service. Used for the Mid Canada Line, they needed factory support. Piasecki moved to Arnprior in February 1954, getting the NRC's shops and equipment as a concession. Frank Piasecki hired Jack Charleson to head his operation, while Ed Ritti managed daily operations. Charleson was one of the famed Twirly Birds, the earliest helicopter pioneers. He flew R-4s during WWII, launched Canadian Coast Guard helicopter operations, and established the DOT's helicopter type certification and operating regulations. The first six H-21s arrived at Arnprior between August-October 1954. Twenty were delivered by 1960. Through this period Max Nebergall, an original USMC helicopter pilot, was the test and instructor pilot.

It was clear that the piston-engine era eventually would give way to turbines. Piasecki flew the turbine YH-16 in the mid-1950s. In 1956 it became Vertol Helicopters; in 1960 Vertol was sold to Boeing. The YH-16 was abandoned for a more practical design, the Model 107, featuring Piasecki's proven twin-rotor concept that dated to the famed HRP "Flying Banana". The 107 flew in April 1958. The first military version (CH-46) flew in October 1962. Canada ordered six (designation CH-113), the first being delivered in October 1963. A dozen more were ordered in 1964. Eight bigger CH-47 Chinooks were ordered in 1974. The CH-113 was being supported by Boeing Canada at Arnprior into the late 1990s. By 1997 Boeing's Canadian operations also manufactured components for jetliners.

The Chinook

By the mid-1950s the US Army wanted bigger, faster types. Boeing Vertol of Philadelphia designed such a helicopter; the prototype (YHC-47A), flew in September 1961. The first production CH-47A (AUW 33,000 pounds) flew in August 1962. Soon CH-47s were in Vietnam. Besides moving troops, casualties, refugees, ammunition, supplies, artillery and vehicles, they salvaged hundreds of downed aircraft, including Caribous. 354 CH-47As and 108 CH-47Bs were delivered before production gave way to the CH-47C (AUW 46,000 pounds), which flew in October 1967.

In August 1973 Canada ordered eight C-models for 10 Tactical Air Group (formed that May at St. Hubert). Designated CH-147s, they were needed for bridge building, moving 155-mm field pieces, bringing troops quickly to a battle line, putting in Arctic fuel caches, medevacs, etc. With its 28,000-pound sling capacity and 20,000-pound internal load the Chinook was ideal for such work. To improve performance, Canada specified the Lycoming T55-L-11C, which was more powerful than the standard -11. The CH-147 also came with a higher-capacity transmission. This upped the all-up weight to 50,000 pounds. Seating for troops (in summer kit) was increased from 33 to 42 by adding a bench down the cabin centre. There was a cargo hook at the front cabin entrance. The CH-147s also had skis, a long-range internal fuel system, and a water dam at the rear ramp for emergency water ops. With five 500-Imp. gallon bladders a CH-147 could ferry 1,300 nm. Also unique was instrumentation (cruise guide indicator) that provided the pilots information about in-flight stresses on their machine on a moment-by-moment basis. Thus could they be flown without fear of exceeding stress limits.

The first of eight CH-147s (147001) set off on delivery to 450 Squadron at Uplands on October 18, 1974. In flight the transmission system failed. This threw the fore and aft rotor blades out of synch. Blade strikes on the fuselage led to a crash that killed the crew of five. The air force pushed ahead and soon had seven Chinooks. No. 450 Squadron, formed on March 29, 1968, was the core of the operation. It traced its lineage to No.1 Transport Helicopter Platoon, formed at St. Hubert in 1964 with Voyageurs. 450 sent a detachment of three Chinooks to CFB Edmonton (Namao); on January 1, 1979 this Det became 447 Squadron. Both units were steadily busy, whether on army co-operation exercises like Ex. Rendezvous at CFB Wainwright, or with a host of other duties such as Op. Morning Light (January-March 1978)—the clean-up of debris scattered in the NWT after re-entry of the Soviet nuclear-powered satellite Cosmos 954. Chinooks served in NATO (e.g. Ex. Brave Lion), supporting Canada's commitment to the defence of Norway. They annually took part in Op Hurricane, servicing remotely-operated DEW Line sites. Several training missions involved recovering historic wrecks, e.g. a Canso in Labrador, a Lancaster from Goderich to Hamilton, and a Ventura from the NWT. Not all such taskings succeeded. When Voodoo 101010 was being slung from Mountain View to Trenton for the base museum on June 25, 1991, it began swinging wildly and had to be jettisoned into the Bay of Quinte.

Discussions about the viability of the

Chinook 147005 of 447 Squadron at Sudbury on April 12, 1991. It was en route from Namao to Mountain View for storage. Then, one of the ex-CanForces Chinooks with the Royal Netherlands Air Force. D-666 was at Deelen Air Base in September 1996. It previously had been CF 147008. (Larry Milberry, Boxman/Van der Mark)

Chinook in 10 TAG went on for years. The air force always could use a medium-lift helicopter, but the army sometimes doubted its usefulness. An argument was that most Chinooks would be destroyed quickly in a war scenario. Another was that they were too expensive to support, e.g. their metal rotors blade were not commonly available (most CH-47s had glass fibre blades). Meanwhile, AIRCOM was gearing for a fleet of new helicopters—it seemed certain that EH101s would be ordered, so the Chinooks could be hurried out of service. In 1990 they were retired and stored at Mountain View, then sold to the Dutch in 1993 for $15.74 million. They flew to Boeing in Philadelphia to be converted to CH-47D standards. The first was delivered by sea to Rotterdam on December 23, 1995 aboard *Atlantic Conveyor II*, (whose namesake had been sunk with several Chinooks during the Falklands War).

By the time Canada's Chinooks were grounded, they had logged 52,500 hours for an average of 6,563 hours per machine. Besides the loss of the first aircraft, 147002 also was destroyed. While it was taxiing at Rankin Inlet on August 17, 1982, its rear rotor severed a light standard. Blade strikes ensued, and 002 flipped into a fuel dump and burned. Three crew died.

Helicopter Support Industries

Canadian manufacturers of helicopter accessories and avionics broke into international markets during the 1980s. Systems perfected for local needs found international markets. The Bambi Bucket from SEI Industries became standard in forest fire fighting. The Richmond, BC manufacturer produced thousands of buckets for customers in about 100 countries. Similarly, Northern Airborne Technologies developed avionics and radio systems. It started by producing radio control units for BC operators, but its products won global acclaim.

Helicopter Welders of Canada dealt internationally almost since it began. In the 1980s its repair business grew with the addition of jigs for 222 and AStar cabins, and tail booms for the 214ST and BO105. Repair jigs and fixtures for the 204, 205, 212 and 412 were sold to customers abroad. Other Canadian repair specialists included Eagle Copters in Calgary, and S&H Helipro, which became a major repair and overhaul business by the mid-1990s. Helicopter Welders of Canada later joined CHL.

Canada's Great Helicopters

Paul Kristapovich started flying helicopters in the 1950s and was still in the business in the mid-1990s. He took these photos early in his career. In the first view Saskatchewan natural resources ranger Anscar Aschim poses in 1959 by an Associated Helicopters Bell 47G2 at Hudson Bay Junction, Saskatchewan. Then, a typical scene as an Alberta government Bell 47J services a forestry lookout tower at Lac la Biche. (Below) Alberta Bell 47J CF-KEY in the Crowsnest Forest in 1962. Then, Associated's Bell 47G CF-NTK is loaded aboard a Wardair Bristol Freighter at Edmonton for a flight north in the spring on 1963.

Ontario Hydro Bell 47G4A CF-OHQ at Malton on September 21, 1971. Ontario Hydro was a pioneer in using helicopters for line patrol, spraying and construction. Besides the Bell 47 its types since the early 1950s included the Alouette, S-55, S-58T, Jet Ranger, AStar and Super Puma. (Larry Milberry)

Radio stations started using helicopters in the 1950s for traffic reporting and news coverage. Bell 47J C-FMFO, owned by Associated, was at Vancouver in September 1979. (Kenneth I. Swartz)

VIH Jet Ranger C-FANC departs Campbell River on an April 9, 1992 charter. Forestry, mining, power and communications, construction, fishing, EMS, film making and tourism all help to keep BC's helicopter industry busy. (Larry Milberry)

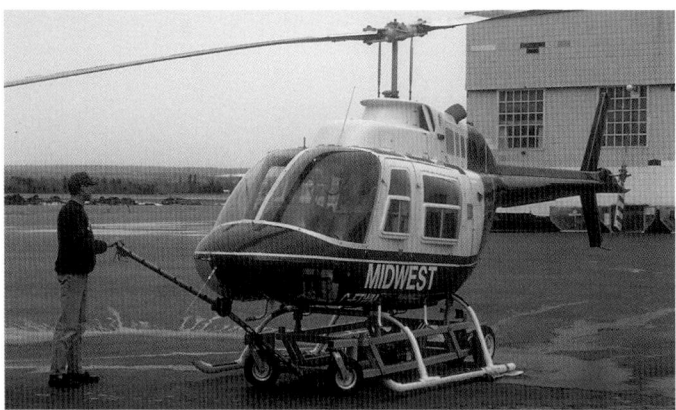

Midwest Jet Ranger C-FTWM goes into its hangar at Thunder Bay on May 1, 1996. The handy battery-powered dolly was designed by George Mannhardt, a local entrepreneur. (Larry Milberry)

(Above) Vern Zelent in CHC Jet Ranger C-GAHR spraying herbicide on a cut near Squamish on June 15, 1989. This mixture of chemicals kills deciduous growth, but causes no damage to coniferous seedlings. (Brent Wallace)

(Right) The light helicopter always has been popular in the sightseeing trade, whether at fall fairs or for viewing spectacular natural vistas. Niagara Helicopters Jet Ranger C-GSEE was over the Niagara Falls with a load of tourists in June 1988. (Kenneth I. Swartz)

(Right) The Bell 205 succeeded the 204. Its 1,400-shp Lycoming T53 gave it greater lifting capacity than the 204 (1,100-shp T53). Basic dimensions remained the same, but all-up weight rose from 8,500 to 10,500 pounds. The first 205s came to Canada in 1968-69 for operators like Dominion. Trans Quebec 205 CF-IBT is seen on July 14, 1977 at Caniapiscau during the James Bay project. Note the sturdy helipad to keep machines out of self-induced dust balls, and clear of moving vehicles, unwary pedestrians, etc. Pilot Roy Heibel was refuelling for a day of slinging. Like many pilots in the northland in this era he was an American—when the US left Vietnam in 1975, thousands of its pilots found civilian work. Roy had more than 6,000 hours, mostly with the CIA. He later lost his life in a Hughes 369 crash. (Larry Milberry)

(Above) Nahanni Bell 205 C-GQLL at Whistler, BC in February 1984 in the colours of its previous owner, Liftair. Then, pilot Ruthanne Page polishing her Campbell Helicopters 205 (C-FFJY) during a forestry contract at Sioux Lookout on July 13, 1991. (Brent Wallace, Larry Milberry)

More BC scenes. Northern Mountain's C-GGAT departs with an IA team for a 1994 fire near Cranbrook. Then, Frontier Bell 212 C-FNSA makes a drop from its belly tank. (Simon Milberry, Frontier Helicopters)

(Above) A typical CanForces CH-135 (Bell212) . This example, flown by Capts Dabros and Tremblay of VU32 Squadron, was at CFB Summerside on January 14, 1987. The CH-135 served 1974-97, the last being replaced by the CH-146. (Larry Milberry)

A CHC Aérospatiale AS350B C-GBBX over canola fields near Grande Prairie, Alberta on July 5, 1989. Then (lower right), a standard helicopter IA (initial attack) fire fighting team ready for action around Boston Bar, BC in August 1994. With their CHL AS350B AStar C-GVHV are Mike, Simon Milberry, Jerry Orlowski and John Walton. (Brent Wallace, via Simon Milberry)

(Above) The Bell 214ST was the biggest of Bell's medium-class family (17,000 pounds AUW). Here Quasar's C-GSTQ, acquired for offshore jobs, was at Abbotsford in August 1982. The 214ST used a 2,930-shp Lycoming T55. (Kenneth I. Swartz)

(Left) Blackcomb Helicopters' AS350B at Whistler carrying a camera pod in August 1996. (Brent Wallace)

(Right) The AS350B is ideal for patrol work. With its greater speed and range it completes such tasks faster than a Jet Ranger. Viking's C-FIOC was near Sept-Îles on November 20, 1992. Its work included patrolling hydro lines and the railroad from the iron ore mines to the north. (Larry Milberry)

The Hughes 369 series have served Canadian operators since the 1960s. Jeffco's C-GTTE (above) departs Harrison Hot Springs, BC in June 1982. Heli-Max's C-FMAI is seen slinging cement, while C-GXON helps offload a Buffalo Airways DC-3 in the NWT. These ubiquitous turbines evolved from the piston 269 of the 1950s (right, C-FCBL of St. Hubert-based Helicraft). Like the Jet Ranger and FH1100 the turbine Hughes started with the 317-shp Allison 250-C18. (Gary Vincent, Heli Max, Jim Smith, Larry Milberry)

(Above) The vintage Hiller UH-12 remained in Canadian service in the late 1990s, some being converted by Soloy to turbine power. This UH-12E (CF-OKF) was with Yukon operator Mayo Helicopters when shot in November 1976. While the 1950s vintage UH-12 returned to production in the 1990s, the FH1100 (above right) disappeared from the Canadian scene. CF-KHB of Kenting-Klondike was at Carp, Ontario in June 1976. (Brent Wallace, Kenneth I. Swartz Col.)

(Right) The Brantly also had all but disappeared from Canada by the mid-1990s. C-FFFB was still operating in BC when shot by Ken Swartz at Delta Air Park on September 2, 1994.

The Enstrom and Robinson series of light piston helicopters filled a niche in the 1970s-90s. The first Enstrom flew in 1962, the Robinson R-22 in 1975 and the R-44 in 1990. All three were popular in Canada, especially the Robinsons, which were ideal for flight training, patrol work, and heli-logging support, e.g. moving loggers around. In 1992 Aerial Recon Surveys of Whitecourt, Alberta had 10 R-22s on oilfield, forestry, mapping and remote sensing contracts. Enstrom C-FBGT is seen at Abbotsford on August 5, 1993. R-22 C-GFIO of CHL was at Science North in Sudbury on June 18, 1994; while R-44 C-FXBB (below) was at Sudbury in October 1996 with R-22 C-GOGL, both Huisson Aviation. (Larry Milberry)

Vintage Sikorskys... CF-FDF (above) was the last S-51 flying in Canada. Here it is at Abbotsford on August 15, 1973, shortly before it was wrecked in a mishap at Langley. Then (left), Eldorado's S-55 C-FJTI at Uranium City in September 1976, and (below) S-55T C-FJTB of MF Air Service at Delta, BC in December 1980. (Kenneth I. Swartz, John Kimberley, Kenneth I. Swartz)

(Left) Campbell's S-58T C-GLOG came into Canada to specialize in logging. Ken Swartz shot it at Chilliwack in August 1996.

(Left) Okanagan's S-61N C-GOKA salvaging a downed BN Trislander from Nighthawk Lake near Timmins on January 7, 1977. Don Mackenzie and Vic Schiebler were the pilots. Both later were killed in helicopter crashes. (Larry Milberry)

(Right) S-61N C-GOKZ aboard the drill rig *Gulftide* near Sable Island on July 4, 1978. It was on a scheduled passenger-freight-mail run from Halifax. (Larry Milberry)

CHC S-61L C-GJDR during a construction job in downtown Vancouver on March 10, 1990. Then (right), it uses a Bambi Bucket on a small fire on Vedder Mountain in the Fraser Valley. The date was July 13, 1989. Forest fire fighting was one of the earliest uses of the helicopter. Perhaps the first trial was in California in early 1946, with six USAAF helicopters working on a US Forest Service project. (Brent Wallace)

(Below) The CH-124 Sea King continued to serve Canada in the late 1990s, about 30 of the original 41 examples having survived the rigors of more than three decades of naval aviation. Here 12419 lands aboard a destroyer escort; then, 12441 is seen at Shearwater on January 13, 1987. (DND SWC86-1253-7, Larry Milberry)

Skycrane N6962R (left) lifting a component for the CN Tower's crowning spire on March 15, 1975; C-GJZK during fire fighting training from Sproat Lake, BC, home of the Martin Mars; then, an Erickson Air-Crane machine raising a hydro pylon near Timmins, following damage from a violent summer storm. When logging, the S-64 can lift 25,000 pounds compared to 11,500 for the Vertol 107 and 28,000 for the Boeing 234. In 1991 Erickson Air-Crane Canada became Canadian Air-Crane. (Larry Milberry, Paul Mavarnak, Ron Trapper)

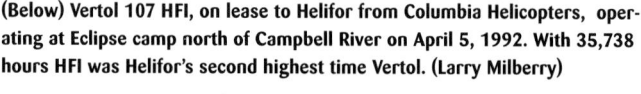

(Below) Vertol 107 HFI, on lease to Helifor from Columbia Helicopters, operating at Eclipse camp north of Campbell River on April 5, 1992. With 35,738 hours HFI was Helifor's second highest time Vertol. (Larry Milberry)

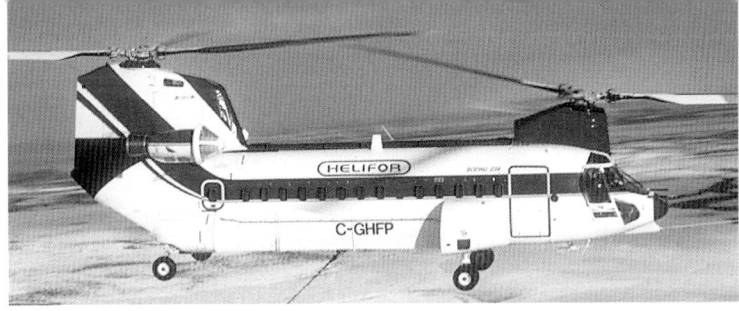

Helifor's Boeing 234 operating from the Salmida Mine on October 5, 1993. It was slinging freight brought in by Buffalo Airways DC-3s and NWT Air Herc. Destination was BHP's diamond exploration camp at Lac le Gras. (Henry Tenby)

Ex-RCAF H-21 CF-GQS (ex-CF-JJQ) was at Saint John, NB on August 9, 1975. Several of these old Piaseckis had brief careers in BC and the Yukon, mainly in logging. (Larry Milberry)

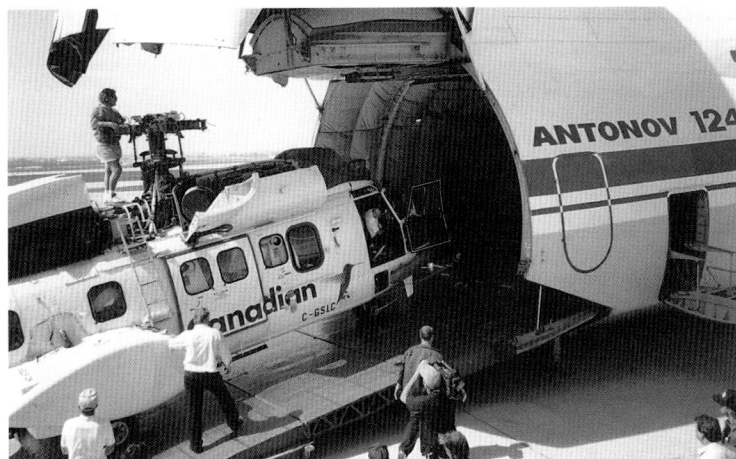

On August 28, 1993 AN-124 UR82008 visited Toronto to collect three AS332s for fire fighting in Greece. Involved were Canadian's C-GSLC, Ontario Hydro's C-GOOH and ERA's N171EH. Here SLC goes aboard in a tight squeeze. Then, a view aboard as the AN-124 gets ready to depart. Next, Ontario Hydro's OOH being readied for flight at Athens the next day, following its 10-hour, non-stop flight. SLC is shown ready to fly. After all this effort little fire fighting was done and the Super Puma fleet soon dispersed. (Larry Milberry)

(Left) One of the newest types in Canada is the US$3.5 million-Kaman K-Max. N134KA was shot at Vancouver on March 15, 1995, while on lease to Erickson. First flown in December 1991, the Kaman K-Max (10,000 pounds AUW) uses an 1,800-shp Lycoming T53. Its design uses inter-meshing main blades that make a tail rotor unnecessary. To Kaman, the 10% of total power used by a tail rotor is better expended on the main blades. Its ability to lift 5,000 pounds at 8,000 feet, while remaining within engine operating limits make the K-Max attractive in forestry. In April 1995 a K-Max had a close encounter in California. Its main blades chewed into some treetops, losing 30 inches from one, 13 from another. The pilot was able to fly to a clearing and land safely. In early 1995 Erickson returned its two leased K-Maxs, stating that it planned to concentrate on heavy-lifters. Later that year Midwest Helicopters became the first Canadian K-Max operator, when it imported C-GMHJ. (Jan Stroomenbergh)

(Left) This brutish Russian Mil Mi-8 visited Canada and the US in 1992 to help celebrate the 50th anniversary of Lend Lease. It was at Winnipeg in October with a Mil MI-24 Hind D attack helicopter. Both machines were crewed by women. In the mid-1990s efforts were being made by Kelowna Flightcraft to market the Mi-8 for duties in the developing world. (R.W. Arnold)

Canada and Space

The aurora borealis (northern lights), comets, eclipses and other heavenly displays figured prominently in the mythology of Canada's prehistoric inhabitants. The early European navigators sailed by the stars. Those coming to Canada's east and north shores, probably as early as the 10th century, kept note of heavenly phenomena, likely using them for steering. Jesuit missionaries in New France wrote of an eclipse as early as 1633. Surveyors, such as those laying out the seigneuries of New France, calculated using astronomy; so did explorers like Champlain, who travelled inland from New France, beginning in the early 1600s. His astrolabe is one of Canada's historical treasures. Canada even attracted international astronomical expeditions, such as one from England led by William Wales. In 1769 his party ventured to forbidding Fort Churchill (as daunting an adventure in those days as a trip to the moon today) to observe a solar eclipse. Wales' party successfully observed the eclipse, which included a transit of Venus.

The first observatory in Canada is thought to have been at the French fortress of Louisbourg on Cape Breton Island about 1750. One was built at the University of Toronto in 1840. In 1904 the U of T established the first university astronomy department in Canada. In 1868 the Toronto Astronomical Club became the first astronomy society in Canada. It became the Royal Astronomical Society of Canada. The federal government built observatories in Ottawa (1905) and Victoria (1918). In 1916 the National Research Council was established to monitor worldwide scientific progress, undertake its own scientific ventures, and encourage new science in Canada. In 1970 all government astronomical research was placed under the NRC.

Radio astronomy got its start in Canada in 1946. Scientist were as drawn to the skies as had been Canada's ancient inhabitants. Now they were seeking answers to how the aurora borealis interfered with radio transmissions. In the late 1950s Canada and the US agreed to cooperate in satellite technology. This led to the design and construction of Canada's first satellite—Alouette I. This work was done in Ottawa at the Defence Research Telecommunications Establishment under communications engineer John H. Chapman. NASA (National Aeronautics and Space Administration) would provide test, launch and tracking requirements. Subcontractors such as Spar of Toronto (a company spun off from de Havilland Canada's Special Products and Research branch) were closely involved. The 145-kg Alouette I went aloft on September 29, 1962 from Vandenburg AFB, California. This made Canada the third nation in space, after the USSR and USA. Alouette I, a science satellite for studying the aurora borealis, had two communications antennae. It was followed by Alouette II (November 29, 1965), ISIS I—International Satellites for Ionospheric Studies (January 30, 1969) and ISIS II (April 1, 1971). Alouette I established a record for satellite longevity, remaining functional till turned off on its 10th

Canada's original astronauts: Ken Money, Roberta Bondar and Bjarni Tryggvason (standing) and Bob Thirsk, Steve MacLean and Marc Garneau. Then (left), a spectacular view as Shuttle mission STS-74 blasts through cloud at the Kennedy Space Centre on November 12, 1995. Aboard was Canadian astronaut Chris Hadfield, Canada's first mission specialist to fly in space, the first Canadian to use the Canadarm operationally, and the first Canadian to visit the Russian space station Mir. (Canadian Space Agency, NASA STS074-(S)-020)

birthday. Canada's initial success in space led to new dimensions. Ottawa would fund research, then let industry determine what it could produce that would be marketable.

While the Avro CF-105 Arrow offered nothing of great value to humanity, its cancellation opened doors for many ex-Avro men. Upon hearing of the Arrow's demise, NASA officials from the Langley Research Center in Hampton, Virginia came to Toronto to recruit former Avro engineers and technicians. In an agreement between the two governments, these Canadians would spend two years working at NASA to gain experience in new technologies. It turned out that many took permanent NASA jobs. Of NASA's initial 140 engineers recruited for its original Space Task Group, 30 were Canadians. Men like Bruce Aikenhead, James

A. Chamberlin, Owen Coons, Bryan Erb, David Ewart, Stanley Galezowski and Owen Maynard worked on the Mercury, Gemini and Apollo programs, helping to put Neil Armstrong on the moon on July 20, 1969. Barry French and Ben Etkin of the University of Toronto Institute for Aerospace Studies helped evaluate whether its heat shield would survive re-entry, after Apollo 13 suffered damage on the way to the moon in April 1970. Apollo 16 and 17 crews trained in Sudbury, Ontario in 1971-72, studying the geology related to impact craters in the Sudbury Basin. Erb headed quarantine labs to examine moon-walking astronauts and the rocks they brought home, checking lest dangerous substances had hitchhiked from moon to Earth.

John H. Chapman was one of Canada's early space visionaries, foreseeing a great future in satellite communications. (Canadian Space Agency)

John Chapman envisioned a Canada with satellites playing a key role. He noted in 1967: "... in the second century of Confederation the fabric of Canadian society will be held together by strands in space just as strongly as the railway and telegraph held together the scattered provinces in the last century." Under his guidance Canada's first telecommunications satellite had its beginning. In 1969 Telesat Canada was formed by Act of Parliament. Through it Canada became the first nation to implement a commercial domestic satellite system, when, on November 10, 1972 Anik-1 (557 kg) was launched into a 36,200-km geostationary orbit over the equator. Although built mainly by Hughes in California, Anik-1 was viewed as a Canadian first. It gave Canadians in the Arctic their first quality telephone, radio and television service. Future Aniks had increasingly greater Canadian content. Anik-2, was launched April 20, 1973; Anik-3 on May 7, 1975. The first practical use of Anik-3 was made by the Toronto *Globe and Mail*, when it relayed its October 23, 1981 edition from Toronto to Montreal. Anik B1 was launched in December 16, 1978; Anik D1—August 26, 1982; Anik D2—November 9, 1984. D1 was the first communications satellite built by a Canadian prime contractor—Spar Aerospace. E2 and E 1 were launched in April and September 1991, respectively; each weighed nearly 3,000 kg. In 1992 a consortium of Canadian telephone companies purchased 52% of Telesat. It then devised a plan to sell 49 of its transponders (4-8 channels per transponder) in the US and 15 in Canada. These transponders would be carried on satellites flying in two Canadian orbital slots. In a solar storm of January 20, 1994 E1 and E2 were damaged, E2 taking five months to return on line. In a later storm E1 again was damaged, losing 50% of its capacity.

In the early 1970s Canada led in development of the Hermes CTS—communications technology satellite. Its partners were NASA and the European Space Research Organization. RCA of Montreal and Spar of Toronto were major subcontractors. Hermes was launched January 17, 1976. It did research into making satellite communications more efficient, e.g. testing an improved antenna-pointing system. Hermes was the first satellite to broadcast radio and television signals to small home receiver antennae. Built to run for two years, Hermes worked for twice that. Canada was involved in the cleanup after the USSR's Cosmos 954, a nuclear-powered reconnaissance satellite, which crashed in the NWT on January 24, 1978. In the early 1980s it became a partner with the US, France and Russia in a satellite-based search and rescue venture. Satellites would receive signals from the ELTs of aircraft in distress. Earth stations then interpreted the data, leading to SAR missions. Known as Search and Rescue Satellite-Aided Tracking (SARSAT), this system saved many lives yearly.

On September 29, 1982 Canada and the US jointly announced the start of the Canadian astronaut program. On November 12-14, 1981 the Canadian-developed robotic "Canadarm" first flew (STS-2, Columbia). It did its first work aboard STS-4 (Columbia) of June 27-July 4, 1982. On November 12, 1982 Anik-C3 (632 kg) was one of two satellites deployed from Shuttle Columbia during STS-5, the first time such a launch had been made. On June 22, 1983 the Canadarm was used during STS-7 (Challenger) for the first time in deploying a satellite—Anik-C2/Telesat-F. Other Aniks were launched, e.g. on November 9, 1984

(D2—STS51A, Discovery) and April 13,1985 (C1—STS51-D, Discovery). Spar provided Brazil with two Anik D-type domestic communications satellites. These went aloft in 1985-86.

There was much other Canadian space activity through the 1980s. Specialized Canadian scientific equipment flew on other nations' satellites. On February 22, 1986, for example, Canada's ultraviolet auroral imager flew on a Swedish Viking satellite launched on an Ariane from Kourou in French Guiana. Another Canadian instrument went up on Japan's Exos-D launched February 21, 1989 from Kagoshima, Japan. This was the first foreign instrument to fly on a Japanese spacecraft. Anik-E2 was boosted into orbit on April 4, 1991 from Kourou on an Ariane; Anik-E1 followed on September 26. On November 4,1995 RADARSAT went aloft. This was Canada's first full non-communications satellite launched since 1971. In March 1989 Ottawa established the Canadian Space Agency with headquarters in Ottawa (later moved to St. Hubert). The CSA has the task of uniting institutes of higher learning with industry; and funding space R&D. One of its prime facilities was the David Florida Laboratory near Ottawa, dedicated to the assembly, integration and testing of Canadian and foreign spacecraft. Capabilities included the quantification of space hardware, which otherwise defies normal testing on account of size, complexity, etc.

Canadarm

Following the signing of a contract in 1975 between Spar Aerospace and the NRC, the Shuttle Remote Manipulator System, or Canadarm, was designed and built as Canada's contribution to NASA's Space Shuttle Program. The concept for such a tool, nothing the likes of which existed, dated to the 1960s when NASA invited international partners to join in the space race. In the early 1970s the small To-

The earliest Shuttle over Canada was the Enterprise, seen aboard the Shuttle Carrier Aircraft while returning to Edwards AFB from a European visit in 1983. As it flew over Quebec towards Toronto, it was escorted by a 433 Squadron CF-5 and a 425 Squadron CF-101. (DND IMC83-69)

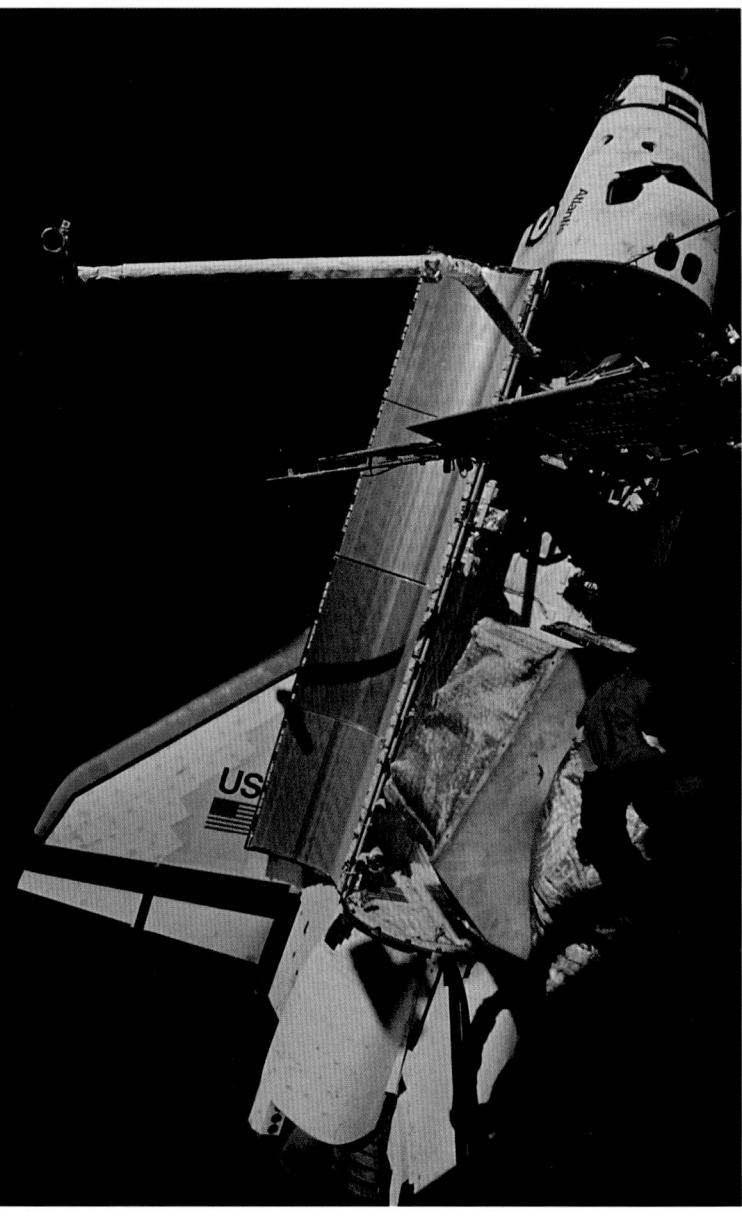

(Above) The Canadarm showing the docking module in the payload bay with Earth as a backdrop. (NASA STS074-321-027)

(Top right) The Canadarm simulator at JSC in April 1995. Since it is not designed to lift loads in gravity, the 50' 3", triple-jointed arm has little capacity on Earth. Astronauts training on it lift full-size inflatables of typical loads. They require at least 25 simulator sessions to qualify on the arm. Hadfield noted that there is no need to focus on the arm's sections and joints, only on the "hand" at the end. Anything between the operator and the hand functions by computer command. The original (SRMA No. 201) spent its 15th birthday in December 1996 being overhauled at SPAR in Brampton, Ontario. Five Canadarms had logged 46 flights in space by May 1997. One was lost when Challenger exploded soon after launch on January 28, 1986. (Larry Milberry)

(Right) Atlantis during STS-74, showing the Canadarm deployed. On this mission its job was to lift a docking module from the payload bay, then position it for attachment to space station Mir. (NASA STS074-324-016)

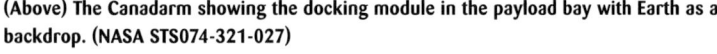

ronto firm DSMA-Atcon Ltd. saw an opportunity to adapt its robot arms, used in the nuclear industry, to space. It teamed with Spar (Toronto), CAE Electronics (Montreal) and RCA Canada (Montreal) in proposing to NASA a manipulator (run manually or by computer) and simulator. Frank Thurston of the National Aeronautical Establishment in Ottawa helped the consortium win Ottawa's funding support and in July 1975 Ottawa and NASA signed an agreement to develop the robot arm with Spar the primary contractor. NASA's requirements in precision, weight, safety and reliability were stringent. Since Canada agreed to fund development of the arm, NASA agreed that Canadi-

ans would be allotted extra astronaut slots on future flights.

As finally laid out, the SRMS had an upper and lower arm made of lightweight graphite epoxy composite. It had six shoulder, elbow and wrist joints; and an end effector for grappling. The joints allowed six degrees of freedom of motion in roll, pitch and yaw, roughly approximating the motion of the human arm. The arm manoeuvred at 3 cm/sec and could place a load to within 5 cm of target. The complete SRMS was 15.3 m long, 38 cm in diameter and weighed 411 kg. It could lift a payload of 29,510 kg in micro-g. The first arm was turned over to NASA at Spar in Toronto on

February 11, 1981 at which time Larkin Kerwin of the NRC dubbed it "Canadarm". Its first test in space came aboard STS-2 in November 1981. Astronaut Richard Truly flew the arm and found that it lived up to Spar's predictions. Four additional arms now were made. These were serviced, modified and updated by Spar ever since. In January 1991, for example, NASA gave Spar a five-year, C$67-million contract for technical and management support, and for upgrading two arms to handle payload weight up to 122,500 kg (equivalent to the mass of the Shuttle Orbiter with full payloads). In March 1997 Spar won another NASA Canadarm upgrading contract.

Astronauts use the arm to deploy and retrieve satellites and free-flying experiment platforms, and to help astronauts move around in the payload bay. During STS41-C (April 1984) the arm retrieved the unserviceable Solar Maximum Mission satellite. During STS41-D (September 1984) it was used to chip ice from the Shuttle's waste water vent. On STS-61 (December 1993) and STS-82 (February 1997) it was vital during missions to service and modify the Hubble Space Telescope. The arm captured Hubble and moved it into the cargo bay, then later relaunched it. Of the Canadarm's usefulness during this work astronaut Jerry Ross noted: "The Canadian mechanical arm with the foot restraint unit is a marvellous way to do business. If we didn't have [it] we would need foot restraints throughout the payload bay. Crew members would have to get in and out of those and kind of leapfrog the handling of the various instruments to and from the telescope as they did their work." To May 1997 the Canadarm had flown on 46 of 84 Shuttle missions.

From the Canadarm evolved the Mobile Servicing System, which will be essential in construction, maintenance and operation of the International Space Station (ISS). It will have one large arm (Space Station Remote Manipulator System) and another smaller, more adept one with 19 joints (Special Purpose Dexterous Manipulator). The SSRMS is 17.6 m long and has seven motorized joints. It will be used to handle large payloads and assist with Shuttle dockings. With its fine capabilities (including some touch sensitivity) the SPDM will take over some tasks previously done by astronauts on EVAs (extravehicular activity). A three-year C$207 million commitment for the construction of the SPDM was announced on April 8, 1997 during Prime Minister Chrétien's first visit to Washington, D.C. On that occasion, Chrétien gave SPDM its nickname—"Canada Hand". Earth applications of the SSRMS are envisioned in such areas as mining and handling hazardous wastes. The IMP Group of Halifax, FRE Composites of St. André, Quebec, CAE Electronics of Montreal, CAL Corp of Ottawa, SED Systems of Saskatoon, and MacDonald, Dettwiler & Associates of Richmond, BC all participated in developing MSS.

RADARSAT and MSat-1

While Canada was involved gathering data from US weather satellites, ERTS (Earth Resources Technology Satellites), etc. since the 1960s, RADARSAT (3,200 kg), launched on November 4, 1995, was its first venture into Earth observation from space. This satellite sits in a sun-synchronous, near-polar orbit about 500 miles high and has a five-year life expectancy. It uses a synthetic aperture radar (SAR) beam that is steerable for elevation (other earth resource satellites such as ESA's ERS series, have only a single-incidence angle). Imaging is unimpeded by cloud cover or darkness. The C$620 million project, financed by public and private funds, was 15 years in the making. Prime contractor was Spar Aerospace, ground station supplier—MacDonald, Dettwiler and

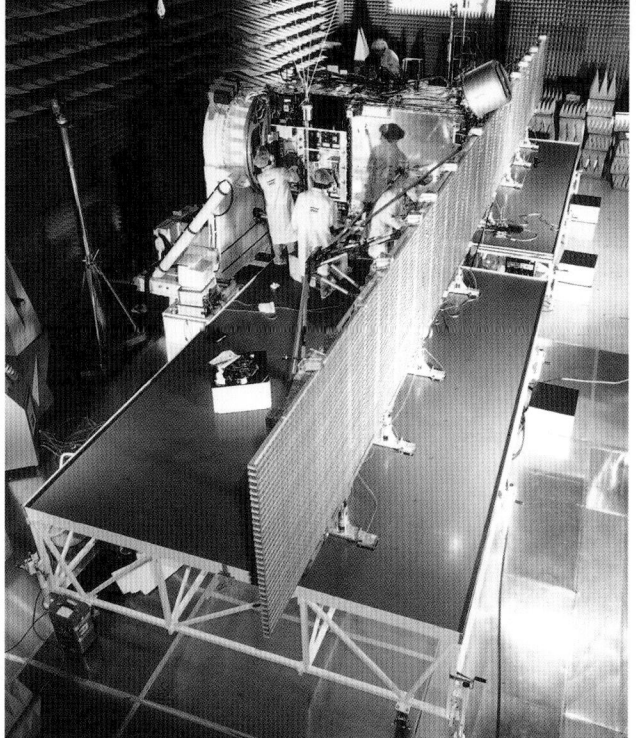

RADARSAT being readied in the integration area of Ottawa's David Florida Labs. Once launched, it occupied a sun-synchronous, near-polar orbit about 500 miles up. While earlier earth resources satellites like ESA's ERS 1 and 2 had a fixed single-incidence angle, RADARSAT used a synthetic aperture radar (SAR) beam that was steerable for elevation. (SPAR)

Associates, satellite payload subsystems supplier—Com Dev, and RADARSAT data distributors—Lockheed Martin Astronautics in the US and RADARSAT International in Canada. The consortium hoped to market to commercial clients 25,000 images yearly for five years, earning up to $400 million. The RADARSAT control centre is at the CSA in St. Hubert, Quebec; and a multi-million dollar data processing centre was built in Gatineau, Quebec. A series of ground receiving stations was established around the world to serve offshore clients. The first RADARSAT image was released on November 16, 1995—a crystal-clear view of Cape Breton Island taken at night through cloud.

In early 1996 RADARSAT began supplying data to commercial customers in 25 countries. The Canadian government has access to 51% of RADARSAT's observation time and initially used much of this for the Canadian Ice Centre, which provided navigation data to ships in icebound waters. RADARSAT can cover the Arctic completely once weekly. It also can monitor oceans (e.g. in locating fisheries), crops, forests, geological features, weather, etc. Any point in Canada will be accessible on a three-day basis, equatorial areas every five days. In return for launching RADARSAT, the US government took a 15% interest in its observation time.

On April 20, 1996 Canada's 6,280-pound MSat-1 was launched by an Ariane 42P rocket to its geosynchronous orbit over the Pacific. MSat-1, developed in Ottawa at the David Florida Laboratory by partners Spar Aerospace, Hughes Space and Communications, the Canadian government Communications Research Centre and the CSA, was the most advanced communications satellite for commercial use and encompassed within its market Canada, the US, Mexico and the Caribbean. One of MSat-1's capabilities would be to facilitate low-cost air-to-ground telephone service.

Canada's Manned Space Program

Canadian interest in the manned space program began in the late 1970s with an invitation from NASA for Canada to take part in its plans. The invitation went to John Chapman, but his untimely death in September 1979 delayed progress. Then, in September 1982, on the occasion of the 20th anniversary of Alouette I, NASA formally invited Canada to send an astronaut on a Shuttle mission. Canada soon established a permanent corps of astronauts to coordinate and conduct experiments in space. This was organized and managed interdepartmentally—through the NRC, DND, EMR, etc. In July 1983 advertisements appeared in newspapers across Canada inviting "applications from Canadian men and women to fly as astronauts on future Space Shuttle missions." They would be payload specialists i.e. non-NASA astronauts added to the Shuttle crew if there were unique activities to perform and when more than the minimum crew size of five was needed. Some 4,300 applied. Roberta Bondar (age 38, assistant professor of neurology and director of the multiple sclerosis clinic at McMaster University in Hamilton), Marc Garneau (34, naval commander stationed in Ottawa), Steve MacLean (28, post-doctoral student, Stanford University in California), Ken Money (48, senior scientist Defence and Civil Institute of Environmental Medicine in Toronto), Bob Thirsk (30, chief resident in family medicine, Queen Elizabeth Hospital in Montreal) and Bjarni Tryggvason (38, aerodynamics research scientist, NRC in Ottawa) were the successful candidates. In 1992 a second recruiting campaign brought some 5,330 applications, resulting in four new candidates:

Marc Garneau, Canada's first man in space. (NASA)

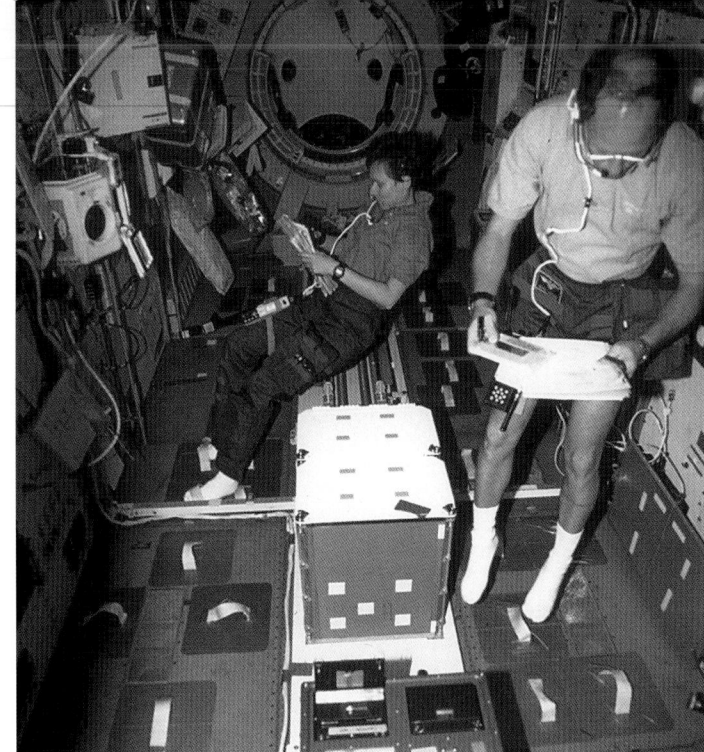

The crew of STS-42. Starting with Roberta Bondar (PS, Canada) at the bottom right. The members (going clockwise) are David C. Hilmers (MS, lowest), Ronald J. Grabe (commander), William F. Readdy (MS, moustache), Ulf D. Merbold (PS, Germany), Norman E. Thagard (MS, top of head obscured) and Steven S. Oswald (pilot, knee by Bondar's shoulder). Except for Bondar and Merbold the crew was American. In the second view, Bondar and Thagard are at work on an experiment. (NASA S42-78-061, S42-11-016)

Chris Hadfield, Julie Payette, Robert Stewart and Dave Williams. Stewart quickly left the program. Michael McKay replaced him. McKay ceased training for medical reasons in 1995, but remained with the CSA as an engineer providing technical support for future Canadian Shuttle flights.

The Original Six:
Roberta Lynn Bondar

Canadian astronaut Roberta Bondar was born in Sault Ste. Marie, Ontario on December 4, 1945. Her early academic achievements included a Bachelor of Science degree in zoology and agriculture in 1968 from the University of Guelph. In 1974 she earned a PhD in neurobiology from the University of Toronto; and in 1977 graduated in medicine from McMaster University in Hamilton. She then pursued research in areas such as neuroscience, receiving various honours for her work.

After a long astronaut training period, and working in such areas as space life sciences, in early 1990 Bondar was designated as a payload specialist for the first International Microgravity Laboratory. As such she flew on STS-42 (Discovery) January 22-30, 1992. Science experiments ranged from plant and animal cell growth (including a leukemia cell experiment), to studies of crew work performance, and lower back pain (a symptom of space flight), to examining a range of human adaptive responses to living in space. Bondar conducted microgravity vestibular experiments, using a rotating/oscillating chair to study a crew member's orientation and balance system in space. During STS-42 Bondar logged eight days, one hour, 15 minutes, travelling 5.4-million kilometres on 129 orbits. Like all astronauts she received a great deal of recognition: Officer of the Order of Canada, NASA Space Flight

Medal, and honorary degrees/appointments galore. Her many interests included scuba diving, ballooning, parachuting and flying. Her special interest in photography and the natural environment led to her book, *Touching the Earth*. Bondar left the space program in 1992. In 1997 she was doing research at the University of Western Ontario in London.

Marc Garneau

The first Canadian to fly in space was Marc Garneau. Born in Quebec City on February 23, 1949, Marc attended Royal Military College in Kingston, graduating in 1970 in engineering physics. In 1973 he added a doctorate in electrical engineering from the Imperial College of Science and Technology in London, England. In 1982-83 he attended the Canadian Forces Command and Staff College in Toronto. Like all astronauts he had many interests—flying, diving, auto mechanics, etc. He sailed the Atlantic twice in a 59-foot yawl.

Garneau served as a combat systems engineer aboard HMCS *Algonquin*. Later he developed a training simulator for the missile system used on Canada's Tribal Class destroyers. While posted in Ottawa and Halifax, he was involved in other research and support projects, including development of a target drogue system for training naval gunners. Selected for astronaut training, he was seconded from the DND to the Canadian Astronaut Program (CAP). On October 5-13, 1984 he flew as a payload specialist on STS41-G (Challenger), managing a series of Canadian experiments known as CANEX-1. He logged eight days, five hours, 23 minutes, 38 seconds in space. In 1986 he was promoted to naval captain. In 1989 he left the military and became deputy head of the CAP, providing technical and program support during preparation of upcoming

Canadian space experiments. Garneau was recipient of numerous honours from the CD (Canadian Decoration) in 1980, to the NASA Space Flight Medal (1984), Officer of the Order of Canada (1984), five honorary doctorates, and was co-recipient in 1985 of the F.W. Casey Baldwin Award.

On June 13, 1995 Garneau was named to crew on STS-77 (Endeavour). He would fly the Canadarm on this mission. STS-77, commanded by Col John H. Casper, was launched May 19, 1996. Its goals included deploying satellites and performing biological and material science experiments. Its deployment of two satellites was associated with demonstrating a low-cost attitude control system and the use of inflatable structures. The inflatable antenna experiment carried on the Spartan satellite was to assist in developing large space antennae that are light and easily storable aboard a small launch vehicle. Marc Garneau had the job of retrieving the Spartan satellite with the Canadarm. Once the antenna experiment was complete, the inflatable was jettisoned for burn-up. In the second satellite experiment a small payload was deployed and rendezvoused with a record four times to study aerodynamic drag and using the Earth's magnetic field in stabilizing a small satellite. Results could lead to less costly spacecraft, i.e. ones not needing expensive and heavy gyroscopic- or propellant-based attitude control systems. For STS-77 science experiments, Spacehab-4 was positioned in the cargo bay.

Marc Garneau took part in other experiments, one being the "Commercial Float Zone Furnace". It used a special furnace to produce ultra-pure semiconductor material. Garneau explained: "We make semiconductors on Earth and try improving on them so we can have faster, more reliable, more powerful computers... but there are serious quality limitations on Earth." In space it is possible to produce better-formed crystals (outside the influence of gravity which deforms them, and outside the usual container, particles from which contaminate crystals) that give improved semiconductors, i.e. allowing a better flow of electrons

Marc Garneau first flew on Shuttle mission 41-G (STS-17) in 1984. In 1996 he became the first Canadian to fly twice, this time on STS-77. Here he appears with his second crew. Standing are Daniel W. Bursch (MS), Mario Runco, Jr. (MS), Marc Garneau (MS) and Andrew S.W. Thomas (MS). In front are Curtis L. Brown, Jr. (pilot) and John H. Casper (commander). (NASA)

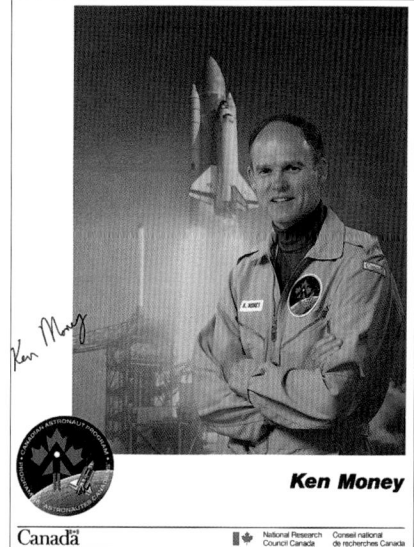

Ken Money

Canadä

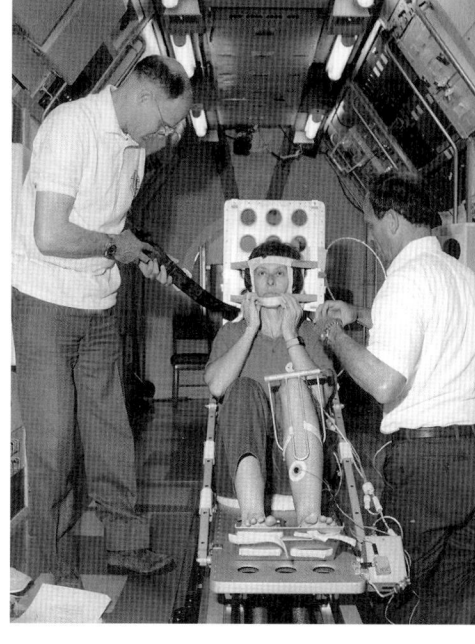

Ken Money autographed this portrait for astronomer Andrew Yee. Then, Money, Bondar and Thagard during training on July 12, 1990 for STS-42 (NASA)

CANADIAN NATIONAL GYMNASTICS TEAM

Payload specialist Dr. Robert Brent Thirsk in an official portrait; then, at work on the WETF during survival training for STS-78. An astronaut tour is indefinite and can be expected to include at least two Shuttle missions. (NASA)

Steve MacLean in his official NASA portrait; then, floating aboard the Columbia during STS-52 in 1992. (NASA, NASA STS052-24-014)

through a computer. In the experiment, rods of semiconducting material float through a small furnace. The rods melt in segments, then resolidify (without dripping) outside gravity's influence. Garneau explained: "In space, surface tension, which is a secondary force on Earth, becomes a primary force. In the zero gravity of space, surface tension will hold that wobbly liquid, which we call a float zone, and it will remain suspended between the two solid portions." With this work, lessons about making better crystals on Earth would be learned, leading to a lower rejection rate in crystal production. Besides its unique experiments, STS-77 made history, being the first to be fully controlled from the new Mission Control Center in Houston, and the first where a full set of three upgraded main Shuttle engines was used. STS-77 returned to KSC on May 29 after 161 orbits, entailing 10 days, 39 minutes and 20 seconds of flight.

Steven Glenwood MacLean

Steve MacLean was born in Ottawa on December 14, 1954. He earned a bachelor's degree in honours physics (1977), then his doctorate in laser physics from York University in Toronto (1983). While at York he was a member of Canada's National Gymnastics Team. In December 1985 he was designated as a payload specialist. From 1987-93 he was astronaut ad-

viser to the Strategic Technologies in Automation and Robotics Program, and program manager of the Advanced Space Vision System. He crewed on STS-52 from October 22-November 1, 1992, when his main task was evaluating the ASVS. This work led to an operational version being fitted to every Shuttle orbiter. His mission entailed nine days, 20 hours, 56 minutes and 13 seconds. In 1997 MacLean was completing training as a mission specialist.

Kenneth Money

Ken Money was born in Toronto on January 4, 1935. He was on Canada's 1956 track and field team during the Olympics in Australia. He earned his RCAF pilot's wings in 1957, and graduated in physiology and biochemistry from the University of Toronto. Next he pursued master's and doctoral degrees there in physiology, finishing in 1961. During the Apollo era Money was a consultant to NASA in his special field of motion sickness, a malady effecting most astronauts in their first few days in space. He served many years with the RCAF/CF Air Reserve, flying such types as the F-86, Beech 18, Otter and Kiowa. In 1972 he graduated from the National Defence College. His

many interests over the years included aerobatics, astronomy, skiing and skydiving.

Money worked for years as a scientist at DCIEM in Toronto, where his fields of interest included motion sickness and aircrew disorientation. He became a worldwide expert in these areas, being honoured for his work by the US National Academy of Sciences, etc. He was involved in evaluating vestibular experiments conducted on Spacelab 1 and other Shuttle missions. Although he never flew in space, Money was Roberta Bondar's back-up on STS-42. He left the CSA in July 1992 to continue research in aviation and space medicine at DCIEM, retiring from this work in April 1995.

Robert Brent Thirsk

Bob Thirsk was born in New Westminster, BC on August 17, 1953. He studied mechanical engineering at the University of Calgary, during which he won the 1976 Association of Profes-

Each Shuttle crew produces its own crew patch, this one representing STS-78, on which Bob Thirsk crewed. (via Andrew Yee)

sional Engineers, Geologists and Geophysicists of Alberta gold medal. He next pursued a master's at MIT, then moved to McGill for a doctorate in medicine. Thirsk was back-up payload specialist for Marc Garneau (Mission 41-G). One of his specialties was studying the effects of weightlessness on the human venous system. This involved many flights on the NASA Zero-g C-135. In 1993-94 he was Chief Astronaut at the CSA. Like many astronauts, one of his interests in the 1990s was learning to speak Russian.

In April 1995 Thirsk was named to STS-78 (Columbia), which began its 17-day mission at 1049 hours EST from KSC on June 20, 1996. He worked till July 7 with his six crewmates in the Life and Microgravity Spacelab (a record to this time for Shuttle mission duration). He conducted 15 experiments investigating the physiological changes and adaptations occurring in microgravity. One experiment was the Canadian-devised Torso Rotation Experiment—studying the relationship of certain body movements to motion sickness. Throughout STS-78 Chris Hadfield was the Capcom (capsule communicator), keeping in touch with the mission on all matters.

Bjarni V. Tryggvason

Payload specialist Bjarni Tryggvason was born in Reykjavik, Iceland on September 21, 1945. He came to Canada at age eight and grew up in Nova Scotia and BC. He graduated in engineering physics from the University of British Columbia in 1972. He later worked in meteorology; and attended the University of Western Ontario to study the effects of wind on structures like Toronto's CN Tower. He also helped in the NRC investigation into the sinking of the oil rig *Ocean Ranger*. As a payload specialist he backed up Steven MacLean on STS-52. He was project engineer for the design of the spacecraft deployed on this mission as a target for the Space Vision System being evaluated by MacLean. In 1995 Tryggvason was involved designing Canadian scientific experiments to be carried aboard NASA's C-135, and, in cooperation with UBC, developed an acceleration monitoring and data acquisition system for this aircraft. He also developed the Microgravity Vibration Isolation Mount (MIM) which would travel in space with him on STS-85 (due for launch in August 1997). The MIM totally

STS-78 launch close-up view. An international crew and an international payload begin their journey as the Space Shuttle Columbia lifts off on Mission STS-78 from Launch Pad 39B at 10:49:00 a.m. EDT, June 20, 1996. On board for Columbia's 20th space flight were mission commander Terence "Tom" Henricks; pilot Kevin R. Kregel; payload commander Susan J. Helms; mission specialists Richard M. Linnehan and Charles E. Brady Jr.; and two payload specialists, Jean-Jacques Favier of the French Space Agency and Robert Brent Thirsk of the Canadian Space Agency. Flying in the payload bay was the Life and Microgravity Spacelab, carrying US and international experiments. Mission duration was planned for 15 days, 22 hours, and 20 minutes. (NASA KSC-96EC-0830)

Bjarni Tryggvason wearing the crew patch for STS-52, for which he was the alternate payload specialist. NASA always provides excellent captions for its photographs and bends over backwards to serve all those enquiring for information. For the second photo it notes: "Steven G. MacLean (left), payload specialist for the STS-52 mission; and Bjarni V. Tryggvason, alternate PS, examine a camera for the Earth observations portion of the scheduled October spaceflight. The two Canadians joined other crew members for this training session in the Crew Compartment Trainer (CCT), part of the Shuttle mockup and integration laboratory at the Johnson Space Centre." (NASA)

Dave Williams (right) with Chris Hadfield at the warbird museum in Galveston, Texas in 1995. (Larry Milberry)

Julie Payette at KSC in November 1995; then, with Chris Hadfield, watching the airshow at CFB Bagotville on June 9, 1996. (Larry Milberry, Andrew Yee)

dampens ambient vibrations, even ones caused by crew members using exercise equipment. Among his many extracurricular interests, Tryggvason was qualified in flying instructing and aerobatics, scuba diving and parachuting.

The Second Group: Julie Payette

Julie Payette was born in Montreal on October 20, 1963. She earned an engineering degree at McGill University in 1986 (graduated as Faculty Scholar), a master's in applied science at the University of Toronto in 1990, then commenced a doctorate at McGill in electrical engineering. In 1991 she joined the IBM research lab in Zurich. The following year she moved to the Speech Research Group at Bell-Northern in Montreal. Payette entered the CSA astronaut program in June 1992. By 1996 she had been involved in such areas as robotic control, human-computer interaction and NASA C-135 activities. Her special accomplishments included piano and vocal expertise (Orpheus Singers, Montreal Symphony Orchestra Chamber Choir, etc.), linguistics, skiing, cross-country running and scuba diving. In 1997 Payette was involved in training to be a mission specialist.

Dave Williams

Dafydd Rhys "Dave" Williams was born in Saskatoon May 16, 1954, but grew up in suburban Montreal. In 1976 he graduated from McGill with a BSc in biology; then attended medical school at McGill, finishing in 1983 with a doctorate, and the title Master of Surgery. After practising family medicine in Ottawa, he studied trauma medicine at the U of T, then worked in the Department of Emergency Services at Toronto's Sunnybrook Hospital and lectured at the Department of Surgery, U of T. Meanwhile, Williams' wife, Cathy Fraser, was

at her own job—flying A320s for Air Canada.

Williams made the astronaut application short list of 20. His course began with a year of general indoctrination at the CSA in Ottawa. The candidates studied disciplines from geology and oceanography, to space design. They did gliding, scuba and sky diving; spent a week in the decompression chamber at DCIEM in Toronto; and visited Russia's Star City to learn about the Mir space station and other Russian accomplishments. Basic training finished in May 1993. Meanwhile, Williams became an assistant professor of surgery at McGill and maintained affiliations with two Montreal hospitals. He arrived in Houston in March 1994 to join the pool of about 100 astronauts always active there. There he was designated manager of the Missions and Space Medicine Group. Training commenced after a few days of orientation. This included land and water survival training, T-38 jet trainer ground school, para-sailing and a high-altitude indoctrination course. All astronaut candidates fly the T-38 as back-seaters. This keeps them in tune with the high speed environment, and with the tight confines of a fighter-type cockpit, ejection seat, peculiarities of "pulling g", etc. One of Williams' early assignments was a seven-day Shuttle mission simulation. He also was a crew member and crew medical officer, and principal investigator of a study evaluating resuscitation skills by non-medical astronauts. In January 1995 Williams joined the class of NASA mission specialist astronaut candidates, graduating in May 1996.

Robert Stewart

At the time he was selected for astronaut training in June 1992, Bob Stewart was a geophysi-

The boys from STS-74 clockwise from the left: Chris Hadfield (MS1), William S. McArthur, Jr. (MS3), Jerry L. Ross (MS2), Kenneth D. Cameron (commander) and James D. Halsell, Jr. (pilot). The photo was taken on November 14, 1995—Day 3 of their mission. (NASA S74E5031)

cist at the University of Calgary. Shortly after beginning his indoctrination at the CSA, he left the program, having decided that his personal long-term goals would be achieved best by returning to the academic world.

Chris Hadfield

Chris A. Hadfield, born in Sarnia, Ontario on August 29, 1959, grew up on a farm near Milton, a few miles west of Toronto. His father, Roger, was a captain with TCA. Chris and his brothers Dave and Phil were introduced to flight in the family plane. As farmers, they learned the meaning of hard work and became adept with mechanical equipment. Chris developed a love for sports and excelled at skiing, sailing and horseback riding. He earned gliding and powered flying licences with 820 Royal Canadian Air Cadet Squadron in Milton (years later he was an 820 reviewing officer).

As a nine-year old Chris Hadfield had watched TV as Neil Armstrong set foot on the moon. That evening, as he and his father walked near the family cottage on Stag Island in the St. Clair River, Hadfield looked into the sky, amazed that there was somebody up there. Then and there he decided that one day he would fly in space. The military seemed to offer the best solution for this goal. In 1977, after graduating from Milton District High School, he joined the Canadian Forces. After basic training he attended Royal Roads Military College in Victoria, then RMC in Kingston. During a field trip to the Johnson Space Center in Houston his keenness for space was piqued, especially after meeting astronaut greats like John Young (Gemini, Apollo and STS-1 commander)

In 1980 Chris Hadfield took basic flight training on Beech Musketeers at CFB Portage la Prairie, finishing top of the class. In 1982 he graduated with honours from RMC in mechanical engineering. In 1982-83 he took basic jet training on Tutors at CFB Moose Jaw, graduating first in September 1983 and winning the Golden Centennaire Award for flying excellence. This entitled him to choose the next step in his career. He opted for fighters and spent the next two years at CFB Cold Lake on the basic CF-5 course, then converted to the CF-18. He was on the second Canadian Hornet course, which included many who would become famous in the fighter world—Brosseau, Comtois, De Koninck, Lang, Major and others. This was a stimulating environment for any young flier.

From Cold Lake Hadfield's course formed the nucleus of 425 Squadron at CFB Bagotville. In June 1985 he and Capt Eric Matheson scrambled from the alert hangar at Bagotville. They staged through Gander for fuel, then flew the first CF-18 intercept on a Soviet Bear bomber. In 1986 Hadfield was part of Canada's elite team at the USAF biennial air weapons meet at Tyndall AFB, Florida. That year the Canadians made their best showing, finishing second overall, and nearly winning the coveted William Tell Trophy. His years as a fighter pilot seasoned Hadfield. Besides the exhilaration came some painful lessons. Wonderful friends were lost, one being Tristan De Koninck. A former Snowbird, he died in a CF-18 crash in PEI on May 24, 1986. Hadfield later recalled: "That taught us that life can be over in a flash, but the rest of us have to carry on."

Next, Hadfield attended the USAF Test Pilot School at Edwards AFB, California (Course 88A). In a year of intense classroom grinding and flying, he flew 28 types from a lightweight Grob glider to the 800,000-pound C-5A Galaxy. He placed first in his class to earn the Liethen-Tittle Award. This was only the third time it had gone to a foreign student. Now Hadfield was posted to NAS Patuxent River to fly the A-7 and F/A-18 on such experimental programs as out-of-control recovery, a vital series of flights to help the US Navy reduce its rate of F/A-18 accidents associated with loss of control. The program required throwing an F/A-18 violently out of control, then seeking the quickest way back to normal flight. In the end Hadfield found the solution—don't fight the situation, set the throttles at idle, rest your

hands on the canopy rails and let the plane self-recover. Hadfield flew 120 of these profiles. This was leading edge work about which he commented: "It's exciting doing things that have never been done before. We were writing the book on testing the F/A-18." Besides test flying, Hadfield did postgraduate studies at the University of Tennessee, writing his thesis on evaluating aircraft handling qualities at high angles of attack. In 1992 he was named US Navy test pilot of the year, the first time a Canadian exchange officer had been so honoured. Meanwhile, Chris and his wife Helene were raising their young children—Kyle, Evan and Kristin.

The Attainable Dream

In June 1992 Chris Hadfield was accepted for astronaut training along with Julie Payette, Robert Stewart and Dave Williams. Stewart soon withdrew and was replaced by Mike McKay. While his mates started a year-long CSA indoctrination program in Ottawa, Hadfield, with his depth of background as a test pilot, etc. was posted in August to JSC for astronaut training as a mission specialist. This included a wide range of activities from missions in the Zero-g C-135; to working in the WETF (weightless environment training facility) on space-like tasks, while wearing a complete EVA suit. About the WETF Hadfield noted: "To me it's a combination of scuba diving, rock climbing and old fashioned farm work. Nothing makes you feel more like an astronaut." He also trained on various simulators, including one for the Canadarm; kept current in the T-38; and visited the Star City space complex near Moscow. There Hadfield viewed Buran—Russia's Shuttle look-alike. De-funded in the post-communist era, Buran flew only once, then became a museum piece.

Hadfield was awarded his silver astronaut's pin August 3, 1993, but training continued. There were C-135 trips on each of which about 40 Zero-g parabolas are flown. On one of these Hadfield conducted electron-beam welding in a vacuum chamber, as if he were welding in space. Through all such programs astronauts pursue a host of other activities, from physical fitness, recreational pastimes, academics, etc. He and several other astronauts took intensive Russian language training in 1995—four hours daily, four days a week. Hadfield, already fluent in French, focused on Russian vocabulary and syntax particular to his mission. He also took up such matters as technical and safety issues for the Astronaut Office, and Shuttle glass-cockpit development. Hadfield became the first non-American in the ASP (astronaut support personnel) category. On September 2, 1994 he was named as the first Canadian mission specialist assigned to fly (STS-74). By this time he had strong views about the benefits of the space program. He saw it as one of humanity's greatest motivators, engendering the best kind of initiative among all involved. To him a mission to Mars (15 months of travel each way) was a realistic personal objective. "We know how to do it," he said. "Life's an adventure, so why not!"

Cost, as Hadfield put it, is "the common en-

emy" of progress in space; but, according to him, nations should combine their resources to have a sustained, gravity-free space lab, where unique experiments can be conducted, e.g. crystal growth for the study of molecular structure in proteins. Chemical gallium arsenide, a material essential in making computer micro chips, can be grown in the pure vacuum of space, but not on Earth. As well, medical science is sure to benefit from research performed without the effects of gravity. Other than the fact that it's there, gravity remains a mystery. In a space station environment more will be learned about it. Of course, with space stations like Mir there is the ongoing requirement for re-supply and crew change missions, upgrades and repairs, so the field of space transportation must steadily evolve.

Simulator Training

Of all JSC equipment the Shuttle static and motion simulators ("sims") occupy most of an astronaut's time. Astronauts must keep current on all seats aboard the Shuttle. In the sims they grow increasingly familiar with the Shuttle's hundreds of instruments, switches, alarms, etc. They must know instinctively what can fail, how it can go wrong, how to get around every emergency and, if necessary, how to survive disaster—if something drastic goes wrong in the initial moments after launch, a crew has seconds to react.

Each crew has its dedicated training personnel and the CAE motion simulator becomes the focus of much activity. Time after time a crew dons space suits to practice scenarios. It might be a simple case of repeating the routines of getting ready for a mission, strapping in and "launching". The Canadian-made sim imitates all the fundamental motions. Once a crew is strapped in, the sim tips 90° to the vertical. For the launch everything seems realistic, from sounds and visuals to the "twang" (a subtle motion as the Shuttle rocks on its mounts a few seconds before lift-off) to post-launch roll, and the "poof" as the SRBs (solid rocket boosters) blow off. The crew performs every function as if it were "for real". It's in touch with Mission Control at JSC and, if the simulation is "integrated", is communicating with Kennedy Space Center (KSC) and perhaps TsUP (Moscow Mission Control). On the other hand a simulation may focus on re-entry and landing. At any moment an emergency can be introduced, anything from engine to electrical failures, even a bird strike. In other JSC crew simulators anything can be replicated, but without motion. A crew sometimes moves into the main fixed simulator to work in isolation for days, although such work is not scheduled till a crew has its flight schedule.

Hadfield the ASP

In January 1994 Chris Hadfield was designated one of NASA's astronaut support personnel. As one of three ASPs his duties included working with Shuttle crews starting three months before launch. This entailed representing them at the launch facility and solving issues like loading cargo, instruments and experiments; window

(Right) The Shuttle program includes three important aircraft types, operated by NASA's Aircraft Operations Division. The Northrop T-38 trainer is used to acclimatize astronauts to the world of high-performance flight. Other important uses include taxiing astronauts between JSC and KSC, and getting them around on PR visits. In this view at Ellington AFB, near JSC, T-38s are seen on NASA's ramp, with a Gulfstream II Shuttle training aircraft and the 747 Shuttle transport (with Columbia atop) in the distance. Astronauts log about 15 T-38 hours monthly. (Larry Milberry)

(Left) NASA's most famous aircraft is its C-135 Zero-g trainer. Based at Ellington AFB near Houston, it flies regularly on missions to create short periods of Zero-g, allowing astronauts to experience this phenomenon. On each mission 40 profiles (called parabolas) are flown—climbing steeply, then pushing over the top and diving, a manoeuvre that creates a few seconds of near-weightlessness. Most trainees suffer air sickness until they get hardened after a few missions. Hence the C-135's nickname—the Vomit Comet.

(Right) One of NASA's four Grumman Gulfstream II Shuttle Training Aircraft (STA) based at Ellington. The STAs are reconfigured to simulate the Shuttle's flight characteristics on re-entry and landing. Pilots fly hundreds of simulated landings with the STA (1,000 to qualify as commander). Then (below), a look at the STA cockpit—standard GII on the right, full Shuttle controls on the left. A typical STA approach is made at 18°-20° nose down with flaps out (3° for a normal airliner approach), main gear down, and engines in full reverse. The pilot flares, drops to within a few feet of the runway, then overshoots. Training takes place at KSC, White Sands, New Mexico and Edwards AFB, California. NASA pilot Dave Finney noted, "Landing the Shuttle is not incredibly demanding. What really counts is being able to handle all the potential situations." Although a qualified test pilot, Canadian astronaut Chris Hadfield didn't fly the STA before STS-74. He had, however, practiced simulated Shuttle landings with a specially-configured T-38 at Edwards. (Larry Milberry)

Columbia at Ellington Field, Houston aboard the Boeing 747 Shuttle Carrier Aircraft (SCA) on April 9, 1995. Columbia had landed earlier at Edwards AFB on account of poor weather at KSC. Edwards was the primary recovery site for many early Shuttle missions. After 1990, once the orbiters had been fitted with upgraded undercarriage and brakes, and runway improvements had been made, KSC was preferred. It was opened for Shuttle recovery in 1976. Including paved overruns, the runway is 17,000 feet long. If Edwards has to be used, this adds five to seven days of preparation for its next mission. (Larry Milberry)

clarity; thermal tile mods; and system anomalies. Six weeks before launch a crew does a strap-in and pre-launch dry run at JSC. Three weeks later comes a full rehearsal at KSC, involving the hundreds of people dedicated to the forthcoming launch. The ASP is involved at every step. A few days before launch, Hadfield would fly with the crew to KSC, brief them on all procedures and follow them through other vital briefings, e.g. weather and safety. Meanwhile, as primary ASP he scheduled and directed the other ASPs. Other ASP tasks included responsibility for EVA suits, photographing stowage and set-up, and overseeing final closure of the airlock hatch. In the few days before launch, one ASP remains aboard the Shuttle to ensure that no unauthorized person enters.

Three hours before a launch Hadfield would enter the "White Room" (the access to the Shuttle), then the Shuttle itself to make check-

list updates and finalize switch configurations. Shortly before launch one of the ASPs flies a weather recce in a T-38. Then the ASPs join the families of the crew to watch the launch and enjoy a traditional post-launch meal of baked beans. When a Shuttle returns, Hadfield would be the first to board, helping the crew unstrap and get on their feet, and taking over switch reconfiguration. With a 1995-96 schedule of a launch every six weeks, ASP Hadfield had plenty to do.

Hadfield in Space—STS-74
STS-74 (Atlantis) was the 73rd Shuttle mission and the second of seven to Mir for 1995-97. It was a dry run—the first true test of space station assembly techniques for the future International Space Station (ISS) when, in 1997, Atlantis would deliver and attach a Russian-made docking module (DM) to the Soyuz FGB space tug. This would be the first time a Shut-

tle had carried a piece of Russian hardware (STS-71, under Capt R.L. "Hoot" Gibson, had been the first Shuttle to dock with Mir). In spite of the demands, Atlantis docked with Mir within three seconds of the scheduled time.

The first Mir building block, its 20.4-ton core module, was launched in February 1986. With life support system, living quarters and science research labs, it soon was manned with operating crews of two or three, as well as visiting crews. The core module had two docking ports—for manned Soyuz-TM transports and automated Progress-M supply ships.; and four spare ports for future modules. One (Kristall) was added in 1990 to facilitate Shuttle dockings. The orbiter docking module would be added by STS-74 in November 1995. It had two solar arrays to augment Mir's power supply. Priroda, a microgravity research and Earth observatory unit, was added in April 1996. With Priroda, Mir weighed more than 100

(Right) In 1996 JSC had 14 simulators on the go, some fixed, some motion. Here is one of the major fixed sims at JSC—a full-size Shuttle. Then a scene inside the Crew Compartment Trainer with Chris Hatfield explaining operations of one of the most important pieces of Shuttle equipment—the toilet. (Mike Valenti)

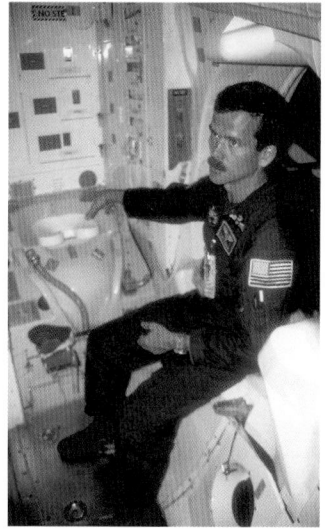

(Left) The Canadian-built full-motion CAE Shuttle simulator at JSC. (Larry Milberry)

(Below) NASA's Sonny Carter Neutral Buoyancy Laboratory (formerly called the Weightless Environment Training Facility) at Ellington in May 1997. Astronauts train in it in full EVA gear. Floating in simulated Zero-g, balanced perfectly by weights, they practice manipulating equipment and operating tools. Included in the NBL is a full-scale Shuttle cargo bay, and a full-scale Canadarm (which Chris Hadfield is explaining during a tour); and a Russian module, which astronauts can use in EVA training. In the second photo Chris Hadfield displays a mock-up tool like those he trained with under water before STS-74. (Larry Milberry)

(Left) International Space Station mock-ups being used at JSC in April 1995. Then, one of JSC's many exotic facilities the NASA Lunar Sample Laboratory Facility in Bldg 31. In this view Chris Hadfield and Andrew Yee of Toronto are observing Moon rocks under the helpful eye of Linda Ann Watts, one of the lab's geology-curatorial staff. (Larry Milberry, Mike Valenti)

(Left) Chris Hadfield in a mid-deck Crew Compartment Trainer position during ascent countdown rehearsal, where a staged emergency was introduced requiring the crew to implement their egress procedures. (NASA S95-07970)

Earth tons. Comprising seven modules, it had been 10 years in the making. Its final phases constituted Phase I of the ISS program.

By 1995 thirteen nations were involved in the joint space program. Phase I of the ISS began when Russian cosmonaut Sergei Krikalev flew aboard Discovery (STS-60) February 3-11, 1994. He conducted scientific research in the Spacehab in Discovery's payload bay. Two years later Discovery (STS-63) flew around Mir, closing to 37 feet. Cosmonaut Vladimir Titov crewed on Discovery for this trip. On June 27, 1995 Atlantis (STS-71) left KSC with five US and two Russian astronauts on the first of seven Mir missions. It docked using a Russian-built Orbiter Docking System, which it carried aloft. (The idea for a docking module to facilitate Shuttle visits to Mir began with talks in 1993. Assembly of the unit began in February 1995 and functional testing was complete by that May). The STS-71 cosmonauts were dropped off at Mir; two others, along with NASA's Norman Thagard, went home on the Shuttle. Thagard had been 115 days aboard Mir.

Crewing STS-74 were Col Kenneth D. Cameron (3rd flight, commander: STS-37, STS-56), LCol James D. Halsell, Jr. (2nd flight, pilot: STS-65), Maj Chris A. Hadfield (1st flight, MS1), LCol Jerry L. Ross (5th flight, MS2: STS-61B, STS-27, STS-37, STS-55) and LCol William S. McArthur, Jr. (2nd flight, MS3:STS-58). Hadfield called this group "the crew of crews" for its immense experience in space—he was the only one who hadn't flown a Shuttle mission. The overall mission included crew from four of five ISS partners—the US, Russia (Mir 20 crew Yuri Pavlovich Gidzenko and flight engineer Sergei Vasilievitch Avdeev), ESA (Thomas Reiter of Germany aboard Mir) and Canada. Only Japan was absent. Chris Hadfield's main task was to use the Canadarm to move the docking module from the cargo bay and berth it to the Shuttle's Orbiter Docking System.

STS-74 departed Launch Complex 39A at Kennedy Space Center at 0730:43 EST on Sunday, November 12, 1995. All went by the book, although Hadfield later joked that he repeated the astronaut's prayer as he waited for engine start: "Lord, please don't let me screw up." At lift-off Atlantis weighed 4,512,395 pounds (orbiter weight: 248,266 pounds, including payload of 14,064). The 8 1/2-minute launch phase was wild and as rough as screaming down a washboard road in a pick-up truck. That was mainly so with the SRBs firing and pushing the launch vehicle to 30 nm and 2,900 mph in two minutes. Then the SRBs drop away and the flight, now heading for 17,000 mph, is uncannily smooth as the Shuttle approaches orbit .

The mission had been slated for 24 hours earlier, when conditions at KSC were good, but the three transoceanic abort landing (TAL) sites in Spain and Morocco, some 4,200 nm eastward, were weathered in. In case of an emergency requiring trans-Atlantic recovery, landing Atlantis at a TAL sites would have been compromised by winds and poor visibility. There was hope that Zaragoza, Spain might

Atlantis on Pad 39A at KSC on November 10, 1995. It had been moved there from the vehicle assembly building on October 12. A Shuttle crew boards two hours before strap-in. A 2 1/2-hour wait ensues, although this can be longer. Thus do the astronauts wear diapers through the pre-launch hours. For STS-74 Chris Hadfield was determined not to use his. The way he saw it, after eight-hour missions in the CF-18 he was well prepared for eventualities.

The STS-74 insignia was designed by the crew. It depicts Atlantis docked to Mir. The central focus is the Russian-built docking module. The rainbow represents Earth's atmosphere and the flags of the tri-nation crew are shown.The sunrise symbolizes the dawn of NASA's space station construction era. (STS-74)

Weighing some 6.75 million pounds, Atlantis departs Pad 39A at 0731 EST on November 12, 1995, heading towards a 162-nm orbit. The launch window was 5 minutes, 10 seconds (for Russia's Soyuz shots it is only 20 seconds, since Soyuz launchers lack the fuel for longer manoeuvring in orbit). On a Shuttle launch the main engines burn for 8.5 minutes and take the Shuttle to 60 nm and a speed of M25. Main engines cut-off for STS-74 was at 0739; 90 minutes after launch the payload bay doors were opened. At this moment Mir was 7,000 nm ahead of Atlantis, in a higher and slower orbit. In the three-day chase that followed, Atlantis pilots Col Kenneth D. Cameron and LCol James D. Halsell, Jr. used their ship's reaction control system in a series of burns to raise or lower the orbit as required to finally close with Mir. (NASA STS074-(S)-016)

clear, but only Runway 30 there was available. It had a Shuttle Orbiter Arrester System—a large net to snag the Shuttle should it over-run. Although winds at Zaragosa were OK for R12, that runway did not have an arrester system. Zaragosa was out, but not till the countdown had ticked down to T-5 minutes. In retrospect, however, some at NASA felt that R12 would have been safe, considering that Atlantis had improved brakes and a bigger drag chute.

STS-74 was Atlantis' 15th flight. It was also the first Shuttle flight to conduct space station assembly. Its primary task was to attach to Mir the 15'4"-DM needed to give room for safer docking—to allow vehicles a safety margin while approaching Mir, with its jumble of appendages. The DM arrived at KSC from Russia on June 7, 1995 aboard an AN-124. The Shuttle crew would install it on the Kristall module, then use it in transferring materiel to and from the space station. Atlantis also would deliver two Russian-designed, 9,000-pound (Earth weight) solar arrays needed to boost power on Mir. These arrays were the first major pieces of ISS hardware co-operatively made and carried into space. STS-74 also carried food, water, experiments and gifts to Mir, about 213 nm above Earth.

STS-74 experiments included GLO-4, Photogrammetric Appendage Structural Dynamics Experiment (Pasde), Mir experiments, and the Shuttle Amateur Radio Experiment (Sarex). Also in the cargo bay was the Canadian-designed and built Orbiter Space Vision System (OSVS, virtual reality equipment to be used by Hadfield in moving and positioning the DM); and a 65mm Imax camera using a 30mm fixed fish-eye lens and 3,500 feet of film. Hadfield would run it from the aft flight deck, shooting footage for NASA feature films. GLO-4 was for studying Earth thermosphere, ionosphere and mesosphere energetics and dynamics, using broadband spectroscopy. It also would observe interactions with the atmosphere of the Shuttle and Mir, study spacecraft glow, Shuttle engine firings, water dumps and fuel cell purges. Using three photogrammetric data-gathering canisters in the cargo bay, Pasde would record structural response data from Mir's solar arrays during docking.

A close-up of Atlantis during STS-74 with the crew hamming it up at the aft windows. Chris Hadfield is at the far right. He was the 200th NASA astronaut to fly. During his flight he also occupied the commander's seat for a short time, after undocking from Mir. This was a chance to "fly" the Shuttle although Hadfield's description is better: "Orbital mechanics make any definition of stick time inaccurate. It's more like nudging the vehicle (doing the occasional thruster firings) than flying it." (NASA STS074-320-019)

Down to Work

About 43 minutes after launch, a two-minute engine firing placed Atlantis in a 162-nm circular orbit. Ninety minutes into the flight Hadfield opened the payload bay doors and mission ops commenced. The pilots fired the reaction control jet to refine the flight path towards Mir. Meanwhile Hadfield activated the DM. Other activity focused on closing with Mir and preparing to position the DM for docking. Hadfield tested the Canadarm, and he and McArthur checked out the OSVS. Cameron got VHF radios ready for Shuttle-Mir

communications, and McArthur and Ross made ready for EVA should difficulties in the payload bay require a space walk. About 24 hours after launch, Atlantis was some 3,500 nm behind Mir. A further engine firing took place on Monday. Another activity was a live press conference. At 1146 CST on Monday, Chris Hadfield and Bill McArthur, with the OSVS powered up, grappled the DM with the Canadarm, ready to lift it the following day. Success at this depended on the Canadian-made Advanced Space Vision System.

Advanced Space Vision System and the Canadarm

On Flight Day 3 of STS-74 (Tuesday, November 14) Hadfield began manoeuvring the DM using the Canadarm in conjunction with the ASVS (other terminology for OSVS). Conceived 20 years earlier by the NRC's Dr. Lloyd Pinckney, the original Space Vision System was designed to enhance astronauts' vision in the difficult viewing conditions of space (alternate periods of extreme darkness and brightness, which make it difficult for astronauts to gauge speed and distance of objects). The SVS could track four targets simultaneously. Canadian astronaut Steve MacLean tested it on STS-52 in 1992. As stated by NASA, the OSVS "consists of a series of large dots on the exterior of the docking module and the ODS (orbiter docking system). Using digitized television camera views of the dots, the OSVS generates a display on a laptop computer aboard the Shuttle that indicates alignment both graphically and digitally." Chris Hadfield described SVS: "What the vision system does is look at targets stuck on the two pieces that I'm trying to assemble. From that it figures out very exactly, in three dimensions, where both of these things are. Once the computers figure out this information, then you can create any computer display you want to show your alignment."

In its 1995 form ASVS could deal with as many as 32 items and function in a greater range of lighting. It obviated a natural shortcoming of the Canadarm in accomplishing such a task. Previously, the operator, in spite of having a variety of camera angles, could not see the station section to which the hardware

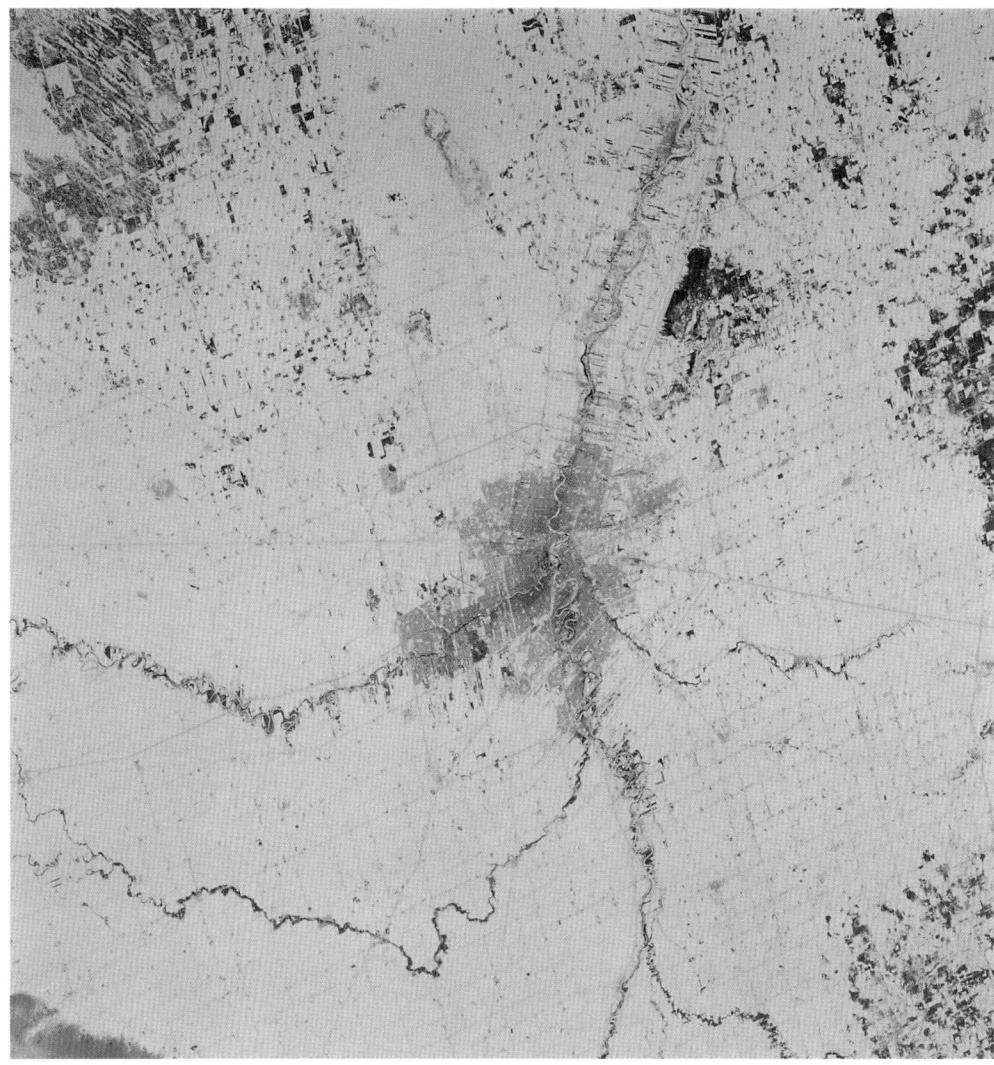

Somalia and Ethiopia viewed by STS-74 during a pass on November 18, 1995. The photo is towards the southeast, while Atlantis was about overhead Djibouti. On the left is the Arabian Sea, while the Indian Ocean is above. The peninsula halfway across the Indian Ocean shore is Ras Hafun. This sweeping view takes in the entire Horn of Africa out to Cape Guardafui. In the early 1990s Canadian Hercules served the area in the bottom middle, taking relief to places like Dire Dawa, Jijiga, Kabridar and Gode. Hargeisa, described in Ch. 48, is also in this parched highland region. (NASA S74E5336)

on the arm's tip was to be attached. ASVS data is updated 30 times per second. With software that included models of equipment to be installed, Shuttle structure, camera locations, etc., it provided information about the exact location (within 1 mm), orientation and motion of its target. Working with Bill McArthur, just before midnight on Monday, Hadfield, began attaching the arm to a grapple fixture on the DM, using ASVS. Then latches holding it horizontal in the payload bay were released. Hadfield lifted the DM above the bay, rotated it vertical to the Shuttle, and aligned it with the ODS, first to a distance of 30", then to 5". The arm was used to hold the DM exactly in place. Cameron fired steering jets to force the Shuttle up into the DM, counting on the flex of the arm to overcome the 5" gap, allowing hooks and latches to engage and lock the DM to the ODS. Mating was confirmed at 0117 CST on November 14.

The Docking

Further flight path corrections were made using manoeuvring jets; the two spaceships met on November 15. Two hours before rendez-vous, they were eight miles apart. The Canadarm was extended so that its elbow-joint camera could assist in formating with Mir. Nav equipment supplemented by a laser-ranging device in Atlantis' payload bay, and a hand-held laser ranger were also used. A half-mile below Mir, Cameron moved from the commander's seat to the aft flight deck controls. On the next

orbit Atlantis closed to a point below Mir called the R-bar—an imaginary line between Mir's centre of gravity and the centre of the Earth. This allowed natural forces to help slow Atlantis more efficiently compared to a more standard Shuttle approach (from directly in front of Mir). Using cockpit cameras and the centreline camera on the module's docking mechanism (ASVS was not used, as there were no target dots on Mir), Cameron centred the module docking device with the one on Mir, continually refining alignment as he approached within 170 feet. There he paused while Mir manoeuvred into docking attitude, then edged in to 30 feet. Cameron inched closer, while keeping Mir updated. At 1238 CST November 15 (orbit 44, and 216 nm above Mongolia) the DM contacted the port on Kristall. It was a textbook success and so satisfying to the STS-74 crew after 14 months of training. In talking about this event, Hadfield described it as "the most tense and exhilarating moment... like a mating of elephants." Once the ships were united and pressures were equalized, the crews moved easily from one to another.

Now STS-74 crew transferred 2,132 pounds of cargo to Mir—food, equipment, oxygen, hydrogen and water. The NASA crew carried gifts. Hadfield's main one was a collapsible electric guitar. He knew that the same old guitar had been aboard Mir since 1980. Cosmonaut Reiter, an accomplished classical guitarist, immediately put the replacement to use. As to his impressions, Hadfield reported to Earth:

"Mir is actually quite large. It's like a long maze-like collection of tunnels. I've turned the wrong way several times already. For us it can be a little disorienting. If you stand up straight, in most places your head will touch the ceiling and your feet will touch the floor. In some places it's about the size of your body... you kind of sneak through a gap. There's quite a bit of room. In fact, there are some places that are very quiet and private on Mir, so it is quite a livable environment."

In preparation for ending its mission, Atlantis loaded experiments and other materiel from Mir. Its payload coming home would total 4,049 pounds. Atlantis and Mir separated on November 18 on their 92nd orbit after three days and 58 minutes together. Now Cameron conducted 90 minutes of Mir fly-arounds, mainly for photography. The cameras were recording the long-term effects of space flight on Mir, e.g. damage from micrometeor strikes.

Atlantis landed uneventfully at KSC at

Mir from Atlantis, first with Earth as backdrop, then, deepest space. The brownish docking module is attached to the Kristall module. The DM gave visiting Shuttles more room to approach clear of Mir's solar panels and other appendages. On June 25, 1997 Mir collided with a Progress supply ship. Its Spektr module was decommissioned and power was halved. Considering previous troubles, Mir was deteriorating. American astronaut Michael Foale decribed his June 1997 Mir mission as similar to "a very dirty and grimy camping trip in an old car." (NASA STS074-716-021, S74E5239)

1201:27 EST on November 20. It had been away eight days, four hours, 20 minutes and 44 seconds; and completed 129 Earth orbits. Distance travelled was 3.4 million miles. Transfers from Mir totalled 816 pounds of science hardware and samples. Included were American biomedical and microgravity science samples and data collected by Mir crews 18, 19 and 20. Water samples (potable, reclaimed hygiene, unprocessed hygiene and humidity condensate water) from Mir were carried home to study their purity and gain further knowledge for the design of water purification units for the ISS. Also returned was a crystal growing experiment, samples of plants grown aboard Mir, and data relating to physiological deterioration suffered by astronauts.

Back Home

Weeks after returning, Chris Hadfield still was "flying high" and getting used to what he had done—he had competed successfully for astronaut training, won a slot on STS-74, become an ASP, learned Russian, flown the Canadarm. He couldn't wait for the next opportunity. "I just see this as the beginning," he said. "My big dream now is to do a space walk." In this period Hadfield was busy with press conferences and celebrity appearances. He talked about some of his impressions. First of all, the launch was not frightening; rather, it was "a magnificent feeling." Once in orbit, he noted how most of the crew had some experience with nausea as their bodies resisted weightless-

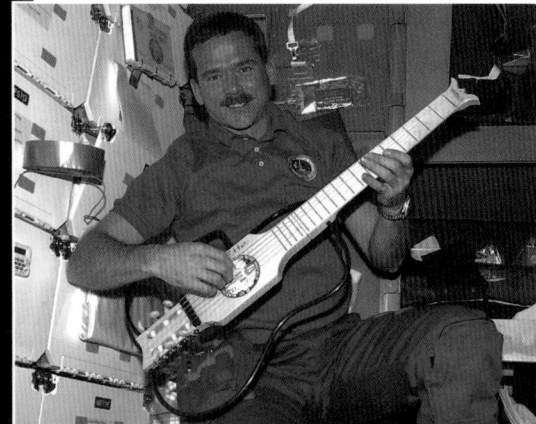

(Right) Chris Hadfield tries out the collapsible guitar that STS-74 brought to Mir. Behind him in this setting aboard Atlantis is a bank of storage lockers containing everything from food to science kits to cameras and batteries. Everything about the space program thrilled Hadfield. He viewed it as "universally cool" and as something important in drawing humanity together in the long run. Then, Hadfield at the Hoagie Ranch—a neighbourhood spot near JSC, where he regularly played guitar and sang with other fun people on Saturday nights. With him here is flutist Cady Coleman, a mission specialist who flew on STS-73, and keyboard man Andy Upchurch. (NASA S74E195, Larry Milberry)

ness, but that this subsided within a day. He noted that after four days, "everything was working normally." Meanwhile, the rigours of adjusting had cost him seven pounds. He had some trouble adjusting to sleep on day one, and needed a pillow to get comfortable. Thereafter he slept floating free and loved the sensa-tion of weightlessness, describing "how right, how natural" it felt. Besides the challenges and excitement of performing his crew duties, Hadfield and his mates spent one enjoyable hour after another spellbound by the ever-changing visions of Earth. On one occasion he watched as a huge meteorite burned up below the Shuttle over Australia. The thought crossed his mind of the disaster that a meteoroid strike could bring to the Shuttle or that an asteroid strike could bring to humanity.

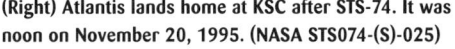

STS Facts
Length: system 184.2'; orbiter 122.17'; external tank 153.8'; solid rocket boosters 149.16'
Wingspan: 78.06'
Cargo bay: 60' x 15'
Gross weight lift-off: 4.5 million pounds
Dry weight including three Shuttle main engines:
Columbia – 181,365 pounds
Discovery – 173,504
Atlantis – 172,778
Endeavour – 173,746
External tank (full): 1,655,600 pounds
SRBs (at launch): 1,292,000 pounds

(Above) The Atlantis-Mir crew on November 15, 1995. In front are Sergei Avdeev, Thomas Reiter, Yuri Gidzenko, Bill McArthur and Jerry Ross. Behind are Ken Cameron, Jim Halsell and Chris Hadfield. In June 1997 Hadfield was assigned to STS-99, the eighth of 40 missions to assemble the ISS. His main work would be to install the 17-meter Spar Aerospace Space Station Remote Manupulator System—the next-generation Canadarm. It will be crucial in constructing the station, using components from Europe, Japan, Russia and the USA. Hadfield would walk in space (three times), the first Canadian to do so. The ISS was scheduled to be complete in 2003. (NASA S74E5162)

(Right) Atlantis lands home at KSC after STS-74. It was noon on November 20, 1995. (NASA STS074-(S)-025)

Canadian Astronaut Missions Flown and Scheduled to 1998

Astronaut	Mission	Shuttle	Dates	Mission Duration	Distance Travelled
Marc Garneau	STS 41-G	Challenger	Oct. 5-13, 1984	8 days 5 hours 23 minutes 38 seconds	5,526,020 km
Roberta Bondar	STS-42	Discovery	Jan. 22-30, 1992	8 days 1 hour 14 minutes 44 seconds	5,405,967 km
Steve MacLean	STS-52	Columbia	Oct. 22 - Nov. 1, 1992	9 days 20 hours 56 minutes 13 seconds	6,643,606 km
Chris Hadfield	STS-74	Atlantis	Nov. 12-20, 1995	8 days 4 hours 30 minutes 45 seconds	Approx. 5,500,000 km
Marc Garneau	STS-77	Endeavour	May 19-29, 1996	10 days 0 hours 39 minutes 20 seconds	Approx. 6,600,000 km
Robert Thirsk	STS-78	Columbia	Jun. 20 - Jul. 7, 1996	16 days 21 hours 47 minutes 36 seconds	Approx. 11,000,000 km
Bjarni Tryggvason	STS-85	Discovery	Targeted for Aug. 1997	Approx. 11 days	N/A
Dafydd Williams	STS-90	Columbia	Targeted for Apr. 1998	Approx. 16 days	N/A
Chris Hadfield	STS-99	Atlantis	Targeted for June 1999	Approx. 11 days	N/A

At Century's End

North America's first Airbus A330 at Toronto on May 14, 1997. The Airbus and Boeing families of airliners were the transport industry's way into the 21st Century. (Larry Milberry)

The Airlines Today

By the late 1990s the world airline industry was on the upswing. In an atmosphere of economic growth people were travelling more. This pumped cash into airline coffers. Frequencies increased, destinations multiplied, and new fleets were ordered. Canada's airlines swept along on these waves. Now Air Canada was making money and its fleet was being renewed at an unprecedented rate—A319s, A340s, CRJs. Good for Air Canada, although Canadian's future remained doubtful—all its top people could claim in 1997 was that losses were not as bad a last year's. Even so, it stayed competitive by downsizing in some domestic and European markets, while taking an aggressive stance with the burgeoning trans-Pacific trade and with Open Skies opportunities in the US. When Air Canada established Toronto-Raleigh/Durham CRJ service, Canadian Regional followed, placing F.28s on the same run in June 1997, offering an introductory return fare of $249 (plus double air miles). Meanwhile, charter carriers were on a roll. Canada increasingly was a vacation destination, while Canadians were holidaying abroad more. Air Transat, Canada 3000 and Royalair were growing. Older fleets were fading—the 757, A310 and A320 were taking hold. For the summer of 1997 Canada 3000 was operating six A320s and six 757s. For its purposes Air Transat still found the L.1011 profitable and had 13 in service over the summer of 1997.

New carriers still were appearing, one being Skyservice. It had begun as an FBO business in 1986 under Russell Payson. It chartered types like the Learjet, then entered the airline business in Toronto in 1992, being closely affiliated with Sunquest Tours. Equipment was the 180-seat A320. In May 1997 Skyservice introduced the Airbus A330 (363 seats) to North America. This was a surprise, since Canada 3000 earlier had announced that it would lead with the A330, taking delivery in the spring of 1998. Skyservice, however, needed a widebody for Sunquest markets from Ireland to Spain, Italy and Greece. A310s were available, but fell short in range for nonstop return flights from Mediterranean destinations with at least

A ir transport in Canada, which started so tentatively, became integral to society. Each day Canadians reap the benefits it brings, whether by travelling swiftly, comfortably, economically and safely from city to city in modern airliners, receiving the day's mail or an overnight package brought to the door by courier, or enjoying foods and manufactured products from around the world that arrive by plane. Many hi-tech companies depend on the sophistication of global aviation. Bombardier, for example, flies the wings of its Global Express and the fuselage for its Dash 8-400 to Toronto from Japan aboard AN-124s. Pratt & Whitney Canada flies in components

from subcontractors abroad. Helicopters made in Mirabel go by air freight to overseas customers. Almost every night, auto makers air freight JIT parts to their factories—components made in Kitchener, Oshawa or Peterborough can be rushed by air to a plant in Michigan, while US-made parts cross into Canada.

Air transport is indispensable to mega projects. Without bushplanes and helicopters it would have been impossible to begin ventures like the James Bay hydro development, the diamond fields of the NWT or the great ore properties around Voisey Bay, Labrador. Meanwhile, air transport serves Canadians in all the usual ways, letting them conduct business in Winnipeg and be home in Toronto the same day, visit family wherever, flee winter aboard charter sun flights, etc. Head offices are connected by corporate jets to far-flung plants, mining properties, sources of supply and markets. Helicopters support offshore oil projects, rush accident victims to hospital, patrol pipelines, log, fight forest fires and monitor highway traffic. Hercules and Airbuses back the military at home and the UN overseas, while helicopters support the army and ships at sea, and conduct search and rescue.

Canada's Helicopters by Region in 1994	
	No. machines
Alberta	210 (98,845 hours flown)
British Columbia	483 (234,146)
Manitoba	53 (20,578)
New Brunswick	5 (1,178)
Newfoundland	90 (36,173)
Nova Scotia	17 (5,375)
Northwest Territories	55 (25,161)
Ontario	173 (59,332)
Prince Edward Island	2 (158)
Quebec	194 (71,843)
Saskatchewan	18 (4,777)
Yukon	18 (9,902)

Some Canadian Helicopter Fleets in 1994	
Aérospatiale AS350 AStar	147
Aérospatiale AS355 TwinStar	6
Aérospatiale Puma/Super Puma	3
Bell 47	29
Bell 204/205/212	142
Bell 206	622
Bell 222	14
Enstrom	28
Hiller	26
McDonnell Douglas (Hughes)	118
MMB BO105	18
Robinson	96
Sikorsky S-58/S-58T	8
Sikorsky S-61	11
Sikorsky S-76	21

The Highland Helicopters ramp at Vancouver International in January 1981. A Bell 212 lifts off, as a Jet Ranger waits. Highland, founded in 1959, was a leading BC operator in the 1990s. Its fleet in 1997 numbered 24 Jet Rangers, six Long Rangers, three AS350Bs and two Bell 204s. (Gary Vincent)

(Above) A CAIL Boeing 767-375ER lands at Toronto on September 2, 1989. For 1997 CAIL's future remained uncertain, although a new restructuring plan was in force. Through cutting overhead, and agreements with employees, government and AMR Corp. it planned to save $560 million in four years. While retreating in some markets, it offered new services, especially in the Pacific, where it had a half-century of experience, eight destinations, strong partners, and its newly-renovated hub in Vancouver. In 1996 CAIL had 160 destinations in 17 nations; it carried 11.6 million passengers. Including its partnership with American Airlines, it had 1,000 daily flights to 90 centres in its Canada-US market. (Larry Milberry)

A330 C-FBUS during a turnaround at Belfast on May 28, 1997. (Larry Milberry)

Keith Levia, Larry Fischer and Ivan Morrell during their Toronto-Belfast crossing. Note the EFIS cockpit design. (Larry Milberry)

265 passengers. The B.767-200 had the range and was available, but the A330 was enormously more capable. An opportunity for an A330 arose when Germany's LTU freed a leased machine. Skyservice capitalized on this. Aircraft C-FBUS (ex-D-AERJ) was delivered to Toronto on May 6, operating its first service (Toronto-Barcelona) on May 9. Certification was facilitated by Transport Canada under inspectors Ivan Morrell and Brian Campbell.

The Airbus family began in the 1960s with design of the A300 (first flight: October 1972). Slow to win a place in the market, the series eventually caught on, especially with the A310, which Wardair introduced to Canada in 1987. The versatile A320 appeared on the Canadian registry with Air Canada in 1990, CAIL in 1991 and Canada 3000 in 1993. Air Canada brought the first intercontinental, four-engine A340 to North America in 1995, then added the A319 in 1996.

The A330, basically a twin-engine A340, flew on November 2, 1992. First delivery was to Air Inter of France in December 1993. The A330 is categorized as a medium extended-range widebody with high-density seating for as many as 440. Like all modern airliners it comes with the customer's choice of engines— GE CF6-80E series (67,500 pounds of thrust), Pratt & Whitney PW4000 series (64,000) or Rolls-Royce Trent series (67,500); more powerful versions were available for higher-gross-weight A330s. Typical of the Airbus family the A330 has electronic flight controls with fly-by-wire control for all wing/tail surfaces. The pilots' sidesticks and the autopilot are linked to three flight control primary computers, which serve ailerons, elevators, rudder and spoilers. There also are two flight control secondary computers. Due to computer control the A330 has immunity from stall, overspeeding and airframe overstressing, as well as windsheer protection. The flight deck is an A320 derivative and there is cross-crew qualification with the A320, A321 and A340. This reduces training

costs and allows a pilot to move from type to type (once an airline is certified for this). The avionics package includes a central maintenance system. It has a central maintenance computer (CMC) that interfaces among aircraft systems, built-in test equipment, and a multipurpose control and display unit on the pedestal between the pilots. The CMC reports on such things as air conditioning, autopilot, electrical power, fire protection, flight controls, fuel, hydraulics, and the undercarriage.

A typical Skyservice A330 flight (F674) operated on May 28. C-FBUS was at its gate at Toronto's Terminal 1 late on the 27th after arriving from Italy. Once it was turned around by the fuel, ground servicing, catering and grooming contractors, it was ready to go again. The crew was on hand two hours prior, the pilots to go over the flight plan and discuss technical matters. The eight cabin crew reviewed their duties and procedures, to be ready to serve 244 Belfast-bound passengers.

On the flight deck for F674 were Capt Keith Levia, FO Larry Fischer and TC's Ivan Morrell. All had started in aviation as young men in the RCAF. Capt Levia had joined the

RCAF in 1962. He started on the Dakota, then spent years flying Buffalos, on which he logged about 4,500 hours. Hercs came next, then (1986) Levia moved to the airline world, flying for Worldways. When it folded on October 11, 1990 he migrated to Royal Air, then Skyservice. Larry Fischer had started as a navigator in 1961, his first posting being on SAR Lancasters at Torbay. In 1970 he cross-trained to pilot, flying Twin Hueys with 427 Squadron. Later he had tours on Hercs and 707s. He left the air force in 1985, flying with Worldways, Nationair, Royal Air, then Skyservice.

Ivan Morrell joined the RCAF in 1966. At the time he was a teenager playing hockey for the Oshawa Generals and went into the air force on a dare from some buddies. He was on one of the few "all-through" jet training programs, where recruits straight off the street took their first flight training in a Tudor jet, bypassing the usual primary course on the Chipmunk. Nearly every student failed and the program was scrapped. Morrell was one of the few course survivors and was posted to fighters. After advance training on T-33s he went to Voodoos, flying them for 1,700 hours. In 1974 then joined Quebecair to fly the F.27, BAC111, 707, etc. In 1986 he started with ACS, a Montreal freight company with DC-8s. This sent him around the world on such jobs as freighting during the Falklands and Gulf wars. In 1992 Morrell went to Transport Canada, where he became an Airbus inspector. When Air Canada inaugurated A340 service on the Pacific on September 20, 1994 (Vancouver-Osaka), he was aboard as check pilot with Air Canada Capts Ray Hauser, Al Burch and Mal MacDonald.

At 0449 GMT BUS was cleared to start (a fully automatic process). It was carrying 40.7 tons of fuel for the trip. As the engines spooled up, a peek at the logbook showed that BUS already had 5,645.49 flying hours and 2,302 cycles since new in April 1995. Now a tug pushed it from the gate. Capt Levia taxied for R33R, from which BUS lifted off at 141 knots at 0502. Initial clearance was for 5,000 feet, but the crew quickly was given 23,000, then 37,000. BUS climbed steadily at about 250 KIAS. Passing 10,000 it was cleared for North American exit at 56°19'N, 58°5'W, a spot on the charts named Porgy. Soon the flight deck was settling down for the trip and the always-

Under Tom Syme, formerly of Air Ontario, Vistajet started operations in May 1997 with a blockbuster $26 fare on any leg in its small system. It hoped to add planes and destinations like Thunder Bay and Calgary by summer's end.

interesting conversations began. Capt Levia opened with some comments about the A330. He had flown the DC-8 and L.1011, at this stage jokingly referred to as "Jurassic jets". None rivalled the A330 as far as ergonomics went. It was incredibly quiet. Pilots no longer had to raise their voices to talk cross-cockpit. Levia also found the ventilation system outstanding. Such features eased a crew's job, and they arrived at destination refreshed instead of tired.

As BUS passed 25,500 feet a computer began an automatic transfer of fuel from the wing tanks to the tail. This would ensure optimal trim for best performance and fuel economy for the rest of the flight. For descent, fuel would transfer the other way. Further conversation was about EFIS displays. Airbus had one layout, Boeing another, McDonnell Douglas another. The crew thought that manufacturers should get together to standardize layout. BUS cruised along. At 0551, by when the sun was up, a glance at the nav display showed Porgy 792 nm in front on a heading of 061°. Ground speed and true air speed were 441 and 464 knots, and winds were 74 knots at 327°. ETA at Porgy was 0739. The Mach number was .820 and altitude FL370. Passing Porgy the flight climbed to FL410. Now it arced north to 60°N latitude on an ETOPS trial—extended twin operating procedures, referring to twin-engine airliners on long over-water flights being able to reach alternate airports on one engine. On this flight the A330 would not be more than 531 nm or 75 minutes from its alternates— Goose Bay, Sondrestrom and Keflavik. Skyservice eventually would be certified for 90 minute ETOPS.

At 1044 BUS started descent for Belfast, 151 nm away. The weather was fine, Ireland being an inviting sight as it came in view. BUS

soon was lined up at 150 KIAS two miles back for R25. A Virgin Air A340 was ahead, outbound, doing touch-and-goes. BUS came across the threshold at 131 knots, touched down at 1116 GMT and shut down at the gate five minutes later. Virgin Air was touching down again as Skyservice's passengers started to fill the aisles. Plastered on the A340 was a typical Virgin Air slogan — "No way BA/AA", referring to its arch rivals on the Atlantic, British Airways and American Airlines. Now the quick turnaround started. BUS had burned 34 tons of fuel— refuelling started quickly. A full house was expected for Toronto, so the caterers began replenishing the galleys. Fresh water was pumped aboard and the potties were emptied. The groomers set to work cleaning the cabin. On the ramp techs did their inspections, checking fluid levels, watching for leaks, etc. In Belfast's international lounge 321 Toronto-bound passengers waited patiently—chatting, snoozing, pacing, sipping teas and quaffing Harps. Meanwhile, a new Skyservice crew arrived. Capt Levia's was off to its hotel. Tomorrow they would deadhead to Barcelona, where they'd rendezvous with BUS in three days to take it home.

At 1314 BUS leaped into the air and turned west for Canada. Soon it was sitting at FL390. In time Greenland passed below, then the Torngats of Labrador. Glacier- and logging-scarred Northern Quebec came next, followed by Ontario's placid Kawartha lakes. The drumlin fields around Rice Lake and Lake Scugog sailed by, then BUS was again over urban Ontario, with the CN Tower in the distance. It flew directly over LBPIA, took a long downwind, turned 180° and landed on R06R 6:40 hours after leaving Belfast. Its load of jolly charter passengers gave the crew a rousing applause. After all, they had been well treated by a team of topnotch flight attendants. Soon they disappeared to Customs, to get their bags, then to do battle with Toronto at the peak of afternoon rush hour. BUS, which had crossed the Atlantic twice in the same day, now was quickly readied for five hours of crew training.

By the time it landed again it would have spent 18 hours aloft for May 28.

While the major carriers moved ahead, others were busy—the upstart inter-urban lines. Such operators started appearing in the 1980s. In 1989 Intair entered the Montreal-Toronto market, offering low prices and high frequency. It invested heavily in Fokker F.100s and ATR42s, but lasted only into 1991. In 1994 Atlantic Island Airlines began F.28 service from Canada's Maritime provinces to Toronto, but it folded in less than a year. Competing with "the big guys" had never been easy, but investors were available to try again. Thus, it was no surprise when two western-based companies appeared— WestJet of Calgary in 1995 and Greyhound Air of Kamloops in 1996. WestJet looked for business between Victoria and Winnipeg with 737s, Greyhound between Vancouver and Toronto with 727s. Astoria entered the market in this period, operating a 737 Toronto-Ottawa-Montreal, but it quickly failed. In May 1997 Vistajet began service from Windsor to Toronto and Ottawa using a 737. Such operators sought to tap business that traditionally belonged to Air Canada and CAIL. The majors conceded thousands of seats daily to the upstarts and to holiday charter companies like Canada 3000, which also were eroding the Air Canada/CAIL domestic hegemony with frequent trans-Canada service at economy rates. This forced Air Canada and CAIL to focus elsewhere. They expanded internationally, especially on the Pacific; and poured millions into new trans-border services that were theirs to grab under the Canada-US Open Skies agreement. For 1995-96 the majors added 60 new US destinations. For June 1997 Air Canada and Canadian were operating 87 Open Skies-related services, many being to new US destinations like Kansas City and Denver. Air Canada dominated among all operators, the secret to this success being its fleet of new Canadair CRJs.

While Canada's leading airlines were on the move, there were reasonably good times among the smaller ones. Companies like BC's Central Mountain, which barely had survived the last recession, were growing. In 1997 CMA took over several Air BC routes, now impractical for Dash 8s. It added a large fleet of Beech 1900Cs and D. Elsewhere, aviation on the BC coast seemed steady, with heli-logging especially busy. In 1997 Helifor added Boeing 234

In the late 1990s the 19-seat Embraer Bandeirante commuter started appearing on the West Coast, years after it was popular in Ontario and Quebec. This one of Kenn Borek was at Vancouver in May 1997. Pacific Coastal of Vancouver adopted the Bandeirante about the same time. (Jan Stroomenbergh)

A Helifor Vertol 107 deposits logs in a BC inlet. In 1997 this Vancouver-based company added a Boeing 234 Chinook to its fleet of three 107s and three Hughes 500Ds. (Larry Milberry)

Chinook C-FHFH. This business, however, remained risky. On April 19, 1997 Coulson Aircrane S-61N C-GBRF crashed while logging at Stave Lake, BC.

New helicopters appeared each month on the Canadian Civil Aircraft Register. In the spring of 1997 these included Bell 407 C-FALA for Alpine Helicopters, Bell 412 C-GUAY for Air Alma, Eurocopter AS350B-2s C-FZTA for Questral Helicopters and C-GMAN for Heli-Manicouagan, AS350BA C-GZPT for West Coast Helicopters, Robinson R22 C-FMNC for Domac Equipment. In June 1997 Bell Boeing announced that 29 of its 275-knot, 750-nm, 21st-century "609" tiltrotors had been reserved, one customer being Canadian Helicopters Ltd.

In the NWT, besides routine work, development of Canada's first producing diamond mines kept the NWT active. Statistics, however, showed the aircraft movements had waned at Yellowknife. At the height of diamond exploration in 1994 there were 83,281. Movements plummeted to 65,340 in 1995, to 62,881 in 1996. In Labrador, Voisey Bay allowed local carriers to expand. Frank Kelner of Pickle Lake set up an operation at Goose Bay using Pilatus PC-12s. Thanks to the reliable

PT6, this type was cleared by Transport Canada for full commercial IFR operations, the first such single-engine approval in North America. The tow-out from St. John's in May 1997 of the 700-foot high, 600,000-ton *Hibernia* drill rig promised renewed offshore helicopter business.

Through the 1990s Canada's smaller air carriers continued as always. For most it was the age-old status quo, although there was occasional growth. A bush or coast operator might add a Cessna or Beaver. It might just as well shrink by a plane. Cessna Caravans were becoming more common on the used plane market, although they were unattainable for most small operators. Their niche seemed more with the courier market than the bush. Each year through the 1990s two or three new turbo Otter conversions appeared among the larger bush operators. In northern BC Bronson Creek remained a focus of air transport. Vintage planes seemed best for the work, even though some supplies were flown in during the spring of 1997 by Southern Air Hercules. Bronson Creek also kept its reputation as a graveyard for old clunkers. On February 1, 1997 Brooks Fuel's DC-4 N44909 was arriving with groceries. It touched down short and was wrecked (without injury). Meanwhile, Hawkair of Terrace soldiered on with the Bristol Freighter. After C-FDFC was sold in the UK (where it crashed), C-FTPA was on its own to do the work at Bronson Creek. On April 24, 1997 it was landing there when the undercarriage collapsed. Within moments TPA was a total wreck, but nobody was hurt. Now Hawkair rushed to bring its last Bristol (C-GYQS), stored at Terrace, back on line. This looked feasible, since many useful spares were salvaged from TPA. At the same time Hawkair purchased a Carvair, long dormant in the US.

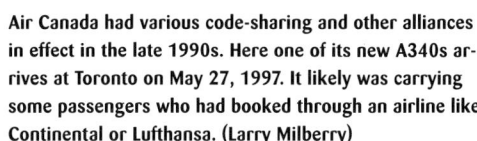

Air Canada had various code-sharing and other alliances in effect in the late 1990s. Here one of its new A340s arrives at Toronto on May 27, 1997. It likely was carrying some passengers who had booked through an airline like Continental or Lufthansa. (Larry Milberry)

Global Alliances

With growing airline competitiveness, and the high costs of fuel and of financing new equipment, airlines found it increasingly difficult to function independently. Thus began a trend of two or more companies forming alliances. The simplest approach was code-sharing, by which one airline could book passengers on another, listing its partner's flights as if they were its own. In 1996 CAIL had code-sharing arrangements with Air New Zealand, Japan Airlines, Mandarin Airlines, Malaysia Airlines and Qantas on the Pacific; British Airways and Alitalia on the Atlantic; and Varig in South America. Other arrangements were more intricate, e.g. in 1992 Washington approved an Air Canada share purchase in Continental Airlines. This gave access to Continental's markets in Mexico, South America, Africa, Asia and the Middle East. The benefits from such arrangements included coordination of schedules and pricing, joint marketing, as in merged frequent flyer incentives, unified travel agency commission programs, and shared revenues. Another benefit was that one partner would feed business to another's hubs. In the long run, alliances could save billions by coordinating fleet purchases, maintenance, training and computer systems. In July 1993 Air Canada, Continental and Air France formed an alliance including 600 aircraft serving 75 million passengers and 400 destinations. In this triangle arrangement, the partners began feeding each other's hubs in Toronto, Houston and Paris. Air Canada's involvement with Continental proved wise—within a few years it sold most of its shares for huge profits. In May 1997 the Star Alliance was announced—a partnership among Air Canada, Lufthansa, SAS, Thai Airways and United Airlines. Under Star Alliance, passengers could earn mileage points from any member but apply them later on other members' flights, check in with any member, make connections from one member to another with ease, and use any member's VIP lounge.

Airport Developments

Growth at airports like Calgary, Toronto and Vancouver came largely from international

The Carvair, which disappeared from the Canadian scene in 1973, reappeared a generation later in the form of Hawkair's C-GAAH. It had served in Australia/New Zealand 1965-78. After years in storage it had gone to Hawaii as N5459X, returned to dormancy, then found salvation in Canada. Jan Stroomenbergh shot it at Abbotsford on June 3, 1997, days after it arrived for duties at Bronson Creek.

Canada's International Charter Carriers*			
Airline	Main Tour Operator	Passengers	Market Share
Air Transat	Air Transat Holidays	1,021,429	36.3%
Canada 3000	Signature (UK)	631,020	22.4%
Royal Aviation	Sunquest (UK)	596,709	21.2%
Air Club	Tours Mont Royal	222,434	7.9%
Canadian	Canadian Holidays	195,491	7.0%
Air Canada	Air Canada Vacation	23,407	0.8%
Others		123,223	4.4%

*Year ending September 30, 1995

Some Canadian Fixed-wing Fleets in 1994	
Type	No. Listed in CCAR
Airbus A310/A320	53
ATR42	14
Beech 18	36
Beech 99	22
Beech King Air 90/100	88
Beech King Air 200/300	49
Boeing 727	25
Boeing 737	53
Boeing 747	14
Boeing 757	9
Boeing 767	32
BAe125	24
BAe146	13
BAe748	33
BAe Jetstream 31	19
BN Islander	27
Canadair CL600	16
CL-215	45
Cessna 180/185	1,367
Cessna 206/207	314
Cessna Caravan	32
Cessna 400 series	76
Cessna Citation	65
Consolidated PBY	13
Convair 240/340/440	6
Convair 580	34
Curtiss C-46	6
Dassault Falcon	20
DHC Beaver	353
DHC Turbo Beaver	21
DHC Otter	116
DHC Twin Otter	104
DHC Dash 7	2
DHC Dash 8	97
Douglas A-26	18
Douglas DC-3	33
Douglas DC-4	4
Douglas DC-6	4
Douglas DC-8	4
Douglas DC-9	35
Douglas DC-10	10
Embraer Bandeirante	8
Fairchild F.27	7
Fokker F.28	10
Gates Learjet	28
Grumman G.I	10
Grumman G.II/III	6
Grumman Goose	8
Grumman Tracker	19
Helio	18
IAI Westwind	6
Lockheed L.1011	10
Martin Mars	2
Mitsubishi MU-2	20
Noorduyn Norseman	22
Piper Apache/Aztec	171
Piper Navajo/Chieftain	231
Short 330/360/Skyvan	13
Swearingen Metro/Merlin	60
Ted Smith Aerostar	16

Canadian Civil Aviation 1994	
Total aircraft	17,940
Total hours flown	3,775,735
Total fixed-wing aircraft	16,622 (hours flown: 3,208,267)
Total rotary-wing aircraft	1,318 (hours flown: 567,468)
Private aircraft	12,681 (hours flown: 835,449)
Commercial aircraft	4,979 (hours flown: 2,830,683)
State aircraft	280 (hours flown: 109,603)
No. accidents	384
Accidents/10,000 hours	1.02

(Above) A Royal Air L.1011 arrives at Toronto on May 27, 1997. By this time Canada's major charter airlines were solidly established, unlike the 1980s-90s, when companies like Nationair, Wardair and Worldways went out of business. (Larry Milberry)

(Right) Canada has a large fleet used in agriculture and forestry. These Ayers Thrush sprayers fought an infestation of budworms in Alberta in 1992. (R.S. Petite)

(Left) Canada's overall aircraft fleet included a wide range of types by century's end. The Boeing 727, although long since phased out by Air Canada and CAIL, remained in service, mainly as a charter plane. This example, at Toronto on September 5, 1996, was with cargo carrier All Canada Express. The Northwestern Jetstream 31 was at Yellowknife on August 3, 1996. This type also had faded from the big fleets, but had found a second wind with smaller airlines and medevac operators. In 1997 Canada had the largest fleet of commercial A-26 Invaders. Eighteen were in service as water bombers with Calgary-based Airspray. This attractive example, flown for many seasons by Turbo Tarling, was at Calgary on May 27, 1988. In the late 1990s Airspray was building a fleet of Electra water bombers, but the versatile and economic A-26 still managed to hold its own. (Larry Milberry, Turbo Tarling)

travel in the 1980s-90s. In Toronto's case, for 1993 growth was 6% internationally and 4.5% trans-border, while domestic travel dropped 1%. International growth was due mainly to the success of Canadian charter carriers at Terminal 1—Air Transat, Canada 3000, Royal Air., etc. Ratty T-1 handled 3.7 million passengers in 1993. In the 1990s Vancouver and Toronto, now ruled by local airport authorities, each added a runway, control tower and other major infrastructure. Vancouver's new terminal, Canada's first designed with anti-earthquake features, opened on May 1, 1996. It was expected to handle more than 12 million passengers in year one. Toronto boomed, its local authority buying privately-held Terminal 3 for $720 million. Its new runway (R33L-15R) opened late in 1997. The authority looked forward to long term (11-17 years) growth costing $2.8 billion that would include a new terminal to replace obsolete Terminals 1 and 2. About 50 new gates would be added.

In May 1993 Vancouver International Airport's LAA began collecting user fees. This tax was needed to help finance upgrades. Passengers for flights within BC paid $5, those for out-of-province domestic $10, and those on international flights $15. Since the banks considered any airline too high a risk, they wanted repayments on $225 million in airport improvement loans paid directly by the LAA. Initially the user fees at Vancouver brought $30 million yearly. Travellers were incensed at what they considered highway robbery. Toronto's LAA, perhaps after seeing the furor at Vancouver, decided against user fees.

Meanwhile, Montreal struggled to make sense of its air transport infrastructure. Although Dorval (capacity for 8 million passengers yearly and potential for expansion) was adequate for Montreal, in the mid-1960s Ottawa had decided on a new international airport, arguing that Dorval could not meet forecast growth. Twenty sites were studied, the final choice being farmland near St. Jerome,

north of Montreal. The go-head for the new airport (Mirabel International) came in July 1968. Mirabel was touted as an airport to encompass intermodal transportation and distribution, as well as a manufacturing zone.

In Canada's largest land expropriation Ottawa took over 88,000 acres for Mirabel. Hundreds of families were pushed off farms and 800 buildings were destroyed, many by fire. In 1984 Ottawa established a process (through its Canada Lands Co.) whereby much expropriated land was sold back to the original owners. In September 1970 the contract was let to build the first runway. The airport, dubbed "the largest aviation facility in the world", opened on October 4, 1975. Prime Minister Trudeau stated at the event that Mirabel, which had cost more than $300 million, was ready for the forthcoming boom in air travel—it could handle 10 million passengers yearly. Almost 20 years later, only 2.5 million were arriving (compared to Dorval's 6.5). A million of those were flying to other destinations, so never left the terminal. Meanwhile, Dorval lost its overseas business and became Montreal's hub for domestic and trans-border air travel. For 1996

Mirabel recorded 55,800 aircraft movements compared to Dorval's 202,220.

Although Mirabel had superb passenger and cargo facilities, and good highway connections to Montreal, there was stubborn resistance to using it. Travellers from the rest of Canada objected when their international flights were obliged to stop at Mirabel, when most had Toronto as their point of departure or destination. As all domestic flights arrived at Dorval, that meant transferring to a bus for the drive to Mirabel. This was not a brilliant setup; it's astounding that Ottawa thought that it could foist such a sham on the public. European carriers all preferred Toronto, which was convenient to use and was the heart of Canada's business world. They objected when forced to use Mirabel. Olympic Airlines of Greece cancelled daily service to Mirabel in July 1977, and increased pressure for Toronto rights. One by one the foreign carriers won. This was facilitated by Ottawa's obligation to negotiate European gateways for Canada's air carriers. If Canada wanted in to new European centres, it would have to trade gates at Toronto. Ottawa must have known that LBPIA and Dorval were the airports which made sense for global operations. Mirabel made little sense, although its day may come in years ahead. After all, in 1938 Torontonians scoffed at Malton Airport and

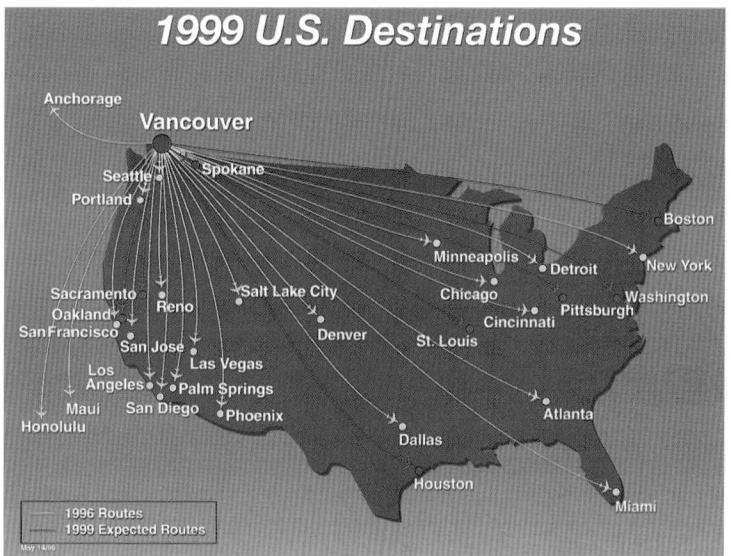

1999 U.S. Destinations

Anchorage
Vancouver
Seattle
Spokane
Portland
Boston
Minneapolis
Detroit
New York
Sacramento
Salt Lake City
Chicago
Washington
Oakland
Reno
Cincinnati
Pittsburgh
San Francisco
Denver
St. Louis
San Jose
Los Angeles
Las Vegas
Palm Springs
Maui
San Diego
Phoenix
Atlanta
Honolulu
Dallas
Houston
Miami

1996 Routes
1999 Expected Routes

Twenty Busiest Transport Canada Airports in December 1996	
Airport	*No. Movements*
Toronto/LBPIA	29,648
Vancouver International	22,800
Calgary International	18,147
Montreal/Dorval	15,037
Winnipeg International	11,365
Montreal/St. Hubert	10,017
Ottawa International	9,911
Victoria International	9,144
Halifax International	8,628
Edmonton International	8,222
Boundary Bay, BC	7,284
Quebec City	6,860
Calgary/Springbank	6,578
London	6,426
Toronto City Centre	6,369
Abbotsford, BC	5,962
Toronto/Buttonville	5,801
Saskatoon	5,208
Hamilton	5,013
Thunder Bay	4,914

Canada's Five Busiest Tower-Controlled Airports in 1996	
Airport	*No. Movements*
Toronto/LBPIA	372,418
Vancouver International	329,960
Calgary International	235,167
Montreal/Dorval	202,220
Montreal/St. Hubert	185,245

wondered, "Who will travel 20 miles to catch an airplane?"

ATC

In July 1994 Ottawa announced that it would lease all its major airports and terminate funding to 100 smaller, money-losing ones. This was in accordance with plans to privatize, commercialize and modernize transportation in Canada. As far as leasing was concerned, Vancouver, Calgary, Edmonton and Montreal already had been turned over to self-sufficient local airport authorities (1992). In early 1996 Ottawa signed a $1.125 billion agreement to transfer the national air navigation system to Nav Canada, a not-for-profit Ottawa-based company scheduled for July 1, 1996 start-up. Nav Canada management included people from Canada's airline, corporate aviation and air traffic control sectors. It took over seven area control centres, 86 flight service stations and national radar, communications and ground-based air navigation systems, plus related equipment. Some 6,400 Transport Canada employees transferred to Nav Canada for which Transport Canada would monitor operations.

Nav Canada planned to operate on a fee-for-service basis, including levying overflight fees. This was an increasingly important source of revenue in the burgeoning world of "user pay". Hundreds of millions in overflight fees were waiting to be raked in. In the United States user fees for overflights were to be implemented in May 1997 at the rates of US$78.90 per 100 nm over US land, and US$69.50 per 100 nm over US-controlled waters. In April 1997 Vancouver-based Signature Vacations, a

major Canadian tour operator, calculated that this would add $20.00 to the price of each ticket it sold between Vancouver and Mexico, even though no aircraft would land in the US. For 1996 Signature sent 87,000 holidayers along this route.

Flight Safety—The Case of Arrow Air

In the 1990s flight safety was a greater concern than it ever had been. Technology had given us safer airplanes, engines and systems. Those in aviation had never been so thoroughly informed, and accident investigation was highly advanced. It had been many years since there was a major accident in Canada. The loss in Saudi Arabia of the Nationair DC-8 was the worst incident involving a Canadian airliner to 1997. Nonetheless, accidents still happened, generally being related to human failure at at least one level. One infamous crash occurred at Gander on December 12, 1985— Arrow Air DC-8 N950JW. Following a brief refuelling stop, Capt John Griffin of N950JW had taxied from the terminal in a pre-dawn drizzle. The flight already had operated Cairo-Cologne and was leaving on its final leg to Fort Campbell, Kentucky with 248 members of the 101st Airborne Division and eight crew. The soldiers had been on UN duty in the Sinai. Within moments the DC-8 was rolling on Runway 22. It rotated, got airborne, but failed to climb. It struck trees on a ridge overlooking Gander Lake less than a mile from the departure end of R22, crashed on the down-slope and exploded, instantly killing all aboard.

Within hours the site had been scoured by Canadian Forces and other personnel. A thorough effort was made to understand what had gone wrong. The area was photo-mapped the day after the crash, enabling such things as measuring the tops of 378 trees clipped, to tell their heights before impact. A 1:100 scale model of the site was built showing the flight path. This helped to establish the yaw, roll and pitch attitude of the aircraft within a few degrees. The cockpit voice recorder was not functioning

The hillside devastation the day after Arrow Air crashed at Gander, then a view up the same hill in the winter of 1992 showing the emotive sculpture that commemorates those killed in the disaster. (Falk Foto, Larry Milberry)

for the take-off. Its outmoded flight data recorder measured only speed, heading, altitude, vertical acceleration, and time (Canadian safety standards listed 17 parameters to be measured). These deficiencies made it harder for investigators to find a precise cause for the accident. A public enquiry in April 1986 included representatives from the Canadian Aviation Safety Board, Gander airport, Canadian Forces, RCMP, Newfoundland government, Health and Welfare Canada, US Army, US Department of Defence Institute of Pathology, and US National Transportation Safety Board. Their efforts resulted in controversy. One faction came to a simple explanation—icing alone has caused hundreds of crashes at Gander, and conditions on December 12, 1985 were perfect for this phenomenon. For some reason Capt Griffin had not had his plane de-iced. His wings could have carried enough ice, imperceptible though it may have been in the darkness and drizzle, to bring him down, especially in his heavily-laden (if not overloaded) state. Some also wondered about the sharpness of the crew, who had been on duty for so long that day.

Another faction argued that one of four thrust reversers had deployed during takeoff. Had such an event occurred, the DC-8 would have been in trouble. Although physical evidence indicated that a thrust reverser *might* have been deployed at the time of the accident, this was iffy. Conspiracy is the best of all theories as far as lawyers are concerned, and the media love a good conspiracy theory far more than anything factual. As with many disasters, the conspiracy theorists soon identified themselves at the Arrow Air hearings. There were suggestions of clandestine cargo and clandestine passengers. There was a list of supposedly weird events after the crash, such as unauthor-

ized burying of wreckage. There were alleged comings and goings at Gander of "suspicious" US military flights delivering shady characters. But those who had to investigate and work with facts and concrete evidence had to keep focused. They did their best but, more than 10 years after the crash, there still was no certainty why Arrow Air crashed at Gander.

Another long-standing mystery involved Air India Flight 182. On June 23, 1985 it departed Toronto for Bombay with 329 aboard. While approaching the Irish coast it was destroyed by a bomb, killing all aboard. On the same day a bomb, destined to destroy a CPAir DC-8, exploded in the baggage area at Tokyo's Narita airport. Both bombings were ascribed to terrorists seeking independent homelands in India. After 12 years the Air India bombing still was unsolved, even though the RCMP had spent $20 million, and a $1-million reward was offered. It was learned in 1997 that the chief suspect secretly had been executed years earlier by the Indian government.

The Industry: Bombardier

While Bombardier was busy rejuvenating Canadair in the late 1980s, it also became the salvation of three other floundering companies. It purchased 51% of de Havilland Canada for $100 million (Ontario took the other 49%). Like Canadair, DHC had spent more than 40 years on the boom-bust roller coaster. It had turned out superb designs, but always struggled for profitability and suffered unhappy labour-management relations. Ownership had bounced around—de Havilland in the UK, Ottawa, Boeing. Now Bombardier acquired Boeing assets in a deferred cash payment of $70 million, while assuming $190 million in Boeing liabili-

The Bombardier Global Express takes off for the first time. The occasion was at Downsview on October 13, 1996. Here was the ultimate in corporate jets—all-new with a 6,500-nm cruise range. By mid-1997 three examples were busy, doing all the test flying required for certification. The Global Express made its international debut at the 1997 Paris airshow. It arrived there on June 12 with the crew of Pete Reynolds, Alain Lacharité, Jeff Kirdikis and Geoff Foster. Their flight covered 4,185 nm in a record 8:28 hours. (Bombardier)

The overall scene at Canadair's Dorval facility in 1994. To the right is the administrative centre. Beyond the flightline is the main Challenger and CRJ production plant. To the right of the main bay is the preflight hangar. Beyond are Air Canada's maintenance base, the runways and the St. Lawrence River. (Canadair Ltd 71143)

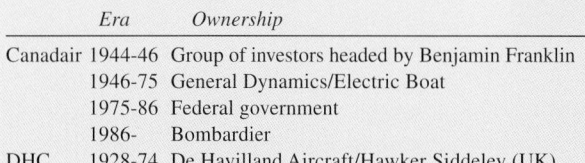

Canadair and de Havilland Canada over the Decades		
	Era	Ownership
Canadair	1944-46	Group of investors headed by Benjamin Franklin
	1946-75	General Dynamics/Electric Boat
	1975-86	Federal government
	1986-	Bombardier
DHC	1928-74	De Havilland Aircraft/Hawker Siddeley (UK)
	1974-86	Federal government
	1986-92	Boeing Aircraft Co.
	1992-97	Bombardier/Ontario government
	1997	Bombardier

ties. The deal became official on March 10, 1992. Within a few days Bombardier gave the go-ahead for the 39-seat Dash 8-200, the "hot and high" version. Using the PW123E, it would have 2,380 shp at an ambient temperature of 41°C, compared to the same output at 35°C for the -100. A new company, de Havilland Inc. came into being at this time.

In the mid-1990s Bombardier comprised four airframers (1997 employment: 24,000. These were Canadair (Cartierville, Dorval, Mirabel, 8,000 employees), de Havilland Canada (Toronto, 3,400), Learjet (Wichita, Tucson, 3,700) and Short Brothers (Belfast, 1,000). There were four main product branches—business aircraft, amphibious aircraft, defence systems and manufacturing. The Challenger, Regional Jet, Dash 8, Learjet, CL-415 and RPV (remote pilotless vehicles)

Bombardier chairman and CEO Laurent Beaudoin, and Canadair president Bob Brown. (Bombardier)

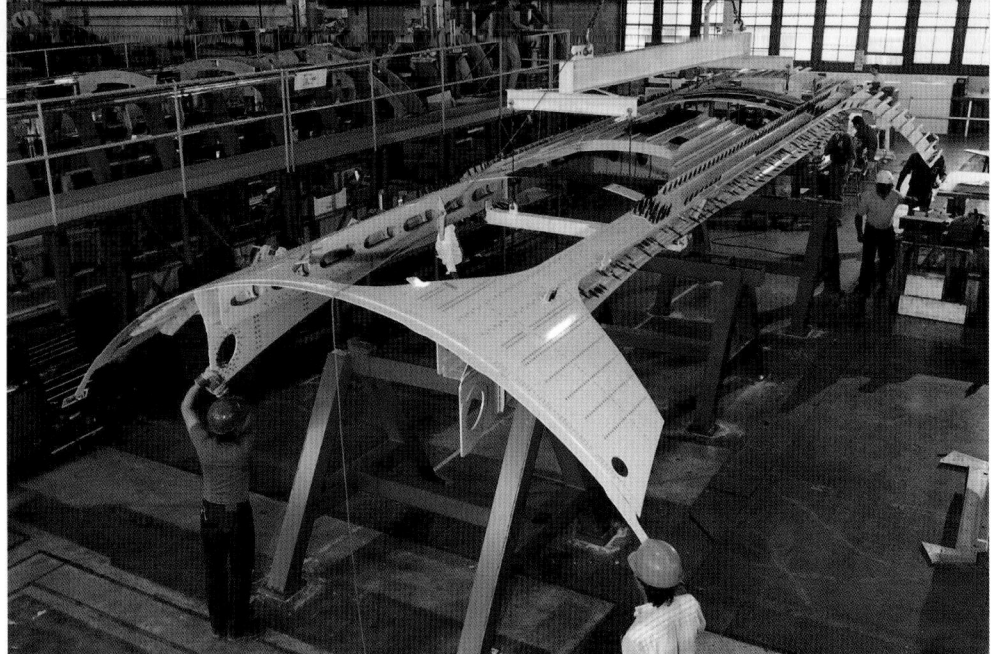

One of the major A330/A340 components (centre fuselage keel beam) manufactured at Canadair in Montreal. Then (below), one of the Boeing 767 rear fuselage bulkheads. About 700 shipsets had been delivered by the end of 1997. (Canadair Ltd.)

families were in production; the Global Express bizjet and Dash 8-400 were under development; and a range of advanced components was being manufactured under contract to major companies like Aérospatiale (A330/340) and Boeing (737, 747, 767). The biggest Boeing project was production of the 31-foot long, 15-foot diameter rear fuselage section for the 767. This included the pressure dome bulkhead, vertical stabilizer mounting structure, horizontal stabilizer pivot bulkheads, screwjack mounting structure and the APU mounting. In all, the unit had 3,400 parts and 69,000 fasteners. As of June 1997 672 rear fuselage sections had been delivered for 767s ordered by 84 airlines

The 100th Regional Jet was delivered on January 25, 1996 and the Global Express was into its flight test and certification programs. Dash 8 production was booming. In the summer of 1996 BRAD (Bombardier Regional Aircraft Division, formed to market Dash 8s and CRJs) received the largest Dash 8 order to date—25 -200s for Horizon Air, with options for an additional 45. Tavaj of Brazil ordered four Dash 8Q-200 combis for areas of the Amazon previously served mainly by river. At Canadair's Mirabel centre the CF-18 system engineering support was in full swing. Never before could Canada boast such a diversified, technically advanced and stable aerospace company.

When Fokker collapsed early in 1996, several companies considered a takeover. Aérospatiale, Alenia, British Aerospace, and Samsung all looked at the possibilities; but so did Bombardier. It was already a master at rescuing down-and-out companies. Besides, it had a special interest in Fokker—its Short Brothers subsidiary employed about 1,000 people making components for Fokker (25% of Short's output). In the end Bombardier decided against a Fokker takeover. The loss of business was partially offset in the late 1990s by placing new CRJ700 and Dash 8-400 work in Belfast.

Flight Test Centre

Canadair's flight test centre in Wichita opened in June 1991. It included a 40,000-sq. ft. hangar,

and almost as much office space. Its first certification programs were the CRJ and CL-415 and through 1995 it was busy with Challenger 604 and Lear 45 flying. Certification of the Global Express began in September 1996 with completion of the program slated for May 1998. Dash 8-400 flying was to begin in late 1997, followed by the stretched CRJ. Heading flight test activities in 1996 was Peter Reynolds, who had joined Learjet in 1973. He made the Lear 45 inaugural flight in October 1995, and was chief project pilot for the Global Express.

Some 200 people worked on Global Express test flying and certification—2,000 flight test hours would be needed for this work, compared to 1,500 for the Regional Jet and 850 for the Challenger 604. Besides certification programs, various test bed aircraft also were based at the Bombardier centre—CRJ, "604" and Lear 31A, 45 and 60, usually on product improvement programs. One such was testing a head-up display for the CRJ capable of Category 3 landings. The system was initially adopted by Lufthansa CityLine. Key test equipment at the centre included telemetry rooms used to provide real-time data from airborne test aircraft. Data were fed down from lightweight Sony data recorders capable of recording 3,000 parameters for three hours. The flight test centre also housed offices for various suppliers and risk sharers like BMW Rolls-Royce (power plant), Dowty (undercarriage), Honeywell (avionics) and Sextant Avionique (flight controls). In February 1997 a 100,000-sq. ft. addition to the flight test centre was opened. Included was a hangar to take any three large Bombardier aircraft. Meanwhile, further Bombardier expansion was heralded by a ground-breaking ceremony in Mississauga in February 1997. This was for an 84,000-sq. ft. BRAD spares centre.

For its fiscal year ending January 31, 1995 Bombardier delivered 123 new aircraft. In June 1995 Canadair Defence Systems Division won a $75 million contract to convert four

CanForces A310s to combi transports. Most of the work was subcontracted to an Aérospatiale subsidiary in Bordeaux. The modified aircraft were to have three configurations: 194-seats for passengers; 16-pallet capacity for the all-cargo version; and the combi setup for 12 pallets and 48 passengers. Canadair, meanwhile, continued with its major involvement with Airbus Industrie—the design and manufacture of six major A330 fuselage components. By 1996 Canadair's parent company, Bombardier had some 37,000 employees worldwide and sales exceeding $6 billion annually.

Global Express

In the mid-1990s the worldwide business climate was hot in almost every area. More than ever, business decisions were being made in rapid fire style. Satellite communications, faxes, E-mail, etc. were universal. While airline service was infinitely better than in the postwar years, corporations relied on their own fleets more than ever. Faster, longer range and bigger planes were needed. While types like the Falcon 50, Gulfstream II and Jetstar had served well, they were limited, mainly in range and spaciousness. With this in view, Bombardier decided to build the most exotic purpose-built corporate jet. In 1991 it announced the Global Express bizjet. It would have a range of 5,650 nm, allowing it to carry passengers nonstop between such centres as Los Angeles and Tokyo, or London and Los Angeles. To power the new plane Bombardier chose the BMW-Rolls-Royce BR710-48-C2. Some were sceptical about Bombardier entering this new field, but the Global Express was officially launched with an announcement on December 20, 1993. At that time Canadair president Robert Brown mentioned that there were firm orders for 30 aircraft. There already were 120 staff dedicated to the project. Meanwhile, Gulfstream was rushing ahead with its GV with similar specs. It was an upgraded version of the earlier GIV, while the Global Express was all-new. The Global Express was designed using IBM/Dassault Systèmes Catia computer-aided design work stations. Data were routinely exchanged via satellite communications among project partners Canadair, de Havilland Canada, Mitsubishi and Short Brothers. Design was to the standards used for the Boeing 777, the first airliner designed fully using computers.

In December 1995 the Global Express forward fuselage barrel arrived in Montreal from Belfast for mating with the nose at Canadair. This assembly then was trucked to Toronto. In February 1996 the wing and wing box/centre fuselage reached DHC from Mitsubishi aboard an AN-124, and the first engines arrived at Downsview. All airframe elements (the rear fuselage and tail were made at DHC) were assembled by April. Meanwhile, John O'Meara had flown the first GV on November 28, 1995, giving Gulfstream a one-year jump on Bombardier. By June 1996 Gulfstream had four GVs in the air and anticipated certification by November.

The prototype Global Express rolled out at DHC on August 26, 1996 before company em-

The No. 1 Global Express (c/n 9001) in final assembly at Downsview. Then (below), one of the structural test articles used in the pre-flight period to prove structural integrity. (Bombardier)

The grandest rollout held at de Havilland Canada took place on August 26, 1996 with the unveiling of the Global Express. Retirees Doug Adkins (Canadair chief pilot) and R.D. Richmond (executive vice-president) attended. (Larry Milberry)

ployees and 2,100 guests. It flew initially on October 13. The crew of Pete Reynolds, Ron Haughton and John Holding were aloft in C-FBGX for 2:46 hours. On its second flight (October 25) BGX visited Canadair headquarters at Dorval. BGX attended the 1996 NBAA convention in Orlando, Florida. There Bombardier announced that range for the new plane would be 6,700 nm (with eight passengers),

200 nm greater than the rival GV. Global Express certification was scheduled for mid-1998, following a 2,000-hour program involving the first four aircraft at Wichita. Meanwhile, the GV won provisional certification early in April 1997. Now Gulfstream could fly nine GVs to completion centres. At this time the Global Express was about 18 months from service. One of Gulfstream's advertisements was an effort to win customers by suggesting that it might be a risky gamble to order a Global Express: "This is not the time or the place to experiment. Go with a proven winner." By June 1997 Gulfstream was claiming about 72 sales, Bombardier 40, in a 500-800 plane market.

The third Global Express flew on April 22, 1997. Its main role in the program was in testing the Honeywell Primus 2000 XP avionics installation. This included a six-tube electronic flight instrument system (EFIS) and an automated engine indicating and crew alerting system using 8 x 7-inch CRT displays, Laseref 3 inertial guidance system, TCAS-2 traffic-alert and collision avoidance system, and Honeywell Primus weather radar. By mid-April the first two aircraft had logged 227 flying hours on 90 flights and reached M0.995 (630.4 mph) and a maximum altitude of 51,000 feet. Aircraft one was involved in flutter tests. Aircraft two was evaluating electrical, hydraulic, fuel, landing gear and brake systems, as well as the 14,690-pound/thrust BR710-48-C2 engines.

The media made much of Gulfstream's head start over Bombardier, but a year is a small thing in aviation. Customers are not concerned about a year, when spending US$35 million on a corporate jet. They are more likely to focus on the value they are getting for their dollar. Product loyalty is the second element influencing sales. Companies well served by Canadair over the years could be expected to replace a Challenger with a Global Express. By the same token, it would be difficult for Bombardier to woo away a satisfied Gulfstream customer. Thus did Bombardier occupy itself not in worrying about the GVs head start, which received its type certificate in April 1997. Instead, it focused on the strengths of the Global Express over the GV— its larger cabin, high cruise speed over its 6,500-nm range (M.85 compared to the G5s M.80), etc. In late 1997 Bombardier opened a Global Express completion centre at Dorval, aircraft No. 5 being the first aircraft sent there.

Regional Jet Update

The Canadair Regional Jet (CRJ-200) came quietly onto the market in 1989. Operators were cautious, for theirs was the world of the Dash 8, ATR and Saab 340—the turboprop world. Commuter jets were something else; the regional airlines were not ready for them. They were expensive and required a new corporate mentality. Nonetheless, a few European carriers committed to small orders. A break came in October 1991, when Comair of Cincinnati became the first North American customer, ordering 20. The CRJ entered service a year later with Lufthansa CityLine of Berlin. Before long, CityLine was serving 10 German and 34 other European centres with the CRJ, with av-

Global Express No. 4 takes flight at Downsview on April 25, 1997. (Kenneth I. Swartz)

erage flight times of 80 minutes. In August 1993 Air Canada ordered 24 CRJs. The market was coming alive. Once travellers got a taste of the CRJ, they were spoiled, and no longer cared to fly on slower turboprops, especially on long runs. A snowballing effect set in and soon operators were scurrying to add CRJs. The situation couldn't have been better for Bombardier. Rarely is a manufacturer in the "predicament" of having so many orders that it barely can meet demand. By late 1994 new versions were in the works. First came the CRJ-LR—the long-range version capable of flying nearly 2,200 nm. Other improvements reduced takeoff roll and increased payload. In September 1996 CityLine accepted its 25th CRJ. In early 1997 Comair, with 50 CRJs, took delivery of the 100th off the line. It added to this in May by ordering a further 30 CRJs. Meanwhile, in January 1997 Atlantic Southeast

An early view of the CRJ production line. Aircraft 7003 flew in November 1991 and remained with Bombardier through the type certification trials. In November 1993 it was delivered to Air Littoral of France as F-GNMN. Then, the prototype (C-FCRJ) taking flight for the first time on May 10, 1991 with pilots Doug Adkins and Don Stephen. In July this aircraft went to Wichita to begin certification trials. During sideslip tests on July 26, 1993 it stalled and crashed near Wichita with the loss of pilots David Martin and Robert Normand, and engineer Roger Booth (Canadair Ltd.)

CRJ No. 100 in a special colour scheme before its delivery to Comair in Cincinnati. Another spiffy-looking Comair CRJ honoured the company's 20th anniversary. Canadair photographer Lucio Anodal captured the view below on March 24, 1997. (Bombardier)

Lufthansa CityLine had a fleet of 30 CRJs in early 1997. These were ideal on longer European runs, where turboprops were too slow and traffic insufficient for economic use of bigger jets. In mid-1997 average stage length for the CRJ was 481 nm, while that for the Dash 8 was 183. The jet thrived on longer runs, the turboprop on shorter runs. For 1996 CityLine earned US$3.8 million. (Bombardier)

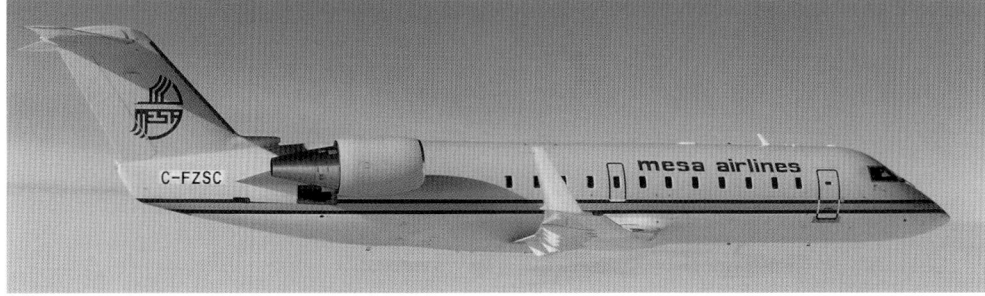

Mesa of Farmington, New Mexico accepted its first CRJs in 1997. This was a major leap for the commuter line that, in the spring of 1997, had a fleet of 33 Beech 1900s and five Brasilias. (Bombardier)

(Below) A Brit Air CRJ over the ski runs of Mont Tremblant north of Montreal on March 17, 1997. Canadair photographer Cliff Symons took this view. (Bombardier)

Airlines of Atlanta, a regional carrier with ATRs, BAe146s and Brasilias, ordered 90 CRJs, options included. Delivery began in June 1997, by when Canadair was turning out five CRJs monthly.

Comair was founded in 1975 by David R. Mueller with a loan from his father. Operations began with a Navajo, but soon jumped ahead. Mueller realized that in the rapidly evolving commuter world fleets had to be tailored to routes. Planes had to be cost-effective and reliable, and the travelling public had to enjoy flying on them. Comair soon moved into Short 330s, Metros, Saab 340s and Brasilias; but Mueller sensed that only jets would really please his customers. As he put it, "No matter how technologically advanced the turboprop is, passengers see propellers out there and they think the plane is old."

Being in the commuter business assumes a willingness by the CEO to gamble. That is what Mueller did when he committed to the CRJ. His first was delivered in April 1993 and by mid-1997 the fleet was the largest in the world. Winning this order was a great moment for Bombardier, for Comair had seemed committed to Embraer. It had 40 Brasilias and Embraer was trying hard to sell it on its forthcoming EMB145 regional jet. But the CRJ already was flying, while the EMB145 was still being designed (the CRJ had four years of operations by the time the first EMB145 entered service). Mueller did not want to wait, so gave the nod to the CRJ. In 1997 he predicted that within 10 years Comair would be an all-jet operation. Its early 1997 fleet included two Bandeirantes, eight Metro IIIs, 39 Brasilias and 45 CRJs.

While new products like the CRJ and Dash 8-400 were making such an impression, the 19-seat commuter plane held its own where stage lengths were short and/or markets thin. The B.1900D replaced the Dash 8 on several BC routes. In the US, Mesa had a fleet of 139 B.1900Ds in mid-1997. These served 53 cities daily, carrying 14,000 passengers. In early 1997 the 500th Beech 1900 was delivered to Impulse Airlines in Australia. Overall, the commuter airlines remained sophisticated, the successful carrier having a special plane for each market.

In February 1997 Bombardier announced Brit Air of France as lead customer for the 72- to 78-seat CRJ-700, that was due to fly in 1999. To Brit Air and other CRJ customers this was a logical move. Business was expanding, so more seats would be needed; there would be crew crossover between the two types; and spares, maintenance and ground support equipment would be common. For Great China Airways of Taiwan, which ordered six -700s, the picture was different. It had no CRJs, but 50-seat Dash 8s; now it needed a bigger, faster plane. The greatest boost to the CRJ-700 came on June 17, 1997 when Bombardier annnounced an order for 25 aircraft for American Eagle, American Airlines' feeder operation. At the same time, American Eagle ordered 42 EMB145s, so this was a disappointment to Bombardier, which had offered the CRJ-200 in

On November 8, 1978 Doug Adkins, Norm Ronaasen, Bill Greening and Jim Martin crewed the first Challenger 600 on its maiden flight. The Challenger faced many a storm, but those who believed in it persevered. In time the Challenger brought honour to Canada as one of the world's greatest business jets. On September 18, 1994 Doug Adkins, Bruce Robinson and Ted Squelch flew the first Challenger 604, the ultimate in its class. Challenger production continued to the turn of the century with nearly 400 delivered to 30 nations. This view shows the first 604. (Canadair C73040-3)

CRJ-200 and CRJ-700 Specs

	CRJ-200 LR	CRJ-700 ER
Length	87'10"	106'4"
Wingspan	69'7"	75'6"
Wing area	520.4 sq. ft.	738.7 sq. ft.
Max. Takeoff Weight	53,000 pounds	75,000 pounds
Max. range	2,005 nm	2,032 nm
Normal cruise speed	Mach 0.74	Mach 0.77
Engines	GE CF34-3B1*	GE CF34-8C1**

*8,729 pounds/thrust   **12,670 pounds/thrust

TAG of Saudi Arabia shows off its CRJ-SE and one of its Challenger 601-3Rs. The SE is the corporate edition of the CRJ. (Bombardier)

the competition. American Eagle would use its new fleets to replace or complement turboprops, as well as F.100s and MD-80s. The planes also would open new services.

The stretch in the CRJ-700 included two fuselage plugs totalling 186 inches. Other changes included a 72-inch wing root plug,

improved leading edge devices and a bigger horizontal tail. Moving the aft pressure bulkhead back 51 inches and moving the APU to the tail cone provided five extra feet of cabin space in addition to the length provided by the plugs. Meanwhile, headroom was increased an inch to 6' 2" by lowering the floor. Windows were raised by 4.5" to improve the view for passengers, and baggage space was created under the floor. The 12,670-pound thrust GE CF34-8C1 engine was selected. It offered 45% more take-off thrust, while there were 30% fewer parts than in the CRJ version

of the same engine. Unlike the CRJ-200, the -700 would have interchangeable left and right engine nacelles and thrust reversers.

Following an industry practice that became common in the 1980s, Bombardier found several partners to risk-share in the -700. These were Mitsubishi (aft fuselage), Short Brothers (mid fuselage), Canadair (forward fuselage, wing, control surfaces), Sundstrand Aerospace of the US (flaps/leading edge slats control system, and parts for the electric generator system), Menasco Aerospace of Ontario (landing gear), Intertechnique of France (fuel system), Collins Commercial Avionics of the US (avionics) and Liebherr-Aerospace of France (air management system). Final assembly would be at Dorval. Flight testing was scheduled to begin in mid-1999, with first deliveries in late 2000.

Another important event at Bombardier was its January 1997 buy-out of the Ontario government's 49% share of DHC. Ontario had been a partner with Bombardier since they purchased DHC from Boeing. Once it paid $49 million to Ontario, Bombardier became sole owner of DHC.

Dash 8 Developments

For 1997 production at DHC continued for the Dash 8 series: the 37-seat Dash 8-100, 37-seat Dash 8-200, and 56-seat Dash 8-300. First flown in May 1983, the Dash 8 was recognized as the world's finest turboprop regional airliner. The -200, a higher performance version than the -100, was the latest version. The first -200 went to Tyrolean Airways of Austria in January 1997. It joined Tyrolean's fleet of one Dash 7, eight Dash 8-100s, 13 Dash 8-300s, five CRJ-200s and four Fokker 70s. Dash 8 deliveries and firm orders totalled 460 and 57 respectively by June 1, 1997. There were orders from 64 operators in 23 countries.

Design of the Dash 8-400 was complete by late 1995, parts fabrication began the following March, and the first airplane was to fly in late 1997. The -400 takes the turboprop regional airliner to new heights. Based on the -300, but 184 inches longer, the 70-seater uses the PW150 of 4,830 shp and cruises at 350 knots. Specifically designed for 200 to 500 nm sectors, it offers the best direct operating costs of any regional turboprop or jet. Launch customer for the -400 was Great China Airlines of Taipei. Widerøe of Norway soon signed for four -400s, becoming the first European customer. Tyrolean of Austria and Rheintalflug of Germany followed. GCA's Peter Szu noted of his company's order: "These aircraft are long-term replacements for our fleet of 12 Dash 8-300s. Their additional passenger capacity will help us overcome slot restrictions at Taipei's downtown airport. We will be able to replace frequency with capacity." In 1994 GCA, which had been operating two 40-seat Dash 8-100s since 1988, had about 10% of Taipei's domestic market. With delivery of the first of twelve 56-seat -300s in 1991, business soared. For 1995 GCA had 36% of the domestic market. This gave credence to BRAD's slogan, "We make aircraft. We build business."

The Dash 8 is noted as the finest of the large commuter turboprops. More than 500 were in service by the turn of the century. Here -100s of Air Atlantic (CAIL connector) and Air Nova (Air Canada connector) overnight at Fredericton on November 28, 1990. (Larry Milberry)

The Dash 8 Family

Series	Pax	Length	Wing-span	Engine	shp	AUW (lb)	Cruise (knots)	Max Range (nm)
100	37-40	73'	85'	PW121A	2,150	36,300	270	1,020
200	37-40	73'	85'	PW123D	2,150	36,300	295	830
300	50-56	84.25'	90'	PW123B	2,500	43,000	285	1,180
400	70-78	107'9"	93'3"	PW150	5,071	63,250	350	1,296

Georges Van Belleghem of Belgium, one of Europe's many airliner fans, provided these Kodachromes of DHC types. CanForces 132002 (right), seen at Moorsele, Belgium, was the 12th Dash 7. It served mainly with the 412 Squadron Det at Lahr. A short tour with Air Atlantic followed, then it went to Markair of Alaska as N678MA. Dash 8-103A OE-LLI of Tyrolean was on lease to Albanian Airlines, when seen at Tirana in January 1993. D-BOBO, a GPA Propjet -102, was at Antwerp in September 1989 on lease to Hamburg Airlines. Tyrolean's -311 OE-LEC was in Sabena colours at Antwerp in April 1996.

Dash 8: Largest Fleet Orders to June 1997

US Air Express (USA)	66
Horizon Air (USA)	46
Air Ontario (Canada)	36
GPA Jetprop (Ireland)	30
Tyrolean (Austria)	27
Northwest Airlines (USA)	25
Great China Airlines (Taipei)	21
Mesa Air (USA)	20
Air BC (Canada)	18
Widerøe (Norway)	16
Air Atlantic (Canada)	15
Time Air (Canada)	14
Air Wisconsin (USA)	12
America West (USA)	12
SA Express (South Africa)	12

(Left) The fuselage of the first Dash 8-400 arrives at Toronto from Japan aboard an AN-124 on April 30, 1997. Then (above), an artist's impression of the sleek finished product. (BRAD)

CL-415

In October 1991 Canadair got the green light to build the CL-415. First flight was at Dorval on December 6, 1993. The ultimate development of the piston-engine CL-215 and the turboprop CL-215T, the CL-415 features PW123F engines (2,380 shp), fully-powered flight controls and EFIS instrumentation. Tankage is 1,350 Imp. gallons. France and Italy took the first production CL-415s in 1995, followed by Quebec. From October 23, 1995 to March 11, 1996 a CL-415 toured the world, visiting 25 nations and putting on demonstrations at 80 different locations, including Australia, Japan and Malaysia. In the summer of 1996 Turkey leased two CL-215s and a CL-415. This was a growing trend at Canadair—since sales were slow, especially for the expensive new CL-415, leases were sought. California, Portugal and Sweden were other water bomber lessees in this period. By early 1997 CL-415 deliveries totalled 26. A typical CL-415 can complete nine drops per hour, where the mission totals six miles between pick-up and drop points.

Lear

When Bombardier acquired Learjet of Tucson, Arizona in 1990 it began revitalizing sales of proven bizjets like the Lear 31 and 35; while launching the 6- to 10-seat transcontinental Lear 60. For 1993-94 it sold 35 new planes, seven more than the previous year. Leading was the Lear 60, which entered service in April 1993, the first going to Herman Miller, Inc. of New Zealand. The Lear 45, which flew in October 1995, fills a niche between the Challenger and small jets. It was the first collaboration among the various Bombardier aerospace companies. Short built the fuselage and

CL-415 Specs

Length	65' 1/2"
Wingspan	93' 11"
Height (land)	29' 5 1/2"
Empty weight	27,190 pounds
Max takeoff weight	43,850 pounds
Max weight, full water	46,000 pounds
Max cruise speed	203 knots
Scooping time	12 seconds
Ferry range	1,310 nm

A CL-415 in Quebec colours scoops water during a training flight. Then, the CL-415 used on the 1996 world demo tour. While new CL-415s were available in the late 1990s, many old CL-215s remained in use, and turboprop conversions (CL-215T) were available. In this period Canadair adopted a new marketing approach to the water bomber, leasing to customers like the Los Angeles Fire Department. (Bombardier)

A Spanish government CL-215T works on a fire near Valencia. This view, taken by Canadair's Cliff Symons, was a prize winner in *Aviation Week*'s1994 photo contest. Bombardier's total deliveries for 1996-97 were: Challenger 604—30, Lear 31—12, Lear 60—23, CRJ-200—56, Dash 8—39 and CL-215/415—8. (Bombardier)

empennage; DHC the wing (first one completed in September 1994); Lear headed design and program management, did final assembly and flight test to certification, followed by marketing and product support. Design, which was described as "a paperless airplane" because of the exclusive use of CAD/CAM technology, was facilitated by computer links among the three companies. Meanwhile, a system was devised whereby data from design computers was transmitted to manufacturing equipment, allowing closer manufacturing tolerances and faster completion of tasks than previously available. Overall, such technology

N610TM—one of the pre-production Lear 45s. (Bombardier)

The first Global Express leads a Lear 45 in formation during a 1997 photo shoot from Wichita. (Bombardier)

reduced costs by as much as 25%. Bombardier explained its new technologies: "Current engineering is a way of organizing work to allow team members to make the best use of their skills, while minimizing repetition and waste."

The Lear 45 uses the Allied Signal TFE731-20. The throttles have auto settings for takeoff, climb and cruise. In any setting a digital control sets power precisely for the ambient conditions. By the fall of 1996 there were five Lear 45s in the 1,610-hour certification program at Wichita. The test aircraft already had logged 325 flights, 650 hours and 3,100 stalls. Aircraft No. 1 had been tested to its maximum Mach number (0.87) and a maximum speed at lower levels of 380 knots. Operationally, the Lear 45 cruises at Mach 0.81, or 330 KIAS.

Other Happenings

In March 1997 Bombardier formed a joint venture with Lufthansa CityLine— European Business Jet Services, offering charters between North American and European centres lacking adequate scheduled airline service. EBJS would use under-utilized corporate aircraft. EBJS represented an expansion of Bombardier's charter operations, which included BusinessJet Solutions in the US and Global Aviation in Singapore. Bombardier also paired with Lufthansa Technik in 1997 to establish a service centre in Berlin for the Challenger, Lear and Global Express.

Pratt & Whitney Canada in the 1990s

In the late 1990s Pratt & Whitney Canada, a subsidiary of United Technologies of Hartford, Connecticut, held its status as one of the world's great hi-tech aero engine manufacturers. Having begun in sales, service and overhaul in the late 1920s, it grew, manufacturing American engines after WWII, then venturing into design with the JT12 and PT6 in the 1950s. While production JT12s were built in the US, the PT6 became P&WC's mainstay at home. Early success with types like the King Air and Twin Otter led to a PT6 dynasty—after 40 years the series remained in production. More than 60 versions, ranging from 550 eshp, were certified; 5,000 operators in 149 countries were flying 11,000 PT6-powered aircraft. PT6s had logged more than 184 million flying hours.

For 1996 P&WC had sales of nearly $2 billion (Cdn). It employed 8,900 in Alberta, Ontario, Quebec, Nova Scotia, and in several US and offshore operations. As it had since the 1950s, the company nurtured R&D, involving 1,800 employees with ex-

A Corporate Aviation Snapshot

Business flying was in a steady climb in the late 1990s. After several economic downturns that had seen corporate aircraft sales fall and flight departments close, the picture was bright. Indications of activity were provided over the years by the Vancouver-based newsletter *Westflight*. For March-April 1997 it listed typical corporate movements in Canada. These show the variety of types and owners. Statistics usually show US aircraft in far greater numbers than Canadian:

Type	Registration	Owner	Type	Registration	Owner
BAe125	C-GJBJ	Sears Canada, Vancouver	Beechjet 400A	N11GE	Mid American Energy Corp
Cessna 500	N161WC	Washington Corp	Beechjet 400A	N8146J	St. Jude Medical Inc.
Falcon 50	N450K	Kimball Aviation	Cessna 550	N95Q	Captain Bly Inc.
Gulfstream 1159SP	N1823D	Air Group	Cessna 560	N560DC	Dow Chemical Co.
King Air 350	N4S	Weyerhaeuser	Challenger CL601/3A	N818TH	Kimberly-Clark Corp
Learjet 35A	N386CM	Apple Jet	Challenger CL601/3R	N306FX	FlexJets
PA-31T	N500MY	Coldwater Veneer	Falcon 10	N97TJ	GE Capital Corp
Starship 2000A	N8149S	Raytheon	Falcon 50	N84HP	Hewlett Packard Co.
BAe125	N291H	Digital Equipment Co., Ottawa	Falcon 50	N500KJ	Sony Trading Corp
Beechjet 400A	N1124Z	Banking Consultants of America	Falcon 900B	N57EL	Enterprise Rent-A-Car
Cessna 650	N141M	Motorola	Gulfstream 1159B	N776MA	Dennison Enterprises Inc.
Challenger CL600S	N19HF	Hershey Foods	Gulfstream 1159C	N2WL	Warner Lambert Co.
Falcon 900	N906WK	Kellogg Co.	King Air 200	N26G	Marathon Electric
Gulfstream 1000	C-GJEI	Irving Oil Transport	King Air 350	N8048U	Archer Daniels Midland Co.
Learjet 35A	C-FZQP	Sky Charter FBO	Learjet 35A	N650LR	Wal Mart Stores
Learjet 55	N58CG	Corning Enterprises	Saab 2000	N5125	General Motors Corp
Pilatus PC-12	C-FMPA	RCMP			

P&WC president and COO Gilles Ouimet and chairman and CEO David Caplan. (P&WC)

(Right) A PW900-series APU, the type used on all B.747-400s. Before a "400" gets into the air, its APU has systems like the hydraulics and air conditioning on line. (P&WC)

penditures exceeding $300 million yearly. In years gone by P&WC had kept this side of its operation together, even in the worst of economic times. This proved wise, just as it had for Canadair—when the chips were down and thousands were laid off, Canadair somehow held on to its R&D people.

In the mid-1980s the PT6 reached above 1,400-shp for planes like the Short 3-60, Turbo DC-3 and Turbo Tracker. Each year saw new applications. In August 1993 Tec Avia of Germany selected the PT6 for the 24-seat MM-1 utility plane being developed in partnership with Myasischechev in the CIS. In September that year the PT6A-114A (twinned) was chosen by Soloy of Olympia, Washington for a Caravan conversion; and a new version of the 1,800-shp PT6AT-3D (TwinPac) was announced for the Bell 412HP. By this time there were 2,200 rotorcraft (368 operators in 87 nations) using the TwinPac. Agusta selected the new PT6B-37 for its A119 Koala helicopter; PZL-Swidnik the PT6A-65/67 for the W-3 Sokol helicopter; and BHT/Boeing the 1,850-shp PT6C-67B for its radical "609" civil tiltrotor. In 1996 yet another use appeared for the PT6. Daewoo Heavy Industries of South Korea was developing the KTX-1 advanced trainer to replace its T-37s and Ce.172s. By mid-1996 four prototypes had flown with the 550-shp PT6A-25; the fifth adopted the 950-shp PT6A-62. By far the plum of 1996 orders was one announced in February—the US selected the Beech Mk.II basic trainer for the USAF and USN. The requirement was for 711 aircraft powered by the PT6A-68. Engine deliveries began in mid-1997, by when there were some 33,200 PT6 turboprops and turboshafts.

The PW100 Family

Realizing that the PT6 turboprop could not grow forever, in the mid-1970s P&WC began researching an engine of 1,500-2,000 eshp. It ran a prototype in December 1977; less than two years later it gave the green light for development. By this time P&WC could see a market—the US had enacted the Airline Deregulation Act in 1978, so bigger commuter planes would be needed. An engine around 2,000 eshp would be best, since the next generation of commuters would carry 30+ passengers, up from the 19 of airline regulation days, when the Beech 99, Bandeirante and Twin Otter were in vogue. The 2,000-shp range historically was dominated in the West by the Rolls-Royce Dart, but it was old technology and burned too much fuel. Whichever company manufactured the first new such turboprop, would have the market. P&WC's design team for its engine, designated PT7, formed in February 1979. In September Embraer chose a 1,500 shp PT7 for its forthcoming EMB120 Brasilia. DHC, needing power for its imminent Dash 8, followed. In 1980 a new designation system was adopted, the PT7 becoming the PW100 series. For a typical example, the PW115, the "1" indicated 100 series, the "15" indicated 1,500 shp.

The recession of the early 1980s hurt P&WC—2,300 lost their jobs in 1982—but it endured. The PT7 took flight on February 27, 1982 aboard P&WC's Viscount. The first PW100-powered airliner to win FAA certification (July 1985) was the Brasilia. Avions de Transport Régional of France picked the PW100 for its ATR42. So did British Aerospace (ATP), Canadair (CL-215T), Catic/XAC (Y7), Dornier (Do.328) and Fokker (F.50). Of the commuter airliners only the Saab 340 and CASA CN-235 used another engine—the GE CT7. Soon stretched ATRs and Dash 8s appeared. For them P&WC offered bigger PW100s. The CL-215T evolved into the CL-415, which also had more power. The PW100 peaked in versions like the 2,500-shp PW123B in the Dash 8-300B.

By June 1997 some 1,600 aircraft were flying worldwide (about half in North America) with 25 versions of the PW100. The series had logged more than 36 million hours with 234 operators in 81 countries. Using a concept called "on condition", operators now were getting as many as 10,000 flying hours on certain versions; e.g. PW118/120. P&WC introduced the term "threshold inspection interval" (TII)

instead of "time between overhaul" (TBO). The latest and most powerful PW100 (2,750 shp) was the PW127. It was aimed at higher performance versions of existing aircraft (e.g. ATR72-210, ATR42-500, Fokker 50 and 60, Jetstream 61), and at new aircraft in the 40- to 70-passenger range. The go-ahead for the Dash 8-400, announced in June 1995 at the Paris Airshow, was music to P&WC's ears. A few weeks earlier Bombardier had announced that the -400 would have a new engine—the 6,500-7,500-eshp PW150. There would be commonality in parts, servicing, and training with PW100s. Ottawa agreed to pump $100 million into the project (it also invested in the -400). Plans for the PW150 included initial test runs at Longueuil in mid-1996. The first ran in the test cell in June 1996. First flight was on P&WC's Boeing 720 on January 30, 1997.

PW200/300/500 Series

Several new engine families parallelled the PT6 and PW100. The PW200 turboshaft was for the light helicopter market, launch customer being McDonnell Douglas' MD900 Explorer (PW206A). Eurocopter followed, ordering the 566-shp PW206B for its EC135. The PW200 was certified in 1991, deliveries beginning in late 1994. Next, Agusta chose the 640-shp PW206C for its A109 Power helicopter; and BHTC selected the PW206C for the Bell 427.

The PW300 turbofan series was launched in the mid-1980s in collaboration with Motoren-und Turbinen Union (MTU) of Germany. The market for the 4,500-7,000-pound/thrust engine was light bizjets. The PW305 soon was adopted for the Lear 60, Raytheon Hawker 1000 and IAI Galaxy bizjets. In November 1996 the 6,575-pound/thrust PW308 was selected for the Raytheon Hawker Horizon, a new mid-size bizjet. Another PW308 was chosen in January 1997 for the Fairchild Dornier 328JET, a 32+-seat regional airliner scheduled to fly in January 1998.

Meanwhile, P&WC/MTU were developing the 3,000- 4,500-pounds/thrust PW500. Design started in August 1992, core testing of a demonstrator beginning in October 1993. After 300 hours of test cell running, the demonstrator flew on the B.720B on May 28, 1994. By then five development engines were involved. MTU held 25%, being responsible for development, manufacturing, testing and assembly of the low

P&WC Boeing 720 flying testbed C-FETB. It can carry a large engine on the No. 3 pylon (as seen here), something like a PW150 on the nose, and a small turbofan like a PW300 on the mount near the forward (starboard) crew door. This Boeing started in 1961 with American Airlines (Flagship Idaho). A decade later it went to Middle East Airlines, then joined P&WC in 1986, where it replaced a Viscount. (P&WC)

pressure turbines, exhaust case and combustion chamber. The PW530 was selected for the Cessna Bravo. It had 10% more power than the JT15D, but a 15% lower fuel burn. TBO was noted as 4,000 hours compared to 3,000 for the JT15D. With four passengers and three crew the Bravo could fly 1,900 nm at M.7. Cessna also picked the PW545 (4,500 pounds/thrust) for its new Excel light bizjets. By June 1997 sixteen development PW500s had logged 8,000 running hours, including 540 aboard the B.720 and more than 2,300 on Cessna prototypes. While P&WC's many new projects went ahead, production continued of its tried-and-true JT15D, and PW900-series APUs. The first JT15D had been certified in 1971. By June 1997 more than 5,300 JT15Ds had logged in excess of 20 million flying hours, mainly on versions of the Cessna Citation. Introduced in 1989, by late 1997 nearly 500 APUs had been delivered for the Boeing 747-400 series.

Overseas Partners

In the mid-1990s P&WC was moving to strengthen global connections. After all, most of the 44,000+ turbine engines (1997) it had produced were outside Canada—in 168 different nations. In the period 1983-93, of $12 billion (Cdn) in sales, $10.5 billion were to foreign customers. Also in this period P&WC re-focused from general aviation to the regional airlines. To extend markets, it began seeking foreign partners to manufacture its products locally. Thus, in a place like the CIS, engines from Canada could find new applications at affordable prices. A natural development would be for partners to work with P&WC improving existing engines, then bring forward new engines. In the first such venture P&WC aligned with the Klimov engine company in 1993. It opened an office in St. Petersburg, owning a 51% interest in Pratt & Whitney/Klimov Ltd. (later termed P&W-Rus). Each company placed three members on the board. P&WC granted P&W-Rus licences for the 600-shp PW200 turboshaft and the 1,272-shp PT6A-67 turboprop. CIS engines were studied as potential P&W-Rus products, as were various airframes and ground vehicles, to see how adaptable they would be for P&W-Rus

power. The first engines were assembled in 1995 from kits shipped from Canada.

While setting up in the CIS, P&WC also looked to China for a partnership, especially with Xian Aero Engines. P&WC had a long relationship with the Chinese, dating to early PT6 days; some of the earliest exported JT15Ds had gone to China.

Other Directions

In order to enhance customer service, P&WC opened a service base in Singapore in 1994. It was handling some 120 engines yearly by 1996. Meanwhile, a new service centre was opened in Dallas early in 1995. A full-service operation, it offered spares, fly-in service, on-line assistance, mobile repair teams, hot section component rework, power section repair and rental engine support. P&WC's other main service bases were in Calgary, Atlanta, Burlington (Vermont), Chicago and Long Beach.

In 1996 P&WC purchased a 51% share in Anglo-African Airmotive of Johannesburg, renaming it Pratt & Whitney Canada CSG Africa. The new company focused on product support in South Africa, where many PT6s and PW100s were in use. Also in 1996 Transport Canada granted certification to the 621-shp

PW206B turboshaft for the EC135. In the same year P&WC and SNECMA of France signed an agreement to jointly develop at 12,000-to-16,000 pound/thrust turbofan engine for the anticipated regional airliner of tomorrow. In late 1997 P&WC opened a major worldwide parts distribution centre in Mississauga, Ontario.

Found Brothers Aviation

A dream that began late in WWII appeared ready for fulfilment more than 50 years later—the Found Brothers Aviation FBA-2. The latest phase of the story began with N.K. "Bud" Found in the early 1990s. By then he had been retired for several years and two of his brothers had passed away. Conversations over the years with pilots like John Vandene of BC indicated that the FBA-2C might be placed back into production. After all, its ruggedness, reliability and economy were well known. Then, Canadian bush and coast operators were facing a shortage of four- and five-seat planes—years earlier, US production in this category (Ce.185, etc.) had ceased.

Bud Found agreed that this would be a good time to bring back the FBA-2C. He was encouraged when Transport Canada agreed to re-instate the FBA-2C Type Certificate with minor engineering changes. This would save the millions required to certify a new aircraft. A surplus -2C airframe (C-FSVD) was loaned to Found by John Blackwell of Fawnie Mountain Outfitters in BC. With additional components supplied by Glenn Tudhope of Hudson, Ontario, Found rebuilt SVD with a Lycoming fuel-injected engine. Other mods planned for the production aircraft were incorporated. Elton Townsend of Muskoka-based Lake Central Air Service (which would be manufacturing tail surfaces for new FBA-2s) flew SVD from Parry Sound on November 27, 1996. SVD now began a series of flights to develop on-going versions under the designation FBA-2E. On February 6 Doug Holtby, an MNR pilot, flew SVD to McCauley in Dayton, Ohio for a factory-conducted propeller vibration study. Holtby later played a key roll in demonstrating

The FBA-2C proof-of-concept aircraft on a test flight at Parry Sound, Ontario in November 1996. (Found Brothers Canada Inc.)

Found Brothers Aviation FBA-2E Bush Hawk

Length – 25.5 ft.
Wing span – 36 ft.
Wing area – 180 sq. ft.
Cabin volume – 120 cu. ft.
Empty weight – 1,750 pounds
Gross weight – 3,200 pounds
Disposable load (wheels) – 1,450 pounds
Disposable load (Edo 2960 floats*) – 1,200 pounds
Take off over 50' obstacle, gross weight – 1,000 feet
Rate of climb – 1,200 fpm
Normal cruise speed (floats) – 120 mph
Stall (full flap) – 63 mph
Endurance (75 Imp. gallons) – 6.5 hrs.
Engine – Lycoming IO-540-D4A5 260 hp, or
    O-540-A1C5 250 hp

* Optional: Edo 2790 amphibious floats

The C-130 Hercules remained Canada's leading military air transport in th late 1990s. This example was at CFB Edmonton on May 7, 1994. (Larry Milberry)

Canada's military helicopters in the late 1990s included the new Griffon as well as old types like the Labrador. Here a Griffon from 439 Combat Support Squadron (Bagotville) lowers SARTech Sgt Normand Boutin during training near Trenton on May 27, 1997. Boutin's crewmates this day were Capts Donald Nault and Mario Boily (pilots) and MCpl Mario Giasson (FE). Then, a 424 Squadron Labrador on exercise near Trenton on May 27, 1997 with the crew of Maj Terry Swanson and Capt John Edwards (pilots), MCpl Brad Edwards (FE), and WO Bill Barber and MCpl Gerry Wile (SarTechs). (Tony Cassanova)

the value of the FBA-2 series for MNR resource management operations.

Contact with the Canadian market showed that there was considerable interest in the FBA-2 and letters of purchase intent were received from operators in various regions. Several, who had been flying the -2C for years, contacted Found, asking how soon they could get a new machine. Enquiries from abroad suggested that there also was an export market worth exploring. Found received funding from private investors as well as several federal government departments. An important hurdle was crossed when Transport Canada renewed the FBA-2C type certificate on April 22. SVD started water trials in June, using Edo 2960 floats. By this time plans for the production FBA-2E Bush Hawk were starting to take shape. It would be a refined -2C with the 260-hp Lycoming, to be followed by the -2E-300 with a 300-hp Lycoming.

The Air Force

In 1997 Air Transport Group was dissolved and air force transport operations devolved to the individual wings. After more than 25 years the Boeing 707 was retired, the last two of five leaving Trenton in May 1997 for the US to be rebuilt as USAF Joint-STARS advanced radar ground surveillance aircraft. Meanwhile the first CC-150 (A310) returned from France with major mods—strengthened floor and a new cargo door. At this time Canada's air force no longer had a strategic aerial refueller, although five C-130 tankers were operated by 435 Squadron in Winnipeg. In 1997 the DND was considering an option whereby a commercial company would provide Canada's strategic AR using 707s.

In May ATG honoured its oldest C-130, aircraft 130315. Delivered in 1965, by May 10, 1997 it had accumulated 40,000 hours in 17,183 cycles, the most hours of any military

Canada's Military 707s

Aircraft	Flying Hours	Cycles	Last Flying
137701	38,415	22,205	27-8-93
13702	38,762	22,197	1-3-93
13703	38,512	20,329	31-3-97
13704	37,418	18,987	31-3-97
13705	38,038	21,554	29-8-95

C-130 worldwide (several civil Hercs had passed 75,000 hours). In its career 130315 had some 600 mods, was configured 568 times, had 40 engine changes, 77 prop changes, and was awaiting an avionics update.

Preserving Canada's Air Transport History

Each region of Canada preserves some aspect of the nation's air transport heritage. Of course there are museums, but sometimes the venue is a dock where a Beech 18, Norseman or other vintage plane still earns its keep. Such living history also might be enjoyed at an airport like Yellowknife, where Buffalo Airways' fleet of C-46s, DC-3s and DC-4s remained busy into the late 1990s. The sight, sound and, even, smell of a vintage plane "doing its thing" is what really gets the aviation fan's blood pumping.

Aviation museums display most of Canada's leading transport aircraft. The National Aviation Museum in Ottawa is the key focus. There one may view everything from a meticulously-built HS-2L replica to Beaver No. 1, a Bell 47, Bellanca Pacemaker, Curtiss JN-4, DC-3, Goose, Junkers W.34, Lockheed 10, Norseman and Viscount. Other transports on display include a Canso and Jetstar with the Atlantic Canada Aviation Museum in Halifax; a Dakota and Kiowa at the RCAF Memorial Museum in Trenton; a Beaver, Fox Moth, Otter and Seabee at the Canadian Bushplane Heritage in Sault Ste. Marie; a Bellanca Airbus, Bristol Freighter, Canadian Vickers Vedette, Fairchild Husky, Froebe brothers helicopter, Junkers "Flying Boxcar" and Stinson Reliant at the Western Canada Aviation Museum in Winnipeg; Bellanca Skyrocket, DC-3 and Hornet Moth in the Reynolds Museum in Wetaskiwin, Alberta; and a DC-3, H-21, Lockheed 18, Norseman and S-55 in the Canadian Museum of Flight and Transportation at Langley, BC.

Besides seeing the airplanes, anyone interested enough can belong to one of the many important Canadian aviation heritage groups, premier among them being the Canadian Aviation Historical Society. Through such groups a member meets fellow enthusiasts, including old timers from bush flying, the airlines and the air force. The leading sources of research material also are available—the National Archives of Canada, the Directorate of History in Ottawa, and all the provincial, territorial and local archives. Certainly anyone following Canadian aviation history should also belong to the local public library, where many of the standard reference volumes are on hand.

Canadians have a richness in aviation history, whether in museums, archives or at the local airport. This view shows part of the collection in the Reynolds Alberta Museum in Wetaskiwin. Nearest is D.H.60X Moth G-CYYG. It started with the Post Office Department in 1928, then spent many years with the Edmonton and Northern Alberta Aero Club. Also shown are Curtiss Robin CF-ALZ, Hornet Moth CF-AYG and J-2 Cub CF-BEE. (Larry Milberry)

History often is a matter of just sitting there, sometimes for decades. This ex-TCA (CF-TGE) and ex-World Wide Airways (CF-RNR) Super Constellation sits weathering on a farm near St-Jean Port Joli, Quebec. In time it was rescued by aviation buff Phil Yull of Mississauga. He arranged to have it shipped to Toronto, where it was restored in Carl Millard's hangar. In August 1996 it was assembled in front of the Regal Constellation Hotel near LBPIA, and opened as a bar and meeting place. In this June 1985 view by Richard Beaudet the Super Connie awaits its rescuer.

Other airplanes are flying museum pieces. Typical is this Buffalo Airways Canso water bomber operating from Yellowknife in 1997. It appears in the modified colours of its previous owner, the Newfoundland and Labrador Forest Service. (Henry Tenby)

Many Canadians are involved in preserving Canada's aviation heritage. Here Ken Molson (right) and Bob Bradford, the first and second curators of the National Aviation Museum, chat at the 1991 CAHS convention. (Larry Milberry)

Since the 1920s keen photographers and artists have played a vital role in preserving Canadian aviation history. Jack McNulty (right) led the way in photography and in recording the minute historic details aircraft by aircraft. Jack's with his cohort, artist Bob Finlayson. They were attending the August 18, 1984 *Sixty Years* book launch at CANAV in Toronto. (Larry Milberry)

For more than 30 years historian Fred Hotson was a leading light in the CAHS. He wrote dozens of learned articles and authored *The de Havilland Canada Story* and *The Bremen*. Here he was at work at home in Mississauga in 1997. (Larry Milberry)

Photographers covering the Canadian scene included John Kimberley of Vancouver, at the controls of a Buffalo Airways C-46 in May 1996; and Wilf White (above) of Glasgow, who often visited Canada. Here he is with Larry Milberry (right) aboard aviator Mike Mushet's sailboat *Mary Jenny*. Stephen Piercey (left) founded *Propliner* magazine, a respected aviation journal. Stephen (on the right in the group of four) was visiting Malton in October 1982 to do an air-to-air session with Goodyear's DC-3 C-FTDJ. He's with TDJ's crew: Capt Fred Livermore, engineer Doug Rock and FO Andy Paulionis. Not long afterward Stephen was killed in a midair collision taking air-to-air photos in Germany. (Larry Milberry, Mike Mushet)

(Right) Many groups meet regularly or on-and-off to promote aviation history. Here a quartet of ex-TCA stewardesses get together. Inez (Dodge) Webster, Betty (Grant) Slade, Hazel (Warrum) Belyea and Joyce (Campbell) Keith met on course at Dorval in January 1951. They had get-togethers ever since. This one was in Port Colborne, Ontario in September 1996. (Robert M. Webster)

(Left) Besides photography and writing, modellers help maintain the overall interest in aviation history. Here Des McGill of Montreal, an Air Canada retiree, checks out his model of a TCA North Star. (CANAV Col.)

In the spring of 1997 a group of aviation history enthusiasts from Bremen, Germany visited the Ford Museum in Dearborn, Michigan. Their trip was to reclaim the Junkers "Bremen", which had made the first east-to-west, non-stop Atlantic crossing. Fred Hotson spent years researching the saga, eventually publishing his award-winning book—*The Bremen*. Because of Fred's efforts the Bremen group came together and started the move to get the Bremen back to Germany. In this candid snap, taken by a friendly Dearborn cab driver, the Bremen fans pose: Fred Hotson, Dr. Hilmar Rauschert, Gunter Strangemann, Wilhelm Lebens, Christian Plato, Harald Claasen, Heinz Jurgen Duwe, Dr. Jens Petersen, Prof. Bernd Hamacher, Volker Schmidt and Larry Milberry.

Glossary

A/C/M – air chief marshal
A/M – air marshal
A/V/M – air vice marshal
AB – Alberta
AC – air commodore
AEA -- Aerial Experiment Association
AFB – air force base
AFC – Air Force Cross
AFM – Air Force Medal
AGGTS – Air Gunner Ground Training School
AITA – Air Industries and Transport
  Association of Canada
aka – also known as
ALCE – airlift control element
AN – Antonov
APC – armoured personnel carrier
APU – auxiliary power unit
ASC – Aero Service Corp.
ASP – astronaut support personnel
ASR – air-sea rescue
ASVS – Advanced Space Vision System
ASW – anti-submarine warfare
ATFERO – Atlantic Ferry Organization
ATG – Air Transport Group
ATR – Avions de Transport Régional
B&GS – Bombing and Gunnery School
BA Oil – British American Oil
BC – British Columbia
BCATP – British Commonwealth Air Training
  Plan
BCIT – British Columbia Institute of
  Technology
BHP – Broken Hill Proprietary Co.
BHTC – Bell Helicopter Textron Canada
BOAC – British Overseas Airways Corporation
BRAD – Bombardier Regional Aircraft
  Division
BRINCO – British Newfoundland Co.
CAB – Civil Aeronautics Board
CADC – Crown Assets Disposal Corporation
CAF – Canadian Air Force
CAHS – Canadian Aviation Historical Society
CAL – Canadian Aeroplanes Ltd.
CAL – Canadian Airways Ltd.
CALPA – Canadian Airline Pilots Association
CanForces – Canadian Forces
CAP – Canadian Astronaut Program
Capcom – capsule communicator
CAR – civil aviation regulations
CARDE – Canadian Armament Research and
  Development Establishment
CASARA – Civil Air Search and Rescue
  Association
CATDS – Canadian Air Transport Detachment
  Somalia
CBC – Canadian Broadcasting Corporation
CBCA – Central British Columbia Airlines
CBD – central business district

CCA – Controller of Civil Aviation
CCAR – Canadian Civil Aircraft Register
CCF – Canadian Car and Foundry
CCGS – Canadian Coast Guard Ship
CDIC – Canadian Development Investment
  Corp.
CF, C-F, C-G – prefixes for Canadian civil
  aircraft registrations
CF – Canadian Forces
CFB – Canadian Forces Base
CFE – Canadian Forces Europe
CGTAS – Canadian Government Trans
  Atlantic Air Service
CIA – Central Intelligence Agency
CIBC – Canadian Imperial Bank of Commerce
CIDA – Canadian International Development
  Agency
CIS – Commonwealth of Independent States
CJFC – Canadian Joint Forces Somalia
Cmdr – commander
CMMMK – Cartier, McNamara, Mannix,
  Morrison and Knudsen
CNR – Canadian National Railways
CNS – Central Navigation School
CO – commanding officer
CPA – Canadian Pacific Airlines
Cpl – corporal
CPR – Canadian Pacific Railway
CSA – Canadian Space Agency
CWDS – clean wing detection system
D-Day – Invasion of France on June 6, 1944
DCIEM – Defence and Civil Institute of
  Environmental Medicine
DEA – Drug Enforcement Agency
DFC – Distinguished Flying Cross
DHC – de Havilland Canada
DM – docking module
DME – distance measuring equipment
DNCO – duty not carried out
DND – Department of National Defence
DOT – Department of Transport
DSO – Distinguished Service Order
DZ – drop zone
ECM – electronic counter measures
ECCM – electronic counter-counter measures
EFIS – electronic flight instrumentation system
EFTS – Elementary Flying Training School
ELT – emergency location transmitter
EMR – Department of Energy, Mines and
  Resources
EMS – emergency medical services
EPA – Eastern Provincial Airways
EPRDF – Eretrian People's Revolutionary
  Democratic Front
ERO – engine running offload
ESA – European Space Agency
ETA – estimated time of arrival
ETOPS – extended twin operation procedures

EVA – extravehicular activity
F/L – flight lieutenant
F/O – flying officer
F/Sgt – flight sergeant
FAA – Federal Aviation Administration
FAI – Fédération Aéronautique Internationale
FAMA – Flota Aera Mercante Argentina
FedEx – Federal Express
FIS -- Flying Instructors School
FLIR – forward-looking infrared
FMS – flight management system
FOL – forward operating location
FSS – flight service station
FY – fiscal year
G/C – group captain
GCA – ground controlled approach
GMT – Greenwich mean time
GNA – Great Northern Airways
GNWT -- Government of the Northwest
  Territories
GPS – Global Positioning System
GPWS – ground proximity warning ystem
HAS – hardened aircraft shelter
HEPC – Hydro Electric Power Commission of
  Ontario
HMS – His/Her Majesty's Ship
HRS – humanitarian relief sector
IAI – Israeli Aircraft Industry
IATA – International Air Transport Association
ICAO – International Civil Aviation
  Organization
ICBM – inter-continental ballistic missile
ICRC – International Commission for the Red
  Cross
IFR – instrument flight rules
IMP – International Marine Products
IPMS – International Plastic Modellers'
  Society
IR – infrared
ISIS – International Satellites for Ionospheric
  Studies
ISS – International Space Station
IT&T – International Telegraph and Telephone
JATC – Joint Air Training Centre
JATO – jet assisted takeoff
JIT – just in time
JKIA – Jomo Kenyata International Airport
JSC – Johnson Space Center
JSTARS – Joint Surveillance and Target Attack
  Radar System
kg – kilogram
KIAS – knots indicated air speed
KSC – Kennedy Space Center
LAA – local airport authority
Lab – Labrador helicopter
LAC – aircraftsman
LAPES – low altitude parachute extraction
  system

LAS – Laurentide Air Service
LAV – light armoured vehicle
LBPIA – Lester B. Pearson International
  Airport
LCmdr – lieutenant commander
LM&E – Labrador Mining and Exploration
LOLEX – low level extraction system
LORAN – long range navigation equipment
Lt – lieutenant
LTU – Lufttransport-Unternehmen
LZ – landing zone
Mach 1 – the speed of sound, noted as 763
  mph at sea level at 59°F; named for
  Austrian physicist Ernst Mach 1838-1916
MAJAID – major air disaster
MAMS – mobile air movements section
  MATS – Military Air Transport Service
MB – Manitoba
MCA – Maritime Central Airlines
MD – McDonnell Douglas
METO – maximum emergency take-off power
MGAS – Manitoba Government Air Service
MGen – major general
MiD – Mention in Despatches
MSS – Mobile Servicing System
MTU – Motoren-und Turbinen Union
N – prefix for US civil aircraft registrations
NAFTA – North American Free Trade
  Agreement
NAME – Northern Aerial Mineral Exploration
NASA – National Aeronautics and Space
  Administration
NB – New Brunswick
NBAA – National Business Aircraft
  Association
NDB – non-directional beacon
NF – Newfoundland
NGO – non-government organization
nm – nautical mile
NOE – nap of the earth
NORAD – North American Air Defence
  System
NRC – National Research Council
NS – Nova Scotia
NTA – National Transportation Agency
NTSB – National Transportation Safety Board
NVG – night vision goggles
NWI – Northwest Industries
NWSR – North West Staging Route
NWT – Northwest Territories
OCA – Ontario Central Airlines
ON – Ontario
OPAS – Ontario Provincial Air Service
ops – operations
OSVS – Orbiter Space Vision System

OTF – operational training flight
OTU – operational training unit
P&WC – Pratt & Whitney Canada
P/O – pilot officer
PAT – Patricia Airways and Transportation
PE – Prince Edward Island
PEP – Provincial Emergency Program
PHI – Petroleum Helicopters Inc.
PR – Puerto Rico
PWA – Pacific Western Airlines
QC – Quebec
R&D – research and development
R&R – rest and recreation
RAAF – Royal Australian Air Force
RAF – Royal Air Force
RANSA – Rutas Aereas Nacionales
RCAF – Royal Canadian Air Force
RCC – rescue co-ordination centre
RCN – Royal Navy
RCNVR – RCN Volunteer Reserve
RN – Royal Navy
RNAS – Royal Naval Air Service
RNHS – Royal Navy Helicopter School
RNZAF – Royal New Zealand Air Force
S/L – squadron leader
SAIT – Southern Alberta Institute of
  Technology
SAM – surface to air missile
SAR – search and rescue
SAREX – search and rescue exercise
SARSAT – Search and Rescue Satellite-Aided
  Tracking
SAS – Special Air Service
SCA – Shuttle Carrier Aircraft
SEBJ – Société d'Energie de la Baie James
SFTS – Service Flying Training School
Sgt – sergeant
SHORAN – short range aid to navigation
SK – Saskatchewan
SKAD – survival kit air dropable
SLt – sub-lieutenant
Snowbird – a Canadian senior citizen who
  spends the winter in the southern US
SOE -- Special Operations Executive
Soo – Sault Ste. Marie
Spar – Special Products and Research
SRB – solid rocket booster
SRMA – Shuttle Remote Manipulator System
SSRMS – Space Station Remote Manipulator
  System
SST – supersonic transport
STA – Shuttle Training Aircraft
STARS – Shock Trauma Air Rescue Society
TAG – Taxi Air Group
TALEX – tactical airlift exercise

TAMS – temporary air movements section
TBO – time between overhaul
TCA – Trans-Canada Air Lines
TCAS – traffic-alert and collision avoidance
  system
THP – Transport Helicopter Platoon
TIA – Toronto Island Airport
TII – threshold inspection interval
TPA – Trans-Provincial Airlines
TSB – Transportation Safety Board
U-boat – German submarine
UK – United Kingdom
uk – unknown
UN – United Nations
UNAMIR – United Nations Assistance
  Mission in Rwanda
UNHCR – United Nations High Commissioner
  for Refugees
UNITAF – Unified Task Force
UNOSOM – United Nations Operations in
  Somalia
UNWHO – United Nations World Health
  Organization
UPS – United Parcel Service
US DEA – US Drug Enforcement Agency
USAAC – US Army Air Corps
USAAF – US Army Air Forces
USAF – US Air Force
USCAR – US Civil Aircraft Register
USCG – United States Coast Guard
USN – United States Navy
UTTH – utility tactical transport aircraft
VE-Day – Victory in Europe Day
VIA – Vancouver Island Airlines
VIH – Vancouver Island Helicopters
VJ-Day – Victory over Japan Day
W/C – wing commander
WAAF – Womens Auxiliary Air Force
WAG wireless air gunner
WASP – Women Airforce Service Pilots
WCAL – Western Canada Airways Ltd.
WCAM – Western Canada Aviation Museum
WETF – weightless environment training
  facility
WFP – World Food Program
WO – warrant officer
WopAG – wireless operator-air gunner
WOXOF – indefinite ceiling, zero feet, sky
  obscured, visibility zero, fog
WSPS – wire strike protection system
WWI – World War I
WWII – World War II
YK – Yukon
YSAT – Yukon Southern Air Transport

Bibliography

Airways: A Global Review of Commercial Flight, publisher John Wegg, Sandpoint, Idaho.

Aviation Week and Space Technology, New York, NY.

Badcock, Capt T.C., *A Broken Arrow: The Story of the Arrow Air Disaster in Gander, Newfoundland*, St. John'n, Nfld., Al Clouston Publications, 1988.

Bain, Donald M., *Canadian Pacific Air Lines: Its History and Aircraft*, Calgary, Kishorn Publications, 1987.

Blatherwick, John, *A History of Airlines in Canada*, Toronto, Unitrade Press, 1989.

CAHS Journal, Canadian Aviation Historical Society, Willowdale, Ontario.

Canada Flight Supplement: Canada and North Atlantic Terminal and Enroute Data, Ottawa, Transport Canada, 1993.

Canadian Aircraft Operator, publisher Robert Halford, Streetsview, Ontario.

Christie, Carl A., *Ocean Bridge: The History of RAF Ferry Command*, Toronto, University of Toronto Press, 1995.

Condit, John, *Wings over the West: Russ Baker and the Rise of Pacific Western Airlines*, Madeira Park, BC, Harbour Publishing, 1984.

Corley-Smith, Peter, *Barnstorming to Bush Flying 1910-1930*, Victoria, BC, Sono Nis Press, 1989.

Corley-Smith, Peter, *Bush Flying to Blind Flying 1930-1940*, Victoria, BC, Sono Nis Press, 1993.

Corley-Smith, Peter and Parker, David N., *Helicopters: The British Columbia Story*, Toronto, CANAV Books, 1985.

Countryman, Barry, *R100 in Canada*, Erin, Ontario, Boston Mills Press, 1982.

Creed, Roscoe, *PBY: The Catalina Flying Boat*, Annapolis, Maryland, Naval Institute Press, 1985.

Davies, R.E.G. and Quastler, I.E., *Commuter Airlines of the United States*, Washington, DC, Smithsonian Institution Press, 1995.

Davis, John M., Martin, Harold G. and Whittle, John A., *The Curtiss C-46 Commando*, Tonbridge, England, Air Britain, 1978.

Delve, Ken, *The Source Book of the RAF*, Shrewsbury, England, Airlife, 1994.

Dotto, Lydia, *A Heritage of Excellence: 25 Years at Spar Aerospace Limited*, Toronto, Spar Aerospace Limited, 1992.

Dotto, Lydia, *The Astronauts: Canada's Voyageurs in Space*, Toronto, Stoddart, 1993.

Douglas, W.A.B., *The Creation of a National Air Force: The Official History of the Royal Canadian Air Force, Volume II*, Toronto, University of Toronto Press, 1986.

Eastwood, A.B. and Roach, J.R., *Piston Engine Airliner Production List*, West Drayton, England, The Aviation Hobby Shop, 1991.

Ellis, Frank H., *Canada's Flying Heritage*, Toronto, University of Toronto Press, 1954.

Floyd, Jim, *The Avro Canada C.102 Jetliner*, Erin, Ontario, Boston Mills Press, 1986.

Fuller, G.A., Griffin, J.A., Molson, K.M., *125 Years of Canadian Aeronautics: A Chronology 1840-1965*, Willowdale, Ontario, Canadian Aviation Historical Society, 1983.

Gandt, Robert L., *China Clipper: The Age of the Great Flying Boats*, Annapolis, Maryland, Naval Institute Press, 1991.

Geren, Richard and McCullogh, Blake, *Cain's Legacy: The Building of Iron Ore Company of Canada*, Sept-Îles, Quebec, Iron Ore Company of Canada, 1990.

Gomersall, Bryce G., *The Stirling File*, Tonbridge, England, Air Britain, 1987.

Gradidge, J.M.G., *The Douglas DC-3 and Its Predecessors*, Tonbridge, England, Air Britain, 1984.

Granthan, A. Kevin, *P-Screamers: The History of the Surviving Lockheed P-38 Lightnings*, Missoula, Montana, Pictorial Histories Publishing Co. Ltd., 1994.

Green, William and Swanborough, Gordon, *Observer's Directory of Military Aircraft*, London, England, Frederick Warren, 1982.

Greenhous, Brereton, Harris, Stephen J., Johnston, William C. and Rawling, William G.P., *The Crucible of War 1939-1945: The Official History of the Royal Canadian Air Force, Volume III*, Toronto, University of Toronto Press, 1994.

Griffin, J.A., *Canadian Military Aircraft Serials and Photographs*, Ottawa, Canadian War Museum, 1969.

Hatch, F.J., *The Aerodrome of Democracy: Canada and the British Commonwealth Air Training Plan 1939-1945*, Ottawa, Directorate of History (Department of National Defence), 1983.

Hayes, Karl E., *De Havilland Canada DHC-3 Otter*, Dublin, Ireland, Irish Air Letter, 1982.

Hoare, Robert J., *Wings over the Atlantic*, London, England, Phoenix House Ltd., 1956.

Horrall, S.W., *The Pictorial History of the Royal Canadian Mounted Police*, Toronto, McGraw-Hill Ryerson, 1973.

Hotson, Fred W., *The Bremen*, Toronto, CANAV Books, 1989.

Hotson, Fred W., *The De Havilland Canada Story*, Toronto, CANAV Books, 1983.

Jefford, W/C C.G., MBE, *RAF Squadrons: A Comprehensive Record of the Movement and Equipment of All RAF Squadrons and Their Antecedents Since 1912*, Shrewsbury, England, Airlife, 1988.

Juptner, Joseph P., *U.S. Civil Aircraft Series*, (Vols. 1-9), Blue Ridge Summit, Pennsylvania, TAB Aero, 1981.

Kostenuk, S. and Griffin, J.A., *RCAF Squadrons and Aircraft*, Toronto, Samuel Stevens Hakkert & Co., 1977.

Leigh, Z. L., *And I Shall Fly: The Flying Memoirs of Z. Lewis Leigh*, Toronto, CANAV Books, 1985.

Life at the Crossroads of the World: A History of Gander, Newfoundland 1936-1988, Gander Seniors Club, 1988.

Main, J.R.K., *Voyageurs of the Air*, Ottawa, Queen's Printer, 1967.

Marson, Peter J., *The Lockheed Constellation Series*, Tonbridge, England, Air Britain, 1982.

McDaniel, William H., *The History of Beechcraft*, Wichita, McCormick-Armstrong, 1982.

McGrath, Thomas M., *The History of Canadian Airports*, Toronto, Lugus Publications, 1992.

McGregor, Gordon R., *The Adolescence of an Airline*, Montreal, Air Canada, 1970.

McLaren, Duncan D., *Bush to Boardroom: A Personal View of Five Decades of Aviation History*, Winnipeg, Watson & Dwyer Ltd., 1992.

McVicar, Donald M., *A Railroad from the Sky*, Dorval, Quebec, Ad Astra Books, 1992.

McVicar, Donald M., *Distant Early Warning*, Dorval, Quebec, Ad Astra Books, 1992.

McVicar, Donald M., *Through Cuba to Oblivion*, Dorval, Quebec, Ad Astra Books, 1994.

Milberry, Larry, *Austin Airways: Canada's Oldest Airline*, Toronto, CANAV Books, 1985.

Milberry, Larry, *Aviation in Canada*, Toronto, McGraw-Hill Ryerson, 1979.

Milberry, Larry, *Sixty Years: The RCAF and CF Air Command 1924-84*, Toronto, CANAV Books, 1984.

Milberry, Larry, *The Canadair North Star*, Toronto, CANAV Books, 1982.

Molson, K.M and Taylor, H.A., *Canadian Aircraft since 1909*, Stittsville, Ontario, Canada's Wings, 1982.

Molson, K.M. and Short, A.J.,*The Curtiss HS Flying Boats*, Ottawa, National Aviation Museum, 1995.

Molson, K.M., *Canada's National Aviation Museum: Its History and Collections*, Ottawa, National Aviation Museum, 1988.

Molson, K.M., *Pioneering in Canadian Air Transport*, Winnipeg, James Richardson and Sons, Ltd., 1974.

Murray, Charles, and Cox, Catherine Bly, *Apollo: The Race to the Moon*, New York, Simon and Schuster, 1989.

Parsons, Bill, *The Challenge of the Atlantic: A Photo-Illustrated History of Early Aviation in Harbour Grace, Newfoundland*, Newfoundland, 1983.

Pelletier, A.J., *Beech Aircraft and their Predecessors*, London, England, Putnam Aeronautical Books, 1995.

Pelletier, A.J., *Bell Aircraft since 1935*, London, England, Putnam Aeronautical Books, 1992.

Phillips, Edward H., *Piper: A Legend Aloft*, Eagan, Minnesota, Flying Books International, 1993.

Phillips, Edward H., *Beechcraft: Pursuit of Excellence*, Eagan, Minnesota, Flying Books, 1992.

Phillips, Edward H., *Wings of Cessna: Model 120 to the Citation III*, Eagan, Minnesota, Flying Books, 1986.

Pickler, Ron and Milberry, Larry, *Canadair: The First 50 Years*, Toronto, CANAV Books, 1995.

Pittet, Richard G. and Kostyniuk, Brent, *Determined on Delivery: The History of 435 (T) Squadron*, Edmonton, 435 (T) Squadron Anniversary Committee, 1994.

Propliner, founder Stephen Piercey, editor Tony Merton Jones, Redlynch, Wilts., England.

Quarrie, Bruce, *Airborne Assault: Parachute Forces in Action 1940-91*, Somerset, UK, Patrick Stephens Ltd., 1991.

Rankin-Lowe, Jeff, and Cline, Andy, *Aircraft of the Canadian Armed Forces: A Checklist of Current Aircraft and Disposals*, London, Ontario, Sirius Productions, 1995.

Roach, J.R. and Eastwood, A.B., *Jet Airliner Production List*, West Drayton, England, The Aviation Hobby Shop, 1992.

Ryan, Richard W., *From Boxkite to Boardroom*, Moose Jaw, Saskatchewan, Moose Jaw Publications.

Smith, Philip, *It Seems Like Only Yesterday: Air Canada, The First 50 Years*, Toronto, McClelland and Stewart, 1986.

Spenser, Jay P., *Vertical Challenge: The Hiller Aircraft Story*, Seattle, University of Washington Press, 1992.

Stevens, Robert W., *Alaskan Aviation History*, Vols. I and II, Desmoines, Washington, Polynyas Press, 1990.

Stevenson, Garth, *The Politics of Canada's Airlines from Diefenbaker to Mulroney*, Toronto, University of Toronto Press, 1987.

Sullivan, Kenneth H. and Milberry, Larry, *Power: The Pratt & Whitney Canada Story*, Toronto, CANAV Books, 1989.

Sutherland, Alice Gibson, *Canada's Aviation Pioneers: 50 Years of McKee Trophy Winners*, Toronto, McGraw-Hill Ryerson, 1978.

Swanborough, Gordon, *Civil Aircraft of the World*, New York, Charles Scribner's Sons, 1980.

Taylor, Michael J.H., *The Aerospace Chronology*, London, England, Tri-Service Press Ltd., 1989.

Tibbo, Frank, *Charlie Baker George: The Story of Sabena OO-CBG*, St. John's, Newfoundland, Jesperson Press Ltd., 1993.

Time Air: 25 Years of People, Service and Memories, Calgary, 1991.

Wegg, John, *General Dynamics Aircraft and their Predecessors*, London, England, Putnam Aeronautical Books, 1990.

West, Bruce, *The Firebirds: How Bush Flying Won Its Wings*, Toronto, Queen's Printer for Ontario, 1974.

West Coast Aviator, publisher Jack Scofield, Sidney, British Columbia.

Westflight: The Canadian Airline Journal, Richmond, British Columbia.

White, Howard and Spilsbury, Jim, *The Accidental Airline: Spilsbury's QCA*, Madeira Park, BC, Harbour Publishing, 1988.

Wilmot, Chester, *The Struggle for Europe*, London, England, Collins, 1952.

Wings, publisher Paul Skinner, Calgary.

Wise, S.F., *Canadian Airmen and the First World War: The Official History of the Royal Canadian Air Force, Volume I*, Toronto, University of Toronto Press, 1980.

World Aircraft & Systems Directory, editor Michael Taylor, Brassey's, London, England, 1996.

50 Years of Technology 1930-1980, Winnipeg, Bristol Aerospace Ltd., 1980.

Grumman Mallard CF-FLC of Hudson Bay Mining and Smelting came to Canada in April 1947. Based at Schist Lake, near Flin Flon, it was flown by Alex More, Ross Lennox, Bob Ross and Pat Donahay. On April 4, 1963 the hangar burned at Flin Flon, destroying CF-FLC. (Bristol Aerospace)

Index

BAC 1-11 391, 393, 395, 777, 784, 790, 861, 869
Bachman, Charles 310
Baddeck, NS 14, 15, 17
Baden-Soellingen, Germany 898, 932, 933
BAe: BAe146 11, 680, 681, 688, 702, 703, 705, 709, 790, 815, 869, 878; BAe748 363, 364, 389, 488, 577, 655, 666, 676, 759-762, 807, 813, 814, 831, 833, 834, 837-840, 845, 846; Harrier 913; Jetstream 11, 680, 780, 841, 856, 864, 865, 869, 878, 992
Baert, Ray 569
Baffin Island 77, 79, 316, 637
Baggett, P/O W.R. 345
Bagotville, QC 197, 348, 393, 396, 512, 789, 928, 933, 934, 979
Bahamas Airways 487
Bahrain 430, 899
Baie Comeau, QC 30, 88, 123, 125, 265, 297, 395, 405, 526, 806, 856
Baier, Hartwig 602
Bailey, A.K. 120
Bailey, Joan 273
Bailley-Maxwell Ltd. 134
Baily, Sgt Francis 159
Bain, James T. 142, 174, 223, 224, 229, 230, 232, 308, 868
Baird, J.N. 223, 684
Baird, Maj P.D. 413
Baker Lake, NT 73, 81-83, 361, 369, 646
Baker, A.W. "Bill" 229, 231, 323, 605
Baker, Al 538, 600
Baker, Capt D. 435
Baker, Harry 327
Baker, Madge 279
Baker, Maj R.F. 81
Baker, Ron 225
Baker, Russell 64, 126, 274-276, 331, 388, 702, 778
Bakhtar Afghan Airlines 548-550, 577
Bakkeli, Kjell 945
Balbo, Gen Italo 99, 119
Balchen, Bernt 71, 97, 242
Baldwin, Frederick W. 14, 15, 18, 24, 976
Baldwin, J.R 266
Baldwin, John 120
Baldwin, ON 617, 621
Bales, G. 319, 325
Ball, R.J. 645
Ballantyne, Alec 383
Ballard Aircraft 247
Ballentine, C.G. 75
balloon flight 12-14
Baltimore, MD 613
Bambi Bucket 363, 693, 958, 965, 970
Bandera, Al 633
Banfe, Chuck 626
Banff Oil 484
Bangkok 145
Bangladesh 672, 674
Bank of Commerce 206
Bank of Montreal 893, 895
Banks, Doug 695
Bannock Aerospace 545, 575
Bannock, Russell 238, 241, 242, 329, 387, 402, 528, 543, 553, 566, 836, 893, 895
Banting, C.S. 131
Banting, Dr. F. 314, 378
Bar XH Aviation 680
Barager, F.B. 74, 398
Baratta, Rick 844
Barbaro, Ron 819
Barber, Bill 1006
Barber, Jim 758

Barber, Sgt 441
Barbery, John C. 244
Barclay, M.B. 174, 176, 216, 372, 881
Barilkow, Bill 357, 850
Barker Field, Toronto 199, 254, 336, 400
Barker, F/O E.A. 429, 434
Barker, William 24, 26, 91
Barkley-Grow T8P-1 115, 155, 197, 277, 306, 329, 331, 371, 378, 478
Barling NB-3 359
Barnes, Professor 125
Barnes, Rodney 451
Barrass, John 901, 903
Barrault, Cpl J.S. 434
Barrett, Bob 800
Barrett, Harry 378
Barrow, Elgin 330
Bartch, Dawn and Gordon 752
Bartel, Hugo 602
Barter Island 316
Barton, Richard 675
Basler, Warren 803, 893
Bastien, Irenée 216
Batchawana Bay Air Service 845
Batchawana Bay, ON 257
Bate, F/O Harry 211, 212
Bates, Louisa 19
Bath, F/O C.L. 76
Bathurst Inlet 81, 82, 803, 804
Baudoux, E.L. 165
Bauer, Col A.J. 424
Baxter, LAC 117
Baxter, Les 894
Bay Aviation 626
Bay Maud 82, 818
Baydala, Allan 715
Bayley, Ray 656
Bayview Air Service 630
BC Air Lines 669
BC Air Transport 127
BC Airlines (new) 277, 279, 280, 652, 618, 619, 682, 684
BC Airlines (old) 133
BC Airways 76
BC Aviator 280
BC Development Corp 629
BCIT/BC Vocational Institute 351, 693, 964
Beachey, Hillery and Lincoln 19, 20
Beadle, F.P.H. 91
Beall, Molly 400
Beamsville, ON 23, 68, 940
Beardmore, W/C 158
Bearskin Airlines 630, 698, 700, 758, 764, 816, 828, 838, 840, 844-846, 851, 856, 904
Bearskin Lake, ON 844
Beasley, Bert 530
Beaton, S/L R.W. "Bob" 398, 426, 915
Beatty, Sir Edward 126, 263
Beaty, Chris 932
Beaulieu, Al 70-1
Beauchene, George 843
Beauchêne, QC 126
Beaudet, Richard 611, 640, 1007
Beaudoin, Laurent 604, 995
Beaudoin, Yves 451
Beaudry, L. 12
Beaufort Sea 796, 941, 949, 953, 954
Beaugrand, Jim 906
Beauregard, Jack 540
Beaver, The 888
Bechtel, Price and Callahan 164
Becker, Clem 807
Becker, Cy 84, 750
Bedard, Jeannine 723
Beech: AT-11 458, 459, 490, 491, 647, 811; Baron 403; Beech Jet 400 897; Bonanza 259, 324, 626, 927; Duke

896; King Air 539-543, 686-688, 845, 865, 889, 896, 918, 927; Model 17 115, 130, 195, 255, 256, 402, 452, 703, 886, 895; Model 18 (Expeditor) 123, 127-129, 183, 197, 250, 251, 258, 288, 295, 303, 363, 367, 368, 373, 377, 392, 394, 395, 397, 403, 404, 406, 436, 442, 460, 470, 492, 506, 521, 539, 541, 542, 627, 641-647, 675, 683, 686-691, 696, 703-705, 720, 723, 734, 753, 763-765, 767, 768, 795, 798, 820, 848, 849, 870, 880, 882, 883, 888, 890-892; Model 99 544, 701, 757, 845, 846, 849, 850, 858; Model 1900 686-688, 715, 755-758, 834, 835, 837, 840, 865, 999; NU-8F 541; Queen Air 257, 461, 541, 542, 763, 896; Starship 897; T-34 251; Travel Air 763, 887; Twin Bonanza 492, 647
Beech, Olive Ann 539, 542, 882
Beeching, Charles 90, 963
Beeman, Jack 354
Beeman, LCdr J.H. 423
Beer, Roy 340
Beery, Wallace 367
Beirut 435
Belair, Bix 398
Belem, Brazil 169
Belet Uen, Somalia 906, 908, 909
Belfast, Ireland 165, 989, 990
Belgium 314, 315
Bell Boeing 609 991
Bell Canada/Bell Telephone/Bell Northern 216, 546, 798, 846, 897, 979
Bell Helicopters/BHTC 120, 626, 960-963
Bell, A.N. 319
Bell, Alexander Graham 14, 15
Bell, Art 478
Bell, Bob 671, 673
Bell, F/L Jack 441
Bell, F/O John 440
Bell, Harry 517
Bell, James C. 146, 257, 394, 398, 408, 613, 797, 828-832, 835, 836
Bell, Joel 602-604, 895
Bell, Larry 350
Bell, N.H. 95
Bell, Ralph 190
Bell, W.R. 176
Bell: Model 47 241, 299, 307, 309, 316, 329-331, 347, 350-354, 363, 367, 370, 388, 407, 443, 456, 459, 513, 526, 720, 756, 841, 850, 940, 944, 950, 953, 965, 966; Model Bell 204/205/ UH-1 Huey 331, 363, 540, 590, 717, 928, 939, 941-943, 948, 950, 952, 961, 966, 967; Model 206/OH-58 Jet Ranger/Long Ranger 11, 331, 540, 648, 692, 717, 826, 847, 859, 939, 943, 949, 950, 953, 956, 960-963, 966, 988; Model 212/412/CH-146 837, 906, 909, 910, 921, 928, 937, 939, 949, 952, 954, 958-960, 962, 963, 966, 988, 1006; Model 214, 941, 954, 966; Model 222 700, 959, 961; Model 230 961-963; Model 400 Twin Ranger 960, 961; Model 407 963; P-39 161, 162
Bella Bella, BC 73, 160, 561
Bellanca Aircraft 93, 97, 99, 831
Bellanca: Model 31-55 Skyrocket 251, 262, 295, 329, 374, 470, 792; Model 66-70/66-75 126, 136, 136, 195, 197, 275, 359, 366, 373, 470, 751, 771; Model CH-300 Pacemaker 84, 100, 101, 108, 117-119, 121, 123, 131, 134, 140, 198, 291, 373

Bellanger, A. 421
Bellis, F/L Jack 258
Belyea, Hazel 1008
Ben Ocean Lancer 949
Benavidez, Hope 693
Bendall, Eric 176
Bendix Trophy 324, 445, 721
Benedik, George 473
Benkie, Donald 448-450
Benner, Sheldon 811
Bennett Cup 13, 14
Bennett, Floyd 97
Bennett, Geoff 725, 744, 746
Bennett, Gil 602-604
Bennett, Hon. R.B. 83, 109, 118, 138
Bennett, Vic 892
Benoist biplane 20
Benoit, E.M. 445, 446, 450
Bensen, Igor 356
Benson, Kevin 679
Bent, Barney 352
Bentham, Dick 624, 626
Berens River, MB 44, 63
Bergren, Don 639
Berland River, AB 513
Berlin Airlift 193, 218
Berlyn, Martin J. 95
Bennett, Donald C.T. 164, 165, 311
Bermuda 31, 168, 169, 170, 179, 218, 226, 350
Bernard, Ray and René 834
Bernatchez, MGen 424
Berry, Arthur Massey "Matt" 59, 74, 102, 124, 329
Berry, Cy 294
Berry, Oscar 56
Berthelet, Mike 805
Bertrand, Jim 902, 920, 921
Best, Bob 926
Best, Bruce 93, 195, 241, 354, 943
Bethlehem Steel 520
Beveridge, Les 58
Beveridge, W.H. 525
Bews, Ken 805
BHP 822, 971
Biafra 333-335, 489, 505
Bibby, R.H. 107, 145
Bickell, J.P. 157, 228, 280, 651, 886
Biddle, W/C W.H. 166
Bieck, A.H. 260
Biehl, Jim 921
Big Bird Sky Farmers 848, 849
Big Delta, Alaska 161
Big Indian Drilling 942
Big Trout Lake Air 680, 698
Big Trout Lake, ON 360, 368, 637, 844, 846, 851
Biggar, O.M. 63
Biggin Hill, UK 182
Billington, Eric 61, 62
Bing Mah, Stanley 37
Binns, Stan 921
Birch Lake Lodge 854
Birch Lake, ON 569
Birch, Tony 360
Birchall, A/C L.J. 423
Bird, F/O Jim 182
Bird, MB 348, 944
Bird, W.H. 229, 230
Bishop, W.A. "Billy" 24, 26, 123, 155, 346
Bisley Trophy 116
Bison APC 907
Bissett, MB 180, 375
Bissette, André 391
Bisson, Johnny 253
Bisson, Louis 134, 170, 180, 323, 324
Bisson, Sgt L.J. 424, 506
Bittle, Ken 844

(Left) The line of classic Stinson Reliants has graced Canadian skies since Konnie Johannesson purchased SR-7A CF-AUS in 1933. About 50 examples served, mainly in the bush, and a few remained flying and in museums at the turn of the century. Here SR-9 CF-BIM of Wheeler-Northland Airways appears in a Quebec setting in the mid-1960s. It had come to Canada in 1933 for McIntyre Porcupine Mines. (PAC/McNulty Col. PA191805)

(Right) In 1996 Ontario's Ministry of Natural Resources began putting its Twin Otters on million-dollar Wipline amphibious, water-bombing floats. Richard Hulina shot recently-fitted C-GOGB at Sioux Lookout.

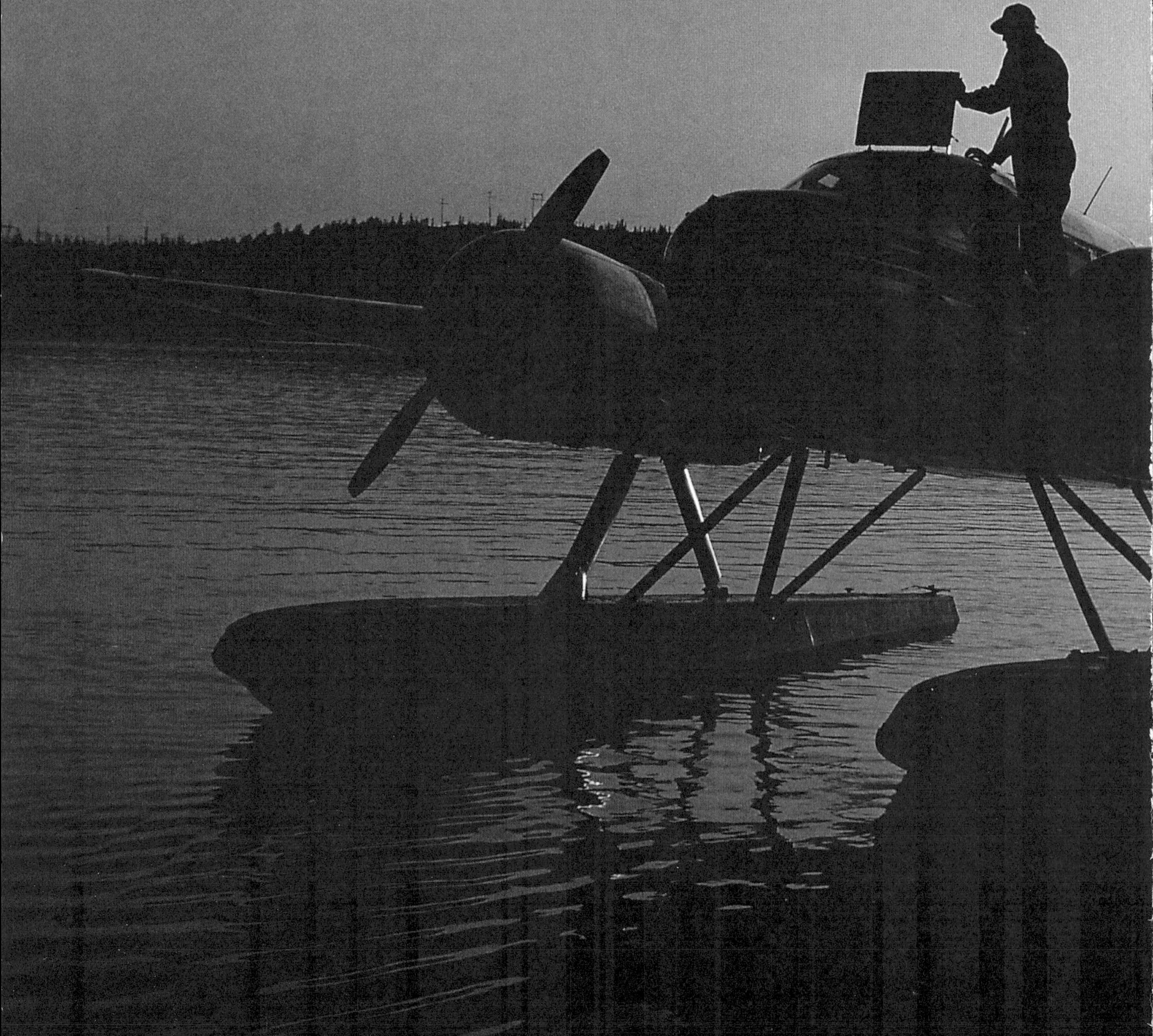